Lecture Notes in Bioinformatics   9043

Subseries of Lecture Notes in (

T0234564

Francisco Ortuño   Ignacio Rojas (Eds.)

Bioinformatics and Biomedical Engineering

Third International Conference, IWBBIO 2015
Granada, Spain, April 15-17, 2015
Proceedings, Part I

 Springer

Volume Editors

Francisco Ortuño
Ignacio Rojas
Universidad de Granada
Dpto. de Arquitectura y Tecnología de Computadores (ATC)
E.T.S. de Ingenierías en Informática y Telecomunicación, CITIC-UGR
Granada, Spain
E-mail: {fortuno, irojas}@ugr.es

ISSN 0302-9743   e-ISSN 1611-3349
ISBN 978-3-319-16482-3   e-ISBN 978-3-319-16483-0
DOI 10.1007/978-3-319-16483-0
Springer Cham Heidelberg New York Dordrecht London

Library of Congress Control Number: 2015934926

LNCS Sublibrary: SL 8 – Bioinformatics

Typesetting: Camera-ready by author, data conversion by Scientific Publishing Services, Chennai, India

Printed on acid-free paper

Springer is part of Springer Science+Business Media (www.springer.com)

Preface

We are proud to present the set of final accepted full papers for the third edition of the IWBBIO conference "International Work-Conference on Bioinformatics and Biomedical Engineering" held in Granada (Spain) during April 15–17, 2015.

The IWBBIO 2015 (International Work-Conference on Bioinformatics and Biomedical Engineering) seeks to provide a discussion forum for scientists, engineers, educators, and students about the latest ideas and realizations in the foundations, theory, models, and applications for interdisciplinary and multidisciplinary research encompassing disciplines of computer science, mathematics, statistics, biology, bioinformatics, and biomedicine.

The aims of IWBBIO 2015 is to create a friendly environment that could lead to the establishment or strengthening of scientific collaborations and exchanges among attendees, and therefore, IWBBIO 2015 solicited high-quality original research papers (including significant work-in-progress) on any aspect of Bioinformatics, Biomedicine, and Biomedical Engineering.

New computational techniques and methods in machine learning; data mining; text analysis; pattern recognition; data integration; genomics and evolution; next generation sequencing data; protein and RNA structure; protein function and proteomics; medical informatics and translational bioinformatics; computational systems biology; modeling and simulation and their application in life science domain, biomedicine, and biomedical engineering were especially encouraged. The list of topics in the successive Call for Papers has also evolved, resulting in the following list for the present edition:

1. **Computational proteomics.** Analysis of protein–protein interactions. Protein structure modeling. Analysis of protein functionality. Quantitative proteomics and PTMs. Clinical proteomics. Protein annotation. Data mining in proteomics.
2. **Next generation sequencing and sequence analysis**. De novo sequencing, re-sequencing, and assembly. Expression estimation. Alternative splicing discovery. Pathway Analysis. Chip-seq and RNA-Seq analysis. Metagenomics. SNPs prediction.
3. **High performance in Bioinformatics**. Parallelization for biomedical analysis. Biomedical and biological databases. Data mining and biological text processing. Large-scale biomedical data integration. Biological and medical ontologies. Novel architecture and technologies (GPU, P2P, Grid,...) for Bioinformatics.
4. **Biomedicine**. Biomedical Computing. Personalized medicine. Nanomedicine. Medical education. Collaborative medicine. Biomedical signal analysis. Biomedicine in industry and society. Electrotherapy and radiotherapy.
5. **Biomedical Engineering**. Computer-assisted surgery. Therapeutic engineering. Interactive 3D modeling. Clinical engineering. Telemedicine.

Biosensors and data acquisition. Intelligent instrumentation. Patient Monitoring. Biomedical robotics. Bio-nanotechnology. Genetic engineering.
6. **Computational systems for modeling biological processes**. Inference of biological networks. Machine learning in Bioinformatics. Classification for biomedical data. Microarray Data Analysis. Simulation and visualization of biological systems. Molecular evolution and phylogenetic modeling.
7. **Healthcare and diseases**. Computational support for clinical decisions. Image visualization and signal analysis. Disease control and diagnosis. Genome-phenome analysis. Biomarker identification. Drug design. Computational immunology.
8. **E-Health**. E-Health technology and devices. E-Health information processing. Telemedicine/E-Health application and services. Medical Image Processing. Video techniques for medical images. Integration of classical medicine and E-Health.

After a careful peer review and evaluation process (268 submission were submitted and each submission was reviewed by at least 2, and on the average 2.7, Program Committee members or Additional Reviewer), 134 papers were accepted to be included in LNBI proceedings.

During IWBBIO 2015 several Special Sessions will be carried out. Special Sessions will be a very useful tool to complement the regular program with new and emerging topics of particular interest for the participating community. Special Sessions that emphasize on multidisciplinary and transversal aspects, as well as cutting-edge topics are especially encouraged and welcome, and in this edition of IWBBIO 2015 are the following:

1. **SS1: Expanding Concept of Chaperone Therapy for Inherited Brain Diseases**
   Chaperone therapy is a new concept of molecular therapeutic approach, first developed for lysosomal diseases, utilizing small molecular competitive inhibitors of lysosomal enzymes. This concept has been gradually targeted to many diseases of other categories, utilizing various compounds not necessarily competitive inhibitors but also non-competitive inhibitors or endogenous protein chaperones (heat-shock proteins).
   In this session, we discuss current trends of chaperone therapy targeting various types of neurological and non-neurological diseases caused by misfolded mutant proteins. This molecular approach will open a new therapeutic view for a wide variety of diseases, genetic and non-genetic, and neurological and non-neurological, in the near future.

   Organizer: Dr. Prof. Yaping Tian, Department of Clinical Biochemistry, Chinese PLA General Hospital, Beijing (China).
2. **SS2: Quantitative and Systems Pharmacology: Thinking in a wider "systems-level" context accelerates drug discovery and enlightens our understanding of drug action**
   "Quantitative and Systems Pharmacology (QSP) is an emerging discipline focused on identifying and validating drug targets, understanding existing

therapeutics and discovering new ones. The goal of QSP is to understand, in a precise, predictive manner, how drugs modulate cellular networks in space and time and how they impact human pathophysiology." (QSP White Paper - October, 2011)

Over the past three decades, the predominant paradigm in drug discovery was designing selective ligands for a specific target to avoid unwanted side effects. However, in the current postgenomic era, the aim is to design drugs that perturb biological networks rather than individual targets. The challenge is to be able to consider the complexity of physiological responses to treatments at very early stages of the drug development. In this way, current effort has been put into combining 0 chemogenomics with network biology to implement new network-pharmacology approaches to drug discovery; i.e., polypharmacology approaches combined with systems biology information, which advance further in both improving efficacy and predicting unwanted off-target effects. Furthermore, the use of network biology to understand drug action outputs treasured information, i.e., for pharmaceutical companies, such as alternative therapeutic indications for approved drugs, associations between proteins and drug side effects, drug–drug interactions, or pathways, and gene associations which provide leads for new drug targets that may drive drug development.

Following the line of QSP Workshops I and II (2008, 2010), the QSP White Paper (2011), or QSP Pittsburgh Workshop (2013), the goal of this symposium is to bring together interdisciplinary experts to help advance the understanding of how drugs act, with regard to their beneficial and toxic effects, by sharing new integrative, systems-based computational, or experimental approaches/tools/ideas which allow to increase the probability that the newly discovered drugs will prove therapeutically beneficial, together with a reduction in the risk of serious adverse events.

Organizer: Violeta I. Perez-Nueno, Ph.D., Senior Scientist, Harmonic Pharma, Nancy (France).

3. **SS3: Hidden Markov Model (HMM) for Biological Sequence Modeling** Sequence Modeling is one of the most important problems in bioinformatics. In the sequential data modeling, Hidden Markov Models(HMMs) have been widely used to find similarity between sequences. Some of the most important topics in this session are:

   (a) Modeling of biological sequences in bioinformatics;
   (b) The application of Hidden Markov Models(HMM);
   (c) HMM in modeling of sequential data;
   (d) The advantages of HMM in biological sequence modeling compared to other algorithms;
   (e) The new algorithms of training HMM;
   (f) Gene sequence modeling with HMM;

Organizer: Mohammad Soruri, Department of Electrical and Computer Engineering, University of Birjand, Birjand (Iran).

4. **SS4: Advances in Computational Intelligence for Bioinformatics and Biomedicine** Biomedicine and, particularly, Bioinformatics are increasingly and rapidly becoming data-based sciences, an evolution driven by technological advances in image and signal non-invasive data acquisition (exemplified by the 2014 Nobel Prize in Chemistry for the development of super-resolved fluorescence microscopy). In the Biomedical field, the large amount of data generated from a wide range of devices and patients is creating challenging scenarios for researchers, related to storing, processing, and even just transferring information in its electronic form, all these compounded by privacy and anonymity legal issues. This can equally be extended to Bioinformatics, with the burgeoning of the .omics sciences.

New data requirements require new approaches to data analysis, some of the most interesting ones are currently stemming from the fields of Computational Intelligence (CI) and Machine Learning (ML). This session is particularly interested in the proposal of novel CI and ML approaches to problems in the biomedical and bioinformatics domains.

Topics that are of interest in this session include (but are not necessarily limited to):

(a) Novel applications of existing CI and ML methods to biomedicine and bioinformatics.
(b) Novel CI and ML techniques for biomedicine and bioinformatics.
(c) CI and ML-based methods to improve model interpretability in biomedical problems, including data/model visualization techniques.
(d) Novel CI and ML techniques for dealing with nonstructured and heterogeneous data formats.

More information at
http://www.cs.upc.edu/ avellido/research/conferences/
IWBBIO15-CI-BioInfMed.html
Main Organizer: Alfredo Vellido, PhD, Department of Computer Science, Universitat Politécnica de Catalunya, BarcelonaTECH (UPC), Barcelona (Spain).

Co-organizers: Jesus Giraldo, PhD, Institut de Neurociències and Unitat de Bioestadística, Universitat Autònoma de Barcelona (UAB), Cerdanyola del Vallès, Barcelona (Spain).
René Alquézar, PhD, Department of Computer Science, Universitat Politécnica de Catalunya, BarcelonaTECH (UPC), Barcelona (Spain).

5. **SS5: Tools for Next Generation Sequencing data analysis** Next Generation Sequencing (NGS) is the main term used to describe a number of different modern sequencing technologies such as Illumina, Roche 454 Sequencing, Ion torrent, SOLiD sequencing, and Pacific Biosciences. These technologies allow us to sequence DNA and RNA more quickly and cheaply than Sanger sequencing and have opened new ways for the study of genomics, transcriptomics, and molecular biology, among others.

The continuous improvements on those technologies (longer read length, better read quality, greater throughput, etc.), and the broad application of NGS in several research fields, have produced (and still produce) a huge amount of software tools for the analysis of NGS genomic/transcriptomic data.

We invite authors to submit original research, pipelines, and review articles on topics related to software tools for NGS data analysis such as (but not limited to):

(a) Tools for data preprocessing (quality control and filtering).

(b) Tools for sequence alignment.

(c) Tools for de novo assembly.

(d) Tools for the analysis of genomic data: identification and annotation of genomic variants (variant calling, variant annotation).

(e) Tools for functional annotation to describe domains, orthologs, genomic variants, controlled vocabulary (GO, KEGG, InterPro...).

(f) Tools for the analysis of transcriptomic data: RNA-Seq data (quantification, normalization, filtering, differential expression) and transcripts and isoforms finding.

(g) Tools for Chip-Seq data.

(h) Tools for "big-data" analysis of reads and assembled reads.

Organizers: Javier Perez Florido, PhD, Genomics and Bioinformatics Platform of Andalusia (GBPA), Seville, (Spain).

Antonio Rueda Martin, Genomics and Bioinformatics Platform of Andalusia (GBPA), Seville, (Spain).

M. Gonzalo Claros Diaz, PhD, Department of Molecular Biology and Biochemistry, University of Malaga (Spain).

6. **SS6: Dynamics Networks in System Medicine**

Over the past two decades, It Is Increasingly Recognized that a biological function can only rarely be attributed to an individual molecule. Instead, most biological functions arise from signaling and regulatory pathways connecting many constituents, such as proteins and small molecules, allowing them to adapt to environmental changes. "Following on from this principle, a disease phenotype is rarely a consequence of an abnormality in a single effector gene product, but reflects various processes that interact in a complex network." Offering a unifying language to describe relations within such a complex system has made network science a central component of systems biology and recently system medicine. Despite the knowledge that biological networks can change with time and environment, much of the efforts have taken a static view. Time-varying networks support richer dynamics and may better reflect underlying biological changes in abnormal state versus normal state and this provides a powerful motivation and application domain for computational modeling. We introduce this session on the Dynamics Networks in System Medicine to encourage and support the development of computational methods that elucidate the Dynamics Networks and its application in medicine. We will discuss current trends and potential biological and clinical applications of network-based approaches to human disease. We aim to bring together experts in different fields in order to promote cross fertilization between different communities.

Organizer: Narsis Aftab Kiani, PhD, Computational Medicine Unit, Department of Medicine, Karolinska Institute (Sweden).

7. **SS7: Interdisciplinary puzzles of measurements in biological systems**

Natural sciences demand measurements of the subject of interest as a necessary part of the experimental process. Thus, for the proper understanding of the obtained datasets, it is the necessity to take into question all mathematical, biological, chemical, or technical conditions affecting the process of the measurement itself. While assumptions and recommendations within the field itself are usually concerned, some issues, especially discretization, quantization, experiment time, self-organization, and consequent anomalous statistics might cause puzzling behavior.

In this special section we describe particular examples across disciplines with joint systems theory-based approach, including noise and baseline filtration in mass spectrometry, image processing and analysis, and distributed knowledge database. The aim of this section is to present a general overview of the systemic approach.

Organizer: Jan Urban, PhD, Laboratory of Signal and Image Processing, Institute of Complex Systems, Faculty of Fisheries and protection of Waters, University of South Bohemia. (Czech republic).

8. **SS8: Biological Networks: Insight from interactions**

The complete sequencing of the human genome has shown us a new era of Systems Biology (SB) referred to as omics. From genomics to proteomics and furthermore, "Omics"-es existing nowadays integrate many areas of biology. This resulted in an essential ascent from Bioinformatics to Systems Biology leaving room for identifying the number of interactions in a cell. Tools have been developed to utilize evolutionary relationships toward understanding uncharacterized proteins, while there is a need to generate and understand functional interaction networks. A systematic understanding of genes and proteins in a regulatory network has resulted in the birth of Systems Biology (SB), there-by raising several unanswered questions. Through this conference, we will raise some questions on why and how interactions, especially protein–protein interactions (PPI), are useful while discussing methods to remove false positives by validating the data. The conference is aimed at the following two focal themes:

(a) Bioinformatics and systems biology for deciphering the known–unknown regions.

(b) Systems Biology of regulatory networks and machine learning.

Organizers: Prof. Alfredo Benso, PhD, Department of Control and Computer Engineering, Politecnico di Torino (Italy).

Dr. Prashanth Suravajhala, PhD, Founder of Bioclues.org and Director of Bioinformatics.org

9. **SS9: Tissue engineering for biomedical and pharmacological applications**

   The concept of tissues appeared more than 200 years ago, since textures and attendant differences were described within the whole organism components. Instrumental developments in optics and biochemistry subsequently paved the way to transition from classical to molecular histology in order to decipher the molecular contexts associated with physiological or pathological development or function of a tissue. The aim of this special session is to provide an overview of the most cutting edge updates in tissue engineering technologies. This will cover the most recent developments for tissue proteomics, and the applications of the ultimate molecular histology method in pathology and pharmacology: MALDI Mass Spectrometry Imaging. This session will be of great relevance for people willing to have a relevant summary of possibilities in the field of tissue molecular engineering.

   Organizer: Rémi Longuespée, PhD, Laboratoire de Spectrométrie de Masse, University of Liege (Belgium).

10. **SS10: Toward an effective telemedicine: an interdisciplinary approach**

    In the last 20 years many resources have been spent in experimentation and marketing of telemedicine systems, but — as pointed by several researchers — no real product has been fully realized — neither in developed nor in underdeveloped countries. Many factors could be detected:

    (a) lack of a decision support system in analyzing collected data;

    (b) the difficulty of using the specific monitoring devices;

    (c) the caution of patients and/or doctors toward E-health or telemedicine systems;

    (d) the passive role imposed on the patient by the majority of experimented systems;

    (e) the limits of profit-driven outcome measures;

    (f) a lack of involvement of patients and their families as well as an absence of research on the consequences in the patient's life.

    The constant improvement of ICT tools should be taken into account: at-home and mobile monitoring are both possible; virtual visits can be seen as a new way to perform an easier and more accepted style of patient-doctor communication (which is the basis of a new active role of patients in monitoring symptoms and evolution of the disease). The sharing of this new approach could be extended from patients to healthy people, obtaining tools for a real preventive medicine: a large amount of data could be gained, stored, and analyzed outside the sanitary structures, contributing to a low-cost approach to health.

    The goal of this session is to bring together interdisciplinary experts to develop (discuss about) these topics:

    (a) decision support systems for the analysis of collected data;

    (b) customized monitoring based on the acuteness of the disease;

    (c) integration of collected data with E-Health systems;

    (d) attitudes of doctors and sanitary staff;

(e) patient-doctor communication;

(f) involvement of patients and of their relatives and care-givers;

(g) digital divide as an obstacle/hindrance;

(h) alternative measurements on the effectiveness of telemedicine (quality of life of patients and caregivers, etc.)

(i) mobile versus home monitoring (sensors, signal transmissions, etc.)

(j) technology simplification (auto-calibrating systems, patient interface, physician interface, bio-feedback for improving learning)

The session will also have the ambition of constituting a team of interdisciplinary research, spread over various countries, as a possible basis for effective participation in European calls.

Organizers: Maria Francesca Romano, Institute of Economics, Scuola Superiore Sant'Anna, Pisa (Italy).

Giorgio Buttazzo, Institute of Communication, Information and Perception Technologies (TeCIP), Scuola Superiore Sant'Anna, Pisa (Italy).

11. **SS11A: High Performance Computing in Bioinformatics, Computational Biology, and Computational Chemistry**

The goal of this special session is to explore the use of emerging parallel computing architectures as well as High-Performance Computing systems (Supercomputers, Clusters, Grids) for the simulation of relevant biological systems and for applications in Bioinformatics, Computational Biology, and Computational Chemistry. We welcome papers, not submitted elsewhere for review, with a focus on topics of interest ranging from but not limited to:

(a) Parallel stochastic simulation.

(b) Biological and Numerical parallel computing.

(c) Parallel and distributed architectures.

(d) Emerging processing architectures (e.g. GPUs, Intel Xeon Phi, FPGAs, mixed CPU-GPU, or CPU-FPGA, etc).

(e) Parallel Model checking techniques.

(f) Parallel algorithms for biological analysis.

(g) Cluster and Grid Deployment for system biology.

(h) Biologically inspired algorithms.

(i) Application of HPC developments in Bioinformatics, Computational Biology, and Computational Chemistry.

Organizers: Dr. Horacio Perez-Sanchez, Dr. Afshin Fassihi and Dr. Jose M. Cecilia, Universidad Católica San Antonio de Murcia (UCAM), (Spain).

12. **SS11B: High-Performance Computing for Bioinformatics Applications**

This Workshop has a focus on interdisciplinary nature and is designed to attract the participation of several groups including Computational Scientists, Bioscientists, and the fast growing group of Bioinformatics, researchers. It is primarily intended for computational scientists who are interested in Biomedical Research and the impact of high-performance computing in the analysis of biomedical data and in advancing Biomedical Informatics. Bioscientists with some background in computational concepts represent another group of intended participants. The interdisciplinary group of research groups with interests in Biomedical Informatics in general and Bioinformatics in particular will likely be the group attracted the most to the workshop. The Workshop topics include (but are not limited to) the following:

(a) HPC for the Analysis of Biological Data.
(b) Bioinformatics Tools for Health Care.
(c) Parallel Algorithms for Bioinformatics Applications.
(d) Ontologies in biology and medicine.
(e) Integration and analysis of molecular and clinical data.
(f) Parallel bioinformatics algorithms.
(g) Algorithms and Tools for Biomedical Imaging and Medical Signal Processing.
(h) Energy Aware Scheduling Techniques for Large-Scale Biomedical Applications.
(i) HPC for analyzing Biological Networks.
(j) HPC for Gene, Protein/RNA Analysis, and Structure Prediction.

For more information, you can see the Call for Paper for this special session.
Organizers: Prof. Hesham H. Ali, Department of Computer Science, College of Information Science and Technology, University of Nebraska at Omaha (EEUU).
Prof. Mario Cannataro, Informatics and Biomedical Engineering University "Magna Graecia" of Catanzaro (Italy).

13. **SS12: Advances in Drug Discovery**
We welcome papers, not submitted elsewhere for review, with a focus in topics of interest ranging from but not limited to:
(a) Target identification and validation.
(b) Chemoinformatics and Computational Chemistry: Methodological basis and applications to drug discovery of: QSAR, Docking, CoMFA-like methods, Quantum Chemistry and Molecular Mechanics (QM/MM), High-performance Computing (HPC), Cloud Computing, Biostatistics, Artificial Intelligence (AI), Machine Learning (ML), and Bio-inspired Algorithms like Artificial Neural Networks (ANN), Genetic Algorithms, or Swarm Intelligence.
(c) Bioinformatics and Biosystems: Methodological basis and applications to drug design, target or biomarkers discovery of: Alignment tools, Pathway analysis, Complex Networks, Nonlinear methods, Microarray analysis, Software, and Web servers.
(d) High Throughput Screening (HTS) of drugs; Fragment-Based Drug Discovery; Combinatorial chemistry, and synthesis.
Organizers: Dr. Horacio Perez-Sanchez and Dr. Afshin Fassihi, Universidad Católica San Antonio de Murcia (UCAM), (Spain).

14. **SS13: Deciphering the human genome**
Accomplishment of "1000 Genomes Project" revealed immense amount of information about variation, mutation dynamics, and evolution of the human DNA sequences. These genomes have been already used in a number

of bioinformatics studies, which added essential information about human populations, allele frequencies, local haplotype structures, distribution of common and rare genetic variants, and determination of human ancestry and familial relationships. Humans have modest intra-species genetic variations among mammals. Even so, the number of genetic variations between two persons from the same ethnic group is in the range of 3.4–5.2 million. This gigantic pool of nucleotide variations is constantly updating by 40–100 novel mutations arriving in each person. Closely located mutations on the same DNA molecule are linked together forming haplotypes that are inherited as whole units and span over a considerable portion of a gene or several neighboring gene. An intense intermixture of millions of mutations occurs in every individual due to frequent meiotic recombinations during gametogenesis. Scientists and doctors are overwhelmed with this incredible amount of information revealed by new-generation sequencing techniques. Due to this complexity, we encountered significant challenges in deciphering genomic information and interpretation of genome-wide association studies.

The goal of this session is to discuss novel approaches and algorithms for processing of whole-genome SNP datasets in order to understand human health, history, and evolution.

Organizer: Alexei Fedorov, Ph.D, Department of Medicine, Health Science Campus, The University of Toledo (EEUU).

15. **SS14: Ambient Intelligence for Bioemotional Computing**

    Emotions have a strong influence on our vital signs and on our behavior. Systems that take our emotions and vital signs into account can improve our quality of life. The World Health Organization (WHO) characterizes a healthy life first of all with the prevention of diseases and secondly, in the case of the presence of disease, with the ability to adapt and self-manage. Smart measurement of vital signs and of behavior can help to prevent diseases or to detect them before they become persistent. These signs are key to obtain individual data relevant to contribute to this understanding of healthy life.

    The objective of this session is to present and discuss smart and unobtrusive methods to measure vital signs and capture emotions of the users and methods to process these data to improve their behavior and health.

    Organizers: Prof. Dr. Natividad Martinez, Internet of Things Laboratory, Reutlingen University (Germany).
    Prof. Dr. Juan Antonio Ortega, University of Seville (Spain).
    Prof. Dr. Ralf Seepold, Ubiquitous Computing Lab, HTWG Konstanz (Germany).

In this edition of IWBBIO, we are honored to have the following invited speakers:

1. Prof. Xavier Estivill, Genomics and Disease group, Centre for Genomic Regulation (CRG), Barcelona (SPAIN).
2. Prof. Alfonso , Structural Computational Biology group, Spanish National Cancer Research Center (CNIO), Madrid (SPAIN).

3. Prof. Patrick Aloy, Structural Bioinformatics and Network Biology group, Institute for Research in Biomedicine (IRB), Barcelona (SPAIN).

It is important to note that for the sake of consistency and readability of the book, the presented papers are classified under 21 chapters. The organization of the papers is in two volumes arranged basically following the topics list included in the call for papers. The first volume (LNBI 9043), entitled "Advances on Computational Intelligence. Part I" is divided into seven main parts and includes the contributions on:

1. Bioinformatics for healthcare and diseases.
2. Biomedical Engineering.
3. Biomedical image analysis.
4. Biomedical signal analysis.
5. Computational genomics.
6. Computational proteomics.
7. Computational systems for modeling biological processes.

In the second volume (LNBI 9044), entitled "Advances on Computational Intelligence. Part II" is divided into 14 main parts and includes the contributions on:

1. E-Health.
2. Next generation sequencing and sequence analysis.
3. Quantitative and Systems Pharmacology.
4. Hidden Markov Model (HMM) for Biological Sequence Modeling.
5. Biological and bio-inspired dynamical systems for computational intelligence.
6. Advances in Computational Intelligence for Bioinformatics and Biomedicine.
7. Tools for Next Generation Sequencing data analysis.
8. Dynamics networks in system medicine.
9. Interdisciplinary puzzles of measurements in biological systems.
10. Biological Networks: Insight from interactions.
11. Toward an effective telemedicine: an interdisciplinary approach.
12. High-Performance Computing in Bioinformatics, Computational Biology, and Computational Chemistry.
13. Advances in Drug Discovery.
14. Ambient Intelligence for Bioemotional Computing.

This third edition of IWBBIO was organized by the Universidad de Granada together with the Spanish Chapter of the IEEE Computational Intelligence Society. We wish to thank our main sponsor BioMed Central, E-Health Business Development BULL (España) S.A., and the institutions Faculty of Science, Dept. Computer Architecture and Computer Technology and CITIC-UGR from the University of Granada for their support and grants. We also wish to thank the Editor-in-Chief of different international journals for their interest in editing special issues from the best papers of IWBBIO.

We would also like to express our gratitude to the members of the different committees for their support, collaboration, and good work. We especially thank the Local Committee, Program Committee, the Reviewers, and Special Session Organizers. Finally, we want to thank Springer, and especially Alfred Hoffman and Anna Kramer for their continuous support and cooperation.

April 2015   Francisco Ortuño
                                                              Ignacio Rojas

Organization

Program Committee

Jesus S. Aguilar	Universidad Pablo de Olavide, Spain
Carlos Alberola	Universidad de Valladolid, Spain
Hesham H. Ali	University of Nebraska at Omaha, USA
René Alquézar	Universitat Politécnica de Catalunya, Barcelona TECH, Spain
Rui Carlos Alves	University of Lleida, Spain
Eduardo Andrés León	Spanish National Cancer Center, Spain
Antonia Aránega	University of Granada, Spain
Saúl Ares	Spanish National Center for Biotechnology, Spain
Rubén Armañanzas	Universidad Politécnica de Madrid, Spain
Joel P. Arrais	University of Coimbra, Portugal
O. Bamidele Awojoyogbe	Federal University of Technology, Minna, Nigeria
Jaume Bacardit	University of Newcastle, Australia
Hazem Bahig	University of Haíl, Saudi Arabia
Pedro Ballester	Inserm, France
Oresti Baños	Kyung Hee University, Korea
Ugo Bastolla	Center of Molecular Biology "Severo Ochoa", Spain
Steffanny A. Bennett	University of Ottawa, Canada
Alfredo Benso	Politecnico di Torino, Italy
Concha Bielza	Universidad Politécnica de Madrid, Spain
Armando Blanco	University of Granada, Spain
Ignacio Blanquer	Universidad Politécnica de Valencia, Spain
Giorgio Buttazzo	Scuola Superiore Sant'Anna, Italy
Gabriel Caffarena	Universidad Politécnica de Madrid, Spain
Mario Cannataro	University Magna Graecia of Catanzaro, Italy
Carlos Cano	University of Granada, Spain
Rita Casadio	University of Bologna, Italy
Jose M. Cecilia	Universidad Católica San Antonio de Murcia, Spain
M. Gonzalo Claros	University of Málaga, Spain

Darrell Conklin	Universidad del País Vasco/Euskal Herriko Unibertsitatea, Spain
Clare Coveney	Nottingham Trent University, UK
Miguel Damas	University of Granada, Spain
Guillermo de La Calle	Universidad Politécnica de Madrid, Spain
Javier De Las Rivas	CSIC, CIC, USAL, Salamanca, Spain
Joaquin Dopazo	Centro de Investigación Principe Felipe Valencia, Spain
Hernán Dopazo	CIPF, Spain
Werner Dubitzky	University of Ulster, UK
Khaled El-Sayed	Modern University for Technology and Information, Egypt
Christian Esposito	ICAR-CNR, Italy
Afshin Fassihi	Universidad Católica San Antonio de Murcia, Spain
Jose Jesús Fernandez	University of Almeria, Spain
Jean-Fred Fontaine	Max Delbrück Center for Molecular Medicine, Germany
Xiaoyong Fu	Western Reserve University, USA
Razvan Ghinea	University of Granada, Sapin
Jesus Giraldo	Universitat Autònoma de Barcelona, Spain
Humberto Gonzalez	University of Basque Country, Spain
Daniel Gonzalez Peña	Bioinformatics and Evolutionary Computing, Spain
Concettina Guerra	Georgia Institute of Technology, USA
Christophe Guyeux	IUT de Belfort-Montbéliard, France
Michael Hackenberg	University of Granada, Spain
Luis Javier Herrera	University of Granada, Spain
Lynette Hirschman	MITRE Corporation, USA
Michelle Hussain	University of Salford, UK
Andy Jenkinson	European Bioinformatics Institute, UK
Craig E. Kapfer	KAUST, Saudi Arabia
Narsis Aftab Kiani	Karolinska Institute, Sweden
Ekaterina Kldiashvili	New Vision University/Georgian Telemedicine Union, Georgia
Tomas Koutny	University of West Bohemia, Czech Republic
Natalio Krasnogor	University of Newcastle, Australia
Marija Krstic-Demonacos	University of Salford, UK
Sajeesh Kumar	University of Tennessee, USA
Pedro Larrañaga	Universidad Politecnica de Madrid, Spain
Jose Luis Lavin	CIC bioGUNE, Spain
Rémi Longuespée	University of Liège, Belgium

Omer F. Rana	Cardiff University, UK
Jairo Rocha	University of the Balearic Islands, Spain
Fernando Rojas	University of Granada, Spain
Ignacio Rojas	University of Granada, Spain
Maria Francesca Romano	Scuola Superiore Sant'Anna, Italy
Gregorio Rubio	Universitat Politècnica de València, Spain
Antonio Rueda	Genomics and Bioinformatics Platform of Andalusia, Spain
Michael Sadovsky	Siberian Federal University, Russia
Yvan Saeys	Ghent University, Belgium
Maria Jose Saez	University of Granada, Spain
José Salavert	European Bioinformatics Institute, UK
Carla Sancho Mestre	Universitat Politècnica de València, Spain
Vicky Schneider	The Genome Analysis Centre, UK
Jean-Marc Schwartz	University of Manchester, UK
Ralf Seepold	HTWG Konstanz, Germany
Jose Antonio Seoane	University of Bristol, UK
Istvan Simon	Research Centre for Natural Sciences, Hungary
Richard Sinnott	University of Glasgow, UK
Mohammad Soruri	University of Birjand, Iran
Prashanth Suravajhala	BioClues.org, India
Yoshiyuki Suzuki	Tokyo Metropolitan Institute of Medical Science, Tokyo
Li Teng	University of Iowa, USA
Pedro Tomas	INESC-ID, Portugal
Carolina Torres	University of Granada, Spain
Oswaldo Trelles	University of Málaga, Spain
Paolo Trunfio	University of Calabria, Italy
Olga Valenzuela	University of Granada, Spain
Alfredo Vellido	Universidad Politécnica de Cataluña, Spain
Julio Vera	University of Rostock, Germany
Renato Umeton	CytoSolve Inc., USA
Jan Urban	University of South Bohemia, Czech Republic

Additional Reviewers

Agapito, Giuseppe	Gonzalez-Abril, Luis
Alquezar, Rene	Ielpo, Nicola
Asencio Cortés, Gualberto	Julia-Sape, Margarida
Belanche, Lluís	König, Caroline Leonore
Calabrese, Barbara	Milano, Marianna
Caruso, Maria Vittoria	Mir, Arnau
Cárdenas, Martha Ivón	Mirto, Maria
Fernández-Montes, Alejandro	Navarro, Carmen

Ortega-Martorell, Sandra
Politano, Gianfranco Michele Maria
Ribas, Vicent
Rychtarikova, Renata

Sarica, Alessia
Tosi, Alessandra
Vellido, Alfredo
Vilamala, Albert

Table of Contents – Part I

Bioinformatics for Healthcare and Diseases

Biomedical Engineering

Biomedical Image Analysis

Biomedical Signal Analysis

Computational Genomics

Computational Systems for Modelling Biological Processes

Table of Contents – Part II

eHealth

Next Generation Sequencing and Sequence Analysis

Quantitative and Systems Pharmacology

Hidden Markov Model (HMM) for Biological Sequence Modeling

Advances in Computational Intelligence for Bioinformatics and Biomedicine

Tools for Next Generation Sequencing Data Analysis

Dynamics Networks in System Medicine

High Performance Computing in Bioinformatics, Computational Biology and Computational Chemistry

Advances in Drug Discovery

Ambient Intelligence for Bioemotional Computing

A Segmentation-Free Model for Heart Sound Feature Extraction

Hai-Yang Wang, Guang-Pei Li, Bin-Bin Fu, Hao-Dong Yao,
and Ming-Chui Dong

Electrical and Computer Department, Faculty of Science and Technology,
University of Macau, Macau S.A.R., China
dylanchriswang@hotmail.com, lee.gp@foxmail.com,
ariespleo51@gmail.com, yaohaodong0108@126.com, mcdong@umac.mo

Abstract. Currently, the fatality of cardiovascular diseases (CVDs) represents one of the global primary healthcare challenges and necessitates broader population checking for earlier intervention. The traditional auscultation is cost-effective and time-saving for broader population to diagnose CVDs early. While many approaches in analyzing heart sound (HS) signal from auscultation have been utilized successfully, few studies are focused on acoustic perspective to interpret the HS signal. This paper proposes a segmentation-free model that can interpret HS effectively, which aligns engineering with clinical diagnostic basis and medical knowledge much more. The presented model stems from timbre analysis model but is adapted for HS signal. The relevant theoretical analysis and simulation experiments indicate that the proposed method has good performance in HS analysis.

Keywords: Acoustics, Feature Extraction, Spectral Centroid, Temporal Centroid.

1 Introduction

The American Heart Association studies predict that 40.5% of US population will have some form of CVD by 2030 with associated indirect cost reaching $ 276 billion [1]. CVD affects individuals in their peak mid-life years disrupting the future of the families dependent on them and undermining the development of nations by depriving valuable human resources in their most productive years. An early detection and prevention care is much significant and is a big challenge on the global scale. Clinical check of CVDs is costly and prolonged, such as echocardiogram (ECHO), electrocardiography (ECG), etc. Besides the cost, they are inconvenient and patients themselves cannot often be checked away from clinics. Traditional auscultation provides a possible solution. Based on HSs, auscultation could also find signs of pathologic conditions, like many cardiac abnormalities including valvular heart disease, congestive heart failure and congenital heart lesions etc. Unfortunately, the auscultation requires extensive training and experience to perform effectively. This leads the research on HS signal interpretation and analysis in order to provide faster, better and more cost-effective healthcare support for the victims of CVDs.

F. Ortuño and I. Rojas (Eds.): IWBBIO 2015, Part I, LNCS 9043, pp. 1–7, 2015.
© Springer International Publishing Switzerland 2015

HS signals are typically dynamic, complex and non-stationary. For analysis of HS signal, many approaches, such as wavelet decomposition and reconstruction method [2-4], short time Fourier transform (STFT) method [5, 6], and S-transform method [7, 8], have been proposed in literatures Most will solely analyze the time frequency domain for feature extractions. This leads to a conclusion that feature extractions are less aligned with medical knowledge.

From acoustic perspective, a doctor diagnoses the CVDs generally through a stethoscope. Common descriptive terms about what it sounds like in auscultation include rumble, blowing, machinery, scratchy, harsh, gallop, ejection, click, drum, cymbal, etc. From acoustic perspective to analyze such a bio-signal, its main advantage is that engineering is aligned with clinical diagnosis. For instance, it is described by a doctor that continuous machinery sound is heard on the left sternal border between the second and third ribs in auscultation indicating the patient's ductus arteriosus. The melfrequency cepstral coefficient (MFCC) method has been utilized for HS feature extraction [9-13], which is based on the theory that human audition spaces linearly at low frequency band and logarithmically at high frequency. Unfortunately, it is a segmentation-based feature extraction technique and its effect suffers from segmentation error greatly.

Instead, a segmentation-free model stems from classical timbre analysis model and is creatively proposed for HS analysis. Timbre, which embodies the texture of acoustic source, is a significant attribute among timbre, loudness and pitch in acoustics. When the pathology changes in heart or blood vessels, different timbre information can be sampled, from which can trace back the reason causing such a timbre. However, current timbre analysis model mainly aims at recognizing different music instruments, and only few literature reports their explorations on HS feature detection. In this paper, the proposed model is elaborated and its performance for characterizing features with diagnostic significance from HS is evaluated through experiments.

2   Methodology

2.1   Feature Extraction

Psychoacoustic literatures conclude that the dimensions of timbre analysis model can be categorized as fundamental frequency, log attack time *(LT)*, temporal centroid *(TC)*, spectral centroid *(SC)*, harmonics, etc. [13-17].

In this paper, *SC*, *TC* and *LT* are selected to construct feature set for HS signal. *SC* reflects the signal power distribution in frequency domain. It is computed as the power spectrum weighted average of the frequency in the power spectrum shown in Eq. (1) and (2).

$$S(k) = \sqrt{\sum_{i=1}^{M} P_i(k)/M} \tag{1}$$

$$SC = \frac{\sum_{K=1}^{NFFT} f(k) \cdot S(k)}{\sum_{K=1}^{NFFT} S(k)} \tag{2}$$

where M is the total number of frames in a sound segment, $P_i(k)$ is the kth power spectrum coefficient in the ith frame, $f(K)$ is the kth frequency bin.

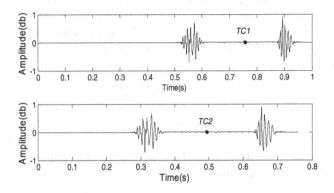

Fig. 1. The same type of HS signal with different TC values caused by different locations of starting point

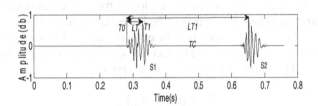

Fig. 2. False $LT1$ is mistaken as the LT while S2 is sometimes stronger

TC reflects the strength of the signal distribution in time domain. It is the geometric centroid of the signal in temporal domain. However, for the traditional timbre analysis model, TC is affected by different locations of starting point (Fig. 1). HS signals A and B are regarded as the same type of HS although a delay in time dominion exists. To solve this problem, TC is defined as:

$$TC = \frac{\sum_1^N \frac{n}{s_r} A(n)}{\sum_1^N A(n)} - T_0 \tag{3}$$

where s_r is the sample rate, T_0 is the time when the signal starts.

LT called rising time in other literatures is of great use in feature extraction on morphology. It indicates the very short period of time that elapses before the sound has formed its spectrum. How timbres are identified usually depends on log attack time. It is defined as the logarithm of duration between the time where signal starts and the time where it reaches its stable part. Conventional computation of LT is on basis of HS signal's envelope, which is typically computed through Hilbert transform. Virtually this will lead to heavy computational load so as to impair the proposed model's promising of pervasive computing. Instead, according to physical definition of *LT*, it can be achieved by computing the period between the minimum and the maximum of the HS signal amplitude. However, when the HS signal is processed as a whole without envelope, the situation appears that the false *LT1* is mistaken as the *LT*, while S2 is sometimes stronger (Fig. 2). As *TC* of HS signal is known as aforementioned, the *T1* value is the supermum of set {*0<t<TC*}. Then the mistaken *LT* is avoided as formula (4) and (5) show.

$$LT = log_{10}(T_1 - T_0) \tag{4}$$

$$T_1 = sup\{0 < t < TC\} \tag{5}$$

where T_0 is the time when signal starts. Here we use the time while the signal amplitude is 0.001 db. T_1 is the supermum of the set of time between 0 and *TC*.

2.2   Classification

The *k*-nearest neighbor algorithm (KNN) is a non-parametric method used for classification [18]. The input consists of the *k* closest training examples in the feature space. The output is a class membership. The HS is classified by a majority vote of its neighbors based on distances, with the object being assigned to the class most common among its *k* nearest neighbors. Distance comes in many flavors. The distance d_{ij} between any two HSs *i* and *j* thus takes the following form:

$$d_{ij} = ||x_{in} - x_{jn}||_p \tag{6}$$

where x_i is the coordinate value of HS *i* on dimension *n*, *n* is the total number of dimensions in the model. If $p = 3$, the distance is the Minkowski metric. Here for the HS, $p = 2$, a simple Euclidean distance is utilized and this is what usually used in timbre studies. As the model has 3 different physics dimensions, so the data are normalized before classification.

3   Test Results

3.1   Setting of Experiments

The experiment is built on a set of HSs obtained from on-line benchmark database provided by eGeneral Medical Inc., USA [19] as well as the datasets recorded at hospital with an electronic stethoscope. The sounds are digitized using a sampling frequency of 44,100 Hz. The spectrograms are computed with Fast Fourier Transform (FFT). The sound types are comprised of ejection, rumble, gallop, and normal HS. Furthermore, the origins of all recorded HSs are from subjects of different ages, genders and physical characteristics at the time of recording. The heart rates in the dataset vary from 60 to 120 beats per minute.

3.2 Results and Discussion

The extracted features are showed as 3-D diagram with each dimension representing individual selected timbre attribute respectively (Fig. 3). It is easily observed that the 4 types of HS demonstrate great divergence spatially.

Table 1. Confusion Matrix of Classification Results

		Ejection	Rumble	Gallop	Normal
I	Ejection	3		3	
	Rumble		6		
	Gallop	1	1	4	
	Normal				6
II	Ejection	5	1		
	Rumble		6		
	Gallop			6	
	Normal				6

Table 2. Comparison Results of Two Models' Classification Accuracy

	I		II	
	TPR (%)	FPR (%)	TPR (%)	FPR (%)
Ejection	50	5	83.3	0
Rumble	100	5	100	5.6
Gallop	66.7	16.7	100	0
Normal	100	0	100	0
ACC(%)	79.1		95.8	

To evaluate the performance of feature extraction, both timbre analysis model [20] (model-I) and this proposed model (model-II) are applied to the same classifier by utilizing the real HS datasets. The results are illustrated as confusion matrix (Table 1). Table 2 shows the comparison results of two models' classification accuracy. Judged by the data, model-I shows poor performance on distinguishing ejection type and gallop type from other types of HS. Comparatively speaking, the results of model-II are quite encouraging. All of the test results are well categorized by model-II except one case in which one ejection HS signal is mistaken as rumble HS signal.

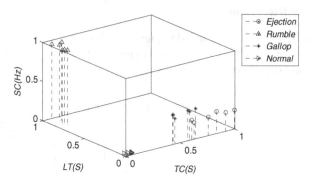

Fig. 3. Feature extraction results of 4 types of HS

Concisely, for ejection type and gallop type HS signal, the *SC* value extracted by model-I show high discrimination index however there is overlap between *LT* and *TC* values bringing about misclassification of these two types. By contrast, the algorithm of *LT* and *TC* is enhanced in model-II by taking HS's intrinsic character into account, and solve the problem that there exists a delay in time domain for real HS signal. It is believed to be the reason why the proposed model is superior to the conventional timbre analysis model.

4   Conclusion

A new segmentation-free model for the HS feature extraction is depicted, which adapts three timbre attributes to modeling HS signal acoustically so as to align engineering with medical knowledge much more. Furthermore, it voids computation complexity and mistakes introduced by segmentation. Experimental results indicate that it has better performance in feature extraction and robustness for HS classification than conventional timbre analysis model. Hereby, the proposed model shows great potential for pervasive health monitoring with intelligent auscultation function.

Acknowledgements. This work was supported in part by the Research Committee of University of Macau under Grant No. MYRG2014-00060-FST, and in part by the Science and Technology Development Fund (FDCT) of Macau under Grant No. 016/2012/A1, respectively.

References

1. Heidenreich, P.A.: Forecasting the Future of Cardiovascular Disease in the United States (2011)
2. Dokur, Z.: Dimension Reduction by a Novel Unified Scheme Using Divergence Analysis and Genetic Search. Digital Signal Processing 19, 521–531 (2009)

3. Liang, H., Hartimo, I.: Feature extraction algorithm based on wavelet packet decomposition for heart sound signals. Presented at the Proceedings of the IEEE-SP International Symposium (1998)
4. Vikhe, P., Hamde, S., Nehe, N.: Wavelet Transform Based Abnormality Analysis of Heart Sound. In: Proc. ACT 9th. International Conf. Advances in Computing, Control, & Telecommunication Technologies, pp. 367–371 (2009)
5. Kao, W., Wei, C., Liu, J., Hsiao, P.: Automatic Heart Sound Analysis with Short-time Fourier Transforms and Support Vector Machines. In: Proc. MWSCAS 9th IEEE International Midwest Symp. Circuits and Systems, August 2-5, pp. 188–191 (2009)
6. Djebbari, A., Bereksi, F.: Short-time Fourier Transform Analysis of the Phonocardiogram Signal. In: Proc. 7th IEEE International Conf. Electronics, Circuits and Systems, pp. 844–847 (2000)
7. Hadi, H.M., Mashor, M.Y., Suboh, M.Z., Mohamed, M.S.: Classification of Heart Sound Based on S-transform and Neural Network. Presented in 10th International Conference, 10-13 (2010)
8. Varanini, M., De Paolis, G., Emdin, M., Macerata, A., Pola, S., Cipriani, M., Marchesi, C.: Spectral Analysis of Cardiovascular Tme Series by theS-Transform. In: Computers in Cardiology, pp. 383–386 (1997)
9. Sunita, C., et al.: A computer-aided MFCC-based HMM system for automatic auscultation. Computers in Biology and Medicine 38, 221–233 (2008)
10. Wenjie, F., Yang, X., Wang, Y.: Heart sound diagnosis based on DTW and MFCC. In: Proc. 3rd International Congress, vol. 6, pp. 2920–2923 (2010)
11. Kamarulafiza, I., et al.: Heart Sound Analysis Using MFCC and Time Frequency Distribution. In: World Congress on Medical Physics and Biomedical Engineering 2006. Springer, Heidelberg (2007)
12. Grey, J.: An exploration of musical timbre. Ph.D. dissertation (Dept. of Music Report No. STAN-M-2, Stanford University, CA) (1975) (unpublished)
13. Lakatos, S.: A Common Perceptual Space for Harmonic and Percussive Timbres. Perception & Psychophysics 62, 1426–1439 (2000)
14. McAdams, S.: Perspectives on the Contribution of Timbre to Music Structure. Computer Music Journal 23, 85–102 (1999)
15. Gianpaolo Evangelista, H.V.: Separation of Harmonic Instruments with Overlapping Partialsin Muti-channel Mixtures. In: IEEE Workshop an Applications of Signal Prwessinp to Audio and Acoustics (2003)
16. Jensen, K.: Timbre Models of Musical Sounds. Nr. 99/7 (1999) ISSN 0107-8283
17. Caclin, A., McAdams, S., Bennett, K.: Smith, S.Winsberg.: Acoustic Correlates of Timbre Space Dimensions: A Confirmatory Study Using Synthetic Tones. Acoustical of Society Amecia 118, 471 (2005)
18. Altman, N.S.: An Introduction to Kernel and Nearest-Neighbor Nonparametric Regression. The American Statistician 46(3), 175–185 (1992)
19. Egeneral Medical, http://www.egeneralmedical.com/listohearmur.html
20. Wang, H.Y., Li, G.P., Fu, B.B., Huang, J., Dong, M.C.: Multidimensional Feature Extraction Based on Timbre Model for Heart Sound Analysis. International Journal of Bioscience, Biochemistry and Bioinformatics 4(5), 318–321 (2014) ISSN: 2010-3638

A Semantic Layer for Unifying and Exploring Biomedical Document Curation Results

Pedro Sernadela[1,*], Pedro Lopes[1], David Campos[2], Sérgio Matos[1],
and José Luís Oliveira[1]

[1] DETI/IEETA, Universidade de Aveiro,
Campus Universitário de Santiago, 3810-193 Aveiro, Portugal
{sernadela,pedrolopes,aleixomatos,jlo}@ua.pt
[2] BMD Software,
Rua Calouste Gulbenkian n. 1, 3810-074 Aveiro, Portugal
david.campos@bmd-software.com

Abstract. Tackling the ever-growing amount of specialized literature in the life sciences domain is a paramount challenge. Various scientific workflows depend on using domain knowledge from resources that summarize, in structured form, validated information extracted from scientific publications. Manual curation of these data is a demanding task, and latest strategies use computerized solutions to aid in the analysis, extraction and storage of relevant concepts and their respective attributes and relationships. The outcome of these complex document curation workflows provides valuable insights into the overwhelming amount of biomedical information being produced. Yet, the majority of automated and interactive annotation tools are not open, limiting access to knowledge and reducing the potential scope of the manually curated information. In this manuscript, we propose an interoperable semantic layer to unify document curation results and enable their proper exploration through multiple interfaces geared towards bioinformatics developers and general life sciences researchers. This enables a unique scenario where results from computational annotation tools are harmonized and further integrated into rich semantic knowledge bases, providing a solid foundation for discovering knowledge.

Keywords: Document curation, text-mining, semantic web, knowledge discovery, data integration.

1 Introduction

The publication of articles, books and technical reports, particularly concerning the scientific biomedical literature, is growing at a high rate. In the MEDLINE bibliographic database alone, the continued exponential growth [1] resulted in a total of 21 million references to journal articles in fields related to the life sciences, until 2013. This evolution has made it harder than ever for researchers to find and assimilate all the information relevant to their research, and promoted various efforts to summarize

---

* Corresponding author.

F. Ortuño and I. Rojas (Eds.): IWBBIO 2015, Part I, LNCS 9043, pp. 8–17, 2015.
© Springer International Publishing Switzerland 2015

the knowledge scattered across multiple publications and store it in structured form. However, creating and updating these valuable resources is a time-demanding and expensive task that requires highly trained data curators.

Computerized text-mining solutions have been increasingly applied to assist bio-curators, allowing the extraction of relevant information regarding biomedical concepts such as genes, proteins, chemical compounds or diseases [2], and thus reducing curation times and cost [3]. However, few information extraction systems carry out a complete analysis to take advantage of the produced results. On the one hand, most of the available text-mining tools successfully identify relevant data in documents, usually providing these results through a text file using a specific format, but do not have the capabilities to further organize or exploit this information. Assisted curation tools, on the other hand, typically follow an internal data structure to acquire, organize and store the curated information. However, since these tools are usually developed for specific curation requirements, their data structure is usually not interoperable, hindering integration with external knowledge bases and full exploitation of the curated knowledge. In this perspective, novel mechanisms are required to explore and sustain such text-mining outcomes.

Nowadays, the Semantic Web (SW) paradigm [4] encompasses a broad set of modern technologies that are a perfect fit for the intrinsic interrelationships within the life sciences field. Technologies such as RDF (Resource Description Framework), OWL (Web Ontology Language) and SPARQL (SPARQL Protocol and RDF Query Language) are technical recommendations of the World Wide Web Consortium that facilitate the deployment of advanced algorithms for searching and mining large integrated datasets [5]. SW standards are used to tackle traditional data issues such as heterogeneity, distribution and interoperability, providing an interconnected network of knowledge. Furthermore, ontologies provide an excellent way to represent biomedical resources, especially considering the complexity of the domain. For instance, Gene Ontology [6] demonstrates that ability by describing gene products in any organism with structured vocabularies. For these reasons, Semantic Web technologies are increasingly being adopted to link, exploit and deliver knowledge for both machine and human consumption, creating large and complex networks.

In this document, we propose an interoperable semantic layer to unify text-mining results originated from different tools, information extracted by curators, and baseline data already available in reference knowledge bases, to enable their proper exploration.

2 Background

The huge amount of information currently available in the biomedical scientific literature demands the application of Information Extraction (IE) tools to automatically extract information from the data, induce new knowledge and understand its meaning. To accomplish that, text-mining techniques are needed to convert unstructured data into a unambiguous representation of concepts and relations [7], targeting a specific goal and domain.

Usually, state-of-the-art solutions follow a combination of pre-defined and sequential processes in order to apply and perform biomedical information extraction. Natural Language Processing (NLP) techniques [8] are commonly applied as pre-processing tasks to split documents text into meaningful components, such as sentences and tokens, assign grammatical categories (a process named part-of-speech tagging), or even apply linguistic parsing to identify the structure of each sentence. Next, concept recognition methods are employed, which involve Named Entity Recognition (NER) [2] to detect the concept mentions, and normalization processes [9] to distinguish and attribute unique identifiers to each detected entity name. Complete biomedical text-mining solutions also apply relation-mining techniques to identify the events and entity relations that make up complex biological networks. Conventional solutions are focused on investigating and extracting direct associations between two concepts (e.g. genes, proteins, drugs, etc.) [10]. The study of these associations has generated plenty of interest, especially in respect to protein-protein interactions (PPIs) [11] [12], drug-drug interactions (DDIs) [13], and relations between chemicals and target genes. Other solutions, such as FACTA [14], are targeted at uncovering implicit heterogeneous connections between different types of concepts.

Recently, interactive text-mining solutions have gained more attention due to the added benefits of including automatically extracted information in the manual curation processes. With these solutions, the curation time is improved and possible mistakes from computational information extraction results are minimized. Brat [15], MyMiner [16], Argo [17] and Egas [18] are state-of-the-art interactive solutions, each providing different features but with the same goal: simplify the annotation process.

Due to the complexity and diverse challenges addressed by such systems, different data models are adopted. This fragmentation of formats is not desirable, and text-mining research should encourage the use of modern data exchange standards, allowing researchers to leverage a common layer of interoperability. A large number of projects have been undertaken with the purpose of enabling or enhancing the prospects for interoperability and reusability of text-mined information. Recently, BioC [19] has emerged has a community-supported format for encoding and sharing textual data and annotations. This minimalist approach envisages to simplify data reuse and sharing, and to achieve interoperability for the different text processing tasks in biomedical text-mining. However, BioC is still a verbose format, and data exchange methods are simply based on sharing a common XML file.

In recent years, Semantic Web became the *de facto* paradigm for data integration at a web-scale, focused on the semantics and the context of data [20]. It enables the creation of rich networks of linked data, establishing new possibilities to retrieve and discover knowledge (e.g. reasoning). Moreover, Semantic Web makes data integration and interoperability a standard feature, enabling the representation of data elements in the web as real-world entities and links containing logical relations between those entities. Currently, there are several systems, such as COEUS [21], Sesame [22] or SADI [23], that explore the potential behind Semantic Web technology, enabling the quick creation of new knowledge bases to store and/or deliver information to end-users. For instance, the COEUS framework includes advanced data integration tools, base ontologies, a web-oriented engine with interoperability features, such as REST (Representational State Transfer) services, a SPARQL endpoint, Linked Data publication interface, as well as an optional nanopublication extension [24].

Strategies that combine the benefits of text-mining methods with Semantic Web technologies represent a growing trend that allows the establishment of curated databases with improved availability [25]. Coulet et al. [26] provide an overview of such strategies and propose a specific approach to make the integration and publication of heterogeneous text-mined pharmacogenomic relationships on the Semantic Web. Mendes et al. [27] provide another example, describing a system that automatically annotates text documents and translates the outputs into the DBpedia namespace. These strategies represent ongoing efforts striving towards the integration of knowledge present in text documents into the Linked Open Data network. Ciccarese et al. [28] provide another interesting integration system by presenting DOMEO, a software framework that allows to run text-mining services on the web, translating the outcomes into the Open Annotation Ontology model [29]. The system stores the results as RDF annotations that can be queried with SPARQL. The DOMEO architecture can rely on external text-mining services to enhance the curation process. However, integration of state-of-the-art algorithms and services needs to be additionally developed.

In this way, such combination represents an attractive strategy that allows curation outcomes to be discoverable and shared across multiple research institutions. However, this process is still a challenge for developers, requiring extensive efforts to integrate and enhance the outputs of their solutions.

3   Methods

Computational annotation solutions focus on strategies to mine textual data, producing large amounts of information in different formats. Generally, these annotations include entity names, entity identifiers from the normalization step, and events or relations between those entities. To fully exploit the potential behind these discovered associations, modern knowledge paradigms need to be adopted in order to streamline the workflow from data acquisition to knowledge delivery. We propose an integrative solution for facilitating the publishing of these valuable data on the Semantic Web ecosystem.

3.1   Case Study

The notion of unifying current text-mining tools with Semantic Web technologies emerged from the need to better explore and share, in an efficient way, the mined results. Currently, publishing or sharing the outcomes of a computational annotation tool across multiple institutions is hindered by serious difficulties and the usual option is to simply share the resulting output file. However, consuming this type of annotation outputs is sometimes not a trivial task and interoperability issues still arise. To better describe this problem, we consider the interaction between Egas[1], a collaborative annotation tool, and COEUS[2], a Semantic Web Framework. Egas is a web-based platform for biomedical text-mining and collaborative curation, supporting manual and

---

[1] https://demo.bmd-software.com/egas/
[2] http://bioinformatics.ua.pt/coeus/

automatic annotation of concepts and relations. COEUS includes several interfaces to integrate, abstract and publish information. The resulting interaction between these tools results in a clear improvement of curated data availability, enabling a multitude of new exploration opportunities. By connecting these tools, we aim to create a seamless semantic layer (Fig. 1), unifying access to the rich curated knowledge.

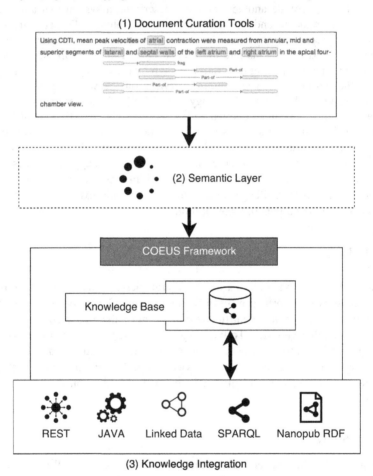

Fig. 1. Example workflow showing the resulting exploration tools from unifying (2) document curation tools such Egas (1) and Semantic Web technologies available on COEUS (3)

3.2   Architecture

The approach proposed in this paper aims to integrate, harmonize and improve existing curation results. In Fig. 2, we present the proposed architecture focused on the transition process from current annotation formats to the Semantic Web paradigm. In order to handle existing tools, we consume different formats from text-mining outputs making an integration parser for each one. The ability to acquire data from several and miscellaneous annotation formats benefits developers, allowing each one to implement and integrate their format in a common interface. With this shared interface, entities and relations extracted from input data files can be processed (Fig. 2, block 1). Next, ontology-based annotation techniques [30] [31] are used to automatically attribute each identified name or relation to a specific object or data property from selected ontologies (Fig. 2, block 2). This mechanism will perform improved relation mapping methods capable to provide an answer to the question: *What is the best ontology property to describe a specific biomedical relation?* Then, the triples generation process (Fig. 2, block 3) results from an advanced Extract-Transform-and-Load process that is responsible for performing an adequate linkage between each entity and related properties. Finally, the integrated information can be combined with existing and related knowledge, providing additional inference and reasoning capabilities (Fig. 2, block 4). This combination takes advantage of existing knowledge bases, associating and discovering related information to support query and retrieval services.

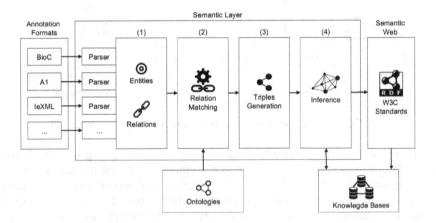

Fig. 2. Semantic layer architecture: annotation results from external text-mining tools integrated into Semantic Web standards: (1) extraction of available entities and relations; (2) Ontologies describe the susceptible relations; (3) ETL processes generate triples; (4) Inference and publication mechanisms to combine, store and share data

This process translates the resulting annotations into the Annotation Ontology[3] (AO), an open representation model for representing interoperable annotations in RDF (Fig. 3). The model provides a robust set of methods for connecting web resources, for instance, textual information in scientific publications, to ontological elements, with full representation of annotation provenance. Through this model, existing domain ontologies and vocabularies can be used, creating extremely rich stores of metadata on web resources. In the biomedical domain, subjects for ontological structuring can include biological processes, molecular functions, anatomical and cellular structures, tissue and cell types, chemical compounds, and biological entities such as genes and

---

[3] http://www.openannotation.org/spec/core/

proteins [29]. By linking new scientific content to computationally defined terms and entity descriptors, AO helps to establish semantic interoperability across the biomedical field, facilitating pathway analysis, disease modeling, cross-species comparisons, and the generation of new hypotheses.

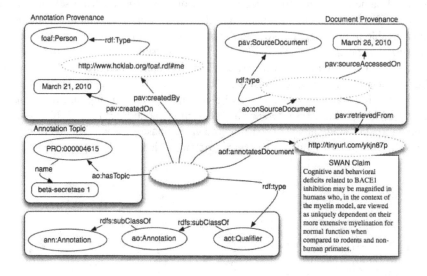

Fig. 3. Example of a document annotation using Annotation Ontology [29]

A prototype implementation of this architecture involved adding features to both Egas and COEUS. At the annotator level, API methods enable and automate the transition of curation results to COEUS. In COEUS, parsers read several annotation formats to an abstraction engine. The engine interacts with the mentioned ontology term matching methods, generating rich triples for integration. The resulting annotations are appended as *coeus:concept* items creating a Linked Data network on COEUS. With a COEUS instance deployed for our case study, Egas regularly pushes new curation data, triggering the integration process. Once this finishes, the semantically enhanced curation outcomes are immediately available for publishing and exploration.

4   Discussion

Dealing with information outputs resulting from several and different annotation tools is an ongoing challenge. Several strategies promote their own rules to deal with annotated data, establishing a barrier to efficiently store and exchange this type of information. With the Semantic Web paradigm, integration became an additional challenge, while at the same time representing the perfect solution for scenarios where the main issue is interoperability. Hence, several research groups are migrating their curated information to promote better features and services using this prominent paradigm.

The architecture proposed in this manuscript adopts a hybrid strategy to connect information extraction tools with Semantic Web technologies, providing a unified layer between the platforms' communication level. To facilitate the emergence of interoperable annotations on the web, our architecture makes use of the Annotation Ontology (AO), a recent and suitable model to perform versioning, sharing and viewing of both manual and automatic or semi-automatic annotations, such as those created by text-mining solutions. However, AO only provides the model for encoding and sharing the annotation, being necessary to additionally develop tools to create, publish and share the resulting annotations. For this reason, our architecture makes use of the COEUS framework to share annotations through several services such as a SPARQL endpoint and Linked Data publication interfaces.

Our strategy does not aim to provide text-mining services to users (e.g. similar to the DOMEO system). In contrast, we are focused on reusing existing curated data from external text-mining tools to improve their availability through an open representation model on the web. This results in a more suitable transition process, in which researchers can optionally select the annotations (from external text-mining tools) to share and obtain credit for the shared content. However, dealing with this type of transitions is not a trivial task and some limitations are expected. For instance, the whole process depends not only on the curated data in each file format, but also on the extra inference mechanisms required to discover related information. Yet, we believe that the described mechanism will result in an improvement of the annotated data, providing adequate ontology-based annotation mechanisms, enabling triples generation and creating new exploration opportunities.

5   Conclusions

Despite the latest efforts, there is no current unified strategy to enable the comprehensive exploration of curation results from external annotations tools. In this manuscript, we propose an architecture to tackle this problem. We introduce a unified layer to harmonize and integrate such outcomes into rich semantic knowledge bases. The system deals with several annotations formats by using an abstraction engine. This approach promotes new strategies to formalize and enrich the knowledge resulting from text-mining applications, enabling the creation of new semantically enhanced systems to aid the exploration of biomedical relationships and derived facts. At last, this architecture simplifies the delivery of curated data in state-of-the-art formats, empowering modern knowledge discovery techniques.

Acknowledgments. The research leading to these results has received funding from the European Community (FP7/2007-2013) under ref. no. 305444 – the RD-Connect project, and the CloudThinking project (funded by the QREN "MaisCentro" program ref. CENTRO-07-ST24-FEDER-002031). Pedro Sernadela is funded by Fundação para a Ciência e Tecnologia (FCT) under grant agreement SFRH/BD/52484/2014. Sérgio Matos is funded by a FCT Investigator grant.

References

1. Hunter, L., Cohen, K.B.: Biomedical language processing: what's beyond PubMed? Mol. Cell. 21, 589–594 (2006)
2. Campos, D., Matos, S., Oliveira, J.: Current Methodologies for Biomedical Named Entity Recognition. Biol. Knowl. Discov. Handb. Preprocessing, Mining, Postprocessing Biol. Data. 839–868 (2012)
3. Alex, B., Grover, C., Haddow, B.: Assisted Curation: Does Text Mining Really Help? In: Pacific Symp. Biocomput., vol. 13 (2008)
4. Berners-Lee, T., Hendler, J., Lassila, O.: The semantic web. Sci. Am. 284(5), 28–37 (2001)
5. Wild, D.J., Ding, Y., Sheth, A.P., Harland, L., Gifford, E.M., Lajiness, M.S.: Systems chemical biology and the Semantic Web: what they mean for the future of drug discovery research. Drug Discov. Today 17, 469–474 (2012)
6. Ashburner, M., Ball, C.A., Blake, J.A., Botstein, D., Butler, H., Cherry, J.M., Davis, A.P., Dolinski, K., Dwight, S.S., Eppig, J.T., Harris, M.A., Hill, D.P., Issel-Tarver, L., Kasarskis, A., Lewis, S., Matese, J.C., Richardson, J.E., Ringwald, M., Rubin, G.M., Sherlock, G.: Gene ontology: tool for the unification of biology. The Gene Ontology Consortium. Nat. Genet. 25, 25–29 (2000)
7. Franzén, K., Eriksson, G., Olsson, F.: Protein names and how to find them. Int. J. Med. Inform. 67(1), 49–61 (2002)
8. Nadkarni, P.M., Ohno-Machado, L., Chapman, W.W.: Natural language processing: an introduction. J. Am. Med. Inform. Assoc. 18, 544–551 (2011)
9. Schuemie, M., Kors, J., Mons, B.: Word sense disambiguation in the biomedical domain: an overview. J. Comput. Biol. (2005)
10. Zhu, F., Patumcharoenpol, P., Zhang, C., Yang, Y., Chan, J., Meechai, A., Vongsangnak, W., Shen, B.: Biomedical text mining and its applications in cancer research. J. Biomed. Inform. 46, 200–211 (2013)
11. Bui, Q.-C., Katrenko, S., Sloot, P.M.A.: A hybrid approach to extract protein-protein interactions. Bioinformatics 27, 259–265 (2011)
12. Hoffmann, R., Valencia, A.: Implementing the iHOP concept for navigation of biomedical literature. Bioinformatics. 21(suppl. 2), ii252–ii258 (2005)
13. Tari, L., Anwar, S., Liang, S., Cai, J., Baral, C.: Discovering drug-drug interactions: a text-mining and reasoning approach based on properties of drug metabolism. Bioinformatics 26, i547–i553 (2010)
14. Tsuruoka, Y., Tsujii, J., Ananiadou, S.: FACTA: a text search engine for finding associated biomedical concepts. Bioinformatics 24, 2559–2560 (2008)
15. Stenetorp, P., Pyysalo, S., Topić, G.: BRAT: a web-based tool for NLP-assisted text annotation. In: Proc. Demonstr. 13th Conf. Eur. Chapter Assoc. Comput. Linguist., pp. 102–107 (2012)
16. Salgado, D., Krallinger, M., Depaule, M., Drula, E., Tendulkar, A.V., Leitner, F., Valencia, A., Marcelle, C.: MyMiner: a web application for computer-assisted biocuration and text annotation. Bioinformatics 28, 2285–2287 (2012)
17. Rak, R., Rowley, A., Black, W., Ananiadou, S.: Argo: an integrative, interactive, text mining-based workbench supporting curation. Database (Oxford), bas010 (2012)
18. Campos, D., Lourenco, J., Matos, S., Oliveira, J.L.: Egas: a collaborative and interactive document curation platform. Database 2014 (2014)
19. Comeau, D.C., Islamaj Doğan, R., Ciccarese, P., Cohen, K.B., Krallinger, M., Leitner, F., Lu, Z., Peng, Y., Rinaldi, F., Torii, M., Valencia, A., Verspoor, K., Wiegers, T.C.,

Wu, C.H., Wilbur, W.J.: BioC: a minimalist approach to interoperability for biomedical text processing. Database (Oxford), bat064 (2013)

20. Machado, C.M., Rebholz-Schuhmann, D., Freitas, A.T., Couto, F.M.: The semantic web in translational medicine: current applications and future directions. Brief. Bioinform. (2013)

21. Lopes, P., Oliveira, J.L.: COEUS: "semantic web in a box" for biomedical applications. J. Biomed. Semantics. 3, 11 (2012)

22. Broekstra, J., Kampman, A., van Harmelen, F.: Sesame: A generic architecture for storing and querying RDF and RDF schema. In: Horrocks, I., Hendler, J. (eds.) ISWC 2002. LNCS, vol. 2342, pp. 54–68. Springer, Heidelberg (2002)

23. Wilkinson, M.D., Vandervalk, B., McCarthy, L.: The Semantic Automated Discovery and Integration (SADI) Web service Design-Pattern, API and Reference Implementation. J. Biomed. Semantics. 2, 8 (2011)

24. Sernadela, P., van der Horst, E., Thompson, M., Lopes, P., Roos, M., Oliveira, J.L.: A Nanopublishing Architecture for Biomedical Data. In: Saez-Rodriguez, J., Rocha, M.P., Fdez-Riverola, F., De Paz Santana, J.F. (eds.) 8th International Conference on Practical Applications of Computational Biology & Bioinformatics (PACBB 2014). AISC, vol. 294, pp. 277–284. Springer, Heidelberg (2014)

25. Laurila, J., Naderi, N., Witte, R.: Algorithms and semantic infrastructure for mutation impact extraction and grounding. BMC Genomics, S24 (2010)

26. Coulet, A., Garten, Y., Dumontier, M., Altman, R.B., Musen, M.A., Shah, N.H.: Integration and publication of heterogeneous text-mined relationships on the Semantic Web. J. Biomed. Semantics 2(suppl. 2), S10 (2011)

27. Mendes, P.N., Jakob, M., García-Silva, A., Bizer, C.: DBpedia spotlight. In: Proceedings of the 7th International Conference on Semantic Systems, I-Semantics 2011, pp. 1–8. ACM Press, New York (2011)

28. Ciccarese, P., Ocana, M., Clark, T.: Open semantic annotation of scientific publications using DOMEO. J. Biomed. Semantics 3(suppl. 1), S1 (2012)

29. Ciccarese, P., Ocana, M., Garcia Castro, L.J., Das, S., Clark, T.: An open annotation ontology for science on web 3.0. J. Biomed. Semantics 2(suppl. 2), S4 (2011)

30. Spasic, I., Ananiadou, S., McNaught, J., Kumar, A.: Text mining and ontologies in biomedicine: Making sense of raw text. Brief. Bioinform. 6, 239–251 (2005)

31. Khelif, K., Dieng-Kuntz, R., Barbry, P.: An Ontology-based Approach to Support Text Mining and Information Retrieval in the Biological Domain. J. UCS (2007)

Blind and Visually Impaired Students Can Perform Computer-Aided Molecular Design with an Assistive Molecular Fabricator

Valère Lounnas[1,*], Henry B. Wedler[2], Timothy Newman[2], Jon Black[1], and Gert Vriend[1]

[1] CMBI Radboudumc, Geert Grooteplein 26-28, 6525 GA Nijmegen, The Netherlands
v-lounnas@unicancer.fr
[2] Department of Chemistry, University of Davis, Davis, CA 95616, USA

Abstract. Life science in general and chemistry in particular are inaccessible to blind and visually impaired (BVI) students at the exception of very few individuals who have overcome, in a seemingly miraculous way, the hurdles that pave the way to higher education and professional competency. AsteriX-BVI a publicly accessible web server, developed at the Radboud University in the Netherlands already allows BVI scientists to perform a complete series of tasks to automatically manage results of quantum chemical calculations and produce a 3D representation of the optimized structures into a 3D printable, haptic-enhanced format that includes Braille annotations.[1]

We report here the implementation of Molecular Fabricator 1.0, a molecular editor which is a new assistive feature of AsteriX-BVI. This molecular editor allows BVI scientists to conceptualize complex organic molecules in their mind and subsequently create them via the server. It was developed around the concept that molecules are composed of chemical fragments connected together. Fragments from either a predefined dictionary or defined as short SMILES strings can be used, ensuring that almost any structure can be generated. A fragment-based relative atom-numbering scheme that can be mentally tracked is used together with the support of an automatically generated report in Braille providing topological information and a clean 2D sketch of the constructed molecule, that can be unambiguously recognized tactilely. The R or S stereo-chemical configuration of asymmetric centers is controlled via feedback provided in the Braille report. The molecular fabricator creates output files in a variety of formats currently used in drug design and theoretical chemistry. It was tested and refined with the help of HBW, the blind co-author of this article, who pursues a PhD in chemistry at UC Davis, California. With Molecular Fabricator 1.0 HBW can now create molecules very fast and verify their structures without external help as was the case before. Molecular Fabricator 1.0 and its tutorial are accessible at: http://swift.cmbi.ru.nl/bitmapb/access/runit_bvi.html.

Keywords: Blind, visually impaired, chemistry, molecular editor, assistive software, computer-aided molecular design, Braille, science, higher education.

---

* Corresponding author.

F. Ortuño and I. Rojas (Eds.): IWBBIO 2015, Part I, LNCS 9043, pp. 18–29, 2015.

1 Introduction

Acquiring the knowledge and understanding the physical rules governing chemistry do no cause problems to blind and visually impaired (BVI) students.[1,2] However, applied chemistry remains operationally inaccessible to them because sight is indispensable to laboratory bench experimentation and to the reading of outputs from measurement devices. Attempts in that direction are still experimental and cannot be achieved without the strict and permanent supervision of a mentor closely monitoring BVI students.[3] This has been the main hindrance that BVI students aspiring to engage in chemistry have encountered until now, and only a very few of them have seemingly miraculously accessed to professorship in the past.[4,5]

Nowadays, the same impossibility is observed despite the fact that modern chemistry has the potential to offer BVI students considerable opportunities to express their talent and join, on an equal footing with their sighted schoolmates, the community of highly qualified professionals.[6] Since the end of the 20th century, with the elaboration of new chemical compounds in a large variety of industrial and research domains chemistry is omnipresent. It involves the systematic use of computers to help select and design compounds via a series of computer-assisted processes that we hereby refer to as in silico chemistry also known as chemo-informatics or computer aided molecular design according to the specific field of activity involved.

For this reason life sciences in general and chemistry in particular are still inaccessible to BVI students at the exception of very few individuals who have overcome, in a seemingly miraculous way, the hurdles that pave the way to higher education and professional competency.[7,8]

Very recently Navmol 2., the first molecular editor for BVI high-school and college students was developed.[9] It allows the construction and navigation of molecules on an atom-based approach controlled via the keyboard with a vocal feedback. However, NavMol presents important drawbacks that preclude its use for higher education: (1) the atom-based approach is not adequate for constructing and navigating complex molecules while memorizing the connections and relative orientation between their components; (2) it is a Java application that need to be installed and run locally, in most case preventing its immediate availability.

We present here the elaboration of Molecular Fabricator 1.0, a simple but efficient software that was conceived with the feedback from HBW, the blind coauthor of this article who is a PhD student in chemistry. Molecular editing is controlled via a fragment-based approach with a feedback provided by a report in Braille. It allows BVI users to elaborate and mentally keep track of complex organic molecules. Molecular Fabricator 1.0 is embedded in AsteriX-BVI a publicly accessible web server that we have developed to bridge the gap between BVI users' aspiration for proficiency and the highly assistive potential of computer-related technologies that could allow them to realize high standard in silico chemistry.[1]

2   Methods

2.1   Extending the Assistive Capacity of AsteriX-BVI

The AsteriX web server was initially designed to fully automatically decomposes articles in the .pdf format, detects 2D plots of low molecular weight molecules, removes meta data and annotations from these plots, and converts them into 3D atomic coordinates.[10] Subsequently, AsteriX was converted into AsteriX-BVI, a fully BVI assistive server, that automatically converts 3D representation of molecules into a 3D printable haptic-enhanced format that includes Braille annotations.[1] These Braille-annotated and tactile physical 3D models allow BVI scientists to generate an accurate mental model of the molecule they hold in their hands. The assistive capacity of AsteriX-BVI is now extended further with the Molecular Fabricator 1.0 computer software we developed to allow BVI users to construct in silico organic and inorganic molecules.

The Molecular Fabricator is developed around a per-component (or per fragment) description of molecules. The chemical fragments that compose molecules can be assembled using simple connecting rules described in the next sections. Fragments are represented by SMILES strings.[11] Input to the Molecular Fabricator is made via a list of sequential commands gathered in an input ASCII file that is uploaded to the server.

2.2   A Fragment-Based Approach to Describe Molecules

Structures of organic molecules that a have a molecular weight (M_w) typically < 1000 can be visually analyzed and decomposed in a reduced number of constitutive components or fragments also called chemical groups. Usually chemical groups contain between 1 and 10 atoms and can therefore be easily memorized and recognized. Structures of complete molecules can thus be described as assemblies of fragments connected together (Figure 1) using a small set of possible virtual actions we refer to as the five 'fabricator commands' that are: 'branch', 'fuse', 'replace', 'connect' and 'change bond'. These commands obey specific syntax rules described in Figure 2. It must be noted that while the first three commands (branch, fuse, replace) refers to actions of growing the molecule by branching, fusing or inserting new atoms or groups of atoms the last two commands (connect and change bond) refer to actions that modify the molecule connectivity and/or stereochemical configuration without appending new fragment to it. Any of these five actions defines a component so that a molecule is actually a sum of components that can be defined one after the other in a sequential order (Figures 1 and 2). The order does not necessarily matter and the Molecular Fabricator allows many different ways of constructing a molecule to be envisaged. In addition, the stereochemical configuration of asymmetric centers can be defined locally by indicating the orientation either 'up' or 'down' of the groups substituted on them, with respect to the plane on which the molecule structure can be visually projected (Figures 2a and 2b). These stereochemical indications are not new components but just modifiers of the action of 'branching' and 'change bond'. It must be pointed out

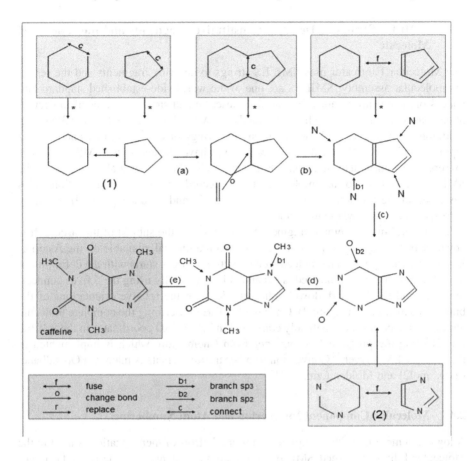

Fig. 1. Shows how components that constitutes the chemical structure of caffeine (M_w= 184.) can be assembled with the Molecular Fabricator. A logical construction process for caffeine is shown in five steps from (a) to (e) starting with two ring structures (1). Thin arrows (green panel) indicate the six different types of possible connection between components. The decomposetion in components is not unique as different starting components can be considered (yellow panels). For instance, cyclic structures with endocyclic nitrogens can also be considered as starting fragments (2) reducing the construction process to 3 steps only instead of 5.

that the absolute stereochemical configuration (R or S) does not need to be determined for each asymmetric center in order to correctly and completely describe the structure of a molecule in its right stereochemical configuration. Only the relative orientation of the substituted group at each asymmetric center must be defined. However, the R or S absolute stereochemical configuration of each asymmetric is an indispensable information that is calculated by the server and provided to the BVI users as a feedback information rather than an input parameter.

2.3   SMILES Strings to Describe Constitutive Fragments and the Assembled Molecule

The Molecular Fabricator uses SMILES strings to describe fragments and the resulting molecular assembly. SMILES are one of the worldwide-established standard languages of chemo-informatics. It is used to describe and store structures of molecules in databases and manipulate them efficiently.[11] SMILES strings replace the 3D representation of molecules with one dimensional strings of characters that encode their topology and stereo-chemical configuration. We have chosen to use SMILES strings to represent individual molecular fragments. The Molecular Fabricator uses the SMILES grammar to assemble fragments according to the combination rules described in the previous section (and in Figure 2) and produce the SMILES string corresponding to the whole molecule.

The algorithm to assemble fragment SMILES is not the subject of this article. It is however briefly described in Figure 2 (see also the tutorial available on the AsteriX-BVI server front page). The molecule construction process starts with the definition of a seed fragment (ic=1). The seed is expanded incrementally using the 5 fives connecting rules and the 'up' and 'down' indicators defining the relative orientations of the branched substituents (Figure 2). Once the SMILES describing the complete molecule has been created it is automatically converted in 2D and 3D coordinates in the .sdf file format using the program Molconverter from ChemAxon, which is implemented in the AsteriX-BVI server.[12] Conversion to other useful formats is made via Opendbabel (.pdb, .mol2) and Molden (.zmat).[13,14]

2.4   Molecular Component Numbering and Atom Numbering

A logical numbering system implying a minimal effort of memorization is used in the Molecular Editor. The seed SMILES is the component number one (ic = 1). Every time a new molecular component is defined, the Molecular Fabricator increments this number (ic + 1) (Figure 2). The component number and atom numbering relative to each component remain unchanged during the construction. Each individual atom index corresponds to its position within the SMILES string, read from left to right (Figure 2). To keep track easily of atom numbering in ring fusion specific rules were established (Figure 3).

2.5   Using a Library of Predefined Fragments

The Molecular Fabricator contains a small internal library of predefined fragment that can be invoked, via the input file, using short-hand name instead of an explicit SMILES string. In addition, each user can define its own library of predefined SMILES strings associated with short-hand name using the 'add' and 'rem' special commands followed by a 3-letter code that specifies his/her personalized database access and storage. The 'add' command appends the SMILES fragment to the library corresponding to the 3 letters code and the 'rem' command removes it (see tutorial link

(a)

comp. #	SMILES at step: ic *numbering*	Command syntax	resulting SMILES at step: ic +1 *numbering*	
ic = 1	C1CCCCC1 *1 23456*	fuse 1 5 6 with 5ring	C1CCCC(CCC2)C21 *1 2345* ‖‖ 6 *2 3 4*	ic = 1 ic = 2
ic = 1 ic = 2	C1CCCC(CCC2)C21 *1 2345* ‖‖ 6 *2 3 4*	branch 2 4 with smi up OH	C1CCCC(CC[C@@](OH)2)C21 *1 2345* ‖‖‖ *2 3 4* *1 2* 6	ic = 1 ic = 2 ic = 3
ic = 1 ic = 2	C1CCCC(CCC2)C21 *1 2345* ‖‖ 6 *2 3 4*	replace 2 3 with smi OCC(N)=C	C1CCCC(C[OCC(N)=C]C2)C21 *1 2345* ‖‖‖‖ *1 2 3 4 5* 6	ic = 1 ic = 2 ic = 3

(b)

comp #	SMILES at step: ic *numbering*	Command syntax	resulting SMILES at step: ic +1 *numbering*	
ic = 1 ic = 2 ic = 3	C1CCCC(C(C2COCC2))C1 *1 2345* ‖‖‖‖ 6 *1 2345*	connect 1 6 , 3 5	C1CCCC(C(C2COCC32=))C31 *1 2345* ‖‖‖‖ 6 *1 2345*	ic = 1 ic = 2 ic = 3
	C1CCCC(C(C2COCC2))C1 *1 2345* ‖‖‖ 6 *1 2345*	change bond 3 4 5 2	C1CCCC(C(C2COCC32=))C31 *1 2345* ‖‖‖ 6 *1 2345*	ic = 1 ic = 2 ic = 3
ic = 1 ic = 2 ic = 3	C1CCCC(CC(C2COCC2))C1 *1 2345* ‖‖ 6 *1 2* *1 2345*	change bond 2 1 2 2 Z	C1CCCC(C\=/C(C2COCC2))C1 *1 2345* ‖‖ 6 *1 2* *1 2345*	ic = 1 ic = 2 ic = 3
ic = 1 ic = 2 ic = 3 ic = 4	C1CCCC(CC(Br)(C2COCC2))C1 *1 2345* ‖‖ 6 *1 2* *1 2345*	change bond 2 1 2 1 U	C1CCCC(C[C@@](Br)(C2COCC2))C1 *1 2345* ‖‖ 6 *1 2* *1 2345*	ic = 1 ic = 2 ic = 3 ic = 4

Fig. 2. Examples that illustrate the syntax for the five commands the Molecular Fabricator uses to (a) combine and/or (b) modify the connectivity or stereochemical configuration of molecular fragments. The modification of the SMILES strings is shown accordingly with the relative atom numbering scheme that is stored by the Molecular Fabricator indicated below. The numbering scheme is used to keep track of the relative position of each atom within the molecule under construction with *ic* values denoting each components. Both component indices and relative atom indices remain invariant during the building process.

on server front page). The 3-letter code defining each user specific fragment library is unique and attributed to the user upon his/her request. The Molecular Fabricator contains a default library of predefined fragments described in the tutorial.

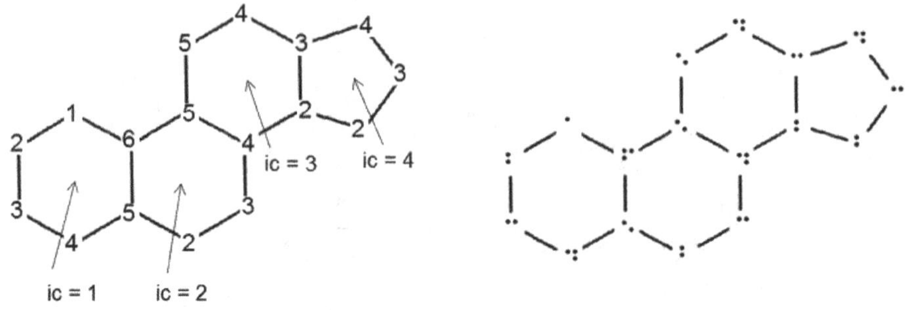

Fig. 3. Illustrates the rules established for relative atom numbering to mentally control ring fusion, in Arabic ciphers (left view) and in Braille (right view). The right view is an extract from the Molecular Fabricator tutorial. Numbering on cycles are visualized anticlockwise (1st rule) with atom 1 positioned at noon on the seed cycle ($ic=1$). The seed component is a 6-member cycle whereon a second 6-member cycle ($ic=2$) is fused at user selected positions 5 and 6. The Molecular Fabricator automatically allocates position 1 and 6 of the second cycle to position 5 and 6 on the first cycle (2nd rule). The numbering on the previous component has the priority (3rd rule) and thus atom 1 and 6 of the second cycles, that are equivalent to 5 and 6 of the first cycle, will not exist in the resulting SMILES string. The same rules are successively applied to the relative numbering of the third ($ic=3$) and fourth ($ic=4$) cycle subsequently fused.

2.6   Reporting the Molecule Topology and Stereochemistry in Braille

One essential component of the Molecular Fabricator is its reporting capacity. A complete document is produced in Braille, summarizing the construction steps fragment by fragment and providing a clean 2D sketch of the molecule with Braille annotations for atom type and absolute numbering scheme. 2D sketches are generated using the optimal 2D coordinates generated via the 'beautify' command of the publicly available Cactvs molecular editor, which allows optimal representation of difficult-to-draw molecules using the Craig Shelley algorithm which provides a representation particularly well suited for tactile recognition when printed in relief.[15] Classical Braille signs are combined with line segments that can be tactilely recognized to provides a schematic 2D representation which is topologically and stereochemically exhaustive similarly to an image generated by a graphical interface (Figures 4 and 5).

The absolute numbering scheme that is provided in the Braille report corresponds to the numbering used in the different file formats produced by the server. The topology report, at the exception of the mixed Braille and segment molecular representation, is also produced in a non-Braille format that allows reading by a voice synthesizer.

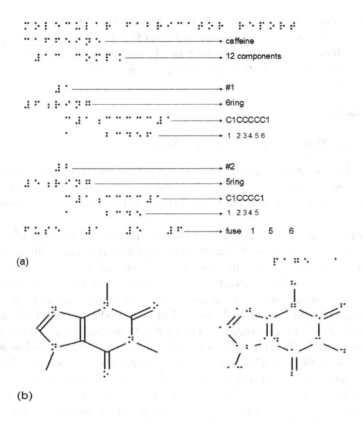

Fig. 4. The first page of the Braille report automatically generated for the building of caffeine (a). The report includes the tactile 2D representation of caffeine (b); combining atom labels in Braille (left) and absolute numbering (right).

2.7   Inputting and Outputting the Molecular Fabricator via the AsteriX-BVI Web Server

Input to and output from the Molecular Fabricator are exchanged over internet via the AsteriX-BVI server. Molecular Fabricator input files are simple text (flat) files that must be prepared locally by the BVI users on their computer and given the .mfh file-name extension. They must be uploaded on the server via the general input text area using the 'Browse' and 'Submit job' buttons. Calculation are performed in the back-ground and results are dispatched after about half a minute. Result are stored for 8 days and can be accessed via hyper links using the 'Browse completed jobs' button. Result files are provided in the form a tarball file (.tgz) that can be retrieved from the server via the 'retrieve work' links on the result html page. Result files include 2D and 3D representations of the constructed molecules in a variety of standard formats suitable for chemo-informatics and computer aided molecular design (.sdf, .pdb, .mol2, .smi, .zmat) as well as a Braille report in .pdf format. The report can be printed on a regular office printer with a special type of thermal paper that creates a relief tactilely readable once it has been heated by a special device.[16]

3   Results

3.1   Learning How to Use the Molecular Fabricator

HBW, the blind co-author of this article, has learned over a few days how to use the Molecular Fabricator and master the syntax of the five connecting rules as well as the 'up' and 'down' rules that define the stereochemical configuration of the asymmetric centers.

HBW is a graduate student engaged in a PhD program at the University of Davis in California. Prior to learning the Molecular Fabricator he was already fully acquainted with organic chemistry and the process of visualizing molecular structures in his mind. He also knew the basic rules for generating molecules with the SMILES language. However, using SMILES he could never really manage the elaboration of a large molecule such as cycloartanol, the molecule he studies for his PhD research project (Figure 5). This impossibility was partly due to the overwhelming complexity of generating a SMILES string for a complex molecule like cycloartanol as well as the absence of feedback mechanism to control the construction process. The fragment-based approach together with the system of reporting in Braille, and the fact that the Molecular Fabricator is embedded in a BVI assistive web server, has provided a synergy that allowed him to take control of the process of constructing entirely by himself the molecules of his choice.

3.2   Flexibility of Use

The choice of the seed (the initial component) and order of the subsequent commands necessary to construct a molecule is not unique. Selecting the fragment SMILES to be used for the construction is the decision of each user according to his/her own logic and preferred way to visualize and envisage how to build the molecule. The construction is a multistage process that involves: (a) elaborating the molecule scaffold, (b) decorating the scaffold with chemical substituents and (c) orienting the substituents. If necessary, a Braille report can be printed at every stage of the process and work can be resumed by editing and submitting again the modified input file to the Molecular Fabricator via AsteriX-BVI.

According to HBW, one extremely useful and beneficial aspect of Molecular Fabricator 1.0 is that the generated structures can be converted straight to z-matrices and then immediately converted to input files for Gaussian, the computational software used by HBW to perform quantum chemical calculations.

3.3   Constructing Cycloartanol - A Large and Complex Molecule

Cycloartanol (Figure 5) is a complex steroidal triterpenoid found in most olive oils.[17] HBW investigates the mechanism of isomerization of this molecule with quantum chemical calculation to find a plausible rearrangement route, which may be of relevance to provide a simple test to determine if olive oil has been refined by a non-mechanical process.[18] Despite its apparent complexity and the presence of 10 asymmetric centers, this molecules can be produced simply with the Molecular Fabricator. First, the steroid scaffold is constructed by assembling 5 ring-components fused together. Second, the substituents are branched on the scaffold and their orientations with respect to the visualization plane indicated using the 'up' and 'down' modifiers. Third, the stereo-chemical configurations of all remaining asymmetric centers are defined by branching hydrogen atoms with the up or down orientation.

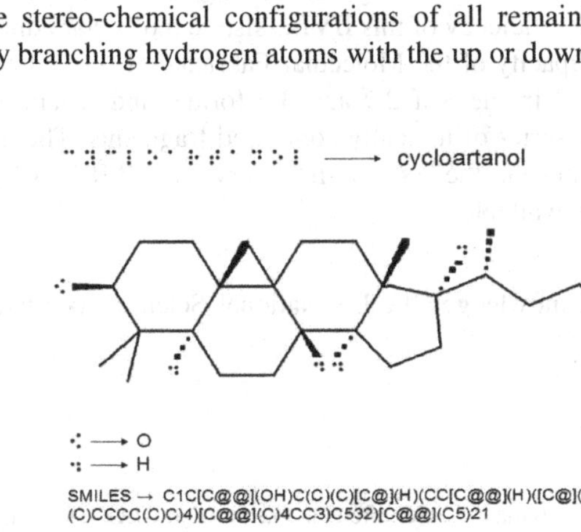

SMILES → C1C[C@@](OH)C(C)(C)[C@](H)(CC[C@@](H)([C@](H)(CC[C@](H)([C@](C)CCCC(C)C)4)[C@@](C)4CC3)C532)[C@@](C5)21

Fig. 5. Tactile 2D representation of cycloartanol extracted from the Braille report generated automatically by the Molecular Fabricator. To have a size suitable for tactile recognition bonds in the 2D sketch must be between 2.5 and 3. cm long. Large molecules are automatically split and printed on two (or 4) A4 sheets that can be attached side by side. The corresponding SMILES string assembled by the Molecular Fabricator is reported here below the structure for the purpose of illustrating the impossibility of generating it without the help of a dedicated algorithm.

The absolute R or S configuration of asymmetric centers is provided as a feedback in the Braille report. Correcting a R configuration into a S configuration can be achieved simply by replacing the 'up' and 'down' modifiers in the input command file and resubmit it to the Molecular Fabricator; complete description of how cycloartanol was produced is in the tutorial at http://swift.cmbi.ru.nl/bitmapb/access/runit_bvi.html.

4   Discussion

With the Molecular Fabricator BVI users can accomplish high-standard tasks afferent to graduate-level chemistry projects. The Molecular Fabricator extends the AsteriX-BVI server capabilities toward allowing BVI students to independently access higher education.

With the Molecular Fabricator complex organic chemistry molecules can be tackled by BVI students. An intuitive fragment-based approach is used and stereo-chemistry is controlled in an adequate and secured manner using a feedback mechanism.

Using SMILES strings was determinant for the efficiency of the Molecular Fabricator because they are a worldwide standard used already for many years by institutional and professional researchers. SMILES strings comprehensively cover all organic and inorganic chemistry and the rules governing their syntax are very intuitive which make them easily learned. This represents a considerable advantage for BVI students who can be trained to use the Molecular Fabricator very rapidly. Creating a Braille report to support the memorization and visualization effort required for the elaboration of complex molecules, when sight is lacking, is certainly one essential component contributing to the efficiency of this BVI-assistive molecular editor.

We plan to augment the capacity of the Molecular Fabricator 1.0 to allow the uploading of molecules provided in the .sdf 2D and 3D format and reverse engineer them to decompose them in a series of logically connected fragments. The molecular fabricator can be used remotely via the AsteriX-BVI server at CMBI. All AsteriX-BVI server codes are publicly available.

Acknowledgments. HBW acknowledges the US National Science Foundation for a research fellowship.

References

1. Lounnas V., Wedler H.B., Newman T., Schaftenaar G., Harrison J.G., Nepomuceno G., Pemberton R., Tantillo D.J., Vriend G.: Visually impaired researchers get their hands on quantum chemistry: application to a computational study on the isomerization of a sterol. J. Comput. Aided Mol. Des. (August 5, 2014) (Epub. ahead of print)
2. Wedler, H.B., Boyes, L., Davis, R.L., Flynn, D., Franz, A.K., Hamann, C.S., Harrison, J.G., Lodewyk, M.W., Milinkevich, K.R., Shaw, J.T., Tantillo, D.J., Wang, S.C.: Nobody Can See Atoms: Science Camps Highlighting Approaches for Making Chemistry Accessible to Blind and Visually-Impaired Students. J. Chem. Educ. 91, 188–194 (2014)
3. Albright professor helps blind students discover chemistry at summer camps, http://readingeagle.com/news/article/albright-professor-helps-blind-students-discover-chemistry-at-summer-camps#.VEFIsb5OKpp
4. History of Analytical Chemistry by Ferenc Szabadvary, Gordon and Breach Sciences Publishers, Yverdon, Switzerland Horstmann, August (1843 – 1929) studied at the University of Heidelberg, and later became a professor there. He became blind quite young as the result of a disease
5. Blind students confront chemistry lab, http://www.kpbs.org/news/2009/aug/18/blind-students-confront-chemistry-lab/
6. Which problems would a blind person have when applying for a professorship, http://academia.stackexchange.com/questions/23421/which-problems-would-a-blind-person-have-when-applying-for-a-professorship

7. Skawinski, W.J., Busanic, T.J., Ofsievich, A.D., Luzhkov, V.B., Venanzi, T.J., Venanzi, C.A.: The Use of Laser Stereolithography to Produce Three-Dimensional Tactile Molecular Models for Blind and Visually-Impaired Scientists and Students. Inf. Technol. Disabil. 1(4), Article 6 (1994)
8. Chemistry by touch: blind scientist fashions new models of molecules, http://www.thefreelibrary.com/Chemistry+by+touch%3A+blind+ scientist+fashions+new+models+of+molecules.-a016723790 (accessed April 30, 2014)
9. Fartaria, R.P.S., Pereira, F., Bonifácio, V.B.D., Mata, P., Aires-de-Sousa, J., Lobo, A.M.: NavMol 2.0: a molecular structure navigator/editor for blind and visually impaired users. Eur. J. Org. Chem. 1415–1419 (2013)
10. Lounnas, V., Vriend, G.: AsteriX: A Web Server To Automatically Extract Ligand Coordinates from Figures in PDF Articles. J. Chem. Inf. Model. 52(2), 568–576 (2012), http://www.ncbi.nlm.nih.gov/pubmed?term=Lounnas%20AsteriX, doi:10.1021/ci2004303
11. SMILES – A Simplified Chemical Language, http://www.daylight.com/dayhtml/doc/theory/theory.smiles. html
12. ChemAxon Home Page, http://www.chemaxon.com/ (accessed April 30, 2014)
13. O'Boyle, N.M., Banck, M., James, C.A., Morley, C., Vandermeersch, T., Hutchison, G.R.: Open Babel: An open chemical toolbox. Journal of Cheminformatics 3, 33 (2011), doi:10.1186/1758-2946-3-33
14. Schaftenaar, G., Noordik, J.H.: Molden: a pre- and post-processing program for molecular and electronic structures. J. Comput-Aided Mol. Design. 14, 123–134 (2000)
15. Cactvs editor, http://www2.chemie.uni-erlangen.de/software/cactvs/
16. Humanware, developers of PIAF, http://www.humanware.com/ (accessed April 23, 2014)
17. Antonopoulos, K., Valet, N., Spiratos, D., Siragakis, G.: Olive oil and pomace olive oil processing. Grasas Y Aceites 57, 56–67 (2006)
18. Wedler, H.B., Pemberton, R.P., Lounnas, V., Vriend, G., Tantillo, D.J., Wang, S.C.: Quantum Chemical Study of the Isomerization of 24-Methylenecycloartanol, A Potential Marker of Olive Oil Refining. J. Agric. Food. Chem. (submitted)

Characterization of Pneumonia Incidence Supported by a Business Intelligence System

Maribel Yasmina Santos[1], Vera Leite[1], António Carvalheira[2],
Artur Teles de Araújo[2], and Jorge Cruz[2,3]

[1] ALGORITMI Research Centre, University of Minho, Portugal
maribel@dsi.uminho.pt, a55604@alunos.uminho.pt
[2] Portuguese Lung Foundation, Portugal
antonio.carvalheira@gmail.com, artur@telesdearaujo.pt
[3] Champalimaud Centre for the Unknown, Portugal
jorge.cruz@fundacaochampalimaud.pt

Abstract. Pneumonia, when talking about respiratory diseases, is the leading cause of death and hospital admissions in Portugal, following the global trend, as described by the World Health Organization, which state that is the leading infectious cause of death in children worldwide, accounting for 15% of all deaths of children under 5 years old and, also, that the lower respiratory infections are among the 10 leading causes of death at a Mundial level. If at a worldwide level it is a serious concern, at a local level, country size, pneumonia has also shown an increase in its incidence over the last past decade. This paper presents the overall characterization of pneumonia in Portugal from 2002 to 2011, being possible to study its evolution, degree of fatality and, also, the geo-spatial distribution of the disease over the country. In this decade, a total of 369 160 patients were assisted in hospitals, being the corresponding data analyzed with the help of a Business Intelligence system implemented for the integration, storage and analysis of all the data collected in this study. Besides the information collected in the hospital units, demographic data collected in the 2011 Portugal census were also stored in the data warehouse, allowing the overall characterization of pneumonia and its incidence in the population.

Keywords: Pneumonia, Business Intelligence, Data Warehouse, Dashboards.

1 Introduction

Nowadays, organizations collect large amounts of data that need to be stored and processed in order to extract knowledge about a specific application domain and support the decision making process. However, as the volume of collected data increases, also increases the difficulty of organizations to process such data.

Business intelligence systems emerged aiming to help organizations to integrate, store and analyze the data in order to extract useful information that can support the decision-making process [1].

Although numerous areas can benefit from the application of business intelligence systems, this paper focuses on the healthcare domain. In the area of respiratory

F. Ortuño and I. Rojas (Eds.): IWBBIO 2015, Part I, LNCS 9043, pp. 30–41, 2015.

diseases, pneumonia is the disease that is the leading cause of death and hospital admissions in Portugal [2], following the global trend, as described by the World Health Organization, which state that is the leading infectious cause of death in children worldwide, accounting for 15% of all deaths of children under 5 years old and, also, that the lower respiratory infections are among the 10 leading causes of death at a Mundial level [3].

For the overall characterization of pneumonia in Portugal, this paper presents the implementation of a Business Intelligence system that intends to integrate, store and analyze data collected in hospital units[1] as well as demographic data that allows the analysis of the incidence of the disease in the population.

The proposed business intelligence system integrates a data warehouse [4], which data model was designed to store the mentioned data, making available a wide range of indicators that can be analyzed in specific analytic tools, as dashboards, providing valuable insights on data.

In terms of data, ten years of data concerning the pneumonia cases assisted in Portuguese hospitals from 2002 to 2011, totalizing 369 160 cases, and the demographic data collected in the 2011 Portugal census [5], were effectively integrated. Specific ETL routines (Extraction, Transformation and Loading) ensure data quality and make available the collected data for data analysis.

This paper is organized as follows: Section 2 presents the related work. Section 3 presents the data model for the data warehouse that is responsible for the integration of the available data. Section 4 shows a set of dashboards that allow the interactive analysis of the data. Section 5 concludes with some remarks about the presented work and some proposals of future work.

2   Related Work

There are several studies in the literature mentioning data analytics tasks associated with respiratory diseases in which pneumonia is included.

The authors in [6] performed a study of the prevalence of pneumonia in children under 12 year old in the Tawau General Hospital, Malaysia. The purpose of the performed analyses was identifying the profile of the patients who were admitted to the hospital. The authors report that there are several factors that may have caused the pneumonia, such as family background, or genetic and environmental factors, alerting the government and doctors for taking appropriate actions. All the patients were asked to fill out a form with specific information, such as age, area of origin, parent's smoking background, parent's medical background, and patient's medical background, among other information. In total, data from 102 patients were collected. As main results, the authors point that 86.27% of the patients are from rural areas, reinforcing poor hygiene as an important factor in the origin of pneumonia in Malaysia, a fact stressed in several studies in this field.

---

[1] These data were extracted from the HDGs database (Homogeneous Diagnosis Groups) of the Central Administration of Health Services - ACSS (*Administração Central dos Serviços de Saúde*).

In terms of technological infrastructure, and given the limited scope of the sample, data from 102 patients, the authors do not mention any special needs.

Another study [7] reported that pneumonia is a disease most often fatal, usually acquired by patients during their stay in intensive care units. In this study, data from patients admitted to the intensive care unit at Friedrich Schiller University Jena were collected and stored in a real-time database. Based on the data collected during two years (11 726 cases), these authors developed an early warning system for the onset of pneumonia that combines Alternating Decision Trees for supervised learning and Sequential Pattern Mining. This detection system estimates a prognosis of pneumonia every 12 hours for each patient. In case of a positive prognosis, an alert is generated. In this case, data mining algorithms [8], one of the data analytics techniques used by business intelligence systems, showed to be useful in the analysis of the collected data.

In [9], pneumonia and other respiratory infections are identified as the leading cause of mortality and morbidity in children. These authors conducted a study on five diseases: common cold, sore throat, croup, viral acute bronchiolitis and pneumonia, warning about the importance of an effective diagnosis.

In [10], the authors conducted a study that allowed the development and validation of an ALI (Acute Lung Injury) prediction score in a population-based sample of patients at risk. For the prediction score the authors used a logistic regression analysis. Patients at risk of acquiring an acute respiratory distress syndrome, the most severe form of ALI, were first identified in an electronic alert system that uses a Microsoft SQL-based database and a data mart for storing data about patients in an intensive care unit. A total of 876 records were analyzed, divided in 409 patients for the retrospective derivation cohort and 467 for the validation cohort.

In [11], the authors conducted a study about ventilator-associated pneumonia (VAP) where state that this is the second most common hospital-acquired infection in pediatric intensive care units. Despite the high volume of existing information in the literature on VAP in adults, the amount of these relating to children is limited. These authors reported that prevention is the most appropriate intervention, although few studies have been done in children to identify necessary skills and strategies on how to proceed. This work provided background evidence to support the use of intervention and prevention strategies, with some reliance on data from adults, and explain its application in children.

The several works mentioned in previous paragraphs show the interest of the scientific community in the analysis of data related with respiratory diseases and, in some cases, with pneumonia. In global terms, the work presented in this paper presents a general overview of pneumonia not restricting the sample or group of people under analysis. It is worth to mention that, to the best of our knowledge, no other study uses a dataset with a decade of pneumonia cases. Moreover, this work follows the recent advances in data analytics, proposing a specific business intelligence approach for such study.

3   Data Model and Warehouse

For the analysis of the pneumonia incidence in Portugal, along 10 years with a total of 369 160 cases, this work proposes the implementation of a data warehouse dedicated to store and allow the analysis of the available data. The proposed decision support data model integrates 3 fact tables dedicated to store data related with pneumonias, other pathologies verified by the patients and, also, demographic data relevant to characterize the incidence of the disease in the population.

The work presented in this paper intends to integrate several data sources in a data warehouse with the aim of providing contextual information to the study of pneumonias in Portugal. In a first stage, the incidence of this respiratory disease in the population is studied considering the demographic distribution of the population in Portugal. With this aim in mind, the data collected in the previous census exercise in 2011[2] is integrated, providing a clear overview of the regions were a higher incidence of the disease is verified.

Figure 1 presents the data warehouse model, in which three fact tables allow the integration of the information previously mentioned. The Pneumonias Incidence fact table is used to store the information about each patient with pneumonia, verified in continental Portugal, from 2002 to 2011. This fact table in supported by four dimensions tables, which provide the context about the Patient, the Hospital, the Date of admission and the geographic Location where the patient lives. Moreover, it integrates four indicators, used to study the incidence of this disease. For each patient, the Number of readmissions indicates how many times the patient was readmitted in the hospital as consequence of pneumonia. The Number of days in hospital represents the total number of days the patient was in the hospital. The Flag mortal victim indicates if a specific patient died, or not, as consequence of the pneumonia and related problems. The Pneumonias counter if an event-tracking attribute used to summarize the gathered information.

For each patient with pneumonia, information about other pathologies is recorded in the hospital database, characterizing the overall physical condition of the individual. For this reason, the data warehouse integrates the Pathologies Incidence fact table. This is a factless fact table that integrates a Pathology counter attribute, used to summarize this event-tracking process. The Patient, Pathology and Year dimensions give context to this fact table.

The Census Statistics fact table is used to store the main statistics collected in the 2011 Portugal census and that indicate, at the parish level (Location dimension), the number of individuals living there. Besides the total number (Number of individuals), partial statistics are also known either by gender (F, M) or by age class. Is this study, the age classes ([0,4], [5,9], [10,13], [14,19], [20,24], [25,64], [+65]) proposed by the Portuguese Statistics Institute are used in order to study the incidence of pneumonia by gender and also by age in each location.

---

[2]   Data made available by the Portuguese Statistics Institute, www.ine.pt.

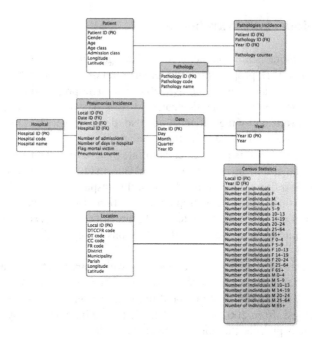

Fig. 1. The Data Warehouse Model

The ETL process started by the extraction of the relevant data and the cleaning of noise and other identified problems in data. Several transformations were also undertaken on data, as the definition of classes or clusters of pathologies. While the data warehouse was implemented in MySQL (www.mysql.pt), the ETL process was implemented using Talend (www.talend.com). All the dashboards presented in the data analytics section were implemented using Tableau (www.tableau.com).

The implementation of the data warehouse proposed in this section allowed the integration of a vast diversity of data, which is fundamental to understand the evolution of pneumonia in Portugal. After the loading of the data warehouse, different dashboards were implemented integrating the several perspectives by which the disease is going to be analyzed.

4   Data Analytics

After the implementation of the data warehouse and its loading with all the available data, several dashboards were implemented to make available a wide range of data analytics instruments that can be used in the characterization of pneumonia and its incidence in the population.

The first dashboard (Figure 2) intends to support the overall analysis of pneumonia and it is divided in four main charts. The first one shows how the number of pneumonias has increased in the analyzed 10 years and, also, how the number of mortal victims has accompanied this increase (Chart 1.Pneumonia incidence). It is worth to mention that in 10 years the number pneumonias increased 33.9% (from 31 257 cases in 2002 to 41 847 cases in 2011) and the number of victims increased 65.3% (from 4 995 cases in 2002 to 8 259 in 2011). In terms of gender and age, almost 70% of the population has 65 or more years, being male more affected than female (Chart 2.Gender incidence). In the [25-64] age class, a relevant difference in terms of the gender incidence is verified, with an average percentage of 65% for males and 35% for females. In no other age class is verified such difference. In terms of mortal victims, it is possible to verify that a significant number of fatalities was verified in individuals that, although presenting the disease, had no admission in a hospital (Chart 3.Admissions). This value was increasable high in 2005 where a variation of 70.24%, with respect to 2004, was verified. Moreover, the variation of mortal victims per year presented an irregular pattern until 2007, with several highs and lows, turning into a more regular pattern in the last analyzed years (Chart 4.Mortal Victims). This last chart also shows this variation regarding gender. Although the incidence of pneumonia is higher in males, it is on females that the variation is more impressive mainly related with the increasing of the number of cases.

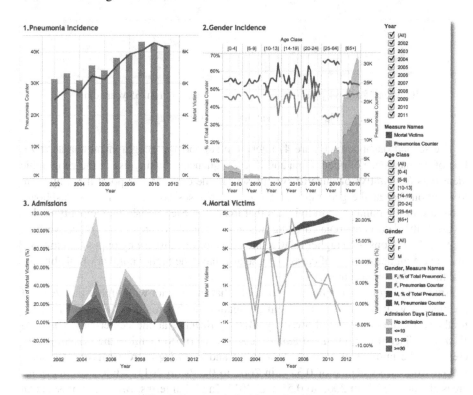

Fig. 2. Pneumonia Incidence Dashboard

To complement the several analyses made with the help of this dashboard, many other charts can be created, mainly to understand or to study in more detail particular patterns. For example, there is a huge difference in the gender incidence in individuals

between 25 and 64 years old. A drill-down by age, Figure 3, shows that this difference is transversal to almost all ages in this class, with the exception of individuals
with 25 to 29 years old. More impressive yet, is the difference between genders in
terms of mortal victims affected by pneumonia in this range of ages. For the age class
[65+], a closer analysis shows the change of behavior in terms of the incidence of
pneumonia in males and females, mainly after the eighties. For 84 years old, the incidence in males is of 49.77% and 50.23% in females, expressing the point of change.
In terms of mortal victims, it is from 86 to 87 years old that we identify the change of
pattern related to genders.

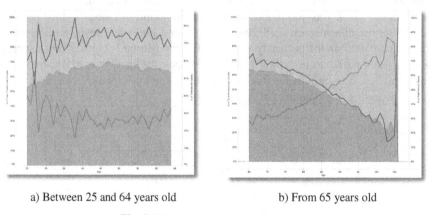

a) Between 25 and 64 years old   b) From 65 years old

Fig. 3. Mortal victims by gender and age

The second dashboard (Figure 4) intends to support the geographical characterization of the incidence of pneumonia and its consequences, making use of the geographic
identifiers (qualitative or quantitative) present in the data. At this point it is important
to mention that at the geographical level, the analyses considered in this dashboard
make use of the number of individuals that inhabit in each parish, municipality or district. All the data is aggregated using this geographical hierarchy. The objective is to
identify the regions with higher relative incidence of this disease by opposition to those
that have a higher number of cases because also have the higher number of inhabitants,
like the metropolitan centers.

Analyzing Figure 4 it is possible to verify that the district with higher incidence of
pneumonia with respect to the global population in that region is *Bragança*, at the
northeast of the country, followed by *Castelo Branco*, at the center-east, and *Vila
Real*, an adjacent district of *Bragança*. It is also worth to mention the systematic increase in the number of cases that *Coimbra* and *Portalegre* are verifying. *Coimbra*
moved from an incidence of 0.38% in 2002 to 0.64% in 2011, while *Portalegre* increased from 0.37% in 2002 to 0.55% in 2011. In global terms, *Bragança* presents an
average incidence of 0.65% of pneumonia in the population, obtaining a maximum of
0.81% in 2009. The presented dashboard also includes a set of filters (District, Year
and Incidence) that can be used to slice the data to be analyzed.

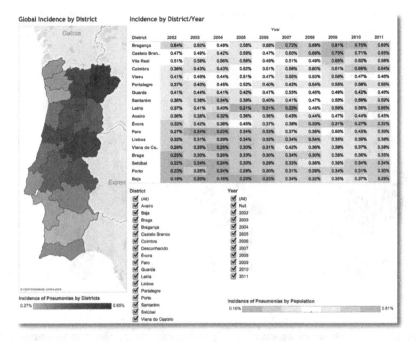

Fig. 4. Geographic incidence of pneumonia

Having in mind the global number of pneumonias in continental Portugal and also its geographic incidence, it is now important to verify if distinct behavioral patterns exist along the country in terms of the number of days patients stay in hospitals, the number of readmissions to the hospital and, also, where fatalities are more frequent attending to the number of pneumonia cases (Figure 5). In terms of the number of days in the hospital, the districts with longer stays are *Guarda*, with an average of 12.3 days, *Lisboa*, with an average of 12.1 days and *Coimbra*, with an average of 11.7 days. For the number of readmissions, the three districts with more readmissions are *Setúbal*, *Faro* and *Portalegre*. Finally, in this dashboard, is presented the incidence of fatalities attending to the number of pneumonias verified in each region. In this case, *Beja* presents an incidence of fatalities of 25.43%, followed by *Setúbal*, with 23.86%, *Portalegre*, with 21.57%, *Santarém* and *Faro* with 20.94%.

While *Setúbal* and *Faro* present higher values in terms of the incidence of mortal victims, these districts were also pointed out as verified significant values in terms of readmissions, and averages values around 11.2 days in terms of staying of the patients in the hospital. Somehow, they seamed to stand out in the undertaken analysis. However, *Beja* presents the lowest incidence of pneumonia regarding the number of inhabitants, but the highest incidence of fatalities considering the number of patients with pneumonia. In order to acutely understand this phenomenon, Figure 6 presents the percentage of pneumonia cases along the several years in analysis and the percentage of victims in this district, considering the age of the individuals. While the maximum incidence of pneumonia in *Beja* is 4.11% for individuals with 82 years old, the maximum incidence of fatalities is of 6.00% for 85 years old. These numbers are calculated considered the total number of cases in the database.

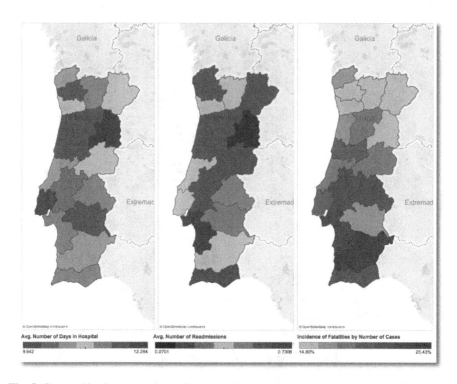

Fig. 5. Geographic characterization of number of days in hospital, number of readmissions and number of fatalities

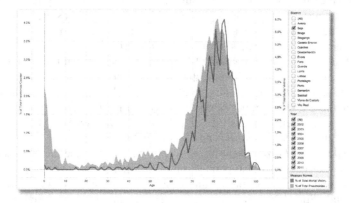

Fig. 6. Distribution by age of the number of pneumonias and fatalities in *Beja*

The previous analyses showed the incidence of pneumonia in the population and how this disease has evolved along the years. It is now important to understand how other pathologies existing in the several patients affect or influence the course of pneumonia and mainly, its consequence, in terms of fatalities.

For the study of the related pathologies, a cluster of pathologies was defined considering the relevance of the several pathologies in the study of pneumonia. The cluster of pathologies includes the chronic pulmonary, cardiac, pancreatic, renal and hepatic diseases, and also, the mellitus diabetes disease. All the other pathologies were classified as *other pathologies*.

For the analysis of the associated pathologies in the fatalities by pneumonia, Figure 7 shows the dashboard implemented to facilitate this analysis. In the first chart (1.Pathologies incidence) it is possible to see the distribution of the mortal victims by pneumonia considering the other pathologies of the patients. Besides presenting other pathologies, the pulmonary disease presents a global incidence of 18.09% in 2002 and a value of 17.91% in 2011, showing a minor decreasing along the years. In the opposite way, the chronic cardiac disease increases its incidence from 16.63% in 2002 to 18.24% in 2011. The incidence of mellitus diabetes also increased from 9.71% in 2002 to 10.03% in 2011. The chronic hepatic disease has an average incidence lower than 0.50%, while the chronic pancreatic disease has an incidence lower than 0.03%. In global terms, it is worth to mention the relevant increase that the chronic renal disease presents, advancing from 0.02% in 2002 to 6.69% in 2011. When we look into the incidence of pathologies excluding the class *other pathologies* (2.Pathologies Incidence (excluding other pathologies)), it is possible to see the relative incidence of each one in the fatalities by pneumonia.

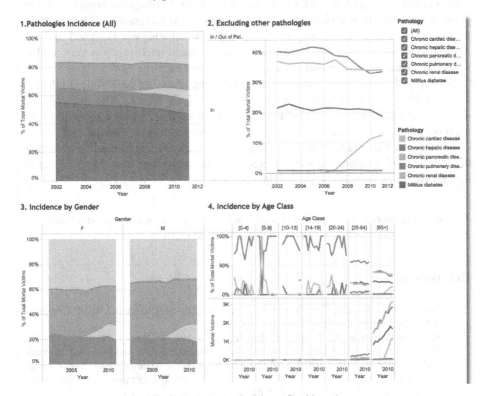

Fig. 7. Pathologies Incidence Dashboard

Analyzing the gender of the patients (3.Incidence by Gender), a different pattern of incidence is verified in terms of mortal victims and the cluster of pathologies. While female present a higher incidence of chronic cardiac disease (ranging from 39.6% in 2002 to 37.39% in 2011), male present a higher incidence of chronic pulmonary disease (ranging from 43.71% in 2002 to 35.65% in 2011). For the other pathologies, female present a higher incidence than male in mellitus diabetes, while male present a higher incidence than female in the chronic hepatic and renal diseases.

Regarding the incidence by age (4.Incidence by Age Class), and considering the defined age classes, for the younger patients, the majority of the mortal victims by pneumonia also presented chronic pulmonary disease. For the [25-64] class, the incidence of the chronic pulmonary disease decreases, increasing the incidence of chronic cardiac disease and mellitus diabetes. For the [65+] age class, the incidence of chronic cardiac disease gains importance falling the incidence of chronic pulmonary disease. This last chart also shows the number of mortal victims with respect to the associated pathologies, and, for the younger patients, it is a residual value.

5   Conclusions

This paper presented the several analyzes made to a data set with 369 160 individuals with pneumonia, aggregating data collected in the several Portuguese hospitals. To be possible the analysis of such volume of data, and also the integration of demographic data collected in the 2011 Portugal census, a data warehouse was designed and implemented, being the principal component of the business intelligence system used in this study to integrate, store and analyze the available data.

Several dashboards were made available providing analytical capabilities to the decision making process, allowing the interactive analysis of data. The implemented data analytics capabilities revealed relevant insights on data, which are useful to the appropriate characterization of pneumonia and its incidence in the population.

As future work, data mining algorithms will be used to identify predictive models that can be used to anticipate specific conditions or events and, also, to go further in the knowledge extracted from the available data.

Acknowledgements. This work was supported by FCT – *Fundação para a Ciência e Tecnologia*, within the Project Scope: UID/CEC/00319/2013.

References

[1]   Watson, H.J., Wixom, B.H.: The current state of business intelligence. IEEE Computer 40(9), 96–99 (2007)
[2]   Carvalheira Santos, A., Gomes, J., Barata, F., Munhã, J., Ravara, S., Rodrigues, F., Pestana, E., Teles de Araújo, A.: Central dos Serviços de Saúde, and Infarmed, Prevenir a doença acompanhar e reabilitar o doente. Observatório Nacional das doenças Respiratórias (2013) (in Portuguese), http://www.fundacaoportuguesadopulmao.org/Relatorio_ONDR_2013.pdf

[3]   W. H. Organization: The top 10 causes of death (2014),
      http://who.int/mediacentre/factsheets/fs310/en/ (accessed: Novem-
      ber 26, 2014)
[4]   Kimball, R., Ross, M.: The Data Warehouse Toolkit, The Definitive Guide to Dimen-
      sional Modeling, p. 600. John Wiley & Sons, Inc. (2013)
[5]   I.N. Estatística: Quadros população (2011), http://censos.ine.pt/xportal/
      xmain?xpid=CENSOS&xpgid=censos_quadros_populacao
[6]   Sufahani, S.F., Razali, S.N.A.M., Mormin, M.F., Khamis, A.: An Analysis of the Preva-
      lence of Pneumonia for Children under 12 Year Old in Tawau General Hospital,
      Malaysia (May 2012)
[7]   Oroszi, F., Ruhland, J.: An early warning system for hospital acquired. In: 18th
      European Conference on Information Systems, ECIS (2010)
[8]   Han, J., Kamber, M., Pei, J.: Data Mining: Concepts and Techniques, 3rd edn. The Mor-
      gan Kaufmann Series in Data Management Systems, p. 744. Morgan Kaufmann (2012)
[9]   Omar, A.H., Zainudin, N.M., Aziz, B.B., Rasid, M.A., Nurani, N.K., Kiong, P.C.,
      Tuan, K.C., Paramjothy, M., Lan, W.S.: Clinical Practice Guidelines on Pneumonia and
      Respiratory Tract Infections in Children, http://www.acadmed.org.my/view_
      file.cfm?fileid=204 (accessed: November 20, 2014)
[10] Trillo-Alvarez, C., Cartin-Ceba, R., Kor, D.J., Kojicic, M., Kashyap, R., Thakur, S.,
      Thakur, L., Herasevich, V., Malinchoc, M., Gajic, O.: Acute lung injury prediction
      score: Derivation and validation in a population-based sample. Eur. Respir. J. 37(3),
      604–609 (2011)
[11] Cooper, V.B., Haut, C.: Preventing ventilator-associated pneumonia in children: an
      evidence-based protocol. Crit. Care Nurse 33(3), 21–29 (2013)

Are Wildfires and Pneumonia Spatially and Temporally Related?

Maribel Yasmina Santos[1], Vera Leite[1], António Carvalheira[2],
Artur Teles de Araújo[2], and Jorge Cruz[2,3]

[1] ALGORITMI Research Centre, University of Minho, Portugal
maribel@dsi.uminho.pt, a55604@alunos.uminho.pt
[2] Portuguese Lung Foundation, Portugal
antonio.carvalheira@gmail.com, artur@telesdearaujo.pt
[3] Champalimaud Centre for the Unknown, Portugal
jorge.cruz@fundacaochampalimaud.pt

Abstract. The global incidence of Pneumonia has increased along the past decade, being the leading cause of death and hospital admissions in Portugal, following the global trend, as described by the World Health Organization. Several studies have emerged trying to study how the consequences of wildfires, namely in what concerns to the generated smoke, can influence and potentiate the emergence of respiratory infections, namely pneumonia. Wildfires are common phenomenon in warmer climates such as Portugal and may have devastating effects, which can get worst with the verified climate changes. As wood smoke contains tiny particles and gases that can have serious effects when breathed, this paper presents a study of the influence of wildfires smoke on the health of the Portuguese population from 2002 to 2011, namely in what concerns to the influence of respiratory infections like pneumonia. In this decade, a total of 369 160 patients were assisted in hospitals as consequence of pneumonia and a total of 338 109 wildfires were registered, being the corresponding data analyzed with the help of a Business Intelligence system implemented for the integration, storage and analysis of the available data. The obtained results showed the emergence of a strong correlation in space and time between these two events in specific municipalities.

Keywords: Pneumonia, Wildfire, Business Intelligence, Dashboards.

1 Introduction

Business Intelligence Systems are being designed and implemented to support data analysis tasks that help organizations in the decision making process. These tasks are accomplished using several technologies aimed to store and analyze data. These systems emerged aiming to help organizations to integrate, store and analyze the data in order to extract useful information that can support specific analytical tasks [1].

F. Ortuño and I. Rojas (Eds.): IWBBIO 2015, Part I, LNCS 9043, pp. 42–53, 2015.
© Springer International Publishing Switzerland 2015

Although numerous areas can benefit from the application of business intelligence systems, this paper focuses on the healthcare domain. In the area of respiratory diseases, pneumonia is the disease that is the leading cause of death and hospital admissions in Portugal [2], following the global trend, as described by the World Health Organization, which state that is the leading infectious cause of death in children worldwide, accounting for 15% of all deaths of children under 5 years old and, also, that the lower respiratory infections are among the 10 leading causes of death at a Mundial level [3]. As Pneumonia has increased its incidence in the past years, and as climate changes also have as consequence the increase of wildfires mainly in warmer countries, it is necessary to study the evolution of these two phenomena and verify if any relation between the two emerges.

In the literature, several studies indicate that wildfires smoke can be the cause of some of the assisted pneumonia in hospitals around the world. As described by the World Health Organization, the chronic health impacts resulting from the exposure to wildfires smoke increased the risk of carcinogenesis, increased the incidence of asthmatic and respiratory diseases and may be responsible for the development of new cases of chronic lung diseases, decreasing life expectancy [4].

The overall study, influence of wildfires smoke on development of pneumonia in Portugal, was supported by a Business Intelligence system that integrates the available data through the implementation of a data warehouse [5], which data model makes available a wide range of indicators that can be analyzed in specific analytic tools, such as dashboards, providing valuables insights on data. This paper gives emphasis to the description of the obtained results and shows how specific dashboards are used to analyze the evolution of both phenomena along the years under analysis. Moreover, and in order to go deeper in the analysis, the mathematical correlation between the two phenomena was calculated, measuring the strength of the relationship between the two events.

In terms of data, ten years of data concerning the pneumonia cases assisted in Portuguese hospitals[1] from 2002 to 2011, totalizing 369 160 cases, and wildfire occurrences from 2002 to 2011, totalizing 338 109 cases, were integrated.

This paper is organized as follows: Section 2 presents the related work. Section 3 describes the available data and analyzes the evolution of both phenomena along the years. Section 4 presents the results obtained when analyzing a cause-effect relationship between these phenomena. Section 5 concludes with some remarks about the presented work and some proposals of future work.

2   Related Work

There are several studies in the literature mentioning that human health can be severely affected by wildfires. In [6], the authors performed a study aimed to verify the impact on health of smoke from wildfires in the United Kingdom standpoint.

---

[1] These data were extracted from the HDGs database (Homogeneous Diagnosis Groups) of the Central Administration of Health Services - ACSS (*Administração Central dos Serviços de Saúde*).

These authors report that wildfires smoke is constituted by tiny particles and gases that can have serious effects when breathed, including an increase in daily mortality.

The tiny particle is the pollutant that prevails in air resulting from bushfire smoke, caused mostly by the burning of vegetation and wood. Some particles are able to pass through the upper respiratory tract and are deposited in the airways and some may reach deeper areas of the lungs and be deposited in the gaseous exchange region of terminal bronchi and alveoli. The wildfires also release gaseous emissions including carbon monoxide, nitrous oxides and benzene, as well as carcinogens including polycyclic aromatic hydrocarbons, aldehydes, and volatile organic compounds.

Certain population groups are at particular risk of respiratory effects from bushfire smoke, including small children, people with pre-existing cardiopulmonary conditions and smokers.

The same authors mention a study based on symptoms of 21 patients with chronic obstructive pulmonary disease (COPD), after two months of the occurrence of wildfires in Denver in 2002, and refer that these patients revealed dyspnea, cough, chest tightness, wheezing and increased sputum production. All these symptoms increased in the days in which the tiny particles levels and carbon monoxide in the atmosphere also increased. The same authors also report the a peak in respiratory consultations was verified five weeks after the Canada wildfires in 2003 and that the long term exposure to the particles may help to explain an increase in pneumonia and acute bronchiolitis after the California wildfires in 2003.

Another study [7] makes several recommendations and gives orientations about this matter. The authors mention several episodes caused by wildfires, like in Indonesia between July and October 1997, in that the Malaysia Ministry of Health reported an increase from two to three times the number of outpatient visits for respiratory diseases during high tiny particle concentration. In south Sumatra, in 1997, pneumonia cases increased by 1.5-5 times compared to 1996. Also, in September 1997, in the province of Jambi, Indonesia, an increase of 50% of upper respiratory tract infections, compared to the observed in the previous month, or in Southeast Kalimantan, Malaysia, in 1997, where pneumonia cases increased from 5 to 25 times, compared with 1995/1996.

In [8], the authors indicate that the wood smoke is very similar to cigarette smoke containing hundreds of air pollutants that can cause cancer and other health problems. Once again, the tiny particles are mentioned and it is mention that this is the factor of more concern with regard to wood smoke. In the document, classes of vulnerable individuals are pointed out, infants and children, the elderly and adults with existing heart or lung conditions.

Through the analysis of the previous paragraphs it is possible to see the concern that this subject raised in the academic community. In this paper, it is intended to enrich this knowledge presenting the analysis of a decade of data, for pneumonia and wildfires.

3   Available Data

For the analysis of the influence of wildfires smoke on the development of pneumonia in Portugal, ten years of data collected in Portuguese hospitals are used, a total of 369 160 cases, and ten years of wildfires, a total of 338 109, made available by the National Authority of Civil Protection[2], are used. These two data sets were integrated, stored and analyzed to understand how the two phenomena evolved along the ten years and to identify if any relationship exist among them.

As can be seen in Figure 1, the number of pneumonias has increased during the period in analysis (blue line), while the number of wildfires as well as the burnt area have decreased along the years (red bars). Despite the relevant decrease in the burnt area, wildfires continue to have a strong impact in the environment and in the society.

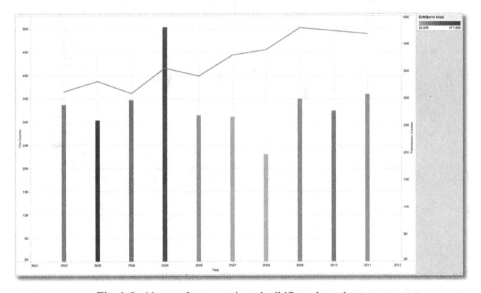

Fig. 1. Incidence of pneumonia and wildfires along the years

To support the analysis of such vast amount of data and allow the implementation of an ad-hoc query environment, a business intelligence system was used. A data warehouse model that allowed the integration of, among other data, the information about the pneumonias and the fires, supports this system.

This paper will not give emphasis to the implemented business intelligence system, as its objective is to show the results of its implementation and its usefulness. Before continuing with the presentation of the main findings, the spatial and temporal characterization of these two phenomena is shown.

---

[2] Available at http://incendios.pt/pt.

Figure 2 presents the characterization of the occurrences in the fires table, where it is possible to see that in terms of the number of occurrences, the *Porto* district is the one with higher incidence, with a total of 69 113 in ten years, while *Guarda* is the district with more burnt area, a total of 190 500 hec (hectares) for 13 619 occurrences. It is worth to mention that it is in the North part of the country that more fires as well as higher burnt areas are verified.

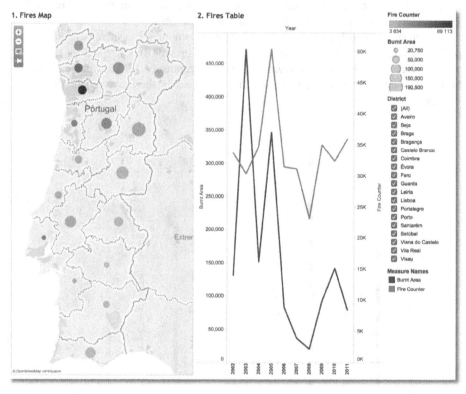

Fig. 2. Spatio-temporal characterization of the wildfires and distribution by year of the number of wildfires and burnt area

For pneumonias, the districts with more cases are obviously the main metropolitan areas like *Lisboa* or *Porto*. Figure 3 shows the distribution of cases along the country, also pointing the number of fatalities (mortal victims) and the distribution of these cases by the age of the patients. In 10 years, the number pneumonias increased 33.9% (from 31 257 cases in 2002 to 41 847 cases in 2011) and the number of victims increased 65.3% (from 4 995 cases in 2002 to 8 259 in 2011). In terms of age, almost 70% of the population with pneumonia has 65 or more years, being this group also the more affected in terms of fatalities.

After the initial data characterization at the district level, and looking forward to a deeper understanding of these two phenomena, next section presents the main findings in the analysis of wildfires and pneumonias.

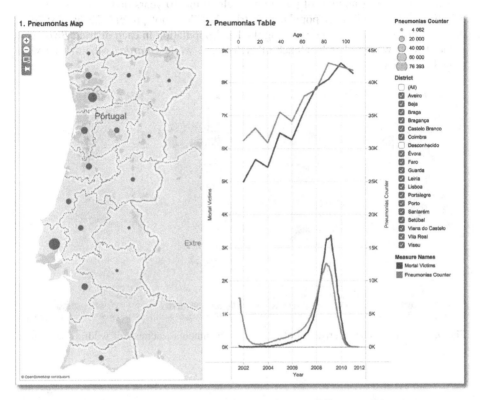

Fig. 3. Geographic characterization of the pneumonias and distribution by year of the number of pneumonias and mortal victims

4   Analyzing Forest Fires and Pneumonia

Figure 4 presents the spatial characterization of the number of pneumonias in the population, the incidence of pneumonia in each district attending to the local population, the number of occurrences in the fires database and the burnt area, for all the years under analysis. In terms of number (either pneumonias or fires), the metropolitan areas like *Lisboa* and *Porto* present a high concentration of events, while the incidence of pneumonia (considering the local population) as well as the extension in terms of burnt areas, are mainly verified in the interior of the country, namely at the Northeast and Central East parts of the country.

Looking at the spatial distribution, we can see that districts like *Guarda*, *Castelo Branco*, *Vila Real* and *Santarém* present higher incidences of burnt area, which in some cases matches the districts with higher incidence of pneumonia considering the local population, which are the districts of *Bragança*, *Castelo Branco*, *Vila Real* and *Coimbra*. However, although coincident in space, both phenomena present a different temporal behavior, as can be seen in Figure 5, where the number of pneumonias along the several months and the burnt area are depicted for some of the districts. *Lisboa* presents the highest number of pneumonias along the 10 years under analysis, as the country's capital and most populated district, but does not present the highest incidence of pneumonia when considering the local population in each district. *Bragança* is the district with higher incidence of the disease while *Guarda* is the district with more burnt area.

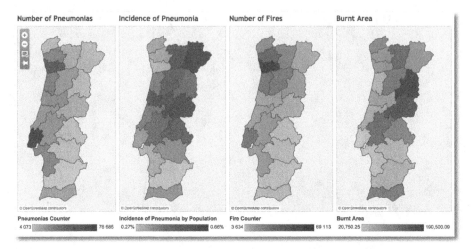

Fig. 4. Geographic characterization of number of pneumonias, number of wildfires and burnt area

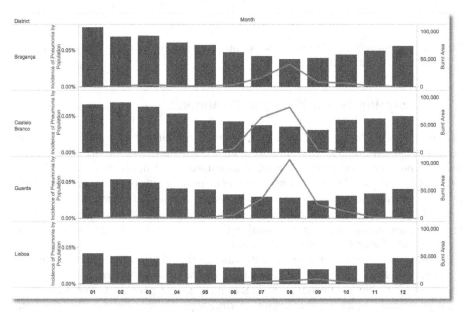

Fig. 5. Incidence of pneumonia by the population of the districts and burnt area

In temporal terms, it is difficult to analyze the data along a year and find a relationship between the phenomena, as their distribution is very different, having fires more cases in summer, and pneumonia in autumn and winter. Even if we drill-down the available data, having for example a quarter or monthly analysis base, the visual analysis of ten years of data is very difficult considering that at the spatial level with have 18 districts, more than 270 municipalities and more than 4 000 parishes.

Given this difficulty to analyze the available data, and in order to allow a global analysis of the available data at the municipality level, for the 10 years of data, the mathematical correlation between the two phenomena was calculated, as it allows the quantification of the strength of the relationship between the two variables.

A positive correlation indicates that the two phenomena are evolving the same way, when one increases the other also increases and when one decreases the other also decreases. A correlation of 0.7 or higher is considered a strong correlation. When the correlation is negative, it means that we have an inverse behavior, that is, when one increases, the other event decreases.

For the pneumonias data, the selected indicator was the number of pneumonias while in the case of wildfires the selected indicator was the burnt area, as the number of fires is not proportional to the burnt area, as shown in Figure 4, and small fires may not produce the quantity of tiny particles and gases that can affect breathing causing serious effects in lungs. Assuming this premise, for each year under analysis and for each municipality, from 2002 to 2011, the correlation between the two phenomena was calculated.

One of the aspects considered when calculating the correlation was the need to establish a time gap between the two phenomena, which is the incubation time of pneumonia. For this, the scientific committee of the Portuguese Lung Foundation pointed from one to two months as the incubation time required for the appearance of the disease, which agrees with the work of [6] already mentioned, pointing that after the forest fires in Canada in 2003 there was a peak in respiratory consultations five weeks after the fires.

As result of the mentioned analysis, this section presents some cases of municipalities for which a strong relationship between the two phenomena was identified. Besides the identified strong correlation, it is worth to mention that these correlations were identified in municipalities with a high incidence of pneumonias during summer, which corresponds to the time of the year with less incidence of the disease, as shown in Figure 6.

In a total of 278 municipalities and 10 years of data, 5 strong correlations were identified, considering a delay of one month between the events, as can be seen in Table 1.

Month	2002	2003	2004	2005	2006	2007	2008	2009	2010	2011
January	4.468	3.208	3.371	4.556	3.840	4.629	4.336	5.396	4.247	5.224
February	3.686	3.057	3.251	5.176	3.682	4.801	4.173	4.128	4.209	4.616
March	3.125	3.287	3.388	4.437	3.513	3.729	3.764	4.252	4.377	4.513
April	2.636	2.852	2.788	2.632	2.781	3.128	3.091	3.675	3.738	3.856
May	2.394	2.575	2.552	2.646	2.795	2.734	2.812	3.529	3.461	3.370
June	2.101	2.345	2.105	2.233	2.274	2.483	2.729	2.805	3.085	2.951
July	1.994	2.036	1.997	1.934	2.412	2.340	2.358	2.415	3.041	2.823
August	1.747	2.287	1.819	1.902	2.234	2.114	2.138	2.644	2.849	2.620
September	1.711	1.854	1.883	1.808	2.066	2.063	2.300	2.803	2.552	2.583
October	2.110	2.402	2.250	2.463	2.505	2.820	2.956	3.423	3.366	3.196
November	2.240	3.533	2.476	2.460	2.469	3.358	3.384	3.649	3.224	3.289
December	3.045	3.707	3.036	3.319	3.568	3.763	4.919	4.255	4.247	2.806

Year

Number of pneumonias

1.711 5.396

Fig. 6. Distribution by year and month of the number of pneumonias

Table 1. Correlations identified considering a delay of one month between the events

Municipality	Year	Correlation
Barrancos	2010	0.719
Alandroal	2002	0.706
Portel	2007	0.836
Borba	2010	0.709
Oliveira de Frades	2010	0.700

Presenting one of these cases, related with the municipality of Oliveira de Frades, a correlation of 0.700 was obtained, considering the burnt area of the fires occurred between January and December of 2010 and the pneumonia cases between February of 2010 and January of 2011. Considering this delay of one month, Table 2 presents relevant values of the burnt area during July and August, and the incidence of pneumonia in August and September with unusual values for this time of the year.

Table 2. Correlation in 2010 for the municipality of Oliveira de Frades

Year	Month	Burnt area (Hec)	Number Pneumonias
2010	January	0	-
	February	0	3
	March	0,001	0
	April	0	2
	May	0	1
	June	0,00030	0
	July	272,515	2
	August	77,3066	7
	September	0,05	5
	October	0,1	2
	November	0	3
	December	0	4
2011	January	-	5

Looking now at the results obtained considering a period of incubation of two months, Table 3 presents the obtained strong correlations, being worth to mention that this number more than duplicated when compared with the one-month scenario.

Table 3. Correlations identified considering a delay of two months between the events

Municipality	Year	Correlation
São João da Madeira	2007	0,804
Castanheira de Pêra	2009	0,716
Sobral de Monte Agraço	2007	0,720
Alter do Chão	2005	0,804
Castelo de Vide	2011	0,749
Marvão	2005	0,764
Matosinhos	2007	0,728
Golegã	2011	0,700
Tarouca	2007	0,700
Vila Nova de Cerveira	2007	0,860
Vila Nova de Paiva	2011	0,766

Presenting one of these cases, for the municipality of Marvão, Table 4 presents the highest incidence of pneumonia during the year in this municipality, in October, after a peak of burnt area in August (Figure 7).

Table 4. Correlation in 2005 for the municipality of Marvão

Year	Month	Burnt area (Hec)	Number Pneumonias
2005	January	0	-
	February	0	-
	March	0,001	2
	April	0,001	1
	May	2	0
	June	0,002	2
	July	4,306	2
	August	77,214	0
	September	0	1
	October	0,003	5
	November	0	1
	December	0	1
2006	January	-	3
	February	-	2

Although difficult to measure and relate in terms of a cause-effect, forest fires and pneumonias present similar spatial patterns if the incidence of the disease in the population is analyzed against the burnt area of the forest fires. However, the two phenomena present different patterns of incidence in what time concerns. Being mentioned by several authors that seem to exist a relation between the two phenomena, the previous results point out that the two events are related in specific places and months of the year.

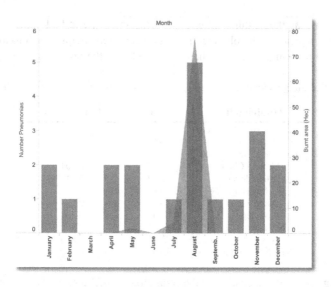

Fig. 7. Incidence of pneumonia and burnt area (Municipality of Marvão)

5 Conclusion

This paper presented the several analyzes made to a data set with 369 160 individuals with pneumonia, data collected in diverse Portuguese hospitals, and 338 109 occurrences of fires, data made available by the Portuguese National Authority of Civil Protection.

To support the analysis of such vast amount of data and to allow the implementation of an ad-hoc query environment, a business intelligence system was used, maintaining a data warehouse for data storage and analytical tools like dashboards for data analysis. Besides the analytical tools used to analyze the available data, and in order to complement the findings, the mathematical correlation between the two phenomena was also calculated. The obtained results point to strong correlations between the two events for some municipalities, in specific years, calling our attention to a possible cause-effect relationship between them. Moreover, these findings are mainly related with peaks of pneumonia during summer or in the beginning of autumn, the periods of the year where the incidence of pneumonia is usually lower.

As future work, and apart from the use of other analytical tools like data mining algorithms, the two phenomena will now be analyzed considering also the duration of the fires, to check if the large exposure to the tiny particles and smoke allow a further comprehension and characterization of the relation between both phenomena.

Acknowledgements. This work was supported by FCT – *Fundação para a Ciência e Tecnologia*, within the Project Scope: UID/CEC/00319/2013.

References

[1] Watson, H.J., Wixom, B.H.: The current state of business intelligence. IEEE Computer 40(9), 96–99 (2007)

[2] Carvalheira Santos, A., Gomes, J., Barata, F., Munhã, J., Ravara, S., Rodrigues, F., Pestana, E., Teles de Araújo, A.: Central dos Serviços de Saúde, and Infarmed, Prevenir a doença acompanhar e reabilitar o doente. Observatório Nacional das doenças Respiratórias (2013) (in Portuguese),
http://www.fundacaoportuguesadopulmao.org/Relatorio_ONDR_2013.pdf

[3] W. H. Organization, The top 10 causes of death (2014),
http://who.int/mediacentre/factsheets/fs310/en/
(accessed: November 26, 2014)

[4] W. H. Organization: Vegetation Fires - Technical Hazard Sheet - Natural disaster profiles,
http://www.who.int/hac/techguidance/ems/vegetation_fires/en/
(accessed: November 26, 2014)

[5] Kimball, R., Ross, M.: The Data Warehouse Toolkit, The Definitive Guide to Dimensional Modeling, p. 600. John Wiley & Sons, Inc. (2013)

[6] Finlay, S.E., Moffat, A., Gazzard, R., Baker, D., Murray, V.: Health impacts of wildfires. PLoS Currents (November 2012)

[7] Schwela, D.: The WHO/UNEP/WMO Health Guidelines for Vegetation Fire Events - An Update. International Forest Fire News, IFFN (2004),
http://www.fire.uni-freiburg.de/iffn/specials/specials.htm
(accessed: November 27, 2014)

[8] Dahlgren, Duerr, Kadlec: How Wood Smoke Harms Your Health (1992),
https://fortress.wa.gov/ecy/publications/summarypages/91br023.html (accessed: November 22, 2014)

Health Technology Assessment Models Utilized in the Chronic Care Management

Ondřej Gajdoš, Ivana Juřičková, and Radka Otawova

CzechHTA, Faculty of Biomedical Engineering, Czech Technical University in Prague,
Kladno, Czech Republic
{ondrej.gajdos,ivana.jurickova,radka.otawova}@fbmi.cvut.cz

Abstract. High prevalence chronic illnesses consume a substantial part of the healthcare budget. We present a cost-effectiveness comparison of the current standard treatment of type 2 diabetes mellitus patients in the Czech Republic with the PCMH (Patient-Centered Medical Home) concept applied in the USA. A randomized sample of 100 probands was chosen out of 1825 Prague diabetological outpatient facility patients. The real-world costs of treatment were calculated for this sample; simultaneously, the costs of the PCMH concept were modelled for the same sample of patients. Then, the outcomes of both technologies (concepts) were figured out using value engineering and multiple-criteria decision making methods (Saaty's matrix and TOPSIS method). Although more expensive, the PCMH concept shows much higher outcome values, which reflects in a significantly higher cost-effectiveness. Preventive, focused and complex care similar to the PCMH concept can be beneficial due to decreasing complications and enhancing patient's quality of life.

Keywords: chronic illness, diabetes mellitus, patient-centered medical home, cost of illness, cost effectiveness analysis.

1 Introduction

A chronic disease is a long-term health disorder often accompanying a person throughout the whole life. A large proportion of the funds covering health care is allocated to the treatment of chronically ill people. The problem is that these funds are not used effectively, which results in the reduction of the health care quality and a waste of money. The trend in the treatment of chronically ill patients is in the non-coordinated application of the same, unnecessary medications and in repeating examinations that have already been done. All is due to the lack of information and non-cooperation among the medical staff providing health care to the respective patient. Another problem is the lack of prevention leading to late detection of the disease and the emergence of potential complications. A related issue is the large number of hospitalizations, which are very costly. The objective of the study was to propose a concept of the chronic care management which will increase the effectiveness and reduce the cost of the health care provided.

F. Ortuño and I. Rojas (Eds.): IWBBIO 2015, Part I, LNCS 9043, pp. 54–65, 2015.
© Springer International Publishing Switzerland 2015

2   Analysis of Current State

According to the European selective health survey of 2009, the most common chronic diseases in the Czech Republic are pains in the lumbar and cervical spine, hypertension and allergies. The most restrictive diseases identified were arthritis and mental disorders [1]. Another very severe chronic disease is diabetes mellitus (DM). Over 347 million people are suffering from this widespread chronic disease worldwide [2]. In Europe, its prevalence accounts for around 60 million patients, of whom 10.3 % of men and 9.6 % of women aged 25 years and older. The DM prevalence within the entire European territory is growing, primarily as a consequence of the growth in overweight and obesity, unhealthy diet and the lack of physical activity [3].

In 2012, the total of 841 227 persons with diagnosed DM were registered by diabetologist's and general practitioner's surgeries. The number of patients with DM grew by 1.9 % against 2011. Thus, the prevalence of this disease affected over 8 % of the population of the Czech Republic.

Type 2 diabetes mellitus (non-insulin dependent) is a disease typical of middle-aged and elderly people. In 2012, 772 585 people were treated with type 2 DM in the Czech Republic. It follows that every seventh Czech citizen suffers from type 2 DM.

Diabetes mellitus represents a well documented example of a disease with prevailing high treatment costs and insufficient health care quality. The ever growing prevalence of the disease may cause significant economic burden to society [5].

2.1   Cost of Illness

Several studies from different countries were selected for the elaboration of studies on the cost of illness with a focus on DM. When calculating the cost of illness both direct and indirect costs are taken into account. One study did not include indirect costs at all due to the lack of respective data. Other differences were found in the very breakdown of direct costs, which should consist of direct medical and non-medical costs. This was only considered in two studies.

Each of the methods was based on a different method of data collection, data processing, had a different focus, but all the studies arrived at the conclusion that DM largely relies on health care resources, its prevalence is increasing and it will constitute a major health care problem in the future. Therefore, it is necessary to conduct studies investigating the cost of illness and, based on it, to focus on the ways of reducing the cost of illness.

2.2   Chronic Care Management

The objective of this study was to find a new model solution for the chronic disease management which would save the cost of treatment being effective, raising the quality of health care provided and reducing hospitalized cases. These criteria may be met by the interconnection and communication within the medical staff team, by eliminating duplicate procedures and over-prescription of medications to reduce costs, and by a complex focus on patients. The Patient-Centered Medical Home (PCMH) concept was selected as a possible solution for use in the Czech Republic. This concept designed in the USA was used as an option for solving the chronic care management as it brings highly positive results in patient care, but it is also beneficial for health care providers and, last but not least, it saves costs.

3   Methods

Methods from the Health Technology Assessment area were selected to compare the health care management systems for patients with type 2 DM [6]. First, cost models were used to determine the cost of illness. The result was then used in the model to identify the cost effectiveness using value engineering and multi-criteria analysis methods. These methods allowed comparing the PCMH concept with the adherence to recommended standards against the system currently applied in the Czech Republic.

3.1   Cost of Illness

Sampling. The cost of illness was calculated in a diabetology outpatient clinic in Prague where a randomized selection of 100 patients was performed. The target group consisted of patients aged over 18 years with a confirmed diagnosis of type 2 DM. A random selection of numbers was generated from a total of 1825 patients in 2013 with the online random number generator. The percentage of men and women represented in the whole group for selection was almost identical.

Perspective. The analysis worked with inputs and outputs from the all-society perspective, from the patient's perspective and from the payer's perspective. The above perspective was essential for the processing of the cost effectiveness analysis (CEA) and the benefits potentially resulting from PCMH.

Costs. Direct medical costs relating to examinations and treatment per patient were identified for a randomized sample of patients. In data collection, access was provided to data on the number of patient visits, the codes assigned for the insurance company, the examinations undertaken by other specialists in connection with the treatment of type 2 DM and the number of laboratory tests.

One of the biggest items in direct costs are drugs. The medications permanently taken by respective patients were available for data collection. These include drugs used primarily for the treatment of type 2 DM, drugs secondarily used for the treatment of type 2 DM and other permanently taken medications required for the treatment of related disorders.

Approach. The calculation of the direct cost of illness was made using the bottom-up procedure.

3.2   Cost Effectiveness Analysis

Cost effectiveness analysis was used to compare the standard treatment of patients with type 2 DM in the Czech Republic with the Patient-Centered Medical Home concept of treatment. The result of the comparison is the identification of the more effective solution for the health care management for patients with type 2 DM.

Perspective. CEA was considered from the payer's point of view.

Costs. The costs again included only direct medical costs, which is adequate for the payer's perspective. Only the cost of procedures performed by outpatient doctors was used for the calculation.

Standard Treatment. The cost of the standard treatment is based on the randomized sample of 100 patients for the cost of illness analysis. The average cost of provided procedures per patient per year is presented in Tab. 2.

Table 1. Average cost of standard treatment per patient (1 year)

total points for registered codes	213 534
point value for 2013	CZK 1.02
total cost of registered procedures	CZK 217 805
average cost per patient	**CZK 2 178**

Patient-Centered Medical Home. The calculation of the cost of procedures which would be provided to patients in the model treatment in PCMH is based on the best practices recommended by the Czech Diabetes Society (CDS) and the professional association's opinion. Patients with type 2 DM were divided into three groups according to the type of treatment. The first group includes patients without any drug treatment, the second group comprises patients taking oral antidiabetic drugs (OAD), and the third group includes patients using insulin. Out of the total number of 772 585 patients with type 2 DM, 116 196 were treated solely by diabetic diet [8]. The procedures performed per patient per year without drug treatment are listed in Tab. 3. The remaining patients had prescribed drug treatment. Of them, 116 181 patients were treated by insulin. The largest percentage of patients corresponding to the number of 540 208 was treated by OAD [4; 8]. The procedures performed per patient taking OAD per year are listed in Tab. 4. The procedures performed per patient taking insulin per year are listed in Tab. 5. The average cost of registered procedures per patient with type 2 DM per year in the PCMH concept is presented in Tab. 6.

Table 2. List of procedures in PCMH per patient without drug treatment

code	name of procedure	points per procedure	No. of procedures	total points
13021	complex examination by a diabetologist	645	1	645
13051	targeted education of a patient with diabetes	327	1	327
75022	targeted examination by an ophthalmologist	232	1	232
75153	biomicroscopic examination of fundus in mydriasis (1 eye)	147	2	294
75161	contact-less tonometry (1 eye)	32	2	64
75163	refraction examination with an autorefractor (1 eye)	36	2	72
points in total				1634

Table 3. List of procedures in PCMH per patient taking OAD

code	name of procedure	points per procedure	No. of procedures	total points
13021	complex examination by a diabetologist	645	1	645
13023	re-examination by a diabetologist	163	3	489
13051	targeted education of a patient with diabetes	327	4	1308
75022	targeted examination by an ophthalmologist	232	1	232
75153	biomicroscopic examination of fundus in mydriasis (1 eye)	147	2	294
75161	contact-less tonometry (1 eye)	32	2	64
75163	refraction examination with an autorefractor (1 eye)	36	2	72
17021	complex examination by a cardiologist	645	1	645
points in total				3749

Table 4. List of procedures in PCMH per patient taking insulin

code	name of procedure	points per procedure	No. of procedures	total points
13021	complex examination by a diabetologist	645	1	645
13023	re-examination by a diabetologist	163	3	489
13051	targeted education of a patient with diabetes	327	4	1308
75022	targeted examination by an ophthalmologist	232	1	232
75153	biomicroscopic examination of fundus in mydriasis (1 eye)	147	2	294
75161	contact-less tonometry (1 eye)	32	2	64
75163	refraction examination with an autorefractor (1 eye)	36	2	72
17021	complex examination by a cardiologist	645	1	645
29021	complex examination by a neurologist	645	1	645
18021	complex examination by a nephrologist	645	1	645
points in total				**5039**

Table 5. Average costs per patient in the PCMH concept (1 year)

total points per patient without drug treatment	1 634
number of patients without drug treatment	116 196
total number of points for patients without drug treatment	**189 864 264**
total points per patient with OAD	3 749
number of patients with OAD	540 208
total number of points for patients with OAD	**2 025 239 792**
total points per patient with insulin	5 039
number of patients with insulin	116 181
total number of points for patients with insulin	**585 436 059**
point value for 2013	1.02
total cost	CZK 2 856 550 917
number of patients with type 2 DM for 2012	772 585
average cost per patient with type 2 DM	**CZK 3 697**

Economic-Clinical Parameters. Among the parameters serving for the comparison of the standard treatment against PCMH, all seven parameters related to the treatment of type 2 DM were selected from individual studies. The monitored parameters were divided into 2 groups. The first group of costs and cost utilization is listed in Tab. 7. The average number of visits for the standard treatment was calculated from the randomized sample, and for PCMH from the estimated number of visits according to the procedures recommended by the Czech Diabetes Society and the professional association's brainstorming. The second group related to the quality of treatment and clinical characteristics is listed in Tab. 8. Both groups of parameters produce improved overall health seen from the patient's perspective [9].

Table 6. List of parameters related to costs and cost utilization

	waiting period length	number of hospitalizations	hospitalization length	average number of visits to outpatient doctors
standard	26 days	10 526	10.5 days	3.880
PCMH	1 day	8 631	6.3 days	5.699

Table 7. List of parameters related to treatment quality and clinical characteristics

	HbA1c level	long-term cholesterol level monitoring	quality of care for patients with DM
standard	0 %	0 %	0 %
PCMH	26 %	25 %	76 %

Assessment of the Weight of Economic-Clinical Parameters and Multi-Criteria Decision Making. Individual assessments of the parameters were made based on the brainstorming of the doctors from a clinic in Prague, which includes the respective diabetes outpatient clinic where the data for the study and the information on the best practices recommended by the Czech Diabetes Society were collected. Saaty's matrix comparing different parameters against each other was used to identify the weights of the criteria. The obtained weights of the criteria were used for multi-criteria decision-making. The TOPSIS method was used for the calculation of the multi-criteria assessment [10].

Cost Effectiveness Analysis. Using multi-criteria analysis and average costs per patient in the standard treatment and the PCMH concept the magnitude of the effect obtained per spent monetary unit was calculated. Based on the resulting ratios, we assess which of the approaches is more cost-effective.

4   Results

4.1   Cost of Illness

Results from All-Society Perspective. From the all-society perspective, all direct costs registered by the randomized sample were added up. The values necessary for the calculation of direct costs are listed in Tab. 9.

Table 8. List of direct costs from all-society perspective

total points for registered codes	213 536
point value for 2013	CZK 1.02
total cost of registered procedures	**CZK 217 807**
number of regulatory fees	332
regulatory fee value	CZK 30
total cost of regulatory fees	**CZK 9 960**
total number of laboratory tests	230
costs per laboratory test	CZK 1 076.32
total cost of laboratory tests	**CZK 247 554**
cost of covered drugs	CZK 2 461 140
cost of non-covered drugs	CZK 16 592
total cost of drugs	**CZK 2 477 732**
total	**CZK 2 953 053**
median	CZK 24 318
minimum	CZK 1 773
maximum	CZK 194 529
standard deviation	CZK 25 055
average costs per patient	**CZK 29 531**
right-hand side confidence interval	**CZK 34 517**

By adding up all direct cost items, the value of CZK 2 953 053 was obtained. The average direct costs per patient in 2013 amount to CZK 29 531. Using the prevalence of type 2 DM for 2012 the value of the total cost of illness was obtained amounting to CZK 22 815 207 635. Here, however, a deviation arises due to the permanently growing prevalence of the disease. If the prevalence for 2013 was used, the costs would be still higher.

Results from Patient's Perspective. From the patient's perspective, all regulatory fees for visits to doctors were added up[1]. For medications, only surcharges for drugs covered by health insurance and average prices of drugs not covered by health insurance obtained from three selected pharmacies were included in the calculation. The values necessary for the calculation of direct costs from the patient's perspective are listed in Tab. 10.

---

[1] Regulatory fees for visits to doctors will be cancelled from 1. 1. 2015.

Table 9. List of direct costs from patient's perspective

number of regulatory fees	332
regulatory fee value	CZK 30
total cost of regulatory fees	**CZK 9 960**
cost of covered drugs	CZK 893 058
cost of non-covered drugs	CZK 16 592
total cost of drugs	**CZK 909 650**
total	**CZK 919 610**
median	CZK 4 040
minimum	CZK 30
maximum	CZK 41 148
standard deviation	CZK 9 860
average costs per patient	**CZK 9 196**
right-hand side confidence interval	**CZK 11 158**

By adding up all direct cost items necessary from the patient's perspective, the value of CZK 919 610 was obtained. Thus, each patient paid on average CZK 9 196 for health care in 2013.

Results from Payer's Perspective. From the payer's perspective, all costs related to the registered treatment were included in the calculation. The values necessary for the calculation of direct costs from the payer's perspective are listed in Tab. 11.

Table 10. List of direct costs from payer's perspective

total points for registered codes	213 536
point value for 2013	CZK 1.02
total cost of registered procedures	**CZK 217 807**
total number of laboratory tests	230
costs per laboratory test	CZK 1 076.32
total cost of laboratory tests	**CZK 247 554**
total cost of covered drugs	**CZK 1 632 269**
total	**CZK 2 097 630**
median	CZK 16 675
minimum	CZK 1 743
maximum	CZK 192 664
standard deviation	CZK 22 147
average costs per patient	**CZK 20 976**
right-hand side confidence interval	**CZK 25 383**

By adding up all direct cost items necessary for reaching the total cost from the payer's perspective, the value of CZK 2 097 630 was obtained. Thus, the health insurance company paid on average CZK 20 976 per patient in 2013.

4.2   Cost Effectiveness Analysis

One monetary unit spent by the payer within the PCMH concept, which adheres to recommended standards, will produce the 25.7x10-5 effect. This value is by 77% higher than the value calculated for the standard treatment and, therefore, the adherence to recommended standards within PCMH is more cost-effective. The results of multi-criteria analysis, which identified the respective effects, and the results of cost-effectiveness analysis are presented in Tab. 12.

Table 11. Result of cost-effectiveness analysis

	effect	price (see Tab. 2, 6)	CEA $(x10^{-5})$	order of variants
standard	0.3160	2178	14.50766397	second
PCMH	0.9492	3697	25.67476062	first

5   Discussion

Our model confirmed that the treatment of chronic diseases accounts for a large portion of the funds spent in health care [5]. The flow of these funds must be coordinated and exploited effectively to avoid wasting. Type 2 diabetes mellitus was chosen as an example for solving the objectives of our study. Although its percentage among chronic diseases is not the highest, it largely threatens the population by its ever increasing prevalence, and it may become a major health and economic issue in many countries around the world in the future. It is a disease that needs a complex approach by many experts being, therefore, a good example to illustrate the importance of communication and cooperation among doctors and other medical staff. This was confirmed by the results of the comparison of the standard treatment in the Czech Republic with the Patient-Centered Medical Home concept used in the USA. It is obvious that the latter approach to treatment is really worth supporting. It improves, above all, the health of patients and the overall quality of diabetological care. Despite the higher costs due to more frequent visits to outpatient doctors, the cost-effectiveness analysis has manifested that this method is more cost-effective.

An important phenomenon in the treatment of diabetes are frequent complications which reduce the quality of patient's life and raise the cost of treatment. The Czech Republic spends larger amounts of money on the direct treatment of diabetes than the most of developed countries in the world [11]. This is mainly due to the emergence of subsequent complications arising as a consequence of poorly tailored outpatient care. The care is neither sufficiently complex nor intensive [11]. As was demonstrated in the USA, Western Europe and the Czech Republic, direct costs for the treatment of diabetes are substantially lower than those accounting for the treatment of complications [11]. For type 2 diabetes mellitus, it would be highly desirable to combine the PCMH characteristics with the best practices recommended by the CDS [11, 12, 13, 14, 15] and the opinions held by the professional association of diabetologists. The results of our model imply that prevention considered as a crucial measure by the National Diabetes Programme will subsequently improve. In the case of ensuring high-quality prevention and professionally controlled therapy, complications whose treatment is so expensive will not arise. Another benefit will be improved quality of life for the patients whose complications were prevented.

Supposing that preventive, targeted and, particularly, complex treatment is provided, the above benefits may also be expected for other chronic diseases. According to our results, the combination of the PCMH concept with adherence to standards produces a greater effect, despite its increased costs. This resulting effect is primarily an advantage for patients as it reduces the risk of the appearance of later complications decreasing their quality of life. The decision still rests with the payer, who must consider if allocating more funds to medical care providers will produce overall savings in the treatment of complications typical of patients with diabetes in the future.

Acknowledgement. The authors acknowledge financial support for the development of HTA methods for medical devices by the Ministry of Health of the Czech Republic under the Grant No. NT11532 "Medical Technology Assessment" and the Grant No. NT14473 "Information system for medical devices purchase monitoring". Radka Otawová was supported by the European social fund within the framework of realizing the project "Support of inter-sectoral mobility and quality enhancement of research teams at Czech Technical University in Prague", CZ.1.07/2.3.00/30.0034.

References

1. Daňková, Š.: European Health Interview Survey in CR - EHIS CR. Institute of Health Information and Statistics of the Czech Republic, http://uzis.cz/system/files/43_09.pdf
2. World Health Organization. Media Centre: Diabetes. World Health Organization, http://www.who.int/mediacentre/factsheets/fs312/en/index.html
3. World Health Organization Europe. Health and topics: Diabetes. World Health Organization Europe, http://www.euro.who.int/en/health-topics/noncommunicable-diseases/diabetes/data-and-statistics
4. Zvolský, M.: Activity in the Field of Diabetology, Care for Diabetics in 2012. Institute of Health Information and Statistics of the Czech Republic, http://www.uzis.cz
5. Bojadzievski, T., Gabbay, R.A.: Patient-Centered Medical Home and Diabetes. Diabetes Care, 1047–1053 (2011)
6. Rosina, J., et al.: Health Technology Assessement for Medical Devices. The Clinician and Technology Journal 44 (2014)
7. Ministry of Health of the Czech Republic. Decree no. 428/2013 Coll., On the point values, the amount of reimbursement paid services and regulatory constraints for the year, Ministry of Health of the Czech Republic (2014), http://www.mzcr.cz/Legislativa/dokumenty/vyhlaska-c428/2013-sb-o-stanoveni-hodnot-boduvyse-uhrad-hrazenych-sluzeb-a_8581_11.html

8. Institute Of Health Information And Statistics of The Czech Republic. Care of Diabetes in 2012, Institute of Health Information and Statistics of the Czech Republic, http://uzis.cz/system/files/diab2012.pdf ISBN 978-80-7472-082-6

9. Patient-Centered Primary Care Collaborative. Summary of Patient-Centered Medical Home Cost and Quality Results, 2010-2013. Patient-Centered Primary Care Collaborative, http://www.pcpcc.net/sites/default/files/PCPCC%20Medical%20Home%20Cost%20and%20Quality%202013.pdf

10. Rogalewicz, V., Juřlčková, I.: Multiple-Criteria Decision Making: Application to Medical Devices. In: Proc. IWBBIO 2014: 2nd int. Work-Conference on Bioinformatics And Biomedical Engineering, pp. 1359–1372 (2014)

11. Czech Diabetes Society. National Diabetes Program, -, Czech Diabetes Society (2012), http://www.diab.cz/dokumenty/NDP_2012_2022_PDF.pdf

12. Czech Diabetes Society. Guidelines on Diabetes Mellitus 2 Type Treatment. Czech Diabetes Society, http://www.diab.cz/dokumenty/dm2_12.pdf

13. Czech Diabetes Society; Czech Ophthalmological Society; Czech Vitreoretinal Society. Guidelines for Diagnostics and Treatment of Diabetic Retinopathy. Czech Diabetes Society, http://www.diab.cz/dokumenty/standard_oci.pdf

14. Czech Diabetes Society. Guidelines on Diagnostics and Treatment of Diabetic Neuropathy. Czech Diabetes Society, http://www.diab.cz/dokumenty/Standardy_DN_2011.pdf

15. Czech Diabetes Society; Czech Nephrological Society. Guidelines in Diabetic Nephropathy. Czech Diabetes Society, http://www.diab.cz/dokumenty/standard_ledviny_12.pdf

16. Epperly, T.(ed.): The Patient-Centred Medical Home in the USA. Journal of Evaluation in Clinical Practice, 373–375 (2011)

17. Rogers, J.C.: The Patient-Centered Medical Home Movement - Promise and Peril for Family Medicine. Journal of the American Board of Family Medicine (2008)

18. Jaén, C.R., Další.: Patient Outcomes at 26 Months in the Patient-Centered Medical Home national Demonstration Project. Annals of family medicine (2010)

19. Rossenthal, T.C.: The Medical Home: Growing Evidence to Support a New Approach to Primary Care. Journal of the American Board of Family Medicine, 427–440 (2008)

20. Steiner, B.D., et al.: Community Care of North Carolina: improving care through community health networks. Ann. Fam. Med., 361–367 (2008)

21. Merrell, K., Berenson, R.A.: Structuring Payment For Medical Home. Health Affairs, pp. 852-858 (2010)

22. Doležal, T.: The Cost of Treating Type 2 Diabetes Mellitus. Postgraduate Medicine (2011)

23. OECD. Health at a Glance: Europe (2012) OECD, http://www.oecd.org/health/healthataglance/europe ISBN 978-92-64-18389-6

24. EEIP; Faculty of Social Sciences cu in Prague. Investment in Health in the Czech Republic (2014)

Prediction of Human Gene - Phenotype Associations by Exploiting the Hierarchical Structure of the Human Phenotype Ontology

Giorgio Valentini[1], Sebastian Köhler[2],
Matteo Re[1], Marco Notaro[3], and Peter N. Robinson[2,4]

[1] AnacletoLab - DI, Dipartimento di Informatica,
Università degli Studi di Milano, Italy
{valentini,re}@di.unimi.it
[2] Institut fur Medizinische Genetik und Humangenetik,
Charité - Universitatsmedizin Berlin, Germany
{peter.robinson,sebastian.koehler}@charite.de
[3] Dipartimento di Bioscienze,
Università degli Studi di Milano, Italy
marco.notaro@studenti.unimi.it
[4] Institute of Bioinformatics, Department of Mathematics and Computer Science,
Freie Universitat Berlin, Germany

Abstract. The Human Phenotype Ontology (HPO) provides a conceptualization of phenotype information and a tool for the computational analysis of human diseases. It covers a wide range of phenotypic abnormalities encountered in human diseases and its terms (classes) are structured according to a directed acyclic graph. In this context the prediction of the phenotypic abnormalities associated to human genes is a key tool to stratify patients into disease subclasses that share a common biological or pathophisiological basis. Methods are being developed to predict the HPO terms that are associated for a given disease or disease gene, but most such methods adopt a simple "flat" approach, that is they do not take into account the hierarchical relationships of the HPO, thus loosing important a priori information about HPO terms. In this contribution we propose a novel Hierarchical Top-Down (HTD) algorithm that associates a specific learner to each HPO term and then corrects the predictions according to the hierarchical structure of the underlying DAG. Genome-wide experimental results relative to a complex HPO DAG including more than 4000 HPO terms show that the proposed hierarchical-aware approach significantly improves predictions obtained with flat methods, especially in terms of precision/recall results.

Keywords: Human Phenotype Ontology term prediction, Ensemble methods, Hierarchical classification methods, Disease gene prioritization.

1   Introduction

The characterization of human diseases through detailed phenotypic data and the ever increasing amount of genomic data available through high-throughput

F. Ortuño and I. Rojas (Eds.): IWBBIO 2015, Part I, LNCS 9043, pp. 66–77, 2015.

technologies can improve our understanding of the bio-molecular mechanisms underlying human diseases. Indeed phenotypic analysis is fundamental for our understanding of the pathophysiology of cellular networks and plays a central role in the mapping of disease genes [1].

To this end the Human Phenotype Ontology (HPO) project [2] provides a comprehensive and well-structured set of more than 10000 terms (classes) that represent human phenotypic abnormalities annotated to more than 7000 hereditary syndromes listed in OMIM, Orphanet and DECIPHER databases [3]. This resource offers an ontology, that is, a conceptualization of the human phenotypes that can be processed by computational methods, and provides a translational bridge from genome-scale biology to a disease-centered view of human patho-biology [4]. The HPO provides also hierarchical relationships between terms, representing the *is_a* relation between them, whereby each term may have more than one parent, thus resulting in a Directed-Acyclic-Graph (DAG) structure of the overall ontology.

In this context, the prediction or ranking of genes with respect to HPO terms is an important computational task . This task is related but different from the classical disease-gene prioritization problem, in which genes are prioritized with respect to specific diseases [5]. Indeed we rank genes with respect to HPO terms. Note that HPO terms do not themselves represent diseases, but rather they denote the individual signs and symptoms and other clinical abnormalities that characterize diseases. Thus, one disease is characterized by ≥ 1 HPO term, and many HPO terms are associated with multiple distinct diseases.

Several computational methods have been applied to predict gene - phenotype associations [6, 7, 8, 9], but they do not take into account the hierarchical relationships that characterize phenotypes both in human and model organisms. The resulting "flat" predictions, i.e. predictions unaware of the relationships between the different phenotypes, may provide inconsistent results. For instance, if we adopt the HPO to catalogue human phenotypes and we try to predict HPO terms independently of each other, we could associate to some human gene the HPO term "Atrial septal defect" but not the term "Abnormality of the cardiac septa", thus introducing an inconsistency since "Atrial septal defect" is obviously a subclass of "Abnormality of the cardiac septa". Besides inconsistency, flat predictions loose the available "a priori" knowledge about the hierarchical relationships between HPO terms, thus suggesting that hierarchy-aware methods could at least in principle introduce improvements in the gene-phenotype predictions. To overcome the limitations of "flat" approaches, we could apply computational methods for hierarchically structured output spaces, but most of them have focused on tree-structured ontologies [10, 11, 12, 13] and only a few on DAG-structured taxonomies [14, 15] and, even if they have been applied in computational biology, e.g. to the prediction of protein functions [16], to our knowledge no hierarchy-aware methods have been applied to the prediction of HPO terms associated to human genes.

To fill this gap, we propose a simple and novel hierarchical method, i.e. the *Hierarchical Top-Down (HTD)* ensemble method conceived to deal with the DAG structure of the HPO. At first a base learner associated with each considered HPO term is applied to provide "flat" gene-phenotype associations. Then the algorithm gradually visits the HPO DAG level by level from the root (top) to the leaves (bottom), and modifies the flat predictions to assure their hierarchical consistency. One of the main advantages of the proposed approach is that it always provides consistent predictions, that is predictions that respect the hierarchical structure of the HPO. Moreover, by exploiting the parent-child relationships between HPO terms, the proposed hierarchical approach can significantly improve HPO flat predictions, as shown by the large set of experiments involving more than 20, 000 human genes and more than 4, 000 HPO terms. The *HTD* method is simple, fast and can be applied by using in principle any base learner for both hierarchical multi-label phenotypic classification and ranking of human disease genes.

2   Hierarchical Top-Down (HTD) Ensembles for the HPO Taxonomy

Let $G = <V, E>$ be a Directed Acyclic Graph (DAG) with vertices $V = \{1, 2, \ldots, |V|\}$ and edges $e = (i, j) \in E, i, j \in V$. G represents a taxonomy structured as a DAG, whose nodes $i \in V$ represent classes of the taxonomy and a directed edge $(i, j) \in E$ the hierarchical relationships between i and j: i is the parent class and j is the child class. In our experimental setting the unique root node $root(G)$ is represented by the top HPO term "HP:0000001": all the other HPO terms are its descendants. The set of children of a node i is denoted $child(i)$, and the set of its parents $par(i)$.

To each HPO term i is associated a "flat" classifier $f_i : X \rightarrow [0, 1]$ that provides a score $\hat{y}_i \in [0, 1]$ for a given gene $x \in X$. Ideally $\hat{y}_i = 1$ if gene x is associated to the HPO term i, and $\hat{y}_i = 0$ if it is not, but intermediate scores are allowed. The ensemble of the $|V|$ flat classifiers provides a score for each node/class $i \in V$ of the DAG G:

$$\hat{\boldsymbol{y}} = <\hat{y}_1, \hat{y}_2, \ldots, \hat{y}_{|V|}> \tag{1}$$

We say that the multi-label scoring $\boldsymbol{y}$ is valid if it obeys the *true path rule* (also called the *annotation propagation rule*) that holds also for other DAG-structured ontologies, such as the Gene Ontology (when restricted to subclass relations) [17]:

$$\boldsymbol{y} \text{ is valid} \iff \forall i \in V, j \in par(i) \Rightarrow y_j \geq y_i \tag{2}$$

According to this rule, if we assign a HPO term i to a gene, then also its parent HPO terms must be assigned to the same gene: in other words an assignment to a node must be recursively extended to all its ancestors. Note that this implies that a score for a parent HPO term must be larger or equal than that of its children. Consequently, if a certain HPO term is classified as a negative example because its score is below threshold, then all of its descendents must also be classified negative.

In real cases it is very unlikely that a flat classifier satisfies the true path rule, since by definition the predictions are performed without considering the hierarchy of the classes. Nevertheless by adding a further label/score modification step, i.e. by taking into account the hierarchy of the classes, we can modify the labeling or the scores of the flat classifiers to obtain a hierarchical classifier that obeys the true path rule.

To this end we propose a Hierarchical top-down algorithm *(HTD)*, that modifies the flat scores according to the hierarchy of a DAG through a unique run across the nodes of the graph. It adopts this simple rule by per-level visiting the nodes from top to bottom:

$$
\bar{y}_i := \begin{cases} \hat{y}_i & \text{if} \quad i \in root(G) \\ \min_{j \in par(i)} \bar{y}_j & \text{if} \quad \min_{j \in par(i)} \bar{y}_j < \hat{y}_i \\ \hat{y}_i & \text{otherwise} \end{cases} \tag{3}
$$

Note that $\bar{\boldsymbol{y}} = < \bar{y}_1, \bar{y}_2, \ldots, \bar{y}_{|V|} >$ represents the set of the predictions obtained by the *(HTD)* algorithm from the flat predictions $\hat{\boldsymbol{y}} = < \hat{y}_1, \hat{y}_2, \ldots, \hat{y}_{|V|} >$.

The node levels correspond to their maximum path length from the root. More precisely, having $\mathcal{L} = \{0, 1, \ldots, \xi\}$ levels in the HPO taxonomy, $\psi : V \longrightarrow \mathcal{L}$ is a level function which assigns to each HPO term $i \in V$ a level, i.e. its maximum distance from the root. For instance, nodes $\{i | \psi(i) = 0\}$ correspond to the root node, $\{i | \psi(i) = 1\}$ is the set of nodes with a maximum path length from the root (distance) equal to 1, and $\{i | \psi(i) = \xi\}$ are nodes that lie at a maximum distance ξ from the root.

Fig 1 shows that we need to visit the HPO hierarchy per level in the sense of the maximum and not of the minimum distance from the root: this is necessary to preserve the consistency of the predictions. Indeed looking at the *HTD* scores obtained respectively with minimum and maximum distance from the root (bottom-left of Fig. 1), we see that only the maximum distance preserves the consistency of the predictions. Indeed, focusing on node 5, by traversing the DAG levels according to the minimum distance from the root, we have that the level of node 5 is 1 ($\psi^{min}(5) = 1$) and in this case by applying the *HTD* rule (3) the flat score $\hat{y}_5 = 0.8$ is wrongly modified to the *HTD* ensemble score $\bar{y}_5 = 0.7$. If we instead traverse the DAG levels according to the maximum distance from the root, we have $\psi^{max}(5) = 3$ and the *HTD* ensemble score is correctly set to $\bar{y}_5 = 0.4$. In other words at the end of the *HTD*, by traversing the levels according to the minimum distance we have $\bar{y}_5 = 0.7 > \bar{y}_4 = 0.4$, that is a child node has a score larger than that of its parent, and the true path rule is not preserved; on the contrary by traversing the levels according to the maximum distance we achieve $\bar{y}_5 = 0.4 \leq \bar{y}_4 = 0.4$ and the true path rule consistency is assured. This is due to the fact that by adopting the minimum distance when we visit node 5, node 4 has not just been visited, and hence the value 0.4 has not been transmitted by node 2 to node 4; on the contrary if we visit the DAG according to the maximum distance all the ancestors of node 5 (including node 4) have just been visited and the score 0.4 is correctly transmitted to node 5 along the path $2 \rightarrow 4 \rightarrow 5$.

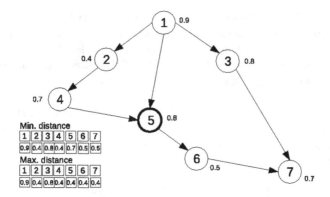

Fig. 1. Levels of the hierarchy must be defined in terms of the maximum distance from the root (node 1). Small numbers close to nodes correspond to the scores of the flat predictions. The Hierarchical top-down scores obtained respectively by crossing the levels according to the minimum and the maximum distance from the root are shown in the bottom-left.

More precisely, given a DAG $G =< V, E >$, the level function ψ, a set of flat predictions $\hat{\boldsymbol{y}} =< \hat{y}_1, \hat{y}_2, \ldots, \hat{y}_{|V|} >$ for each class associated to each node $i \in \{1, \ldots, |V|\}$, the $HTD\text{-}DAG$ algorithm assures that for the set of ensemble predictions $\bar{\boldsymbol{y}} =< \bar{y}_1, \bar{y}_2, \ldots, \bar{y}_{|V|} >$ the following property holds:

$$\forall i \in V, \ j \in par(i) \Rightarrow \bar{y}_j \geq \bar{y}_i \qquad (4)$$

Indeed, by applying the rule (3) from the top to the bottom of the hierarchy we assure that the scores of the parents are larger or equal than those of its children. Moreover by visiting "per level" the hierarchy according to the ψ function (levels are defined in the sense of the maximum distance) we assure that each parent has just been visited before their children and by observing that each node is visited only once it cannot be changed by the successive top-down steps of the algorithm, thus assuring that $\forall i \in V, \ j \in par(i) \Rightarrow \bar{y}_j \geq \bar{y}_i$.

There are several ways to implement the function ψ that computes the maximum distance of each node from the root. We applied the classical Bellman-Ford algorithm [18]: by recalling that it finds the shortest paths from a source node to all the other nodes of a weighted digraph, it is sufficient to invert the sign of each edge weight to obtain the maximum distance (longest path) from the root. We outline that other methods (e.g. procedures based on the topological sort of graphs) are more efficient, but considering that the levels should be computed only once, on modern computers there are not significant differences in terms of the mean empirical computational time.

Fig. 2 shows the pseudo code of the overall HTD algorithm. Rows $1-4$ provide the maximum distance of each node from the root, whereas the block B of the algorithm implements a per-level top-down visit of the graph (rows $5-16$).

```
Input:
- G =< V, E >
- ŷ =< ŷ₁, ŷ₂, ..., ŷ|V| >,   ŷᵢ ∈ [0, 1]
begin algorithm
01:    A. Compute ∀i ∈ V the max distance from root(G):
02:        E' := {e'|e ∈ E, e' = −e}
03:        G' :=< V, E' >
04:        dist := Bellman.Ford(G', root(G'))
05:    B. Per-level top-down visit of G:
06:        ȳ_root(G) := ŷ_root(G)
07:        for each d from 1 to max(dist) do
08:          N_d := {i|dist(i) = d}
09:          for each i ∈ N_d do
10:            x := min_{j∈par(i)} ȳ_j
11:            if (x < ŷᵢ)
12:              ȳᵢ := x
13:            else
14:              ȳᵢ := ŷᵢ
15:          end for
16:        end for
end algorithm
Output:
- ȳ =< ȳ₁, ȳ₂, ..., ȳ|V| >
```

Fig. 2. Hierarchical Top-Down algorithm for DAGs (HTD)

Starting from the children of the root (level 1) for each level of the graph the nodes are processed and the hierarchical top-down correction of the flat predictions $\hat{y}_i$, $i \in \{1, \ldots, |V|\}$ to the HTD-DAG ensemble prediction $\bar{y}_i$ is performed according to eq. 3. It is easy to see that the complexity of block B (rows $5-16$) is linear in the number of vertices for sparse graphs (and the HPO is just a sparse DAG).

3   Experimental Set-Up

3.1   The Human Phenotype Ontology

Ontologies are high-level representations of knowledge domains based upon controlled vocabularies. The Human Phenotype Ontology (HPO) aims at providing a standardized categorization of the abnormalities associated to human diseases (each represented by an HPO term) and the semantic relationships between them. Each HPO term describes a phenotypic abnormality and is developed using medical literature and cross-references to other biomedical ontologies (e.g. OMIM [3]).

A key feature of HPO is its ability, based upon an equivalence mapping to other publicly available phenotype vocabularies, to allow the integration of existing datasets and to strongly promote the interoperability with multiple biomedical resources [4].

The experiments presented in this manuscript are based on the September 2013 HPO release (10,099 terms and 13,382 between-term relationships). The annotations of the 20,257 human genes were taken from the same HPO release. After pruning the HPO terms having less than 2 annotations we obtained a final HPO DAG composed by 4,847 terms (and 5,925 between-terms relationships).

3.2   Construction and Integration of the Protein Functional Network

The set of human genes considered in the experiments presented here was obtained from the recent critical assessment of protein function annotation (CAFA2) international challenge. Starting from an initial set of 20,257 human genes we constructed, for each gene, different binary profile vectors representing the absence/presence of bio-molecular features in the gene product encoded by the considered gene. More precisely, we constructed for each gene 8 binary vectors containing the features obtained, respectively, from InterPro [19], Pfam [20], PRINTS [21], PROSITE [22], SMART [23], SUPFAM [24], Gene Ontology [17] and OMIM [3]. All these annotations were obtained by parsing the raw text annotation files made available by the Uniprot knowledgebase (release May 2013, considering only its SWISSprot component database). We then obtained a similarity score between each pair of genes simply by computing the Jaccard similarity between the feature vectors associated with the genes.

Following this strategy we obtained 8 gene networks (one for each of the aforementioned data sources). The final functional interaction network used in the presented experiments was constructed using a simple unweighted integration strategy that does not involve any learning phase in the network integration process: the *Unweighted Average (UA)* network integration method[25]. In UA the weight of each edge of the combined networks is computed simply averaging across the available n networks:

$$\bar{w}_{ij} = \frac{1}{n} \sum_{d=1}^{n} w_{ij}^{d} \tag{5}$$

In order to construct a more informative network, we added also two more functional gene networks taken from the literature and previously published in [26, 27].

3.3   Kernelized Score Functions

As base learner we used a semi-supervised network-based learning method recently successfully applied to gene disease prioritization [28], gene function prediction [29] and drug repositioning [30]. Kernelized score functions adopt both a local and a global learning strategy. Local learning is accomplished through a generalization of the classical guilt-by-association approach [31], through the introduction of different functions to quantify the similarity between a gene and its neighbours. A global learning strategy is introduced in form of a kernel that can capture the overall topology of the underlying biomolecular network.

More precisely, by this approach we can derive score functions $S : V \longrightarrow \mathbb{R}^+$ based on properly chosen kernel functions, by which we can directly rank a gene v according to the values of $S(v)$: the higher the score, the higher the likelihood that a gene belongs to a given class [29]. The score functions are built on distance measures defined in a suitable Hilbert space $\mathcal{H}$ and computed using the usual "kernel trick", by which instead of explicitly computing the inner product $< \phi(\cdot), \phi(\cdot) >$ in the Hilbert space, with $\phi : V \longrightarrow \mathcal{H}$, we compute the associated kernel function $K : V \times V \longrightarrow \mathbb{R}^+$ in the original input space V.

For instance, given a vertex v, a set of genes V_C belonging to a specific class C, we can obtain the following *Average score* S_{AV}:

$$S_{AV}(v, V_C) = \frac{1}{|V_C|} \sum_{x \in V_C} K(v, x) \qquad (6)$$

In principle any valid kernel K can be applied to compute the aforementioned kernelized score, but in the context of gene - phenotype association ranking, we used *random walk kernels* [32], since they can capture the similarity between genes, taking into account the topology of the overall functional interaction network.

In our experiments we applied a *1-step random walk kernel*: in this way we explicitly evaluate only the direct neighbors of each gene in the functional interaction network. It is worth noting that other kernels may lead to better results, but here we are mainly interested in verifying whether our proposed *HTD* algorithm can improve upon Flat predictions, and not in fine tuning and achieving the best possible results.

4 Results

We compared our proposed *HTD* ensemble methods with flat predictions obtained with 1-step random walk kernelized score functions, by applying classical leave-one-out techniques.

In terms of the average AUC across the 4846 considered HPO terms, even if the difference in favour of *HTD* is very small (0.7923 vs 0.7897), by looking at the results of the single HPO terms, *HTD* improves over flat in 3346 HPO terms, achieves the same AUC for 554 HPO terms and "looses" in 956 terms. This means that for more than 3/4 HPO terms we obtain an improvement, and this explains also why, according to the Wilcoxon rank sum test the difference between the methods is statistically significant in favour of *HTD* at 10^{-5} significance level.

Also better results are obtained when we consider the precision at a fixed recall level. Indeed in this case the average values across HPO terms are quite consistent: for instance the average precision at 20% recall is 0.1535 vs 0.1278,

Table 1. Average AUC, and precision at 10, 20 and 40% recall (P10R, P20R and P40R). Flat stands for flat ensemble method, HTD for Hierarchical Top-Down, Max for Hierarchical Maximum, And for Hierarchical And and Or for Hierarchical Or ensemble methods. Methods that are significantly better than all the others according to the Wilcoxon rank sum test ($\alpha = 10^{-5}$) are highlighted in bold.

	Flat	HTD	Max	And	Or
AUC	0.7897	0.7923	0.7879	**0.8151**	0.7880
P10R	0.1620	**0.1957**	0.1315	0.1665	0.1352
P20R	0.1278	**0.1535**	0.1081	0.1283	0.1110
P40R	0.0812	**0.0890**	0.0728	0.0758	0.0741

and another time for most HPO terms we obtain a significant increment when the *HTD* hierarchical correction is applied to the flat predictions.

Table 1 summarizes the average results across terms for the *HTD*, *Flat* and three heuristic hierarchical ensemble methods originally proposed for the hierarchical prediction of Gene Ontology terms [14]. *HTD* achieves always significantly better results than the flat approach, both in terms of AUC and precision at a fixed recall. Moreover it obtains significantly better results than all the other compared hierarchical ensemble methods. The only exception is with respect to the AUC where the *And* hierarchical method achieves better results.

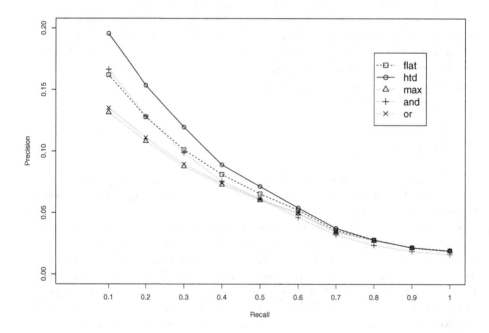

Fig. 3. Precision at different levels of recall, averaged across HPO terms

Fig. 3 compares the precision at different recall levels for all the hierarchical and the flat ensemble methods: the *HTD* solid line marked with circles is consistently above all the other curves, showing that *HTD* achieves on the average better results than all the other competing methods.

Even if the average precision at a fixed recall rate is relatively low with all the methods we presented (Fig. 3), we note that we tried to perform predictions also with terms having only two positive annotations, a very difficult task that likely leads in most case to precision values very close to 0. Moreover by applying score functions with 2-step or more random walk kernels, we could better exploit the overall topology of the network and at least potentially achieve better results, especially with terms having a small number of annotations or with "positive" nodes relatively "far" from each other. In any case, at least for low values of recall, *HTD* shows on the average relative improvements of the precision between 10 and 20%, with respect to the *Flat* approach. On the contrary, the heuristic *Max*, *And* and *Or* methods are not able to outperform the *Flat* approach, confirming previous results in the context of gene function prediction [14].

5   Conclusions

The prediction of human gene–abnormal phenotype associations is an important step toward the discovery of novel disease genes associated with hereditary disorders. Several computational methods that exploit "omics" data can be successfully applied to predict or rank genes with respect to human phenotypes, but usually their predictions are inconsistent, in the sense that do not necessarily obey the parent-child relationships between HPO terms (i.e. a gene may achieve a score for a child term larger than that that of its parent HPO term).

We showed that our proposed method provides predictions that are always consistent, according the "true path rule" that governs the HPO taxonomy. Moreover the *HTD* ensemble method can enhance "flat" predictions by exploiting the hierarchical relationships between HPO terms. Indeed our experimental results showed that *HTD*, by using kernelized score functions as base learner, can significantly improve the precision-recall curves. We obtained a significant increment using also other base learners (e.g. the classical label propagation algorithm described in [33] – data not shown), and in principle our proposed hierarchical method is independent of the base learner used to provide the initial "flat" scores. From this standpoint *HTD* can be applied to improve the performance of any "flat" learning method, and to provide consistent and more reliable predictions for novel gene - phenotype predictions by exploiting the DAG structure of the Human Phenotype Ontology.

Acknowledgments. G.V. and M.R. acknowledge partial support from the PRIN project "Automi e linguaggi formali: aspetti matematici e applicativi", funded by the Italian Ministry of University.

References

[1] Robinson, P., Krawitz, P., Mundlos, S.: Strategies for exome and genome sequence data analysis in disease-gene discovery projects. Cin. Genet. 80, 127–132 (2011)

[2] Robinson, P., Kohler, S., Bauer, S., Seelow, D., Horn, D., Mundlos, S.: The Human Phenotype Ontology: a tool for annotating and analyzing human hereditary disease. Am. J. Hum. Genet. 83, 610–615 (2008)

[3] Amberger, J., Bocchini, C., Amosh, A.: A new face and new challenges for Online Mendelian inheritance in Man (OMIM). Hum. Mutat. 32, 564–567 (2011)

[4] Kohler, S., et al.: The human phenotype ontology project: linking molecular biology and disease through phenotype data. Nucleic Acids Research 42(Database issue), D966–D974 (2014)

[5] Moreau, Y., Tranchevent, L.: Computational tools for prioritizing candidate genes: boosting disease gene discovery. Nature Rev. Genet. 13(8), 523–536 (2012)

[6] McGary, K., Lee, I., Marcotte, E.: Broad network-based predictability of Saccharomyces cerevisiae gene loss-of-function phenotypes. Genome Biology 8(R258) (2007)

[7] Mehan, M., Nunez-Iglesias, J., Dai, C., Waterman, M., Zhou, X.: An integrative modular approach to systematically predict gene-phenotype associations. BMC Bioinformatics 11(suppl. 1) (2010)

[8] Wang, P., et al.: Inference of gene-phenotype associations via protein-protein interaction and orthology. PLoS One 8(10) (2013)

[9] Musso, G., et al.: Novel cardiovascular gene functions revealed via systematic phenotype prediction in zebrafish. Development 141, 224–235 (2014)

[10] Cerri, R., de Carvalho, A.: Hierarchical multilabel protein function prediction using local neural networks. In: Norberto de Souza, O., Telles, G.P., Palakal, M. (eds.) BSB 2011. LNCS, vol. 6832, pp. 10–17. Springer, Heidelberg (2011)

[11] Silla, C., Freitas, A.: A survey of hierarchical classification across different application domains. Data Mining and Knowledge Discovery 22(1-2), 31–72 (2011)

[12] Valentini, G.: True Path Rule hierarchical ensembles for genome-wide gene function prediction. IEEE ACM Transactions on Computational Biology and Bioinformatics 8(3), 832–847 (2011)

[13] Cesa-Bianchi, N., Re, M., Valentini, G.: Synergy of multi-label hierarchical ensembles, data fusion, and cost-sensitive methods for gene functional inference. Machine Learning 88(1), 209–241 (2012)

[14] Obozinski, G., Lanckriet, G., Grant, C., Jordan, M., Noble, W.: Consistent probabilistic output for protein function prediction. Genome Biology 9(S6) (2008)

[15] Schietgat, L., Vens, C., Struyf, J., Blockeel, H., Dzeroski, S.: Predicting gene function using hierarchical multi-label decision tree ensembles. BMC Bioinformatics 11(2) (2010)

[16] Valentini, G.: Hierarchical Ensemble Methods for Protein Function Prediction. ISRN Bioinformatics 2014(Article ID 901419), 34 pages (2014)

[17] Gene Ontology Consortium: Gene Ontology annotations and resources. Nucleic Acids Research 41, D530–D535 (2013)

[18] Cormen, T., Leiserson, C., Rivest, R.: Introduction to Algorithms. MIT Press, Boston (2009)

[19] Apweiler, R., Attwood, T., Bairoch, A., Bateman, A., et al.: The interpro database, an integrated documentation resource for protein families, domains and functional sites. Nucleic Acids Research 29(1), 37–40 (2001)

[20] Finn, R., Tate, J., Mistry, J., Coggill, P., Sammut, J., Hotz, H., Ceric, G., Forslund, K., Eddy, S., Sonnhammer, E., Bateman, A.: The Pfam protein families database. Nucleic Acids Research 36, D281–D288 (2008)

[21] Attwood, T.: The prints database: a resource for identification of protein families. Brief Bioinform. 3(3), 252–263 (2002)

[22] Hulo, N., Bairoch, A., Bulliard, V., Cerutti, L., Cuche, B., De Castro, E., Lachaize, C., Langendijk-Genevaux, P., Sigrist, C.: The 20 years of prosite. Nucleic Acids Research 36, D245–D249 (2008)

[23] Schultz, J., Milpetz, F., Bork, P., Ponting, C.: Smart, a simple modular architecture research tool: identification of signaling domains. Proceedings of the National Academy of Sciences 95(11), 5857–5864 (1998)

[24] Gough, J., Karplus, K., Hughey, R., Chothia, C.: Assignment of homology to genome sequences using a library of hidden markov models that represent all proteins of known structure. Journal of Molecular Biology 313(4), 903–919 (2001)

[25] Valentini, G., Paccanaro, A., Caniza, H., Romero, A., Re, M.: An extensive analysis of disease-gene associations using network integration and fast kernel-based gene prioritization methods. Artificial Intelligence in Medicine 61(2), 63–78 (2014)

[26] Wu, G., Feng, X., Stein, L.: A human functional protein interaction network and its application to cancer data analysis. Genome Biol. 11, R53 (2010)

[27] Lee, I., Blom, U., Wang, P.I., Shim, J., Marcotte, E.: Prioritizing candidate disease genes by network-based boosting of genome-wide association data. Genome Research 21(7), 1109–1121 (2011)

[28] Re, M., Valentini, G.: Cancer module genes ranking using kernelized score functions. BMC Bioinformatics 13(suppl.14/S3) (2012)

[29] Re, M., Mesiti, M., Valentini, G.: A Fast Ranking Algorithm for Predicting Gene Functions in Biomolecular Networks. IEEE ACM Transactions on Computational Biology and Bioinformatics 9(6), 1812–1818 (2012)

[30] Re, M., Valentini, G.: Network-based Drug Ranking and Repositioning with respect to DrugBank Therapeutic Categories. IEEE/ACM Transactions on Computational Biology and Bioinformatics 10(6), 1359–1371 (2013)

[31] Oliver, S.: Guilt-by-association goes global. Nature 403, 601–603 (2000)

[32] Smola, A.J., Kondor, R.: Kernels and regularization on graphs. In: Schölkopf, B., Warmuth, M.K. (eds.) COLT/Kernel 2003. LNCS (LNAI), vol. 2777, pp. 144–158. Springer, Heidelberg (2003)

[33] Zhu, X., et al.: Semi-supervised learning with gaussian fields and harmonic functions. In: Proc. of the 20th Int. Conf. on Machine Learning, Washintgton DC, USA (2003)

Mortality Prediction with Lactate
and Lactate Dehydrogenase

Yasemin Zeynep Engin[1,*], Kemal Turhan[1], Aslı Yazağan[2], and Asım Örem[3]

[1] Department of Biostatistics and Medical Informatics,
Karadeniz Technical University, Trabzon, Turkey
[2] Department of Computer Technologies, Recep Tayyip Erdoğan University, Rize, Turkey
[3] Department of Medical Biochemistry, Karadeniz Technical University, Trabzon, Turkey
{ysmnzynp.engin,kturhan.tr,ayazagan}@gmail.com,
aorem64@yahoo.com

Abstract. It has been proved in many studies that Lactate and Lactate dehydrogenase (LDH) are associated with mortality. In this study lactate test values of inpatients were analyzed with Support Vector Machines (SVM) to identify patients in high risk of death. In the data set containing 686 records with lactate results; 219 patients treated in the pediatric service and 467 of the patients are adults. Lactate levels of 331 patients are normal and levels of 355 patients are high. 89 patients with high lactate levels were recorded as dead. 97%, 96.6% and 92.3% accurate mortality classification rates were recorded with analyzes performed using different data sets and variables. Patient's risk assessment can be assessed with such findings and treatments can be planned. Prediction of patients under high risk can provide opportunities for early intervention and mortality levels can be red.

Keywords: Lactate, Lactate Dehydrogenase, mortality prediction, Support Vector Machine, SVM.

1   Introduction

Lactate is a metabolite that arises as a result of metabolizing glucose with anaerobic respiration. Glucose cannot be converted completely into energy when the oxygen transferred to the muscles is not enough. Lactate dehydrogenase (LDH) is a biomarker which helps to define the cause of tissue damage and the location in the body and to monitor the progression of damage. Sometimes it is used for monitoring some cancers or progressive disorders such as liver disease, kidney disease [1].

It has been proved in many studies that Lactate and LDH are associated with mortality. In a study, LDH levels were evaluated as a prognostic factor for the prediction of lifetime in patients with fatal cancers. Results showed that serum LDH in patients with fatal cancer is a good predictor for survival time [2]. In another study,

---

* Corresponding author.

F. Ortuño and I. Rojas (Eds.): IWBBIO 2015, Part I, LNCS 9043, pp. 78–84, 2015.

association between venous lactate levels and increased risk of death in infected intensive care unit patients was emphasized [3]. In another study on this issue, it was observed that, increased blood lactate levels in critically ill patients are associated with increased morbidity and mortality [4].

2   Methods

In this study, data of patients who treated in various services of Karadeniz Technical University, Faculty of Medicine, Farabi Hospital between 07.01.2010–31.12.2010 was used. 686 records of inpatients containing lactate test were included in the study. 219 patients treated in the pediatric service and 467 of the patients were adults. Lactate levels of 331 patients were normal and levels of 355 patients were high. 89 patients with high lactate levels were recorded as dead.

Survival predictions were made by analyzing lactate test values of patients with Support Vector Machines (SVM). SVM is a data mining method that results highly successful in linear or nonlinear classification problems. The basic principle of the SVM is to determine the optimum decision boundaries between classes in a decision plane. According to these boundaries, it can be determined that the object belongs to which class [5].

3   SVM

Fisher published the first pattern recognition related algorithm in 1936 [6]. Subsequently, in other studies SVMs were further developed out of statistical learning theory [8,9]. Using SVM is a highly successful approach to linear or nonlinear classification problems (Figure 1).

SVM characterize an extension to nonlinear models of the generalized depiction algorithm developed by Vapnik and Lerner [7]. The SVM algorithm is built on the statistical learning theory and the Vapnik–Chervonenkis (VC) dimension [8]. Statistical learning theory describes the properties of learning machines to make reliable estimates, was reviewed [9,10,11]. In the current formulation, the SVM algorithm was developed at AT&T Bell Laboratories by Vapnik et al. [12]

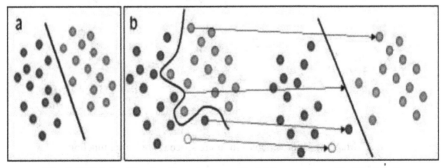

Fig. 1. SVM classification linearly separable (a) or non-separable (b) objects

The basic principle of SVM is Lipkowitz optimal decision boundaries between classes in a decision plane. So it could be easily classify desired objects according to SVM decision plane boundaries. Figure 1 shows how SVM classifies linearly separable or non-separable objects.

SVM finds further distance between two closest objects as shown in Figure 2 - a and 2 - b. Classification problems in real life often are not linear (Figure 1- b). Therefore, for nonlinear classification problems SVM performs mathematical data transformation using kernel functions (Figure 1-b). These kernel functions are listed below:

- Linear
- Polynomial
- Radial Based Function
- Sigmoid

In this study, Radial Based Function (RBF) is used because of its high performance.

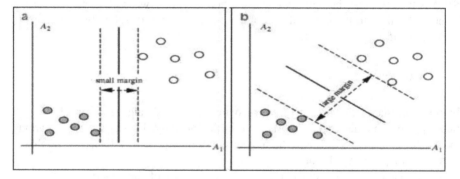

Fig. 2. Finding optimal line between two classes

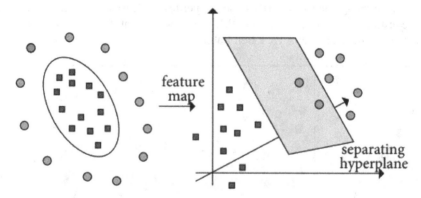

Fig. 3. Mapping input variables to future space using kernel function

The separation surface may be nonlinear in many classification problems, but support vector machines can be extended to deal with nonlinear separation surfaces by using feature functions. The SVM extension to nonlinear datasets is based on mapping the input variables into a feature space of a higher dimension (Figure 3) and then performing a linear classification in that higher dimensional space.

4 SVM Results

Association rules analysis was applied to the data set with 0.2 support level and 0.50 confidence level. First two rules are as follows:

DEATH == No 526 ==> LACTATE == Normal 356 supp (41.2), conf (53.9)

DEATH == Yes 161 ==> LACTATE == High 331 supp (12.9), conf (55.2)

Although derived rules are not very powerful, it may be mentioned that there is a relationship between death and lactate levels. In the first analysis to predict mortality patient's age, lactate level, LDH level and value of sodium (Na) were defined as continuous variables; sex and ICD code were defined as categorical variables. The reason of the addition of Na test to analysis is the association rules we have obtained. According to rules, Na is the most frequently requested test with lactate. SVM results are as follows:

```
Dataset laktat yatan:
    Dependent: OLUM
    Independents: YAS, AC_90225044, AC_90226044,
AC_90367044, CINSIYETI...
    Sample size = 202 (Train), 67 (Test), 269 (Overall)
```

```
 Support Vector machine results:
    SVM type: Classification type 1 (capacity=10,000)
    Kernel type: Radial Basis Function (gamma=0,167)
    Number of support vectors = 71 (17 bounded)
    Support vectors per class: 30 (0), 41 (1)
    Class. accuracy (%) = 96,535(Train), 97,015(Test),
96,654(Overall)
```

According to the results, a total of 269 requests were evaluated in the analysis, 202 of them for training of SVM, 67 of them for testing. The cause of the number of evaluated requests fall to 269 is to evaluate requests that Na, lactate and LDH tests requested together. Radial Basis Function was selected as kernel function. Total of 67 requests used to test, 55 of them were belonging to alive patients and 12 of them were dead. SVM was evaluated 54 of requests who were alive and 11 of requests who were died correctly. Two patients were misclassified. 97% success of the test is thought to be a good value for such an analysis.

Survival prediction analysis was performed on the adult data set. Patient's age and LDH level were defined as continuous variables; sex, ICD code and lactate level were defined as categorical variables. SVM results are as follows:

```
Dataset laktat yatan yetiskin:
    Dependent: OLUM
    Independents: YAS, AC_90226044, POLIKLINIK_KODU,
ICDKOD, LAKTAT
    Sample size = 192 (Train), 59 (Test), 251 (Overall)
```

```
 Support Vector machine results:
    SVM type: Classification type 1 (capacity=10,000)
    Kernel type: Radial Basis Function (gamma=0,167)
    Number of support vectors = 63 (5 bounded)
    Support vectors per class: 34 (0), 29 (1)
    Class. accuracy (%) = 99,479(Train), 96,610(Test),
98,805(Overall)
```

Looking at the results, a total of 251 requests were evaluated in the analysis, 192 of them for training of SVM, 59 of them for testing. All of the 41 alive patients were predicted correctly. 16 of the patients who died were classified correctly and two patients were classified incorrectly. 96.6% success rate was determined in this group.

An analysis was then made to evaluate all of adult data set. Patient's age was used as continuous variables; sex, ICD code and lactate level were used as categorical variables. DVM results are as follows:

```
Dataset laktat yatan yetiskin:
    Dependent: OLUM
    Independents: YAS, POLIKLINIK_KODU, ICDKOD, LAKTAT
    Sample size = 350 (Train), 117 (Test), 467 (Overall)
```

```
 Support Vector machine results:
    SVM type: Classification type 1 (capacity=10,000)
    Kernel type: Radial Basis Function (gamma=0,250)
    Number of support vectors = 142 (24 bounded)
    Support vectors per class: 54 (0), 88 (1)
    Class. accuracy (%) = 96,571(Train), 92,308(Test),
95,503(Overall)
```

In the analysis, all of 467 adult patients requests were used; 350 of them for training of SVM and 117 of them for testing. 80 of 84 alive patients were classified successfully and four patients were misclassified. 28 of 33 deceased patients were predicted correctly, five patients were classified incorrectly. Success rate of the test was 92.3%.

5 Discussion and Conclusion

As a result of many studies, it has proven that lactate and LDH tests to be associated with mortality [2,3,4,13,14]. 97%, 96.6% and 92.3% accurate mortality classification rates were recorded with analyzes performed using different data sets and variables. One of the analyses which Na had analyzed with lactate and LDH, was evaluated with biochemistry field expert. The reason of requesting such a high rate of Na with lactate was, in patients who are constituting the data sets usually be fed through a blood vessel and therefore the Na level was found to be frequently measured.

Patient's risk assessment can be done with such findings and these findings can be useful while treatments are planning. Prediction of patients under high risk can provide opportunities for early intervention and mortality levels can be reduced.

Results obtained with the analysis of medical data were unforeseeable and could not be achieved by conventional methods. But for the signification of these results, professional support of the medical field is essential. Thus, data mining studies that will be performed on medical data should be carried out multidisciplinary. During the planning stage of these studies, both experts from the fields must act together.

Furthermore medical data was subjected to pre-processing and was found to contain many missing records. But for the success of data mining analysis, quality of data is the most important factor. Therefore, a better quality data should be created for future works and missing-false, repetitive records must be cleaned from data.

References

1. Lab Tests Online, http://labtestsonline.org/ (last access: February 6, 2014)
2. Suh, S.Y., Ahn, H.Y.: Lactate dehydrogenase as a prognostic factor for survival time of terminally ill cancer patients: A preliminary study. Eur. J. Cancer 43(6), 1051–1059 (2007)
3. Shapiro, N.I., Howell, M.D., Talmor, D., Nathanson, L.A., Lisbon, A., Wolfe, R.E., Weiss, J.W.: Serum lactate as a predictor of mortality in emergency department patients with infection. Ann. Emerg. Med. 45(5), 524–528 (2005)
4. Bakker, J., Gris, P., Coffernils, M., Kahn, R.J., Vincent, J.L.: Serial blood lactate levels can predict the development of multiple organ failure following septic shock. Am. J. Surg. 171, 221–226 (1996)
5. Han, J., Kamber, M.: Data Mining: Concepts and Techniques, 2nd edn. Morgan Kaufmann Publishers, New York (2006)
6. Fisher, R.A.: The use of multiple measurements in taxonomic problems. Ann. of Eugen. 7, 111–132 (1936)
7. Vapnik, V., Lerner, A.: Pattern recognition using generalizedportrait method. Automat. Rem. Contr. 24, 774–780 (1963)
8. Vapnik, V., Chervonenkis, A.Y.: Teoriya raspoznavaniya obrazov: Statisticheskie problemy obucheniya (Theory of pattern recognition: Statistical problems of learning). Nauka, Moscow, Russia (1974)
9. Vapnik, V.: Estimation of Dependences Based on Empirical Data.Nauka, Moscow, Russia (1979)
10. Vapnik, V.: The Nature of Statistical Learning Theory. Springer, New York (1995)
11. Vapnik, V.: Statistical Learning Theory. Wiley Interscience, NewYork (1998)

12. Lipkowitz, K.B., Cundari, T.R.: Reviews in Computational (2007)
13. Chemistry. Wiley-VCH, John Wiley & Sons Inc. 23, pp. 291–400
14. Von Eyben, F.E., Blaabjerg, O., Hyltoft-Petersen, P., Madsen, E.L., Amato, R., Liu, F., Fritsche, H.: Serum lactate dehydrogenase isoenzyme 1 and prediction of death in patients with metastatic testicular germ cell tumors. Clin. Chem. Lab. Med. 39(1), 38–44 (2001)

Intellectual Property Protection for Bioinformatics and Computational Biology

Dennis Fernandez[1], Antonia Maninang[2], Shumpei Kobayashi[3], Shashank Bhatia[1], and Carina Kraatz[1]

[1] Fernandez & Associates, LLP, Alterton, California
{dennis,shashank}@iploft.com, carinakraatz.iploft@gmail.com
[2] Stanford Hospital, Stanford, California
amaninang@stanfordmed.org
[3] San Diego, California
shkobaya@ucsd.edu

Abstract. Bioinformatics, and computational biology are two ever-growing fields that require careful attention to intellectual property rights (IPR) and strategies. The American patent system is currently going through the biggest reformation since the passage of Patent Act of 1952, and many changes apply directly to the field of biology that utilize computational intelligence. Basic IP definitions, recent IP developments, and advanced protection strategies are discussed in order to better understand the status quo of intellectual property (IP) specifically in the field of evolutionary computation, bioinformatics, and computational biology.

Keywords: Computational Biology, Bioinformatics, Intellectual Property, America Invents Act, Patent Portfolio.

1 Introduction

Bioinformatics is a branch of biological science that aims to use computer science, mathematics, and information theory to analyze and model large database of biochemical or pharmaceutical information. There are three important sub-disciplines within bioinformatics: the development of new algorithms and statistics with which to assess relationships among members of large data sets; the analysis and interpretation of various types of data including nucleotide and amino acid sequences, protein domains, and protein structures; and the development and implementation of tools that enable efficient access and management of different types of information [1]. Computational biology is a closely related discipline to bioinformatics, but concentrates more on the evolutionary, population, and theoretical biology rather than cell and molecular biomedicine.

The field of bioinformatics and computational biology has come a long way since the discovery of the double-helical structure of deoxyribonucleic acid (DNA) by James D. Watson and Francis Crick in 1953. In 2003, the Human Genome Project was completed after 13 years of international research coordinated by the U.S. Department

F. Ortuño and I. Rojas (Eds.): IWBBIO 2015, Part I, LNCS 9043, pp. 85–95, 2015.

of Energy and the National Institutes of Health [2]. The human genome, or our genetic blueprint, is written in four alphabet of chemical compounds called adenine (A), guanine (G), cytosine (C), and thymine (T). With a $3.8 billion federal investment, it is estimated that the human genome project produced nearly $800 billion in economic output and helped drive the bioinformatics and computational biology field move forward with an unprecedented rate [3]. Although the sequencing of the genome is still too expensive for most individuals, Eric Lander, the head of the Broad Institute, in Cambridge, Massachusetts, claims that the cost of DNA sequencing has fallen to a hundred-thousandth of what it was a decade ago [4]. It can be seen from Fig. 1 that the cost of genome sequencing is declining at a speed that surpasses that of Moore's Law for computing hardware [5]. The day is fast approaching when individuals will be able sequence their unique genome sequence at a cost of $1000 or lower. The next generation sequencing techniques such as Illumina and nanopore sequencing are enabling genetic research to be conducted with more accuracy, lower cost, and higher throughput.

Even in the midst of global financial crisis, the biotechnology industry has continued to grow. A recent report from Global Industry Analysts Inc. predicts that the biotechnology market will grow to exceed a value of U.S. $320 billion by 2015 [6]. The United States is the leading country in the industry, with more than 6,800 biotechnology firms and with a share of more than 40% of all biotechnology patents filed under the Patent Cooperation Treaty (PCT) [7].

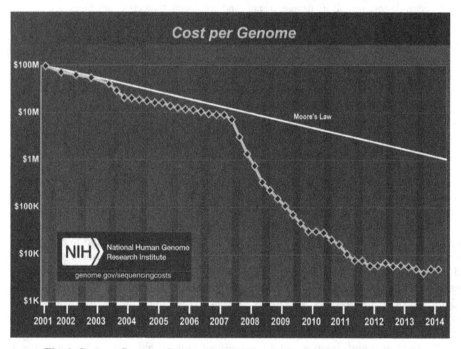

Fig. 1. Cost per Genome (Source: National Human Genome Research Institute)

Within the biotechnology market, the bioinformatics market specifically is projected to reach U.S. $5.0 billion by 2015 [8]. The United States represents the largest regional market for bioinformatics worldwide, and some of the driving factors of the market include the availability of genomic information and the increasing application of genomics in research and development processes. Most pharmaceutical companies have concerns related to their blockbuster drugs going off-patent, narrow product pipelines, and high cost of drug development [9]. Bioinformatics helps companies reduce cost and enhance productivity in many steps of research and development by providing data management tools to simulate biological processes.

With a growing biotechnology market and the recent changes in the United States patent system, careful attention must be given to strategic intellectual property prosecution in the bioinformatics/computational biology field. This paper will give a brief overview of intellectual property rights, its recent developments, and advanced protection strategies in the fields of bioinformatics/computational biology.

2   Basic IPR Definitions

2.1   Patents

In the United States a patent is a property right that protects inventions including any process, machine, manufacture, or composition of matter, or any improvement thereof that are novel, useful, and non-obvious [10]. The three types of patents are utility, design, and plant patents. It is important to note that a patent does not grant any positive or affirmative right to make, use, sell, offer to sell, or import an invention.

In theory, intellectual properties, including patents, help to foster intellectual and economic growth. The disclosure requirement assures the dissemination of innovation to the public, and in return the inventor or the owner is granted legal rights of ownership for duration of 17-20 years. Although some argue that patents are used for anticompetitive purposes that lead to monopolies, economists claim that patents provide important incentives for investment and trade, both of which are vital sources of economic growth.

In order to obtain a patent, an inventor must submit an application to the United States Patent and Trademark Office. The process is lengthy and often very costly. It may take years to get a typical biotechnology patent, and costs an average of ~$15,000 to prosecute such patent application [11].

2.2   Copyrights

Copyrights protect the original expression of an idea [12]. Copyright protection is automatic so you do not have to register. However, it is wise to register the work with the US Copyright office since this will spare you from having to prove actual monetary damages from an infringed activity. Unlike the registration process for patents, copyright registration is as easy as filling out a 1-2 page from, and the fee is dramatically cheaper. Copyrights are usually valid for the author's lifetime plus 70 years or sometimes longer (75-100 years) if the work was created for hire [13].

2.3   Trademark

Trademarks protect the goodwill and branding of one's product or services. It refers to the distinctive signature mark used to protect the company, product, service, name, or symbol. A typical trademark registration takes under two years and costs under $5,000 per registered mark. It can also be re-registered indefinitely.

2.4   Trade Secrets

Trade secrets are any confidential technical or business information that provides an enterprise a competitive advantage [14]. There is no formal filing procedure to register trade secrets. Companies must make efforts to protect their trade secrets through non-compete and non-disclosure agreements. Due to the lack of formal protection, once the information is publicly disseminated, a third party is not prevented from duplicating and using the information.

3   Recent IP Developments

> *"Whoever invents or discovers any new and useful process, machine, manufacture, or composition of matter, or any new and useful improvement thereof, may obtain a patent therefore, subject to the conditions and requirements of this title" (35 U.S.C. § 101) [15].*

The patentability of many biotechnology and bioinformatics inventions has long been a subject of intense debate. Arguments for patentability include the promotion of research and development and public disclosure of innovative technology. On the other hand, many believe that biotechnology and software inventions are fundamentally different from other kinds of patentable inventions. For instance, many argue against the patentability of genetics codes because genetic information is a product of nature. Software patents are often under debate because they are designed solely by their functions without physical features.

Bilski v. Kappos (2010 Supreme Court) and Mayo Collaborative Services v. Prometheus Laboratories (2012 Supreme Court) are two major recent developments regarding patentable subject matter. In Bilski v. Kappos, the Supreme Court affirmed USPTO and Federal Circuit's decision that the patent application describing the process for hedging risk in commodity markets was not a patentable subject matter. However, although all justices supported the Supreme Court's decision, the Court divided 5-4 in holding that under some circumstances some business methods may be patented [16]. Even though the case decision rejected the machine-or-transformation test to be the sole standard when determining patentability of a subject matter, it is crucial to note that the court was one vote away from eliminating an entire category as un-patentable. This implies similar favorable or unfavorable rulings could have happened to any other areas of technology under patentability debate. Overall, the Supreme Court in Bilski v. Kappos showed leniency in forms of biotechnology and bioinformatics inventions that are patent eligible, but demonstrated the strong division in opinion within the court regarding patentability of an invention.

In Mayo Collaborative Services v. Prometheus Laboratories, the Supreme Court reversed the Federal Circuit's decision that the process patent under consideration is patent-eligible under the machine-or-transformation test. The Supreme Court, in light of Bilski v. Kappos, stated that the machine-or-transformation test is not the ultimate test of patentability and that the patent under consideration merely describes a correlation found in law of nature, and thus un-patentable [17]. The Supreme Court's controversial decision received much unenthusiastic reaction from industry leaders. Many believe that if you look close enough, laws of nature can be found in any patent claim and it is the court's responsibility to recognize the innovation and technology embedded in the claims. A number of patent practitioners believes that the court's decision devalued many existing patents, and will reduce entrepreneurship and investment in biotechnology and bioinformatics.

In Alice Corporation v. CLS Bank International, the Supreme Court ruled that a patent application describing generic computer implementation of an abstract idea is not patentable [18]. The Supreme Court asked a two-pronged question to determine if the subject matter of the application is patentable: are the claims directed to a patent-ineligible concept, and, if so, do the elements of the claims transform the nature of the claim into a patentable application that is significantly more than the concept itself [18]? This decision was largely hailed to be the death of software and bioinformatics patents as nearly all of these patents could be drawn down to an abstract idea implemented by a generic computer.

The Supreme Court's decision in Alice Corporation v. CLS Bank International was tested when the Federal Circuit determined that an e-commerce syndication system for generating a web page is patentable in DDR Holdings, LLC v. Hotels.com, LP [19]. Specifically, the patent at issue did not "generically claim 'use of the Internet'" and instead specified how the Internet was manipulated in light of interactions that lead to a desired result when the result is not the same as the mere routine or conventional use of the internet [19]. The decision of the DDR Holdings Court in light of the Supreme Court's decision in Alice Corporation indicates that it is still possible to obtain a software or bioinformatics. Further, obtaining a patent is now more strategically important to guard against competition.

As referenced in Fig. 2, another major recent development in the United States patent system is the Leahy-Smith America Invents Act (AIA), signed into law on September 16, 2011 by President Barack Obama. It is said that the AIA will result in the most significant reform that the United States patent system has gone through since the enactment of Patent Act of 1952. There are several major changes to the patent system within the AIA that will alter patent value, patent acquisition cost, and patent enforcement.

The first notable change is the shift from "First-to-Invent" to "First-to-File" system [20]. Unlike other countries, the U.S. has operated under the first-to-invent system where a patent is entitled to those who first invent the technology. Many have praised the system for honoring the garage inventors who do not always have the monetary resources to apply for a patent at the time of the invention. The first-to-file system on the other hand determines priority of ownership by the earliest date the patent application was filed with the USPTO. This shift to the first-to-file system will eliminate costly and time-consuming interference proceedings with derivation proceedings. As a result of this change, inventors with little monetary resource and legal knowledge will most likely lose the race to attain ownership of an invention to big corporations who file patent applications on a day-to-day basis.

Second major change is the new restriction on the grace period. A grace period is the one-year period in which prior disclosure of an invention by the applicant or any other entity one year prior to the filing date does not count as prior art. AIA will change the grace period so that only the applicant's own disclosure prior to the filing date are exempt as invalidating prior art [21].

AIA will allow new mechanisms for challenging patents as well. These mechanisms include pre-issuance level at the USPTO and the post-issuance level. The pre-issuance level challenges are intended to make USPTO more efficient and lower the cost of post-issuance challenges made in the courts [22]. To many patent practitioners, these changes in AIA are discouraging since acquiring patent is now more difficult, lowering the return for investment in innovation and technology.

So what do these changes in the U.S. patent system mean for the bioinformatics/computational biology sector? First, like technologies in any other field, attaining a patent will be more difficult. AIA substantially expands prior art that can be used to invalidate a patent. Unlike current U.S. patent laws, AIA states that after March 15, 2013, invention will no longer be eligible for patentability if the invention has been patented or published; is in public use or on sale; or is otherwise available to the public *anywhere in the world* prior to the effective filing date of the patent application [23]. The new grace period will no longer protect a patent application with respect to all disclosures, but only those disclosures made by the inventor one year prior to the filing date [24].

Another interesting impact of AIA on biotechnology field is the act's stance on claims directed towards human organism. Historically, the USPTO without statutory or legal authority implemented the policy that human organisms must be rejected under 35 U.S.C. § 101. However, section 33(a) of the AIA specifically states, "[n]otwithstanding any other provision of law, no patent may issue on a claim directed to or encompassing a human organism" [25]. For this reason, those drafting claims in biotechnology patents directed toward cells, nucleic acids, and proteins must make sure that human organism is not captured within the scope of the claims.

Additionally, section 27 of AIA requires the Director of the USPTO to conduct a study on genetic testing that focuses on the following topics: the impact that the current lack of independent second opinion testing has had on the ability to provide the highest level of medical care to patients and recipients of genetic diagnostic testing, and on inhibiting innovation to existing testing and diagnosis; the effect that providing independent second opinion genetic diagnostic testing would have on the existing patent and license holders of an exclusive genetic test; the impact that current exclusive licensing and patents on genetic testing activity have on the practice of medicine, including but not limited to the interpretation of testing results and performance of testing procedures; and the role that cost and insurance coverage have on access to and provision of genetic diagnostic tests [26]. This section of AIA went into effect at the time of enactment, and the study is designed to provide input on how gene patents affect personalized medicine to Congress by June 2012.

| Bilski v. Kappos Supreme Court Decision June 28, 2010 | Leahy-Smith America Invents Act signed into law by President Barack Obama September 16, 2011 | Mayo Collaborative Services v. Prometheus Laboratories, Inc. Supreme Court Decision March 20, 2012 | Alice Corp. v. CLS Bank International Supreme Court June 19, 2014 | DDR Holdings, LLC v. Hotels.com Federal Circuit December 5, 2014 |

Fig. 2. Recent IP Developments (Source: Fernandez & Associates, LLP)

4   Advanced Protection Strategies

The promising market projection of the bioinformatics/computational biology industry and changes in the recent U.S. patent system calls for careful intellectual property protection strategies in order to thrive in the highly competitive market. Table 1 summarizes some protectable applications in biotechnology. Some patentable technologies specifically in the bioinformatics field are data mining and data visualization tools, sequence alignment and pattern recognition tools, molecular modeling tools, and predictive tools.

The initial patent prosecution strategy for biotechnology companies even prior to filing a patent application is to assess and evaluate the current state of the intellectual property owned or licensed by the company. By analyzing the current intellectual property portfolio, companies can align prosecution strategies with its future goals. For instance, separating different technologies into "utilize," "likely not to utilize," or "will not utilize" will allow companies to determine how to create a business plan such as expanding new products or protecting existing products [27].

There are two main ways to utilize patents: offensive and defensive uses. The first and apparent offensive use of patents is suing competitor for damages caused by infringement. Recent infringement case includes the Apple Inc v. Samsung case, in which Apple was awarded $1 billion [28]. Another obvious offensive use is licensing patents to generate royalties. In addition, companies may want to cross-license, or trade patents, in order to obtain technologies that other companies possess.

Patents can also be used as sales and marketing tool. When companies market products as "patented" or "patent pending," it induces the perception of uniqueness and innovation to the consumers. Such marketing strategy will keep away potential competitors in the field that may be infringing your technology. In essence, having a strong patent portfolio gives a company competitive advantage in the market.

Companies use patents defensively to protect investment in research and development. Without protection provided by patents, competitors are free to reverse engineer and steal your technology. Secondly, patents provide a bargaining option

Table 1. Examples of Patentable Biotechnologies

Tools	Software
	Devices and Methods
Healthcare Products	Diagnostics
	Drugs
Composition	Nucleic Acid Sequence
	Protein and Small Molecules

when competitors assert a patent infringement claim on a company. By having a strong intellectual property, you can increase the likelihood of settling a case through cross licensing instead of having to pay significant legal fees to either fight the allegations or pay royalties [29]. In addition, having ample patents may lead to counterclaims that will keep competitors from suing.

There are other various strategic values to patents that make them indispensable. Patents help create leverage in business deals. For instance, by having a patent on innovative technology, it will be substantially easier for you to find joint venture partners who will invest in commercialization of the technology [30]. Patents are considered as part of the overall asset of a business unit or company. Thus it is common for primary value of small technology companies to be its intellectual property when considering mergers and acquisitions. In addition, it will most likely be easier for start-up companies with patents to get funded by venture capitalists and other investors.

The general patent prosecution is outlined in the Fig. 3. Average prosecution time for a US patent is 3.4 years while the average biotechnology patent prosecution time is 4.4 years [31]. However, prosecution time can generally be shortened by submitting a provisional application under 35 U.S.C. § 111(b), making narrower claims, and responding to USPTO office action quickly and facilitating efficient communication with the patent examiner [32].

Some of the specific patent prosecution challenges in the areas of bioinformatics/computational biology include the interdisciplinary nature of the field and variety in business models. In order to successfully prosecute bioinformatics, both the prosecutor and the USPTO must understand the IT and biotechnology aspects of the invention [33]. Thus, it will be wise for applicants to hire patent attorneys from both fields as a team in order to realize the maximum value from the patent portfolio.

The variety of business models in the bioinformatics field poses a unique challenge as well. Examples of bioinformatics business models include licensing data, licensing software, selling and licensing systems, and selling or using test equipment to perform testing [34]. When drafting patent claims, the patent prosecutors must build claims that are tailor-made to the revenue model of the specific business.

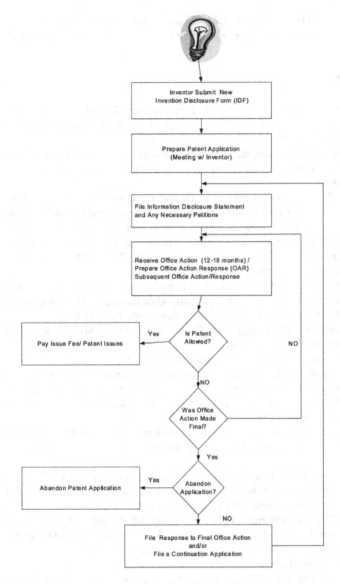

Fig. 3. General Patent Prosecution Procedure (Source: Fernandez & Associates, LLP)

5   Conclusion

Bioinformatics/computational biology is a fast-growing field, and patent prosecution in this field currently faces many difficult challenges and changes. As the United States patent system take on the largest reform since 1952, companies must pay special attention to their business models and how they will build patent portfolios. With efficient and effective patent strategies, bioinformatics/computational biology will continue to grow as an industry and promote technological innovation in healthcare.

References

[1] National Center for Biotechnology Information, Bioinformatics, http://www.ncbi. nlm.nih.gov/Class/MLACourse/Modules/MolBioReview/bioinformat ics.html

[2] Biological and Environmental Research Information System, Human Genome Project Information, http://www.ornl.gov/sci/techresources/Human_Genome/ home.shtml

[3] Battelle, $3.8 Billion Investment in Human Genome Project Drove $796 Billion in Economic Impact Creating 310,000 Jobs and Launching the Genomic Revolution (May 2011), http://www.battelle.org/media/news/2011/05/11/$3.8- billion-investment-in-human-genome-project-drove-$796- billion-in-economic-impact-creating-310-000-jobs-and- launching-the-genomic-revolution

[4] The Economist, A decade after the human-genome project, writes Geoffrey Carr (interviewed here), biological science is poised on the edge of something wonderful (June 17, 2010), http://www.economist.com/node/16349358

[5] Wetterstrand, K.: DNA Sequencing Costs: Data from the NHGRI Genome Sequencing Program (GSP) (January 14, 2015), http://www.genome.gov/sequencingcosts

[6] thepharmaletter, Global biotechnology market set to exceed $320 billion by 2015, says GIA (January 25, 2012), http://www.thepharmaletter.com/file/110532/ global-biotechnology-market-set-to-exceed-320-billion-by- 2015-says-gia.html

[7] Organization for Economic Co-operation and Development, Key Biotechnology Indicators, http://www.oecd.org/innovation/inno/keybiotechnologyindicator s.htm

[8] PRWeb, Global Bioinformatics Market to Reach US$5.0 Billion by 2015, According to a New Report by Global Industry Analysts, Inc. (January 10, 2011), http://www. prweb.com/releases/2011/1/prweb8052230.htm

[9] PRWeb, Global Bioinformatics Market to Reach US$5.0 Billion by 2015, According to a New Report by Global Industry Analysts, Inc. (January 10, 2011), http://www. prweb.com/releases/2011/1/prweb8052230.htm

[10] 35 U.S.C. §§ 101-103

[11] Richards Patent Law, How much does a patent cost? (2012), http://www. richardspatentlaw.com/faq/have-an-idea/how-much-does-a- patent-cost/

[12] Halpern, S.: Copyright Law: Protection of Original Expression (Carolina Academic Press (2002)

[13] Halperon, S.: Copyright Law: Protection of Original Expression. Carolina Academic Press (2002)

[14] Anawalt, H.C., Powers, E.E.: IP Strategy – Complete Intellectual Property Planning, Access, and Protection, West Group, §1.01, at 1-3 (2001)

[15] 35 U.S.C. §101

[16] Simmons, W.J.: Bilski v. Kappos: the US Supreme Court broadens patent subject-matter eligibility. Nature Biotechnology (August 2010), http://www.sughrue.com/files/Publication/5da8dcda-0cd5-4057-8e8e-1215a32ca981/Presentation/PublicationAttachment/a410c397-5170-49c2-a140-163bff055521/Simmons%20Nature%20Biotechnology%20August%2020 10.pdf

[17] Hymei, L.: Mayo v. Prometheus – the emperor's new law of nature? (September 24, 2012), http://www.masshightech.com/stories/2012/09/24/daily1-Mayo-v-Prometheus-the-emperors-new-law-of-nature.html

[18] Alice Corporation v. CLS Bank International, 573 U.S. ____ (2014)

[19] DDR Holdings, LLC v. Hotels.com, L.P., Appeal No. 2013-1505 Fed. Cir. (December 5, 2014)

[20] Waller, P.R.H., Loughran, C.A.: America Invents Act: The Implications of Patent Reform. IP Magazine (November 2011)

[21] Waller, P.R.H., Loughran, C.A.: America Invents Act: The Implications of Patent Reform. IP Magazine (November 2011)

[22] Waller, P.R.H., Loughran, C.A.: America Invents Act: The Implications of Patent Reform. IP Magazine (November 2011)

[23] Fasse, P.J., Baker, E.L., Reiter, T.A.: Patent Updates: The America Invents Act and Its Importance to Patent Prosecution in the Biotech Sector, Industrial Biotechnology (April 2012)

[24] Fasse, P.J., Baker, E.L., Reiter, T.A.: Patent Updates: The America Invents Act and Its Importance to Patent Prosecution in the Biotech Sector, Industrial Biotechnology (April 2012)

[25] Fasse, P.J., Baker, E.L., Reiter, T.A.: Patent Updates: The America Invents Act and Its Importance to Patent Prosecution in the Biotech Sector, Industrial Biotechnology (April 2012)

[26] Fasse, P.J., Baker, E.L., Reiter, T.A.: Patent Updates: The America Invents Act and Its Importance to Patent Prosecution in the Biotech Sector, Industrial Biotechnology (April 2012)

[27] Isacson, J.P.: Maximizing Profits Through Planning and Implementation, 18 Nature Biotechnology 565, 565 (2000)

[28] Wingfield, N.: Jury Awards $1 Billion to Apple in Samsung Patent Case. The New York Times (August 24, 2012), http://www.nytimes.com/2012/08/25/technology/jury-reaches-decision-in-apple-samsung-patent-trial.html?_r=0

[29] Gatto, J.G.: Bioinformatics Patents – Challenges and Opportunities. Mintz Levin (November 2001)

[30] Gatto, J.G.: Bioinformatics Patents – Challenges and Opportunities. Mintz Levin (November 2001)

[31] Huie, J.T., Fernandez, D.S.: Strategic Balancing of Patent and FDA Approval Process to Maximize Market Exclusivity

[32] Huang, W.H., Yeh, J.J., Fernandez, D.: Patent Prosecution Strategies for Biotechnological Inventions. IP Strategy Today (2005)

[33] Gatto, J.G.: Bioinformatics Patents – Challenges and Opportunities. Mintz Levin (November 2001)

[34] Gatto, J.G.: Bioinformatics Patents – Challenges and Opportunities, Mintz Levin (November 2001)

Lupin Allergy: Uncovering Structural Features and Epitopes of β-conglutin Proteins in *Lupinus Angustifolius* L. with a Focus on Cross-allergenic Reactivity to Peanut and Other Legumes

José C. Jimenez-Lopez[1,2,*], Elena Lima-Cabello[2], Su Melser[3], Rhonda C. Foley[3], Karam B. Singh[1,3], and Alché Juan D.[2]

[1] The UWA Institute of Agriculture, The University of Western Australia, 35 Stirling Highway, Crawley, Perth WA 6009 Australia
jose.jimenez-lopez@uwa.edu.au, jcjimenez175@gmail.com
[2] Department of Biochemistry, Cell and Molecular Biology of Plants, Estación Experimental del Zaidín (EEZ), National Council for Scientific Research (CSIC), 1 Profesor Albareda Street, Granada 18008 Spain
[3] Agriculture, CSIRO, 147 Underwood Av., Floreat, Perth WA 6014 Australia

Abstract. The use of sweet lupins as a new food is resulting in an increasing number of cases of allergy reactions, particularly in atopic patients with other pre-existing legume allergies. We performed an extensive *in silico* analysis of seed β-conglutins, a new family of major allergen proteins in lupin, and a comparison to other relevant food allergens such as Ara h 1. We analyzed surface residues involved in conformational epitopes, lineal B- and T-cell epitopes variability, and changes in 2-D structural elements and 3D motives, with the aim to investigate IgE-mediated cross-reactivity among lupin, peanut, and other different legumes.

Our results revealed that considerable structural differences exist, particularly affecting 2-D elements (loops and coils), and numerous micro-heterogeneities are present in fundamental residues directly involved in epitopes variability.

Variability of residues involved in IgE-binding epitopes might be a major contributor to the observed differences in cross-reactivity among legumes.

Keywords: β-conglutins, Computational Biology, Epitopes, Diagnosis, Food Allergy, Legume Seeds, *Lupinus angustifolius* L., Protein Structure Modeling, IgE-binding, Immunotherapy, Recombinant Allergen, Vicilin-Like Proteins.

1   Introduction

Lupin is a popular PULSE (the edible seeds of plants in the legume family) worldwide, which has traditionally been consumed as source of proteins since long ago.

---

[*] Corresponding author.

F. Ortuño and I. Rojas (Eds.): IWBBIO 2015, Part I, LNCS 9043, pp. 96–107, 2015.

From more than 450 species of the *Lupinus* family, only lupin known as "sweet lupins" such as white lupin (*Lupinus albus*), yellow lupin (*Lupinus luteus*), and blue lupin (*Lupinus angustifolius*) are being used in food manufacturing. Flour of raw lupin is increasingly used as food ingredient because of its nutritional value (rich in protein and fibre, poor in fat and gluten-free) [1].

Furthermore, ingestion of lupin-containing foods has been associated with the prevention of obesity, diabetes, and eventually cardiovascular disease. Recently, hypocholesterolaemic properties have been demonstrated for lupin conglutin γ proteins, which may decrease the risk of cardiovascular disease [2].

Lupin belongs to the *Fabaceae* family. As all edible legume seeds, the major protein fraction of lupin seeds is associated with storage proteins, which could be classified in the cupin and prolamin superfamilies, based in structure, solubility and/or sedimentation properties.

The two major lupin storage proteins are α-conglutin (legumin-like or 11S globulin), and ß-conglutin (vicilin-like or 7S globulin). Vilicin proteins are characterized by two cupin (barrel-shaped) domains constituted by α-helices. Another family with a cupin-like structure, γ-conglutin (basic 7S-globulin), displays tetrameric structure integrated by two different disulphide-linked monomers. In contrast, δ-conglutin (2S sulphur-rich albumin) contains 2 disulphide-linked proteins with the typical cysteine-rich prolamin structure [3].

Sweet lupin seeds seem to be particularly promising as a source of innovative food ingredients due to averaged protein content similar to soybean and an adequate composition of essential amino acids. Foods based on sweet lupin proteins include flour for bakery, pasta formulations, gluten-free products and other food items, which are gaining more attention from industry and consumers because the large number of health-promoting benefits described above [2].

On the other hand, with the rapid introduction of novel foods and new ingredients in traditional foods, the number of reports of allergic reactions to lupin proteins is also rising, either as primary lupin allergy or as a result of cross-reactivity to other legume proteins, particularly peanut, soybean, lentil, bean, chickpea, and pea [4]. The most common clinical pattern of lupin allergy is the triggering of an allergic reaction via ingestion of lupin in peanut-allergic individuals, although most commonly triggered via ingestion, inhalation and occupational exposure in individuals without peanut allergy has also been reported. The prevalence varies considerably between studies, but a prevalence of about 1-3% in the general population and 3–8% among childrens is the consensus [5]. Considering the increasing number of clinical cases of lupin allergy reported in the literature, lupin was added in 2008 to the list of foods that must be labelled in pre-packaged foods as advised by the European Food Safety Authority (EFSA) (http://www.efsa.europa.eu/).

Overall, cross-reactivity is the result of IgE-binding to commonly shared epitopes among proteins, i.e. different legume seed proteins, with conserved steric domains (conformational epitopes), and/or amino acid sequences (lineal epitopes).

Given the increase in the number of cases of lupin allergy and the frequency of cross-reactivity with other legume seed proteins, the possible involvement of individual major lupin proteins, i.e. β-conglutins, and their counterparts from other legumes in cross-allergy is of major concern and of great interest to investigate.

In the present study, we add to our results an extensive *in silico* analysis including allergen structure modeling based epitopes (T- and B-cells) identification, aiming to uncover common-shared and specific epitopes, and providing a comprehensive insight of the broad cross-allergy among legume proteins, as well as specific allergic reactions to lupin β-conglutins. This is an important step towards understanding the molecular basis of the allergy phenomenon, particularly cross-reactivity, and towards the development of safe and efficacious diagnosis tools and immunotherapy to lupin-related food allergy.

2   Methods

2.1   Allergen Sequences

We retrieved allergen sequences necessary for the present study from GenBank/ EMBL Database: β-conglutin 1 or Lup an 1 (F5B8V9), β-conglutins 2 to 7 (F5B8W0 to F5BW5), Ara h 1 (P43237, P43238), Gly m 5 (O22120), Len c 1 (Q84UI0, Q84UI1), Gly m β-conglycinin (P25974), Vig r 2 (Q198W3, B1NPN8).

2.2   Phylogenetic Analysis of Food Allergen Sequences

Allergen protein sequences from legumes (lupin, peanut, soybean, Mung bean, lentil, chickpea, pea, mezquite) were retrieved and used to perform a phylogenetic analysis. Sequences alignments were performed by using ClustalW multiple sequence alignment tool (www.ebi.ac.uk/Tools/clustalw) according to Jimenez-Lopez et al. [6]. Trees were visualized using Treedyn (www.treedyn.org).

2.3   Template Assessment

All allergen sequences were searched for homology in the Protein Data Bank (PDB). Suitable homologous templates were selected by using Swiss-Prot database (swiss-model.expasy.org) and BLAST server (ncbi.nlm.nih.gov/) employing fold recognition.

2.4   Proteins Homology Modeling

Sequences were modelled through SWISS-MODEL via the ExPASy web server (swissmodel.expasy.org), by using the top PDB closest template structures previously assessed. Models refinement of 3D structural errors, and structural assessment were performed using stereo-chemical and energy minimization parameters [7].

2.5   Structural Comparison and Evolutionary Conservational Analysis

Allergen proteins structure comparison was performed by superimposition to calculate average distance between their Cα backbones. Protein models were submitted to ConSurf server (consurf.tau.ac.il) to generate evolutionary related conservation scores, in order to identify functional region in the proteins. Functional and structural key residues were confirmed by ConSeq server (conseq.tau.ac.il). 2-D and 3D were visualized and analyzed using PyMol software (www.pymol.org).

2.6 Solvent Accessible Surface Area and Poisson–Boltzmann Electrostatic Potential

Solvent accessible surface area (SASA), defined as the percentage of surface area of a biomolecule that is accessible to a solvent for each residue was calculated by using the GETAREA v1.1. program (curie.utmb.edu/getarea.html). The electrostatic Poisson-Boltzmann (PB) potentials for the built structures were obtained [7,8].

2.7 Allergenicity Profile Assessment

Allergenicity of lupin and other legume allergen sequences was checked by a full FASTA alignment in the Structural Database of Allergenic Proteins (SDAP) (Fermi.utmb.edu/SDAP). Allergenicity profile was assessed by combination of different parameters: hydrophobicity, antigenicity and SASA [9]. Values of absolute surface area (ASA) of each residue were also calculated by DSSP program (swift.cmbi.ru.nl/gv/dssp), and transformed to relative values of ASA and visualized by ASAView (www.netasa.org/asaview).

2.8 Linear and Conformational B-cell Epitopes Analysis

For determination of linear (continuous) epitopes, the allergen proteins sequences were submitted to ABCpred (uses artificial neural networks, www.imtech.res.in/raghava), BepiPred 1.0b (based on hydrophobicity scale with a Hidden Markov Model, www.cbs.dtu.dk), BCPREDS (uses support vector machine, ailab.cs.iastate.edu/ bcpreds), Bcepred (based on a combination of physico-chemical properties, www.imtech.res.in/raghava), and COBEpro (uses support vector machine, scratch.proteomics.ics.uci.edu). Linear and discontinuous antibody epitopes based on a protein antigen's 3D structure were predicted using Ellipro (http://tools.immuneepitope.org/tools/ElliPro/iedb_input/, discontinuous epitopes are defined based on PI values and are clustered based on the distance R (Å) between residue's centers of mass, tools.immuneepitope.org), and Discotope (tools.immuneepitope.org) webservers.

The epitopes identified frequently by most of the tools were selected [9,10].

2.9 T-cell Epitopes Identification and Analysis

The identification of MHC Class-II binding regions for all the allergen sequences was performed by using neuronal networks and quantitative matrices derived from published literature. Promiscuous peptides binding to multiple HLA class II molecules were selected. The analysis was made by using the TEPITOPE software (www.bioinformation.net/ted), with a threshold of 5% for the most common human HLA-DR alleles [DR1, DR3, DR4, DR7, DR8, DR5 and DR2] among Caucasian population [10], and covering a large proportion of the peptides that bind with human HLA.

3   Results

3.1   Searching for Allergen Proteins Templates

We used the Swiss-model server to identify the best possible templates to build allergen structures, finding high scores and very low E-values (ranging 12E–34 to 7E–42) for the templates retrieved from Protein Data Bank (PDB) database and used for homology modeling: lupin β-conglutins (1uijA, 2eaaB), Ara h 1 (3s7i, 3s7e), Gly m 5 (1uijA), Len c 1 (1uijA), Gly m β-conglycinin (1uijA), Vig r 2 (2eaaB).

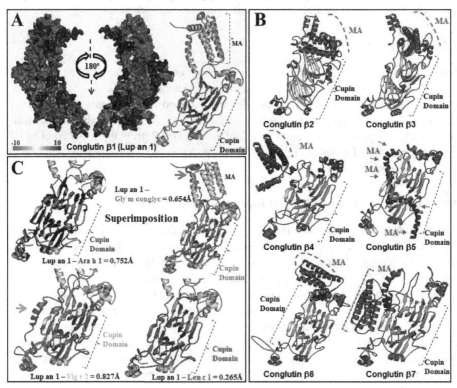

Fig. 1. Lupin β-conglutins structural analysis. A) Cartoon and surface representation views of conglutin β1 rotated 180°, showing the surface electrostatic potential clamped at red (-10) or blue (+10). 2-D elements (α-helices, β-sheets, coils) were depicted in cartoon model, showing main proteins domains (mobile arm and cupin domain). B) 3D structures of conglutins β2 to β7 were depicted as a cartoon diagram. α-helices, β-sheets and coils are depicted in red, yellow and, green respectively, integrating main proteins domains. C) Superimpositions showed the close structural relationship with allergens from other legumes such as peanut (Ara h 1), soybean (β-conglycinin), Mung bean (Vig r 2), and lentils (Len c 1). Å = Armstrong; **MA** = mobile arm.

Figure 1 showed that lupin β-conglutins are characterized by a surface negatively charged, a domain from the Cupin superfamily constituted by 2 barrels of 8-10 α-helices each, and a mobile arm, which position may be different depending of the β-conglutin form. One of these barrels followed the Rossmann fold structure, typically found in oxidoreductase enzymes.

2-D elements comparison by superimposition among allergens showed a comparable low values (< 1Å) of structural differences, when compared Cupin superfamily domain, since the mobile arm is absent in these allergens. Overall, β-conglutins were found structurally close to Len c 1 and most distantly related to the Gly m 5 allergen.

3.2 Structural Assessment of the β-conglutin 1 to 7, Gly m 5, Len c 1, Gly m Conglycinin, and Vig r 2 Structural Models

Different molecular tools (stereochemistry, energy minimization) were used to assess the quality of the models built for this study. A general quality assessment parameter as *QMEAN* displayed adequate values for all models. Most of the residues of the main chain of built models were located in the acceptable regions of the Ramachandran plot shown by *Procheck* analysis. In addition, Z-scores returned from *ProSa* indicated that structures showed negative interaction energy and within the lowest energy range. In addition, the Z-scores were within the range usually found for templates used for allergen structure modeling.

3.3 Phylogenetic Analysis

We analyzed the relationships between lupin β-conglutin proteins and allergens from other species. The data clearly reveal five established groups/clusters. We have identified 5 main groups, where β-conglutins were grouped with allergens of 7S-globulin nature (Fig. 2).

3.4 Identification of Highly Antigenic Regions in Plant Profilins

Physicochemical parameters such as hydrophobicity, accessibility, exposed surface, and antigenic propensity of polypeptide chains have been used to identify continuous epitopes (see methods section). In our analysis, antigenicity determinants were assigned by locating the positive peaks in hydrophilicity plots, and identifying the regions of maximum potential of antigenicity (data not shown).

We identified up to 8 regions in lupin β-conglutins, with high potential of antigenicity, 7 regions in Ara h 1, 7 regions in Gly m 5, 8 regions in β-conglycinin, 7 regions in Vig r 2, 4 in Len c 1, and 5 in Pis s 2 (data not shown). These regions with high antigenicity correlated well with the lineal T- and B-cell and conformational epitopes identified and analyzed in the present study.

The highest differences in terms of antigenicity regions polymorphism correspond to lupin β-conglutins, while the lowest variable allergen was Len c 1 (data not shown).

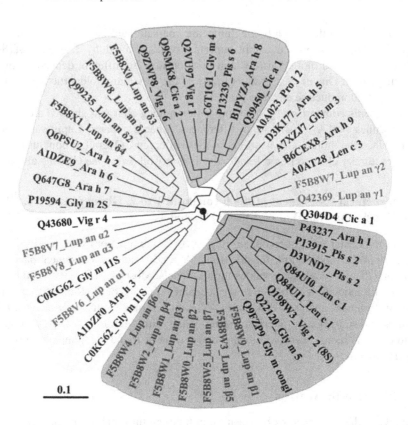

Fig. 2. Phylogenetic analysis of food allergens. Neighbor-joining (NJ) method was used to perform a phylogenetic analysis of 45 legume allergens from lupin (conglutins α1–3, β1–7, γ1–2, and δ1–4), peanut (Ara h 1, 2, 3, 5, 6, 7, 8, and 9), soybean (Gly m 5, and Gly m conglycinin), lentil (Len c 1, and 3), pea (Pis s 2, and 6), Mung bean (Vig r 1, 2, 4, and 6), Mezquite (Pro j 2), and chickpea (Cic a 1, and 2).

3.5   Analysis of B-cell Epitopes

12 antigenic lineal regions prone to B-cell binding were analyzed in conglutins β1, 7 for β 2 and β 7, 5 for β3, 4 for β 4, β 5, β 6. In addition, we identified 6 antigenic regions in Ara h 1, 7 in Gly m 5, 11 in β -conglycinin, 4 in Len c 1, 10 in Pis s 2, and 8 in Vig r 2 (Fig. 3). Comparative analysis of these regions showed that 5 lineal epitopes in conglutin β 1 are located in the mobile arm, 3 of them overlapping with a big conformational epitopic area (black color, Fig. 3) and 2 lineal epitopes independent. Furthermore, β 2 and β 5 present 3 conformational epitopic areas, 1 in β 3, β 6 and β 7, 2 in β 4, related to the differential mobile arm structure.

The biggest difference as structural feature between the β -conglutins and the other legume allergens is the presence of the mobile arm in N-terminal region of the lupin β -conglutins and the epitopes which integrate. Number of epitopes and polymorphism analysis of lineal and conformational B-cell epitopes in other legume allergens showed a wide range of variability in both, the number and the sequence identity of these epitopes (data not shown).

3.6   Identification of T-cell Epitopes

We have identified a variable number of anchor motifs to HLA-DR in the sequences of lupin β-conglutins (8 main T-cell epitopes), and their counterparts in five species of legumes.

Lineal B-cell epitopes   Conformational B-cell epitopes

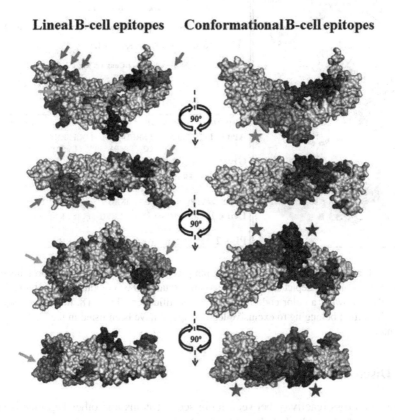

Fig. 3. B-cell epitopes analysis in lupin β-conglutins and their legume proteins counterparts. Cartoon representation of Lup an 1 allergen showing in various colors lineal and conformational B-cell epitopes in its surface. Arrows and stars represent specific lineal and conformational epitopes, respectively, which do not overlap with each other.

T1 was the "solo" epitope in the mobile arm of β-conglutins (Fig. 4), exhibiting a large surface orientation. This epitope is common for other legume allergens such as peanut (Ara h 1), soybean (β-conglycinin), Mung bean (Vig r 2), and pea (Pis s 2). The rest of epitopes identified in β-conglutins were located in the globular (Cupin-like) domain of these proteins. Some of these epitopes were differentially shared with other legume allergens, i.e. T2 is the most commonly shared epitope, and T8 only commonly located in allergens of soybean. In addition, each of the allergen analyzed has specific epitopes not found in other species (Fig. 4). Most of these lineal epitopes displayed 50% or more of their residues not exposed to the surface (T2 to T8).

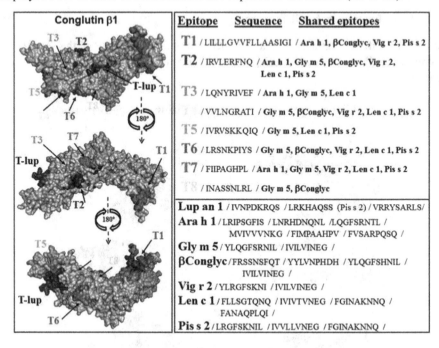

Conglutin β1	Epitope	Sequence	Shared epitopes
	T1	/ LILLLGVVFLLAASIGI /	Ara h 1, βConglyc, Vig r 2, Pis s 2
	T2	/ IRVLERFNQ /	Ara h 1, Gly m 5, βConglyc, Vig r 2, Len c 1, Pis s 2
	T3	/ LQNYRIVEF /	Ara h 1, Gly m 5, Len c 1
		/ VVLNGRATI /	Gly m 5, βConglyc, Vig r 2, Len c 1, Pis s 2
	T5	/ IVRVSKKQIQ /	Gly m 5, Len c 1, Pis s 2
	T6	/ LRSNKPIYS /	Gly m 5, βConglyc, Vig r 2, Len c 1, Pis s 2
	T7	/ FIIPAGHPL /	Ara h 1, Gly m 5, Vig r 2, Len c 1, Pis s 2
	T8	/ INASSNLRL /	Gly m 5, βConglyc

Lup an 1 / IVNPDKRQS /LRKHAQSS (Pis s 2) / VRRYSARLS/
Ara h 1 /LRIPSGFIS /LNRHDNQNL /LQGFSRNTL / MVIVVVNKG / FIMPAAHPV / FVSARPQSQ /
Gly m 5 / YLQGFSRNIL / IVILVINEG /
βConglyc /FRSSNSFQT / YYLVNPHDH / YLQGFSHNIL / IVILVINEG /
Vig r 2 /YLRGFSKNI / IVILVINEG /
Len c 1 /FLLSGTQNQ / IVIVTVNEG / FGINAKNNQ / FANAQPLQI /
Pis s 2 /LRGFSKNIL /IVVLLVNEG / FGINAKNNQ /

Fig. 4. T-cell epitopes comparison between lupin β-conglutins and their legume proteins counterparts. T-cell epitopes depicted on the three views rotated 180° conglutin β1 protein surface, respectively, following a color code for epitopes identification (T1 to T8, upper right square). Epitopes identified belonging to exclusive legume specie have been listed in the figure (bottom right square).

4   Discussion

The immune-cross-reactivity between lupin seed proteins and other legumes is starting to be analyzed, and knowledge at molecular level is scarce.

In *Lupinus angustifolius* L., Lup an 1 (conglutin β 1) has been recently identified as a major allergen by using proteomic analysis and was recognized by serum IgE from most of 12 lupin-allergic patients' sera [11], which matched with a vicilin-like (7S-type) protein.

Knowledge about the linear epitopes of lupin major allergens is of crucial importance to help identify the trigger agent in clinical diagnosis (trials) of lupin allergy and to develop and implement molecular tools in order to identify the presence of lupin allergens as ingredients in food, in order to protect patients with lupin allergy and other cross-allergenic reaction. Sequence homology between lupin major allergens and other legume allergens support cross-reactivity between them. However, in the present study has been identified commonly shared T- and B-cell lineal and conformational epitopes in lupin and allergens from other legumes, which are located in the globular (Cupin Superfamily) characteristic domain. The largest number of epitopes has been identified in conglutin β 1, which may be the reason why Lup an 1 is currently the main allergen among the beta forms. Several of these epitopes are common to other legume proteins. However, others are not well-conserved, finding a noticeable degree of polymorphism. We have identified surface patterns (conformational epitopes), as well as multiple regions (B- and T-cell epitopes) in legume allergens, including lupin, exhibiting differences in length and variability. Furthermore, we have found shared common B- and T-cell epitopes among these legume allergens, as well as epitopes differentially distributed in specific allergens. The variability in their surface residues might contribute to generate areas of the protein enable of being differentially recognized as Th2- inducing antigens. Depending on the location of these polymorphic residues, recognition by IgE/IgG may be also affected [12].

Thus, we propose that the presence of several of these epitopes (T- and B-cell) is the main reason for cross-allergenicity reactions among legume proteins, which however react differentially with lupin β -conglutins forms and between them. The extension of the reactions may be directly linked to the residues variability of these epitopes. It has been reported serological cross-reactivity among legume allergens [4], and Lup an 1 (Ara h1, Len c 1 and Pis s 1). IgE reactivity may not always be related to clinical cross-reactivity (leading to allergy symptoms), which has been observed in lupin, peanut and pea. In this regard, we have found that six T-cell epitopes are shared between Lup an 1 and Len c 1. From these, four epitopes are commonly found in Ara h 1 and Pis s 1 as well. Furthermore, one of these four epitopes is the "T- solo" or T1 located in the mobile arm of β -conglutins. This epitope may play a key role in specific cross-reactivity between legume seeds proteins and lupin β -conglutins as one of the four main families (α, β, γ, δ) of seed storage proteins in lupin.

Molecular modeling of proteins throughout computational biology tools help identifying specific regions, which could be candidates for the development of peptide-based immunotherapeutic reagents for allergy, while conserved regions could be responsible of the cross-reaction between allergens [10]. Epitopes prediction based on knowledge derived from structural surface features such as increased solvent accessibility, backbone flexibility, and hydrophilicity [7,9,10]. Such predictions were found to correlate well with antigenicity in the present study. At structural level, antigenic determinants may be integrated by 2-D structure elements, which protrude from the surface of the protein, such as coils and loops [10]. Our results have shown that conformational epitopes are these more affected by 2-D structure elements, which are mostly integrated by short α-helices and coils (Fig. 1 and 3). Variability in sequence and length of these 2-D elements may additionally increase the differences and the extension of the cross-allergenic reactions between legume allergens [13].

On the other hand, linear B- and/or T- cell epitopes may play most important roles in cross-reactivity between food allergens [14], since food processing or digestion may increase the number or the accessibility of IgE binding epitopes. Thus, some food allergens have been described to lead to a loss of some or all the B-cell epitopes (but not the T-cell epitopes) by denaturalization/digestion [15]. In a similar fashion, vicilin-like allergens such as Ara h 1 and Lup an 1 also share thermal stability. B- and T-cell responses have a defining and differential recognition of antigenic epitopes, and their localization in the allergen does not necessarily coincide. T-cell receptor recognizes only the linear amino acid sequence [16]. In contrast, B-cell epitopes recognized by IgE antibodies are either linear or conformational and are located on the surface of the molecule accessible to antibodies. The extension of the epitope may range from 5 to 8 or longer amino acids for IgE to be able of binding to the epitope [17-20]. However, we have identified lineal B-cell epitopes in lupin β-conglutins and the other legume allergens with a wide range of amino acid lengths, and overlapping with conformational epitopes.

5   Conflict of Interest

The authors confirm that this article content has no conflicts of interest.

Acknowledgement. This research was supported by the European Research Program MARIE CURIE (FP7-PEOPLE-2011-IOF), under the grant reference number PIOF-GA-2011-301550 to JCJ-L, KBS, and JDA, and by ERDF-cofunded projects P2010-AGR-6274 and P2011-CVI-7487 (Junta de Andalucía, Spain).

References

1. Borek, S., Pukacka, S., Michalski, K., Ratajczak, L.: Lipid and protein accumulation in developing seeds of three lupine species: Lupinus luteus L., Lupinus albus L., and Lupinus mutabilis Sweet. J. Exp. Bot. 60(12), 3453–3466 (2009)
2. Bähr, M., Fechner, A., Krämer, J., Kiehntopf, M., Jahreis, G.: Lupin protein positively affects plasma LDL cholesterol and LDL:HDL cholesterol ratio in hypercholesterolemic adults after four weeks of supplementation: a randomized, controlled crossover study. Nutrition J. 12, 107 (2013)
3. Duranti, M., Consonni, A., Magni, C., Sessa, F., Scarafoni, A.: The major proteins of lupin seed: Characterisation and molecular properties for use as functional and nutraceutical ingredients. Trends in Food Sci. Technol. 19(12), 624–633 (2008)
4. Guillamón, E., Rodríguez, J., Burbano, C., Muzquiz, M., Pedrosa, M.M., Cabanillas, B., Crespo, J.F., Sancho, A.I., Mills, E.N., Cuadrado, C.: Characterization of lupin major allergens (Lupinus albus L.). Mol. Nutr. Food Res. 54(11), 1668–1676 (2010)
5. Koplin, J.J., Martin, P., Allen, K.: An update on epidemiology of anaphylaxis in children and adults. Curr. Op. Allergy Clin. Imm. 11(5), 492–496 (2011)
6. Jimenez-Lopez, J.C., Morales, S., Castro, A.J., Volkmann, D., Rodríguez-García, M.I., Alché, J.D.: Characterization of profilin polymorphism in pollen with a focus on multifunctionality. PLoS One 7(2), e30878 (2012)
7. Jimenez-Lopez, J.C., Kotchoni, S.O., Hernandez-Soriano, M.C., Gachomo, E.W., Alché, J.D.: Structural functionality, catalytic mechanism modeling and molecular allergenicity of phenylcoumaran benzylic ether reductase, an olive pollen (Ole e 12) allergen. J. Comput. Aided. Mol. Des. 27(10), 873–895 (2013)

8. Gao, D., Jimenez-Lopez, J.C., Iwata, A., Gill, N., Jackson, S.A.: Functional and structural divergence of an unusual LTR retrotransposon family in plants. PLoS One 10, e48595 (2012)

9. Jimenez-Lopez, J.C., Kotchoni, S.O., Rodríguez-García, M.I., Alché, J.D.: Structure and functional features of olive pollen pectin methylesterase using homology modeling and molecular docking methods. J. Mol. Model. 18(12), 4965–4984 (2012)

10. Jimenez-Lopez, J.C., Rodríguez-García, M.I., Alché, J.D.: Analysis of the effects of polymorphism on pollen profilin structural functionality and the generation of con-formational, T- and B-cell epitopes. PLoS One 8(10), e76066 (2013)

11. Goggin, D.E., Mir, G., Smith, W.B., Stuckey, M., Smith, P.M.: Proteomic analysis of lupin seed proteins to identify conglutin Beta as an allergen, Lup an 1. J. Agric. Food Chem. 56(15), 6370–6377 (2008)

12. Ferreira, F., Hirtenlehner, K., Jilek, A., Godnick-Cvar, J., Breiteneder, H., et al.: Dissection of immunoglobulin E and T lymphocyte reactivity of isoforms of the major birch pollen allergen Bet v 1: potential use of hypoallergenic isoforms for immunotherapy. J. Exp. Med. 183, 599–609 (1996)

13. Valenta, R., Duchene, M., Ebner, C., Valent, P., Sillaber, C., et al.: Profilins constitute a novel family of functional plant pan-allergens. J. Exp. Med. 175(2), 377–385 (1992)

14. Aalberse, R.C., Akkerdaas, J., Van Ree, R.: Cross-reactivity of IgE antibodies to allergens. Allergy 56(6), 478–490 (2001)

15. Schimek, E.M., Zwolfer, B., Briza, P., Jahn-Schmid, B., Vogel, L., et al.: Gas-trointestinal digestion of Bet v 1-homologous food allergens destroys their mediator-releasing, but not T cell-activating, capacity. J. Allergy Clin. Immunol. 116, 1327–1333 (2005)

16. Pomes, A.: Relevant B cell epitopes in allergic disease. Int. Arch. Allergy Immunol. 152, 1–11 (2010)

17. Meno, K.H.: Allergen structures and epitopes. Allergy 66(95), 19–21 (2011)

18. Bannon, G.A., Ogawa, T.: Evaluation of available IgE-binding epitope data and its utility in bioinformatics. Mol. Nutr. Food Res. 50, 638–644 (2006)

19. Tanabe, S.: IgE-binding abilities of pentapeptides, QQPFP and PQQPF, in wheat gliadin. J. Nutr. Sci. Vitaminol. 50, 367–370 (2004)

20. Asturias, J.A., Gomez-Bayon, N., Arilla, M.C., Sanchez-Pulido, L., Valencia, A., et al.: Molecular and structural analysis of the panallergen profilin B cell epitopes defined by monoclonal antibodies. Int. Immunol. 14(9), 993–1001 (2002)

Artificial Neural Networks in Acute Coronary Syndrome Screening

M. Rosário Martins[1], Teresa Mendes[1], José M. Grañeda[2], Rodrigo Gusmão[2],
Henrique Vicente[3] and José Neves[4,*]

[1] Departamento de Química, ICAAM, Escola de Ciências e Tecnologia, Universidade de Évora,
Évora, Portugal
mrm@uevora.pt, teresabmendes@gmail.com
[2] Serviço de Patologia Clínica do Hospital do Espírito Santo de Évora EPE
granedal@sapo.pt, dir.patcli@hevora.min-saude.pt
[3] Departamento de Química, Centro de Química de Évora, Escola de Ciências e Tecnologia,
Universidade de Évora, Évora, Portugal
hvicente@uevora.pt
[4] CCTC, Universidade do Minho, Braga, Portugal
jneves@di.uminho.pt

Abstract. In Acute Coronary Syndrome (ACS), early use of correct therapy plays a key role in altering the thrombotic process resulting from plaque rupture, thereby minimizing patient sequels. Indeed, current quality improvement efforts in acute cardiovascular care are focused on closing treatment gaps, so more patients receive evidence-based therapies. Beyond ensuring that effective therapies are administered, attention should also be directed at ensuring that these therapies are given both correctly and safely. Indeed, this work will focus on the development of a diagnosis support system, in terms of its knowledge representation and reasoning procedures, under a formal framework based on Logic Programming, complemented with an approach to computing centered on Artificial Neural Networks, to evaluate ACS predisposing and the respective Degree-of-Confidence that one has on such a happening.

Keywords: Acute Coronary Syndrome, Healthcare, Logic Programming, Knowledge Representation and Reasoning, Artificial Neuronal Networks.

1   Introduction

The Acute Coronary Syndrome (ACS) stands for a complex medical disorder, associated with high mortality and morbidity, with heterogeneous etiology, characterized by an imbalance between the requirement and the availability of oxygen in the myocardium [1]. In Europe, cardiovascular disease are responsible for more than 2 million deaths per year, representing about 50% of all deaths, and for 23% of the morbidity cases [2]. The clinical presentations of coronary artery disease include

---

* Corresponding author.

F. Ortuño and I. Rojas (Eds.): IWBBIO 2015, Part I, LNCS 9043, pp. 108–119, 2015.

Silent Ischemia, Stable Angina Pectoris, Unstable Angina, Myocardial Infarction (MI), Heart Failure, and Sudden Death [2]. Under a thorough medical profile indicative of ischemia, the ElectroCardioGram (ECG) is a priority after hospital admission, and patients are often grouped in two categories, i.e., patients with acute chest pain and persistent ST-Segment Elevation (STE), and patients with acute chest pain but without persistent ST-Segment Elevation (NSTE) [3]. Besides the higher hospital mortality in STE-ACS patients, the annual incidence of NSTE-ACS patients is higher than STE-ACS ones [4,5]. Furthermore, a long-term follow-up showed the increase of death rates in NSTE-ACS patients, with more co-morbidity, mainly diabetes mellitus, Chronic Renal Diseases (CRD) and anemia [2], [6]. Premature mortality is increased in individuals susceptible to accelerated atherogenesis caused by accumulation of others risk factors, namely age over 65 years, hypertension, obesity, lipid disorders and tobacco habits. Diagnosis of ACS relies, besides clinical symptoms and ECG findings, primarily on biomarker levels. Markers of myocardial necrosis such as cardiac troponins, creatine kinase MB mass (CK-MB mass) and myoglobin reflect different pathophysiological aspects of necrosis and are the gold standard in detection of ACS [7,8]. According to the European Society of Cardiology, troponins (T, I) that reflects myocardial cellular damage, play a central role in the diagnosis establishing and stratifying risk and make possibility to distinguish between NSTEMI and unstable angina [2], [9]. Nevertheless, CK-MB mass is useful in association with cardiac troponin in order to discard false positives diagnosis, namely in pulmonary embolism, renal failure and inflammatory diseases, such as myocarditis or pericarditis [9,10].

Patients with chest pain represent a very substantial proportion of all acute medical hospitalization in Europe. Despite modern treatment, the rates of death, MI and readmission of patients with ACS remain high [2,3].

The stated above shows that it is difficult to make an early diagnosis of ACS since it needs to consider different conditions with intricate relations among them, where the available data may be incomplete, contradictory and/or unknown. In order to overcome these drawbacks, the present work reports the founding of a computational framework that uses knowledge representation and reasoning techniques to set the structure of the information and the associate inference mechanisms. We will centre on a Logic Programming (LP) based approach to knowledge representation and reasoning [11,12], complemented with a computational framework based on Artificial Neural Networks (ANNs) [13].

2   Knowledge Representation and Reasoning

Many approaches to knowledge representation and reasoning have been proposed using the Logic Programming (LP) paradigm, namely in the area of Model Theory [14,15,16], and Proof Theory [11,12]. In this work it is followed the proof theoretical approach in terms of an extension to the LP language. An Extended Logic Program is a finite set of clauses in the form:

$$p \leftarrow p_1, \cdots, p_n, not\ q_1, \cdots, not\ q_m \tag{1}$$

$$? (p_1, \cdots, p_n, not\ q_1, \cdots, not\ q_m)\ (n, m \geq 0) \qquad (2)$$

where $?$ is a domain atom denoting falsity, the p_i, q_j, and p are classical ground literals, i.e., either positive atoms or atoms preceded by the classical negation sign $\neg$ [11]. Under this emblematic formalism, every program is associated with a set of abducibles [14], [16] given here in the form of exceptions to the extensions of the predicates that make the program.

Due to the growing need to offer user support in decision making processes some studies have been presented [17,18] related to the qualitative models and qualitative reasoning in Database Theory and in Artificial Intelligence research. With respect to the problem of knowledge representation and reasoning in LP, a measure of the *Quality-of-Information* (*QoI*) of such programs has been object of some work with promising results [19,20]. The *QoI* with respect to the extension of a predicate i will be given by a truth-value in the interval $[0,1]$.

It is now possible to engender the universe of discourse, according to the information given in the logic programs that endorse the information about the problem under consideration, according to productions of the type:

$$predicate_i - \bigcup_{1 \leq j \leq m} clause_j(x_1, \cdots, x_n) :: QoI_i :: DoC_i \qquad (3)$$

where U and m stand, respectively, for *set union* and the *cardinality* of the extension of *predicate_i*. On the other hand, DoC_i denotes one's confidence on the attribute's values of a particular term of the extension of *predicate_i*, whose evaluation will be illustrated below. In order to advance with a broad-spectrum, let us suppose that the *Universe of Discourse* is described by the extensions of the predicates:

$$f_1(\cdots), f_2(\cdots), \cdots, f_n(\cdots)\ where\ (n \geq 0) \qquad (4)$$

Assuming that a clause denotes a happening, a clause has as argument all the attributes that make the event. The argument values may be of the type unknown or members of a set, or may be in the scope of a given interval, or may qualify a particular observation. Let us consider the following clause where the first argument value may fit into the interval $[20,30]$ with a domain that ranges between 0 (zero) and 50 (fifty), where the second argument stands for itself, with a domain that ranges in the interval $[0,10]$, and the value of the third argument being unknown, being represented by the symbol $\perp$, with a domain that ranges in the interval $[0,100]$. Let us consider that the case data is given by the extension of predicate f_1, given in the form:

$$f_1 : x_1, x_2, x_3 \rightarrow \{0,1\} \qquad (5)$$

where "{" and "}" is one's notation for sets, "0" and "1" denote, respectively, the truth values *false* and *true*. Therefore, one may have:

$\{$

$\quad \neg f_1(x_1, x_2, x_3) \leftarrow not\ f_1(x_1, x_2, x_3)$

$\quad f_1(\underbrace{[20,30],\quad 5,\qquad \perp}) :: 1 :: DoC$
$\qquad\quad \text{\textit{attribute's values for }} x_1, x_2, x_3$

$\qquad\quad \underbrace{[0,50\,][0,10][0,100]}$
$\qquad\quad\ \text{\textit{attribute's domains for }} x_1, x_2, x_3$

. . .

$\}$

Once the clauses or terms of the extension of the predicate are established, the next step is to set all the arguments, of each clause, into continuous intervals. In this phase, it is essential to consider the domain of the arguments. As the third argument is unknown, its interval will cover all the possibilities of the domain. The second argument speaks for itself. Therefore, one may have:

$\{$

$\quad \neg f_1(x_1, x_2, x_3) \leftarrow not\ f_1(x_1, x_2, x_3)$

$\quad f_1\left(\underbrace{[20,30],[5,5],[0,100]}\right) :: 1 :: DoC$
$\qquad \text{\textit{attribute's values ranges for }} x_1, x_2, x_3$

$\qquad\quad \underbrace{[0,50]\ \ [0,10]\ \ [0,100]}$
$\qquad\quad\ \text{\textit{attribute's domains for }} x_1, x_2, x_3$

. . .

$\}$

Now, one is in position to calculate the *Degree of Confidence* for each attribute that makes the term arguments (e.g. for attribute one it denotes one's confidence that the attribute under consideration fits into the interval [20,30]). Next, we set the boundaries of the arguments intervals to be fitted in the interval [0,1] according to the normalization procedure given in the procedural form by $(Y - Y_{min})/(Y_{max} - Y_{min})$, where the Y_s stand for themselves.

$\{$

$\quad \neg f_1(x_1, x_2, x_3) \leftarrow not\ f_1(x_1, x_2, x_3)$

$\quad x_1 = \left[\dfrac{20-0}{50-0}, \dfrac{30-0}{50-0}\right] \quad x_2 = \left[\dfrac{5-0}{10-0}, \dfrac{5-0}{10-0}\right], \quad x_3 = \left[\dfrac{0-0}{100-0}, \dfrac{100-0}{100-0}\right]$

$\quad f_1\quad (\underbrace{[0.4,0.6],[0.5,0.5],[0,1]})\qquad :: 1 :: DoC$
$\qquad\quad\ \text{\textit{attribute's values ranges for }} x_1, x_2, x_3$
$\qquad\qquad\qquad \text{\textit{once normalized}}$

$\qquad\quad \underbrace{[0,1]\qquad [0,1]\quad\ [0,1]}$
$\qquad\quad\ \text{\textit{attribute's domains for }} x_1, x_2, x_3$
$\qquad\qquad\qquad \text{\textit{once normalized}}$

. . .

$\}$

The *Degree of Confidence* (*DoC*) is evaluated using the equation $DoC = \sqrt{1 - \Delta l^2}$, as it is illustrated in Fig. 1. Here Δl stands for the length of the arguments intervals, once normalized. Therefore, one may have:

{

    $\neg f(x_1, x_2, x_3) \leftarrow not\ f_1(x_1, x_2, x_3)$

    $f_1 \underbrace{(0.98, \qquad 1, \qquad 0)}_{\substack{attribute's\ confidence \\ values\ for\ x_1, x_2, x_3}} :: 1 :: 0.66$

    $\underbrace{[0.4, 0.6][0.5, 0.5][0,1]}_{\substack{attribute's\ values\ ranges\ for\ x_1, x_2, x_3 \\ once\ normalized}}$

    $\underbrace{[0, 1] \quad [0, 1] \quad [0, 1]}_{\substack{attribute's\ domains\ for\ x_1, x_2, x_3 \\ once\ normalized}}$

    ...

}

where the *DoC's* for $f_l(0.98, 1, 0)$ is evaluated as $(0.98+1+0)/3 = 0.66$, assuming that all the argument's attributes have the same weight.

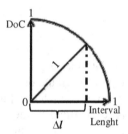

Fig. 1. Evaluation of the Degree of Confidence

3   A Case Study

In order to exemplify the applicability of our problem solving methodology, we will look at the relational database model, since it provides a basic framework that fits into our expectations [21], and is understood as the genesis of the LP approach to Knowledge Representation and Reasoning [11].

    As a case study, consider the scenario where a relational database is given in terms of the extensions of the relations depicted in Fig. 2, which stands for a situation where one has to manage information about ACS predisposing risk detection.

Patients' Information								Cardiac Biomarkers			
#	Age	Gender	Body Mass (Kg)	Height (m)	Previous ACS	Chest Pain	Heredity	#	Troponin	CK-MB	Mioglobin
1	65	F	70	1.63	1	1	$\perp$	1	1	1	0
...	...	...	...	...	...	...	...	...	...	...	...
n	59	M	103	1.71	$\perp$	0	1	n	1	0	0

ECG				ACS Predisposition						
#	STE	Other	#	Age	Previous ACS	Chest Pain	Heredity	Cardiac Biomarkers	STE	Risk Factors
1	0	0	1	65	1	1	$\perp$	2	0	[4,7]
...	...	...	...	...	...	...	...	...	...	...
n	1	0	n	59	$\perp$	0	1	1	1	[3,5]

Risk Factors							
#	Obesity	Hypertension	Tobacco habits	Diabetes mellitus	Lipid disorders	Anaemia	Chronic Renal Diseases
1	1	1	1	1	$\perp$	$\perp$	$\perp$
...	...	...	...	...	...	...	...
n	2	1	0	$\perp$	$\perp$	0	0

Fig. 2. An Extension of the Relational Database model. In *Risk Factors* database and in *ECG* database 0 (zero) and 1 (one) denote, respectively, *no* and *yes*. In *Cardiac Biomarker* database 0 (zero), 1 (one) and 2 (two) denote, respectively, *normal*, *high* and *very high* values.

Under this scenario some incomplete and/or unknown data is also available. For instance, in the *ACS Predisposition* database, the *Heredity* in case 1 is unknown, while the *Risk Factors* ranges in the interval [4,7].

The *Obesity* column in *Risk Factors* database is populated with 0 (zero), 1 (one) or 2 (two) according to patient' Body Mass Index (BMI), evaluated using the equation $BMI = Body\ Mass/Height^2$ [22]. Thus, 0 (zero) denotes $BMI < 25$; 1 (one) stands for a *BMI* ranging in interval [25,35[; and 2 (two) denotes a $BMI \geq 35$. The values presented in the *Cardiac Biomarkers* and *Risk Factors* columns of *ACS Predisposition* database are the sum of the correspondent databases, ranging between [0,6] and [0,8], respectively. Now, we may consider the relations given in Fig. 2, in terms of the *acs* predicate, depicted in the form:

$$acs: Age, Prev_{ious\ ACS}, C_{hest}P_{ain}, Her_{edity}, C_{ardiac}B_{iomarkers},$$

$$STE, R_{isk}F_{actors} \rightarrow \{0,1\}$$

where 0 (zero) and 1 (one) denote, respectively, the truth values *false* and *true*. It is now possible to give the extension of the predicate *acs*, in the form:

{

$\neg acs\ (Age, Prev, CP, Her, CB, STE, RF)$

$\leftarrow not\ acs\ (Age, Prev, CP, Her, CB, STE, RF)$

$$acs \left(\underbrace{65,\quad 1,\quad 1,\quad \perp,\quad 2,\quad 0,\quad [4,7]}_{attribute\text{'}s\ values} \right) :: 1 :: DoC$$

$$\underbrace{[37,91][0,1][0,1][0,1][0,6][0,1][0,8]}_{attribute\text{'}s\ domains}$$

...

}

In this program, the former clause denotes the closure of predicate *acs*, and the next, taken from the extension of the *ACS* relation shown in Fig. 2, presents symptoms with respect to patient 1 (one). Moving on, the next step is to transform all the argument values into continuous intervals, and then move to normalize the predicate's arguments. One may have:

{

$\neg acs\ (Age, Prev, CP, Her, CB, STE, RF)$

$\leftarrow not\ acs\ (Age, Prev, CP, Her, CB, STE, RF)$

$$acs \left(\underbrace{1,\quad 1,\quad 1,\quad 0,\quad 1,\quad 1,\quad 0.93}_{attribute\text{'}s\ confidence\ values} \right) :: 1 :: 0.85$$

$$\underbrace{[0.52,0.52][1,1][1,1][0,1][0.33,0.33][0,0][0.5,0.88]}_{attribute\text{'}s\ values\ ranges\ once\ normalized}$$

$$\underbrace{[0,1]\quad [0,1][0,1][0,1]\quad [0,1]\quad [0,1]\quad [0,1]}_{attribute\text{'}s\ domains\ once\ normalized}$$

...

}

where its terms make the training and test sets of the Artificial Neural Network given in Fig. 3.

4   Artificial Neural Networks

Several studies have shown how Artificial Neural Networks (ANNs) could be successfully used to structure data and capture complex relationships between inputs and outputs [23,24,25]. ANNs simulate the structure of the human brain being populated by multiple layers of neurons. As an example, let us consider the former case presented in Fig. 2, where one may have a situation in which the ACS predisposing is needed. In Fig. 3 it is shown how the normalized values of the interval boundaries and their *DoC* and *QoI* values work as inputs to the ANN. The output translates the ACS predisposing and the confidence that one has on such a happening. In addition, it also contributes to build a database of study cases that may be used to train and test the ANN.

In this study 627 patients admitted on the emergency department were considered, during the period of three months (from February to May of 2013), with suspect of ACS. The gender distribution was 48% and 52% for female and male, respectively. These patients, representing 13% of the total patients admitted to the emergency department in that period, had been submitted to the quantification of serum biomarkers, TnI, CK-MB mass and myoglobin blood, in the first hour after the admission. Among the patients studied, the ACS syndrome was diagnosed in 116 patients, i.e., 18.5% of the analysed population. Data was supplied from the Pathology Department of the Espírito Santo Hospital – Évora, a main health care center in the South of Portugal.

The dataset holds information about risk factors considered critical in the prediction of ACS predisposition. Seventeen variables were selected allowing one to have a multivariable dataset with 627 records. Table 1 shows a brief description of each variable and the data type, i.e., numeric or nominal.

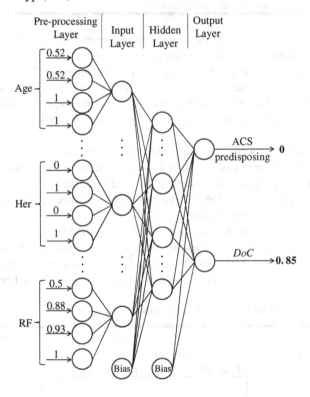

Fig. 3. The Artificial Neural Network topology

Table 1. Variables characterization

Variable	Description	Data type
Age	Patient's age	Numeric
Gender	Patient's gender	Nominal
Body Mass	Patient's body mass	Numeric
Height	Patient's height	Numeric
Previous ACS	Had ACS in the past	Nominal
Chest Pain	Has chest pain	Nominal
Heredity	Has family story	Nominal
Troponin	Troponin values (normal, high or very high)	Nominal
CK-MB	CK-MB values (normal, high or very high)	Nominal
Mioglobin	Mioglobin values (normal, high or very high)	Nominal
STE	ECG with persistent ST-segment elevation	Nominal
Hypertension	Is hypertensive	Nominal
Tobacco habits	Is smoker	Nominal
Diabetes	Has diabetes mellitus	Nominal
Lipid Disorders	Has lipid disorder	Nominal
Anaemia	Has anaemia	Nominal
CRD	Has chronic renal diseases	Nominal

To ensure statistical significance of the attained results, 20 (twenty) experiments were applied in all tests. In each simulation, the available data was randomly divided into two mutually exclusive partitions, i.e., the training set with 67% of the available data and, the test set with the remaining 33% of the cases. The back propagation algorithm was used in the learning process of the ANN. As the output function in the pre-processing layer it was used the identity one. In the other layers we used the sigmoid function.

A common tool to evaluate the results presented by the classification models is the coincidence matrix, a matrix of size $L \times L$, where L denotes the number of possible classes. This matrix is created by matching the predicted and target values. L was set to 2 (two) in the present case. Table 2 present the coincidence matrix (the values denote the average of the 20 experiments).

Table 2 shows that the model accuracy was 96.9% for the training set (410 correctly classified in 423) and 94.6% for test set (193 correctly classified in 204). Thus, the predictions made by the ANN model are satisfactory, attaining accuracies close to 95%. Therefore, the generated model is able to predict ACS predisposition properly.

Table 2. The coincidence matrix for ANN model

Target	Predictive			
	Training set		Test set	
	False (0)	True (1)	False (0)	True (1)
False (0)	333	11	157	10
True (1)	2	77	1	36

5   Conclusions and Future Work

On the one hand early diagnosis and treatment offers the utmost advantage for myocardial recover in the first hours of STEMI, and timely and determined management of unstable angina and NSTEMI reduces adverse events and improves outcome. Thus, it is imperative that healthcare providers recognize patients with potential ACS in order to initiate the evaluation, triage, and management as soon as possible; in the case of STEMI; this recognition also allows for prompt notification of the receiving hospital and preparation for emergent reperfusion therapy. Indeed, delays to therapy occur during 3 (three) timeline intervals, namely from onset of symptoms to patient recognition, during pre-hospital transport or during emergency department evaluation.

On the other hand, once the parameters to assess *ACS Predisposing* are not fully represented by objective data (i.e., are of types unknown or not permitted, taken from a set or even from an interval), the problem was put into the area of problems that must be tackled by Artificial Intelligence based methodologies and techniques for problem solving. Really, the computational framework presented above uses powerful knowledge representation and reasoning methods to set the structure of the information and the associate inference mechanisms. One`s approach may revolutionize prediction tools in all its variants, making it more complete than the existing ones. It enables the use of normalized values of the interval boundaries and their respectives QoI and *DoC* values, as input to the ANN. The output translates the patient`s ACS predisposing and the confidence that one has on such a happening. The last but not the least, involvement of hospital leadership in the process and commitment to support rapid access to STEMI reperfusion therapy are also critical factors associated with successful programs.

Future work may recommend that the same problem must be approached using others computational frameworks like Case Based Reasoning [26], Genetic Programming [12] or Particle Swarm [27], just to name a few.

Acknowledgments. This work has been supported by FCT – Fundação para a Ciência e Tecnologia within the Project Scope UID/CEC/00319/2013.

References

1. Allender, S., Peto, V., Scarborough, P., Kaur, A., Rayner, M.: Coronary heart disease statistics. British Heart Foundation Statistics Database, Oxford (2008)
2. Hamm, C.W., Bassand, J.-P., Agewall, S., Bax, J., Boersma, E., Bueno, H., Caso, P., Dudek, D., Gielen, S., Huber, K., Ohman, M., Petrie, M.C., Sonntag, F., Uva, M.S., Storey, R.F., Wijn, W., Zahger, D.: ESC Guidelines for the management of acute coronary syndromes in patients presenting without persistent ST-segment elevation. European Heart Journal 32, 2999–3054 (2011)
3. Brogan, R.A., Malkin, C.J., Batin, P.D., Simms, A.D., McLenachan, J.M., Gale, C.P.: Risk stratification for ST segment elevation myocardial infarction in the era of primary percutaneous coronary intervention. World Journal of Cardiology 6, 865–873 (2014)

4. Fox, K.A., Eagle, K.A., Gore, J.M., Steg, P.G., Anderson, F.A.: The Global Registry of Acute Coronary Events, 1999 to 2009–GRACE. Heart 96, 1095–1101 (2010)
5. Terkelsen, C.J., Lassen, J.F., Norgaard, B.L., Gerdes, J.C., Jensen, T., Gotzsche, L.B., Nielsen, T.T., Andersen, H.R.: Mortality rates in patients with ST-elevation vs. non-ST-elevation acute myocardial infarction: observations from an unselected cohort. European Heart Journal 26, 18–26 (2005)
6. Anderson, J.L., Adams, C.D., Antman, E.M., Bridges, C.R., Califf, R.M., Casey, J.D.E., Chavey II, W.E., Fesmire, F.M., Hochman, J.S., Levin, T.N., Lincoff, A.M., Peterson, E.D., Theroux, P., Wenger, N.K., Wright, R.S.: ACC/AHA 2007 Guidelines for the Management of Patients With Unstable Angina/Non–ST-Elevation Myocardial Infarction: Executive Summary. A Report of the American College of Cardiology/American Heart Association Task Force on Practice Guidelines. Circulation 116, 803–877 (2007)
7. Antman, E., Bassand, J.-P., Klein, W., Ohman, M., Sendon, J.L.L., Rydén, L., Simoons, M., Tendera, M.: Myocardial infarction redefined–a consensus document of The Joint European Society of Cardiology/American College of Cardiology Committee for the redefinition of myocardial infarction. European Heart Journal 21, 1502–1513 (2000)
8. Hollander, J.E.: The Future of Cardiac Biomarkers, new concepts and emerging technologies for emergency physicians. EMCREG International 4, 1–7 (2005)
9. Thygesen, K., Mair, J., Katus, H., Plebani, M., Venge, P., Collinson, P., Lindahl, B., Giannitsis, E., Hasin, Y., Galvani, M., Tubaro, M., Alpert, J.S., Biasucci, L.M., Koenig, W., Mueller, C., Huber, K., Hamm, C., Jaffe, A.S.: Recommendations for the use of cardiac troponin measurement in acute cardiac care. European Heart Journal 31, 2197–2206 (2010)
10. Agewall, S., Giannitsis, E., Jernberg, T., Katus, H.: Troponin elevation in coronary vs. non-coronary disease. European Heart Journal 32, 404–411 (2011)
11. Neves, J.: A logic interpreter to handle time and negation in logic databases. In: Muller, R.L., Pottmyer, J.J. (eds.) Proceedings of the 1984 Annual Conference of the ACM on The Fifth Generation Challenge, pp. 50–54. Association for Computing Machinery, New York (1984)
12. Neves, J., Machado, J., Analide, C., Abelha, A., Brito, L.: The halt condition in genetic programming. In: Neves, J., Santos, M.F., Machado, J.M. (eds.) EPIA 2007. LNCS (LNAI), vol. 4874, pp. 160–169. Springer, Heidelberg (2007)
13. Cortez, P., Rocha, M., Neves, J.: Evolving Time Series Forecasting ARMA Models. Journal of Heuristics 10, 415–429 (2004)
14. Kakas, A., Kowalski, R., Toni, F.: The role of abduction in logic programming. In: Gabbay, D., Hogger, C., Robinson, I. (eds.) Handbook of Logic in Artificial Intelligence and Logic Programming, vol. 5, pp. 235–324. Oxford University Press, Oxford (1998)
15. Gelfond, M., Lifschitz, V.: The stable model semantics for logic programming. In: Kowalski, R., Bowen, K. (eds.) Logic Programming – Proceedings of the Fifth International Conference and Symposium, pp. 1070–1080 (1988)
16. Pereira, L.M., Anh, H.T.: Evolution prospection. In: Nakamatsu, K., Phillips-Wren, G., Jain, L.C., Howlett, R.J. (eds.) New Advances in Intelligent Decision Technologies. SCI, vol. 199, pp. 51–63. Springer, Heidelberg (2009)
17. Halpern, J.: Reasoning about uncertainty. MIT Press, Massachusetts (2005)
18. Kovalerchuck, B., Resconi, G.: Agent-based uncertainty logic network. In: Proceedings of the IEEE International Conference on Fuzzy Systems, Barcelona, pp. 596–603 (2010)
19. Lucas, P.: Quality checking of medical guidelines through logical abduction. In: Coenen, F., Preece, A., Mackintosh, A. (eds.) Proceedings of AI-2003 (Research and Developments in Intelligent Systems XX), pp. 309–321. Springer, London (2003)

20. Machado, J., Abelha, A., Novais, P., Neves, J., Neves, J.: Quality of Service in healthcare units. International Journal of Computer Aided Engineering and Technology 2, 436–449 (2010)
21. Liu, Y., Sun, M.: Fuzzy optimization BP neural network model for pavement performance assessment. In: 2007 IEEE International Conference on Grey Systems and Intelligent Services, Nanjing, China, pp. 18–20 (2007)
22. World Health Organization: Obesity and overweight.Fact Sheet Number 311, http://www.who.int/mediacentre/factsheets/fs311/en/
23. Caldeira, A.T., Arteiro, J., Roseiro, J., Neves, J., Vicente, H.: An Artificial Intelligence Approach to Bacillus amyloliquefaciens CCMI 1051 Cultures: Application to the Production of Antifungal Compounds. Bioresource Technology 102, 1496–1502 (2011)
24. Vicente, H., Dias, S., Fernandes, A., Abelha, A., Machado, J., Neves, J.: Prediction of the Quality of Public Water Supply using Artificial Neural Networks. Journal of Water Supply: Research and Technology – AQUA 61, 446–459 (2012)
25. Salvador, C., Martins, M.R., Vicente, H., Neves, J., Arteiro, J.M., Caldeira, A.T.: Modelling Molecular and Inorganic Data of Amanita ponderosa Mushrooms using Artificial Neural Networks. Agroforestry Systems 87, 295–302 (2013)
26. Carneiro, D., Novais, P., Andrade, F., Zeleznikow, J., Neves, J.: Using Case-Based Reasoning and Principled Negotiation to provide decision support for dispute resolution. Knowledge and Information Systems 36, 789–826 (2013)
27. Mendes, R., Kennedy, J., Neves, J.: The Fully Informed Particle Swarm: Simpler, Maybe Better. IEEE Transactions on Evolutionary Computation 8, 204–210 (2004)

A Flexible Denormalization Technique for Data Analysis above a Deeply-Structured Relational Database: Biomedical Applications

Stanislav Štefanič* and Matej Lexa

Faculty of Informatics, Masaryk University,
Botanická 68a, 60200 Brno, Czech Republic
{stefanic,lexa}@mail.muni.cz

Abstract. Relational databases are sometimes used to store biomedical and patient data in large clinical or international projects. This data is inherently deeply structured, records for individual patients contain varying number of variables. When ad-hoc access to data subsets is needed, standard database access tools do not allow for rapid command prototyping and variable selection to create flat data tables. In the context of Thalamoss, an international research project on β-thalassemia, we developed and experimented with an interactive variable selection method addressing these needs. Our newly-developed Python library *sqlAutoDenorm.py* automatically generates SQL commands to denormalize a subset of database tables and their relevant records, effectively generating a flat table from arbitrarily structured data. The denormalization process can be controlled by a small number of user-tunable parameters. Python and R/Bioconductor are used for any subsequent data processing steps, including visualization, and Weka is used for machine-learning above the generated data.

Keywords: relational database, PostgreSQL, NoSQL, data flattening, automatic data denormalization

1 Introduction

Relational databases are often used to store biomedical and clinical data in large international projects. For example, [1] recently described their database use in the FINDbase project [2]. The data in this kind of databases is inherently deeply structured, records for individual patients contain varying number of variables, depending on their clinical history and the origin of the data in relation to clinical procedures. At the same time they need to be combined with genetic and phenotypic data of entirely different nature. We are currently working on the Thalamoss project where we collect heterogeneous clinical and molecular data of similar composition.

As in most projects this clinical and genetic data needs to be repeatedly accessed, organized and visualized in many different forms. Standard database

---

* Corresponding author.

F. Ortuño and I. Rojas (Eds.): IWBBIO 2015, Part I, LNCS 9043, pp. 120–133, 2015.
© Springer International Publishing Switzerland 2015

access tools can be used to set up predetermined reports. These can be displayed to the project participants and general public. However, for flexible and universal access to such data, it is desirable to use a method that does not require intimate knowledge of the database structure, enables rapid arbitrary variable selection and provides data in a format that can be readily used in common visualization or machine learning tasks [3] [4]. In other words, variable parts of the database need to be frequently denormalized into a simple table for downstream processing.

While wide and intense studies have covered the problem of normalization in the past, denormalization techniques enjoyed less attention. Most of the interest was in denormalization for read performance increase [5]. However, another type of denormalization is the scenario we assume here, to support generation of frequent ad-hoc views of the data. This has been done, for example, in context of data warehousing [6]. All the examples we have seen assumed carefully prepared JOIN SQL statements for every chunk of data that needed to be denormalized. There seems to be a lack of tools for acheiving this automatically when working on a relatively complex database schema that has not been designed with easy denormalization in mind.

This paper describes an automated method to denormalize an arbitrary subset of data stored in a PostgreSQL database. In Section 2 we briefly describe the algorithm and its implementation. In Section 3 we discuss its use within a biomedical research project support system made of several modules (Apache, Django, Python, R/Bioconductor, Weka).

2   Software and Methods

The denormalization procedure introduced above is based on the idea that rational databases can be represented as graphs, with nodes or vertices representing database tables and edges between the nodes representing relationships between the connected tables. If a relationship between two tables exists, an edge between these two tables is present in the respective graph representation. The relationships may be via primary keys, foreign keys or other constraints in rational databases. Information about them can be extracted from the `information_schema` pertaining to the database. Our script automatically creates a PostgreSQL query which can select data from tables and their columns as chosen by the user.

First, as shown in Figure 3, the script connects to the database and collects the information necessary to create the relationship graph. It subsequently builds the graph and store it as a Python dictionary structure. For PostgreSQL database, the user specified at the beginning of the script is used to launch a database connection. He or she must have access and read permissions to `information_schema`, because all relationships are extracted from there. Next, the script searches for all paths between tables chosen by the user. We built functionality into the script to find the shortest path possible, but it is limited to less than nine nodes. If there are more than eight nodes in the graph, we use a heuristic approach that by definition does not guarantee to find the shortest path, but one that will most probably be of similar length. To find the path between the chosen nodes (tables) we implemented Dijstra's [7] and Floyd-Warshall's [8] algorithms adapted to our purpose. If the path is found, the script creates a complete SQL query based on previous user specifications. The user can optionaly use several types of JOIN statements and various aggregating functions to better control the content of the resulting data table. In the last step the created PostgreSQL query may be executed to get the data table and store it in a file for further analysis as needed.

2.1 Python sqlAutoDenorm Script

To implement the ideas above, we created a Python script which is able to denormalize any part of an arbitrary well-structured rational database. Namely, we follow the *primary_key* and *foreign_key* relationships in the database schema. For practical purposes, we divided the implementation into two parts - a library called *sqlAutoDenormLib.py* and a demo command-line application *runAutoDenorm.py* based on the library. The library contains all necessary functions for database analysis and denormalization. The command-line executable Python script interfaces with the user via a short dialogue. This serves as an example for including the library into other software projects.

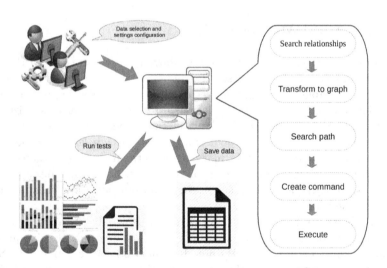

Fig. 1. A simplified schema describing the functioning of the entire web portal. Denormalization library provides key functions of the system. Users select data and set parameters to specify one of the available denormalization modes. A custom-made algorithm generates a PostgreSQL query for requested data (flowchart in the bubble). The query can be executed to obtain data which could be stored in a csv file or forwarded to tests and analyses installed on the portal (such as graph and diagram drawing, etc.).

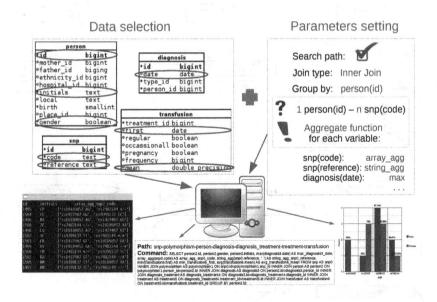

Fig. 2. Similar to the schema on the Figure 1, but presenting a real example with real selection of data and parameter settings. SQL statement is shown in the bottom of the figure.

Package and Requirements. The package `autoDenorm.zip` can be downloaded from authors' website at `http://fi.muni.cz/~lexa/autodenorm`. The archive contains the following files:

runAutoDenorm.py: Dialogue script to select data from a PostgreSQL database in denormalized form using the options set in a dialogue.

sqlAutoDenormLib.py: Library of useful functions for database denormalization designed to be imported in an arbitrary Python script or interpreter.

README.txt: Useful information about the script and the library.

lib directory: Directory with required Python libraries.

__init.py__ and directory __pycache__: Files required for correct import of the library into Python.

The scripts have been written in Python (version 3.3.5) programming language. To run the code, Python 3.x must be installed on your computer. Regarding dependencies, the following Python libraries must also be installed - *psycopg2, ast, csv, json, itertool, random*. The scripts currently only support and work with PostrgreSQL databases.

2.2 sqlAutoDenormLib

This Python library contains all necessary functions for denormalization of the database. Our executable Python script *runAutoDenorm.py* exploits this library

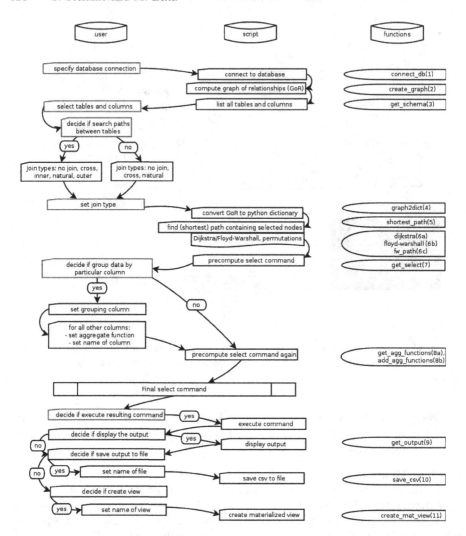

Fig. 3. A block chart showing the database denormalization algorithm and its implementation in *sqlAutoDenormLib.py* and *runAutoDenorm.py*. The first column (user) shows the decisions that have to be made by a user (optional decisions are shown in grey, compulsory are red). The third column (functions) shows the names of *sqlAutoDenorm.py* functions implementing the steps shown in the second column (script).

by importing it and calling several of its functions. The library can be used independently. When using its functions, a working connection to a PostgreSQL database must exist in the program environment.

connect_db()[1]. This function connects to the database using the *psycopg2* library and returns a **cursor** variable which allows us to execute PostgreSQL commands. Database host, name, username and password should be specified as input arguments.

create_graph()[2]. One of the most important functions in the library that creates a materialized view object containting information about relationships between tables in the database. The function executes all PostgreSQL commands necessary to create the view. These commands select information about interacting table names and column names also providing us with the name of constraints for analyzed node pairs. This data is stored as a undirected graph defined by pairs of nodes, where each node represents a table and a relevant column name. Each pair of nodes forms an edge in the graph which is labelled by all relevant constrain names. All this information is selected from the database **information_schema** fields for constraints such primary and foreign keys.

gen_select()[7]. Function *gen_select(chosen, join_type, is_path)* generates an output string containing a PostgreSQL command defined by previous user specifications. Function has three input arguemnts. First is the Python dictionary of chosen tables and columns. The second argument represents one of the following join types:

- cross = CROSS JOIN
- natural = NATURAL JOIN
- inner = INNER JOIN
- left = LEFT OUTER JOIN
- right = RIGHT OUTER JOIN
- full = FULL OUTER JOIN
- nothing - no join, just select all from each column separately

The third argument is a binary value denoting whether we want to connect chosen tables and columns via a common path between them. The path is necessary for proper mapping between values from different tables. An error message is produced if no such path exists. This is probably the most demanding step from the user point of view, since we must have some limited knowledge of the data columns and tables to get proper mapping with the correct number of rows in the denormalized data table.

do_join(). This function is called from function *gen_select()[7]* when user decides to search for a path between tables. The existence of a path allows us then to use any of the seven join types. We do not recommend to use this function explicitly. To generate a select query, we recommend the use function *gen_select[7]*.

shortest_path()[5]. This is the most important function in the whole script. Its input is a list of compulsory (required) nodes. Initially, this function calls *graphs2dict()[4]* (see below) which transforms the materialized view **graphs** into a Python data struncture called dictionary for the purpose of higher efectiveness of computation of the shortest path. Subsequently, the shortest path is calculated in one of two ways depending on the number of selected nodes.

If user specifies only two compulsory nodes (tables), the shortest path is calcuated using function *dijkstra()[6a]*. If more than two nodes are selected, function *floyd_warshall()*[6b] is used to calcuate the path. In addition, if the number of specified compulsory nodes (tables) is between 2 and 8, the algorithm finds the shortest possible path through the compulsory nodes. This is achieved by permuting specified nodes and computing paths for each permutation. Finally, the path which contains the minimum number of nodes is selected and declared to be the shortest path. To achieve this, functions *permutations()*, *fw_path()*[6c] and *extend_without_repeat()* are used in addition to function *floyd_warshall()*[6b].

Because the number of permutations of n nodes grows quickly - $n!$, and computation for $n > 8$ is time consuming, we compute a nearly-minimal path for $n > 8$. To do that, we calculate the minimal length from only a subset of possible paths that can still be calculated in reasonable time (approximately one second). The user can therefore set a bigger n if required.

graphs2dict()[4]. Before starting the computations of the shortest path between tables, it is necesarry to get data about table relationships into Python. We chose dictionary representation of the graph. Function *graphs2dict()*[4] selects the first and second nodes from materialized view called **graphs** which was created by function *create_graph()*[2]. After selection of necessary information from this materialized view, information is inserted into a Python dictionary stored in global variable **G**.

create_mat_view()[11]. If the user wants to reuse the same output (table) several times, there is a function called *create_mat_view()*[11] which takes a name of the view and the final select query on input and creates a materialized view containing the selected information (columns). Then user can use this materialized view and repeatedly select data from it to save time and computing resources. More about benefits of using materialized views can be found in [9].

denormalize_database(). Sometimes it is necessary to denormalize the database with respect to a specific column. This is when we want to keep all data from a selected column in a specific table and assign data from other tables to this column (if a relationship between the selected column and other columns exists). We decided to call this specified column the 'main column'.

In our case of a patient-centric clinical database, the main column often is the person *id* from table **person**. We typically create denormalized tables which include columns in other tables, such as **transfusion, chelation** or column *date* from table **death**. In this case, one person could have more tranfusions and we would only want to have one row for one person id. This is were we use aggregate functions on each column except **person***(id)*. The aggregate function *array_agg()* creates an array from all values in some column (i.e. **transfusion***(date)*) corresponding to one **person***(id)*.

The input arguments of this function are: selected columns and their tables, main column, type of join between tables and the name of materialized view, which will be created. The result is the created materialized view with a chosen name which contains the final denormalized table corresponding to selected settings.

Table 1. Additional functions defined in the *sqlAutoDenorm.py* library

Name	Input	Output	Description
get_schema()[3]	schema name	cursor	Lists all tables and their columns in database given a schema name.
dijkstra()[6a]	graph, start, destination	shortest path	Dijkstra's algorithm which for a given graph, source vertex and destination vertex finds the shortest path between these vertices.
floyd_warshall()[6b]	graph	successors	Floyd-Warshall algorithm which for a given graph computes and returns a dictionary of all vertices in the graph and their successors.
fw_path()[6c]	successors, u, v	shortest path	Builds the shortest path between vertices u and v using dictionary of successors returned by the *floyd_warshall()*[6b] function.
get_agg_functions()[8a]	column type	aggregate functions	For each column and its type returns a list of applicable aggregate functions.
add_agg_functions()[8b]	select cmd, columns with aggregate functions, main column	select cmd	Takes the precomputed select query string and the dictionary of columns and chosen aggregate functions. Returns a SELECT query string with incorporated options.
get_output()[9]	db fetch, head	result of select cmd	Prints results of the select query to the standard output.
save_csv()[10]	db fetch, head, output name	csv file	Saves results of the select query into a csv file.
create_big_table()	main column, join type, view name	view in db	Denormalizes the whole database with respect to chosen column in appropriate table. See *denormalize_database()*.
create_search_paths()	schema name	function in db	Creates a recursive PostgreSQL function in your database which is able to find paths between given start and destination vertices.

2.3   runAutoDenorm

File *runAutoDenorm.py* is a simple Python script which allows us to generate a PostgreSQL select query in accordance to specific requirements. This script uses the *sqlAutoDenormLib.py* library described above. To run the script from command-line enter the directory with scripts and type the following command:

```
$ python3 runAutoDenorm.py
```

As script runs follow the instructions on the display. First, insert information about the desired database connection (database host, database name, user name and user password). After each line press enter to continue. Script connects to the database using function *connect_db()*[1] and creates a materialized view called **graphs**. A function for searching paths in this graph called *shortest_path()*[5] becomes available. Upon successful completion you will be informed by messages on your display.

After successfully connecting to database and creating the view and function, the script selects all tables and columns from the database and prints them for the user to chose in the following form:

table_name: *col1_name, col2_name, ...*

Using this guide type items (table and its columns) which you want to select from your database. After each line (table) press enter to continue. To stop adding items enter **0**. All the chosen items are added to a variable named **chosen** (an ordered dictionary type in Python). This ensures that order of the output will be the same as the order of items inserted in the previous step. After specification of tables and columns continue with decisions on other matters:

Search Path: Decide if you want to search a path between tables - by inserting **1** or **0** for yes or no respectively. It is recommended to use searching path option in order you are not sure if your tables are "close enough" in graph - it means that column in one table is a key in other table.

Join Type: Choose the type of join - by following the instruction on your display insert relevant number from 1 to 7. After this step the script calling function *gen_select()*[7] for generating PostgreSQL select query. In case that user choose own specification of select query algorithm pre-generating select and user will be promted to change this select and insert it on input in one of the following steps.

Group by Column: Decide if you want to group data by one particular column. If you insert **1** for yes you will be prompted for refine additionally informations. From all columns which have been chosen above you must now choose one (*main column*), which will be a major column. That means all data from this column will be on the output and for all other columns must be specified so called aggregate functions. User will be prompted to insert aggregate function for each column. Algorithm get information about type of each column from database `information_schema` and then offer to user set of possible aggregate functions using function *get_agg_functions()*[8a]. In addition user can insert the name of each column for better orientation in the resulting CSV file.

After these steps a PostgreSQL query is generated. The user can copy it and terminate the script or execute the query and then choose one of the following methods for present the results:

Send Results to the Standard Output: Selected results will be printed on the standard output in the terminal emulator using the function *get_output()*[9].

Save Result to File: Selected results will be saved to a CSV file. Before saving it is necessary to specify the output file name. The CSV file will be saved in the current dirrectory using the function *save_csv()*[10]

2.4   Django Variable Selection Interface

The main reason why the scripts were created was our interest to use them within the Thalamoss research project. We implemented the utilities described here within a Django interface at our website. All settings required for correct generation of output are divided into two or three parts. Each part matches a single frame or page on our website. We use the Thalamoss research project database.

Part 1: The vast majority of settings are set on the first page (Figure 4). The left part helps choosing tables and columns. There are two select boxes - the left one contains all table and column names from our database (multiple

Please select tables and columns:

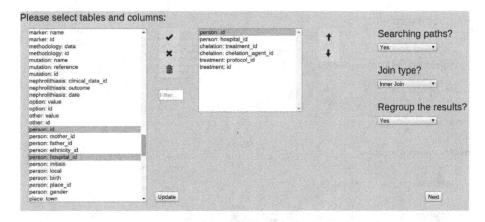

Fig. 4. Screenshot of the portal. The left box lists all database tables and their columns. The right box represents tables already chosen (and their columns). The user can add, remove items or clear the whole box and change order of selected items. The right side provides for parameter settings, from top to bottom - decisions to search for relationships between tables and specification of join type. Data grouping with respect to a specific column and the use of aggregate funcions for other columns can be specified.

rows can be selected). The selected tables with columns are added into the right select box. OK, DELETE and UPDATE buttons are available to manage the selection process. On the right part of the page there are combo boxes to adjust the following:

- searching path between tables - *yes* or *no*
- type of join to use - *cross, natural, inner, left/right/full outer* or *no join*
- grouping result by specific column - *yes* or *no*

Part 2: This page (Figure 5-Left) will be skipped in the case that grouping option on the first page is set to *no*. Grouping by specific column is a feature that can be used in case of one-to-many mappings in the data. For all other columns except the chosen grouping column it is necessary to specify an aggregating function - for example average value or concatenation into a string or an array.

Part 3: This is the first page (Figure 5-Right) with results showing the generated select query. Just press thumbs up button and resulting query will be displayed. COPY TO CLIPBOARD button can be used to copy the query string to clipboard and use it for selecting data from the PostgreSQL database.

Output of the query can be saved to a file. At the moment the script supports output to CSV files.

Last part located at the bottom of this page is dedicated to tests. There are two type of tests - statistical tests in language R and tests written in Python. A tree is available that can be opened by clicking on the relevant buttons. For running tests just mark the checkboxes located next to selected test names and press button RUN TESTS! We are diligently working on creating and adding new tests to both categories.

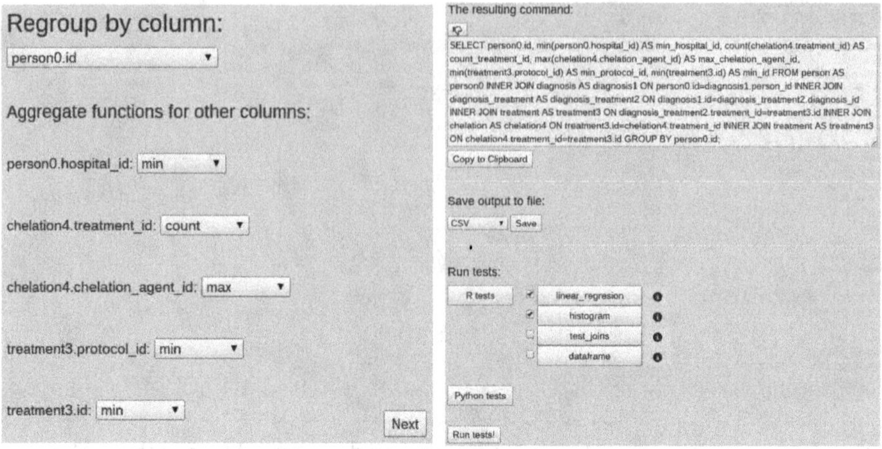

Fig. 5. Portal screenshots. **Left**: In case that user decided to group data by specfic column, it is necessary to specify main column and aggregate functions for all other columns as shown in this figure. **Right**: From top to bottom - final string of a SELECT query which can be copied to the clipboard, options for saving results (data) of this query, options for running further tests with these data.

Part 4: Last page is displayed only in case that some of the tests ware executed. All results from all tests are displayed here one by one. See example of results in Figure 6.

2.5 Post-denormalization Visualization and Analysis

The main motivation behind creating an automated procedure for data denormalization was to allow users perform rapid ad-hoc visualization or machine learning tasks over an arbitrary subset of available data. Commonly used visualization tools often work well with data in single tabular format (e.g. CSV file). A popular machine learning tool Weka [10] has similar requirements (e.g. ARFF file format). By producing data output of this kind, *sqlAutoDenormLib.py* can be easily included in data analysis pipelines and prepare input for a plethora of downstream analysis programs. Machine-learning could be used, for example, to select specific biomarkers for a selected phenotype, diagnosis or condition. We have tested the platform in the context of the Thalamoss research project on clinical and biomedical data for β-thalassemia. An example visualization task is included in the next section of this paper (Figure reffig6).

3 Results

3.1 Performance Tests

We tested out script on two databases. First one is the clinical and genomic research database from the Thalamoss project which has a rich and highly-structured schema but contains few data at the moment. A second database,

Table 2. Scalability tests of the *sqlAutoDenormLib.py* library. We tested the times (shown in **ms**) required to finish key computational steps on Thalamoss database (78 tables, 436 relationships, 12 MB in size) and the Dellstore database with simple layout (8 tables, 8 relationships) but higher data volumes (27 MB). Time of creating a graph of relationships in database (function *create_graph()*[2] was deleted from the table because its time costs are nearly constant. For the *Thalamoss* database it took an average time of **72.75 ms** and for *Dellstore* it was **65.5 ms**.

Database name	number of nodes	is aggregated	searching of shortest path	query generation	query execution	no. of nodes in path	query size
	2	no	3.674	4.710	1.271	3	227
	2	yes	3.769	4.874	1.887	3	301
	3	no	6.925	8.321	1.499	4	297
	3	yes	6.775	8.197	2.164	4	374
	4	no	26.70	30.25	2.407	8	710
	5	no	157.87	161.23	2.307	8	759
	7	no	499.99	505.87	5.26	14	1228
Thalamoss	8	no	1299.67	1307.63	6.85	19	1810
	8	yes	1291.31	1299.14	5.03	19	2201
	9	no	919.83	927.37	9.56	20	1721
	9	yes	818.76	824.24	4.48	19	2088
	10	no	874.68	883.48	5.67	22	2188
	15	no	1125.33	1132.89	15.53	41	3992
	2	no	0.895	1.323	54.56	2	201
Dellstore	2	yes	1.372	1.788	223.81	2	357
	3	no	1.644	2.608	140.00	3	296
	3	yes	1.741	2.602	202.08	3	480

called '*dellstore2*', was downloaded from *pgfoundry.org* - a website with free sample databases for PostgreSQL. Database *dellstore2* contains much more data than *Thalamoss* database but has a simple schema. The results of performance tests are presented in Table 2 below.

The test results show that path search and query generation in a database with complex schema (*Thalamoss*) takes comparable amount of time with query execution in a large but simple database (*Dellstore*). We mostly tuned the path search steps in our implementation. For number of nodes up to 5, we used Dijkstra's algorithm for searching the shortest path. For 6 or more nodes, we used Floyd-Warshall algorithm for searching the shortest paths between all nodes in a graph. All permutations of nodes are explored up to 8 nodes. If the number of nodes is greater or equal than 9, the total number of permutations to evaluate is too high ($n!$ for n nodes), so the algorithm explores only a randomly chosen subset. With increasing number of nodes this heuristic search is not guaranteed to identify the shortest path, but will find good alternatives, at least.

3.2   Sample Application to a Thalamoss Project Visualization Task

The Thalamoss project is a multidisciplinary research and clinical project with the goal of better understanding molecular markers underpinning the various forms of β-thalassemia. The project participants collect a rich spectrum of molecular data characterizing individual patients. A developer or project participant will know the project database contains genomic data for individual patients. When given a task to visualize single nucleotide polymorphisms from HBB genes in the database and show how they differ between sexes, he or she can use *runAutoDenorm.py* or our Django interface to scan variable names. Because they are in alphabetical order, it is relatively easy to spot the following tables and columns relevant to the prescribed task:

- **country**: *name*
- **person**: *initials, gender*
- **snp**: *code, reference*

After clicking through a couple of default selections, the *get_select()*[7] function from *autoDenormLib.py* generates the PostgreSQL query for selecting data from database.

A short R/Bioconductor script was installed on the system that plots the HBB gene cluster using the *genoPlotR* package [11]. This was combined with code that reads a subset of the denormalized data referring to SNP **rs1427407** into a data.frame R object and a set of *ggplot* [12] calls to plot this data into a specialized piechart.

The results can be seen in Figure 6. After installation of this script it becomes available within the Django portal for future use and can be quickly used to visualize SNP distribution data for any SNP from the HBB cluster, for example.

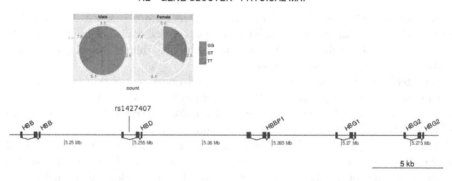

Fig. 6. An example visualization of genomic data for a group of patients, showing the position of a single nucleotide polymorphism (SNP) in a chromosomal physical map (horizontal line) and its variants (colors) in patients of different sex (panels). The map has been drawn with the *genePlotR* package [11], while piecharts come from the deployment of *ggplot* [12].

4   Conclusions

We have created a library in Python that can be used to selectively denormalize subsets of an arbitrarily complex PostgreSQL database. The data can be saved as a new table (materialized view) or a CSV flat file for downstream processing, such as visualization or machine learning. Our implementation is reasonably fast to be practical (see Table 2 for exact times). However, it still provides space for faster execution with the use of database indexing techniques and caches. Our Python code is available for download for immediate inclusion into other projects, especially those using Python as programming environment. We plan to further improve the developed codebase.

Acknowledgements. Financial support provided by the EU 7th Framework Project "Thalassaemia Modular Stratification System for Personalized Therapy THALAMOSS" (FP7-HEALTH-2012-INNOVATION-1 Collaborative Project; http://thalamoss.eu/index.html).

References

1. Viennas, E., Gkantouna, V., Ioannou, M., Georgitsi, M., Rigou, M., Poulas, K., Patrinos, G., Tzimas, G.: Population-ethnic group specific genome variation allele frequency data: A querying and visualization journey. Genomics 100, 93–101 (2012)
2. van Baal, S., Kaimakis, P., Phommarinh, M., Koumbi, D., Cuppens, H., Riccardino, F., Macek Jr., M., Scriver, C., Patrinos, G.: Findbase: a relational database recording frequencies of genetic defects leading to inherited disorders worldwide. Nucleic Acids Res. 35, D690–D695 (2007)
3. Mitropoulou, C., Webb, A., Mitropoulos, K., Brookes, A., Patrinos, G.: Locus-specific database domain and data content analysis: evolution and content maturation toward clinical use. Hum. Mutat. 31(10), 1109–1116 (2010)
4. Smith, T., Cotton, R.: Varivis: a visualisation toolkit for variation databases. BMC Bioinformatics 9, 206 (2008)
5. Zaker, M., Phon-Amnuaisuk, S., Haw, S.-C.: Optimizing the data warehouse design by hierarchical denormalizing, pp. 131–138 (2008)
6. Singh, J., Singh, B., Sriveni, Y.: A convenient way from normalized database to denormalized database. International Journal of Computer Communication and Information System (IJCCIS) 2, 84–87 (2010)
7. Dijkstra, E.: A note on two problems in connexion with graphs. Numerische Mathematik 1(1), 269–271 (1959)
8. Floyd, R.W.: Algorithm 97: Shortest path. Commun. ACM 5, 345 (1962)
9. Chak, D.: Enterprise Rails. O'Reilly, Beijing Farnham (2009)
10. Hall, M., Frank, E., Holmes, G., Pfahringer, B., Reutemann, P., Witten, I.: The weka data mining software: An update. SIGKDD Explorations 11 (2009)
11. Guy, L., Kultima, J.R., Andersson, S.G.E.: genoplotr: comparative gene and genome visualization. Bioinformatics 26(18), 2334–2335 (2010)
12. Yin, T., Cook, D., Lawrence, M.: ggbio: an r package for extending the grammar of graphics for genomic data. Genome Biology 13(8), R77 (2012)

Multilayer Clustering: Biomarker Driven Segmentation of Alzheimer's Disease Patient Population

Dragan Gamberger[1,*], Bernard Ženko[2], Alexis Mitelpunkt[3], and Nada Lavrač[2,4]

[1] Rudjer Bošković Institute, Bijenička 54, 10000 Zagreb, Croatia
dragan.gamberger@irb.hr
[2] Jožef Stefan Institute, Ljubljana, Slovenia
[3] Tel Aviv University, Israel
[4] University of Nova Gorica, Slovenia

Abstract. Identification of biomarkers for the Alzheimer's disease is a challenge and a very difficult task both for medical research and data analysis. In this work we present results obtained by application of a novel clustering tool. The goal is to identify subpopulations of the Alzheimer's disease (AD) patients that are homogeneous in respect of available clinical and biological descriptors. The result presents a segmentation of the Alzheimer's disease patient population and it may be expected that within each subpopulation separately it will be easier to identify connections between clinical and biological descriptors. Through the evaluation of the obtained clusters with AD subpopulations it has been noticed that for two of them relevant biological measurements (whole brain volume and intracerebral volume) change in opposite directions. If this observation is actually true it would mean that the diagnosed severe dementia problems are results of different physiological processes. The observation may have substantial consequences for medical research and clinical trial design. The used clustering methodology may be interesting also for other medical and biological domains.

1 Introduction

Identification of connections between biological and clinical characteristics of Alzheimer's disease patients is a long term goal that could significantly improve the understanding of the Alzheimer's disease (AD) pathophysiology, improve clinical trial design, and help in predicting outcomes of mild cognitive impairment [1]. The difficulty of the task is in the fact that AD is clinically described as a set of signs and symptoms that can be only indirectly measured and that have been integrated into various scoring systems like Clinical Dementia Rating Sum of Boxes, Alzheimer's Disease Assessment Scale, and Montreal Cognitive Assessment [2]. All of these scales as well as everyday cognition problems have proved their usefulness in the diagnostic process but a unique reliable measure does not exist.

---

* Corresponding author.

F. Ortuño and I. Rojas (Eds.): IWBBIO 2015, Part I, LNCS 9043, pp. 134–145, 2015.
© Springer International Publishing Switzerland 2015

On the other side, although relations between some biological descriptors and AD diagnosis have been undoubtedly demonstrated [3,4], currently available biological descriptors are non-specific (e.g., whole brain or hippocampal volume) and their changes may be a consequence of various physiological processes. It means that potentially useful information related to biological causes of the cognitive status of a patient is hidden in the large "noise" of interfering biological processes.

Technically speaking, we are looking for relevant relations in a very noisy data domain (biological descriptors) in which the target function is defined by a large set of imprecise values (clinical descriptors). A simplified approach in which medical AD diagnosis is used as the target function has enabled the detection of some relations, like importance of decreased FDG-PET values for the AD diagnosis, but all the detected relations including those obtained by complex supervised approaches [5] have low predictive quality and did not help significantly in expert understanding of the disease. In line with the approach proposed in [6], our work aims at finding homogeneous subpopulations of AD patients in which it will be easier to identify statistically and logically relevant relations between clinical and biological descriptors. The approach is based on finding subpopulations with clustering algorithms.

Clustering is a well-established machine learning methodology but it still suffers from problems such as definition of the distance measure and optimal selection of the number of resulting clusters. Typically the obtained clustering results are unstable because they significantly depend on user selectable parameters. This is especially true for noisy domains and domains with statistically related descriptors (attributes). Recently we have developed a novel clustering approach called *multilayer clustering* that successfully solves some of the basic problems of data clustering [7]. In this methodology, the quality of the resulting clusters is ensured by the constraint that clusters must be homogeneous at the same time in two or more data layers, i.e., two or more sets of distinct data descriptors. By defining the clinical descriptors as one data layer and biological descriptors as the other layer, we can expect not only more reliable clusters but clusters which will be potentially good candidates for the detection of relevant relations between the clinical and biological descriptors. The AD domain fulfils all the requirements of the ideal setting in which multilayer clustering may demonstrate its advantages.

The rest of the paper is structured as follows. Section 2 presents the multilayer clustering methodology, and Section 3 presents the concrete results for the AD domain. Medical meaning and the results significance are analysed in Section 4.

2   Multilayer Clustering

Clustering is an optimisation task which tries to construct subpopulations of instances so that distances between instances within each subpopulation are small while distances between instances in different subpopulations are as large as possible [8]. The most commonly used distance measure is the Euclidean distance that is well defined for numerical attributes.

Redescription mining is a novel clustering approach in which the quality of the results is ensured by the constraint that the resulting clusters must have meaningful interpretations in at least two independent attribute layers [9]. It is possible, and it occurs often, that some of the training instances remain outside the identified clusters but the detected clusters are more likely really relevant. A very important property of this approach is that the constructed clusters have human interpretable descriptions in all attribute layers.

However, redescription mining has some issues as well. For the approach to be applied, both numerical and nominal attributes have to be transformed into a transactional form [10] or some on–the–fly approaches have to be implemented [11]. Also, selection of the appropriate minimal necessary support level is not a trivial task. Low values may result in unacceptably long execution times of the algorithms and unreliability of the results, while too high values may prevent detection of any useful clusters [12]. An even more serious problem is that in all real life domains some level of attribute noise can be expected. In such cases an error in a single attribute value may prevent the identification of correct descriptions. Such an error does not only cause that the erroneous example is not detected as a member of a cluster, but it causes the descriptions in different attribute layers not to cover all the subsets of examples it would have covered otherwise. As a result, some of the actual clusters may not be detected.

In this work we use an approach to more reliable clustering that reuses the basic idea of redescription mining in a novel setting, proposed in [7]. The first step is to determine the similarity of instances in each attribute layer independently and then to search for clusters that satisfy similarity conditions for *all* layers. The main characteristic of the approach is that the resulting clusters are small but very coherent.

2.1   Single Layer Algorithm

Let us assume a basic clustering task in which we have only one layer of attributes. The clustering approach consists of two steps. In the first step we estimate pair-wise similarity between all examples in the training data set. In the second step we use this similarity estimation in order to construct clusters.

Similarity Estimation. In the first step we compute the so called *example similarity table (EST)*. It is an $N \times N$ symmetric matrix, where N is the number of examples. All its values are in the $[0, 1]$ range, where large values denote large similarity between examples.

We start from the original set of N examples represented by nominal and numerical attributes that may contain unknown values. We define an "artificial" binary classification problem on a data set constructed as follows. The first part consists of examples from the original set, these examples are labelled as positive. The second part consists of the same number of examples which are generated by randomly shuffling attribute values of original examples (within each attribute separately), these examples are labelled as negative.

Next, we use supervised machine learning to build classifiers for discrimination between positive cases (original examples) and negative cases (examples with shuffled attribute values). The goal of learning are not the predictive models themselves, but the resulting information on the similarity of the original (positive) examples. Machine learning approaches in which we can determine if some examples are classified "in the same way" (meaning they are somehow similar) are appropriate for this task. For example, in decision tree learning this means that examples fall in the same leaf node, while in covering rule set induction this means that examples are covered by the same rule. In order to statistically estimate the similarity of the examples, it is necessary to use a sufficiently large number of classifiers. Additionally, a necessary condition for a good result is that the classifiers are as diverse as possible and that each of them is better than random. All these conditions are satisfied by Random Forest [13] and Random Rules algorithms [14]. Here we use the later one with which we construct a large number of rules (100,000) for each EST computation.

Finally, the similarity of examples is estimated so that for each pair of examples we count how many rules cover both examples. The example similarity table presents the statistics for positive examples only. A pair of similar examples will be covered by many rules, while no rules or a very small number of rules will cover pairs that are very different in respect of their attribute values. The final EST values are normalised with the largest detected count value.

Clustering with the CRV Score. In the second step we use the EST values to perform a bottom-up clustering. The agglomeration of clusters is guided by the *Clustering Related Variability (CRV)* score [7]. The score measures the variability of the EST similarity values in a cluster of examples with respect to all other clusters. It is defined as follows.

Let x_{ij} be the similarity between examples i and j from the EST matrix. The CRV score of a single example i from cluster C is the sum of within cluster and outside of cluster components: $\text{CRV}(i) = \text{CRV}_{\text{wc}}(i) + \text{CRV}_{\text{oc}}(i), i \in C$. The two components are sums of squared deviations from the mean value within (or outside of) cluster C: $\text{CRV}_{\text{wc}}(i) = \sum_{j \in C}(x_{ij} - \overline{x_{i,wc}})^2$ and $\text{CRV}_{\text{oc}}(i) = \sum_{j \notin C}(x_{ij} - \overline{x_{i,oc}})^2$. Finally, the *CRV* score of cluster C is the mean value of $C\overline{RV}(i)$ values of all examples in the cluster: $\text{CRV}(C) = \sum_{i \in C} \text{CRV}(i)/|C|$.

The clustering algorithm starts with each example being in a separate cluster and then iteratively tries to merge clusters together. In each iteration for each possible pair of clusters we compute the potential variability reduction that can be obtained by merging the clusters. The variability reduction of joining clusters C_1 and C_2 is computed as: $\text{DIFF}(C_1, C_2) = (\text{CRV}(C_1) + \text{CRV}(C_2))/2 - \text{CRV}(C_2 \cup C_2)$. The pair of clusters with the largest variability reduction is then merged into a single cluster. The iterative process repeats until no pair of clusters exists for which the variability reduction is positive. A more detailed description of the algorithm including some examples can be found in [7].

The algorithm produces a hierarchy of clusters and, in contrast to most other clustering algorithms, it has a very well defined stopping criterion. The algorithm stops when additional merging of clusters cannot further reduce the variability, measured by the CRV score. This means that the algorithm automatically determines the optimal number of clusters and that some examples may stay unclustered, i.e., some clusters may only include a single example.

2.2 Multilayer Algorithm

The single layer approach for clustering presented in the previous section can be easily extended to clustering in multi-layer domains. For each attribute layer we compute the example similarity table independently. Regardless of the number and type of attributes in different layers, the EST tables will always be $N \times N$ matrices, because the number of examples in all layers is the same.

After having all the EST tables, we proceed with the clustering. The clustering procedure for multiple layers is basically the same as for a single layer, except that we merge two clusters only if a variability reduction exists in all layers. For each possible pair of clusters we compute potential variability reduction for all attribute layers. Then we find the minimal variability reduction in all layers, and merge the pair of clusters for which this value is largest. As previously, we only merge two clusters if the (minimal) variability reduction is positive.

3   Data and Results

Data used in the preparation of this article were obtained from the Alzheimer's Disease Neuroimaging Initiative (ADNI) database. ADNI is a long term project aimed at the identification of biomarkers of the disease and understanding of the related pathophysiology processes. The project collects a broad range of clinical and biological data about patients with different cognitive impairment.[1] In our work we started from a set of numerical descriptors extracted from the AD-NIMERGE table, the joined dataset from several ADNI data tables. We have used baseline evaluation data for 916 patients in total with 5 different medical diagnoses: cognitive normal CN (187 patients), significant memory concern SMC (106), early mild cognitive impairment EMCI (311), late mild cognitive impairment LMCI (164), and Alzheimer's disease AD (148). The patients are described by a total of 10 biological and 23 clinical descriptors. Biological descriptors are genetic variations of APOE4 related gene, PET imaging results FDG-PET and AV45, and MRI volumetric data of: Ventricles, Hippocampus, WholeBrain, Entorhinal, Fusiform gyrus, Middle temporal gyrus (MidTemp), and intracerebral volume (ICV). Clinical descriptors are: Clinical Dementia Rating Sum of Boxes (CDRSB), Alzheimer's Disease Assessment Scale (ADAS13), Mini Mental State Examination (MMSE), Rey Auditory Verbal Learning Test (RAVLT immediate,

---

[1] The ADNI was launched in 2003 by the National Institute on Aging (NIA), the National Institute of Biomedical Imaging and Bioengineering (NIBIB), the Food and Drug Administration (FDA), private pharmaceutical companies and non-profit organizations (http://www.adni-info.org and http://adni.loni.usc.edu).

Table 1. Properties of five largest clusters. The clusters are ordered by the decreasing median value of the CDRSB score for patients included in the clusters.

Cluster	Number of patients	Distribution of diagnoses AD	LMCI	EMCI	SMC	CN	CDRSB
A	42	41	1	-	-	-	5.5
B	19	12	4	1	-	2	4.0
C	20	14	2	-	-	4	3.75
D	34	10	6	5	5	8	2.5
E	27	6	3	1	6	11	0

learning, forgetting, percentage of forgetting), Functional Assessment Questionnaire (FAQ), Montreal Cognitive Assessment (MOCA) and Everyday Cognition which are cognitive functions questionnaire filled by the patient (ECogPt) and the patient study partner (ECogSP) (Memory, Language, Visuospatial Abilities, Planning, Organization, Divided Attention, and Total score).

The clustering process started from one table with biological data consisting of 916 rows and 10 columns and one table with clinical data consisting of 916 rows and 23 columns. Each of the two tables represented one attribute layer. The information about medical diagnoses of the patients have not been included into the tables with the intention to use it only for the evaluation of the clustering results. The goal of the clustering process has been to identify as large as possible groups of patients that are similar according to both layers, i.e., biological and clinical characteristics.

The result is a set of five clusters. The largest among them includes 42 patients and the smallest only 19 patients. Table 1 presents the distribution of medical diagnoses for patients included into the clusters. We notice that the largest cluster A is very homogeneous. It has 42 patients and 41 of them have diagnosis AD while only one has diagnosis LMCI. Least homogeneous clusters are D and E.

We have identified the CDRSB score as the clinical characteristic that best discriminates between the constructed clusters. The rightmost column in Table 1 presents median CDRSB values for patients included in each cluster. Values above 3.0 demonstrate that patients in clusters A–C have problems with severe dementia. Patients in cluster D typically have moderate to severe dementia, while the majority of patients in cluster E do not have problems with dementia.

3.1 Distinguishing Properties of Patients in Clusters A and C

For each cluster we have computed median values and standard deviations of all biological and clinical descriptors. The intention is to identify distinguishing properties of patients included in the clusters. It is worth focusing on clusters that have either extremely high or extremely low median values or very low standard deviation of some descriptor. In the former case the patients in the cluster have this property (significantly) increased or decreased, while in the

Table 2. Median values and standard deviations of biological descriptors discriminating clusters A and C. Distinguishing values are typeset in bold. Actual values for ICV, Whole brain, Fusiform, and MidTemp values are 1,000 times larger than presented in the table.

Cluster	FDG-PET	AV45	ICV	Whole Brain	Fusiform	MidTemp
A	**4.37 / 0.52**	**1.45 / 0.13**	1404 / 193	918 /121	**15.4 / 2.7**	**15.6 /3.2**
B	5.67 / 0.67	1.43 / 0.23	1372 / 149	935 /84	16.0 /2.7	17.8 / 2.5
C	5.55 / 1.06	1.35 / 0.25	**1634 /126**	**1107 / 72**	18.2 / **2.0**	19.4 / 2.7
D	6.28 / 0.85	1.20 / 0.26	1453 / 230	1005 / 144	17.2 / 2.4	19.6 / 2.4
E	6.45 / 0.70	1.19 / 0.25	1445 / 174	1022 /93	18.0 / 3.	19.6 / 2.5
AD	5.36 / 0.75	1.42 / 0.21	1490 / 175	986 /115	16.4 / 2.5	17.6 / 3.2
CN	6.57 / 0.54	1.05 / 0.17	1483 /155	1051 /101	18.4 /2.3	20.5 / 2.4

later case most of the patients have very similar values of the descriptor. We can also find a distinguishing descriptor that has both low standard deviation and extreme median value. In all such situations we interpret the descriptor as a distinguishing property.

Clusters A, B, and C are especially interesting because most of the included patients have diagnosis AD and these clusters may be regarded as relatively homogeneous subsets of the AD patient population. The difference between clusters A and C turned out to be especially intriguing. Table 2 presents the values of biological descriptors for which patients in clusters A and C have distinguishing values. Presented are median values and standard deviations for all five clusters as well as for the complete AD population consisting of 148 patients and the complete cognitive normal (CN) population of 187 patients.

These results demonstrate that the patients in cluster A have extremely low values of the FDG-PET descriptor. The median for this cluster is 4.37 while median for the complete AD population is 5.36 and the median for cognitive normal patients is 6.57. Surprisingly, we can notice that the difference between the median of AD patients in cluster A and the median of all AD patients is almost as large as the difference between the median of all AD patients and the median of CN patients. Additionally, it must be noted that the patients in cluster A have very small standard deviation for FDG-PET meaning that the consistency of these small values is high. A similar pattern can be observed for the AV45 descriptor. When compared with other clusters and with the whole AD and CN populations, patients in cluster A have the largest median value 1.45 and the smallest standard deviation of 0.13 . We can conclude that cluster A is characterized by outstandingly low FDG-PET values and outstandingly high AV45 values.

In contrast to cluster A, cluster C is characterized by outstandingly high values for ICV and Whole brain descriptors, which are typeset in bold in the fourth and fifth column of Table 2, respectively. There is a substantial difference between biological descriptors that characterize clusters A and C. For cluster A the extreme values of FDG-PET and AV45 descriptors are relevant, while for cluster C the median values of these descriptors change (compared to CN

Table 3. Median values and standard deviations of clinical descriptors characteristic for patients in cluster A. Distinguishing values are typeset in bold.

Cluster	ADAS13	MOCA	Functional Assessment	RAVLT Perc. Forgetting	Ecog SP Organization
A	**35 / 8.5**	**14.5 / 4.9**	**16 / 6.8**	100 / 10	**3.17 / 0.76**
B	28 / 11.1	17 / 5.2	11 / 8.3	100 / 26	2.83 / 1.07
C	28 / 10.7	19 / 5.0	12 / 9.0	100 / 35	3.08 / 1.07
D	19.5 / 11.8	21.5 / 5.3	6.5 / 7.0	63 / 37	1.83 / 1.03
E	10 / 8.8	25 / 3.6	0 / 4.8	42 / 35	1.00 / 0.94
AD	31 / 8.4	18 / 4.5	13 / 7.1	100 / 20	2.83 / 0.86
CN	9 / 4.5	26 / 2.4	0 / 0.6	31 / 27	1.00 / 0.42

patients) in the same direction but with less intensity. The situation with ICV and Whole brain descriptors is very different. Median values of ICV for CN and all AD patients are very similar: 1483 and 1490, respectively. Patients in cluster C have a high median value of 1634, while patients in cluster A have a low median value of 1404 that is lower than the CN median value. A similar situation is with the Whole brain descriptor. CN patients have a median value of 1051, patients in cluster C have an increased value of 1107, while patients in cluster A have a very low median value equal to 918. Additionally, patients in cluster C have a low standard deviation of ICV and Whole brain descriptors.

Differences between clusters A and C can also be seen in some other biological descriptors. The rightmost two columns in Table 2 demonstrate that patients in cluster A have very low Fusiform and MidTemp volumes. For patients in cluster C these values are almost normal or only slightly decreased.

A good property of the multilayer clustering is that similarity of patients in both attribute layers is needed if they are to be included in the same cluster. Table 3 presents clinical descriptors with distinguishing values for patients in cluster A. CDRSB is a distinguishing descriptor but it is not included in Table 3 because we have already demonstrated in Table 1 that it has very high values for cluster A. The median value for the complete AD population for CDRSB descriptor is 4.5 while the median value for cluster A is 5.5 with a low standard deviation. Clinical descriptors with distinguishing values for patients in cluster C are presented in Table 4.

4 Discussion

Very elaborate medical studies have recently shown that mild and severe cognitive impairment as well as AD diagnosis are correlated with some measurable changes in the human brain. Our results are completely in agreement with these results. Firstly, it has been shown that the progressive reduction in fluorodeoxyglucose positron emission tomography (FDG-PET) measurement of the regional cerebral metabolic rate for glucose is related with cognitive impairment [3]. This follows from a series of increasing values in the second column of Table 2

Table 4. Median values and standard deviations of clinical descriptors characteristic for patients in cluster C. Distinguishing values are typeset in bold.

Cluster	Ecog Pt Memory	Ecog Pt Organization	Ecog SP Divided Attention
A	2.12 / 0.80	1.45 / 0.72	3.00 / 0.74
B	2.38 / 0.89	1.67 / 0.55	3.25 / 0.96
C	**2.62** / 0.85	**1.73** / 0.78	**3.50** / 1.06
D	1.88 / 0.61	1.33 / 0.57	2.00 / 1.08
E	1.75 / 0.65	1.00 / 0.63	1.00 / 0.95
AD	2.38 / 0.75	1.50 / 0.74	3.25 / 0.93
CN	1.50 / 0.44	1.00 / 0.38	1.00 / 0.48

and how it nicely correlates with the decreasing values of the CDRSB score (last column of Table 1) practically for all clusters A–E. Also, there is a recent result presented in [4] which demonstrates that the increased values of Florbetapir F18-AV-45 PET (AV45) are positively correlated with dementia. This effect can be noticed in the third column of Table 2 where we have constantly increasing values for clusters A–E in the order of their cognitive impairment severity.

There is a statistical analysis of ADNI data which demonstrated that whole brain atrophy and ventricular enlargement differ between cognitive impaired patients and healthy controls [15]. The result is very interesting because it is based on the same data source as our analysis. Our analysis detected the same relation for clusters A, B, D, and E. So it is not surprising that in [15] this property has been detected for the complete AD population. In our results the lowest median value for the whole brain volume is for cluster B (1372) but it is also very low for cluster A (1404). The real novelty of our result is cluster C in which the corresponding median value of 1107 is *higher than in all other clusters and also higher in than in the complete cognitive normal population* with median value of 1051.

This result is important because it potentially suggests that we have two different AD related physiological processes. One, that results in significantly decreased IC and whole brain volumes and is characteristic for the majority of AD patients, and second, that results in increased IC and whole brain volumes and appears in only 10-15% of cases. If this hypothesis is actually true, it will have substantial consequences for medical research and treatment of Alzheimer's disease. Namely, it is very likely that each of the two physiological processes will require different research and treatment procedures.

An additional result of our analysis is an observation that decreased FDG-PET values, increased AV45 values and decreased whole brain volume are especially characteristic for cluster A that is the largest and most homogeneous AD subpopulation. So it is not a surprise that these properties have been previously recognized as relevant for the complete AD population [3,4,15]. But if these studies would have concentrated on the AD subpopulation in cluster A only, the statistical significance and usefulness of their results would most probably be even higher.

Especially interesting is the question whether different physiological develop-ment of the AD disease may affect the clinical status of the patient and the patient's clinical prognosis. Figure 1 illustrates the differences between clusters A and C for ICV and ADAS13 values. Black circles denote patients in cluster A while empty squares are patients in cluster C. Signs + and − denote the position of median values for the whole AD and CN populations, respectively. The big black circle is the median value for cluster A patients and the big square is the median value for cluster C patients. Results presented in Table 3 suggest that patients characterized by *decreased ICV, fusiform, whole brain, and MidTemp results* (type A patients) have clinically very intensive general picture of AD iden-tified by high values of CDRSB, ADAS13, MOCA, and Functional Assessment Questioner scores. In contrast, patients characterized by *increased ICV, almost normal fusiform, slightly increased whole brain, and slightly decreased MidTemp* (type C patients) have main problems with Everyday cognition, especially with Memory, Organization, and Divided Attention (see Table 4). This is potentially an interesting issue for further medical expert evaluation. A preliminary obser-vation is that type C patients have self-awareness and insight to the condition. This pattern is not the usual course of classic AD where the level of insight decreases as the disease progresses.

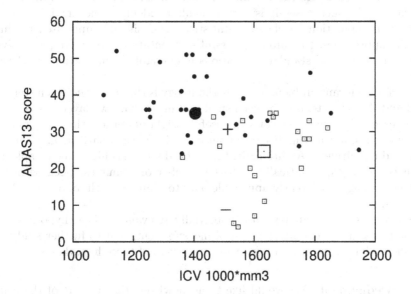

Fig. 1. Alzheimer's disease patients in cluster A (circles) and those in cluster C (squares) presented in the space defined by the ICV (Intracerebral Volume) and ADAS13 (Alzheimer's Disease Assessment Scale)

Four squares at the bottom of Figure 1 denote four patients with cognitive normal diagnosis included into cluster C. This is a potentially interesting fact suggesting that the physiological process of type C (increased IC and whole brain volumes) may in a small number of patients, in spite of significant physiological changes, result in only slightly changed clinical status, which is diagnosed as cognitive normal. In contrast, the significant physiological changes resulting from the process of type A (low FDG-PET, high AV45), according to the available data, practically always results in severe dementia diagnosed as AD.

5 Conclusions

This work presents the application of a novel clustering methodology to discover relations between two distinct sets of descriptors. The result is a set of small but homogeneous clusters of examples (patients). The main advantages of the methodology are that it may be successfully used on instances described by both numeric and nominal attributes and that it has a well-defined stopping criterion. Additionally, its unique property is that the quality of clusters is ensured by the requirement that examples must be similar in at least two different attribute layers. As a consequence the resulting clusters present the segmentation of the complete population where each subpopulation has some homogeneous properties in both layers. AD domain is a good example of a task in which relations between layers can be different or even contradictory for various subpopulations. In such cases it is very difficult to identify relevant relations on a complete population by classical statistical analysis and supervised machine learning approaches. But after successful segmentation even a simple analysis of median values and standard deviations may enable identification of relevant relations.

The most relevant problem of the methodology is that constructed clusters are small and that they tend to be even smaller if more than two layers are used. In the concrete AD domain we got some useful insight about less than a half of AD patients and practically no insight about cognitive normal patients and patients with mild impairment. Additionally, the methodology has high time complexity, which is growing quadratically with the number of examples. Because of this the methodology is currently applicable only to domains with an order of 1,000 examples.

The results of the analysis are potentially relevant medical hypotheses. It will be necessary to extend the clinical measurements and to further study the identified relations in order to confirm correctness of these hypotheses.

Acknowledgement. We would like to acknowledge the support of the European Commission through the Human Brain Project (Grant number 604102) and MAESTRA project (Grant number 612944) as well as the support of the Croatian Science Foundation (under the project number 9623 Machine Learning Algorithms for Insightful Analysis of Complex Data Structures) and Slovenian Research Agency supported grants: program (Knowledge Technologies) and project (Development and Applications of New Semantic Data Mining Methods in Life Sciences).

References

1. Weiner, M.W., et al.: The Alzheimer's Disease Neuroimaging Initiative: A review of papers published since its inception. Alzheimer's & Dementia 9(5), e111–e194 (2013)
2. Smith, G.E., Bondi, M.W.: Mild Cognitive Impairment and Dementia. Oxford University Press (2013)
3. Langbaum, J.B., et al.: Categorical and correlational analyses of baseline fluorodeoxyglucose positron emission tomography images from the Alzheimer's Disease Neuroimaging Initiative (ADNI). Neuroimage 45(4), 1107–1116 (2009)
4. Doraiswarny, P.M., et al.: Florbetapir F 18 amyloid PET and 36-month cognitive decline: a prospective multicenter study. Molecular Psychiatry 19(9), 1044–1051 (2014)
5. Hinrichs, C., et al.: Predictive markers for AD in a multi-modality framework: an analysis of MCI progression in the ADNI population. Neuroimage 55(2), 574–589 (2011)
6. Galili, T., Mitelpunkt, A., Shachar, N., Marcus-Kalish, M., Benjamini, Y.: Categorize, Cluster, and Classify: A 3-C Strategy for Scientific Discovery in the Medical Informatics Platform of the Human Brain Project. In: Džeroski, S., Panov, P., Kocev, D., Todorovski, L. (eds.) DS 2014. LNCS, vol. 8777, pp. 73–86. Springer, Heidelberg (2014)
7. Gamberger, D., Mihelčić, M., Lavrač, N.: Multilayer clustering: A discovery experiment on country level trading data. In: Džeroski, S., Panov, P., Kocev, D., Todorovski, L. (eds.) DS 2014. LNCS, vol. 8777, pp. 87–98. Springer, Heidelberg (2014)
8. Gan, G., Ma, C., Wu, J.: Data Clustering: Theory, Algorithms, and Applications. Society for Industrial and Applied Mathematics (2007)
9. Parida, L., Ramakrishnan, N.: Redescription mining: Structure theory and algorithms. In: Proc.of the Association for the Advancement of Artificial Intelligence, AAAI 2005, pp. 837–844 (2005)
10. Ramakrishnan, N., Kumar, D., Mishra, B., Potts, M., Helm, R.F.: Turning cartwheels: an alternating algorithm for mining redescriptions. In: Proc. of the 10th ACM Intern. Conf. on Knowledge Discovery and Data Mining, pp. 266–275 (2004)
11. Galbrun, E., Miettinen, P.: From black and white to full color: extending redescription mining outside the boolean world. Statistical Analysis and Data Mining, 284–303 (2012)
12. Zaki, M.J., Ramakrishnan, N.: Reasoning about sets using redescription mining. In: Proc. of the 11th ACM SIGKDD International Conference on Knowledge Discovery in Data Mining, KDD 2005, pp. 364–373 (2005)
13. Breiman, L.: Random forests. Machine Learning 45(1), 5–32 (2001)
14. Pfahringer, B., Holmes, G., Wang, C.: Millions of random rules. In: Proc. of the Workshop on Advances in Inductive Rule Learning, 15th European Conference on Machine Learning, ECML 2004 (2004)
15. Evans, M.C.: etal. Volume changes in Alzheimer's disease and mild cognitive impairment: cognitive associations. European Radiology 20(3), 674–680 (2010)

Entropy Analysis of Atrial Activity Morphology to Study Atrial Fibrillation Recurrences after Ablation Procedure

Raquel Cervigón[1,*], Javier Moreno[2], and Francisco Castells[3]

[1] Escuela Politécnica. DIEEAC. UCLM Camino del Pozuelo sn. Cuenca
[2] Arrhythmia Unit. Hospital Ramon y Cajal, Madrid
[3] ITACA Institute, Universitat Politécnica de Valencia, Valencia, Spain
raquel.cervigon@uclm.es

Abstract. Atrial fibrillation (AF) is an abnormal heart rhythm originated in the top chambers of the heart. The goal of pulmonary vein ablation for AF is returning to normal heart rhythm; nevertheless restoration of sinus rhythm is difficult to prognostic. In order to predict AF recurrences regularity of atrial activity morphology was studied. Intracardiac recordings from 43 paroxysmal and persistent AF patients registered previous to ablation procedure were monitored after the intervention. Results showed differences in entropy measurements from dipoles located in the right atrium with lower values of entropy in the recurrent group than in group that maintain sinus rhythm (p=0.004). The same trend was showed by entropy measures from spatial correlation between dipoles located in the right atrium, with lower values in the non-recurrent group than in the group with AF recurrence (p=0.009). Moreover, differences between both atria were found in the non-recurrent group 4.11 ± 0.01 in the left atrium vs. 4.07 ± 0.01 in the right atria (p=0.04). These findings show that atrial activity is more regular in the right atrium in the patients with non recurrences in AF.

Keywords: atrial fibrillation, entropy, correlation, ablation.

1 Introduction

Atrial fibrillation (AF) is the most common sustained arrhythmia. Prevalence increases from 0.1% among adults younger than 55 years to 9.0% in persons aged 80 years or older [1]. AF affects the upper chambers of the heart. It reduces the ability of the atria to pump blood into the ventricles and causes an irregular ventricular response. Since the atria are not emptying properly during fibrillation, blood clots can develop and travel to small vessels in the head and cause a stroke.

---

* Corresponding author.

F. Ortuño and I. Rojas (Eds.): IWBBIO 2015, Part I, LNCS 9043, pp. 146–154, 2015.
© Springer International Publishing Switzerland 2015

Treatment of AF is directed toward controlling underlying causes, slowing the heart rate and/or in restoring and maintaining sinus rhythm in patients to avoid the need for anticoagulation, to reduce thromboembolic risk and prevent tachycardia induced cardiomyopathy and improve survival.

Pharmacology treatments are commonly used for patients suffering from this disease, in the longer term to control or prevent recurrence of AF, but medications may not be effective and may have intolerable side effects. Electrical cardioversion is successful in over 95% of patients with AF, but 75% of patients have a recurrence of AF within 1 to 2 years [2]. Non-medication treatments of AF include pacemakers, atrial defibrillators, and ablation procedure.

Pulmonary vein isolation shows promise for the treatment of AF and has a high rate of success. It is increasingly being used for its efficacy compared with antiarrhythmic drugs. It requires a change in strategy, since recovering the therapeutic ideal of curing a disease that until recently was considered incurable, resulting in improved quality of life.

Pulmonary vein isolation is a catheter ablation technique where radiofrequency energy is applied to destroy this small area of tissue [3]. The use of radiofrequency energy causes scar tissue to form. The resulting tissue blocks the extra electrical signals from the pulmonary veins reaching the left atrium, so the area can no longer generate or conduct the fast, irregular impulses. This process is repeated around the opening of each of the four pulmonary veins. The long-term goal of the pulmonary vein ablation procedure is to eliminate the need for medications to prevent AF.

The success rate for a single pulmonary vein ablation procedure depends on many factors and at the moment it is difficult to predict when AF will be cured. Previous studies have characterized AF by heterogeneous and unstable patterns of activation including wavefronts, transient rotational circuits, and disorganized activity [4]. The goal of this study is to calculate entropy measure from atrial beats correlation sequences derived from intracardiac recordings of paroxysmal and persistent AF subjects in a three months follow-up ablation procedure to determine if it is possible to predict recurrence outcome.

2 Materials

AF intracardiac recordings were registered in 43 patients immediately before AF ablation procedure. Table 1 shows different parameters from recurrent and non-recurrent AF patients where there was no statistically significant difference between groups. A 24-pole catheter (Orbiter, Bard Electrophysiology, 2-9-2 mm electrode spacing) was inserted through the femoral vein and positioned in the right atrium (RA) with the distal dipoles into the coronary sinus (CS) to record left atrial (LA) electrical activity as well. The medium and proximal group of electrodes were located spanning the RA free-wall peritricuspid area, from the coronary sinus ostium to the upper part of the inter-atrial region. Using this catheter, 12 bipolar intracardiac electrograms from the RA (dipoles from 15-16 to 21-22) and LA (dipoles 1-2, 3-4, 5-6 and 7-8), were digitally recorded at

1 kHz sampling rate (16 bit A/D conversion; Polygraph Prucka Cardio-Lab, General Electric). Fifty to 60 seconds recordings from paroxymal and persistent AF patients were analyzed. All patients were monitored after ablation, and were divided in 2 groups according to AF recurrence outcome 3 months after ablation procedure.

Table 1. Patient Clinical Characteristics

Parameters	Recurrent AF	Non-recurrent AF
Paroxysmal AF Patients	10	13
Persistent AF Patients	7	13
Male (%)	11(65%)	24(88%)
Age (years)	51 ± 14	48 ± 14
Structural heart disease (%)	4(29%)	5(22%)
Left atrium size (mm)	44 ± 8	44 ± 6

3 Methods

Our method relies on combining three key insights. First is the fact that the amplitude of atrial peaks can provide information about AF morphology, the second key insight is that maximum correlation from sequential time windows that contain one atrial beat provides information from the variation of signals along the time, and it also allows accurate calculation of the distance from atrial peaks. The third key insight used is the entropy to measure AF irregularity.

3.1 Preprocesing

Electrograms from LA (dipoles from 1-2 to 3-4) and RA (dipoles from 17-18 to 19-20) were preprocessed according to the steps proposed by Botteron [5]. Initially, signals were band-pass filtered between 40 and 250Hz. This filter keep the high frequency oscillations of atrial activations, while damping the low frequency components due to noise and interference from the surroundings that may be captured within silent periods (i.e. the segment between consecutive activations). Subsequently, the signal is rectified. This is the key step of the preprocessing, which makes the fundamental frequency peak arise well over the harmonics. Finally, the absolute value of the filtered waveform was low-pass filtered with a 20Hz cut-off filter. The properties of this preprocessing approach are described in [6].

3.2 Correlation Measurements

Normalized correlation function was applied to examine whether it could be identified differences between wave morphologies along the time in both atrial chambers. This analysis requires two simultaneous signals recorded from closely spaced bipolar endocardial recordings.

The discrete cross-correlation function between discrete time series of N length data x and y can be approximated by:

$$R_{xy}(i) = \frac{1}{N} \sum_{i=1}^{N-k-1} \frac{(x_i - \bar{x})(y_i - \bar{y})}{\sigma_x \sigma_y}, \tag{1}$$

where $\bar{x}$ and $\bar{y}$ are the mean values of the two time series, and σ_x, σ_y are the standard deviations of the corresponding segments x and y, respectively.

Two different applications of normalized cross-correlation were proposed.

First, normalized cross-correlation function was calculated over each electrogram, overlapping segments that contain one atrial beat, where maximum of cross-correlation was calculated. This operation was repeated on sequential windowed segments for the entire data file. As result, it was obtained a time series composed from the correlation coefficients of the segments.

Moreover, spatial correlation analysis was performed between two electrograms registered from very close positions in each patient. The cross-correlation function was calculated over intervals for each electrogram (non-overlapping one atrial beat segments). For each data segment analyzed, the maximum peak was considered as the correlation coefficient of those two signals for that period of time. This operation was repeated on sequential windowed segments for the entire data files and a time series was built with the segment correlation coefficients.

3.3 Shannon Entropy

Due to the huge irregularity of the correlation coefficients time series, it was applied entropy measures to evaluate whether there were differences in the regularity along the atria in the group of patients who maintained sinus rhythm and in the patients with recurrences in AF.

The entropy is usually calculated according to the well-known Shannon definition [7]:

$$SE = - \sum_{i=1}^{M} p(i) \ln p(i), \tag{2}$$

where M is the number of discrete values the considered variable can assume and $p(i)$ is the probability of assuming the ith value.

3.4 Statistical Analysis

As a final step statistical techniques were applied to the extracted parameters. The parameters are expressed in terms of mean and standard deviation values. Independent t-student tests were used for comparison between both groups of results. Results were considered to be statistically significant at $p < 0.05$.

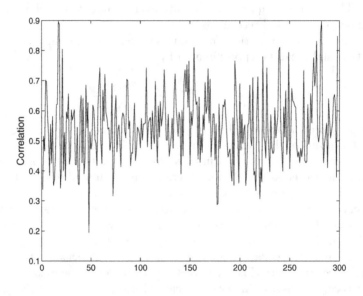

(a) RA correlation dipole (Entropy=3,98)

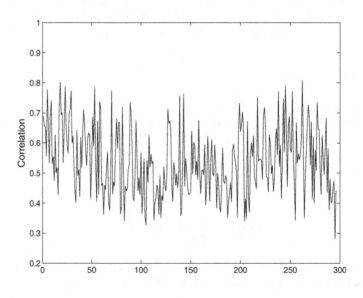

(b) RA correlation dipole (Entropy=4,12)

Fig. 1. Maximum Correlation from sequential time windows along the signal recorded in the RA from a non-recurrent AF and a recurrent AF patients, with a entropy value of 3, 98 in the left and with a entropy value of 4, 12 in the right

4   Results

Figures 1(b) and 1(a) represent the capability of entropy measurements to capture the disorganization of maximum correlation time series. Whereas the entropy of the signal in Figure 1(a) (up) was 3.98 (higher predictability), in the case of Figure 1(b) (down) raised up to 4.12 (lower predictability). These examples illustrate how higher values of entropy correspond to more disorganized signals. In addition, higher correlation values indicate higher organization degrees in the atrial electrical activation and in this case it was correspond with lower entropy values.

Entropy measures from temporal and spatial correlation analysis showed differences between both chambers, shower higher regularity in the RA in patients that did not have AF recurrences.

Entropy measurements from temporal maximum correlation along the right and left atria showed a higher entropy in the measurements from the RA in the patients with recurrences in AF (3.97 ± 0.24) than in patients that maintenance sinus rhythm (3.62 ± 0.59), showing a statistical signification of $p = 0.004$. There was no statistically significant difference between entropy measures in the LA (Figure 2).

In addition, entropy from maximum correlation time series extracted from two dipoles located in very close positions showed higher values in those dipoles located in the RA in patients with recurrences in AF, with 4.08 ± 0.05 for the non-recurrent AF group and 4.11 ± 0.03 for patients associated with AF recurrence ($p=0.009$). Non statistically significant differences were found between entropy measures from correlation between dipoles located int the LA (Figure 3).

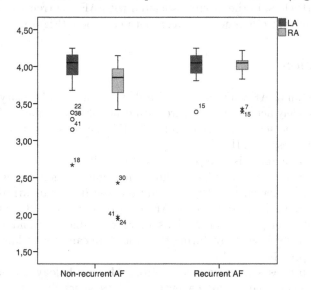

Fig. 2. Entropy from temporal correlation in both atria in recurrent and non recurrent AF groups

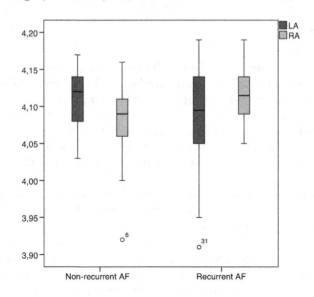

Fig. 3. Entropy from spatial correlation in both atria in recurrent and non recurrent AF groups

Furthermore, comparison between entropy measures from maximum correlation of two dipoles located in the LA and two dipoles sited in the LA showed statistical significant differences in the patients associated with sinus rhythm maintenance. In the LA entropy was 4.11 ± 0.04 vs. 4.08 ± 0.05 measured in the RA. Nevertheless, in the group associated with AF recurrences there was no statistical significant differences between both chambers (Figure 3).

5   Discussion

Catheter ablation of AF with the procedural end point of pulmonary vein isolation is now a widely accepted procedure with generally good efficacy for patients with AF [8,9]. The search for predictors of AF ablation success is currently of high clinical interest [10,11].

On multivariate analysis, the predictors of overall clinical of AF termination by ablation include shorter ablation duration, younger age, male gender, and the presence of hypertension [12]. In addition, the dominant atrial frequency has been described as a predictor of AF ablation outcome [13]. Furthermore, combination of other parameters such as a larger LA diameter and the presence of non-pulmonary veins ectopy during the procedure can predict late recurrence during long-term follow-up [14].

This paper proposes entropy measure from AF morphology as potential predictor for AF recurrence, where low entropy values, specially at the RA, are associated with sinus rhythm maintenance. Patients that remained in sinus rhythm presented lower entropy values in the RA compare with those who turned back to AF. Indeed, patients within the recurrent AF group had higher entropy values, i.e. the atrial morphology in this group was more disorganized. This means that patients with a more chaotic atrial electrical activity in the RA will have a higher risk to AF recurrence. With respect to atrial regions, entropy values exhibited a LA to RA gradient, with higher values at the LA, only in patients that were in sinus rhythm, where similar results were found in previous studies [15]. All these results are consistent and suggest that high entropy values could be a pro-arrhythmic indicator and be related to AF drivers or AF maintenance and perpetuation.

It is possible that after successful AF ablation, the electrophysiological properties of the LA, responsible for AF maintenance, are supposed to be dramatically modified. RA electrophysiological properties could be also changed. These results are consistent with previous studies where sample entropy was used as a tool to measure regularity from atrial activity across the electrogram, with higher differences in the [16]. The major conclusion of this study is the differences found between both groups respect to RA and the LA to RA entropy gradient found in the group that maintains sinus rhythm.

Acknowledgments. This work was supported by the Castilla-La Mancha Research Scheme (PPII-2014-024-P) and for the program: Becas Iberoamérica Jóvenes Profesores e Investigadores. Santander Universidades. Convocatoria 2014. España.

References

1. Go, A.S., Hylek, E.M., Phillips, K.A., Chang, Y., Henault, L.E., Selby, J.V., Singer, D.E.: Prevalence of diagnosed atrial fibrillation in adults: national implications for rhythm management and stroke prevention: the anticoagulation and risk factors in atrial fibrillation (atria) study. JAMA 285(18), 2370–2375 (2001)
2. Grönberg, T., Hartikainen, J.K., Nuotio, I., Biancari, F., Vasankari, T., Nikkinen, M., Ylitalo, A., Juhani Airaksinen, K.E.: Can we predict the failure of electrical cardioversion of acute atrial fibrillation? the fincv study. Pacing Clin. Electrophysiol. (December 2014)
3. Haissaguerre, M., Shah, D.C., Jais, P.: Electrophysiological breakthroughs from the left atrium to the pulmonary veins. Circulation 102, 2463–2465 (2000)
4. Lee, G., Kumar, S., Teh, A., Madry, A., Spence, S., Larobina, M., Goldblatt, J., Brown, R., Atkinson, V., Moten, S., Morton, J.B., Sanders, P., Kistler, P.M., Kalman, J.M.: Epicardial wave mapping in human long-lasting persistent atrial fibrillation: transient rotational circuits, complex wavefronts, and disorganized activity. Eur. Heart J. 35(2), 86–97 (2014)
5. Botteron, G.W., Smith, J.M.: A technique for measurements of the extent of spatial organization of atrial activation during atrial fibrillation in the intact human heart. IEEE Trans. Biomed. Eng. 42, 579–586 (1995)
6. Castells, F., Cervigón, R., Millet, J.: On the preprocessing of atrial electrograms in atrial fibrillation: Understanding botteron's approach. Pacing Clin. Electrophysiol (November 2013)

7. Shannon, C.E., Weaver, W.: The mathematical theory of communication. University of Illinois Press, Urbana (1998)
8. Brooks, A.G., Stiles, M.K., Laborderie, J., Lau, D.H., Kuklik, P., Shipp, N.J., Hsu, L.-F., Sanders, P.: Outcomes of long-standing persistent atrial fibrillation ablation: a systematic review. Heart Rhythm 7(6), 835–846 (2010)
9. Calkins, H., Brugada, J., Packer, D.L., Cappato, R., Chen, S.-A.: Hrs/ehra/ecas expert consensus statement on catheter and surgical ablation of atrial fibrillation: recommendations for personnel, policy, procedures and follow-up. a report of the heart rhythm society (hrs) task force on catheter and surgical ablation of atrial fibrillation developed in partnership with the european heart rhythm association (ehra) and the european cardiac arrhythmia society (ecas); in collaboration with the american college of cardiology (acc), american heart association (aha), and the society of thoracic surgeons (sts). endorsed and approved by the governing bodies of the american college of cardiology, the american heart association, the european cardiac arrhythmia society, the european heart rhythm association, the society of thoracic surgeons, and the heart rhythm society. Europace 9(6), 335–379 (2007)
10. Montserrat, S., Gabrielli, L., Bijnens, B., Borràs, R., Berruezo, A., Poyatos, S., Brugada, J., Mont, L., Sitges, M.: Left atrial deformation predicts success of first and second percutaneous atrial fibrillation ablation. Heart Rhythm 12(1), 11–18 (2015)
11. Pathak, R.K., Middeldorp, M.E., Lau, D.H., Mehta, A.B., Mahajan, R., Twomey, D., Alasady, M., Hanley, L., Antic, N.A., McEvoy, D., Kalman, J.M., Abhayaratna, W.P., Sanders, P.: Aggressive risk factor reduction study for atrial fibrillation and implications for the outcome of ablation: The arrest-af cohort study. J. Am. Coll. Cardiol. 64(21), 2222–2231 (2014)
12. Kevin Heist, E., Chalhoub, F., Barrett, C., Danik, S., Ruskin, J.N., Mansour, M.: Predictors of atrial fibrillation termination and clinical success of catheter ablation of persistent atrial fibrillation. Am. J. Cardiol. 110(4), 545–551 (2012)
13. Okumura, Y., Watanabe, I., Kofune, M., Nagashima, K., Sonoda, K., Mano, H., Ohkubo, K., Nakai, T., Hirayama, A.: Characteristics and distribution of complex fractionated atrial electrograms and the dominant frequency during atrial fibrillation: relationship to the response and outcome of circumferential pulmonary vein isolation. J. Interv. Card. Electrophysiol. 34(3), 267–275 (2012)
14. Lo, L.-W., Tai, C.-T., Lin, Y.-J., Chang, S.-L., Udyavar, A.R., Hu, Y.-F., Ueng, K.-C., Tsai, W.-C., Tuan, T.-C., Chang, C.-J., Kao, T., Tsao, H.-M., Wongcharoen, W., Higa, S., Chen, S.-A.: Predicting factors for atrial fibrillation acute termination during catheter ablation procedures: implications for catheter ablation strategy and long-term outcome. Heart Rhythm 6(3), 311–318 (2009)
15. Cervigon, R., Moreno, J., Quintanilla, J.G., Perez-Villacastín, J., Millet, J., Castells, F.: Frequency spectrum correlation along atria to study atrial fibrillation recurrence. In: Computing in Cardiology (CinC), pp. 1125–1128 (September 2014)
16. Cervigon, R., Moreno, J., Millet, J., Castells, F.: Predictive value of entropy analysis for atrial fibrillation recurrence after ablation procedures. In: Computing in Cardiology (CinC), pp. 825–828 (September 2012)

Portable Low-Cost Heart Attack Detection System Using ZigBee Wireless Technology

Khaled Sayed Ahmed[1] and Shereen M. El-Metwally[2]

[1] Department of Bio-Electronics, Modern University for Technology and Information, Cairo, Egypt
[2] Faculty of Engineering, Cairo University, Giza, Egypt
khaled.sayed@k-space.org, shereen.elmetwally@yahoo.com

Abstract. People suffering from heart diseases are liable to many accidents due to the occurrence of sudden heart attacks. As ECG signal processing can help to detect any heart malfunction as heart attack, in this paper, an integrated system is designed and implemented to help patients with heart disease by sending their ECG signals to the nearest hospital for a rapid intrusion and to alarm their neighbours or relatives. The system is based on collecting the ECG signals through a small portable device that applies signal processing techniques to identify the signal abnormalities. An alarm is then sent to the nearest hospital in the area based on the wireless ZigBee technology. A receiver module, to be established at the nearby healthcare facilities, is designed to receive these alarm signals. An automatic call centre is combined with the receiver unit to initiate a series of automatic calls with the patient's contacts to help save him. The system has been applied to an ECG simulator programmed to produce abnormal ECG signals at specific time intervals. The system experimental results have shown that a receiver established at distances up to 1.24 km from the transmitter unit could receive correct alarm signals in short time delays. The system could achieve an accuracy of 89% of correct data transfer.

Keywords: ECG, Heart attack, wireless transmission, ZigBee.

1 Introduction

A coronary heart attack is a heart block which occurs due to a blood clot. This may happen from the build-up of fat, cholesterol and other substances as plaque which affects the blood flow. When blood is blocked in the coronary artery, the heart muscle supplied begins to die [1]. Damage increases the longer an artery stays blocked. Once that muscle dies, the result is permanent heart damage. Heart attack is one of the most common causes of death. About 325,000 people-a-year die of coronary attack before they get to a hospital or in the emergency room [2]. The heart attack can be detected based on the ECG ST-segment, which corresponds to the level of damage inflicted on the heart [3]. The ST-segment elevation myocardial infarction (STEMI) is the most serious type of heart attack, where there is a long interruption to the blood supply [2].

F. Ortuño and I. Rojas (Eds.): IWBBIO 2015, Part I, LNCS 9043, pp. 155–162, 2015.

This is caused by a total blockage of the coronary artery, which can cause extensive damage to a large area of the heart.

Since ECG signal processing can help to reveal abnormalities and detect heart attack, a three-lead portable device can be connected to the patient to pick up the ECG signals and perform signals analysis to identify the abnormal ST- wave. Many applications have been performed to transmit the patient acquired signals to a receiving station via wireless communication technology in order to receive the data transmitted and analyse it [4, 5].

Wireless communication is a fast-growing technology as it allows for systems flexibility and mobility [6]. The wireless technologies include Bluetooth, ultra-wideband (UWB), Wi-Fi, ZigBee and GPS. IEEE defines the physical (PHY) and media-access control (MAC) layers for wireless communications over ranges around 10-100 meters. Bluetooth, ultra-wideband, ZigBee over (IEEE 802.15.1/3/4) respectively, and Wi-Fi over (IEEE 802.11) represent four protocol standards for short range wireless communications with low power consumption [7]. The main features for the various wireless technologies have been compared in literature [8]. Various metrics for the Bluetooth and Wi-Fi features have been presented in terms of including capacity, network topology, security, quality of service support, and power consumption [9]. The strengths and weaknesses of ZigBee and Bluetooth for industrial applications have been studied in [10]. ZigBee can meet a wider variety of real industrial needs than Bluetooth due to its long-term battery operation, greater useful range, flexibility in a number of dimensions, and reliability of the mesh networking architecture. For data transmission over long distances, ZigBee devices allow passing data through intermediate devices to reach more distant ones, creating a mesh network; a network with no centralized control suitable for applications where a central node can't be relied upon.

ZigBee is intended to be used in applications that require only a low data rate, long battery life, and secure networking [5]. It has a defined rate of 250 kbit/s, best suited for periodic or intermittent data or a single- signal transmission from a sensor or input device. The ZigBee technology has the advantage of being simpler and less expensive than other wireless personal area networks (WPANs), such as Bluetooth or Wi-Fi.

This paper presents the design and implementation of a low-cost wearable ECG-based heart attack detection system. The system has two main modules which have been integrated and adapted to transmit and receive the patients ECG signals. The first module is a small portable device with the patient. It acquires the raw ECG data from the leads placed in predefined locations of the patient chest. All necessary processing for heart attack detection and formatting of the ECG data for its transmission occur in the microcontroller included in the mobile unit. Upon heart attack, the system sends an alarm and/ or complete signals via a low-cost RF transceiver to the second module which is the receiver placed at the healthcare facility.

2   Methodology

In this paper, we have implemented a prototype hardware system consisting of two main modules as shown in Fig. 1. First, data collection & transmission module: a small-sized portable device mainly responsible for collecting the ECG signals from the patient chest via three leads and applying signal processing techniques to detect heart attack, then transmitting an alarm along with the defected signals to the nearest hospital. Second, the receiver module: is connected to a PC at the hospital to receive the transmitted signals in order to allow for the quick hospital intervention either by making automatic phone calls with the patient's relatives and neighbours contacts or by giving an alarm to the hospital emergency department. The ZigBee-based wireless technology is used for the data wireless transmission. The detailed description for the system components is given in the following sections.

Fig. 1. A wireless transmission/receiver system

2.1   Data Acquisition and Transmission

In ECG data acquisition and transmission, an Arduino shield kit [11] is used to read the three-leads output from the patient chest and process it in order to classify three different categories of ECG signals. These categories are:

(a) Normal (patient-1) which reflects normal ECG signals without any abnormalities with normal heart rate of 60 heart beats per minute.
(b) Abnormal (patient-2) which represents an ECG abnormality due to the ST segment defect probably leading to heart attack.
(c) Abnormal (patient-3) that reflects any abnormality of ECG signals other than heart attack, e.g. atrial flutter.

The system has been applied to an ECG simulator programmed to produce abnormal ECG signals at specific time intervals. The collected signals are analyzed in order to identify the signal category (Heart attack or not) and to give a label to each category. If there is no heart attack detected, the system will ignore the data else, it will generate an alarm/ signal to be transmitted via a ZigBee module [12] connected to the Arduino kit as illustrated in the block diagram in Fig. 2.   This alarm or signal can be an on/ off pulse, a short segment of the ECG signal showing the ST-segment defect, or a longer segment of the ECG signal (3 seconds long) to allow the physician receiving it to review and analyze it visually. The data acquisition and transmission module components are shown in Fig. 3.

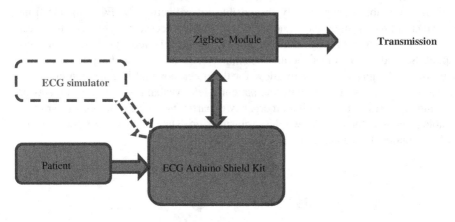

Fig. 2. Data acquisition and transmission block diagram

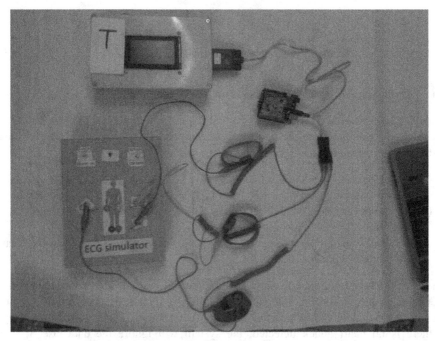

Fig. 3. Data acquisition and transmission setup

2.2   The Receiver

The receiver is composed of a ZigBee module connected via an Arduino microcontroller kit to the PC to be placed at the hospital, where all the patients' contacts are stored. This is illustrated in Fig. 4. According to the received signal class label, the system PC would initiate a series of automatic calls with the patients' relatives or neighbors whose contacts are stored and also would generate an alarm to the emergency department. Sometimes the received alarm is accompanied with a part of the ECG signal indicating the abnormality.

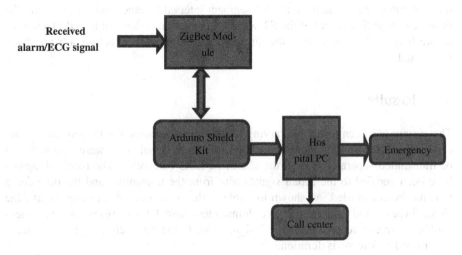

Fig. 4. System receiver block diagram

2.3   Wireless Technology

In our system, XBee RF Modules [12] are used to implement the ZigBee protocol for peer-to-peer wireless communication between the two Arduino systems at the transmitter and receiver, allowing them to communicate over a modified ZigBee protocol. The two Arduino systems should be of the same series (Series 2) as the XBee modules. The XBee receiving kit is connected to the hospital PC at the receiving station, the other transmitting kit is portable with the patient data acquisition and transmitting device.

The system setup can transmit/receive three types of signals along with the ECG class label: Type 1 is only On/ Off pulse alarm, Type 2 is On/ Off pulse alarm in addition to a segment of the ECG signal of one or two ECG cycles length, and Type 3 is On/ Off pulse accompanied with a 3 seconds long ECG signal. The ECG categories labels transmitted are: Category-1 for normal patients, Category-2 for heart attack, and Category-3 for atrial flutter. The different ECG categories and signal types transmitted are tabulated in Table 1. In the system experimental trials, the system transmitter is assumed to be located at a certain position A, while the receiver is placed at four different locations according to the studied distances between the transmitter and receiver: 10 m, 120 m, 1240 m, and more than 2.5 km. Table 2 illustrates the studied transmitter/receiver locations and their separating distances.

2.4   ECG Signal Processing

The signals have been acquired in time domain. A back/forward search algorithm for the ST episodes detection is used. After detecting the R wave of the QRS complex, a timer is initialized to calculate the ST-segment interval together with determining the positive or negative values of the ST wave. This interval is then compared to a specific threshold. If it is longer than the threshold and in negative direction, a heart attack is detected.

3   Results

The system has been tested by varying the distances between the transmitter and receiver. For every studied receiver location, the three signal types were also tested to be transmitted separately for all the classified ECG categories. The received signals have been verified to the actual signals sent from the transmitter and the time delay time has been recorded. As shown in Table 3, the system could correctly receive the signal Types 1 and 2 in the distance domain less than 1.5 km. However, the system couldn't correctly receive the Type 3 signals due to the relatively long ECG segment transmitted of 3 seconds duration.

Table 1. Transmitted signal types and ECG categories

Signal	Description
Type 1	On/ Off pulse alarm
Type 2	On/ Off + segment of ECG signal
Type 3	3 seconds long ECG signal

ECG Category	Description
1	Normal
2	Heart Attack
3	Atrial flutter

Table 2. Transmitter/receiver locations and separating distances

Point	Description	Distance between source and receiver
A	Source	Base point
B	Building B	10 m
C	Building C	120 m
D	Building D	1240 m
E	Open Space	More than 2.5 km

Table 3. The system performance and calculated time delay between signals transmitting and receive

Signal	Time Delay (s)			
	Building B	*Building C*	*Building D*	*E*
Type 1	0.5	0.52	0.6	not received
Type 2	1	1.2	1.4	not received
Type 3	1.3	1.8	received incorrectly	not received

4   Discussion and Conclusion

An integrated system has been introduced to help people suffering from heart disease and liable to a heart attack to send their recorded ECG signals to the nearest hospital in the patient area upon detecting a heart attack. The system is based on collecting the ECG signals via a small portable device which can identify the abnormal signals by applying signal processing techniques. An alarm along with a part of the recorded signal can then be transmitted to the nearest hospital based on ZigBee technology. The system performance has been verified by calculating the time delay between the signals transmit and receive. The percentage of the correctly detected alarm signals was also computed. The system could achieve an accuracy of 89% of correct data transfer.

The system was shown to correctly receive the transmitted signals for distances within the ZigBee domain (1.5 km for series-2) [5]. However at a distance of 1.24 km, the 3 seconds long ECG signal couldn't be received correctly. This is due to the signal reflections problem inside concrete buildings, which increases for longer distances and for transmitted signals of longer length. In order to overcome that, the receiver should be replaced by two perpendicular antennas. Also, the signals may be divided into smaller sub-segments to be sent without error. For distances longer than 2.5 km, the system failed to receive all types of transmitted signals (alarm and the 3 seconds long ECG signals).

Because the reachability radius achieved is only about 1.24 km, it is not likely to expect patients being so close to the hospital. As a future work, the communications from the patient's device to the hospital may to be conveyed using cellular networks or other wide-range communication technologies in order to extend the covered distances.

References

1. National Heart, Lung, and Blood Institute, Disease and Condition Index. What Is a Heart Attack? (August. 2003)
2. National Heart, Lung, and Blood Institute, Disease and Condition Index. What Is Coronary Artery Disease? (August 2003)
3. Bulusu, S.C., Faezipour, M., Ng, V., Banerjee, S., Nourani, M., Tamil, L.S.: Early Detection of Myocardial Ischemia Using Transient ST-Segment Episode Analysis of ECG. International Journal of Medical Implants and Devices 5(2), 113 (2011)
4. Daki, J.: Low Cost Solution For Heart Attack Detection With Walking Stick. International Journal for Scientific Research & Development 2(03) (2014)
5. Abinayaa, V., Jayan, A.: Case Study on Comparison of Wireless Technologies in Industrial Applications. International Journal of Scientific and Research Publications 4(2) (February 2014)
6. Willig, A.: An architecture for wireless extension of Profibus. In: Proc. IEEE Int. Conf. Ind. Electron (IECON 2003), Roanoke, VA, pp. 2369–2375 (November 2003)
7. Wang, X., Ren, Y., Zhao, J., Guo, Z., Yao, R.: Comparison of IEEE 802.11e and IEEE 802.15.3 MAC. In: Proc. IEEE CAS Symp. Emerging Technologies: Mobile & Wireless Commun., Shanghai, China, pp. 675–680 (May 2004)
8. Lee, J.-S., Su, Y.-W., Shen, C.-C.: A Comparative Study of Wireless Protocols: Bluetooth, UWB, ZigBee, and Wi-Fi. In: The 33rd Annual Conference of the IEEE Industrial Electronics Society (IECON) Taipei, Taiwan, November 5-8 (2007)
9. Ferro, E., Potorti, F.: Bluetooth and Wi-Fi wireless protocols: A survey and a comparison. IEEE Wireless Commun. 12(1), 12–16 (2005)
10. Baker, N.: ZigBee and Bluetooth: Strengths and weaknesses for industrial applications. IEE Computing & Control Engineering 16(2), 20–25 (2005)
11. https://www.olimex.com/Products/Duino/Shields/SHIELD-EKG-EMG/
12. http://www.digi.com/products/wireless-wired-embedded-solutions/ZigBee-rf-modules/ZigBee-mesh-module/xbee-zb-module.jsp

Cost-Effectiveness Studies on Medical Devices: Application in Cardiology

Radka Otawova, Vojtech Kamensky, Pavla Hasenohrlova, and Vladimir Rogalewicz

CzechHTA, Faculty of Biomedical Engineering, Czech Technical University in Prague, Kladno, Czech Republic
{radka.otawova,kamenvoj,rogalewicz,hasenpa1}@fbmi.cvut.cz

Abstract. Current discussions about cost-effectiveness in the field of medical devices for cardiology are reviewed. In order to identify the trends, we performed a review of studies published in medical journals. After having reviewed 1143 originally identified papers, we finally selected 61 papers to be studied in detail. The selected papers are classified into two categories: diagnostic and therapeutic medical devices, while technology for telemedicine is briefly mentioned. In each category, cost-effectiveness results of chosen modalities are presented. The main trend seems to be avoiding invasive technology during the treatment. This trend is obvious both in the field of diagnostic and therapeutic devices. In CAD diagnostics, some authors try to prove that CT, PET or SPECT are more cost-effective than ICA. As compared with the classical (open chest) surgery in the past, the treatment of cardio-vascular illnesses is moving to less invasive interventions (TAVI versus SAVR is just one example).

Keywords: HTA review, medical devices, cardiology, cost-effectiveness.

1   Introduction

Coronary artery disease (CAD) is the main cause of death in high-income countries [1]. The World Health Organization estimates there will be about 20 million deaths from cardiovascular reasons in 2015, accounting for 30 percent of all deaths worldwide [2]. Moreover, aging of population mirrors in an increase in the incidence and prevalence of cardiovascular diseases in the elderly. Odden et al. [3] constructed a prognosis of the absolute coronary heart disease prevalence from 2010 until 2040 in U.S. (see Figure 1). According to European Cardiovascular Disease Statistics 2012 [4], costs of cardiovascular diseases are estimated at almost €196 billion a year in the EU (54% direct healthcare costs, 24% productivity losses, and 22% opportunity costs of informal care). In the last decade we have witnessed a great progress in cardiology, especially in non-invasive diagnostic methods and in treatment of a wide range of heart diseases.

Limited resources and continuously growing costs in health care have led to the necessity to assess effectiveness, appropriateness and costs of health technologies incl. medical devices. However, apparatus have their specificities that affect significantly all processes, where they are involved [5]. In Health Technology Assessment,

F. Ortuño and I. Rojas (Eds.): IWBBIO 2015, Part I, LNCS 9043, pp. 163–174, 2015.

they lead to special methods different from other technologies [6]. The main challenge remains to be how to calculate and express the effects (outcomes) of a particular device. The generally used QALY concept is not suitable for medical devices, as they frequently do not directly affect the quality of life and/or life years of the patient. The efficiency of a device depends not only on the device itself, but also on how it is used (the skill and experience of the surgeon, organization of work on the clinic, etc.) [7]. Its effect can be lower radiation, more comfort for the clinician or the patient, better image resolution. We recommend to calculate standard CEA, where the effects are evaluated by means of a combination of value engineering methods and multiple-criteria decision analysis, while costs are evaluated directly (ideally by the micro-costing method) [6].

In this paper we briefly review current discussions about cost-effectiveness in the field of devices for cardiology.

2   Methods

In order to identify current trends in the field of medical devices used in cardiology, we performed a review of studies in journals focused on cardiology, medicine, bio-medical engineering and health economy. We focused the review on state-of-the-art instrumentation in cardiology. The following data bases were searched: Web of Science, Science Direct, SpringerLink, Wiley Online Library, PubMed. We systematically selected relevant papers that contained a cost-effectiveness analysis. We included studies regardless of their design, sample size, time span or a particular coronary disease. The electronic and hand searches retrieved 1143 references. Of these, 155 papers were considered potentially relevant to the review of cost-effectiveness in instrumentation for cardiology. Out of them, 61 papers were selected and included in the detailed study. These papers were classified according to their focus on diagnostic and therapeutic devices. Figure 1 shows the flow of studies through the review.

3   Results - Apparatus

3.1   Diagnostic Medical Devices

Computed Tomography (CT). Non-invasive visualization of coronary arteries is a modern trend in cardiac imaging. The majority of studies in cardiac imaging have been focused on CT, only a few of them on other non-invasive or less-invasive modalities like SPECT, PET and/or scintigraphy [8-10]. Cardiac catheterization, the historically most frequently used modality for coronary imaging, is invasive, costly, and often performed unnecessarily [11].

A full substitution of the invasive coronary angiography (ICA) in revascularization assessment is unlikely, particularly as angiography and angioplasty are often done on the same occasion, but there is a trend to avoid excessive interventions [12].

According to the Catalan et al. [13], the initial CT strategy is better than the strategy with initial ICA, because it allows for avoiding ICA and post-ICA morbidity- mortality, with significant savings in CAD patients diagnostic process. DSCT (dual source CT) coronary angiography is cost-effective for diagnosing CAD in patients with an intermediate pre-test likelihood [14, 15]. Dewey et al. [16] and Halpern et al. [17] report CTA being a cost-effective strategy before cardiac catheterization in patients with the pre-test probability up to 50%. Kreisz et al. [18] allows for the pre-test likelihood up to 65%, and Stacul et al. [19] even up to 86%. CT angiography cost effectiveness proved to be better in the large majority of patients [19], which is supported by [20, 21]. The studies show a large heterogeneity. In order to increase the reliability, more studies are necessary that would combine the existing results [22] or use large randomized controlled trials.

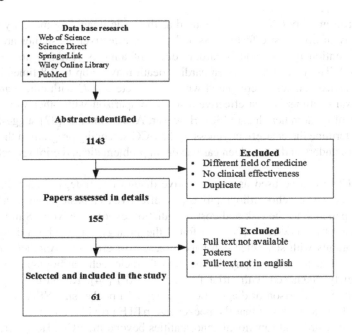

Fig. 1. Flow of studies through the review

Angiography. The invasive coronary angiography (ICA) remains the predominant method for guiding decisions about stent implantation [23]. As mentioned above, CA is most often compared with CT angiography. Dorenkamp et al. and Halpern et al. [14, 17] suggested that the CA is cost-effective in the case of the pre-test probability up to 50% or even more. With the effort to reduce costs of CA, there is a way to transfer some CA examinations to ambulatory conditions, while maintaining clinical effectivity and safety. Matchin et al. [24] showed that ambulatory coronary angiography in patients with coronary heart disease was safe, with a low complication risk, and due to decreased hospitalization days reduced the average procedure cost by 19%.

PET, SPECT. Most studies concerning these diagnostic modalities in cardiology have not been focused on cost-effectiveness. They investigated the clinical effectiveness

[8, 25, 26] and attempted to demonstrate a suitability of these diagnostic modalities for cardiologic examinations and their inclusion into the guidelines [27]. Bengel et al. in his systematic review of the PET diagnostic modality [19] mentioned that some studies supported the overall cost-effectiveness of the technique despite of higher costs of a single examination, because unnecessary follow-up procedures can be avoided.

As concerns SPECT, Shaw et al. [28] reported that SPECT is a cost-effective examination method in intermediate risk patients with a stable chest pain. Meyer et al. [10] compared SPECT with CT and reported a higher cost of the first-line SPECT myocardial perfusion test as compared with CT.

Electrocardiography (ECG). ECG is used as the basic imagining modality in cardiology. Many of both cost-effectiveness and clinical effectiveness studies are focused on the prevention of the sudden cardiac death of athletes [29, 30] and also non-athletes [31]. The risk of the sudden cardiac death may be up to 2.8 times higher in top-performance athletes as compared with non-athletes. ECG with other cardiac and physical examinations is cost-effective also in comparison with other well-accepted procedures of modern health care [30]. However, Anderson et al. [32] suggests that in order to determine the cost-effectiveness of the ECG screening program in the US it is necessary to understand the sudden cardiac death problematic and etiology better.

ECHO. ECHO can be used as an alternative diagnostic strategy. Bedeti et al. [33] concluded that stress echocardiography is cost-effective as a basic strategy in low-risk chest-pain patients, as the risk and cost of radiation exposure is void. Similar results were reported by Shaw et al. [28] that found the strategy using echocardiography in low-risk patients with suspected coronary disease cost-effective. Another cost effective usage of echocardiography is the transthoracic echocardiography (TTE) in patients newly diagnosed with atrial fibrillation (AF) [34]. Kitada et al. [35] were interested in a comparison of diagnostic accuracy of a pocket-size pTTE with a standard TTE. The paper shows that the pocket-size pTTE provides an accurate detection of cardiac structural and functional abnormalities beyond the ECG. However, the cost effectiveness analysis is necessary. Similarly, Trambaiolo et al. [36] was interested in a similar device: hand-carried ultrasound (HCU) device in an outpatient cardiology clinic. His results show that using a simple HCU device in an outpatient cardiology clinic allowed for reliable diagnoses. This modality proved to be cost and time saving.

Another chapter in echocardiography is its application in pediatric diagnostics. Echo screening among siblings suffering from bicuspid aortic valve disease is effective and inexpensive, and it may lower the risk of complications. Hence, it should be incorporated into the clinical guidelines [37].

3.2 Therapeutic Medical Devices

Transcatheter Aortic Valve Implantation (TAVI). Not so long ago, the only option of effective treatment of patients suffering from a heavy form of arterial stenosis was the surgical replacement of the aortal valve (SAVR). Unfortunately the surgery may not be suitable for a part of these patients [38]. For them, the Transcatheter Aortic Valve Implantation (TAVI) represents a less invasive option. The PARTNER high study showed that for the risk patient group, TAVI is approximately as effective as the surgical methods [39].

The results of Neyt et al. [40] and Fairbairn et al. [41] indicated that TAVI shows a better cost efficiency for the non-operable patients. Fairbairn concluded that, in spite of greater procedural costs, TAVI is more cost efficient than the Surgical Aortic Valve Replacement (SAVR). Reynolds   et al. [39] suggested that the transfemoral TAVI is economically more interesting than SAVR, but results do not speak in favour of the transposal TAVI. On the other hand, Doble with his colleagues [42] made a complex analysis comparing TAVI and SAVR. Their results indicate that TAVI is less cost effective than SAVR in a 20-year horizon.

Resynchronization Therapy. Blomstrom et al. [43] indicates that cardiac resynchronization therapy (CRT) is a cost-effective treatment in Scandinavian health care setting as compared to traditional pharmacological therapy. The authors recommended it for routine usage in patients with moderate to severe heart failure and manifestations of dyssynchrony. CRT was shown to be cost-effective in patients with hearth failure also in the Brazilian health care system [35]. The study simultaneously investigated cost-effectiveness of an upgrade of CRT to CRT+ICD (CRT + implantable cardioverter defibrilator). However, this upgrade proved only marginally cost-effective in patients with an implanted ICD [44].

CRT defibrilators (CRT-D) may offer a higher chance of survival from the sudden cardiac death as compared with CRT pacemakers (CRT-P), the increase in clinical effect is significant and justifies the three times higher price of the CRT-D devices [45]. When the cost and effectiveness of CRT and CRT-D are compared with the optimum pharmaceutical therapy, the estimated net benefit from CRT-D is less than in two other strategies, while the ICER threshold exceeds 40 160 GBP/QALY [46].

Implantable Cardioverter Defibrilator (ICD). The implantable cardioverter defibrilator is a machine that is used for the treatment of patients suffering from hearth arrhythmias and a sudden death [47]. This treatment proved to work and it is well received not only for secondary prevention, but also as the primary prevention for patients with a high risk of hearth´s death. The primary prevention can be described as an effort to make an impact on the cause of the disease [48]. In the systematic review by Garcia et al. [49], there are mentioned several studies describing different aspects of ICD for the primary prevention. The cost-effectiveness of ICD as the primary prevention shows some uncertainty.

Recent European surveys describing the primary prevention area brought different conclusions. Smith et al. [50] claim that it is possible to consider ICD as cost effective for patients suffering from both ischemic and non-ischemic heart disease and ejection fraction of the left heart chamber ≤ 0.40. On the other hand, Gandjour et al. [51] conclude that the treatment with ICD is not cost effective for the primary prevention of the sudden cardiac death in patients suffering from heart attack of myocardium and ejection fraction of the heart´s left chamber ≤ 0.3.

Ablation. Vida et al. [52] compared cost-effectiveness of radiofrequency catheter ablation (RFCA) in children and adolescents compared to medical treatment of supraventricular tachycardia (SVT) in a low-income country (Guatemala). The results indicate that radiofrequency catheter ablation of SVT is safe and cost-effective as compared to pharmacological therapy in children and adolescents.

Another study about catheter ablation investigates whether simplified 3-catheter ablation approach is a safe and effective method for ablation of paroxysmal supraventricular tachycardia. However, further studies are needed to assess cost-effectiveness of this simplified procedure [53].

Extracorporeal Membrane Oxygenation (ECMO). ECMO is an expensive intervention and the length of hospital stay and the in-hospital mortality are predictors of total hospital costs [54, 55]. The results of cost-effectiveness for ECMO are variable. Roos et al. [56] claims that it is a cost-effective intervention in patients with high-risk PCI, and the ICER value is below the discussed threshold. Also another study by Graf et al. [57] indicates that the costs are reasonable in the range of provided treatments. In contrast, Clarke et al. [58] in his study presented results that ECMO is not cost-effective (due to too high ICER). Despite variable resulting ICER values, NHS unofficially listed these practices as recommended, though. Some other studies are focused on application of these systems in childcare. In their review study (including four studies) [59], Mugford et al. demonstrated a significant effect of ECMO on mortality. The use of ECMO is cost-effective in a comparable measure as other intensive care technologies in common use. Petrou et al. in their studies [60, 61] provide rigorous evidence of cost effectiveness of ECMO for mature infants with severe respiratory failure based on four-year experience.

4   Conclusions

The main trend in devices for cardiology is to avoid invasive technologies during the treatment. This trend is obvious both in diagnostic and therapeutic devices. In the diagnostics there is above all an effort to replace invasive coronary angiography, the former golden standard of CAD, by less invasive procedures like CT angiography, or to use PET or SPECT [11]. However, the published studies suffer excessive heterogeneity, and other large randomized controlled trials are needed to confirm the cost efficiency [22]. For PET and SPECT, the studies are focused rather on clinical efficiency than on assessing the cost-effectiveness. Shaw et al. [28] and Meyer et al. [10] showed higher costs of PET and SPECT per physical examination as compared with CT. In spite of the effort to replace ICA by non-invasive methods, ICA stays the predominant method when stent implantation is decided [23]. The great advantage of ICA is the possibility of an intervention immediately after the diagnostic phase. A new trend in ICA is an effort to transfer as many ICA examinations as possible to ambulatory conditions. There is space for a decrease in ICA costs due to zero hospitalization costs, while the safety level is maintained as showed Matchin et al. [24].

ECHO is applied to diagnose low-risk chest-pain patients and patients in a low risk pacing with suspected coronary artery disease (Bedetti et al. [33], Shaw et al. [28]). A great advantage of ECHO diagnostics is avoiding radiation exposure and the danger of suffering during a more stressful examination [33]. Presenting no radiation stress, ECHO is the ideal modality to diagnose valve disease in children. Screening among siblings suffering from bicuspid aortic valve disease can serve as an example. Hales et al. [37] proved that ECHO is both clinically and cost effective. Another stress-free examination in cardiology is the basic ECG. It is widely used in the prevention of the sudden cardiac death both in athletes and non-athletes. However, Anderson et al. [32] suggests that in order to determine the cost-effectiveness of ECG screening programs, it is necessary to better understand the mechanism of the sudden cardiac death.

As compared to classic (open chest) surgery, the treatment of cardio-vascular diseases is moving to less invasive interventions in case of the ischemic heart disease, heart valve disease, birth defects, as well as heart rhythm disorders. New interventions are careful and considerate to patients, and are less time-consuming both during the intervention and in the phase of recovery. Accurate imaging methods are important for an execution of these complex interventional procedures in catheterization rooms [39-41]. Quite recent papers raise the cost-effectiveness question of TAVI versus SAVR. Neyt et al. [40] and Fairbairn et al. [41] suggested that TAVI shows better cost efficiency for non-operable patients. In the resynchronization therapy, the main interest is to prove its better cost efficiency as compared with the pharmacological treatment. Blomstrom et al. [43] proved cost-effectiveness superiority of the resynchronization therapy. Also CRT-D offers a relatively higher chance of survival and an increase in clinical-effectiveness, which might justify a three times higher price of the apparatus, when cost-effectiveness is figured out [45].

In the field of ICD, the trend is to use it in the primary care [48, 49]. The indication of a primary prophylactic ICD was expanded to patients suffering from non-ischemic cardiomyopathy with left ventricular ejection fraction less than 30%. The results of studies evaluating cost-effectiveness of ICD implementation in the primary prevention are contradictory [50, 51]. It is necessary to carry out larger studies and gain more evidence to decide. ECMO can serve as an example that despite a high ICER, the clinical effectiveness and irreplaceability by other therapies justify the high costs [54-56]. Other authors prove the cost-effectiveness of ECMO in children [59-61].

Advances in telecommunication technology allow implementing new and more effective strategies to monitoring, screening and treatment of a wide range of disorders in cardiology. De Waure et al. [62] concluded in their systematic review that telemedicine can improve the health outcomes in patients with CAD. Mistry et al. [63] suggest efficiency of teleconsultations in screening of the congenital heart disease. However, De Waure et al. pointed on heterogeneity in study designs and endpoints of most studies and a limited number of papers focused on the topic. There is for sure a great potential for the development of these technologies; their cost-effectiveness shall depend on process management and should be carefully established.

Acknowledgement. This research was supported by research grants NT11532 "Medical Technology Assessment" and NT14473 "Information system for medical devices purchase monitoring". Radka Otawova was supported by the European social fund within the framework of realizing the project "Support of inter-sectoral mobility and quality enhancement of research teams at Czech Technical University in Prague", CZ.1.07/2.3.00/30.0034.

References

1. Go, A.S., Mozaffarian, D., Roger, V.L., Benjamin, E.J., Berry, J.D., Blaha, M.J., et al.: Heart disease and stroke statistics–2014 update: a report from the American Heart Association. Circulation 129, e28–e292 (2014)
2. WHO: World Health Organization, http://www.who.int
3. Odden, M.C., Coxson, P.G., Moran, A., Lightwood, J.M., Goldman, L., Bibbins-Domingo, K.: The impact of the Aging Population on Coronary Heart Disease in the U.S. The American Journal of Medicine 124,827–833 (2012)
4. Nichols, M., Townsend, N., Luengo-Fernandez, R., Leal, J., Gray, A., Scarborough, P., Rayner, M.: European Cardiovascular Disease Statistics 2012. European Heart Network and European Society of Cardiology, Brussels and Sophia Antipolis (2012)
5. Drummond, M., Griffin, A., Tarricone, R.: Economic evaluation for devices and drugs, same or different? Value in Health 12, 402–404 (2009)
6. Rosina, J., Rogalewicz, V., Ivlev, I., Juřičková, I., Donin, G., Jantosová, N., Vacek, J., Otawová, R., Kneppo, P.: Health Technology Assessment for Medical Devices. Lekar a Technika - Clinician and Technology 44, 23–36 (2014)
7. Rogalewicz, V., Jurickova, I.: Specificities of Medical Devices Affecting Health Technology Assessment Methodology. Proceedings Iwbbio 2014: International Work-Conference on Bioinformatics and Biomedical Engineering, vols. 1 and 2, pp. 1229–1234 (2014)
8. Giraldes, M.D.R.: Evaluation of New Technologies PET/CT Nuclear Imaging. Acta Medica Portuguesa 23, 291–310 (2010)
9. Hachamovitch, R., Johnson, J.R., Hlatky, M.A., Cantagallo, L., Johnson, B.H., Coughlan, M., Hainer, J., Gierbolini, J., di Carli, M.F., Investigators, S.: The study of myocardial perfusion and coronary anatomy imaging roles in CAD (SPARC): design, rationale, and baseline patient characteristics of a prospective, multicenter observational registry comparing PET, SPECT, and CTA for resource utilization and cl. Journal of Nuclear Cardiology 16, 935–948 (2009)
10. Meyer, M., Nance, J.W.J., Schoepf, U.J., Moscariello, A., Weininger, M., Rowe, G.W., Ruzsics, B., Kang, D.K., Chiaramida, S.A., Schoenberg, S.O., Fink, C., Henzler, T.: Cost-effectiveness of substituting dual-energy CT for SPECT in the assessment of myocardial perfusion for the workup of coronary artery disease. European Journal of Radiology 81, 3719–3725 (2012)
11. Pelberg, R.A., Mazur, W., Clarke, G., Szawaluk, J.: The What and Why of Cardiac CT Angiography: Data Interpretation and Clinical Practice Integration. Reviews in Cardiovascular Medicine 10, 152–163 (2009)
12. Mowatt, G., Cummins, E., Waugh, N., Walker, S., Cook, J., Jia, X., Hillis, G., Fraser, C.: Systematic review of the clinical effectiveness and cost-effectiveness of 64-slice or higher computed tomography angiography as an alternative to invasive coronary angiography in the investigation of coronary artery disease. Health Technology Assessment 12(17) (2008)

13. Catalan, P., Callejo, D., Juan, A.B.: Cost-effectiveness analysis of 64-slice computed tomography vs. cardiac catheterization to rule out coronary artery disease before noncoronary cardiovascular surgery. European Heart Journal-Cardiovascular Imaging 14, 149–156 (2013)
14. Dorenkamp, M., Bonaventura, K., Sohns, C., Becker, C.R., Leber, A.W.: Direct costs and cost-effectiveness of dual-source computed tomography and invasive coronary angiography in patients with an intermediate pretest likelihood for coronary artery disease. Heart 98, 460–467 (2012)
15. Ladapo, J.A., Jaffer, F.A., Hoffmann, U., Thomson, C.C., Bamberg, F., Dec, W., Cutler, D.M., Weinstein, M.C., Gazelle, G.S.: Clinical outcomes and cost-effectiveness of coronary computed tomography angiography in the evaluation of patients with chest pain. Journal of the American College of Cardiology 54, 2409–2422 (2009)
16. Dewey, M., Hamm, B.: Cost effectiveness of coronary angiography and calcium scoring using CT and stress MRI for diagnosis of coronary artery disease. European Radiology 17, 1301–1309 (2007)
17. Halpern, E.J., Savage, M.P., Fischman, D.L., Levin, D.C.: Cost-effectiveness of coronary CT angiography in evaluation of patients without symptoms who have positive stress test results. American Journal of Roentgenology 194, 1257–1262 (2010)
18. Kreisz, F.P., Merlin, T., Moss, J., Atherton, J., Hiller, J.E., Gericke, C.A.: The pre-test risk stratified cost-effectiveness of 64-slice computed tomography coronary angiography in the detection of significant obstructive coronary artery disease in patients otherwise referred to invasive coronary angiography. Heart, Lung & Circulation 18, 200–207 (2009)
19. Stacul, F., Sironi, D., Grisi, G., Belgrano, M., Salvi, A., Cova, M.: 64-Slice CT coronary angiography versus conventional coronary angiography: activity-based cost analysis. Radiologia Medica 114, 239–252 (2009)
20. Min, J.K., Gilmore, A., Budoff, M.J., Berman, D.S., O'Day, K.: Cost-effectiveness of coronary CT angiography versus myocardial perfusion SPECT for evaluation of patients with chest pain and no known coronary artery disease. Radiology 254, 801–808 (2010)
21. Genders, T.S.S., Ferket, B.S., Dedic, A., Galema, T.W., Mollet, N.R.A., de Feyter, P.J., Fleischmann, K.E., Nieman, K., Hunink, M.G.M.: Coronary computed tomography versus exercise testing in patients with stable chest pain: comparative effectiveness and costs. International Journal of Cardiology 167, 1268–1275 (2013)
22. Nance, J.W., Bamberg, F., Schoepf, U.J.: Coronary computed tomography angiography in patients with chronic chest pain: systematic review of evidence base and cost-effectiveness. Journal of Thoracic Imaging 27, 277–288 (2012)
23. Fearon, W.F., Tonino, P.A.L., de Bruyne, B., Siebert, U., Pijls, N.H.J.: Rationale and design of the fractional flow reserve versus angiography for multivessel evaluation (FAME) study. American Heart Journal 154, 632–636 (2007)
24. Matchin, Y.G., Privalova, O.B., Privalov, D.V., Zateyshchikov, D.A., Boytsov, S.A.: New approach towards diagnostic coronarography at hospitals without their own angiography laboratories. Cardiovascular Therapy and Prevention 7, 51–57 (2008)
25. Miller, T.D., Hodge, D.O., Milavetz, J.J., Gibbons, R.J.: A normal stress SPECT scan is an effective gatekeeper for coronary angiography. Journal of Nuclear Cardiology 14, 187–193 (2007)
26. Le Guludec, D., Lautamaki, R., Knuuti, J., Bax, J., Bengel, F., Cardiology, E.C.N.: Present and future of clinical cardiovascular PET imaging in Europe - a position statement by the European Council of Nuclear Cardiology (ECNC). European Journal of Nuclear Medicine and Molecular Imaging 35, 1709–1724 (2008)

27. Marcassa, C., Bax, J.J., Bengel, F., Hesse, B., Petersen, C.L., Reyes, E., Underwood, R.: Clinical value, cost-effectiveness, and safety of myocardial perfusion scintigraphy: a position statement. European Heart Journal 29, 557–563 (2008)

28. Shaw, L.J., Marwick, T.H., Berman, D.S., Sawada, S., Heller, G.V., Vasey, C., Miller, D.D.: Incremental cost-effectiveness of exercise echocardiography vs. SPECT imaging for the evaluation of stable chest pain. European Heart Journal 27, 2448–2458 (2006)

29. Papadakis, M., Sharma, S.: Electrocardiographic screening in athletes: the time is now for universal screening. British Journal of Sports Medicine 43, 663–668 (2009)

30. Borjesson, M., Dellborg, M.: Is There Evidence for Mandating Electrocardiogram as Part of the Pre-Participation Examination? Clinical Journal of Sport Medicine 21, 13–17 (2011)

31. Chandra, N., Rachel, B., Papadakis, M.F., Panoulas, V.F., Ghani, S., Dusche, J., Folders, D., Raju, H., Osborne, R., Sharma, S.: Prevalence of Electrocardiographic Anomalies in Young Individuals. Journal of the American College of Cardiology 63, 2028–2034 (2014)

32. Anderson, B.R., McElligott, S., Polsky, D., Vetter, V.L.: Electrocardiographic Screening for Hypertrophic Cardiomyopathy and Long QT Syndrome: The Drivers of Cost-Effectiveness for the Prevention of Sudden Cardiac Death. Death. Pediatric Cardiology 35, 323–331 (2014)

33. Bedetti, G., Pasanisi, E.M., Pizzi, C., Turchetti, G., Lore, C.: Economic analysis including long-term risks and costs of alternative diagnostic strategies to evaluate patients with chest pain. Cardiovascular Ultrasound 6(21) (2008)

34. Simpson, E.L., Stevenson, M.D., Scope, A., Poku, E., Minton, J., Evans, P.: Echocardiography in newly diagnosed atrial fibrillation patients: a systematic review and economic evaluation. Health Technology Assessment 17(36) (2013)

35. Kitada, R., Fukuda, S., Watanabe, H., Oe, H., Abe, Y., Yoshiyama, M., Song, J.-M., Sitges, M., Shiota, T., Ito, H., Yoshikawa, J.: Diagnostic Accuracy and Cost-Effectiveness of a Pocket-Sized Transthoracic Echocardiographic Imaging Device. Clinical Cardiology 36, 603–610 (2013)

36. Trambaiolo, P., Papetti, F., Posteraro, A., Amici, E., Piccoli, M., Cerquetani, E., Pastena, G., Gambelli, G., Salustri, A.: A hand-carried cardiac ultrasound device in the outpatient cardiology clinic reduces the need for standard echocardiography. Heart 93, 470–475 (2007)

37. Hales, A.R., Mahle, W.T.: Echocardiography Screening of Siblings of Children With Bicuspid Aortic Valve. Pediatrics 133,E1212–E1217 (2014)

38. Reardon, M.J., Reynolds, M.R.: Cost-effectiveness of TAVR in the non-surgical population. Journal of Medical Economics 16, 575–579 (2013)

39. Reynolds, M.R., Magnuson, E.A., Lei, Y., Wang, K., Vilain, K., Li, H., Walczak, J., Pinto, D.S., Thourani, V.H., Svensson, L.G., Mack, M.J., Miller, D.C., Satler, L.E., Bavaria, J., Smith, C.R., Leon, M.B., Cohen, D.J.: Cost-effectiveness of transcatheter aortic valve replacement compared with surgical aortic valve replacement in high-risk patients with severe aortic stenosis: results of the PARTNER (Placement of Aortic Transcatheter Valves) trial (Cohort A). Journal of the American College of Cardiology 60, 2683–2692 (2012)

40. Neyt, M., van Brabandt, H., Devriese, S., van de Sande, S.: A cost-utility analysis of transcatheter aortic valve implantation in Belgium: focusing on a well-defined and identifiable population. BMJ Open 2 (2012)

41. Fairbairn, T.A., Meads, D.M., Hulme, C., Mather, A.N., Plein, S., Blackman, D.J., Greenwood, J.P.: The cost-effectiveness of transcatheter aortic valve implantation versus surgical aortic valve replacement in patients with severe aortic stenosis at high operative risk. Heart 99, 914–920 (2013)

42. Doble, B., Blackhouse, G., Goeree, R., Xie, F.: Cost-effectiveness of the Edwards SAPIEN transcatheter heart valve compared with standard management and surgical aortic valve replacement in patients with severe symptomatic aortic stenosis: a Canadian perspective. Journal of Thoracic and Cardiovascular Surgery 146, 52–60.e53 (2013)
43. Blomstrom, P., Ekman, M., Lundqvist, C., Calvert, M.J., Freemantle, N., Lonnerholm, S., Wikstrom, G., Jonsson, B.: Cost effectiveness of cardiac resynchronization therapy in the Nordic region: An analysis based on the CARE-HF trial. European Journal of Heart Failure 10, 869–877 (2008)
44. Bertoldi, E.G., Rohde, L.E., Zimerman, L.I., Pimentel, M., Polanczyk, C.A.: Cost-effectiveness of cardiac resynchronization therapy in patients with heart failure: the perspective of a middle-income country's public health system. International Journal of Cardiology 163, 309–315 (2013)
45. Neyt, M., Stroobandt, S., Obyn, C., Camberlin, C., Devriese, S.: se Laet, C., van Brabandt, H.: Cost-effectiveness of cardiac resynchronisation therapy for patients with moderate-to-severe heart failure: a lifetime Markov model. BMJ Open 1,e000276 (2011)
46. Fox, M., Mealing, S., Anderson, R., Dean, J., Stein, K., Price, A., Taylor, R.: The clinical effectiveness and cost-effectiveness of cardiac resynchronisation (biventricular pacing) for heart failure: systematic review and economic model. Health Technology Assessment 11(47) (2007)
47. IKEM: Implantation of cardioverter-defibrilator (ICD) (in Czech), http://www.ikem.cz/www?docid=1004024
48. Boriani, G., Cimaglia, P., Biffi, M., Martignani, C., Ziacchi, M., Valzania, C., Diemberger, I.: Cost-effectiveness of implantable cardioverter-defibrillator in today's world. Indian Heart Journal 66(suppl. 1), S101–104 (2014)
49. Garcia-Perez, L., Pinilla-Dominguez, P., Garcia-Quintana, A., Caballero-Dorta, E., Garcia-Garca, F.J., Linertova, R., Imaz-Iglesia, I.: Economic evaluations of implantable cardioverter defibrillators: a systematic review. European Journal of Health Economics (October 2014)
50. Smith, T., Jordaens, L., Theuns, D.A.M.J., van Dessel, P.F., Wilde, A.A., Hunink, M.G.M.: The cost-effectiveness of primary prophylactic implantable defibrillator therapy in patients with ischaemic or non-ischaemic heart disease: a European analysis. European Heart Journal 34, 211–219 (2013)
51. Gandjour, A., Holler, A., Adarkwah, C.C.: Cost-effectiveness of implantable defibrillators after myocardial infarction based on 8-year follow-up data (MADIT II). Value in Health 14, 812–817 (2011)
52. Vida, V.L., Calvimontes, G.S., Macs, M.O., Aparicio, P., Barnoya, J., Castaneda, A.R.: Radiofrequency catheter ablation of supraventricular tachycardia in children and adolescents - Feasibility and cost-effectiveness in a low-income country. Pediatric Cardiology 27, 434–439 (2006)
53. Wang, L.X., Li, J.T., Yao, R.G., Song, S.K.: Simplified approach to radiofrequency catheter ablation of paroxysmal supraventricular tachycardia. Croatian Medical Journal 45, 167–170 (2004)
54. Mishra, V., Svennevig, J.L., Bugge, J.F., Andresen, S., Mathisen, A., Karlsen, H., Khushi, I., Hagen, T.P.: Cost of extracorporeal membrane oxygenation: evidence from the Rikshospitalet University Hospital, Oslo, Norway. European Journal of Cardio-thoracic Surgery 37, 339–342 (2010)
55. Tseng, Y.H., Wu, M.Y., Tsai, F.C., Chen, H.J., Lin, P.J.: Costs Associated with Extracorporeal Life Support Used in Adults: A Single-Center Study. Acta Cardiologica Sinica 27, 221–228 (2011)

56. Roos, J.B., Doshi, S.N., Konorza, T., Palacios, I., Schreiber, T., Borisenko, O.V., Henriques, J.P.S.: The cost-effectiveness of a new percutaneous ventricular assist device for high-risk PCI patients: mid-stage evaluation from the European perspective. Journal of Medical Economics 16, 381–390 (2013)

57. Graf, J., Muhlhoff, C., Doig, G.S., Reinartz, S., Bode, K., Dujardin, R., Koch, K.-C., Roeb, E., Janssens, U.: Health care costs, long-term survival, and quality of life following intensive care unit admission after cardiac arrest. Critical Care 12,R92 (2008)

58. Clarke, A., Pulikottil-Jacob, R., Connock, M., Suri, G., Kandala, N.-B., Maheswaran, H., Banner, N.R., Sutcliffe, P.: Cost-effectiveness of left ventricular assist devices (LVADs) for patients with advanced heart failure: analysis of the British NHS bridge to transplant (BTT)program. International Journal of Cardiology 171, 338–345 (2014)

59. Mugford, M., Elbourne, D., Field, D.: Extracorporeal membrane oxygenation for severe respiratory failure in newborn infants. The Cochrane Database of Systematic Reviews CD001340 (2008)

60. Petrou, S., Edwards, L.: Cost effectiveness analysis of neonatal extracorporeal membrane oxygenation based on four year results from the UK Collaborative ECMO Trial. Archives of Disease in Childhood. Fetal and Neonatal Edition 89, F263–F268 (2004)

61. Petrou, S., Bischof, M., Bennett, C., Elbourne, D., Field, D., McNally, H.: Cost-effectiveness of neonatal extracorporeal membrane oxygenation based on 7-year results from the United Kingdom Collaborative ECMO Trial. Pediatrics 117, 1640–1649 (2006)

62. de Waure, C., Cadeddu, C., Gualano, M.R., Ricciardi, W.: Telemedicine for the reduction of myocardial infarction mortality: a systematic review and a meta-analysis of published studies. Telemedicine and e-Health 18, 323–328 (2012)

63. Mistry, H., Gardiner, H.M.: The cost-effectiveness of prenatal detection for congenital heart disease using telemedicine screening. Journal of Telemedicine and Telecare 19, 190–196 (2013)

Preliminary Research on Combination of Exponential Wavelet and FISTA for CS-MRI

Yudong Zhang[1,3], Shuihua Wang[1,2], Genlin Ji[1], Zhengchao Dong[3], and Jie Yan[4],

[1] School of Computer Science and Technology, Nanjing Normal University, Nanjing,
Jiangsu 210023, China
[2] School of Electronic Science and Engineering, Nanjing University, Nanjing,
Jiangsu 210046, China
[3] Translational Imaging Division & MRI Unit, Columbia University and NYSPI,
New York, NY 10032, USA
[4] Department of Applied Physics, Stanford University, Stanford, CA 94305, USA
zhangyudong@njnu.edu.cn

Abstract. Compressed sensing magnetic resonance imaging (CS-MRI) is a hot topic in the field of medical signal processing. However, it suffers from low-quality reconstruction and long computation time. In this preliminary research, we took a study of applying exponential wavelet as a sparse representation to the conventional fast iterative shrinkage/threshold algorithm (FISTA) for the reconstruction of CSMRI scans. The proposed method was termed exponential wavelet iterative shrinkage/threshold algorithm (EWISTA). Simulation results demonstrated EWISTA was superior to existing algorithms w.r.t. reconstruction quality and computation time.

Keywords: Fast Iterative Shrinkage/Threshold Algorithm, Exponential Wavelet Transform, Magnetic Resonance Imaging, Compressed Sensing.

1 Introduction

Magnetic resonance imaging (MRI) is a medical imaging technique to investigate the anatomy and physiology of the body in both health and disease [1-4]. MRI scanners use strong magnetic fields to form images of the body [5-8]. The technique is widely used in hospitals for medical diagnosis, staging of disease and for follow-up without exposure to ionizing radiation [9-12].

Recently, fast MRI technique [13] based on compressed sensing (CS) [14-16] was developed that permits the acquisition of a small number of k-space data points at random rates lower than Nyquist sampling rate [17-19]. The random undersampling employed in compressed sensing magnetic resonance imaging (CS-MRI) produces massive and random aliasing, and thus the reconstruction of the CS-MRI is in essence a procedure.

Various algorithms were proposed for solving CS-MRI problem in last decade. Total Variation (TV) employs the sum of Euclidean norms of the gradient [20], choosing finite difference as sparsifying transform. TV is thus only suitable for

F. Ortuño and I. Rojas (Eds.): IWBBIO 2015, Part I, LNCS 9043, pp. 175–182, 2015.

piecewise-constant objects. The iterative shrinkage thresholding algorithm (ISTA) is a nonparametric method that is proven to converge mathematically. However, the slow convergence prevents its widespread use [21]. Fast ISTA (FISTA) exploits the result of past iterations to speed up convergence [22]. Subband adaptive ISTA (SISTA) optimizes wavelet-subband-dependent parameters [23].

In this study, we employed the exponent of wavelet transform (EWT) to replace conventional wavelet transform (WT) as an more efficient way for sparse representation [24]. Subsequently, we proposed a novel exponential wavelet iterative shrinkage/threshold algorithm (EWISTA), taking advantages of both the simplicity of FISTA and the sparse representation of EWT, aimed at improving the reconstruction quality and accelerating computation.

2   Methodology

The EWT is obtained as

$$\Psi_{EWT}(x \mid n) = \left(\left(\frac{\exp(\omega)-1}{e-1} \right)_{n} \ldots \right) \tag{1}$$

where x denotes the spatial image to be reconstructed, ω denotes the sparsity coefficients. The conventional WT is written as

$$x = W\omega \tag{2}$$

where W denotes the sparsity transform. The formula above show the sparsity transform W bijectively maps the spatial image x to the sparsity domain ω. Afterwards, the data formation model of MRI scanner is

$$y = Fx + e \tag{3}$$

where F denotes the Fourier transform corresponding to the k-space undersampling scheme, y denotes the measured k-space data from the MRI scanner, and e denotes the effect on measurements of noise and scanner imprecisions. In another form,

$$y = M\omega + e \tag{4}$$

where $M = FW$. The reconstruction of CS-MRI is obtained by solving the following constrained optimization problem [25]:

$$\omega^* = \arg\min_{\omega} C(\omega) = \arg\min_{\omega} \left(\|y - M\omega\|_2^2 + \lambda \|\omega\|_1 \right) \tag{5}$$

where λ controls the fidelity of the reconstruction to the measured data.

We define Γ_τ the shrinkage operator with threshold τ

$$\Gamma_\tau(u) = \mathrm{sgn}(u)\left(|u| - \min(\tau/2, |u|) \right) \tag{6}$$

The pseudocodes of the procedure of the proposed EWISTA are described in Table 1.

Table 1. Pseudocodes of EWISTA

Input: ω_0

$\Lambda^{-1} = \text{diag}(\tau)$, $M = FW_{EWT}$

Initialization: $n = 0$, $v_0 = \omega_0$, $t_0 = 1$

Repeat

$$\omega_{n+1} = \mathcal{T}_{\lambda\tau}\left(v_n + \Lambda^{-1}\left(a - Av_n\right)\right)$$

$$t_{n+1} = (1 + \sqrt{1 + 4t_n^2})/2$$

$$v_{n+1} = \omega_{n+1} + (t_n - 1)(\omega_{n+1} - \omega_n)/t_{n+1}$$

$$n = n + 1$$

Until termination criteria is reached

Output: ω_{n+1}

---

3   Experiments and Results

We compared the proposed EWISTA with Fast composite splitting algorithm (FCSA), ISTA, FISTA, and SISTA. The coefficients normalization is added to all methods for fair comparison. In addition, we analyze the convergence performance of the proposed EWISTA. All the algorithms are in-house developed using Matlab 2013a on a HP laptop.

We used a MR brain image diagnosed as Meningioma, and an ankle MR image. Their sizes are identical of 256x256. Both acceleration factors are set as 5. We chose bior4.4 wavelet with 5 decomposition level. Parameter n of EWISTA is chosen as 6. Gaussian white noise with standard deviation 0.01 was added to the k-space measurements. The maximum iteration number of each algorithm was set as 100.

We used the median square error (MSE), median absolute error (MAE), and peak signal to noise ratio (PSNR) to evaluate the quality between the reconstructed image and the ground-truth image. Less MSE, less MAE, or larger PSNR indicates a better reconstruction [26, 27].

Figure 1 and 2 showed the reconstruction quality by different algorithms for the brain and ankle images, respectively. They indicated clearly that EWISTA was more effective in terms of preserving edges and textures, and suppressing noises. Table 2 gave the quantitative measure, which demonstrated the proposed EWISTA was superior to existing algorithms with regard to reconstruction quality. In addition, the computation time of EWISTA was faster than ISTA, and slightly slower than FISTA and SISTA.

We applied the proposed EWISTA algorithm to a 256-by-256 brain MR image diagnosed as ADHD. The acceleration factor was set to 3. We recorded the reconstruction and the error map every 9 steps in Figure 3.

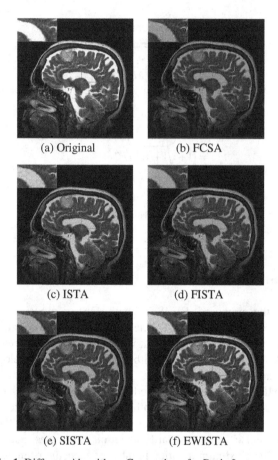

(a) Original   (b) FCSA

(c) ISTA   (d) FISTA

(e) SISTA   (f) EWISTA

Fig. 1. Different Algorithms Comparison for Brain Image

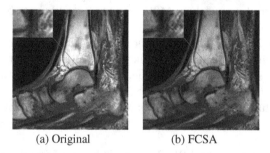

(a) Original   (b) FCSA

Fig. 2. Different Algorithms Comparison for Ankle Image

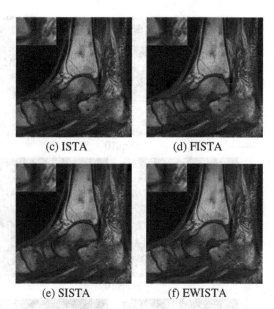

(c) ISTA   (d) FISTA

(e) SISTA   (f) EWISTA

Fig. 2. *(Continued)*

Table 2. Comparisons of different algorithms

Brain	FCSA	ISTA	FISTA	SISTA	EWISTA
MAE	6.53	5.01	4.65	4.73	4.49
MSE	103.86	49.96	45.48	45.70	37.19
PSNR (dB)	27.97	31.14	31.55	31.53	32.43
Time (s)	7.17	10.85	6.78	6.92	8.37
Ankle	FCSA	ISTA	FISTA	SISTA	EWISTA
MAE	4.02	3.02	3.00	2.97	2.74
MSE	32.40	19.53	19.20	18.99	14.34
PSNR (dB)	33.03	35.22	35.30	35.35	36.57
Time (s)	8.24	12.21	9.18	8.84	9.94

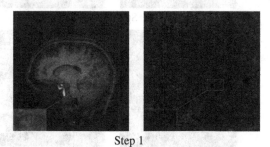

Step 1

Fig. 3. Reconstructions and error maps of the proposed EWISTA method

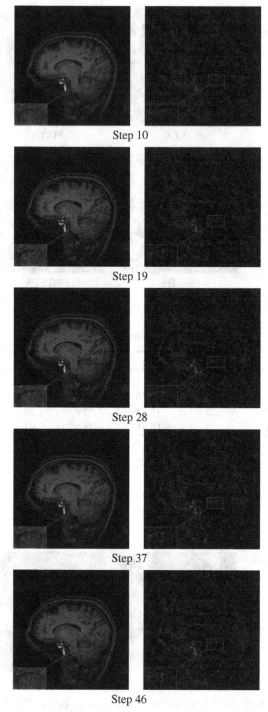

Step 10

Step 19

Step 28

Step 37

Step 46

Fig. 3. *(Continued)*

We observed in Figure 3 that the proposed EWISTA reconstructed the ADHD MR image gradually, and the quality increase was reducing. The enlarged area covered various brain parts that were entangled with adjacent parts, which made them not easily distinguished. The error map appeared to have a tendency that decreased to its minimum. Moreover, the artifacts remained at the early steps were hardly perceived in the late steps.

4   Conclusion

We proposed a novel EWISTA algorithm that took advantage of the simplicity of FISTA and the sparse representation of EWT. The experimental results showed that the proposed method EWISTA noticeably improved the reconstruction quality with somewhat compromised computation time compared with existing methods. In addition, the proposed EWISTA was convergent.

Acknowledgment. This work was supported from NSFC (No. 610011024), Program of Natural Science Research of Jiangsu Higher Education Institutions of China (No. 14KJB520021) and Nanjing Normal University Research Foundation for Talented Scholars (No. 2013119XGQ0061).

References

1. Pérez-Palacios, T., Caballero, D., Caro, A., Rodríguez, P.G., Antequera, T.: Applying data mining and Computer Vision Techniques to MRI to estimate quality traits in Iberian hams. Journal of Food Engineering 131, 82–88 (2014)
2. Hamy, V., Dikaios, N., Punwani, S., Melbourne, A., Latifoltojar, A., Makanyanga, J., Chouhan, M., Helbren, E., Menys, A., Taylor, S., Atkinson, D.: Respiratory motion correction in dynamic MRI using robust data decomposition registration – Application to DCE-MRI. Medical Image Analysis 18, 301–313 (2014)
3. Zhang, Y., Wang, S., Ji, G., Dong, Z.: An improved quality guided phase unwrapping method and its applications to MRI. Progres. Electromagnetics Research 145, 273–286 (2014)
4. Goh, S., Dong, Z., Zhang, Y., DiMauro, S., Peterson, B.S.: Mitochondrial dysfunction as a neurobiological subtype of autism spectrum disorder: Evidence from brain imaging. JAMA psychiatry 71, 665–671 (2014)
5. Sotiropoulos, S.N., Jbabdi, S., Andersson, J.L., Woolrich, M.W., Ugurbil, K., Behrens, T.E.J.: RubiX: Combining Spatial Resolutions for Bayesian Inference of Crossing Fibers in Diffusion MRI. IEEE Transactions on Medical Imaging 32, 969–982 (2013)
6. Lau, D., Zhihao, C., Ju Teng, T., Soon Huat, N., Rumpel, H., Yong, L., Hui, Y., Pin Lin, K.: Intensity-Modulated Microbend Fiber Optic Sensor for Respiratory Monitoring and Gating During MRI. IEEE Transactions on Biomedical Engineering 60, 2655–2662 (2013)
7. Zhang, Y., Wang, S., Dong, Z.: Classification of Alzheimer Disease Based on Structural Magnetic Resonance Imaging by Kernel Support Vector Machine Decision Tree. Progress In Electromagnetics Research 144, 171–184 (2014)
8. Dong, Z., Zhang, Y., Liu, F., Duan, Y., Kangarlu, A., Peterson, B.S.: Improving the spectral resolution and spectral fitting of 1H MRSI data from human calf muscle by the SPREAD technique. NMR in Biomedicine 27, 1325–1332 (2014)

9. Cho, Z.-H., Han, J.-Y., Hwang, S.-I., Kim, D.-S., Kim, K.-N., Kim, N.-B., Kim, S.J., Chi, J.-G., Park, C.-W., Kim, Y.-B.: Quantitative analysis of the hippocampus using images obtained from 7.0. T MRI. NeuroImage 49, 2134–2140 (2010)

10. Renard, D., Castelnovo, G., Bousquet, P.-J., de Champfleur, N., de Seze, J., Vermersch, P., Labauge, P.: Brain MRI findings in long-standing and disabling multiple sclerosis in 84 patients. Clinical Neurology and Neurosurgery 112, 286–290 (2010)

11. Tetzlaff, R., Schwarz, T., Kauczor, H.-U., Meinzer, H.-P., Puderbach, M., Eichinger, M.: Lung Function Measurement of Single Lungs by Lung Area Segmentation on 2D Dynamic MRI. Academic Radiology 17, 496–503 (2010)

12. Zhang, Y., Dong, Z., Wu, L., Wang, S.: A hybrid method for MRI brain image classification. Expert Systems with Applications 38, 10049–10053 (2011)

13. Kumamoto, M., Kida, M., Hirayama, R., Kajikawa, Y., Tani, T., Kurumi, Y.: Active Noise Control System for Reducing MR Noise. Ieice Transactions on Fundamentals of Electronics Communications and Computer Sciences E94A,1479–1486 (2011)

14. Hirabayashi, A., Sugimoto, J., Mimura, K.: Complex Approximate Message Passing Algorithm for Two-Dimensional Compressed Sensing. Ieice Transactions on Fundamentals of Electronics Communications and Computer Sciences E96A, 2391–2397 (2013)

15. Blumensath, T.: Compressed Sensing With Nonlinear Observations and Related Nonlinear Optimization Problems. IEEE Transactions on Information Theory 59, 3466–3474 (2013)

16. Zhang, Y., Peterson, B.S., Ji, G., Dong, Z.: Energy Preserved Sampling for Compressed Sensing MRI. Computational and Mathematical Methods in Medicine 2014,12 (2014)

17. Lustig, M., Donoho, D., Pauly, J.M.: Sparse MRI: The application of compressed sensing for rapid MR imaging. Magn. Reson. Med. 58, 1182–1195 (2007)

18. Smith, D.S., Xia, L., Gambrell, J.V., Arlinghaus, L.R., Quarles, C.C., Yankeelov, T.E., Welch, E.B.: Robustness of Quantitative Compressive Sensing MRI: The Effect of Random Undersampling Patterns on Derived Parameters for DCE- and DSC-MRI. IEEE Transactions on Medical Imaging 31, 504–511 (2012)

19. Zhang, Y., Wu, L., Peterson, B.S., Dong, Z.: A Two-Level Iterative Reconstruction Method for Compressed Sensing MRI. Journal of Electromagnetic Waves and Applications 25, 1081–1091 (2011)

20. Michailovich, O., Rathi, Y., Dolui, S.: Spatially Regularized Compressed Sensing for High Angular Resolution Diffusion Imaging. IEEE Transactions on Medical Imaging 30, 1100–1115 (2011)

21. Daubechies, I., Defrise, M., De Mol, C.: An iterative thresholding algorithm for linear inverse problems with a sparsity constraint. Communications on Pure and Applied Mathematics 57, 1413–1457 (2004)

22. Beck, A., Teboulle, M.: A Fast Iterative Shrinkage-Thresholding Algorithm for Linear Inverse Problems. SIAM J. Img. Sci. 2, 183–202 (2009)

23. Bayram, I., Selesnick, I.W.: A Subband Adaptive Iterative Shrinkage/Thresholding Algorithm. IEEE Transactions on Signal Processing 58, 1131–1143 (2010)

24. Zhang, Y., Dong, Z., Ji, G., Wang, S.: An improved reconstruction method for CS-MRI based on exponential wavelet transform and iterative shrinkage/thresholding algorithm. Journal of Electromagnetic Waves and Applications 1-12 (2014)

25. Menzel, M.I., Tan, E.T., Khare, K., Sperl, J.I., King, K.F., Tao, X., Hardy, C.J., Marinelli, L.: Accelerated diffusion spectrum imaging in the human brain using compressed sensing. Magn. Reson. Med. 66, 1226–1233 (2011)

26. Ravishankar, S., Bresler, Y.: Sparsifying transform learning for Compressed Sensing MRI. In: 2013 IEEE 10th International Symposium on Biomedical Imaging (ISBI), pp. 17–20 (2013)

27. Zhang, Y., Peterson, B., Dong, Z.: A support-based reconstruction for SENSE MRI. Sensors 13, 4029–4040 (2013)

Accurate Microscopic Red Blood Cell Image Enhancement and Segmentation

Syed Hamad Shirazi[1], Arif Iqbal Umar[1], Nuhman Ul Haq[2], Saeeda Naz[1],
and Muhammad Imran Razzak[3]

[1] Hazara University Mansehra, Pakistan
[2] COMSATS Institute of Information Technology, Abbottabad, Pakistan
[3] King Saud bin Abdulaziz University for Health Sciences, Riyadh, Saudi Arabia
razzakmu@ngha.med.sa

Abstract. Erythrocytes (RBC) are the most common type of blood cell. These cells are responsible for the delivery of oxygen to body tissues. The abnormality in erythrocyte cell affects the physical properties of red cell. It may also decrease the life span of red blood cells which may lead to stroke, anemia and other fatal diseases. Until now, Manual techniques are in practiced for diagnosis of blood cell's diseases. However, this traditional method is tedious, time consuming and subject to sampling error. The accuracy of manual method depends on the expertise of the expert, while the accuracy of automated analyzer depends on the segmentation of objects in microscopic image of blood cell. Despite numerous efforts made for accurate blood cells image segmentation and cell counting in the literature. Still accurate segmentation is difficult due to the complexity of overlapping objects and shapes in microscopic images of blood cells. In this paper we have proposed a novel method for the segmentation of blood cells. We have used wiener filter along with Curvelet transform for image enhancement and noise removal. The snake algorithm and Gram-Schmidt orthogonalization have applied for boundary detection and image segmentation, respectively.

Keywords: RBC, SEM, Segmentation, Wiener filter, Curvelet.

1 Introduction

Computer based tools are becoming vital for the quantitative analysis of peripheral blood samples to help experts in diagnosis of blood cells disorder such as malaria, leukemia, cancer. The segmentation stage is initiated where red blood cells are separated from the background and other objects and cells present in the image. The success of these systems depends on the accuracy of segmentation. That is why different studies focus on the development of accurate image segmentation algorithm. Although previous works give better results for segmentation of sparsely distributed normal red blood cells, only few of these techniques focus the segmentation of touching and overlapping cells. The first step of the blood cells image analysis is the image pre-processing where unwanted noise is removed from the image. For accurate

F. Ortuño and I. Rojas (Eds.): IWBBIO 2015, Part I, LNCS 9043, pp. 183–192, 2015.

segmentation image enhancement is vital. After the blood cells image enhancement, image segmentation process is initiated, in which individual cells are detached from each other and as well as from its background. The segmentation makes possible to extract features of the segmented cells and to discriminate among the normal and the infected blood cells. The traditional techniques for blood cell analysis are tedious and time consuming and also involve human intervention. The results of manual techniques are also dependent on the skills of an expert. The blood cell counting is a blood test that counts number of cells. The low or high cell counts as compare to given threshold helps the doctor or physician in diagnosing the disease related to blood cells. It is also called complete blood cell count, complete blood count (CBC), full blood exam (FBE) or full blood count (FBC). There are two methods in practice: manual blood count and automated blood count.

1.1   Manual Blood Count

It is performed by trained physician who visually asses the blood smears by means of light microscope and in practice since 1950s [3]. This process of blood staining is time consuming and required more human effort. It is also require dedicated infrastructure including specialized instruments, dyes and well trained personnel.

1.2   Automated Blood Count

Automated Image segmentation and pattern recognition techniques are used to analyze different cells by means of microscopic images. Quantifying these cells is very helpful in diagnosis and detection of various diseases like leukemia, cancer and many other fatal diseases. This method has been got attention of researcher in 1980's [1]. The automated methods are more flexible and provide accurate results as compared to the orthodox manual methods. However, still image segmentation of overlapping cells and complex shapes is challenging task for accurate blood cell analysis.

A variety of segmentation algorithms are presented in the literature. When the erythrocytes cells are sparsely distributed and the overlapping objects do not exist then the segmentation is quite easy. Methods like thresholding [2],region growing, and edge detection are usually sufficient for delineating cell boundaries. The results achieved from these straight forward methods can also be improved by the use of mathematical morphology. However in the presence of overlapping objects the automation of the cell segmentation is challenging. The erythrocyte image segmentation is challenging because of several reasons i.e. these cells images having low contrast with variations in shapes. Sometimes the boundaries of cells are missing and vague which makes the segmentation more complex.

In cell-based diagnosis, the hematologists or pathologist always make a decision based on the number of cell count, their distribution and their geometrical features. Abundant of computer vision methods are developed to extract useful information from medical images. Mostly due to the complexity of cells structure and overlapping between the cell boundaries and the background it is very challenging task to analyze it manually. The same problem is also faced during the automation of such systems. Because segmented images are used as input for computer aided systems for detection and diagnosis of disease. The precision of the results depend on the accurately segmented regions of the blood cells. The microscopic blood cell counting is still performed by the hematologists, which is essential for diagnostics of suspicious. The automation of this task is very crucial in order to improve the hematological procedures. The development of such systems not only accelerates the diagnosis process but it also improves the accuracy of detection of the disease.

The segmentation phase requires more attention because it is vital for accurate feature extraction and classification. In microscopic blood cell images segmentation is a challenging task because of the complex nature of the cells and uncertainty in microscopic images. Image segmentation is mostly used as a pre-processing step to locate objects in SEM images. It is the process of partitioning an image to objects, shapes and regions. To elaborate more, image segmentation not only focuses on the discrimination between objects and their background but also on the separation between different regions [4]. The image segmentation techniques can be classified as region based and contour based. For region based segmentation [5], morphological operators and watershed[3] approaches are used.

The process of automated recognition of blood cells in microscopic images usually consists of four major steps include preprocessing, image segmentation, feature extraction and classification as shown in figure 1. In this paper our work is confined to pre-processing and image segmentation.

2   Related Work

In the literature, numerous state of the art techniques have been presented for image segmentation date back 1970s [6]. But unfortunately the studies on microscopic image segmentation are rare. Accurate segmentation is an essential for preserving the shape in order to detect the disease, which depends on the number of cells count and its type exists in the blood cell. The key tasks for blood cell analysis involve extracting all the targeted cells from the complicated background and then segmenting them into their morphological components, such as the nuclei and cytoplasm.

Various types of cells are available in human blood including white blood cell (WBC), red blood cells (RBC), Platelets, transmigrated cells (tissue specific cells) and various combinations of these cells. The white blood cells (leukocytes) defend the body against diseases and viral attacks. The quantification of leukocytes is essential for the detection of disease. Granular (polymorph-nuclear cells) and non-granular (mononuclear cells).The granulocytes contain three types of granules i.e. basophiles, nutrophils and eosinophiles, while the non-granular cells contain lymphocytes and monocytes [7]. The leukocyte cells have no color, by using the process of staining with chemicals leukocyte are made colorful to be visible under the microscope. During the staining process variation in color intensity is produced. Similarly during the acquisition of blood smear images from microscopes the quality of these images may suffer from various types of illumination. These microscopic images may also be affected by camera lenses, exposure of time.

To resolve the illumination issue image enhancement techniques are used as pre-processing. Several illumination and contrast enhancement techniques have been applied in literature. Local histogram equalization for contrast enhancement of parasite and RBC is used in [8]. The authors in [9] used adaptive histogram equalization for image enhancement [10-11]. In [12] the authors performed illumination correction using pre-defined illumination factor. Paraboloid modal for illumination correction is used in [13], to highlight the issues related to the features visibility under white light; polarized filters are used in light sources. While the issues related to color illumination still exist and require preprocessing.

The most important phase in the blood cell analysis is segmentation. Blood cell segmentation and morphological analysis is a challenging problem due to both the complex cell nature uncertainty in microscopic videos. Blood cell segmentation yields a number of single cell images and each cell image is segmented into three regions, cell nucleus, cytoplasm and background. Huge amount of literature focused on the segmentation of white blood cells and differential counting. Edge detection is based on HIS (Hue Saturation Intensity) model is used for segmentation [14]. Color features selection and histogram based thresholding are used for segmentation blood cells cytoplasm and nucleus [15].Non supervised nucleus region detection is performed in [16] before nucleus color segmentation using the G channel from RGB color coordinates.

Blood cell segmentation algorithms can be divided into three categories: traditional segmentation, graph cut based segmentation, and active contour model. Traditional segmentation algorithms are based on water shed methods, thresholding and edge detection. Leukocyte segmentation is a difficult task because the cells are often over-laid with each other. In addition, the color and the intensity of an image often change due to instability of staining. Meanwhile, other factors also results in the difficulty of segmentation, such as: five classes of leukocyte and their different morphologies, light variation and noise. A new idea of leukocyte segmentation and classification in blood smear images is presented in [17]. They have utilized features related to cytoplasm and nucleus. These features are color and shape of the nucleus. They have used SVM [18] for classification.

In automated blood cells analysis the major challenge is segmentation for blood cell, because it affects highly the performances of the classification. Various methods have been presented in literature but still a lot of improvements are required. The major challenge to the microscopic image segmentation is the accurate segmentation of overlapping complex objects and shape variations present in image. In this paper our focus is on the blood cell image enhancement and image segmentation.

3   Methodology

The blood automated blood cells analysis consists of image, acquisition, image pre-processing, image segmentation, classification and feature extraction. The scope of our work is confined to the image pre-processing and image segmentation in this paper. In figure1: the block diagram shows our proposed methodology for automated blood cells analysis.

3.1   Image Smoothening

The blood smear images are widely acquired using bright field microscopy. The image quality is affected by the use of different illuminators like HBO, XBO and LED. Illumination variations degrade the efficiency of both the manual and automated system which may lead to biased analysis of blood smear. Therefore, image preprocessing is required to minimize these variations which will facilitate the image segmentation and it will also improve the accuracy of classification.

Our method starts with the removal of unwanted particles or noise present in the image using wiener filter. The purpose of wiener filter is to reduce noise but preserves the edges. Wiener filter is statistical in nature as it adopts a least square (LS) approach for signal recovery in the presence of noise. We have to use the wiener filter which cannot be directly applied to 3D images. Therefore, separation of the RGB channels is required and then wiener filter is applied separately to each channel. The wiener filter measures the mean and variance of each pixel around.

To get the finest detailed coefficients of noise free image, a Forward Discrete Curvelet Transform (FDCT) is applied to the input image. It is a multi-dimensional transform which can sense both the contours as well as curvy edges of the overlapping objects in the image. The FDCT has high directional sensitivity along with the capability to capture the singularities. Edge and singularity details are processed to extract the feature points. After obtaining the highest detailed coefficients Inverse Discrete Curvelet Transform is applied to high frequency band to obtain the detailed image. This detailed image is now having the stronger edges than the original and would perform better in lending edge details to the segmentation step.

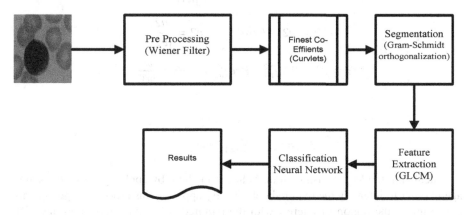

Fig. 1. Proposed Solution

3.2   Image Segmentation

The image segmentation is the most crucial phase in hematological image analysis. Accurate image segmentation produces accurate results in subsequent stages. Our proposed image segmentation technique is based on Gram-Schmidt orthogonalization and snakes algorithm.

For segmentation, we have used Gram-Schmidt process. This method is used in mathematics and above all in numerical analysis for orthogonizing a set of vectors in the inner product space. The Gram-Schmidt process each feature is considered as vector. The pixel intensities in RGB space are the elements of each vector.

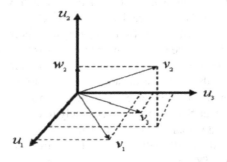

Fig. 2. Relation between w and v in 3D space

The Gram-Schmidt method takes linearly independent set S={v1,v2,v3...Vn} and generate orthogonal set S={u1,u2,u3...un}. Where u and v denotes the vectors having inner product <u ,v>.

$$u1 = v1 \qquad e1 = \frac{u1}{|u1|}$$

$$u2 = v2 - Proj_{u1}^{v2} \qquad e2 = \frac{u2}{|u2|}$$

$$u3 = v3 - Proj_{u1}^{v2} - Proj_{u2}^{v3} \qquad e3 = \frac{u3}{|u3|}$$

$$\cdot$$
$$\cdot$$

$$W_k = v_{vk} - \sum_{j=1}^{k-1} Proj_{uj}^{vk}$$

Calculating the W_k by using Gram-Schmidt method, by applying it to a color image in order to intensify the required color v_k. By applying this process we get a composite image the region of interest with the required color have maximum intensity while the remaining have minimum color intensity. It requires proper thresholding which produce the desired segmentation. This process amplifies the desired color vector while reduce or weakening the undesired vectors.

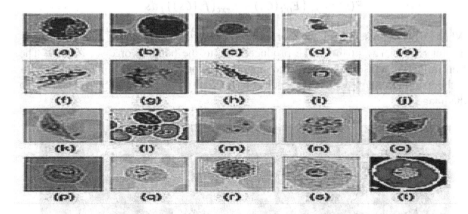

Fig. 3. Step by step segmentation process

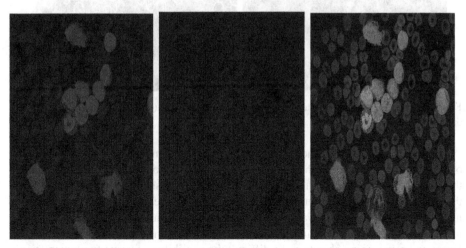

Fig. 4. Image Enhancement (**a**) Input Image (**b**) Curvelet Finest Coefficients (**c**) Enhanced Image

The snake algorithm is useful for the segmentation of objects where the edges are not well defined. The snake segmentation technique has a number of attractive characteristics that makes it useful for shape detection. The snake model can be parameterized   contour defined within an image domain. The internal forces within the image domain control the bending characteristics of the line, while forces like image gradient act to push the snake toward image features. As the snakes coordinate vector can be defined as C(s)=(x(s),y(s)) so the total energy can be represented as.

$$E(s) = \int_0^1 \left(E_{int}(Cs) + E_{ext}\left(f,(Cs)\right)\right) ds$$

Where $E(s)$ represent the image forces, while f represents image intensity and Eint is the internal energy of the snake produced due to the bending or discontinuity. The Eint inflict the snake to be small and smooth. It avoids the wrong solution. The Eext is the external energy which is responsible for finding the objects and boundaries in the image.

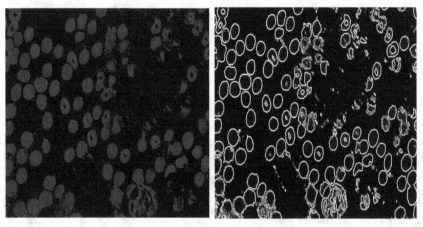

Fig. 5. Erythrocyte Boundary detection and image segmentation (a) Red Blood Cells (b) Boundary detection

In literature different methods are existing for the segmentation and classification of the white blood cells but these methods having limitations. Mostly used method is thresholding approaches including region growing, watershed segmentation and otsu threshholding all of these methods suffers from inconsistencies especially where the images having considerable variations. Genetic algorithm is used for leukocyte segmentation in literature the main limitation of this method is that the presence of irrelevant and overlapping objects make difficult the possibility of accurate segmentation. Similarly methods like artificial neural networks (ANN), Support Vector machine (SVM), K-means, and Fuzzy c-means produce poor results for those images having complex background. Our method provides better results in the presence of complex background as well as the method produce better results for images having complex shapes and overlapping objects.

4   Conclusion

It is concluded that accurate image segmentation is a crucial part for feature extraction and classification of blood cells. The accurate segmentation of blood cells is highly challenging in presence of complex and overlapping objects in microscopic images of blood cells. In this paper, we have proposed a hybrid segmentation technique which utilizes the capabilities of Gram-Schmidt orthogonalization method along with the snake algorithm to tackle inaccurate blood cell segmentation due to complex and overlap objects. The combination of these techniques leads to fruitful results. We have also exploited the Fast Discrete Curvelet Transforms (FDCT) Curvelet transform for image enhancement in order remove the noise and to obtain the finest coefficients which ensures the accurate image segmentation. In our future work will utilize these results for feature extraction and classification purpose.

References

1. Meijering, E.: Cell Segmentation: 50 Years Down the Road 29(5), 140–145 (2012)
2. Savkare, S.S., Narote, S.P.: Automatic System for Classification of Erythrocytes Infected with Malaria and Identification of Parasite's Life Stage. Procedia Technol. 6, 405–410 (2012)
3. Sharif, J.M., Miswan, M.F., Ngadi, M.A., Hj, S., Mahadi, M.: Red Blood Cell Segmentation Using Masking and Watershed Algorithm: A Preliminary Study pp. 27–28 (February 2012)
4. Shirazi, S.H., Haq, N., Hayat, K., Naz, S.: Curvelet Based Offline Analysis of SEM Images. PLoS ONE 9(8), e103942 (2014)
5. Xiao, X., Li, P.: An unsupervised marker image generation method for watershed segmentation of multiespectral imagery. Geoscience Journal 8(3), 325–331 (2004)
6. Duncan, J., Ayache, N.: Medical Image Analysis: Progress over two decades and the challenges ahead. IEEE Transactions on Pattern Analysis and Machine Intelligence, Institute of Electrical and Electronics Engineers (IEEE) 22(1), 85–106 (2000)
7. Kumar, V., Abbas, A.K., Fausto, N., Aster, J.: Robbins and Cotran Pathologic Basis of Disease. Saunders, Philadelphia, PA (2010)
8. Suradkar, P.T.: Detection of Malarial Parasite in Blood Using Image Processing. International Journal of Engineering and Innovative Technology (IJEIT) 2(10) (April 2013)
9. Karel, Z.: Contrast limited adaptive histogram equalization. Graphics Gems IV, 474–485, code: 479–484 (1994)
10. Khan, M.I., Acharya, B., Singh, B.K., Soni, J.: Content Based Image Retrieval Approaches for Detection of Malarial Parasite in Blood Images. International Journal of Biometrics and Bioinformatics (IJBB) 5(2) (2011)
11. Ross, N.E., Pritchard, C.J., Rubin, D.M.: Duse, Automated image processing method for the diagnosis and classification of malaria on thin blood smears. Medical & Biological Engineering & Computing 44, 427–436 (2006)
12. Tek, F.B., Dempster, K.: Parasite detection and identification for automated thin blood film malaria diagnosis. Computer Vision and Image Understanding 114, 21–32 (2010)
13. Ruberto, C.D., Dempster, A., Khan, S., Jarra, B.: Analysis of Infected Blood Cell Images using Morphological Operators. Image and Computer Vision 20 (2002)
14. Angulo, J., Flandrin, G.: Automated detection of working area of peripheral blood smears using mathematical morphology. Analytical Cellular Pathology 25, 37–49 (2003)
15. Trivedi, M., Bezedek, J.C.: Low-level segmentation of Zacrial images with fuzzy clustering. IEEE Trans. on System Man and Cybernetics 16(4), 589–598 (1986)

16. ChulKo, B., Gim, J.W., Nam, J.Y.: Automatic white blood cell segmentation using step-wise merging rules and gradient vector flow snake. Micron 42, 695–705 (2011)
17. Sabino, D.M.U., da Fontoura Costa, L., Gil Rizzatti, E., Antonio Zago, M.: A texture approach to leukocyte recognition. Real-Time Imaging 10, 205–216 (2004)
18. Foran, D., Meer, P., Comaniciu: Image guided decision support system for pathology, machine vision and applications. Machine Vision and Applications 11(4), 213–224 (2000)

A Hyperanalytic Wavelet Based Denoising Technique for Ultrasound Images

Cristina Stolojescu-Crisan

Politehnica University of Timisoara
V. Parvan 2, 300223 Timisoara, Romania
cristina.stolojescu-crisan@upt.ro

Abstract. Medical ultrasonography offers important information about patients health thus physicians are able to recognize different diseases. During acquisition, ultrasound images may be affected by a multiplicative noise called speckle which significantly degrades the image quality. Removing speckle noise (despeckling) plays a key role in medical ultrasonography. In this paper, we propose a denoising algorithm in the wavelets domain which associates the Hyperanalytic Wavelet Transform (HWT) with a Maximum a Posteriori (MAP) filter named bishrink for medical ultrasound images. Several common spatial speckle reduction techniques are also used and their performances are compared in terms of three evaluation parameters: the Mean Square Error (MSE), the Peak Signal to Noise Ratio (PSNR), and the Structural SIMilarity (SSIM) index.

Keywords: Denoising, speckle noise, ultrasound images, wavelets.

1 Introduction

Ultrasound imaging is one of the most non-invasive, less-expensive and widely used medical diagnostic technique [1] for soft-tissue structures of the human body like kidney, liver, spleen, etc. However, medical ultrasound images are of poor quality, resulting from the presence of speckle noise, a multiplicative noise that significantly influences the visual interpretation of the image and complicates diagnostic decisions. Speckle noise also limits the efficient application of intelligent image processing algorithms, for example segmentation. Thus, speckle noise reduction is an essential pre-processing step, but it should be done without affecting the image features.

The ultrasound technology is similar to that used by radar and SONAR. Thus, a transducer that emits high-frequency ultrasound waves through the human body is used. During ultrasound wave propagation, the waves are reflected by obstacles (such as an organ or bone), creating echoes. The reflected waves are picked up by the transducer and transmitted to the device. Ultrasound images generation is done by measuring the intensity of the reflected waves and the time of each echo's return. The machine displays on the screen the distances and the echoes intensities, forming a two dimensional image [2].

F. Ortuño and I. Rojas (Eds.): IWBBIO 2015, Part I, LNCS 9043, pp. 193–200, 2015.

In ultrasound images, the noise content is multiplicative and non-Gaussian, and it is more difficult to be removed than the additive noise since its intensity varies with the intensity of the image. Speckle noise can be modeled as:

$$y_{i,j} = u_{i,j} sp_{i,j}, \qquad (1)$$

where $y_{i,j}$ is the noisy image, $u_{i,j}$ is the image without speckle, and $sp_{i,j}$ is the speckle noise.

Thus, it is necessary to convert the multiplicative noise into an additive noise. This can be done by two methods: the first method consists of applying the logarithm of the acquired image and the second is based on the definition of multiplication as a sequence of repeated addition operations, obtaining a signal-dependent additive noise.

Speckle-reducing filters have been originally used by the Synthetic Aperture Radar (SAR) community. They have been applied to ultrasound imaging since the early 1980s [3]. Filters that are widely used in both SAR and ultrasound imaging were originally proposed by Lee [6], Kuan et al. [7], and Frost et al. [8]. But, during the years, various other techniques have been proposed for denoising in medical ultrasound images [2,3,9].

In this paper, we propose a speckle reduction method in the wavelets domain. This method is based on the Hyperanalytic Wavelet Transform (HWT) associated with a Maximum a Posteriori (MAP) filter named bishrink. We will compare the proposed method with several common spatial filtering methods: Lee, Frost, Kuan, Gamma, Wiener, Median, and Hibrid Median.

The rest of this paper is organized as follows. Section 2 is dedicated to the description of the proposed denoising method based on the association of the HWT with the bishrink filter. In Section 3, the results of the proposed denoising algorithm are presented. The results are further compared with the results obtained using seven spatial filters. The last section is dedicated to conclusions and future work.

2   Image Denoising

The field of images denoising methods is very large. Over the years, various denoising techniques for ultrasound images have been proposed in literature, each of them being based on particular assumptions and having advantages and limitations. The removal of speckle noise is a particularly difficult task, due to the intrinsic properties of speckle noise. More, ultrasound images fine details that integrate diagnostic information should not be lost during denoising.

Despeckling techniques can be classified in the following categories: filters acting in the spatial domain, including linear and nonlinear filtering, and filters acting in the wavelets domain [3].

Spatial filtering methods use the ratio of local statistics. The most used types of spatial filters include: the mean filter, Lee filter, Frost filter, Kuan filter, the Median filter, or Wiener filter.

An alternative to the spatial domain filtering is to apply the filter in the transform domain. The role of the transform is to give a better representation of the acquired image. Generally, a better representation has a higher sparsity then the spatial domain representation. The resulting image will have a reduced number of pixels with high values, so its filtering will be more easy. Wavelets based methods have three steps: the computation of a wavelet transform, the filtering of the detail wavelet coefficients and the computation of the corresponding inverse wavelet transform.

There are a few variants of wavelet transforms (WT) that can be used for images denoising: the Discrete Wavelet Transform (DWT), the Dual Tree Complex Wavelet Transform (DT-CWT), or the Hyperanalytic Wavelet Transform (HWT). All transforms have two parameters: the mother wavelets used and the primary resolution (the number of iterations). In this paper, we will use the HWT which has a high shift-invariance degree and an enhanced directional selectivity.

The HWT of the image can be obtained using the 2D DWT of its associated hypercomplex image [12], as it can be seen in Fig. 1.

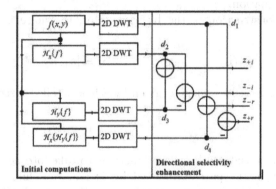

Fig. 1. HWT implementation

The HWT of an image $f(x, y)$ can be computed as:

$$HWT\{f(x,y)\} = \langle f(x,y), \psi_a(x,y) \rangle, \tag{2}$$

where $\psi_a(x,y)$ represents the hypercomplex mother wavelet associated to the real mother wavelets $\psi(x,y)$ and defined as:

$$\psi_a(x,y) = \psi(x,y) + i\mathcal{H}_x\{\psi(x,y)\} + j\mathcal{H}_y\{\psi(x,y)\} + k\mathcal{H}_x\{\mathcal{H}_y\{psi(x,y)\}, \tag{3}$$

where $i^2 = j^2 = -k^2 = -1$, $ij = ji = k$, and $\mathcal{H}$ represents the Hilbert transform. Thus:

$$HWT\{f(x,y)\} = DWT\{f(x,y)\} + iDWT\{\mathcal{H}_x\{\psi(x,y)\}\} + \\ + jDWT\{\mathcal{H}_y\{\psi(x,y)\}\} + kDWT\{\mathcal{H}_x\{\mathcal{H}_y\{\psi(x,y)\}\}\}. \tag{4}$$

In the end we obtain:

$$HWT\{f(x,y)\} = \langle f_a(x,y), \psi(x,y) \rangle = DWT\{f_a(x,y)\}. \qquad (5)$$

Concerning the second choice, various non-linear filter types can be used in the wavelet domain. One of the most efficient parametric denoising methods implies the use of MAP filters.

The bishrink filter is a MAP filter that takes into account the interscale dependency of the wavelet coefficients. Based on the observation $y = w + n$, where n represents the wavelet transform of the noise n_i, obtained as the logarithm of the speckle $n_i = log(sp)$, and w represents the wavelet transform of the useful component of the input image s, obtained as the logarithm of the noiseless component of the acquired image $s = logu$, the MAP estimation of w is given by:

$$\hat{w}(y) = \arg\max_w \{ln(p_n(y - w)p_w(w))\}, \qquad (6)$$

where p_a is the probability density function (*pdf*) of a. Equation (6) represents the MAP filter equation.

For the construction of the bishrink filter, the noise is assumed i.i.d. Gaussian [11]:

$$p_n(\mathbf{n}) = \frac{1}{2\pi\sigma_n^2} e^{-\frac{n_1^2+n_2^2}{2\sigma_n^2}}, \quad \mathbf{n} = [n_1, n_2]. \qquad (7)$$

The model of a noiseless image is given by a heavy tailed distribution:

$$p_w(\mathbf{w}) = \frac{3}{2\pi\sigma^2} e^{-\frac{\sqrt{3}}{\sigma}\sqrt{w_1^2+w_2^2}}, \quad \mathbf{w} = [w_1, w_2]. \qquad (8)$$

These *pdf*s are considered bivariate functions to take into account the inter-scale dependency of the parent (indexed by 1) and the child (indexed by 2) wavelet coefficients. The input-output relation of the bishrink filter is expressed by:

$$\hat{w}_1 = \frac{\left(\sqrt{y_1^2 + y_2^2} - \frac{\sqrt{3}\hat{\sigma}_n^2}{\hat{\sigma}}\right)_+}{\sqrt{y_1^2 + y_2^2}} y_1, \qquad (9)$$

where $\hat{\sigma}_n^2$ is the estimate of variance of the noise with:

$$\hat{\sigma}_n^2 = median(|y_i|), \qquad (10)$$

with $y_i \in$ sub-band HH and $\hat{\sigma}$ is the estimate of the standard deviation of the noiseless coefficients:

$$\hat{\sigma} = \begin{cases} \sqrt{\frac{1}{M}\sum\limits_{y_i \in N(k)} y_i^2 - \hat{\sigma}_n^2} & \text{if } \frac{1}{M}\sum\limits_{y_i \in N(k)} y_i^2 - \hat{\sigma}_n^2 > 0 \\ 0 & \text{if not,} \end{cases} \qquad (11)$$

where M is the size of the moving window $N(k)$ centered on the kth pixel of the acquired image and $(X)_+ = X$, for $X > 0$ and 0, otherwise.

3 Simulation Results

To investigate the performance of the proposed denoising method, we use real noise-free ultrasound images (as reference images) and we added artificial speckle noise obtaining the test images. We compared the performance of the considered denoising technique based on the association of HWT with the bishrink MAP filter with common spatial denoising filters such as: Lee [6], Frost [7], Kuan [8], Wiener filter [5], Gamma filter [4], Median filter [4,5], and Hibrid Median filter [3]. The window size used for these filters was set to 5 × 5.

The performance of the proposed denoising techniques is compared using three evaluation parameters: the Mean Square Error (MSE), the Peak Signal to Noise Ratio (PSNR) and the Structural SIMilarity (SSIM) index. The MSE measures the quality change between the original image and denoised image and is given by:

$$MSE = \frac{1}{mn} \sum_{i=1}^{m-1} \sum_{j=1}^{n-1} [f_{i,j} - \hat{S}_{i,j}]^2, \tag{12}$$

where $f_{i,j}$ is the noiseless image, $\hat{S}_{i,j}$ represents the estimation of the noiseless component of the acquired image (realized by denoising), and m and n are the dimensions of both images (in pixels).

The PSNR is one of simplest and most widely used full-reference quality metric. The PSNR measures the image fidelity (how closely the denoised image resembles the original image) and is computed as:

$$PSNR = 10 log_{10}(255^2 / MSE). \tag{13}$$

A higher value of the PSNR shows a greater similarity between the noiseless component of the acquired image and the image obtained after denoising. However, the PSNR is not very well matched to perceived visual quality, meaning that two distorted images with the same PSNR may have very different types of errors, some of them more visible than others [10].

The Structural SIMilarity (SSIM) gives a much better indication of image quality [10]. The SSIM of two images, x and y, can be computed with the following formula:

$$SSIM(x, y) = \frac{(2\mu_x \mu_y + C_1)(2\sigma_{xy} + C_2)}{(\mu_x^2 + \mu_y^2 + C_1)(\sigma_x^2 + \sigma_y^2 + C_2)}, \tag{14}$$

where μ_x and μ_y represent the mean of x and y respectively, σ_x and σ_y the standard deviations (the square root of variance) of x and y, σ_{xy} represents the covariance of x and y, and C_1 and C_2 are constants used to avoid instability in certain conditions. A value of 1 of the SSIM indicates that the two images are identical.

An example of visual comparison for a test image randomly selected from the database, after applying the denoising methods considered in this paper, is shown in Fig 2. The noisy image is shown in Fig. 2a). The results obtained by applying the proposed denoising method based on the association of the HWT with the bishrink filter is shown in Fig. 2i), while the results obtained by applying the

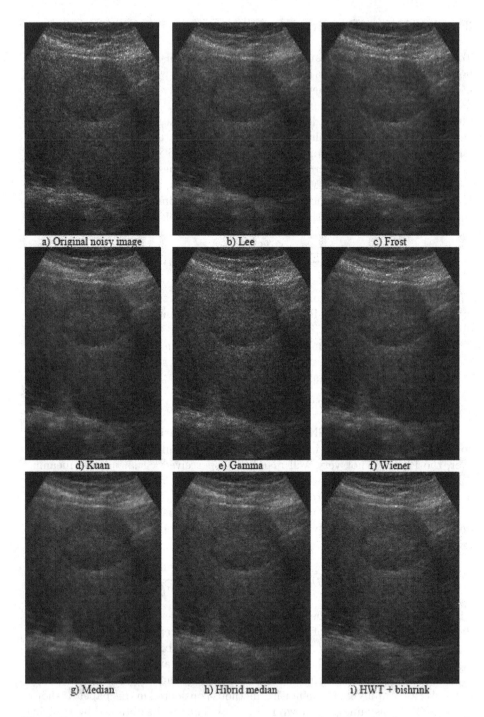

Fig. 2. Performance comparison of various denoising methods by visual inspection of an ultrasound image of the liver corrupted by speckle noise

eight spatial filtering techniques are shown in Fig. 2b)-h). By visual inspection of these resulting images it seems that the wavelet based denoising method is the one that removes most of the noise.

The MSE, PSNR and SSIM mean values obtained by applying the eight denoising techniques on twenty test images are shown in Table 1.

Table 1. MSE, PSNR, and SSIM values for various despeckling techniques

Denoising method	MSE	PSNR	SSIM
Lee	75.80	29.33	0.8516
Frost	67.94	29.80	0.8763
Kuan	73.04	29.49	0.8551
Gamma	144.37	26.53	0.7851
Wiener	65.21	29.98	0.8781
Median	51.86	30.98	0.8864
Hibrid median	48.55	31.26	0.9010
Proposed(HWT+bishrink)	**28.52**	**33.57**	**0.9459**

We can observe that the proposed denoising algorithm that combines the HWT with the bishrink filter gives the best results for all the three image quality measures. It should be pointed out that for all the twenty ultrasound test images considered in the experimental part, the best results were obtained using the proposed denoising method.

4   Conclusions

Removing speckle noise plays a key role in medical image processing, since the speckle noise significantly degrades the image quality and complicates the diagnostic decisions and the study of various illnesses. This paper has focused on the removal of speckle noise in medical ultrasound images of the liver and proposed a new denoising method that associates one of the best wavelet transforms, the HWT with a very good MAP filter, namely the bishrink filter. We have presented a comparative study between the proposed denoising algorithm with seven common spatial denoising filters used in the literature for despeckeling purpose: Lee, Kuan, Frost, Wiener, Gamma, Median and Hibrid Median. The comparative tests of the eight techniques considered for despecklization have shown that the proposed HWT+bishrink gives the best MSE, PSNR and SSIM values. It also outperforms the visual aspect of the other denoising methods considered for comparison. All the denoising methods have been implemented in MATLAB.

A future research direction consists in increasing the diversity of the denoising methods class by including representatives of the class of multiscale denoising methods, for example Wavelet thresholding techniques such as SURE Shrink, Visu Shrink or Bayes Shrink, the use of contourlet transform or the use of MAP filter applied in the Double Tree Complex Wavelet Transform (DTCWT). The investigation will be extended for ultrasound images of the kidney as well.

The results will be used as a pre-processing step for effective segmentation of the organs in ultrasound images.

Acknowledgments. This work was partially supported by the strategic grant POSDRU/159/1.5/S/137070 (2014) of the Ministry of National Education, Romania, co-financed by the European Social Fund Investing in People, within the Sectoral Operational Programme Human Resources Development 2007-2013.

References

1. Narouze, S.N. (ed.): Atlas of Ultrasound-Guided Procedures in Interventional Pain Management. Springer Science and Business Media (2011)
2. Hiremath, P.S., Akkasaligar, P.T., Badiger, S.: Speckle Noise Reduction in Medical Ultrasound Images. In: Gunarathne G. (ed.) Advancements and Breakthroughs in Ultrasound Imaging. InTech (2011)
3. Loizou, C.P., Pattichis, C.S.: Despeckle Filtering Algorithms and Software for Ultrasound Imaging. Andreas Spanias, Arizona State University, Morgan and Claypool Publisher (2008)
4. Gonzalez, R.C., Woods, R.E.: Digital Image Processing, 2nd edn. Prentice-Hall, Englewood Cliffs (2002)
5. Bovik, A. (ed.): Handbook of image and video processing (Communications, Networking and Multimedia), 1st edn. Academic Press (2000)
6. Lee, J.S.: Speckle analysis and smoothing of synthetic aperture radar images. Comput. Graph. Image Processing 17, 24–32 (1981)
7. Frost, V.S., Stiles, J.A., Shanmuggam, K.S., Holtzman, J.C.: A model for radar images and its application for adaptive digital filtering of multiplicative noise. IEEE Trans. Pattern Anal. Machine Intell. 4(2), 157–165 (1982)
8. Kuan, D.T., Sawchuk, A.A., Strand, T.C., Chavel, P.: Adaptive restoration of images with speckle. IEEE Trans. Acoust. 35, 373–383 (1987)
9. Adamo, F., Andria, G., Attivissimo, F., Lanzolla, A.M.L., Spadavecchia, M.: A comparative study on mother wavelet selection in ultrasound image denoising. Elsevier Measurement 46, 2447–2456 (2013)
10. Wang, Z., Bovik, A.C., Sheikh, H.R., Simoncelli, E.P.: Image Quality Assessment: From Error Visibility to Structural Similarity. IEEE Trans. on Image Processing 13(4), 600–612 (2004)
11. Sendur, L., Selesnick, I.W.: Bivariate shrinkage functions for wavelet-based denoising exploiting interscale dependency. IEEE Trans. on Signal Processing 50(11), 2744–2756 (2002)
12. Firoiu, I., Nafornita, C., Isar, D., Isar, A.: Bayesian hyperanalytic denoising of SONAR images. IEEE Geosci. Remote Sens. Let. 8(6), 1065–1069 (2011)

Detection of Pathological Brain in MRI Scanning Based on Wavelet-Entropy and Naive Bayes Classifier

Xingxing Zhou[1], Shuihua Wang[2], Wei Xu[3], Genlin Ji[1], Preetha Phillips[4],
Ping Sun[5], and Yudong Zhang[1,*]

[1] School of Computer Science and Technology, Nanjing Normal University, Nanjing,
Jiangsu 210023, China
[2] School of Electronic Science and Engineering, Nanjing University, Nanjing,
Jiangsu 210046, China
[3] Student Affairs Office, Nanjing Institute of Industry Technology, Nanjing,
Jiangsu 210023, China
[4] School of Natural Sciences and Mathematics, Shepherd University, Shepherdstown,
West Virginia, 25443, USA
[5] Department of Electrical Engineering, The City College of New York, CUNY, New York,
NY 10031, USA
zhangyudong@njnu.edu.cn

Abstract. An accurate diagnosis is important for the medical treatment of patients suffered from brain disease. Nuclear magnetic resonance images are
commonly used by technicians to assist the pre-clinical diagnosis, rating them
by visual evaluations. The classification of NMR images of normal and pathological brains poses a challenge from technological point of view, since NMR
imaging generates a large information set that reflects the conditions of the
brain. In this work, we present a computer assisted diagnosis method based on a
wavelet-entropy (In this paper 2D-discrete wavelet transform has been used, in
that it can extract more information) of the feature space approach and a Naive
Bayes classifier classification method for improving the brain diagnosis accuracy by means of NMR images. The most relevant image feature is selected as the
wavelet entropy, which is used to train a Naive Bayes classifier. The results
over 64 images show that the sensitivity of the classifier is as high as 94.50%,
the specificity 91.70%, the overall accuracy 92.60%. It is easily observed from
the data that the proposed classifier can detect abnormal brains from normal
controls within excellent performance, which is competitive with latest existing
methods.

Keywords: Wavelet transform, Entropy, Naïve Bayes classifier, Classification.

1 Introduction

Finding accurate and appropriate technologies for noninvasive observation and early
detection of the disease are of fundamental importance to develop early treatments for

---

* Corresponding author.

brain disease. Magnetic resonance imaging (MRI) [1, 2], is a medical imaging technique used in radiology to investigate the anatomy and physiology of the body in both health and disease. MRI scanners use strong magnetic fields and radio-waves to form images of the body [3-5], which does not cause any radiation damage to the patient's tissues because it is not using any injurious ionizing radiation to the patients.

Classification of normal/pathological brain conditions from MRIs is important in clinical medicine since MRI focuses on soft tissue anatomy and generates a large information set and details about the subject's brain conditions. However, high volume of data makes manual interpretation difficult and necessitates the development of automated image classification tools [6]. To solve the problem, numerous feature extraction methods are proposed, such as Fourier transform based techniques, wavelet transform based techniques, etc. Among those features abovementioned, the WT is a series of image descriptor and it can analyze images at any resolution desired [7]. The WT has become a choice for many image analysis and classification problems because of following two points [8]: (1) The WT has been found to be good for extracting frequency space information from non-stationary signals. (2) Owing to its time-scale representation, it is intrinsically well-suited to non-stationary signal analysis. However, the major disadvantage of WT is that it requires large storage and may cause curse of dimensionality [9]. Therefore, a novel parameter, namely wavelet-entropy [10, 11], is extracted from the wavelet approximation coefficients.

Classification problem is the next important issue in pattern recognition. Many algorithms can solve classification problem, such as neural network, support vector machine, Naive Bayes classifier (NBC) and decision trees. NBC is widely recognized as a simple and effective probabilistic classification method [12], and its performance is comparable with or higher than those of the decision tree [13] and neural network [14].

The proposed methodology for the MRI image classification is a combination of wavelet entropy and probabilistic NBC to perform a robust and accurate automated magnetic resonance normal/abnormal brain images classification. Those individual techniques were already proven successful, so we expect the proposed method (which is the combination of them) can achieve good results.

2   Discrete Wavelet Transform

The main feature of discrete wavelet transform (DWT) is a multi-scale representation of the function. By using wavelets, the given function can be analyzed at various levels of resolution. 1D discrete wavelet transform (1D-DWT) can be applied to 2D discrete wavelet transform (2D-DWT) easily [15]. The original image is processed along the x and y directions by low pass filters and high pass filters which is the row representation of the image [16]. DWT is applied to each dimension separately. After the 1-level wavelet decomposition there are four sub-band images (DD_1, VD_1, HD_1, A_1) at each scale [17]. The A_1 sub-band can be regarded as the approximation component of the image, while the LH, HL, and HH sub-bands can be regarded as the detailed components of the image [18].

As the level of decomposition increases, compacter but coarser approximation component is obtained. The sub-band A_1 is used for the next 2D-DWT. After the 2-level wavelet decomposition there are four sub-band images (DD_2, VD_2, HD_2, A_2) at each scale [19]. Due to page limit, the readers can refer to literature [8] to get the full concept of DWT.

3   Proposed Feature

However, the discrete wavelet transform allows the image decomposition with different kinds of coefficients preserving the image information. The major disadvantage of this technique is that its excessive features increase computation times and storage memory. It is required to reduce the number of features. Therefore, an additional parameter, namely entropy, is an efficient tool to reduce the dimension of a data set consisting of a large number of interrelated variables while retaining most of the variations.

Entropy is the average amount of information contained in each message received [20]. Here, message stands for an event, sample or character drawn from a distribution or data stream. Entropy thus characterizes our uncertainty about our source of information [21]. Then Entropy is also a statistical measure of randomness that can be used to characterize the texture of the input image. The entropy of an image can be determined approximately from the histogram of the image. The histogram shows different grey level probabilities in the image.

Named after Boltzmann's H-theorem, Shannon defined the entropy H of a discrete random variable X with possible values $\{x_1, x_2, ..., x_n\}$ and probability mass function $P(X)$ as:

$$H(X) = E(I(X)) = E[-\ln(P(X))] \tag{1}$$

here E is the expected value operator, and I is the information content of X. $I(X)$ is itself a random variable. When taken from a finite sample, the entropy can explicitly be written as

$$H(X) = \sum_i P(x_i)I(x_i) = -\sum_i P(x_i)\log_b p(x_i) \tag{2}$$

As is known, the iterative subband decomposition is the core process of WT, so the coding stage has to be performed on some layers, in the case of 1D Mallat-tree decomposition, where only approximation signals (L_1) are recursively decompose into two sub-signals (L_2, H_2). Along the tree decomposition, it can be noticed that in intermediate layers some data are just temporary. For example, in layer 1, L_1 signal data is produced while coding input signal and is successively decompose into L_2 and H_2.

The two-dimensional discrete wavelet transform (2D-DWT) coding is usually based on separable basic scaling functions and wavelet bases so that it can be performed iterating two orthogonal 1D-DWT.

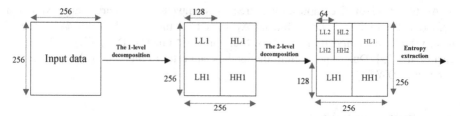

(a) Original MR image; (b) 1-level decomposition; (c) 2-level decomposition; (d) entropy vector

Fig. 1. 2D-DWT with Mallat-tree decomposition (The size of intermediate layers decreases twice as fast as the 1D case)

After 2-level 2D-DWT, we will get seven coefficient matrices (Fig. 1c) from a MR image (Fig. 1a), four of them are 64-by-64 in size, and the rest 128-by-128. Then, we extract entropy from the seven matrices (LL2, HL2, LH2, HH2, HL1, LH1, and HH1) [22, 23]. The above procedure reduces the features from original 256x256=65,536 to an entropy vector of 7 values, which is used as the input for Naïve Bayes classifier.

4   Naive Bayes Classifier

What kind of classification can we choose is one of the important issues in pattern recognition. Many algorithms can solve classification problem, such as decision trees, support vector machine and neural network. Naive Bayes classifier is widely recognized as a simple probabilistic classifier based on the application of the Bayesian theorem with strong (naive) independence assumptions, and its performance is comparable with the decision tree and neural network.

4.1   Probabilistic Model

In simple terms, a NBC assumes that the value of a particular feature is unrelated to the presence or absence of any other feature, given the class variable. A NBC considers each of these features to contribute independently to the probability, regardless of the presence or absence of the other features. In spite of their naive design and apparently oversimplified assumptions, NBCs work quite well in various complicated real-world situations. For some types of probability models, NBCs can be trained very efficiently in a supervised learning setting. In many practical applications, parameter estimation for NBC uses the method of maximum likelihood; in other words, one can work with the NBC without accepting Bayesian probability or using any Bayesian methods [24].

The probability model for a classifier is a conditional model over a dependent class variable C with a small number of outcomes or classes, conditional on several feature variables X_1 through X_n.

$$p(C \mid X_1, \ldots, X_n) \tag{3}$$

However, if a feature can take on a large number of values or if the number of features n is large, then basing such a model on probability tables is infeasible. We therefore reformulate the model to make it more tractable.

$$p(C \mid X_1,...,X_n) = \frac{p(C)\,p(X_1,...,X_n \mid C)}{p(X_1,...,X_n)} \tag{4}$$

In the Bayesian analysis, the final classification is produced by combining both sources of information, the prior and the likelihood, to form a posterior probability using the so-called Bayes' rule, so the above equation can be written as

$$posterior = \frac{prior \times likelihood}{evidence} \tag{5}$$

In practice, there is interest only in the numerator of that fraction, because the denominator does not depend on C and the values of the features X_i are given, so that the denominator is effectively constant, so we just need to maximize the value of $p(C)p(X1,.... Xn|C)$

Now the "naive" conditional independence assumptions come into play: assume that each feature X_i is conditionally independent of every other feature X_j for $j \neq i$, given the category C.

$$p(X_1,...,X_n \mid C) = \prod_{i=1}^{n} p(X_i \mid C) \tag{6}$$

where probability $P(X_1|C)$, $P(X_2|C)$...$P(Xn|C)$ can be estimated by the training sample. Through these calculations, we can get the posterior probabilities of sample belonging to each class, then based on Bayesian maximum a posteriori criteria, select the class with largest posterior probability as class label.

5   Experiment, Result, and Discussion

All the programs were developed in-house, using Matlab2013a and were run on an IBM desktop with 3G Hz Intel Core i3 processor and 2GB RAM. This section only reports objective results.

5.1   Data Source

The dataset consist of T2 weighted MR brain images in axial plane and 256x256 in plane resolution. We used 64 images, 18 of which were normal and the rest abnormal (consisting of Glioma, Alzheimer disease, Sarcoma disease, Huntington disease, Meningioma disease). The two-dimensional discrete wavelet transform (2D-DWT) decomposes an image into several sub-bands according to a recursive process. The 1-level decomposition obtains two kinds of coefficients. One contains the three sub-bands (LH1, HL1, and HH1), which represent details of the original images. The other is LL1 subband that corresponds to the approximation of original image, as is show in Fig. 2(b). The approximation LL1 is then decomposed into second-level approximation and detail images, and the process is repeated to achieve the desired level of resolution. The obtained coefficients for the approximation and detail sub-band images are useful features for texture categorization, as is show in Fig. 2(c).

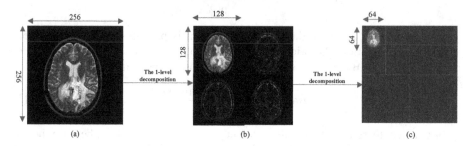

Fig. 2. Illustration of the image decomposition by 2D-DWT: (a) Original image; (b) 1-level decomposition; (c) 2-level decomposition

The 10-fold CV was repeated 10 times, i.e., we carried out a 10x10-fold CV. For each time, we used three established measures: accuracy, sensitivity, and specificity. The order of the class was pathological brain and normal brain following common convention.

5.2   Classification Comparison

We compared the proposed wavelet-entropy with other reported features: including DWT+SVM [25], DWT+PCA +FP-ANN [26], DWT+PCA+SVM [27], BPNN [28], and RBFNN [28]. For the proposed NBC, Gaussian distribution and empirical probability were assumed.

Table 1. Classification Comparison

Feature	Accuracy	Sensitivity	Specificity
DWT+SVM [25]	96.15%	N/A	N/A
DWT+PCA+FP-ANN [26]	97.00%	N/A	N/A
DWT+PCA+SVM [27]	95.00%	96.42%	85.00%
BPNN [28]	86.22%	88.00%	56.00%
RBFNN [28]	91.33%	92.47%	72.00%
Wavelet-Entropy+NBC (Proposed)	92.60%	94.50%	91.70%

Results in Table 1 shows that the sensitivity is 94.50%, the specificity 91.70%, and accuracy 92.60%. The accuracy of the proposed method is higher than RBFNN and BPNN. The sensitivity of the proposed method is higher than BPNN and RBFNN. The specificity of the proposed method is higher than BPNN, RBFNN, and DWT+PCA+SVM. On the other hand, the proposed method performs marginally worse w.r.t. average accuracy than PSO-KSVM, DWT+SVM, and DWT+PCA+FP-ANN, however, it offered an alternative method for MR image classification, and can be used in combination with other classification methods. The standard deviation of our method was relatively small, which demonstrated the robustness of the proposed method.

6 Conclusion and Discussion

This paper proposed a new approach, by combining wavelet-entropy and NBC, for the classification of normal and pathological brains from the MR images. The approach consisted of two steps: feature extraction and classification. The accuracy achieved is 92.60% in detecting abnormal images from normal controls. Sensitivity and specificity of the proposed method are quite good.

The contribution of this paper contained following two aspects: (i) It could help physicians/technicians detect abnormal brains in MR images with good accuracy. (ii) The application of DWT greatly reduced the calculation complexity.

However, although it was difficult to interpret the entropy values or the weights/biases of NBC, the proposed method has satisfying classification results. This is the future research direction.

Another limitation was that we treated different types of brain pathologies in a single "abnormal" label; hence, multi-disease classification was the future research direction.

Finally, the dataset is small and may introduce error to the reported classification accuracy; we shall create larger dataset to re-test our proposed method.

Acknowledgment. This work was supported from NSFC (No. 610011024) and Nanjing Normal University Research Foundation for Talented Scholars (No. 2013119XGQ0061).

Conflict of Interest. We have no conflicts of interest to disclose with regard to the subject matter of this paper.

References

1. Goh, S., Dong, Z., Zhang, Y., DiMauro, S., Peterson, B.S.: Mitochondrial dysfunction as a neurobiological subtype of autism spectrum disorder: Evidence from brain imaging. JAMA Psychiatry 71, 665–671 (2014)
2. Hou, X.S., Han, M., Gong, C., Qian, X.M.: SAR complex image data compression based on quadtree and zerotree Coding in Discrete Wavelet Transform Domain: A Comparative Study. Neurocomputing 148, 561–568 (2015)
3. Lingala, S.G., Jacob, M.: Blind Compressive Sensing Dynamic MRI. IEEE Transactions on Medical Imaging 32, 1132–1145 (2013)

4. Dong, Z., Zhang, Y., Liu, F., Duan, Y., Kangarlu, A., Peterson, B.S.: Improving the spectral resolution and spectral fitting of 1H MRSI data from human calf muscle by the SPREAD technique. NMR in Biomedicine 27, 1325–1332 (2014)
5. Zhang, Y., Wang, S., Ji, G., Dong, Z.: Exponential wavelet iterative shrinkage thresholding algorithm with random shift for compressed sensing magnetic resonance imaging. IEEJ. Transactions on Electrical and Electronic Engineering 10, 116–117 (2015)
6. Schneider, M.F., Krick, C.M., Retz, W., Hengesch, G., Retz-Junginger, P., Reith, W., Rösler, M.: Impairment of fronto-striatal and parietal cerebral networks correlates with attention deficit hyperactivity disorder (ADHD) psychopathology in adults – A functional magnetic resonance imaging (fMRI) study. Psychiatry Research: Neuroimaging 183, 75–84 (2010)
7. Arjmandi, M.K., Pooyan, M.: An optimum algorithm in pathological voice quality assessment using wavelet-packet-based features, linear discriminant analysis and support vector machine. Biomedical Signal Processing and Control 7, 3–19 (2012)
8. Jero, S.E., Ramu, P., Ramakrishnan, S.: Discrete Wavelet Transform and Singular Value Decomposition Based ECG Steganography for Secured Patient Information Transmission. Journal of Medical Systems 38 (2014)
9. Lustig, M., Donoho, D., Pauly, J.M.: Sparse MRI: The application of compressed sensing for rapid MR imaging. Magn. Reson. Med. 58, 1182–1195 (2007)
10. Chen, J.K., Li, G.Q.: Tsallis Wavelet Entropy and Its Application in Power Signal Analysis. Entropy 16, 3009–3025 (2014)
11. Frantzidis, C.A., Vivas, A.B., Tsolaki, A., Klados, M.A., Tsolaki, M., Bamidis, P.D.: Functional disorganization of small-world brain networks in mild Alzheimer's Disease and amnestic Mild Cognitive Impairment: an EEG study using Relative Wavelet Entropy (RWE). Frontiers in aging neuroscience 6 (2014)
12. Anderson, M.P., Dubnicka, S.R.: A sequential naive Bayes classifier for DNA barcodes. Statistical Applications in Genetics and Molecular Biology 13, 423–434 (2014)
13. Zhang, Y., Wang, S., Phillips, P., Ji, G.: Binary PSO with mutation operator for feature selection using decision tree applied to spam detection. Knowledge-Based Systems 64, 22–31 (2014)
14. Kotsiantis, S.: Integrating Global and Local Application of Naive Bayes Classifier. International Arab Journal of Information Technology 11, 300–307 (2014)
15. Ravasi, M., Tenze, L., Mattavelli, M.: A scalable and programmable architecture for 2-D DWT decoding. IEEE Transactions on Circuits and Systems for Video Technology 12, 671–677 (2002)
16. Zhang, Y., Wang, S., Dong, Z.: Classification of Alzheimer Disease Based on Structural Magnetic Resonance Imaging by Kernel Support Vector Machine Decision Tree. Progres. Electromagnetics Research 144, 171–184 (2014)
17. Chavez-Roman, H., Ponomaryov, V.: Super Resolution Image Generation Using Wavelet Domain Interpolation With Edge Extraction via a Sparse Representation. IEEE Geosci. Remote Sens. Lett. 11, 1777–1781 (2014)
18. Darji, A.D., Kushwah, S.S., Merchant, S.N., Chandorkar, A.N.: High-performance hardware architectures for multi-level lifting-based discrete wavelet transform. Eurasip Journal on Image and Video Processing (2014)
19. Ganesan, K., Acharya, U.R., Chua, C.K., Min, L.C., Abraham, T.K.: Automated Diagnosis of Mammogram Images of Breast Cancer Using Discrete Wavelet Transform and Spherical Wavelet Transform Features: A Comparative Study. Technology in Cancer Research & Treatment 13, 605–615 (2014)

20. Zhang, Y.D., Dong, Z.C., Ji, G.L., Wang, S.H.: An improved reconstruction method for CS-MRI based on exponential wavelet transform and iterative shrinkage/thresholding algorithm. Journal of Electromagnetic Waves and Applications 28, 2327–2338 (2014)
21. Nicolaou, N., Georgiou, J.: Detection of epileptic electroencephalogram based on Permutation Entropy and Support Vector Machines. Expert Systems with Applications 39, 202–209 (2012)
22. Lee, S.G., Yun, G.J., Shang, S.: Reference-free damage detection for truss bridge structures by continuous relative wavelet entropy method. Structural Health Monitoring-an International Journal 13, 307–320 (2014)
23. Bakhshi, A.D., Bashir, S., Loan, A., Maud, M.A.: Application of continuous-time wavelet entropy for detection of cardiac repolarisation alternans. IET Signal Processing 7, 783–790 (2013)
24. Sheppard, S., Nathan, D., Zhu, L.J.: Accurate identification of polyadenylation sites from 3′ end deep sequencing using a naive Bayes classifier 29, 2564 (2013), Bioinformatics (Oxford, England) 30, 596–596 (2014)
25. Chaplot, S., Patnaik, L.M., Jagannathan, N.R.: Classification of magnetic resonance brain images using wavelets as input to support vector machine and neural network. Biomedical Signal Processing and Control 1, 86–92 (2006)
26. El-Dahshan, E.S.A., Hosny, T., Salem, A.B.M.: Hybrid intelligent techniques for MRI brain images classification. Digital Signal Processing 20, 433–441 (2010)
27. Zhang, Y., Wu, L.: An Mr Brain Images Classifier via Principal Component Analysis and Kernel Support Vector Machine. Progres. Electromagnetics Research 130, 369–388 (2012)
28. Zhang, Y., Wang, S., Ji, G., Dong, Z.: An MR Brain Images Classifier System via Particle Swarm Optimization and Kernel Support Vector Machine. The Scientific World Journal 2013, 9 (2013)

PloidyQuantX: A Quantitative Microscopy Imaging Tool for Ploidy Quantification at Cell and Organ Level in Arabidopsis Root

Xavier Sevillano[1], Marc Ferrer[1], Mary-Paz González-García[2], Irina Pavelescu[2], and Ana I. Caño-Delgado[2]

[1] Grup de Recerca en Tecnologies Mèdia, La Salle - Universitat Ramon Llull, Barcelona, Spain
[2] Department of Molecular Genetics, Centre for Research in Agricultural Genomics (CRAG) CSIC-IRTA-UAB-UB, Barcelona, Spain
xavis@salleurl.edu, ana.cano@cragenomica.es

Abstract. Mapping centromere distribution in complete organs is a key step towards establishing how this couples to organ development and cell functions. In this context, quantitative microscopy tools play a key role, as they allow a precise measurement of centromere presence at both individual cell and whole organ levels. This work introduces PloidyQuantX, an imaging tool that operates on confocal microscopy image stacks. Tested on imagery obtained from whole-mount centromere Q-FISH of the Arabidopsis thaliana primary root, Ploidy-QuantX incorporates interactive segmentation and image analysis modules that allow quantifying the number of centromeres present in each cell nucleus, thus creating maps of centromere distribution in cells across the whole organ. Moreover, the presented tool also allows rendering three-dimensional models of each individual root cell, which makes it possible to relate their internal topology to specific cell functions.

Keywords: quantitative microscopy, centromere quantification, image processing, biomedical imaging.

1 Introduction

The centromere is a region of highly specialized chromatin found within each chromosome that serves as a site for the junction of sister chromatids. They play a key role in cell division processes, as they ensure accurate chromosomal segregation during mitosis and thus, the successful transfer of genetic information between cell and organismal generations [1]. In fact, abnormal centromere function can end up in aberrant cell division or chromosomal instability [2].

For this reason, centromeres have become a central focus in the study of processes in which cell division plays a primary role, from species and chromosome evolution to cancer, aging and stem cell biology in mammals [3,4].

From a structural and functional perspective, there exists a high similarity between the role of centromeres in animals and in plants. Particularly, in *Arabidopsis thaliana*

F. Ortuño and I. Rojas (Eds.): IWBBIO 2015, Part I, LNCS 9043, pp. 210–215, 2015.

(Arabidopsis) –a popular model organism in plant biology and genetics–, several studies have led to the identification and characterization of the centromere regions [5], as well as to the better understanding of centromere functions [6]. However, the role of centromeres in key aspects of plant growth and development remain largely unexplored.

In this context, a key element of plant morphogenesis is the balance among cell proliferation, expansion, and differentiation to produce organs of characteristic sizes and shapes. During the development of many plants, certain cell types undergo extensive endoreduplication, a modified cell cycle that results in DNA replication without subsequent mitosis [7]. For this reason, cells undergoing endoreduplication contain a higher number of chromosomes in their nuclei, thus presenting higher ploidy levels. Therefore, ploidy quantification is a key issue for studying endoreduplication, the function of which is unknown, although some of the proposed roles include gene amplification, radiation resistance and cell differentiation.

Ploidy level of plant cells has traditionally been determined by using conventional cytophotometric methods. However, these techniques do not allow precise quantification of individual centromeres within a tissue or specific organ.

As an alternative, confocal centromere Quantitative-Fluorescence in situ Hybridization (Q-FISH) applied in tissues [8] or flow-sorted [9] Arabidopsis nuclei allowed for study of a large number of nuclei with a good accessibility to the target DNA for the labeled probes. Additionally, flow sorting allowed to distinguish nuclei according to their ploidy level (C-value). The disadvantages of both preparation techniques are insufficient 3D information due to flattening of nuclei and loss of the spatial context given within native tissues or in particular cell types.

In this context, this work introduces PloidyQuantX, an interactive image processing tool to separate cells nuclei and to identify the number of centromeres inside each cell, creating maps of centromere distribution from confocal microscopy image stacks of the Arabidopsis primary root. Moreover, PloidyQuantX allows conducting centromere quantification from organ level to single-cell level, and rendering virtual 3D models of a specific cell of choice to visualize its internal topology.

The remaining of this paper is organized as follows. Section 2 describes the biological data on which the proposed system operates, while Section 3 provides the reader with an overview of PloidyQuantX's architecture and a description of its constituting modules.

2   Imaging Data

The Arabidopsis thaliana root displays a simple structure, easily traceable cell lineage, distinctly separate zones of growth activities, and transparency. For this reason, the Arabidopsis root has emerged as a powerful model for studying cell division, cell elongation, and cell differentiation in plants).

In this work, a centromeric cytomolecular map was developed using whole-mount immunolocalization in 6-day-old Arabidopsis roots with a PNA Cy3-labelled centromeric probe as described in [10]. On the other hand, cells nuclei were counterstained with 4',6-diamidino-2-phenyl indole (DAPI) stain.

An Olympus FV 1000 confocal microscope was employed to perform the image stack acquisition. The acquired images were exported and saved as 8-bit and 16-bit TIFF color image files, presenting the Cy3-labeled probes (corresponding to centromeres), the fluorescent quiescent center cells, and the DAPI staining (corresponding to cells nuclei) in the red, green and blue channels of the images.

3   PloidyQuantX Architecture

As mentioned earlier, PloidyQuantX is an interactive imaging tool for centromere distribution mapping, and it is implemented in Matlab. Its architecture is depicted in the block diagram in Fig. 1. It can be observed that PloidyQuantX comprises two primary modules, labelled as "Interactive segmentation" and "Interactive quantification". The following paragraphs provide an overview of these two modules.

Notice that both modules employ, as input data, a stack of confocal microscopy images with 8-bit and 16-bit depth. The former are used for displaying and user interaction purposes, while the latter are used for computation, due to their higher precision in gray level quantification. As for the output, PloidyQuantX generates several types of files that allow the analysis of the quantification process results from different perspectives, ranging from statistical analysis to visualization purposes.

Finally, it is important to notice that PloidyQuantX is designed as a highly interactive tool, as user intervention is required at both the segmentation and the quantification stages.

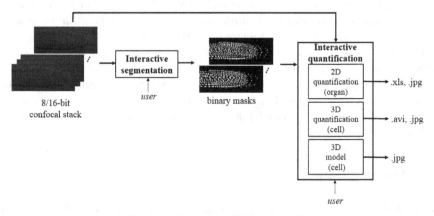

Fig. 1. Architecture of PloidyQuantX

3.1   Interactive Segmentation

The goal of the "Interactive segmentation" module is to allow the individual analysis of cells, which is a basic issue in centromere mapping. To that end, this module aims at creating one or several binary masks that isolate each cell nucleus appearing in one or several image(s) of the confocal stack, determining their limits accurately and reliably. In its current implementation, the binary mask creation process is semi-automatic and semi-supervised by the user, in the sense that a first segmentation approximation is automatically obtained –following the procedure described in the next paragraphs-, and it is refined manually at a later stage.

The proposed segmentation procedure is a simple and adaptable two-step segmentation strategy that combines average filtering based adaptive thresholding segmentation (for obtaining a first version of the binary mask), followed by denoising via morphological operators to improve its quality.

On one hand, average filtering based adaptive thresholding segmentation allows setting a segmentation threshold that is highly local, thus being fairly immune to the large illumination variations found across the image [11,12]. On the other hand, the majority and horizontal breaking morphological operators are sequentially applied to remove isolated white pixels that do not belong to cells nuclei, and order to separate cells nuclei connected by one pixel, respectively [13]. The final version of the binary mask is created by manually separating those cells nuclei that may remain connected and filling the holes of the empty nuclei.

3.2   Interactive Quantification

The "Interactive quantification" module is responsible for the quantification of the centromeres contained in the cells nuclei. Currently, PloidyQuantX supports three types of quantification, namely: quantification at organ level, quantification at cell level, and 3D cell model rendering.

All three quantification modes are based on centromere detection, which is tackled through the detection of local brightness maxima on the 16-bit images red channel by comparing the original image with the image resulting of applying a morpholigical dilation.

As for the quantification at single-cell level, PloidyQuantX offers the user the chance of selecting a cell nucleus through its graphical user interface. Moreover, the user has the chance of selecting the upper and lower limits of the selected cell by means of an animation showing the cell of choice at different stack layers. Next, centromere detection is performed and refined across the selected layers, obtaining the final centromere quantification results for the cell of choice.

As regards the quantification at organ level, it is conducted cellwise using the binary mask obtained from the "Interactive segmentation" module. As a result, Ploidy-QuantX generates an image with a color-coded centromere count for each cell, and also an Excel spreadsheet indicating the centromere count corresponding to each cell at every selected layer of the confocal stack. Moreover, PloidyQuantX also allows focusing the quantification in an arbitrary number of regions of interest selected by the user through the program's graphical user interface.

Finally, the three-dimensional cell model rendering functionality of PloidyQuantX allows providing the user with a visual representation of a cell nucleus of choice, showing valuable information regarding cellular internal morphology. The contour limits of the nucleus of the cell of choice at each layer of the stack are obtained by performing cell nucleus segmentation layer-wise, which constitutes the basis for the cell 3D model rendering.

4 Conclusions

This work has introduced PloidyQuantX as a tool for counting centromeres maps in intact organs such as the primary root of Arabidopsis, allowing to measure ploidy at single-cell and organ levels, plus the generation of 3D virtual models of cells. In its current status, PloidyQuantX provides centromere count, but we plan expanding its capabilities to obtain centromere fluorescence measurements. Moreover, given the importance of the segmentation step, further research efforts will be mainly directed towards the obtainment of accurate binary masks for cell nuclei segmentation in a maximally automatic manner. In this sense, we will investigate alternative algorithms, such as watershed segmentation [13], or exploiting the three-dimensionality of the confocal image stack to refine the obtained masks [14].

References

1. Sullivan, B.A., Blower, M.D., Karpen, G.H.: Determining Centromere Identity: Cyclical Stories and Forking Paths. Nature Reviews Genetics 2, 584–596 (2001)
2. O'Connor, C.: Chromosome Segregation in Mitosis: The Role of Centromeres. Nature Education 1(1), 28 (2008)
3. Tomonaga, T., Matsushita, K., Yamaguchi, S., Oohashi, T., Shimada, H., Ochiai, T., Yoda, K., Nomura, F.: Overexpression and Mistargeting of Centromere Protein-A in Human Primary Colorectal Cancer. Cancer Res. 63(13), 3511–3516 (2003)
4. McGovern, S.L., Qi, Y., Pusztai, L., Symmans, W.F., Buchholz, T.A.: Centromere Protein-A, an Essential Centromere Protein, is a Prognostic Marker for Relapse in Estrogen Receptor-Positive Breast Cancer. Breast Cancer Res. 14(3), R72 (2012)
5. Round, E.K., Flowers, S.K., Richards, E.J.: Arabidopsis thaliana Centromere Regions: Genetic Map Positions and Repetitive DNA Structure. Genome Res. 7, 1045–1053 (1997)
6. Copenhaver, G.P.: Using Arabidopsis to Understand Centromere Function: Progress and Prospects. Chromosome Res. 11, 255–262 (2003)
7. Barlow, P.: Endopoliploidy, Towards an Understanding of Its Biological Significance. Acta Biotheor. 27, 1–18 (1978)
8. Fransz, P., de Jong, H., Lysak, M., Castiglione, M.R., Schubert, I.: Interphase Chromosomes in Arabidopsis are Organized As Well Defined Chromocenters From Which Euchromatin Loops Emanate. Proc. Natl. Acad. Sci. USA 99, 14584–14589 (2002)
9. Pecinka, A., Schubert, V., Meister, A., Kreth, G., Klatte, M., Lysak, M.A., Fuchs, J., Schubert, I.: Chromosome Territory Arrangement and Homologous Pairing in Nuclei of Arabidopsis Thaliana are Predominantly Random Except for NOR-Bearing Chromosomes. Chromosoma 113, 258–269 (2004)
10. Gonzalez-Garcia, M.P., Vilarrasa-Blasi, J., Zhiponova, M., Divol, F., Mora-Garcia, S., Russinova, E., Cano-Delgado, A.I.: Brassinosteroids control meristem size by promoting cell cycle progression in Arabidopsis roots. Development 138, 849–859 (2011)

11. Wellner, P.: Adaptive Thresholding for the DigitalDesk. Techical Report EPC-93-110. Cambridge, UK: Rank Xerox Research Center (1993)
12. Otsu, N.: A threshold selection method from gray-level histogram. IEEE Trans. Syst. Man Cybern. 9, 62–66 (1979)
13. Gonzalez, R.C., Woods, R.E.: Digital Image Processing, 3rd edn. Prentice-Hall (2008)
14. Lin, G., Adiga, U., Olson, K., Guzowski, J.F., Barnes, C.A., Roysam, B.: A Hybrid 3D-Watershed Algorithm Incorporating Gradient Cues and Object Models for Automatic Segmentation of Nuclei in Confocal Image Stacks. Cytometry A 56(1), 23–36 (2003)

Study of the Histogram of the Hippocampus in MRI Using the α-stable Distribution

Diego Salas-Gonzalez[1,*], Oliver Horeth[1], Elmar W. Lang[1],
Juan M. Górriz[2], and Javier Ramírez[2]

[1] CIML Group, University of Regensburg, Germany
[2] SiPBA Group, University of Granada, Spain

Abstract. The hippocampus is a grey matter region of the brain which is known to be affected by Alzheimer's disease at the earliest stage. Its segmentation is important in order to measure its degree of atrophy.

We study the histogram of the intensity values of the hippocampus for 18 magnetic resonance images from the Internet Brain Segmentation Repository. In this dataset, manually-guided segmentation results are also provided. We use this database and the manual segmentation information to select the hippocampus of each of the images for the study of its histogram.

The histogram of intensity values of the left and right hippocampus for each image in the database are unimodal, heavy tailed and lightly skewed, for that reason, they can be fitted in a parsimonious way using the alpha-stable distribution. This results can be used to design a procedure to perform the segmentation of the hippocampus.

Keywords: Magnetic Resonance Image, brain processing, hippocampus, alpha-stable distribution.

1 Introduction

Alzheimer's disease (AD) is the most common cause of dementia. Nowadays, many efforts are being made in the diagnosis in early stage of the disease using magnetic resonance images [1] or functional brain images [2]. The hippocampus is a grey matter region of the brain which is affected at the earliest stage. Hippocampal atrophy rate, has been proven to discriminate AD from controls or other dementia [3]; and mild cognitive impairment (MCI) from controls [4]. Furthermore, regional measures of hippocampal atrophy are found to be useful predictors of progression to AD [5].

Segmentation of the hippocampus in magnetic resonance images is a useful tool for Alzheimer's diagnosis, nevertheless, this is a very challenging and time consuming task which is usually performed manually by experts. For this reason, it is useful to develop automatic segmentation methods.

In this work, we outline the similarities between the histogram of the intensity values of hippocampus in T1 magnetic resonance images (MRI) and the α-stable

---

* Corresponding author.

F. Ortuño and I. Rojas (Eds.): IWBBIO 2015, Part I, LNCS 9043, pp. 216–221, 2015.
© Springer International Publishing Switzerland 2015

distribution. In order to do that, we fit an α-stable distribution to the histogram of the hippocampus of 18 MRI images using a maximum likelihoood method [6].

This paper is organized as follows: Section 2 introduce some properties of the α-stable distribution. Section 3 presents the MRI database. The study of the hippocampus of the brain magnetic resonance images is performed in Section 4 showing the predicted α-stable density fitting the histogram of the hippocampus.Lastly, conclusions are given in Section 5.

2 Properties of the α-stable Distribution

The α-stable probability density function $f_{\alpha,\beta}(y|\gamma,\mu)$ has four parameters: $\alpha \in (0,2]$ is the characteristic exponent which sets the level of impulsiveness, $\beta \in [-1,+1]$ is the skewness parameter, ($\beta = 0$, for symmetric distributions and $\beta = \pm 1$ for the positive/negative stable family respectively), $\gamma > 0$ is the scale parameter, also called dispersion, and μ is the location parameter.

The α-stable distribution only has analytical closed expression in a few particular cases: when $\alpha = 2$ we get the Gaussian distribution and then $\gamma = \sigma/\sqrt{2}$, where σ is the standard deviation. Furthermore, for $\alpha = 1$ and $\beta = 0$ the distribution reduces to a Cauchy distribution and when $\alpha = 1/2$ and $\beta = 1$ to a Lévy distribution.

Figure 1 shows the α-stable probability density function for different values of the parameters. We use the distribution with parameters $\alpha = 1.5$, $\beta = 0$, $\gamma = 1$ and $\delta = 0$ as reference. This figure also explain the name of the parameters: α controls the degree of impulsiveness. When α decreases, the degree of impulsiveness increases. β controls the degree of asymmetry. γ controls the concentration of the samples along the bulk of the distribution. Lower values of γ correspond with higher concentration of the samples. Lastly, different values of δ produce the same density but shifted in the x-axis.

3 IBSR Database

The images were obtained from the Internet Brain Segmentation Repository (IBSR) [1] which provides manually-guided expert segmentation results along with magnetic resonance brain image data. The MR brain data sets and their manual segmentations were provided by the Center for Morphometric Analysis at Massachusetts General Hospital.

In this dataset, manually-guided segmentation results are also provided. The MR image data is a T1-weighted 3-dimensional coronal brain scan after it has been positionally normalized into the Talairach orientation (rotation only).

The segmentation are the result of semi-automated segmentation techniques which require many hours of effort by a trained expert. Even though manual segmentations cannot be considered to be 100% "ground truth", they provide a useful information for the study of the histogram of regions of interests in the

---

[1] http://www.nitrc.org/projects/ibsr

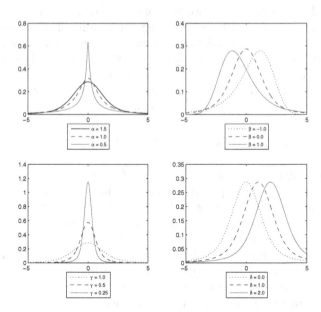

Fig. 1. α-stable probability density function with reference parameters $\alpha = 1.5$, $\beta = 0$, $\gamma = 1$ and $\delta = 0$ with changing: (a) Characteristic exponent α. (b) Skewness parameter β. (c) Dispersion γ. (d) Location parameter δ.

brain. This database also includes segmentation results for of the principle gray and white matter structures of the brain including the hippocampus.

4   Results

We study the histogram of the intensity values of the hippocampus for 18 magnetic resonance images from the Internet Brain Segmentation Repository. We use this database and the manual segmentation information to select the hippocampus of each of the images for the study of its histogram. The shape of the histogram of intensity values in the region of interest is unimodal, heavy tailed and lightly skewed. We fit them accurately using an α-stable distribution. Figure 2 shows the histogram of intensity values of the left and right hippocampus for each image in the database along with the predicted α-stable density.

Unfortunatelly, in real world problem we do not have access to manual segmented regions of the brain. Therefore, when we establish a narrow region of interest including the right and/or left hippocampus, the histogram will be affected by the contribution of adjacent regions of the hippocampus.

We plot the histogram of a dilated region of interest (the striatum and neighbour voxels) for the image 1 in Figure 3. This figure depicts how the contribution of the grey matter of the hippocampus is mixed with white matter voxels from the adjacent voxels. Left to the histogram of the hippocampus, this figure also

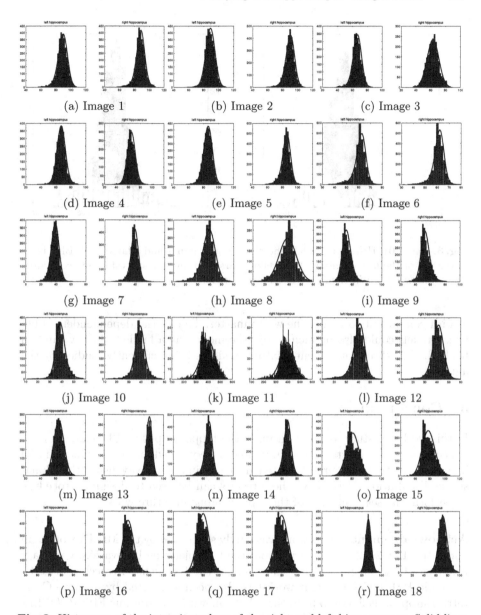

Fig. 2. Histogram of the intensity values of the right and left hippocampus. Solid line: predicted α-stable density.

shows additional grey matter voxels with spreading histogram and some voxels labelled as cerebro spinal fluid in the manual segmentation procedure. These features suggest that a tight region of interest including the hippocampus can be mathematically modelled as a mixture of three distributions: one for the main region of interest comprising the grey matter in hipppocampus, a second component for the cerebro spinal fluid and grey matter of the adjacent area and, lastly,

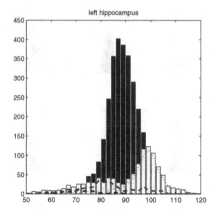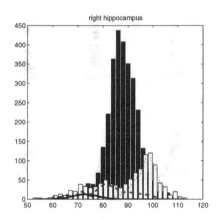

Fig. 3. Image 01. Histogram of left and right hippocampus and dilated ROI. Blue bars: histogram of the hippocampus. Left yellow bars: grey matter and csf of adjacent area. Right yellow bars: White matter of adjacent area.

a third component to model the white matter. The spatial dependendences between voxels could also be included in the model via a hidden Markov random field [7]. This three component mixture model is being currently studied by the authors.

5   Conclusion

In this work, we study the histogram of the hippocampus in T1 magnetic resonance images. This histogram is unimodal, heavy tailed and lightly skewed and it was found to be accurately fitted in a parsimonious way using the alpha-stable distribution. We believe these results can be used to design a procedure to perform the segmentation of the hippocampus in future work.

Acknowledgements. This work was partly supported by the MICINN of Spain under the TEC2012-34306 project and the Consejería de Innovación, Ciencia y Empresa (Junta de Andalucía, Spain) under the Excellence Projects P09-TIC-4530 and P11-TIC-7103. This work has also been supported by a Marie Curie Intra-European Fellowship from the 7th Framework Programme FP7-PEOPLE-2013-IEF (Project: 624453 ALPHA-BRAIN-IMAGING).

References

1. Cuingnet, R., Gerardin, E., Tessieras, J., Auzias, G., Lehricy, S., Habert, M.O., Chupin, M., Benali, H., Colliot, O.: Automatic classification of patients with alzheimer's disease from structural mri: A comparison of ten methods using the ADNI database. NeuroImage 56(2), 766–781 (2011); Multivariate Decoding and Brain Reading

2. Padilla, P., López, M., Górriz, J., Ramírez, J., Salas-Gonzalez, D., Álvarez, I.: Nmf-svm based cad tool applied to functional brain images for the diagnosis of alzheimer's disease. IEEE Transactions on Medical Imaging 31(2), 207–216 (2012)
3. Barnes, J., Scahill, R.I., Boyes, R.G., Frost, C., Lewis, E.B., Rossor, C.L., Rossor, M.N., Fox, N.C.: Differentiating AD from aging using semiautomated measurement of hippocampal atrophy rates. NeuroImage 23(2), 574–581 (2004)
4. Henneman, W., Sluimer, J.D., Barnes, J., van der Flier, W.M., Sluimer, I.C., Fox, N.C., Scheltens, P., Vrenken, H., Barkhof, F.: Hippocampal atrophy rates in alzheimer disease: Added value over whole brain volume measures. Neurology 72(11), 999–1007 (2009)
5. Jack, C.R., Petersen, R.C., Xu, Y.C., O'Brien, P.C., Smith, G.E., Ivnik, R.J., Boeve, B.F., Waring, S.C., Tangalos, E.G., Kokmen, E.: Prediction of ad with mri-based hippocampal volume in mild cognitive impairment. Neurology 52(7), 1397–1403 (1999)
6. Nolan, J.P.: Numerical calculation of stable densities and distribution functions. Communications in Statistics. Stochastic Models 13(4), 759–774 (1997)
7. Zhang, Y., Brady, M., Smith, S.: Segmentation of brain mr images through a hidden markov random field model and the expectation-maximization algorithm. IEEE Transactions on Medical Imaging 20(1), 45–57 (2001)

A Novel Algorithm for Segmentation of Suspicious Microcalcification Regions on Mammograms

Burçin Kurt[1], Vasif V. Nabiyev[2], and Kemal Turhan[1]

[1] Department of Biostatistics and Medical Informatics, Karadeniz Technical University, Trabzon, Turkey
[2] Department of Computer Engineering, Karadeniz Technical University, Trabzon, Turkey
{burcinnkurt,kturhan.tr}@gmail.com, vasif@ktu.edu.tr

Abstract. Microcalcifications can be defined as the earliest sign of breast cancer and the early detection is very important. However, detection process is difficult because of their small size. Computer-based systems can assist the radiologist to increase the diagnostic accuracy. In this paper, we presented an automatic suspicious microcalcification regions segmentation system which can be used as a preprocessing step for microcalcifications detection. Our proposed system includes two main steps; preprocessing and segmentation. In the first step, we have implemented mammography image enhancement using top-hat transform and breast region segmentation using 3x3 median filtering, morphological opening and connected component labeling (CCL) methods. In the second step, a novel algorithm has been improved for segmentation of suspicious microcalcification regions. In the proposed segmentation algorithm, first Otsu's N=3 thresholding, then dilation process and CCL methods have been applied on preprocessed mammography image. After this process, we took the upper region from the biggest two regions and if the pixels number of the taken region was greater than the limit value, that means the upper region was the pectoral muscle region and should be removed from the image. The limit value was determined according to the database results and prevented the false region segmentation for mammography images which have no pectoral muscle region. Successful results have been obtained on MIAS database.

Keywords: Mammography image, microcalcification, image enhancement, segmentation, Otsu's N thresholding.

1 Introduction

Microcalcifications are one of the important signs of the breast cancer and detection of them is very difficult because of their small size. Furthermore, low contrast and noise features of mammograms complicate the detection process. In addition to these, approximately 10%-30% of breast cancer cases are missed by radiologists [1]. Computer-aided detection or diagnosis (CAD) systems can help radiologists as a second reader to increase the diagnosis accuracy. An automatic CAD system includes many processing steps such as image enhancement, segmentation, suspicious regions

F. Ortuño and I. Rojas (Eds.): IWBBIO 2015, Part I, LNCS 9043, pp. 222–230, 2015.
© Springer International Publishing Switzerland 2015

identification and detection. In this study, we aim to develop an automatic system for segmentation of suspicious microcalcification regions and for this purpose; we have implemented image enhancement, breast region segmentation and suspicious regions identification processes. The developed system can be used to attract the attention of the radiologist of suspicious microcalcification regions and also as a processing step for detection of microcalcifications.

For segmentation of suspicious microcalcification (MC) regions, Guan et al. [2] used scale-invariant feature transformation (SIFT) method and with this method key points were computed to segment MCs. However, many points were found initially with SIFT, which makes the segmentation process difficult. For this reason, then they set an appropriate threshold to determine the correct key points by analyzing the suspicious regions marking results of radiologists. However, the determined threshold value was static for all images which can be a disadvantage. In another study [3], the markers were determined for gradient images obtained by multiscale morphological reconstruction and then the MC clusters were identified using the watershed algorithm. However, here the process complexity is more. Balakumar et al [4], primarily decomposed the enhanced mammography image with undecimated wavelet transform and identified the suspicious MC regions using skewness and kurtosis statistical parameters which were obtained from the horizontal and vertical detail images. The mammography image was scanned with size of 32x32 overlapping windows where skewness and kurtosis values for each window were computed. If the skewness and kurtosis values are >0.2 and >4 respectively, this region is identified as suspicious MC region. Using static values for each mammography images may be a disadvantage because of different characteristics of breast tissues. Mohanalin et al. [5] proposed Havrda&Charvat entropy based thresholding method. For parameters which change in the range of (0,2] , in the optimum threshold calculation formula; a fuzzy value corresponding to each parameter is calculated and the threshold value corresponds to the maximum fuzziness is selected. Then fuzzy based enhancement was implemented using the selected threshold value to improve the visibility of microcalcifications and they were segmented by iterative thresholding.

In this study, we have used top-hat transform and 3x3 median filtering, morphological opening, connected component labeling (CCL) methods for image enhancement and breast region segmentation respectively as a preprocessing step. Then for segmentation of suspicious MC regions, Otsu's N=3 thresholding, dilation process and CCL methods have been applied on preprocessed mammography image. For this study, we have used Mammographic Image Analysis Society (MIAS) database [6].

The scheme of the proposed system is given in the following figure.

As seen in Fig. 1., enhancement and breast region segmentation processes are given in the following section and the segmentation of suspicious MC regions is given after that.

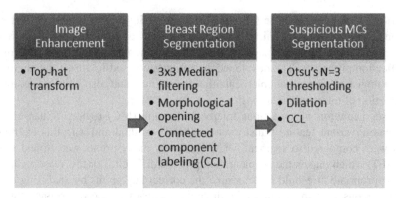

Fig. 1. The scheme of the proposed system

2   Preprocessing

Mammography images usually have low image quality and breast tissue differences also makes detection of abnormalities difficult. In addition to these, there can be noises and tags except breast region. Therefore, image enhancement and breast region segmentation processes are needed.

2.1   Image Enhancement

For enhancement of MCs in mammography image, we have used top-hat transform which is a morphological contrast enhancement method proposed by Meyer [7]. The top-hat filter protects features that conform to the structural element in the image and removes the others [8]. The opening process with top-hat can be defined by the following formula.

$$h_o = f - (f \circ A) \tag{1}$$

Where f and A show the original image and structural element respectively. Similarly, the closing process with top-hat is expressed by the formula below.

$$h_c = (f \odot A) - f \tag{2}$$

With opening process, the hill and back features are obtained and by adding these features to the original image, brighter structures are highlighted. With closing process, features such as valleys and gutters are obtained and by subtracting it from the resulting image, dark structures are highlighted [8].

$$M = f + h_o - h_c \tag{3}$$

The enhancement results for some mammograms are given in Fig. 2.

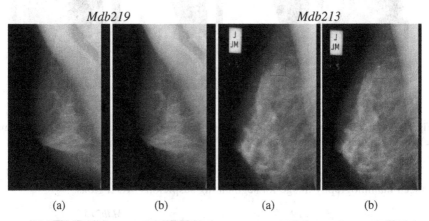

Fig. 2. (a) Original image (b) Enhanced image with top-hat transform

After the enhancement process, we have implemented the breast region segmentation to remove the unwanted parts such as labels and etc.

2.2 Breast Region Segmentation

In this section, by segmenting breast region, we aim to remove the unnecessary parts in the mammography image. Thus, the computational complexity is reduced and the success performance can be increased. The proposed algorithm for breast region segmentation is given in the following figure.

Fig. 3. The proposed algorithm for breast region segmentation

As seen in Fig. 3., while the edges were protected, the noise in the mammography image was reduced by using a 3x3 median filter, then a binary image was obtained with mathematical morphology. The objects in the image were labeled by connected component labeling method and the biggest region was segmented as the breast region. The outputs of the proposed algorithm are given in Fig. 4.

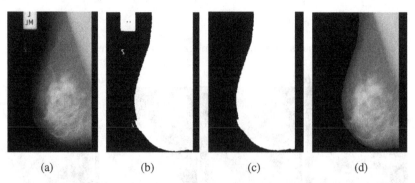

Fig. 4. (a) Original image (b) 3x3 median filter and mathematical morphology applied image (c) The biggest region obtained with CCL (d) The breast region segmented image

As a result of this section, we have obtained the breast region segmented and enhanced mammography image for segmenting suspicious MC regions.

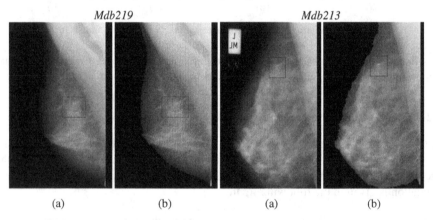

Fig. 5. (a) Original image (b) Enhanced and breast region segmented image

Thus, the preprocessing section was completed. After this section, we focused on the segmentation of suspicious MC regions on the preprocessed image.

3   Segmentation of Suspicious MC Regions

In this section, the suspicious MC regions are identified and for this purpose, the proposed algorithm is given below.

As seen in Fig. 6., the upper region identifies the pectoral muscle region and the limit value is used to control if the pectoral muscle is seen in mammography image or not. Because there is no pectoral muscle in some mammograms or it is just a bit visible. Therefore, if the pectoral muscle is visible in the mammography image it can be removed because it is an unnecessary area for MC regions.

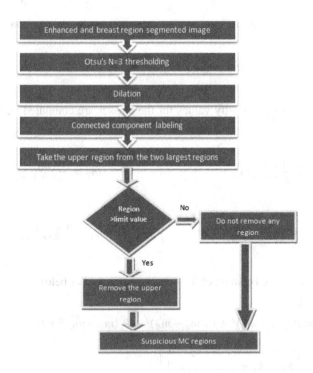

Fig. 6. The proposed algorithm for segmentation of suspicious MC regions

3.1 Otsu's N=3 Thresholding

Otsu's N thresholding is an automatic threshold selection method which separates the image into classes [9]. The steps of Otsu's N=3 thresholding algorithm are given below.

The grey levels of the image are identified as [0,1,2,…,L-1] where L is the number of grey levels.

Step 1. The normalized histogram of the input image is computed and the histogram elements are shown with p_i where $i=0, 1, 2, …, L-1$.

Step 2. The threshold values k_1, k_2, k_3 separates the image into 4 classes and the cumulative sums P_1, P_2, P_3 and P_4 are computed for 4 classes with the following formulas.

$$P_1 = \sum_{i=0}^{k_1} P_i \qquad\qquad P_2 = \sum_{i=k_1+1}^{k_2} P_i$$

$$\text{(4)}$$

$$P_3 = \sum_{i=k_2+1}^{k_3} P_i \qquad\qquad P_4 = \sum_{i=k_3+1}^{L-1} P_i$$

Step 3. The global intensity average, m_G, is computed by the formula below.

$$m_G = \sum_{i=0}^{L-1} i\, P_i \tag{5}$$

Step 4. The average intensity values for the classes are computed with the following formulas.

$$m_1 = \frac{1}{P_1} \sum_{i=0}^{k_1} i\, P_i \qquad\qquad m_2 = \frac{1}{P_2} \sum_{i=k_1+1}^{k_2} i\, P_i$$

$$\tag{6}$$

$$m_3 = \frac{1}{P_3} \sum_{i=k_2+1}^{k_3} i\, P_i \qquad\qquad m_4 = \frac{1}{P_4} \sum_{i=k_3+1}^{L-1} i\, P_i$$

Step 5. The variance between classes can be computed as below.

$$\sigma_\beta^2(k_1, k_2, k_3) = P_1(m_1 - m_G)^2 + P_2(m_2 - m_G)^2 + P_3(m_3 - m_G)^2 + P_4$$

$$\sigma_\beta^2(k_1, k_2, k_3), \begin{cases} k_1 = 1 & , \dots , \ k_2 - 1 \\ k_2 = k_1 + 1, \dots , \ k_3 - 1 \\ k_3 = k_2 + 1, \dots , \ L - 2 \end{cases} \tag{7}$$

Step 6. The optimum threshold values correspond to the values which maximizes the variance between classes.

$$\sigma_\beta^2(k_1, k_2, k_3), max\ \sigma_\beta^2(k_1, k_2, k_3), \ 0 < k_1 < k_2 < k_3 < L - 1 \tag{8}$$

3.2   Segmentation Results

In the proposed algorithm, after the implementation of Otsu's N=3 thresholding to strengthen the visibilities of MCs, dilation process has been used. Then, the number of pixels of the upper region which shows the pectoral muscle was compared with the limit value and if it is higher than the limit value, the upper region was removed from the image. In this study, the limit value was taken as 22.000 pixels according to the experimental results on the database. The outputs of the proposed algorithm are given in the following figure.

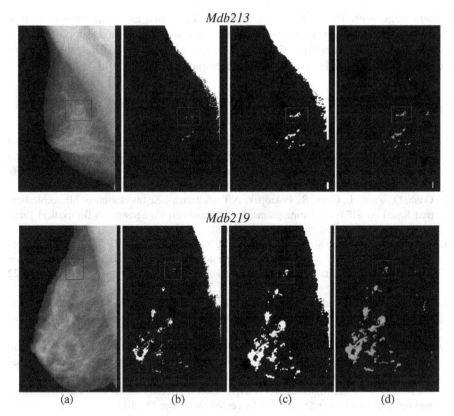

Mdb213

Mdb219

(a)   (b)   (c)   (d)

Fig. 7. (a) Enhanced and breast region segmented image (b) Otsu's N=3 thresholding (c) +Dilation (d) Suspicious MC regions segmented image

As seen in Fig. 7., windows show the MC clusters and after the segmentation of suspicious MC regions, the MC clusters can be seen and detected easily.

There are 23 mammograms which contain 28 MC clusters in MIAS database and as a result, MC clusters were never missed by the obtained suspicious MC regions.

4   Conclusion

In this paper, a novel algorithm has been proposed for segmentation of suspicious MC regions. The developed system can be used to take attention of radiologists to these regions and to simplify the detection of MC clusters. Furthermore, it can be used as a processing step for an automatic computer-aided MCs detection system. The developed algorithm for segmentation of suspicious MC regions has been never used in the literature.

Consequently, all MC clusters in MIAS database have been catched by the developed system.

Acknowledgement. The proposed algorithm has been used as a preprocessing step in developing study which is breast cancer diagnosis system and supported as a SANTEZ project by Republic Of Turkey Science, Technology and Industry Ministry and AKGÜN Computer Programs and Services Industry Tic. Ltd. Şti.

References

1. Jasmine, J.S.L., Govardhan, A., Baskaran, S.: Microcalcification Detection in Digital Mammograms based on Wavelet Analysis and Neural Networks. In: International Conference on Control, Automation, Communication and Energy Conservation, June 4-6, pp. 1–6 (2009)
2. Guan, Q., Zang, J., Chen, S., Pokropek, A.T.: Automatic Segmentation of Microcaicification Based on SIFT in Mammograms. In: International Conference on Biomedical Engineering and Informatics, pp. 13–17 (May 2008)
3. Kumar, S.V., Lazarus, M.N., Nagaraju, C.: A Novel Method for The Detection of Microcalcifications Based on Multi-Scale Morphological Gradient Watershed Segmentation Algorithm. International Journal of Engineering Science and Technology 2, 2616–2622 (2010)
4. Balakumaran, T., Vennila, I.L.A., Shankar, C.G.: Detection of Microcalcification in Mammograms Using Wavelet Transform and Fuzzy Shell Clustering. International Journal of Computer Science and Information Security 7, 121–125 (2010)
5. Mohanalin, J., Beenamol, M., Kalra, P.K., Kumar, N.: A Novel Automatic Microcalcification Detection Technique Using Tsallis Entropy & Type II Fuzzy Index. Computers and Mathematics with Applications 60, 2426–2432 (2010)
6. University of Essec, Mamographic Image Analysis Society (2003), http://peipa.essex.ac.uk/ipa/pix/mias/ (accessed December 15, 2014)
7. Meyer, F.: Iterative Image Transformations for An Automatic Screening of Cervical Smears. Journal of Histochemistry and Cytochemistry 27(1), 128–135 (1979)
8. Dabour, W.: Improved Wavelet Based Thresholding for Contrast Enhancement of Digital Mammograms. In: International Conference on Computer Science and Software Engineering, China, pp. 948–951 (December 2008)
9. Deepa, S., Bharathi, S.: Efficient ROI Segmentation of Digital Mammogram Images Using Otsu's N Thresholding Method. International Journal of Engineering Research & Technology 2(1), 1–6 (2013)

A 3D Voxel Neighborhood Classification Approach within a Multiparametric MRI Classifier for Prostate Cancer Detection

Francesco Rossi[1], Alessandro Savino[1], Valentina Giannini[2],
Anna Vignati[2], Simone Mazzetti[2], Alfredo Benso[1], Stefano Di Carlo[1],
Gianfranco Politano[1], and Daniele Regge[2]

[1] Politecnico di Torino, Control and Comp. Engineering Department, Torino, Italy
{firstname,lastname}@polito.it
http://www.sysbio.polito.it
[2] Radiology Unit, Candiolo Cancer Institute FPO, IRCCS, Candiolo (Torino), Italy
{firstname,lastname}@ircc.it

Abstract. Prostate Magnetic Resonance Imaging (MRI) is one of the most promising approaches to facilitate prostate cancer diagnosis. The effort of research community is focused on classification techniques of MR images in order to predict the cancer position and its aggressiveness. The reduction of False Negatives (FNs) is a key aspect to reduce mispredictions and to increase sensitivity. In order to deal with this issue, the most common approaches add extra filtering algorithms after the classification step; unfortunately, this solution increases the prediction time and it may introduce errors. The aim of this study is to present a methodology implementing a 3D voxel-wise neighborhood features evaluation within a Support Vector Machine (SVM) classification model. When compared with a common single-voxel-wise classification, the presented technique increases both specificity and sensitivity of the classifier, without impacting on its performances. Different neighborhood sizes have been tested to prove the overall good performance of the classification.

Keywords: Prostate cancer, magnetic resonance imaging, support vector machine, MRI classification.

1 Introduction

Prostate Cancer (PCa) is one of the most frequent cancer in males, and it is the third leading cause of cancer-related death among European men [1]. According to clinical guidelines, one of the most commonly used methods to detect prostate cancer is a Transrectal Ultrasound (TRUS) guided biopsy that, unfortunately, has been proven to provide limited efficacy to differentiate malignant from benign tissues [2]. Another accepted screening method is the antigen (PSA) blood test, which has been linked to over diagnosis and over treatments [3]. Recently, diagnostic improvements have been made by evaluating the information extracted from magnetic resonance image (MRI) sequences such as conventional

F. Ortuño and I. Rojas (Eds.): IWBBIO 2015, Part I, LNCS 9043, pp. 231–239, 2015.

morphological T1-weighted (T1-w) and T2-weighted (T2-w) imaging, diffusion-weighted MRI (DW-MRI) and dynamic contrast-enhanced MRI (DCE-MRI). These Multiparametric MRI techniques are promising alternatives for the detection of prostate cancer, as well as the evaluation of its aggressiveness [4–6].

Research studies have shown that Support Vector Machine (SVM) classifiers provide good results for classification [7–9]. Nevertheless, they have to cope with False Positives (FPs) and False Negatives (FNs) that affect the final results. From a clinical point of view FNs may lead to underestimating the cancer by detecting only portions of it, whereas FPs may lead to an extra care. Both scenarios are not acceptable and should be avoided. The reduction of FPs and FNs is therefore still a challenge that needs to be solved to effectively use Computer Aided Detection (CAD) tools or Decision Support System (DSS) for PCa. In particular, physicians look at FNs reduction (and related sensitivity increment) to avoid misprediction by detecting regions that cannot be easily seen with the naked eye.

On the computational side, these aspects are usually addressed either by implementing extra filter steps, trained to increase the specificity by reducing FPs, or by modifying the classification method. Sometimes both approaches are taken, negligently forgetting that procedures adding post-prediction filters may also decrement True Positives (TPs) or may negatively impact on future improvements of the classification step.

In this article, we propose a new methodology that relies on a 3D voxel-wise neighborhood features evaluation instead of single voxel one. We implemented the classification pipeline on top of a SVM supervised machine learning classification technique. The tool is mainly written resorting to Insight Toolkit (ITK) libraries [10] to provide a modular and cross-platform implementation of the flow. Moreover, actual implementation based ITK algorithms may take advantage of the multiple processors present in most common systems and ensure faster classification time.

In terms of classification, preliminary results show interesting classification improvements w.r.t. the single voxel classification process.

2   Materials and Methods

In this section, we first describe the available multiparametric MRI dataset, then we present the idea of the 3D-voxel neighborhoods classifier approach, together with its implementation based on ITK and LIBSVM tool [11].

2.1   Dataset

The available dataset consists of 28 patients, who underwent MRI before prostatectomy. The mean age is 64 years and they were selected among patient of the same hospital. Since personal data are confidential and removed from the MRI sequences, we have no other information about them. A pathologist contoured tumors on histological sections and a radiologist outlined regions of interest

(ROI) on the T2-w images in areas corresponding to each PCa. Non tumor ROI were also outlined in each patient to define a balanced dataset useful in the training stage. A total of 28 tumors with size bigger than 0.5 cc (median: 1,64 cc; 1st-3rd quartile: 0,75-2,25 cc) were included in the dataset. DW-MRI and DCE-MRI sequences were automatically aligned to T2-w images and finally each pixel, belonging to the prostate automatically segmented, was represented as a vector of quantitative parameters (i.e. T2-w signal intensity, the ADC value and K trans)[12, 13].

2.2   Classifier

Before going into methods and implementation details, we briefly describe the architecture of the classification tool. It consists of two different stages: a training and a testing stage. Both stages are summarized in Figure 1. Since the ITK libraries follows the Objected Oriented Programming (OOP) paradigm, it can be noticed that the architecture is fully modular. Together with the usage of the LIBSVM library as a reliable implementation for a SVM classifier [11], this choice aims at guaranteeing compatibility and flexibility towards further functionality improvements.

The very first step of both stages is a data adaptation provided by means of an ITK filter (called DataProcess in the schema of Figure 1). This step is merely required by the discrepancy between the actual dataset format and the one imposed by LIBSVM. It consists on vector permutations and aggregations; actual values are not modified during this process. If the DataProcess filter, due to further changes on the dataset format, will be no longer necessary, it will be avoided.

Once adapted, all the data are normalized and standardized as required by the SVM classification methodology [14]. The Normalization class filter implements this step. It provides both the normalized data and the normalization parameters. Normalization parameters are necessary to apply the normalization process to new data, i.e., when any further prediction is needed.

The normalized data are then forwarded to the training class, where the final SVM classifier model is trained and created. The Training class filter is also able to output a final model description that is very useful to set up the classifier whenever needed, without training the model from scratch again. Avoiding re-training time waste hugely impacts to the classifier timing performances.

When the training is completed and we need to employ it for prediction, the testing flow is very similar to the training one: data coming from the feature extraction are formatted and subsequently normalized resorting to the parameters saved during the training phase. The prediction is also performed restoring the model previously saved during the training phase. The outputs is an ITK compliant image, representing a probability map where, for each pixel of the original morphological sequence, a cancer probability value is computed.

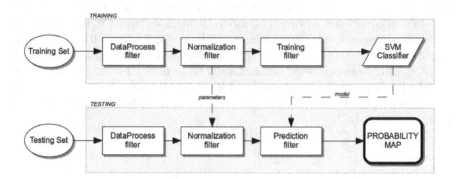

Fig. 1. The Classifier Training and Testing Flows

2.3 The 3D Voxel Neighborhood Approach

At the morphological level, tumors are clusters of voxels, spreading along three dimensions. Even on MRI sequences, all of them start as a two-dimensions artifact on one slice and then progress slice by slice until the end of their mass. Thus, it makes sense to consider at the same time not only on the set of features extracted for the voxel under-evaluation but also on the pixels surrounding it, both on the same slice and on the previous and following slices. Under these assumptions, when the classifier deals with potential FP voxels surrounded by True Negative (TN) voxels, we may increase the chance of classifying them as TN. The same way potential FN voxels within a TP voxels neighborhood likely results in a TP classification.

To formalize this idea, we first define a radius concept as the distance from a central voxel (see Figure 2). This distance can be evaluated as a voxel-wise one (R_{vx}) or using a common length metrics, i.e. millimeters (R_{mm}), resorting to the spacing information ($S_{x,y,z}$) available within the header of any MRI sequence compliant with the DICOM standard [15]. This last type of evaluation should allow better results since spacing is known to be bigger on Z-axis than on X- and Y-axis (e.g., 3 millimeters from one slice to another versus 0.3 millimeters between two adjacent voxels within a slice). The radius identifies a 3D box where all belonging voxels can be processed together.

Assuming that for each axial direction (x, y, z) we can relate a voxel-wise radius with common length metrics radius (millimeters in the equation) as:

$$R_{vx} = \left\lceil \frac{R_{mm}}{S_{(x,y,z)}} \right\rceil \tag{1}$$

we are able to express the number of voxel selected with respect to each direction as:

$$\begin{cases} NumVox_x = ((2R_{vx}) + 1) \\ NumVox_y = ((2R_{vx}) + 1) \\ NumVox_z = ((2R_{vx}) + 1) \end{cases} \tag{2}$$

Eventually, equation 2 allows to evaluate the total number of voxels processed in the 3D neighborhood as:

$$NumVox_{tot} = (NumVox_x NumVox_y NumVox_z)$$
$$= ((2R_{vx}) + 1)^3 \tag{3}$$

Figure 2 shows quantitatively how many voxel will be considered when the radius is set to 1 voxel. Both the Training and the Prediction class filters are able to process 3D voxel-wise neighborhoods as well as single voxels.

Once a radius is defined, each feature belonging to the voxels included in the 3D box is evaluated by averaging its value among all single voxel values. This averaged value is employed as final value of the feature. Currently, studies on different evaluation approaches are under analysis.

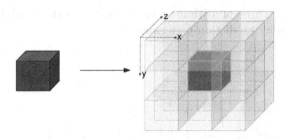

Fig. 2. 3D Voxel Neighborhood with a $R_{vx} = 1$. A total of 27 voxels are then considered.

It has to be emphasized in here that, during the developed supervised training, where the dataset is originated by user-selected malignant and benignant ROI, the Training class filter averages the feature value only of voxels of the same ROI target type. This strategy avoids the training class filter to break the supervised training rules. Moreover, this way, volumetric information could be weighted in the malignant voxels set described by $NumVox_{tot}$ neighborhood voxels; thereby FP isolated voxels could be filtered and the borders of malignant ROI may result more accurate. In line with these assumption physicians consider and highlight tumors with a minimum volume size that have diagnostic relevance (i.e. 0,5cc). It is also important to mention that features are collected voxel by voxel. Any other form of feature extraction introduces losses in the original data.

Experiments were performed with different radius values in order to investigate the classifier performances against single voxel-wise classification and will be provided on Section 3. In technical terms, the neighborhood selection is implemented making use of *itkConstNeighborhoodIterator* and *itkImageRegionConstIterator* ITK class templates.

3   Results

In this section we present the results for 3D voxel neighborhood classifier and we discuss about obtained performances.

3.1   Data Evaluation

The training set was built upon 28 patients and a leave-one-out cross-validation (LOOCV) has been implemented to validate the model. We performed different experiments varying the radius (R_{vx}): six different radius (R_i) size were tested: $R_{i=\{0,1,2,3,4,5\}}$, where $R_{i=0}$ means a standard single voxel-wise classification. Figure 3 compares results by means of the following statistical functions:

- Area Under the ROC Curve (AUC) [16]
- Sensitivity (Se): $Se = \frac{TP}{TP+FN}$
- Specificity (Sp): $Sp = \frac{TN}{TN+FP}$
- Accuracy (Acc): $Acc = \frac{TP+TN}{TP+TN+FP+FN}$

In particular, we present arithmetic mean (AM) and standard deviation (SD) to compare classifier prediction performances.

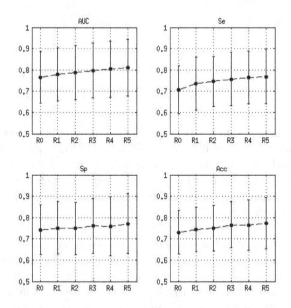

Fig. 3. AUC, Se, Sp and Acc average (dot) and standard deviation (whiskers) values, with relation to radius size (R_i), for testing set

Since the final decision has to be binary (tumor vs healthy tissue), a thresholds cutoff was set to 0.5 value during the experimental setup. This is a very weak threshold, which commonly leads to worst results comparing with best cutoff search algorithms, such as the Yoden one [17, 18]. The meaning of our choice is strongly related to the way physicians exploit the classification results. Within a diagnosis path, the CAD software is usually able to show the physician the classification results as a colored overlay map on the T2-w sequence [19]. Colors help focusing on the tumor areas but their relation to the classifier outcome is changed as physician commodity.

3.2   Discussion

Figure 3 reveals that our 3D voxel-wise neighborhood classifier provide an improvement of the classification performances. We highlight the AM improvement in term of AUC, Se, Sp and Acc. In particular, when comparing R_0 to R_5, the results seem very promising.

Generally speaking, the progressive increase of the radius relates to a continuous improvement in results. The Se reveals a significant improvement ranging its average from 0.71 to 0.78 (R_0 to R_5), as well as the improvement of the AUC value is significant when single voxel-wise classifier performances are compared with R_5 neighborhood classifier. In this case the average rises from 0.76 to 0.81. Nevertheless, the Sp seems to do not benefit from our proposed technique (R_0 to R_5 delta is less than 0.03), suggesting that the way features on different voxel are considered may need further investigation to refine the methodology. In particular, the AM variation expressed along the radius, may suggest that more considerations on the morphological characteristic of tumors among the dataset are needed. We expected such kind of impact on final results thus further investigations will be planned on that.

Eventually, SD values are generally not negligible but this may depend on the actual available dataset size.

4   Conclusion

We present a 3D voxel neighborhood SVM classifier methodology, implemented using ITK libraries, based on MRI sequences to discriminate prostate cancer lesion from healthy tissue.

Obtained results indicate improvements if compared against traditional single voxel-wise classifier; especially FNs take advantage from the proposed approach. Some minor drawbacks suggest a further analysis involving an extended dataset to confirm the validity volumetric neighborhood approach performance.

References

1. European Cancer Observatory (ECO), http://eu-cancer.iarc.fr/
2. Hegde, J.V., Mulkern, R.V., Panych, L.P., Fennessy, F.M., Fedorov, A., Maier, S.E., Tempany, C.M.C.: Multiparametric MRI of Prostate Cancer: An Update on State-Of-The-Art Techniques and their Performance in Detecting and Localizing Prostate Cancer. Journal of Magnetic Resonance Imaging 37, 1035–1054 (2013)
3. Ongun, S., Celik, S., Gul-Niflioglu, G., Aslan, G., Tuna, B., Mungan, U., Uner, S., Yorukoglu, K.: Are Active Surveillance Criteria Sufficient for Predicting Advanced Wtage Prostate Cancer Patients? Actas Urologicas Espanolas 38, 499–505 (2014)
4. Tan, C.H., Wang, J.H., Kundra, V.: Diffusion Weighted Imaging in Prostate Cancer. European Radiology 21, 593–603 (2011)
5. Desouza, N.M., Riches, S.F., VanAs, N.J., Morgan, V.A., Ashley, S.A., Fisher, C., Payne, G.S., Parker, C.: Diffusion-Weighted Magnetic Resonance Imaging: a Potential Non-Invasive Marker of Tumor Aggressiveness in Localized Prostate Cancer. Clinical Radiology 63, 774–782 (2008)
6. Turkbey, B., Bernardo, M., Merino, M.J., Wood, B.J., Pinto, P.A., Choyke, P.L.: MRI of Localized Prostate Cancer: Coming of Age in the PSA Era. Diagnostic and Interventional Radiology 18, 34–45 (2012)
7. Artan, Y., Haider, M.A., Langer, D.L., van der Kwast, T.H., Evans, A.J., Yang, Y.Y., Wernick, M.N., Trachtenberg, J., Yetik, I.S.: Prostate Cancer Localization With Multispectral MRI Using Cost-Sensitive Support Vector Machines and Conditional Random Fields. IEEE Transactions on Image Processing 19, 2444–2455 (2010)
8. Zhou, T., Lu, H.L.: Multi-Features Prostate Tumor Aided Diagnoses Based on Ensemble-SVM. In: 2013 IEEE International Conference on Granular Computing (Grc), pp. 297–302 (2013)
9. Shah, V., Turkbey, B., Mani, H., Pang, Y.X., Pohida, T., Merino, M.J., Pinto, P.A., Choyke, P.L., Bernardo, M.: Decision Support System for Localizing Prostate Cancer Based on Multiparametric Magnetic Resonance Imaging. Medical Physics 39, 4093–4103 (2012)
10. The Insight Segmentation and Registration Toolkit, http://www.itk.org
11. Chang, C.C., Lin, C.J.: LIBSVM: A Library for Support Vector Machines. ACM Transactions on Intelligent Systems and Technology 2, 27 (2011), Software available at http://www.csie.ntu.edu.tw/~cjlin/libsvm
12. Peng, Y.H., Jiang, Y.L., Yang, C., Brown, J.B., Antic, T., Sethi, I., Schmid-Tannwald, C., Giger, M.L., Eggener, S.E., Oto, A.: Quantitative Analysis of Multiparametric Prostate MR Images: Differentiation between Prostate Cancer and Normal Tissue and Correlation with Gleason Score-A Computer-aided Diagnosis Development Study. Radiology 267, 787–796 (2013)
13. Tamada, T., Sone, T., Jo, Y., Yamamoto, A., Ito, K.: Diffusion-Weighted MRI and its Role in Prostate Cancer. Nmr in Biomedicine 27, 25–38 (2014)
14. Duda, R.O., Hart, P.E., Stork, D.G.: Pattern Classification. Wiley, New York (2001) ISBN:0-471-05669-3
15. Pianykh, O.S.: Digital Imaging and Communications in Medicine (DICOM). Springer, Heidelberg (2012) ISBN:978-3-642-10849-5
16. Hanley, J.A., McNeil, B.J.: The Meaning and Use of the Area Under a Receiver Operating Characteristic (ROC) Curve. Radiology 143, 29–36 (1982)
17. Martnez-Camblor, P.: Nonparametric Cutoff Point Estimation for Diagnostic Decisions with Weighted Errors. Revista Colombiana de Estadstica 34(1), 133–146 (2011)

18. Fluss, R., Faraggi, D.: FAU - Reiser, Benjamin, Estimation of the Youden Index and its associated cutoff point. Biometrical Journal. Biometrische Zeitschrift 47(4), 458–472 (2005)
19. Savino, A., Benso, A., Di Carlo, S., Giannini, V., Vignati, A., Politano, G., Mazzetti, S., Regge, D.: A Prostate Cancer Computer Aided Diagnosis Software including Malignancy Tumor Probabilistic Classification. In: International Conference on Bioimaging (BIOIMAGING), Eseo, Angers, FR, March 3-6, pp. 49–54 (2014)

Towards Precise Segmentation
of Corneal Endothelial Cells

Adam Piórkowski[1] and Jolanta Gronkowska–Serafin[2,3]

[1] Department of Geoinfomatics and Applied Computer Science
AGH University of Science and Technology,
A. Mickiewicza 30 Av., 30–059 Cracow, Poland
pioro@agh.edu.pl
[2] Department of Ophthalmology
Pomeranian Medical University
Powstańców Wielkopolskich 72 Av., 70–111 Szczecin, Poland
[3] Oejenafdelingen, Regionshospitalet Holstebro
Laegaardvej 12a, 7500 Holstebro, Denmark

Abstract. This article describes an algorithm for defining the precise, objective, repeatable and unambiguous segmentation of cells in images of the corneal endothelium. This issue is important for clinical purposes, because the quality of the grid cells is assessed on the basis of segmentation. Other solutions, including commercial software, do not always mark cell boundaries along lines of lowest brightness.

The proposed algorithm is comprised of two parts. The first part determines the number of neighbors of less than or equal brightness to each image point in the input image, then a custom-made segmentation of the binary image is performed on the basis of the constructed map. Each of the 9 iterations of the segmentation considers a number of neighboring points equal to the iteration index, thinning them if they have equal or lower value than the analyzed input point, which allows the boundaries to be routed between cells through the darkest points, thus defining an objective and unambiguous selection of cells.

Keywords: image processing, cell counting, segmentation, thinning, corneal endothelium, endothelial cell.

1 Introduction

The corneal endothelium is a monolayer of mostly hexagonal cells which are responsible for stable corneal hydration rate. Unlike in other species, corneal endothelial cells in humans and primates do not present mitotic activity in vivo, which results in diminishing corneal endothelial cell density with increased age. Age related density loss is exacerbated by many diseases as well as surgery on the anterior segment of the eye, which can lead to irreversible corneal swelling leading to functional sight loss. Corneal endothelial cells can be assessed in vivo in a specular microscope. Corneal cell density (CD, cells/mm^2), percentage of hexagonal cells (H,%), and the variation coefficient of the cell's size (CV, %)

F. Ortuño and I. Rojas (Eds.): IWBBIO 2015, Part I, LNCS 9043, pp. 240–249, 2015.

can be obtained. All these parameters are sensitive and dependent on accurate segmentation [3,4]. The proposed new $CVSL$ parameter is very sensitive, as it is based on triple points in a cell's grid [6,17].

2   Features of Corneal Endothelial Images and Preprocessing

Corneal endothelium images obtained with confocal microscopy are characterized by moderately large ranges of distribution variability and brightness dynamics a lot of noise [15]. These features make it impossible to use standard segmentation algorithms such as Watershed, which is described in detail in [16]. For this reason, a custom solution for appropriate cell segmentation in images should be developed.

2.1   Image Preprocessing

The first step of image processing and analysis is usually preprocessing. If the degree of noise in images of the corneal endothelium is significant, this can cause local minima and maxima to appear in unexpected places, however this can be reduced by smoothing with a convolution filter mask. Tests have proved that a 5 × 5 mask filled with values of 1 positioned in the shape of a diamond is sufficient. The issue of noise removal from corneal endothelial images will be the subject of future work [5]. Alignment of dynamics and ranges of brightness can be achieved in other ways, if this is required for a binarization algorithm. Detailed preprocessing of corneal endothelial images is described in [15,24].

2.2   Binarization

Images of the corneal endothelium can be binarized using various algorithms.

In [21] the author describes the binarization process as including the following stages: removal of the lowest frequency, gap closing, contrast enhancement, thresholding, skeletonization, and finally improvements to the output image (pruning and elimination of spurious structures).

An algorithm which uses shape masks for convolution is presented in [10]. The authors use three kinds of mask: 'tricorn', orthogonal (vertical and horizontal) and diagonal.

An interesting approach is to use pyramid methods in image processing [1]. In [8] the authors used wavelet pyramidal decomposition to carry out the binarization of corneal endothelial cells, followed by morphological operations to assess the thinned boundaries.

A very interesting method is presented in [11]. The authors propose a scissoring operator that separates cells in the binary image, where the standard solution is to use a kind of watershed algorithm.

One of the standard approaches in the identification of cells is the top-hat algorithm. The authors also propose a binarization method that consists of adjusting masks to captured parts of the valleys. In [14] a mask is proposed for detecting local reductions. The KH Algorithm [7] improves this algorithm by indicating the course of the valleys between cells and a filter which makes it less susceptible to local defects. This algorithm uses four directional edge masks and an object size filter to remove small objects. Fig. 1 shows an example portion of an image of a corneal endothelium (Fig. 1(a)), the effect of smoothing (Fig. 1(b)), and the binarization results of the top-hat (Fig. 1(c)) and KH (Fig. 1(d)) algorithms at a manually selected optimum threshold.

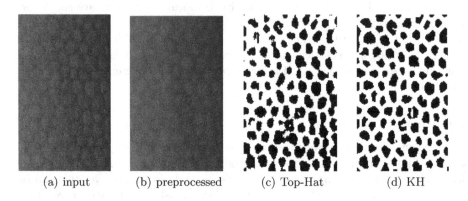

| (a) input | (b) preprocessed | (c) Top-Hat | (d) KH |

Fig. 1. Corneal endothelial image sample - original, after preprocessing, top-hat and KH algorithm

2.3   Thinning

Thinning is the next step, which allows the contour of cells to be extracted from a binarized image. There are a several approaches to implementing thinning [23,13], a summary of which is in [20]. The authors of this article performed thinning by carrying out a series of iterations of morphological operations for one of three sets of masks. These masks are as follows (2(a), 2(b) and 2(c)), with orientation 0°, 90° 180° and 270°.

Set A performs simpler segmentation (3(c)), while set C extracts a more precise course of the skeleton (3(d)).

3   The Problem of Imprecise Segmentation

The issue of determining precise cell boundaries is quite important. For example, Fig. 4 shows a sample of segmentation produced by a commercial program, containing several places where the boundaries of the division between cells do not run strictly in the valleys between cells.

The presented software does not provide a precise and fully correct description of grid cells. It should be also noted that the use of thinning in the aforementioned binarization algorithms draws a grid segmentation in the middle of cells. Therefore, it is worth considering the creation of an algorithm which makes precise (repeatable and objective) segmentation lines that run in the valleys.

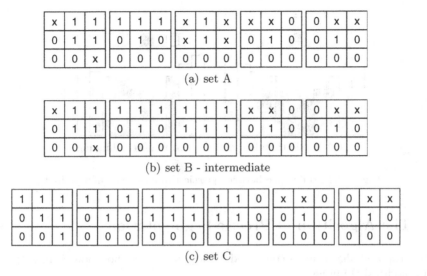

(a) set A

(b) set B - intermediate

(c) set C

Fig. 2. The sets of mask for thinning

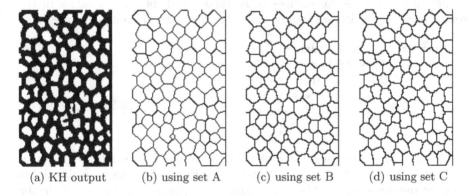

(a) KH output   (b) using set A   (c) using set B   (d) using set C

Fig. 3. Thinning outputs

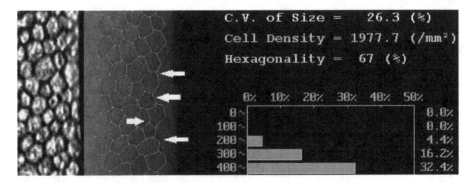

Fig. 4. An example of commercial software

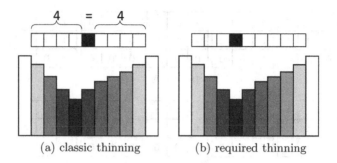

(a) classic thinning   (b) required thinning

Fig. 5. The difference between classic thinning and the required

4   The Algorithm of Precise Segmentation

The proposed algorithm is composed of two parts: neighborhood point mapping and modified thinning.

The first part of the algorithm creates a neighborhood map (n) which, after preprocessing, selects the number of neighbors with brightness values in the source image (p) which are greater than or equal to the given input image pixel.

```
for (x = 1; x < Width -1; x++)
    for (y = 1; y < Height - 1; y++)
        for (i = -1; i <= 1; i++)
            for (j = -1; j <= 1; j++)
                if (p[x + i, y + j] >= p[x, y])
                    n[x, y]++;
```

This (3×3) neighborhood map shows the top of cells as smaller values (min = 1) and valleys as higher values (max = 9). Figure 6 shows a neighborhood map for the input image (Fig. 1(a)), normalized (range of 1-9 to 0-255), in the negative.

Fig. 6. The neighborhood map (normalized to 0-255, inverted)

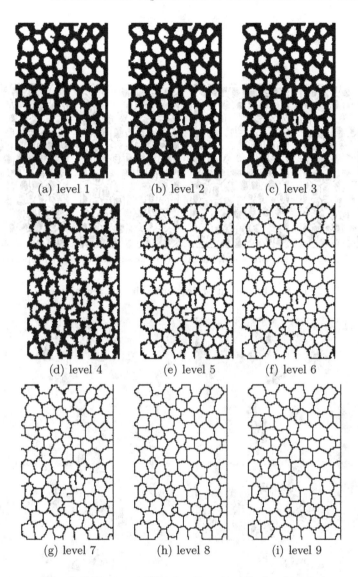

(a) level 1   (b) level 2   (c) level 3

(d) level 4   (e) level 5   (f) level 6

(g) level 7   (h) level 8   (i) level 9

Fig. 7. Subsequent thinning iterations for levels 1-9

The second stage is modified thinning. This consists of performing 9 iterations of classical thinning. Each iteration processes thinning for points that have values in the neighborhood map corresponding to a value that is no greater than the number of the current iteration (1-9). Taking the second iteration as an example, if there are 2 pixels in the mask that have a brightness value equal to or higher than the central pixel then the value of the map for this pixel is of 2.

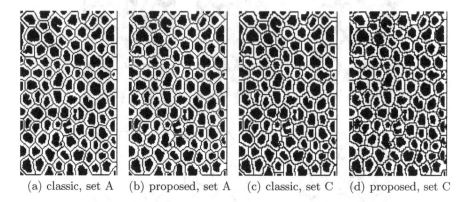

(a) classic, set A   (b) proposed, set A   (c) classic, set C   (d) proposed, set C

Fig. 8. The output of descibed thinning techniques

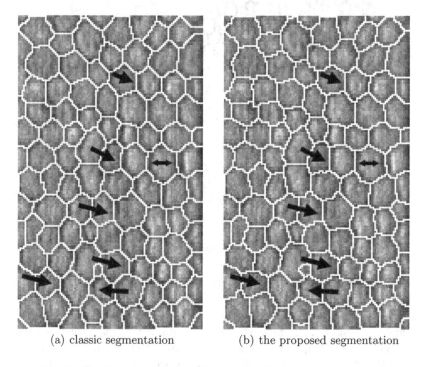

(a) classic segmentation   (b) the proposed segmentation

Fig. 9. The output of classic segmentation (a) and the proposed (b)

5   Results

Fig. 8 shows examples of binarization algorithm KH with overlapped classic thinning using A and C masks, and thinning using the proposed algorithm (also for thinning A and C). It can be seen that while the segmentation lines in Fig. 8(a) and Fig. 8(c) run in the middle, in the case of the proposed algorithm the lines adapt to the valleys.

The difference between classic segmentation (Fig. 9(a)) and the proposed (Fig. 9(b)) is shown on real images (magnified, using mask B).

6   Conclusion

The article presents the problem of imprecise segmentation in corneal endothelium images. This problem also affects commercial software. One of the reasons for this error is the use of binarization. The next step after binarization is thinning, which creates a result dependent on the selection of masks which are not related to input image. A method which overcomes this problem is presented in the article.

The authors' method of determining cell boundaries allows the precise and representative segmentation of the corneal endothelium. The first part of the algorithm maps neighboring pixels The second part performs thinning by referring to the neighborhood map. As indicated in the resulting images, this method offers better segmentation.

Further studies should involve full testing of the proposed algorithm, its impact on the quality of the grid, as well as estimation of approximation depending on a grid based on triple points with regards to the different methods of segmentation. Other neighborhood sizes and binarization methods will be considered [19,22]. The most promising approach seems to involve the use of an active contour technique [2]. The result of segmentation in some cases can be improved using intelligent methods such as neural networks or rough sets [12,18,9].

Acknowledgement. This work was financed by the AGH - University of Science and Technology, Faculty of Geology, Geophysics and Environmental Protection, Department of Geoinformatics and Applied Computer Science as a part of statutory project.

References

1. Adelson, E.H., Anderson, C.H., Bergen, J.R., Burt, P.J., Ogden, J.M.: Pyramid methods in image processing. RCA Engineer 29(6), 33–41 (1984)
2. Charłampowicz, K., Reska, D., Boldak, C.: Automatic segmentation of corneal endothelial cells using active contours. Advances in Computer Science Research 11, 47–60 (2014)
3. Doughty, M.: The ambiguous coefficient of variation: Polymegethism of the corneal endothelium and central corneal thickness. International Contact Lens Clinic 17(9-10) (1990)

4. Doughty, M.: Concerning the symmetry of the 'hexagonal' cells of the corneal endothelium. Experimental Eye Research 55(1), 145–154 (1992)
5. Fabijańska, A., Sankowski, D.: Noise adaptive switching median-based filter for impulse noise removal from extremely corrupted images. IET Image Processing 5(5), 472–480 (2011)
6. Gronkowska-Serafin, J., Piórkowski, A.: Corneal endothelial grid structure factor based on coefficient of variation of the cell sides lengths. In: Choras, R.S. (ed.) Image Processing and Communications Challenges 5. AISC, vol. 233, pp. 13–20. Springer, Heidelberg (2014)
7. Habrat, K.: Binarization of corneal endothelial digital images. Master's thesis, AGH University of Science and Technology (2012)
8. Khan, M.A.U., Niazi, M.K.K., Khan, M.A., Ibrahim, M.T.: Endothelial cell image enhancement using non-subsampled image pyramid. Information Technology Journal 6(7), 1057–1062 (2007)
9. Korkosz, M., Bielecka, M., Bielecki, A., Skomorowski, M., Wojciechowski, W., Wójtowicz, T.: Improved fuzzy entropy algorithm for x-ray pictures preprocessing. In: Rutkowski, L., Korytkowski, M., Scherer, R., Tadeusiewicz, R., Zadeh, L.A., Zurada, J.M. (eds.) ICAISC 2012, Part II. LNCS, vol. 7268, pp. 268–275. Springer, Heidelberg (2012)
10. Mahzoun, M., Okazaki, K., Mitsumoto, H., Kawai, H., Sato, Y., Tamura, S., Kani, K.: Detection and complement of hexagonal borders in corneal endothelial cell image. Medical Imaging Technology 14(1), 56–69 (1996)
11. Nadachi, R., Nunokawa, K.: Automated corneal endothelial cell analysis. In: Proceedings of Fifth Annual IEEE Symposium on Computer-Based Medical Systems, pp. 450–457. IEEE (1992)
12. Ogiela, M., Tadeusiewicz, R.: Artificial intelligence methods in shape feature analysis of selected organs in medical images. Image Processing and Communications 6, 3–11 (2000)
13. Pavlidis, T.: Algorithms for graphics and image processing. Computer science press (1982)
14. Piorkowski, A., Gronkowska-Serafin, J.: Analysis of corneal endothelial image using classic image processing methods. In: KOWBAN - XVIII The Computer-Aided Scientific Research. The Works of Wroclaw Scientific Society, B, vol. 217, pp. 283–290. Wroclawskie Towarzystwo Naukowe (2011)
15. Piorkowski, A., Gronkowska-Serafin, J.: Selected issues of corneal endothelial image segmentation. Journal of Medical Informatics & Technologies 17, 239–245 (2011)
16. Piórkowski, A., Gronkowska–Serafin, J.: Towards automated cell segmentation in corneal endothelium images. In: Choraś, R.S. (ed.) Image Processing & Communications Challenges 6. AISC, vol. 313, pp. 181–189. Springer, Heidelberg (2015)
17. Piórkowski, A., Mazurek, P., Gronkowska–Serafin, J.: Comparison of assessment regularity methods dedicated to isotropic cells structures analysis. In: Choraś, R.S. (ed.) Image Processing & Communications Challenges 6. AISC, vol. 313, pp. 171–180. Springer, Heidelberg (2015)
18. Placzek, B.: Rough sets in identification of cellular automata for medical image processing. Journal of Medical Informatics & Technologies 22, 161–168 (2013)
19. Ridler, T., Calvard, S.: Picture thresholding using an iterative selection method. IEEE Transactions on Systems, Man and Cybernetics 8(8), 630–632 (1978)
20. Saeed, K., Tabędzki, M., Rybnik, M., Adamski, M.: K3M: A universal algorithm for image skeletonization and a review of thinning techniques. International Journal of Applied Mathematics and Computer Science 20(2), 317–335 (2010)

21. Sanchez-Marin, F.J.: Automatic segmentation of contours of corneal cells. Computers in Biology and Medicine 29(4), 243–258 (1999)
22. Sauvola, J., Pietikäinen, M.: Adaptive document image binarization. Pattern Recognition 33(2), 225–236 (2000)
23. Serra, J., Mlynarczuk, M.: Morphological merging of multidimensional data. In: Proceedings of STERMAT 2000, pp. 385–390 (2000)
24. Szostek, K., Gronkowska-Serafin, J., Piorkowski, A.: Problems of corneal endothelial image binarization. Schedae Informaticae 20, 211–218 (2011)

Sliding Box Method for Automated Detection of the Optic Disc and Macula in Retinal Images

Dan Popescu, Loretta Ichim, and Radu Dobrescu

Faculty of Automatic Control and Computers, University Politehnica of Bucharest, Romania
dan.popescu@upb.ro, loretta.ichim@aii.pub.ro,
rd_dobrescu@yahoo.ro

Abstract. In this paper we propose two simple and efficient algorithms for automated detection and localization of optic disc and macula with high accuracy. In the learning phase a set of statistical and fractal features were tested on 40 images from STARE database. The selected features combine spatial and spectral properties of the retinal images and are based on minimum or maximum criteria. In the first step of the algorithm, a sliding box method is used for primary detection of optic disc and macula. For the second phase, a non overlapping box method is proposed for accurate localization of the center of the optic disc. The features are different for the two phases and also for the two cases optic disc and macula. The algorithms were tested on a set of 100 retinal images from the same database. By accurate determination position of the optic disc and macula, the results confirm the efficiency of the proposed method.

Keywords: retinal images, image processing, features, sliding box method, optic disc detection, macula detection.

1   Introduction

Loss of vision affects quality of life and creates difficulty carrying out daily activities. It can be prevented by early detection through regular screenings and appropriate medical treatment. Therefore, the analysis of the main retinal components is fundamental in ophthalmological care. The detection and localization of the optic disc (OD) and macula (MA) can be considered as a starting point in the diagnosis of ophthalmic diseases such as: diabetic retinopathy, glaucoma and macular degeneration, which are major causes of a loss of vision.

The main features of eye fundus images include OD, MA and blood vessels (BV). The OD has an approximately round form and is a brighter region than the rest of the ocular fundus interrupted by the outgoing blood vessels. The MA is a highly sensitive part of the retina and is located roughly in the center of it.

Previously, several teams of researchers have studied a variety of automated methods to segment and quantify the OD and MA from digital retinal images [1-9].

Carmona et al. introduced an automatic system to locate and segment the OD in eye fundus images using genetic algorithms [1]. However, this method has a potential

F. Ortuño and I. Rojas (Eds.): IWBBIO 2015, Part I, LNCS 9043, pp. 250–261, 2015.

error if the shape of the OD is not a perfect ellipse, and the method did not segment the OD cup.

Xu et al. developed an automated system for assessment of the OD on stereo disc photographs [2]. They used a deformable model technique, which deforms an initial contour to minimize the energy function defined from contour shape.

Youssif et al. [3] proposed an algorithm for OD detection based on matching the expected directional pattern of the retinal blood vessels. The retinal vessels are segmented using a simple and standard 2-D Gaussian matched filter. The segmented vessels are then thinned, and filtered using local intensity, to represent finally the OD center candidates.

Dehghani et al. [4] detected the OD in retinal images by extracting the histograms of each color component. Then, the average of the histogram for each color was calculated as a template for localizing the center of OD. Another model-based (template matching) approach was employed by Osareh et al. to approximately locate the OD [5]. The images were normalized by applying histogram specification, and then the OD region from color-normalized images was averaged to produce a gray-level template.

In [6], Qureshi et al. suggested an approach to automatically combine different OD and MA detectors, to benefit from their strengths while overcoming their weaknesses. The criterion for the selection of the candidate algorithms to be combined is based either on good detection accuracy or low computation time.

More recently, Fuente-Arriaga et al. [7] presented a methodology for glaucoma detection based on measuring displacements of blood vessels within the OD (vascular bundle) in human retinal images.

Pereira et al. [8] applied an algorithm using ant colony optimization inspired by the foraging behavior of some ant species that has been used in image processing with different purposes. This algorithm, preceded by anisotropic diffusion, was used for OD detection in color fundus images.

Li and Chutatape [9] proposed a new method to localize the optic disc based on principal component analysis, which, using the advantage of a top-down strategy, can extract the common characteristics among the training images. Also, the authors propose to extract other main features in color retinal images, such as the shape of the OD, fovea localization, set up fundus coordinate system, and detection of exudates.

In this paper we propose a new approach for automated detection of the OD and MA using the sliding box method, the non overlapping box method and adaptive texture analysis. The localization of these regions of interest (ROIs) in fundus images is an important problem in retinal image processing in order to improve the diagnostic accuracy in eye diseases.

2   Methodology and Algorithms

We considered images from the STARE database (700×605 pixels), with dimension $M \times M$ and RGB representation on 24 bits (8 bits for each color component) [10]. Because the dominant color is red-brown, we used the green channel to emphasize the areas with blood vessels and the blue channel to reduce the background influence and to emphasize the bright areas, like the OD. The red channel was used to extract complementary features. In order to select some efficient features for discriminating and localizing the regions of interest (OD and MA), we analyzed a set of 40 images.

The proposed algorithm has two phases. The first is based on the sliding box method with a window of 132×132 pixels. This allows dividing the window in 16 small windows (33×33) with a central point. The dimension of the window and also the sliding step length (5 pixels) were conveniently chosen so that the OD and MA are completely framed in one window. The output of this phase is the box which fits best the OD or MA, and is determined by optimizing the selected features described below (different for OD and MA). For ODs, the procedure continues the second phase of the algorithm with a fine localization of the OD center inside the selected box by using analysis based on non overlapping small boxes with dimensions of 33×33 pixels. The selected features also differ from those of the first step.

The retinal image has regions with different intensities and also contains pixels or small patterns in the background which can be considered as noise. Therefore, preprocessing operations, like noise rejection and contrast enhancement, are needed. The noise rejection is the first operation which must be done. On each color channel, two operations achieve this goal: the median filtering and the morphological erosion and dilation on a 3×3 window.

Because the threshold operator generates errors [11], direct binarization of retinal images is not recommended, so we propose a morphological combined operation: erosion and dilation directly on gray images.

The retinal images have a relatively low local contrast. Therefore, histogram equalization is carried out.

2.1 Selection of Features

In order to detect and localize the regions of interest in the retinal image, we tested both statistical and fractal features on the R, G and B color channels. By combining the spectral information (color channels), statistical and spatial information (fractal and textural features) the performance of the algorithm is increased. During the learning process, a set of simple and efficient features from different color channels was tested and then selected. The tested features were the most frequently used statistical and fractal types: mean intensity, energy, contrast, homogeneity, variance, entropy, different estimations of fractal dimension and lacunarity. For our application we selected the features by the following criterion: inside the pattern class (e.g. DO and MA) the differences must be small and between the classes they must be high. In order to select the better features we used a set of 40 images from STARE database. Finally we chose the following features: mass fractal dimension (Dm), lacunarity (L), contrast (C), energy (En) and mean intensity (Im). These features ensure proper detection and localization of the OD and MA. Texture and color components inside the OD differ from those of the retina. Therefore, different features depending of the color channel for the two phases of the algorithm were used.

Mean intensity. The mean intensity *Im* [11], a simple statistical feature, can be computed as in (1):

$$Im = \frac{1}{M \times M} \sum_{i=1}^{M} \sum_{j=1}^{M} I(i, j) \tag{1}$$

Energy and contrast. Aiming to obtain a statistical feature relatively insensitive to texture rotation and translation, we introduce the *mean co-occurrence matrix* notion [12]. For each pixel we consider symmetric neighborhoods with dimension of $(2h+1) \times (2h+1)$, $h = 1, 2, 3, \ldots$ Inside each neighborhood there are eight principal directions: 1, 2, 3, 4, 5, 6, 7, 8 (corresponding to $0°$, $45°$, ...,$315°$) and we evaluate the normalized co-occurrence matrices $N_{d,k}$ corresponding to the displacement d determined by the central point and the neighborhood edge point in the k direction ($k = 1,2, \ldots,8$). For each neighborhood type, the mean normalized co-occurrence matrix N_d is calculated by averaging the eight co-occurrence matrices $N_{d,k}$, (2):

$$N_d = \sum_{k=1}^{8} N_{d,k}, \quad d = 1, 2, \ldots 10 \tag{2}$$

The energy and the contrast can be extracted from the mean co-occurrence matrix N_d.

Mass fractal dimension. For gray images, Sarker and Chaudhuri [13] proposed an extension of the basic box counting algorithm (BC), named Differential Box-Counting (DBC) algorithm. It is proved that DBC has a good efficiency in texture classification of monochromatic images [14]. In order to calculate the fractal dimension for gray level images (mass fractal dimension – *Dm*), a 3D relief is created to be covered with boxes of sizes $r \times r \times s$, where r is the division factor, s is the box height $s = r\, I_{max}/M$, where I_{max} represents the maximum value of the intensity and $M \times M$ is the image dimension.

For each square $S_r(u,v)$ of dimension $r \times r$, at the point (u,v) in the grid, exists a stick of boxes with height s. Each pixel (i,j) from $S_r(u,v)$ has an intensity value (gray level) $I(i,j)$ which belongs to a box from the stick. The maximum value of $I(i,j)$, for $(i, j) \in S_r(u,v)$, is denoted by $p(u,v)$. Similarly, the minimum value is denoted by $q(u,v)$. It can be considered that $n_r(u,v)$ from (3) covers the 3D relief of the gray image created by $S_r(u,v)$ and $I(i,j)$, for $(i, j) \in S_r(u,v)$.

$$n_r(u,v) = p(u,v) - q(u,v) + 1 \tag{3}$$

In order to calculate *Dm* with DBC, the relation (4) is used:

$$Dm = \frac{\log(\sum_u \sum_v n_r(u,v))}{\log r} \tag{4}$$

The sum $\sum_u \sum_v n_r(u,v)$ covers the entire relief, which was created by the gray image I. Otherwise the DBC algorithm behaves the same as the standard BC. For ease of calculation, it is required that M and r are powers of 2.

Lacunarity. Another selected feature, lacunarity, characterizes the homogeneity and gaps in textured images. It is a powerful feature in medical and satellite image analysis [15]. Lacunarity can distinguish between two images of the same fractal dimension; therefore we can say that the lacunarity and fractal dimension are complementary. A great value of the lacunarity means that the texture has variable holes and high heterogeneity. For monochrome images, the lacunarity is based on DBC (5):

$$L(r) = \frac{\sum_N N^2 \cdot P(N,r)}{[\sum_N N \cdot P(N,r)]^2}, \quad N = \sum_u \sum_v n_r(u,v) \tag{5}$$

2.2   Detection of Optic Disc

In the learning phase a set of 5 features was selected to indicate if the OD belongs to a window (132×132 pixels): Dm_G (mass fractal dimension, on the color channel G), Dm_R (mass fractal dimension, on the color channel R), L_G (lacunarity, on the color channel G), L_R (lacunarity, on the color channel R) and C_G (normalized contrast, extracted from co-occurrence matrix, on the color channel G). The origin of the box B is considered the upper left corner and corresponds to the point (i_B, j_B) in the whole image. The OD is localized in the box that satisfies the following optimal criterion over all the investigated boxes: $Dm_G = \min$, $Dm_R = \min$, $L_R = \max$, $L_G = \max$, $C_G = \max$.

In order to localize and mark the center of the OD, the selected box is investigated in a similar way, but using the non overlapping boxes method and a window with a size of 33×33 pixels (16 boxes). The features selected for this phase are the following: Dm_B (mass fractal dimension, on the color channel B), L_R (lacunarity, on the color channel R), C_G (normalized contrast, extracted from the co-occurrence matrix, on the color channel G), En_B (normalized energy, extracted from the co-occurrence matrix, on the color channel B) and Im_B (normalized average intensity, on the color channel B). As one can observe 3 of 5 features refer to the blue channel. Therefore, the blue channel is important and histogram analysis can produce valuable information concerning the validity of the algorithm. In most cases, the blue channel histogram is concentrated in the low values domain and, moreover, the mode is 0. For this reason, the test of histogram is considered useful to correctly apply the algorithm. The OD is localized in the box that satisfies the following optimal criterion over all the investigated boxes: $Dm_B = \min$, $L_R = \max$, $En_B = \min$, $C_G = \max$, $Im_B = \max$. In most of the cases, the above ideal situation is not possible and therefore for each box and each feature, a weight is established by considering 15 points for the extreme (minimum or maximum, as the case), 14 points for the following and so on (0 for the last). The total weight for a box is calculated by a normalized sum of the five weighting values (6).

$$w(u,v) = \frac{w(Dm_B(u,v)) + w(L_R(u,v)) + w(En_B(u,v)) + w(C_G(u,v)) + w(Im_B(u,v))}{75} \quad (6)$$

where $u,v \in \{1,2,3,4\}$ and $w(F(u,v))$ represents the weight associated with feature F in the position (u, v) from the grid of boxes.

Only the boxes with $w(u,v) \geq 0.67$ are selected and, from their coordinates, the position of the OD (x_o, y_o) is calculated. The positions of these boxes are noted by (u_i, v_i), $i = 1,...,k$, $k \leq 5$. By considering the associated weights, the parameters, u and v are obtained by the formula of the mass center (7):

$$u = \frac{\sum w_i u_i}{\sum w_i}, \qquad v = \frac{\sum w_i v_i}{\sum w_i} \quad (7)$$

The (x_o, y_o) coordinates are obtained, approximating the values x,y to integers (8):

$$x = i_B + u \cdot 33 - 17, \quad y = j_B + v \cdot 33 - 17 \quad (8)$$

where (i_B, j_B) represents the origin coordinates of B.

The input of the algorithm is image I.

The output of the algorithm is the position of the optic disc (x_o, y_o).

The algorithm consists of the following steps:

1. Image decomposition on color channels: R, G, B.
2. Testing the algorithm applicability: Histogram analysis for blue channel: if mode is 0, the algorithm is aborted.
3. Noise rejection by median filter 3×3 on each color channel.
4. Morphological noise rejection by erosion and dilation grey level method.
5. For each window with size 132×132 created by sliding box algorithm, the mentioned features and its extremes are calculated: $Dm_G = \min$, $Dm_R = \min$, $L_R = \max$, $L_G = \max$, $C_G = \max$.
6. Testing the algorithm applicability and the first step localization: The box with the five optimal criteria satisfied is localized and selected. If not, the box with the four optimal criteria satisfied is localized and selected. If not, the algorithm is aborted.
7. Inside of the selected window, the non overlapping box algorithm with box size 33×33 pixels starts and the extremes are calculated: $Dm_B = \min$, $L_R = \max$, $En_B = \min$, $C_G = \max$, $Im_B = \max$.
8. Testing the algorithm applicability and the second step localization: For each of the non overlapping boxes the weight w is calculated. The boxes with $w(u,v) \geq 0.67$ are localized and selected. If none are found, the algorithm is aborted.
9. The third step localization: The mass center (u,v) of the selected box is calculated.
10. The center of the optic disc is considered as (x_o, y_o).

2.3   Macula Detection

For MA detection and localization we propose a similar method to the like optic disc detection, but without the second step – with small boxes. The difference consists in the fact that in the learning phase another set of features was selected to indicate if the MA belongs to a window: L_B (lacunarity, on the color channel B), L_G (lacunarity, on the color channel G), En_G (normalized energy, extracted from the co-occurrence matrix, on the color channel G) and C_G (normalized contrast, extracted from the co-occurrence matrix, on the color channel G). The MA is considered to belong to the box that satisfies the following optimal criterion over all the investigated boxes: L_B = min, L_G = min, En_G = max, C_G = min. The center of the MA is considered as the center of the selected box (9):

$$x_M = i_B + 66, \quad y_M = j_B + 66 \tag{9}$$

The input of the algorithm is image I.
The output of the algorithm is position of MA (x_M, y_M).
The algorithm consists of the following steps:

1. Image decomposition on color channels: G and B.
2. Testing the algorithm applicability: Histogram analysis for blue channel - if mode is 0, then the image is aborted.
3. Noise rejection by median filter 3×3 on each color channel.
4. Morphological noise rejection by erosion and dilation grey level method.
5. For each window with size 132×132 created by the sliding box algorithm, the following features and their extremes are calculated: L_B = min, L_G = min, En_G = max, C_G = max.
6. Testing the algorithm applicability: If it exists, the box which satisfies the four optimal criteria is localized and selected. If not, the box which satisfies the three optimal criteria is localized and selected. If this does not exist, the algorithm is aborted.
7. Calculating the coordinates of the MA center (x_M, y_M).

3   Experimental Results

The algorithm was tested on a set of 100 retinal images from the STARE database. From these we chose, for exemplification, 9 images, normal (healthy) as well as abnormal (images affected by disease): Im_1, ..., Im_9 (fig. 1). The images were decomposed in the components RGB both on global images and on the boxes (132×132 pixels). For example, in figure 1 the decomposition of the Im_1 is presented (Im_1_G, Im_1_R, for the global image, and Im_1_DO_G, Im_1_DO_R, Im_1_DO_B, for the box selected for the optic disc). The image Im_1_W represents samples from the sliding box method containing three ROIs: OD, MA and BV. The last image Im_1_L marks the position of the OD and MA obtained by our method.

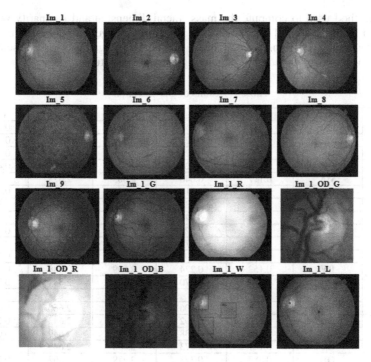

Fig. 1. Test images

The input images were tested by the histogram requirement. For example, the image Im_6 does not fulfill the condition mode $\neq 0$ (fig. 2). It can be seen, from table 2, that the result for MA detection in Im_6 is faulty on the blue channel.

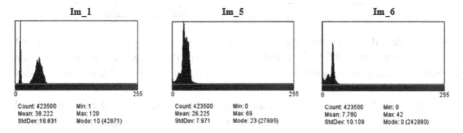

Fig. 2. Examples of a histogram on the blue channel (Im_1 and Im_5 fulfill the test criterion and Im_6 does not)

The results for the first step of OD detection and MA detection are presented in table 1 and table 2 respectively. For presentation, only three box types are considered: one containing OD, one containing MA and another containing BV.

From table 1 it can be seen that the conditions: $Dm_G = \min$, $Dm_R = \min$, $L_R = \max$, $L_G = \max$ and $C_G = \max$ were satisfied for all tested images. In the case of Im_5 a box with MA was not considered.

The method used to compute the mass fractal dimension and lacunarity was implemented using FracLac [16], which is a plug-in of ImageJ, a public domain Java image processing program. Several other MATLAB programs were subsequently used for image processing and for computing the texture features.

Table 1. Results of OD detection

Image	Features	L_R	C_G	Dm_R	Dm_G	L_G
	OD	0.481	0.356	2.514	2.565	0.299
Im_1	MA	0.190	0.022	2.635	2.596	0.069
	BV	0.166	0.123	2.569	2.597	0.227
	OD	0.558	0.526	2.520	2.580	0.485
Im_2	MA	0.117	0.040	2.605	2.625	0.085
	BV	0.115	0.212	2.586	2.632	0.192
	OD	0.682	0.688	2.575	2.592	0.359
Im_3	MA	0.244	0.047	2.633	2.611	0.096
	BV	0.229	0.428	2.620	2.624	0.254
	OD	0.247	0.370	2.554	2.573	0.247
Im_4	MA	0.138	0.104	2.579	2.596	0.125
	BV	0.152	0.274	2.609	2.588	0.240
	OD	0.610	0.370	2.508	2.581	0.579
Im_5	MA	-	-	-	-	-
	BV	0.218	0.006	2.633	2.660	0.075
	OD	0.262	0.392	2.556	2.643	0.245
Im_6	MA	0.255	0.138	2.616	2.649	0.083
	BV	0.141	0.173	2.656	2.649	0.176
	OD	0.883	0.328	2.519	2.597	0.293
Im_7	MA	0.264	0.098	2.587	2.623	0.079
	BV	0.167	0.267	2.616	2.620	0.208
	OD	1.181	0.517	2.506	2,558	0.362
Im_8	MA	0.202	0.088	2.652	2.643	0.073
	BV	0.367	0.274	2.553	2.619	0.183
	OD	0.486	0.508	2.527	2.581	0.306
Im_9	MA	0.114	0.133	2.595	2.611	0.094
	BV	0.142	0.212	2.599	2.632	0.156

From table 2 it can be seen that the conditions: L_B = min, L_G = min, En_G = max and C_G = min were satisfied for all tested images except the images Im_6 and Im_7 (on the blue channel, marked with *). In the case of Im_5 a box with MA was not considered.

The position detection of the OD, after the first step (detection of i_B,j_B), is refined inside of the selected box by the second step analyzing all 16 images (33×33 pixels). Table 3 presents the results of the features selected for image Im_1 and the corresponding weights. It can be seen that only images in grid positions (2,3), (3,2) and (3,3) meet the condition $w(u,v) \geq 0.67$. The ideal condition for Dm_B is Dm_B = min. The corresponding weight of the minimum value is 15. The weights decrease if Dm_B grows, taking the value 0 for the maximum value of Dm_B. Similarly, for the minimum value of Em_B the correspondent weight is 15.

For L_R, C_G and I_{mB} the weight corresponding to the maximum value is 15 and the weight corresponding to the minimum value is 0.

Table 2. Results of features selected for MA detection (note that * represents the values which do not meet the extreme criteria)

Image	Features	L_B	L_G	En_G	C_G
	OD	0.295	0.300	0.145	0.356
Im_1	MA	0.054	0.069	0.946	0.022
	BV	0.170	0.227	0.396	0.123
	OD	0.516	0.485	0.165	0.526
Im_2	MA	0.056	0.085	0.830	0.040
	BV	0.126	0.192	0.391	0.212
	OD	0.456	0.921	0.111	0.688
Im_3	MA	0.095	0.096	0.809	0.047
	BV	0.170	0.253	0.215	0.428
	OD	0.303	0.247	0.200	0.370
Im_4	MA	0.075	0.125	0.464	0.104
	BV	0.116	0.249	0.293	0.274
	OD	0.366	0.574	0.244	0.370
Im_5	MA	-	-	-	-
	BV	0.043	0.075	0.986	0.006
	OD	0.301	0.245	0.248	0.392
Im_6	MA	1.947*	0.083	0.579	0.138
	BV	1.386	0.145	0.550	0.173
	OD	0.091	0.193	0.204	0.328
Im_7	MA	0.146*	0.079	0.499	0.098
	BV	0.034	0.208	0.295	0.267
	OD	0.171	0.362	0.268	0.093
Im_8	MA	0.073	0.073	0.672	0.088
	BV	0.104	0.183	0.337	0.274
	OD	0.324	0.306	0.117	0.508
Im_9	MA	0.038	0.094	0.571	0.133
	BV	0.100	0.156	0.337	0.212

Table 3. Results of the fine localization of OD for Im_1

Im_1	w_D	Dm_B	w_L	L_R	w_C	C_G	w_E	En_B	w_I	Im_B	w
1,1	2	2.49	13	0.94	0	0.05	2	0.97	0	0.14	17/75=0.23
1,2	4	2.49	11	0.55	8	0.31	0	1	1	0.16	25/75=0.33
1,3	12	2.45	7	0.28	10	0.42	8	0.56	2	0.18	29/75=0.39
1,4	5	2.49	1	0.14	4	0.23	10	0.48	8	0.20	28/75=0.37
2,1	1	2.50	12	0.74	2	0.18	6	0.65	3	0.18	24/75=0.32
2,2	7	2.47	4	0.19	14	0.57	7	0.61	6	0.19	38/75=0.51
2,3	15	2.42	14	0.99	15	0.94	14	0.32	13	0.28	**71/75=0.95**
2,4	11	2.46	8	0.37	5	0.25	3	0.96	12	0.27	34/75=0.45
3,1	8	2.47	6	0.23	7	0.29	12	0.34	7	0.19	40/75=0.53
3,2	13	2.44	5	0.22	13	0.53	15	0.31	9	0.22	**55/75=0.73**
3,3	14	2.43	15	2.03	12	0.46	13	0.33	15	0.36	**69/75=0.92**
3,4	10	2.46	9	0.40	3	0.23	9	0.56	14	0.30	45/75=0.60
4,1	0	2.51	3	0.16	1	0.12	4	0.85	4	0.18	12/75=0.16
4,2	6	2.47	2	0.15	11	0.43	11	0.46	5	0.19	35/75=0.47
4,3	9	2.46	10	0.45	9	0.37	5	0.65	11	0.26	44/75=0.59
4,4	3	2.49	0	0.12	6	0.27	1	0.96	10	0.25	20/75=0.27

Equations (7) and (8) lead to: $u = 2.63$, $v = 2.72$,

$$u_1 = 2, u_2 = 3, u_3 = 3, v_1 = 3, v_2 = 3, v_3 = 2, w_1 = 0.95, w_2 = 0.92, w_3 = 0.73$$

By considering the coarse localization (i_B, j_B), the approximate position of the OD (x_o, y_o) are found from equation (9). The results are presented together with the approximate position of the MA in table 4.

Table 4. Coordinates of the OD and MA

Image	(i_B, j_B)	(u, v)	(x_o, y_o)	(x_M, y_M)
Im_1	(171, 34)	(2.63, 2.72)	(243, 106)	(305, 357)
Im_2	(253, 539)	(2.12, 2.39)	(306, 601)	(335, 371)
Im_3	(212, 462)	(2.21, 2.48)	(268, 527)	(291, 315)
Im_4	(182, 119)	(2.36, 2.45)	(243, 183)	(273, 405)
Im_5	(191, 524)	(2.60, 3.51)	(260, 623)	-
Im_6	(193, 60)	(2.45, 2.33)	(257, 120)	(286, 350)
Im_7	(164, 27)	(2.33, 1.78)	(224, 69)	(277, 347)
Im_8	(224, 550)	(2.18, 3.09)	(279, 635)	(319, 320)
Im_9	(220, 81)	(2.39, 2.48)	(282, 146)	(314, 378)

Testing on 100 images, the center of the OD (manually marked by an optometrist) was localized inside a square of 5×5 pixels found with an accuracy of 98%, while an accuracy of 94% was reached for localization of the MA. The number of failures was 2 for OD and 6 for MA (table 5).

Table 5. Failure classification criteria

Algorithm	OD test (failures)	MA test (failures)
Step 2 (not fulfill histogram)	1	1
Step 6 (under the minimum number of matches)	1	5

In table 6, the accuracy of our results comparing with other works is presented.

Table 6. Comparison with other similar works

Similar works	[3]	[4]	[6]	[8]	Our method
Success rate OD / MA	98.77 / -	91.36 / -	96.34 / 95.53	95.75 / -	98 / 94

4   Conclusions

The objective of this paper was the accurate localization of the OD and MA in retinal images. We considered a simple and efficient method based on fractal and statistical features for textured images, considering mass fractal dimension, lacunarity, energy, contrast and mean intensity. This method combined spatial, statistical and spectral information. The corresponding features were selected in a learning phase from a large set of different characteristics based on extreme criteria. The method avoids some difficult problems like blood vessel extraction and binarization for OD and MA detection. For accurate localization of the OD, we proposed an algorithm with two distinct steps: the first is a sliding box type, based on large windows, and the second is a non overlapping type based on partitioning the previous windows into small boxes. For a high resolution, a weighting scheme of the results from the second step was applied.

References

1. Carmona, E.J., Rincon, M., Garcıa-Feijoo, J., Martinezde-la Casa, J.M.: Identification of the optic nerve head with genetic algorithms. Artificial Intelligence in Medicine 43, 243–259 (2008)
2. Xu, J., Ishikawa, H., Wollstein, G., Bilonick, R.A., Sung, K.R., Kagemann, L., Townsend, K.A., Schuman, J.S.: Automated assessment of the optic nerve head on stereo disc photographs. Investigative Ophthalmology & Visual Science 49, 2512–2517 (2008)
3. Youssif, A.A., Ghalwash, A.Z., Ghoneim, A.S.: Optic disc detection from normalized digital fundus images by means of a vessels' direction matched filter. IEEE Trans. Med. Imag. 27, 11–18 (2008)
4. Dehghani, A., Moghaddam, H.A., Moin, M.-S.: Optic disc localization in retinal images using histogram mathing. EURASIP Journal on Image and Video Processing, 1–11 (2012)
5. Osareh, A., Mirmehdi, M., Thomas, B., Markham, R.: Classification and localisation of diabetic-related eye disease. In: Heyden, A., Sparr, G., Nielsen, M., Johansen, P. (eds.) ECCV 2002, Part IV. LNCS, vol. 2353, pp. 502–516. Springer, Heidelberg (2002)
6. Qureshi, R.J., Kovacs, L., Harangi, B., Nagy, B., Peto, T., Hajdu, A.: Combining algorithms for automatic detection of optic disc and macula in fundus images. Computing Vision and Image Understanding 116, 138–145 (2012)
7. de la Fuente-Arriaga, J.A., Felipe-Riverón, E.M., Garduño-Calderón, E.: Application of vascular bundle displacement in the optic disc for glaucoma detection using fundus images. Computers in Biology and Medicine 47, 27–35 (2014)
8. Pereira, C., Gonçalves, L., Ferreira, M.: Optic disc detection in color fundus images using ant colony optimization. Med. Biol. Eng. Comput. 51, 295–303 (2013)
9. Li, H., Chutatape, O.: Automated feature extraction in color retinal images by a model based approach. IEEE Trans. Biomed. Eng. 51, 246–254 (2004)
10. Structured Analysis of the Retina, http://www.ces.clemson.edu/~ahoover/stare/
11. Pratt, W.: Digital Image Processing, PIKS Scientific Inside, 4th edn. Wiley (2006)
12. Popescu, D., Dobrescu, R., Angelescu, N.: Statistical texture analysis of road for moving objectives. U.P.B. Sci. Bull. Series C. 70, 75–84 (2008)
13. Sarker, N., Chaudhuri, B.B.: An efficient differential box-counting approach to compute fractal dimension of image. IEEE Transactions on Systems, Man, and Cybernetics 24, 115–120 (1994)
14. Chaudhuri, B.B., Sarker, N.: Texture segmentation using fractal dimension. IEEE Transactions on Pattern Analysis and Machine Intelligence 17, 72–77 (1995)
15. Barros Filho, M.N., Sobreira, F.J.A.: Accuracy of lacunarity algorithms in texture classification of high spatial resolution images from urban areas. In: XXI Congress of International Society of Photogrammetry and Remote Sensing (ISPRS 2008), Beijing, China, 417–422 (2008)
16. Karperien, A.: FracLac for ImageJ, http://rsb.info.nih.gov/ij/plugins/fraclac/FLHelp/Introduction.htm

A Simple Hair Removal Algorithm from Dermoscopic Images

Damian Borys[1], Paulina Kowalska[1], Mariusz Frackiewicz[1], and Ziemowit Ostrowski[2]

[1] Institute of Automatic Control, Silesian University of Technology
ul. Akademicka 16, 44-100 Gliwice, Poland
[2] Institute of Thermal Technology, Silesian University of Technology
ul. Konarskiego 22, 44-100 Gliwice, Poland
{damian.borys,mariusz.frackiewicz,ziemowit.ostrowski}@polsl.pl

Abstract. The main goal of our work is initial preprocessing of dermoscopic images by hair removal. Dermoscopy is a basic technique in skin melanoma diagnostics. One of the main problems in dermoscopy images analysis are hairs objects in the image. Hairs partially shade the main region of interest that's why it needs special treatment. We have developed a simple and fast hair removal algorithm based on basic image processing algorithms. The algorithm was tested on available online test database PH2 [8]. Primary results of proposed algorithm show that even if hair contamination in the image is significant algorithm can find those objects. There is still a place for improvements as long as some air bubbles are marked as a region of interest.

Keywords: dermatoscopy image, hair removal algorithm, melanoma.

1 Introduction

Melanoma, the most frequent malignant skin cancer, characterizes an aggressive growth with early and numerous metastases that are very difficult in pharmacological treatment and have rather a poor prognosis. The most frequent localisation is in the region of pigmental change. However, it can also arise in the unchanged skin area. Early diagnosis, followed by immediate therapy (surgery procedure), allows to save about 80% of patients with melanomas. Dermatoscopy and videodermatoscopy are the major methods used in a diagnosis of skin cancer [10]. In the case of the disease risk identification, physician orders further specialistic treatment. Next step is surgery procedure to remove suspicious lesion with a safe margin and perform the precise study of removed skin. Image processing is an important part of the dermatoscopy diagnosis of malignant melanoma. It allows to quickly diagnose the changes in nevus, is non-invasive and allows to determine the probability of melanoma. During the diagnostic procedure, using the videodermatoscopy device, skin lesions are acquired. Specific features help in melanoma recognition from the image using the so-called "ABCD rule" [6] : A- asymmetry, B- border, C- color, D- differential structures. But before we can

F. Ortuño and I. Rojas (Eds.): IWBBIO 2015, Part I, LNCS 9043, pp. 262–273, 2015.

analyse dermoscopic image towards lesion malignancy, at first the preprocessing step has to be done to mark the region of interest in the image. Preprocessing is the initial step toward more advanced analysis. For example to discretize images of malignant skin lesions and benign changes based on features calculated from region of interest, like texture parameters that will indicate porosity or directionality. To accurately calculate those features hairs registered in the image have to be removed, and then region of interest can be found. It is needed to prepare dermoscopic images to further processing by removing unwanted contaminations in the image like hairs, air bubbles etc..

One of the main problems in dermatoscopy images analysis are hairs objects in the image. Hairs partially shade the main region of interest that's why it needs special treatment. In the literature one can find a lot of works, that uses some processing methods, for example, special filtration methods, thresholding and morphological operations [5],[9]. In other, authors use morphological top-hat operation or filtration together with morphological operations and lesion filling step [1,2]. In this work, we have proposed our algorithm for hair removal from dermoscopic images.

2   Materials and Methods

To test the algorithm an open-access database PH2 with dermoscopic images was used [8]. Images have been acquired with a videodermatoscope, and each of them has been manually contoured. All contours are stored in the database. Examples of the images are shown in Fig. 1, 2 and 3. On the first of them one can see almost clear image with lesion in the center whilst on the second large image contamination with the hairs can be observed. Third test image contains few hairs in colour similar to the lesion. In addition, some air bubbles are also present in Fig. 2 and 3.

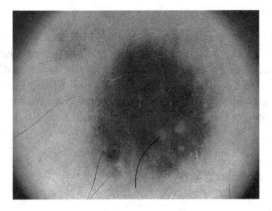

Fig. 1. Example of dermoscopic image. Image almost without contaminations. Source [8]

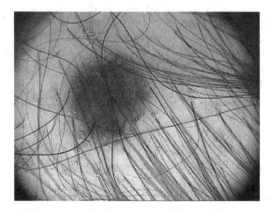

Fig. 2. Example of dermoscopic image. Image contaminated with hairs and air bubbles. Source [8]

Fig. 3. Example of dermoscopic image. Image contaminated with hairs and air bubbles. Source [8]

The proposed algorithm includes a transformation from color image to grayscale, sharpening the image by stretching the histogram, finding the edge of dark and light hair (using Canny edge detector [3,4]), using mathematical morphology methods to filter inappropriate objects and finally hair removal. After processing the image, the next step is to analyze the structure of the image. Presented algorithm contains the following steps:

- Data acquisition
- RGB to Grayscale conversion
- Contrast Enhancement
- Binarization
- Edge Detection
- Removal non-hair edges
- Hair mask creation

- Hair mask with RGB image merge
- Lesion (region of interest) mask creation

and is presented on a block diagram in Fig. 4.

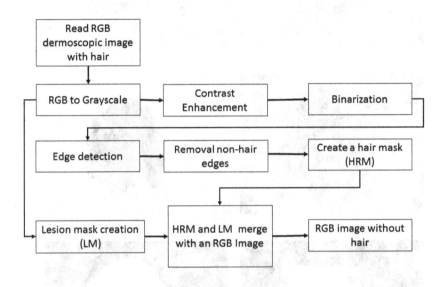

Fig. 4. Block diagram of the algorithm

Results of proposed method have been compared with DullRazor algorithm [7].

3   Results

We present results of each step of the algorithm that was performed on exemplary dermoscopic images (Fig. 1, 2 and Fig. 3). Results for Fig. 1 are presented on Figures: Fig.5 (a) - (h), Fig.8 (a) - (b). Results for Fig. 2 are presented on Figures: Fig.6 (a) - (h), Fig.8 (c) - (d) and results for Fig 3 are presented on Figures: Fig.7 (a) - (h), Fig.8 (e) -(f). Results presented in this section show each step of the algorithm that was presented in Fig. 4. Conversion from RGB space to grayscale is in b), contrast enhancement result in c), lesion ROI segmentation d). Point e) is d) negation, next this ROI is enhanced by morphological operations f). Final steps for hair edge detection are g) and h) where non-hair objects are removed.

For a better representation of the final results on Fig. 9 hairs are marked in red, ROI (lesion) is marked in green contour.

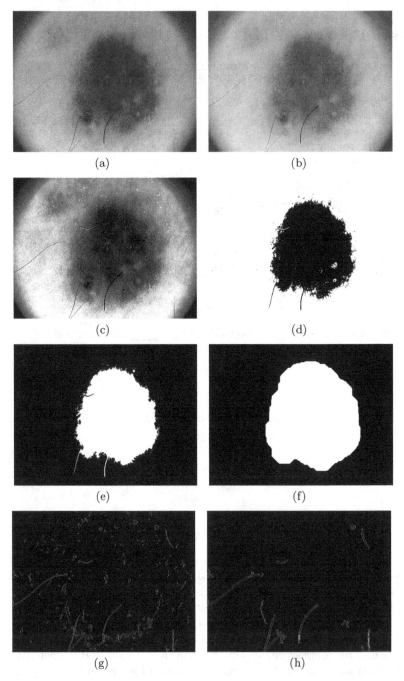

Fig. 5. Image processing shown in Fig. 1: (a) input image, (b) RGB to Grayscale, (c) contrast enhancement, (d) lesion ROI segmentation LM, (e) negation, (f) LR enhancement by morphological operations, (g) hair edge detection, (h) remove non-hair edges and create hair mask HRM

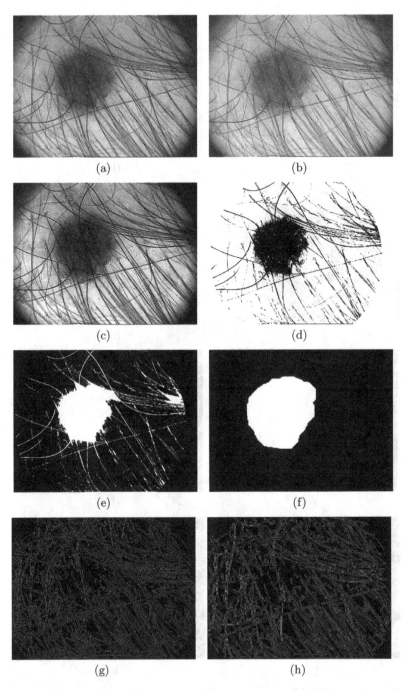

Fig. 6. Image processing shown in Fig. 2: (a) input image, (b) RGB to Grayscale, (c) contrast enhancement, (d) lesion ROI segmentation LM, (e) negation, (f) LR enhancement by morphological operations, (g) hair edge detection, (h) remove non-hair edges and create hair mask HRM

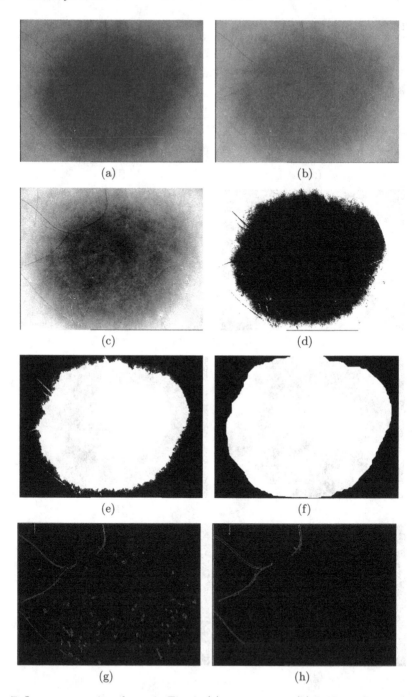

(a)

(b)

(c)

(d)

(e)

(f)

(g)

(h)

Fig. 7. Image processing shown in Fig. 3: (a) input image, (b) RGB to Grayscale, (c) contrast enhancement, (d) lesion ROI segmentation LM, (e) negation, (f) LR enhancement by morphological operations, (g) hair edge detection, (h) remove non-hair edges and create hair mask HRM

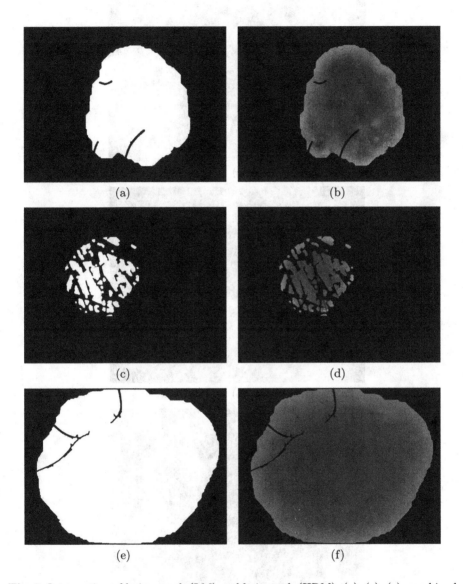

(a)   (b)

(c)   (d)

(e)   (f)

Fig. 8. Intersection of lesion mask (LM) and hair mask (HRM): (a), (c), (e), combined with input images: (b), (d), (f)

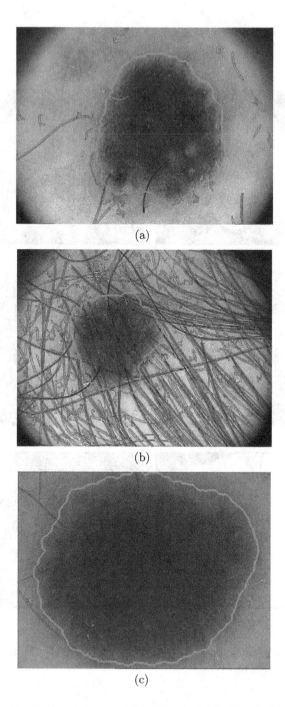

Fig. 9. Final result - for input image: (a) in Fig. 1, (a) in Fig. 2, (a) in Fig. 3. Hairs are marked in red, ROI (lesion) is marked in green contour

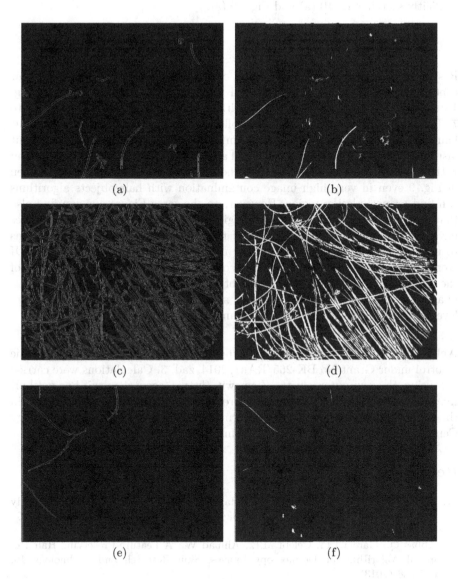

Fig. 10. Comparison of hair masks obtained by our algorithm (a, c, e) and DullRazor software [7] (b, d, f)

We compare obtained hair masks with results generated by DullRazor software [7]. Comparison is shown in Fig. 10 and as we can see there is some differences between this two approaches. Our algorithm generates less hair areas which explicitly seen in Fig. 10 (a) and Fig. 10 (e).

4 Discussion

Results of the algorithm were presented on three test images with variable hair contamination level (Fig. 1 and 2) and different hair to lesion contrast (Fig. 1 and 3). All crucial steps of the algorithm have been presented in Figs. 5 to 7. Processed final images that were collected in Fig. 8 shows lesion without hair objects. In most cases, the algorithm removes hairs from the image without disrupting the structure of the lesion and allows us to create contour of the lesion without unnecessary (hair) objects in the region of interest. As it can be seen in Fig. 9 even in very high image contamination with hair objects algorithms can find properly lesion region. However, on the same Figure one can find also that not only lesion is classified as a region of interest. Also, some air bubbles have been classified as ROI. Results of this algorithm, after some corrections to improve air bubbles removal, will be used for verification with experts ROI creation, and further for deeper analysis of this images. For example, analysis of the texture will be performed by means of entropy, standard deviation, grayscale image, the average value of brightness, skewness of the histogram and energy level and will be the subject of further analysis.

Acknowledgment. This work was supported by the Institute of Automatic Control under Grant No BK-265/RAu1/2014, zad. 3. Calculations were carried out using the computer cluster Ziemowit (http://www.ziemowit.hpc.polsl.pl) funded by the Silesian BIO-FARMA project No. POIG.02.01.00-00-166/08 in the Computational Biology and Bioinformatics Laboratory of the Biotechnology Centre at the Silesian University of Technology.

References

1. Abbas, Q., Celebi, M.E., Garcia, I.F.: Hair removal methods: A comparative study for dermoscopy images. Biomedical Signal Processing and Control 6, 395–404 (2011)
2. Abbas Q., Garcia I.F., Celebi M.E., Ahmad W.: A Feature-Preserving Hair Removal Algorithm for Dermoscopy Images. Skin Research and Technology 19, e27–e36 (2013)
3. Canny, J.: A Computational Approach To Edge Detection. IEEE Trans. Pattern Analysis and Machine Intelligence 8, 679–714 (1986)
4. Deriche, R.: Using Canny's criteria to derive a recursively implemented optimal edge detector. The International Journal of Computer Vision 1(2), 167–187 (1987)
5. Heijmans Henk, J.A.M., Roerdink Jos, B.T.M. (eds.): Mathematical Morphology and its Applications to Image and Signal Processing. In: Proceedings of the 4th International Symposium on Mathematical Morphology (1998)

6. Johr, R.H.: Dermoscopy: Alternative melanocytic algorithms. The ABCD rule of dermatoscopy, menzies scoring method, and 7-point checklist. Clin. Dermatol. 20, 240–247 (2002)
7. Lee, T., Ng, V., Gallagher, R., Coldman, A., McLean, D.: DullRazor: A software approach to hair removal from images. Computers in Biology and Medicine 27, 533–554 (1997)
8. Mendoca, T., et al.: PH2 - A dermoscopic image database for research and benchmarking. In: 35th Annual International Conference of the IEEE EMBS, Osaka, Japan, July 3 - 7 (2013)
9. Nguyen, N.H., Lee, T.K., Atkins, M.S.: Segmentation of light and dark hair in dermoscopic images: a hybrid approach using a universal kernel. In: Dawant, B.M., Haynor, D.R. (eds.) Medical Imaging 2010: Image Processing. Proc. of SPIE, vol. 7623, p. 76234N (2010)
10. Soyer, H.P., Argenziano, G., Hofmann-Wellenhof, R., Zalaudek, I.: Dermoscopy: The Essentials, 2nd edn. Elsevier Saunders (2011)

Automatic Segmentation System of Emission Tomography Data Based on Classification System

Sebastian Student[1], Marta Danch-Wierzchowska[1], Kamil Gorczewski[2], and Damian Borys[1]

[1] Institute of Automatic Control, Silesian University of Technology
ul. Akademicka 16, 44-100 Gliwice, Poland
[2] Department of PET Diagnostics, Maria Sklodowska-Curie Memorial Cancer Center and Institute of Oncology, Gliwice Branch, ul. Wybrzeże AK 15, 44-100 Gliwice, Poland
{sebastian.student,marta.danch-wierzchowska,damian.borys}@polsl.pl,
kgorczewski@io.gliwice.pl

Abstract. Segmentation and delineation of tumour boundaries are important and difficult step in emission tomography imaging, where acquired and reconstructed images presents large noise and a blurring level. Several methods have been previously proposed and can be used in single photon emission tomography (SPECT) or positron emission tomography (PET) imaging. Some of them relies on the standard uptake value (SUV) used in PET imaging. Presented approach can be used in both (SPECT and PET) modalities and it is based on support vector machines (SVM) classification system. System has been tested on standard phantom, widely used for testing the emission tomography devices. Results are presented for two classifiers SVM and DLDA.

1 Introduction

Segmentation of three-dimensional data from emission tomography devices is important, difficult and also a time-consuming task, especially, when performing manually by medical experts. The problem of structures delineation with real uptake have been studied in the literature [5,14] and few methods have been proposed. Some of the proposed approaches are based on fixed threshold (for example 70 % of maximum value) [1] or even simpler, based on fixed absolute SUV value threshold (for example SUV 2.5) [10]. Others proposed methods that are based on a contrast, calculated from phantoms with spheres [11]. Unfortunately, all of those methods are acquisition parameters and parameters of reconstruction methods dependent. The importance and the influence of the method for the system matrix calculation for ordered-subset expectation maximization (OSEM) reconstruction algorithm have been studied by our group earlier [2].

Due to high noise in obtained images and high similarity between tumor and normal tissue, automating segmentation process is a challenging task. In this

F. Ortuño and I. Rojas (Eds.): IWBBIO 2015, Part I, LNCS 9043, pp. 274–281, 2015.
© Springer International Publishing Switzerland 2015

paper, we propose new, automatic segmentation system, based on extraction of different image features, which are used to train the classifiers such as SVM and Bayesian classifier. In our experiment, we have compared the performance of the classifiers to widely used in clinical practice method based on fixed on the level 2.5 threshold of standardized uptake value (SUV) segmentation [5].

2 Methodology

Experiment Design. Data have been taken with Philips Gemini PET device at Department of PET Diagnostics. Phantom was a standard cylinder with five spheres of different radius and volume (diameter from 1.3 cm up to 4 cm). CT scan of the phantom is presented on Fig. 1. Spheres, visible on the CT slice, have been filled with F-18 with constant concentration. The rest of the phantom was filled with water. PET images were acquired with standard settings for iterative reconstruction algorithm with CT-attenuation correction. Values of counts reconstructed in PET images were scaled to SUV values to take into account object's mass and injected activity.

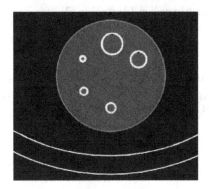

Fig. 1. Slice CT of a phantom

Classifier Design. Optimal design of the classification system for 3D image data described in this article constitute challenging, computational and conceptional problem. To the most important and crucial steps of classifier design, we can include data preprocessing, choice of proper classification and feature selection methods. For that reason, we compare results from different selection methods and different number of selected features. Our classifier is based on the classification system proposed in [12] and is shown in Fig. 2.

First of all we have assigned different image features. All of these features are computed for raw data obtained from PET-CT device and are shown in Tab. 1.

For all features (Tab. 1) we have compared selection algorithms based on method proposed in [13], Recursive Feature Elimination (RFE) [6] and the classical t-statistics. Basing on the estimated train error for training data image, we

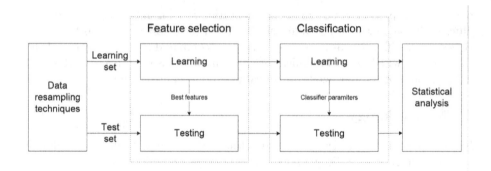

Fig. 2. Classification system scheme

have decided to use the method proposed by Yang. Then, we have selected 15 best features, ranked as shown in Tab. 1. As a classification method we have used the Support Vector Machines method (SVM) [3,4],[8] and compared the results with Diagonal Linear Discriminant Analysis classifier (DLDA), which uses the procedure called Nearest Shrunken Centroids [7],[9].

2.1 SVM Methods

In this article we have used binary SVM. We denote $Z = (z_1, z_2, \ldots, z_l)$ as our dataset, where $z_i = (x_i, y_i)$ and $i = 1, \ldots, l$ and $y_i \in \{1, \ldots, K\}$. The response y_i is the class of predictor vector x_i. The SVM solves the following problem

$$\min_{\mathbf{w}, b, \xi} \frac{1}{2} K(\mathbf{w}, \mathbf{w}) + C \sum_{i=1}^{l} \xi_i, \tag{1}$$

subject to

$$y_i(K(\mathbf{w}, \mathbf{x}) + b) \geq 1 - \xi_i, \qquad \xi_i \geq 0, \qquad i = 1, \ldots, l; \tag{2}$$

where $K(,)$ is the kernel function and C is penalty parameter. When we minimize $\frac{1}{2} K(\mathbf{w}, \mathbf{w})$, we maximize the margin between two groups of data $\frac{2}{||\mathbf{w}||}$. After solving this problem we have decision function:$\mathbf{w} \cdot \mathbf{x} + b$ for linear kernel function $K(,)$. We don't discuss about kernel function, because in our case the best results were obtained for the linear kernel function $K(\mathbf{w}, \mathbf{x}) = \mathbf{w}^T \mathbf{x}$.

3 Results

The numerical experiment includes comparison of selection methods, number of features and classification methods. As shown in Fig. 3 the train error decrease with increased number of features and has the lowest level for SVM classifier and 15 selected features (ranked as shown in Table 1). In Fig. 4 we can see the

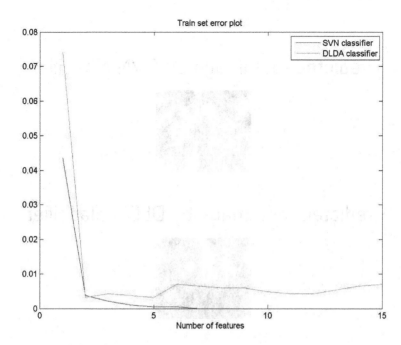

Fig. 3. Train error of classification for different number of image features used

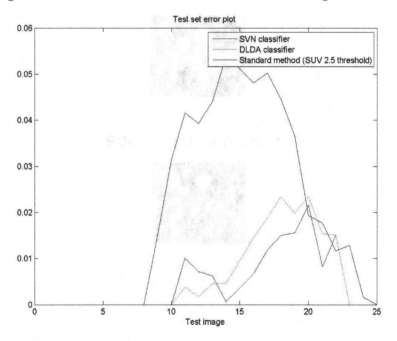

Fig. 4. Error plot for test set data for SPECT stacks

Predicted test image by SVN classifier

Predicted test image by DLDA classifier

Predicted test image by SUV 2.5 method

CT orginal test image

Fig. 5. Example of trained images for SVM and DLDA classifier for 15 best features

Table 1. Table of features that were computed for raw data obtained from PET-CT device

Symbol	description	Selection rank
rd	raw data	6
mean4	mean value for 4-neighbourhood	3
mean8	mean value for 4-neighbourhood	2
mean12	mean value for 4-neighbourhood	0
maxmin4	maximum value-minimum value for 4-neighbourhood	0
maxmin8	maximum value-minimum value for 4-neighbourhood	0
maxmin12	maximum value-minimum value for 4-neighbourhood	0
nbnum4_40	number of 4-neighbours for 40% thresholding	5
nbnum8_40	number of 8-neighbours for 40% thresholding	7
nbnum12_40	number of 12-neighbours for 40% thresholding	8
nbnum4_50	number of 4-neighbours for 50% thresholding	9
nbnum8_50	number of 8-neighbours for 50% thresholding	11
nbnum12_50	number of 12-neighbours for 50% thresholding	12
nbnum4_70	number of 4-neighbours for 70% thresholding	15
nbnum8_70	number of 8-neighbours for 70% thresholding	14
nbnum12_70	number of 12-neighbours for 70% thresholding	13
nbnum4_2.5 SUV	number of 4-neighbours for 2.5% SUV thresholding	0
nbnum8_2.5 SUV	number of 8-neighbours for 2.5% SUV thresholding	0
nbnum12_2.5 SUV	number of 12-neighbours for 2.5% SUV thresholding	0
rdiff4	relative difference in min. and max. values of 4-neighbourhood	10
rdiff8	relative difference in min. and max. values of 8-neighbourhood	4
rdiff12	relative difference in min. and max. values of 12-neighbourhood	1

Table 2. Table of test error indicator results from numerical experiment

Method	Mean error rate	95% confidence interval
SVM	0.0049	(0.0021 , 0.0077)
DLDA	0.0060	(0.0024 , 0.0097)
SUV 2.5	0.02	(0.011 , 0.028)

comparison of test result for the test image not included in train set for SVM and DLDA classifier. We can also observe that for the DLDA method we lose the smallest object in phantom used for this study (images not shown in the manuscript). Consequently, looking at all the classification errors indicators and the 95% confidence interval as shown in Table 2, one general conclusion is that there are no significant differences between SVM and DLDA method. Typically, SVM method outperforms the other methods but we only observe significant differences between SVM and standard method based on SUV 2.5 thresholding (p value¡ 0.002). Also, DLDA method is significantly better that SUV 2.5 method (p value¡ 0.005).

The Fig. 5 illustrate how the test error is changing for all used test images obtained for the phantom. The biggest error we denote for slides where on the slide with all visible spheres. Also for the test images the smallest error we found for increased a number of features (Fig. 6). In this case, the error rate is much higher for feature number smaller than 5.

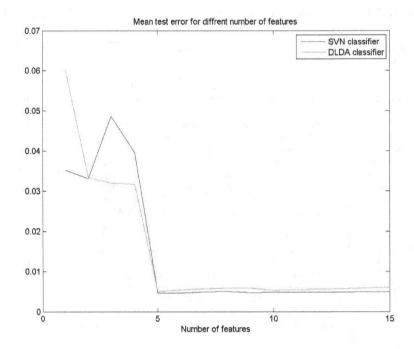

Fig. 6. Mean error for test set data for different number of features

4 Discussion

Emission tomography images are widely used method in clinical practice, although the segmentation step is crucial for proper diagnosis and medical decision tasks. This work presents the segmentation method for emission tomography images, based on feature selection technique and SVM classifier. From the experimental results it is concluded, that SVM gives better classification accuracy than standard method based on simple SUV thresholding (SUV level 2.5 is widely used to differentiate between benign and malignant lesions) and DLDA classifier. For the chosen dataset, it is more effective to use more selected features for classification. The best results we have obtained for 15 selected features. The presented method makes it possible to obtain better segmentation results without manual adjustment. Due to that, medical diagnostics, could be a faster and more effective process. Of course, our methods can be applicable also to other data where we consider segmentation problems such as segmentation of confocal microscope images of cells or tissues.

Acknowledgment. This work was supported by the National Science Centre (NCN) under Grant No. 2011/03/B/ST6/04384 (DB) and by the Institute of Automatic Control under Grant No. BKM-524/RAU1/2014/ t.30 (SS) and

BK-265/RAu1/2014, t3. and Grant No. BKM-524/RAU1/2014/ t.9 (MDW). Calculations were performed using the infrastructure supported by the computer cluster Ziemowit (http://www.ziemowit.hpc.polsl.pl) funded by the Silesian BIO-FARMA project No. POIG.02.01.00-00-166/08 in the Computational Biology and Bioinformatics Laboratory of the Biotechnology Centre at the Silesian University of Technology.

References

1. Boellaard, R., Krak, N.C., Hoekstra, O.S., Lammertsma, A.A.: Effects of noise, image resolution, and ROI definition on the accuracy of standard uptake values: a simulation study. J. Nucl. Med. 45, 1519–1527 (2004)
2. Borys, D., Szczucka-Borys, K., Gorczewski, K.: System matrix calculation for iterative reconstruction algorithms in SPECT based on direct measurements. International Journal of Applied Mathematics & Computer Science 21(1), 193–202 (2011)
3. Boser, B.E., Guyon, I.M., Vapnik, V.: A training algorithm for optimal margin classifiers. In: Proceedings of the Fifth Annual Workshop on Computational Learning Theory, pp. 144–152. ACM (1996)
4. Koonin, E.V., Altschul, S.F., Bork, P.: Knowledge based analysis of microarray gene expression data by using support vector machines. Proceedings of the National Academy of Sciences 4 97(1), 262–267 (1996)
5. Cheebsumon, P., et al.: Impact of $[^{18}F]$FDG PET imaging parameters on automatic tumour delineation:need for improved tumour delineation methodology. Eur. J. Nucl. Med. Mol. Imaging 38, 2136–2144 (2011)
6. Guyon, I., Weston, J., Barnhill, S., Vapnik, V.: Gene selection for cancer classification using support vector machines. Machine Learning 46, 389–422 (2006)
7. Huang, D., Quan, Y., He, M., Zhou, B.: Comparison of linear discriminant analysis methods for the classification of cancer based on gene expression data. Oxford University Press (1988)
8. Kumar, M.S., Kumaraswamy, Y.S.: An improved support vector machine kernel for medical image retrieval system. In: Pattern Recognition, Informatics and Medical Engineering (PRIME), pp. 257–260 (2012)
9. Krzanowski, W.J.: Principles of Multivariate Analysis: A User's Perspective. Journal of Experimental and Clinical Cancer Research 28(1), 149 (2009)
10. Paulino, A.C., Koshy, M., Howell, R., Schuster, D., Davis, L.W.: Comparison of CT- and FDG-PET-defined gross tumor volume in intensity-modulated radiotherapy for head-and-neck cancer. Int. J. Radiat. Oncol. Biol. Phys. 61, 1385–1392 (2005)
11. Schaefer, A., Kremp, S., Hellwig, D., Rbe, C., Kirsch, C.M., Nestle, U.: A contrast-oriented algorithm for FDG-PET-based delineation of tumour volumes for the radiotherapy of lung cancer: derivation from phantom measurements and validation in patient data. Eur. J. Nucl. Med. Mol. Imaging 35, 1989–1999 (2008)
12. Student, S., Fujarewicz, K.: Stable feature selection and classification algorithms for multiclass microarray data. Biology Direct 7, article 33 (2012)
13. Yang, K., Cai, Z., Li, J., Lin, G.: A stable gene selection in microarray data analysis. BMC Bioinformatics 7, 228 (2006)
14. Zaidi, H., El Naqa, I.: PET-guided delineation of radiation therapy treatment volumes: a survey of image segmentation techniques. Eur. J. Nucl. Med. Mol. Imaging 37, 2165–2187 (2010)

Alpha Rhythm Dominance in Human Emotional Attention States: An Experimentation with 'Idling' and 'Binding' Rhythms

Mohammed G. Al-Zidi, Jayasree Santhosh[*], and Jamal Rajabi

Department of Biomedical Engineering, Faculty of Engineering
University of Malaya, 50603 Kuala Lumpur, Malaysia
{bingamal,jsanthosh}@um.edu.my, jamal.rajabi@gmail.com

Abstract. The aim of the present study was to investigate the prominence of type of brain oscillations in processes combined with attention voluntarily directed to emotionally significant stimuli. An indigenously designed experiment paradigm was conducted on specific class of healthy subjects using a series of audio and video clippings as stimuli. Each of the clippings was associated with one of the three different types of emotions namely sad, calm and excitation. Changes in peak of power spectrum and Inter hemispheric coherences showed an enhanced alpha rhythm dynamics compared to gamma rhythms. Results from repeated measures using T-test and ANOVA showed that Alpha rhythm dynamics play more possible functional roles as compared with gamma rhythms. The fact that the study conducted on a similar class of subjects increases the chance of nullification of factors such as age, gender, educational and societal status which also influences the presence of alpha.

Keywords: Electroencephalogram (EEG), Alpha, Gamma, Attention, Emotion, Power Spectral Density [PSD], EEG Coherence.

1 Introduction

Exploring brain waves with its role in attentive aspects of human activities has been the interest of numerous studies especially with audio and visual tasks where changes in brain waves always expected to occur [1-3]. Many researchers had varied opinion about the role of alpha and gamma in attentive tasks.

Alpha rhythm as one of the best known EEG phenomenon [4-6] is mostly present during the state of awake relaxation with eyes closed. In adults, it was reported that EEG alpha waves are more dominant and has higher amplitude of about 100 µV in the parietal-occipital region of the brain. However, activity of alpha waves were reportedly decreases in response to all types of motor activities including visual inspection, or even blinking of eyes [7],[8]. Alpha, initially known as 'idling rhythms'

---

[*] Corresponding author.

F. Ortuño and I. Rojas (Eds.): IWBBIO 2015, Part I, LNCS 9043, pp. 282–291, 2015.
© Springer International Publishing Switzerland 2015

started revealing its functional roles lately, when more about its correlates contributing towards cognitive processes and memory were identified.

A study using audio and visual sensory responses from occipital and temporal lobes of normal subjects, showed higher alpha in occipital compared to temporal [9]. The range of alpha and the highest values of Power Spectral Density (PSD) with different types of Sensory Response were reported recently [10]. In another study that was analyzed using graph theoretical tools, authors reported an enhanced synchronization level of alpha [11]. A recent short review gave detailed information about alpha activity in cognitive processes [12],[13], where the significance of alpha activity in both memory and in cognition was explained.

Almost in parallel attempts, similar experiments and results were reported for significance of gamma in cognition [14-16]. While a group of scientists agreed on the importance of alpha [2],[17], many others pointed out the role of gamma for the same. Gamma rhythms called as 'binding' rhythms was reported to be involved in perception, cognition and cognitive task execution [18-21].

Researchers attempted to study processes combined with emotion and attention with experiments using faces or natural scenes in experiments. In one of such attempt, the focus was on the relationship between emotion and attention in terms of processing periods [22],[23]. In one of the latest reports, analysis was done to elucidate alpha and cortical inhibition in affective attention [24] where role of Gamma was not explored. In another latest study, the multimodality aspects of facial and vocal expressions were reported [20]. Thus effectively not much attempts were reported in this area especially about the dominance of type of EEG rhythms.

The present study, made an effort as per the above facts and lacunae. Simple and indigenously designed experiments were conducted using audio and video clippings associated with three different types of emotions. The subject selection was restricted to a class of subjects similar in age, gender, culture, education and societal level, done purposefully to nullify the effect of changes in alpha rhythms with respect to age, gender, education and societal status [25].

2   Method and Material

2.1   Participants

Final year graduate male students of University Malaya, mean age of 22.5 years, right-handed, healthy having normal vision and hearing were selected. The study was approved by the engineering department of Biomedical Engineering, University of Malaya. Procedures were explained to each of the participants and written informed consent, as per prevailing medical ethics, was obtained prior to participation. Table 1 summarizes the information about the participants.

Table 1. Subject details

No. of sub.	Mean age	Gender	ethnicity	Edu, level	handedness
15	22.5 years	Male	Malaysian	Undergraduate	Right-handed

2.2   Materials

EEG was recorded on a 40-channel Medlec Profile EEG Machine in Department of Biomedical engineering, University Malaya. Non-Invasive recording with Ag–Ag/Cl electrodes from 10 scalp positions were done according to the international 10–20 system [26]. Fp1, Fp2, C3, C4, P3, P4, T3, T4, O1 and O2 were used for data extraction, where impedance values were kept below 10 k ohm. Brain vision professional recorder software by Brain Products GmbH and Brain maps were used. Sampling frequency was fixed at 256 Hz and the Sensitivity at 150 μV/mm.

2.3   Procedure

Visual and audio clippings were shown as emotional stimuli. Each clipping, associated with one of the three emotions, lasted for 30 second duration. The type of clippings used is shown in Table 2. Clippings were combined and organized in a power point slide to play them effectively. The participant was required to sit on a comfortable chair and initially, three minutes of awake relaxation with eyes closed, were recorded which was later used as baselines activity.

Table 2. Emotional stimuli used for present study

Task	Type of Emotion	Recording source
	Sad video clip	cartoon animation
	Calm video clip	Mother Nature scene
Visual	Excitement video clip	cultural event
Audio	Sad music	famous television program theme song
	Calm music	religious song
	Excitement music	Cultural event song

Figure 1 show the details of the procedure for one session, which lasted for 6 minute and was repeated for four times from each subject. The entire duration for data collection from one subject lasted for a maximum of 45 minutes to an hour, including preparation of subject.

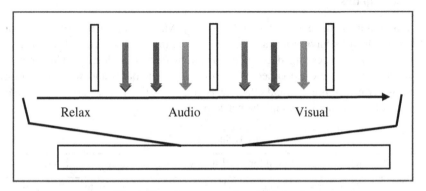

Fig. 1. Experiment procedure for 1 Session

2.4   Data Analysis

Occipital and temporal lobes were focused in all analyses mainly due to the major role played by these lobes in processing sensory inputs related to audio and video emotional attention stimuli [27]. The analysis consisted of normalization of the data, relevant band and feature extraction using discrete wavelet transform. The approach used was as in following steps:

Artifact Removal. Data went through pre-processing for removal of eye blinks and other signal interferences. Band pass filter to get signal between 0.5-100 Hz and a 50 Hz notch filter to remove power-line interference components are used. Eye blink artifacts were removed using visual inspection.

Feature Extraction. Discrete Wavelet Packet Transform (DWPT) method was used for feature extraction in alpha and gamma band. Wavelet packet analysis is a generalization of wavelet decomposition that offers richer range of possibilities of signal analysis [28]. By applying the discrete wavelet packet analysis on the original signal, wavelet coefficients in the 32-64 frequency band in the 2nd level node and 8-16 Hz frequency band at the 4th level node (4, 2) was extracted   for gamma and alpha bands respectively [29]. For analysis, Daubechies (db4) wavelets were used as mother wavelet because of its resemblance in physiological shape, and their near optimal time-frequency localization properties.

Following this, PSD of each signal for the left and right side of the brain was calculated, where the length NFFT (non-equispaced fast Fourier transform or non-uniform fast Fourier transform) was selected at 1024 with a window size of 256. The inter-hemispheric coherence values for occipital and temporal lobes were also obtained and compared between alpha and gamma. The coherence function was obtained by dividing the cross-spectral density of the targeted signals over the auto-spectral density of each signal [30]. MATLAB software version 7.12 was used for all analyses.

3   Results

The wavelet decomposition method was performed and the detail coefficients for alpha and gamma bands were extracted.   The PSD of each signal from occipital and temporal electrodes was calculated for both right hemisphere and left hemisphere of the brain. The PSD of baseline activities for all subject were calculated for the purpose of comparison between the relaxing state and emotional attention state. The mean value for relax state was obtained as 749 μV2. A sample of a two second epoch for a calm video recording scene is shown in Figure 2 A.

A combined plot showing the PSD values for three types of emotional attention using audio stimuli is shown in Figure 2 B, and the same using visual scenes is shown in Figure 2 C.

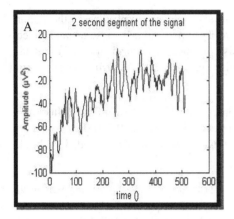
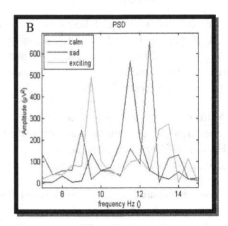
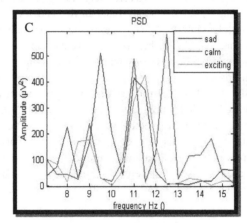

Fig. 2. (A) A two second segment from O1 electrode for calm video clip, (B) PSD of Alpha rhythm from T3 electrode using Audio clipping (C) PSD of Alpha rhythm from O1 electrode for visual clipping

The mean of the PSD values for alpha and gamma were obtained as shown in figure 3 and figure 4. Repeated T-test using Graphic-pad Prism software was conducted to examine the impact of each of the emotionally targeted attention on alpha and gamma. Statistically significant

Differences in mean values were found, where values for alpha was always higher compared to that of gamma. The P-value obtained for alpha, was $0.0001 < 0.05$ for each type of emotional attention states.

Figure 3 and 4 indicated the PSD values for each emotional attention state recorded from respective electrodes. Though the PSD values are quite high for alpha, it can be seen that the mean value of the alpha and gamma tends to be the highest in calm emotions compared with sad and excitement. The difference in mean value for sad and excitement was insignificant in both alpha and gamma, though the values were always higher in alpha.

Graphic-pad Prism software was also utilized to perform a repeated one way ANOVA between the three emotions (sad , calm and exciting) in order to find how significantly the mean values differ from each other. The calculated P-value showed that there is a statistically significant differences in the mean value of PSD between the three emotions in alpha ($F=3.336$, $P=0.0425<0.05$) and there is in-significant difference for the same in gamma ($F=2.0365$, $P=0.1398>0.05$). This result clearly showed that changes in alpha are more observable in attentive states associated with emotion compared with the same in gamma.

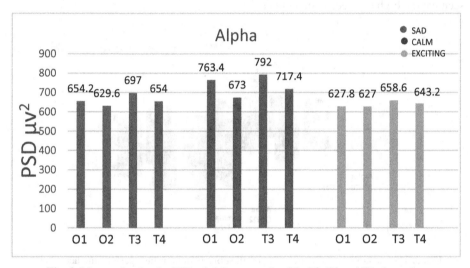

Fig. 3. Mean value for the PSD of alpha waves for O1, O2, T3 and T4 electrodes

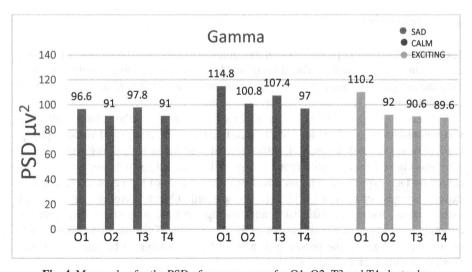

Fig. 4. Mean value for the PSD of gamma waves for O1, O2, T3 and T4 electrodes

The Inter-hemispheric coherence was also calculated in order to quantitatively measure the linear dependency among alpha and gamma between the two brain regions. The calculation was done between T3 and T4 electrodes and between O1 and O2 electrodes for all type of emotional attention states used in the study. The mean value was calculated and is shown in figure 5. The result showed high coherence in Alpha rhythm compared to Gamma in all type of emotional attention states between both left and right temporal and occipital lobes. The range of coherence for O1 and O2 electrodes was between *0.7-0.9* in alpha and gamma where alpha was always shown higher values than gamma. An average coherence between *0.5-0.7* in both alpha and gamma, was obtained between T3 and T4 electrodes where alpha was again seen having higher values than gamma.

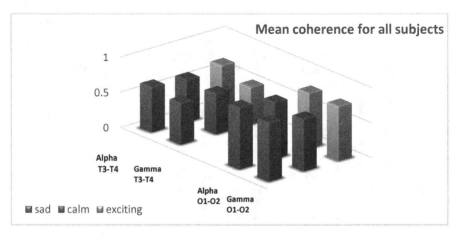

Fig. 5. The mean values of coherence for all subjects in their task performing activities

4   Discussion

The study revealed the prominence in dynamics of Alpha over Gamma in processing human emotional attention. The finding would show that the functionality of alpha dynamics might have better roles to play in its processes other than or associated with cognition and short term memories reported so far.

We recorded ongoing EEG during an experiment in which the subject's attention was voluntarily targeted to emotionally significant stimuli. Audio and video clippings used were scenes and music's of nature or that associated with living environment. This was chosen so as to differentiate from other research, which were mostly conducted on face expressions and on affective image scenes of high arousal, fear, anger, extreme sadness or neutral [17],[31]. The selected subjects were all right handed, male, healthy, having normal vision and hearing and are students of final year undergraduate study who are all known to the department. This was done purposely to nullify the changes in the presence of alpha which is known to be affected by age, gender, education and societal level.

The analyses were done for occipital and temporal lobes due to two reasons. One, because of the audio and video emotional attention stimuli used, where temporal and occipital lobes would be playing the major role in processing the related sensory inputs respectively [27]. Secondly, an increased activation in occipital and temporal cortical regions was reported in emotional attention by various researchers [22],[32].

The PSD values were significantly high for all the three type of emotions in Alpha compared to Gamma. In all three types of emotions, alpha (763.4) and gamma (114.8) gave the highest PSD values for 'calm' compared with sad and excitement. This indicated that calm emotions enhance both alpha and gamma which is a fact in accordance with prior researchers. The statistical analysis reported the p value as (0.0001<0.05). PSD values for both 'sad' (654.2) and 'excitement' (627.8) were shown as having insignificant differences in both alpha and gamma (96.6 for sad; 110.2 for excitement). Values from the left hemispheric electrodes were higher compared with that from right, for both occipital and temporal lobes. The most interesting observation was that, though the pattern of change in PSD values displayed the same characteristics for both Alpha and Gamma for all the three emotional attention analyses, values are quite high (0.0001<0.05) for Alpha. Thus the changes in Alpha were quite distinct in comparison with that in Gamma. The same characteristic pattern shown by alpha and gamma in emotional states are in agreement with previous researchers in their opinion on alpha and gamma, observed separately.

The inter-hemispheric coherence analyses shown high values in occipital as between 0.7-0.9 and a lower value in temporal lobe as between 0.5-0.7.This showed that visual scenes had more effect compared to audio in measuring the performance associated with emotional attentiveness. This established the role of these two lobes in processing emotion attention and is also in agreement with the views from earlier researchers [33],[34]. This fact also adds to the credibility to the present experiment.

The result clearly showed that the changes in Alpha were more observable in all the three emotional attention states used, compared with that in Gamma. The present study is the first time reporting on the prominence of functional alpha dynamics in comparison with Gamma and emphasizes its importance in processing attention associated with human emotional states.

5 Conclusion

Cognitive neuroscience has been very active with number of research in the area of attentive tasks and human emotion. Research reported so far, varied mostly in its opinion about the role played by Alpha and Gamma rhythms in these areas. The present study reported the major role by alpha in its presence and availability during three different emotional attention states in an experiment with similar class of subjects. The most impressive findings in this study was the higher PSD values in 'Calm' as compared to 'Sad' and 'Excitement' emotion and the higher alpha PSD relative to that of Gamma. Since emotional responses are also known to modify perception, alpha rhythms might be a major part of neural correlates that may modulate specialized mechanisms of emotional attention. Authors acknowledge the fact that, repetition of the study with a bigger sample size should pave way for more research attempts to elucidate the role of alpha dynamics governing such mechanisms and in explaining related biological implications.

Acknowledgement. This research was funded by the University Malaya Research Grant UMRG RP016D-13AET. The authors express their gratitude to all volunteers participated and contributed in conducting the experiment. The authors declare no competing interests.

References

1. Klimesch, W.: Alpha-band oscillations, attention, and controlled access to stored information. Trends in Cognitive Sciences 16, 606–617 (2012)
2. Ray, W., Cole, H.: EEG alpha activity reflects attentional demands, and beta activity reflects emotional and cognitive processes. Science 228, 750–752 (1985)
3. Schaefer, R.S., Vlek, R.J., Desain, P.: Music perception and imagery in EEG: Alpha band effects of task and stimulus. International Journal of Psychophysiology 82, 254–259 (2011)
4. Angelakis, E., Lubar, J.F., Stathopoulou, S., Kounios, J.: Peak alpha frequency: an electroencephalographic measure of cognitive preparedness. Clinical Neurophysiology 115, 887–897 (2004)
5. Klimesch, W., Doppelmayr, M., Russegger, H., Pachinger, T., Schwaiger, J.: Induced alpha band power changes in the human EEG and attention. Neuroscience Letters 244, 73–76 (1998)
6. Lutzenberger, W.: EEG alpha dynamics as viewed from EEG dimension dynamics. International Journal of Psychophysiology 26, 273–283 (1997)
7. Bazanova, O.M., Vernon, D.: Interpreting EEG alpha activity. Neuroscience and Biobehavioral Reviews (2013)
8. Klimesch, W., Schimke, H., Pfurtscheller, G.: Alpha frequency, cognitive load and memory performance. Brain Topography 5, 241–251 (1993)
9. Huisheng, L., Mingshi, W., Hongqiang, Y.: EEG Model and Location in Brain when Enjoying Music. In: 27th Annual International Conference of the Engineering in Medicine and Biology Society, IEEE-EMBS 2005, pp. 2695–2698 (2005)
10. Hussin, S.S., Sudirman, R.: Sensory Response through EEG Interpretation on Alpha Wave and Power Spectrum. Procedia Engineering 53, 288–293 (2013)
11. Wu, J., Zhang, J., Ding, X., Li, R., Zhou, C.: The effects of music on brain functional networks: A network analysis. Neuroscience 250, 49–59 (2013)
12. Başar, E., Güntekin, B.: A short review of alpha activity in cognitive processes and in cognitive impairment. International Journal of Psychophysiology 86, 25–38 (2012)
13. Klimesch, W.: EEG alpha and theta oscillations reflect cognitive and memory performance: a review and analysis. Brain Research Reviews 29, 169–195 (1999)
14. Aoki, F., Fetz, E., Shupe, L., Lettich, E., Ojemann, G.: Increased gamma-range activity in human sensorimotor cortex during performance of visuomotor tasks. Clinical Neurophysiology 110, 524–537 (1999)
15. Başar, E., Başar-Eroglu, C., Karakaş, S., Schürmann, M.: Gamma, alpha, delta, and theta oscillations govern cognitive processes. International Journal of Psychophysiology 39, 241–248 (2001)

16. Tallon-Baudry, C., Bertrand, O., Peronnet, F., Pernier, J.: Induced γ-band activity during the delay of a visual short-term memory task in humans. The Journal of Neuroscience 18, 4244–4254 (1998)

17. Güntekin, B., Basar, E.: Emotional face expressions are differentiated with brain oscillations. International Journal of Psychophysiology 64, 91–100 (2007)

18. Balconi, M., Pozzoli, U.: Event-related oscillations (EROs) and event-related potentials (ERPs) comparison in facial expression recognition. Journal of Neuropsychology 1, 283–294 (2007)

19. Engel, A.K., Singer, W.: Temporal binding and the neural correlates of sensory awareness. Trends in Cognitive Sciences 5, 16–25 (2001)

20. Fitzgibbon, S.P., Pope, K.J., Mackenzie, L., Clark, C.R., Willoughby, J.O.: Cognitive tasks augment gamma EEG power. Clinical Neurophysiology 115, 1802–1809 (2004)

21. Martini, N., Menicucci, D., Sebastiani, L., Bedini, R., Pingitore, A., Vanello, N., Milanesi, M., Landini, L., Gemignani, A.: The dynamics of EEG gamma responses to unpleasant visual stimuli: from local activity to functional connectivity. Neuroimage 60, 922–932 (2012)

22. Schupp, H.T., Stockburger, J., Codispoti, M., Junghöfer, M., Weike, A.I., Hamm, A.O.: Selective visual attention to emotion. The Journal of Neuroscience 27, 1082–1089 (2007)

23. Gerdes, A.B., Wieser, M.J., Bublatzky, F., Kusay, A., Plichta, M.M., Alpers, G.W.: Emotional sounds modulate early neural processing of emotional pictures. Frontiers in Psychology 4 (2013)

24. Uusberg, A., Uibo, H., Kreegipuu, K., Allik, J.: EEG alpha and cortical inhibition in affective attention. International Journal of Psychophysiology 89, 26–36 (2013)

25. Kolev, V., Yordanova, J., Basar-Eroglu, C., Basar, E.: Age effects on visual EEG responses reveal distinct frontal alpha networks. Clinical Neurophysiology 113, 901–910 (2002)

26. Jasper, H.H.: The ten twenty electrode system of the international federation. Electroencephalography and Clinical Neurophysiology 10, 371–375 (1958)

27. Goodale, M.A., Milner, A.D.: Separate visual pathways for perception and action. Trends in Neurosciences 15, 20–25 (1992)

28. Vetterli, M., Herley, C.: Wavelets and filter banks: theory and design. IEEE Transactions on Signal Processing 40, 2207–2232 (1992)

29. Nussbaum, P.A., Hargraves, R.H.: Pilot Study: The Use of Electroencephalogram to Measure Attentiveness towards Short Training Videos. International Journal (2013)

30. Weiss, S., Mueller, H.M.: The contribution of EEG coherence to the investigation of language. Brain Lang. 85, 325–343 (2003)

31. Höller, Y., Thomschewski, A., Schmid, E.V., Höller, P., Crone, J.S., Trinka, E.: Individual brain-frequency responses to self-selected music. International Journal of Psychophysiology 86, 206–213 (2012)

32. Kastner, S., Ungerleider, L.G.: Mechanisms of visual attention in the human cortex. Annual Review of Neuroscience 23, 315–341 (2000)

33. Cvetkovic, D., Cosic, I.: EEG inter/intra-hemispheric coherence and asymmetric responses to visual stimulations. Medical & Biological Engineering & Computing 47, 1023–1034 (2009)

34. Knyazeva, M., Kiper, D., Vildavski, V., Despland, P., Maeder-Ingvar, M., Innocenti, G.: Visual stimulus–dependent changes in interhemispheric EEG coherence in humans. Journal of Neurophysiology 82, 3095–3107 (1999)

New Insights in Echocardiography Based Left-Ventricle Dynamics Assessment

Susana Brás[1], Augusto Silva[1], José Ribeiro[2], and José Luís Oliveira[1]

[1] Instituto de Engenharia Electrónica e Telemática de Aveiro (IEETA),
Campus Universitário de Santiago, 3810-193 Aveiro, Portugal
[2] Centro Hospitalar V.N. Gaia, Cardiology Department,
4434-502 V.N. Gaia, Portugal
{susana.bras,augusto.silva,jlo}@ua.pt,
jri@chvng.min-saude.pt

Abstract. Cardiovascular diseases affect a high percentage of people worldwide, being currently a major clinical concern. Echocardiograms are useful exams that allow monitoring the heart dynamics. However, their analysis depends on trained physicians with well-developed skills to recognize pathology from morphological and dynamical cues. Furthermore, these exams are often difficult to interpret due to image quality. Therefore, automatic systems able to analyze echocardiographic quantitative parameters in order to convey useful information will provide a great help in clinical diagnosis. A robust dataset was built, comprising variables associated with left-ventricle dynamics, which were studied in order to build a classifier able to discriminate between pathological and non-pathological records. To accomplish this goal, a network classifier based on decision tree was developed, using as input the left ventricle velocity over a complete cardiac cycle. This classifier revealed both sensitivity and specificity over 90% in discriminating non-pathological records, or pathological records (dilated or hypertrophic).

Keywords: decision tree, echocardiogram, classification.

1 Introduction

The automatic characterization of individual cardiac conditions will allow a faster diagnostic of cardiovascular diseases, a major cause of death worldwide [1]. Therefore, the quantification of the left ventricular activity is of the utmost importance for the objective assessment of the myocardial function [2]. Echocardiography is an imaging modality that has been for long a readily available and effective means to support the diagnosis and follow-up procedures of cardiovascular diseases. Although the echocardiogram is a much-used method of diagnosis, there is a lack of information in the literature on the usage of quantitative parameters since consensus is mostly absent regarding the clinical meaning of standard values or those considered normal [3]. It is our basic assumption that methods that evaluate the left ventricle dynamics throughout a (virtual) cardiac cycle will

F. Ortuño and I. Rojas (Eds.): IWBBIO 2015, Part I, LNCS 9043, pp. 292–302, 2015.
© Springer International Publishing Switzerland 2015

provide, at least, the basis for automatic discrimination between groups (normal versus pathological).

With adequate procedural technique, echocardiographic images may depict the left ventricle dynamics and activity. Using appropriate image processing tools it is therefore possible to obtain quantitative descriptors related with the contour, wall velocity, diameter, area and volume, among others. It must be emphasized that, even with standard image processing tools provided by the scanners manufacturers those descriptors are difficult to obtain accurately. Most of the techniques are still semi-automatic requiring skillful user interaction. There is anyway a vast amount of raw quantitative data that is likely to be fed into a decision support system. Within a large image repository framework this decision support system will be able to combine information extracted from many echocardiograms and produce useful knowledge that will help clinical diagnosis.

There is, so far, no rule for selecting the best data mining method, which is normally evaluated according to each problem [4]. Therefore the aim of this study is to extract useful quantitative information from echocardiograms of pathological cases, in order to characterize the pathology. To accomplish this goal, a network classifier based on a decision tree was implemented allowing a primary classification between non pathological and pathological (dilated and hypertrophic cardiomyopathy).

In recent years, several studies tried to identify motion abnormalities in the left ventricle wall, a goal that has been accomplished through diverse statistical or modeling techniques, e.g. hidden markov models, shape statistical analysis, independent component analysis classifiers, statistical modeling [5] [6] [7]. These works are of utmost importance in the detection of local pathologies. They had as objective to identify heart regions that were characterized with abnormal motion, not identifying the type of pathology. By the opposite, in our work, the left ventricle movement information is studied globally in order to infer possible myocardial pathologies (normal, hypertrophic or dilated cardiomyopathy).

2   Material and Methods

2.1   Dataset

The database was built from clinical records collected in regular procedures at Centro Hospitalar Gaia-Espinho. The local ethics committee approved the procedures and an informed consent was obtained from patients. In all collected data, a four-chamber view was recorded during, at least, a complete cardiac cycle. The data were then anonymized and processed offline using a proprietary software package from Siemens, the Syngo Velocity Vector Imaging technology (VVI) [8]. The left ventricle contour is visually identified and manually marked, at the beginning of the systolic phase. From this point, the software is able to track the evolution of the first landmark through the entire record, extracting variables related with the left ventricle function [9]:

- **Velocity** dynamical measure over the left ventricle. Evaluates velocity over the cardiac cycle at 49 image points, reflecting the ventricle displacement per unit of time. (49 data points over a sequence of time.) Units: cm/sec
- **Strain** describes the ventricle deformation, by the fraction of change on the myocardial deformation. (49 data points over a sequence of time.) Units: %
- **Strainrate** the rate of change in strain. (49 data points over a sequence of time.) Units: 1/sec
- **Tx** left ventricle coordinates in the xx axis. (49 data points over a sequence of time.)
- **Ty** left ventricle coordinates in the yy axis. (49 data points over a sequence of time.)
- **tVx** velocity vector projection on the xx axis. (49 data points over a sequence of time.) Units: cm/sec
- **tVy** velocity vector projection on the yy axis. (49 data points over a sequence of time.) Units: cm/sec
- **Volume** left ventricle volume estimation over time. (a vector.) Units: ml
- **Segment volume** left ventricle volume estimation in its six segments. (6 data points evaluated over time) Units: ml
- **Minimum diameter** left ventricle minimum diameter over time. Units: mm
- **Maximum diameter** left ventricle maximum diameter over time. Units: mm
- **dV/dt** volume over time (derivate of volume) Units: ml/sec

2.2   Data Normalization

The software used for variable extraction has the pre-requisite of starting the echocardiogram analysis in the beginning of systole (i.e. at the maximum volume). This requirement guarantees that the first analyzed frame in each record corresponds to the same physiological point (i.e. comparable frames). However, depending on the machine, or its configured parameters, the sampling frequency may be different between records. The echocardiogram frame rate may vary between 10 and 80 frames/sec [9]. Therefore, in a first phase, it is necessary to standardize the sampling frequency, allowing the comparison of variables and records, indexed to the frame. To meet this requirement, a cubic spline was used to up resample all the records in the dataset to a higher frequency (100 frames/sec).

Briefly, the spline will find a cubic polynomial between two x points and link them, its function being defined as:

$$f(x) = P_i(x), x_i \leq x \leq x_{i+1}, i = 1, ..., k - 1 \qquad (1)$$

where k is the length of the x vector. The spline building is based on a balance between the best data fitting and the best smoothing. Therefore, to obtain the best spline to the data [10], equation (2) is minimized:

$$(1 - q)\frac{1}{k} \sum_{i=1}^{k}(y_i - f(t_i))^2 + q \int_0^1 f''(t)^2 dt \qquad (2)$$

where q is a parameter that varies between 0 and 1, and y is the "real" data.

2.3   Variables Representation

The cardiac cycle is divided in systole and diastole. In the systole the ventricle decreases its volume until reaching the minimum. In the diastole, the ventricle increases its volume to the maximum. Consequently, by inspecting the volume curve, the cardiac phases are identified. After the spline building and data interpolation, all previously presented variables were split in two cardiac phases (systole and diastole). In each one, the mean and standard deviation of each variable cardiac phase interval was calculated, allowing the inference of the variation of a particular descriptor related with the LV dynamics. These variables were used as inputs in the classification system.

Several variables extracted from the echocardiogram were used as inputs of the decision tree. The output will provide a classification for the myocardial status (non-pathological, dilated or hypertrophic cardiomyopathy). For tree split, the Gini's diversity index is used, which evaluates the level of diversity in a sample [11] [12]. Essentially, this index increases when the number of types in which the entities are classified increases also. For the decision tree construction, the data was divided in two groups, the training and testing dataset. In this study, the testing dataset was defined as 50% of each echocardiogram type, chosen randomly. The training dataset corresponds to the other 50%. To evaluate possible overfitting of the tree, it had been trained and tested 100 times using randomly chosen heartbeats in the proportion defined before. This method allows a balanced proportion of echocardiogram types, and also datasets not biased by the records, because the random choice inside each type guarantees data from different patients.

2.4   Method Evaluation

The global classification error (e) is assessed as the ratio between the true positive (correctly classified cases) and the total number of cases.

To evaluate the classification performance, sensitivity (Sen, Eq.1), specificity (Spe, Eq.2) and error (Error, Eq.3) were used, where TP is the true positive, TN the true negative, FN the false negative, FP the false positive, and N the total number of evaluated records. Sensitivity evaluates the probability of the method classifying an echocardiogram in a class and effectively belonging to that class. On the other hand, specificity evaluates the probability of the method not classifying an echocardiogram in a class and not in fact belonging to that class. Error translates the degree of correctly classified records by the classification system.

$$Sen = \frac{TP}{(TP + FN)} \tag{3}$$

$$Spe = \frac{TN}{(FP + TN)} \tag{4}$$

$$Error = 1 - \frac{TP}{(N)} \tag{5}$$

3   Results

The analyzed dataset is composed of 54 echocardiogram records, of which 32 are non-pathological and 22 pathological (12 dilated and 10 hypertrophic cardiomyopathy). All variables were recorded on one complete cardiac cycle and they were split in systole and diastole, allowing comparison of different patient variables in each cardiac phase.

Figure 1 presents a record comprising 2 complete cardiac cycles. The volume is used to mark the transition point between phases, allowing association of the ventricle variation in each cardiac cycle.

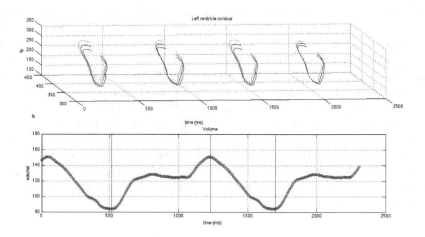

Fig. 1. Left ventricle variation model ($mean \pm 2std$) representation through two complete cardiac cycles, describing systolic (first and third representations) and diastolic (second and fourth representations). The blue line represents the mean, and the read and green lines represent the model standard deviation. Below, the left ventricle volume is represented, with red marks representing the transition between each cardiac phase. tx/ty is the xx/yy left ventricle coordinates, respectively.

Based on the idea that the pathological and non-pathological data will have distant representations, the variables' discriminatory power was studied based on their statistical distribution. All the variables studied had superimposed statistical distributions, especially considering the hypertrophic and non-pathological data. The dilated cardiomyopathy (by physiological definition) is characterized as an abnormal left ventricle movement, which consequently influences the extracted variables. Therefore those variables have intrinsic differences comparing to non-pathological data. Nevertheless, by statistical analysis over the variables, this difference is not enough to split between the considered classes.

The use of classifiers allows extraction of information from data that is not accessible by itself. They may combine data information, extracting characteristics able to discriminate between groups. The chosen classifier in this analysis was the decision tree, due to its characteristics that allow a straightforward reading of the discrimination rules by both technical and clinical staff. The classifiers are able to differentiate between several classes, but if the data is not well defined by a group of characteristics or variables, the differentiation between classes may be lower. Therefore, it is important to select the input variables in the classifier, which reveal the best discriminatory power. The introduction of irrelevant or noisy data may lead the classifier to weak or false results [4].

Considering that the main objective of the study is to differentiate between pathological and non-pathological data, in the first step the intention is to find the variables that are clearly relevant in records discrimination. Since individual evaluation using the empirical statistical distribution does not give enough information to infer such evidence, in this study the variables will be used individually in a decision tree. The variables revealing the high accuracy will be used in combination with each other in order to evaluate if the classifier performance increases.

Considering that the pathological records are divided between dilated and hypertrophic cardiomyopathy, a network classifier was also implemented. That is, a record, in the first step, is analyzed and classified as pathological or non-pathological. If it is classified as pathological, it will enter a new classifier that will differentiate between dilated or hypertrophic. Again, the classifier (in this case decision tree) input variables will be analyzed in order to identify which of them are relevant in pathology discrimination. The final classifier will be the one with the best performance and the lowest number of inputs.

3.1 Relevant Variables in Pathological/ Non-pathological Differentiation

Considering the goal to find the most relevant variables in echocardiogram discrimination, the training and test samples were randomly chosen in a proportion of 50-50. However, to guarantee that the tree structure and the evaluation are comparable, these samples were the same in all the evaluations presented.

In the first approach, all variables were considered as inputs considering different variables combinations. However, tree results interpretation revealed that the tree was only using observations from a few variables. Knowing that the introduction of not useful information in the classifier may lead to reduced performance, it is important to evaluate the individual discriminatory power of each variable, in the separation of both pathological and non-pathological classes.

In the tree implementation, all the extracted variables evaluated in both systole and diastole were used. These variables by themselves do not reveal their total power in the data discrimination, as they may do when combined with other variables. Nevertheless, it is always necessary to omit features that may only contribute noise to the tree, making information evaluation difficult.

Table 1 resumes the classification tree performance in the differentiation between pathological and non-pathological records, considering the different input

Table 1. Performance evaluation of the tested classifiers with the inputs extracted from the Siemens software. Bold symbolizes the best performances.

Input	Sensitivity	Specificity
dV_dt	70.37%	70.37%
$dmax$	68.52%	68.52%
$dmin$	79.63%	79.63%
$segvol$	74.07%	74.07%
$strain$	88.89%	88.89%
$strainrate$	83.33%	83.33%
tVx	75.93%	75.93%
tVy	88.89%	88.89%
tx	79.63%	79.63%
ty	72.22%	72.22%
$velocity$	**90.74%**	**90.74%**
vol	79.63%	79.63%

combinations. From inspection of this table, there is one possible candidate: velocity.

In order to overcome possible bias in the data, the classifier was tested in 100 randomly selected samples (separation between train and test dataset). In that evaluation, velocity was the variable with the best results, with a mean performance of $75.8 \pm 8.8\%$ (in sensitivity and specificity evaluation over the 100 tests).

The first step classifier was built with the left ventricle velocity as input argument.

3.2   Relevant Variables in Dilated/ Hypertrophic Differentiation

In the first step of the network classifier, the record is classified as pathologic or non-pathologic, but the classifier does not have 100% classification accuracy, so there will be an error that will accumulate from the first step to the second.

Some records will be classified in the first step as pathological, when they are actually non-pathological, and consequently, in the second step evaluation, these records will increase the error.

In the second step classifier evaluation, the principle used was the same as in the first step. Each variable will be evaluated as a potential input to the system. The variable or variables that conduce to the lowest error / highest performance will be chosen as the relevant variables in cardiomyopathy differentiation. When the performance is evaluated, it is observed to be similar considering all the variables. Therefore, as previously, the performance is evaluated with 100 randomly selected samples. The results are presented in Table 2. Evaluating sensitivity, the variable with the highest value is the maximum and minimum left ventricle diameter. Considering specificity, the most relevant measure is the yy left ventricle coordinates. However, the difference between these two measures and the other evaluated measures is not significant. Therefore, the classification error is

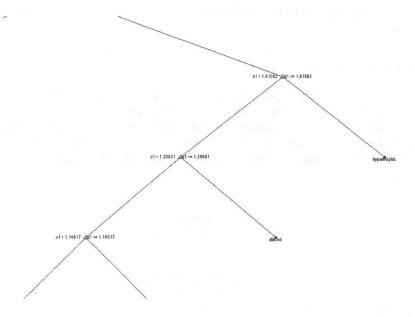

Fig. 2. Part of the pathology decision tree used in the differentiation between dilated and hypertrophic cardiomyopathy

Table 2. Performance evaluation in the second step network classifier. Bold symbolizes better performances/ lower error.

Input	Sensitivity	Specificity	Error
dV_dt	$94, 3 \pm 6, 1\%$	$85, 8 \pm 8, 8\%$	$16, 2 \pm 8, 2\%$
$dmax$	$\mathbf{96,8 \pm 4,7\%}$	$86, 2 \pm 9, 1\%$	$14, 4 \pm 7, 6\%$
$dmin$	$\mathbf{96,8 \pm 4,7\%}$	$86, 2 \pm 9, 1\%$	$14, 5 \pm 8, 5\%$
$segvol$	$93, 7 \pm 5, 2\%$	$83, 6 \pm 8, 7\%$	$15, 8 \pm 7, 9\%$
$strain$	$91, 7 \pm 5, 9\%$	$83, 0 \pm 7, 6\%$	$17, 0 \pm 8, 4\%$
$strainrate$	$94, 1 \pm 6, 3\%$	$82, 5 \pm 8, 6\%$	$16, 1 \pm 1, 0\%$
tVx	$92, 5 \pm 6, 1\%$	$86, 1 \pm 9, 0\%$	$17, 2 \pm 8, 8\%$
tVy	$94, 9 \pm 6, 3\%$	$\mathbf{87,9 \pm 9,2\%}$	$14, 0 \pm 9, 1\%$
tx	$91, 8 \pm 6, 0\%$	$82, 9 \pm 8, 3\%$	$16, 2 \pm 8, 6\%$
ty	$95, 3 \pm 5, 9\%$	$86, 8 \pm 9, 3\%$	$14, 8 \pm 7, 8\%$
$velocity$	$95, 0 \pm 5, 2\%$	$85, 9 \pm 8, 8\%$	$\mathbf{13,3 \pm 9,7\%}$
vol	$92, 8 \pm 6, 5\%$	$84, 3 \pm 8, 9\%$	$16, 5 \pm 8, 5\%$

also inspected. This evaluation leads to one possible variable as potential discriminator between dilated and hypertrophic cardiomyopathy: velocity (of the left ventricle in systole and diastole). Considering that velocity is already used in the first step, this makes the system simpler to explain and analyze. This decision tree is exemplified in figure 2.

The final performance is $95.0 \pm 5.2\%$ considering sensitivity and specificity of the classifier.

3.3   Network Classifier

Velocity was chosen as the input variable, since in terms of performance, it achieves high values, and the lowest mean error with 14.8%.

In summary, a pathological cardiomyopathy classifier was built evaluating as input the velocity of the left ventricle, described by 49 points that delineates the ventricle and evaluates the velocity at each frame. Figure 3 represents the workflow information.

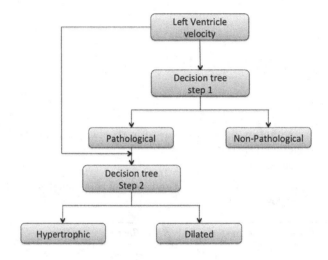

Fig. 3. Classifier workflow information

To help understanding the left ventricle delineation process, whose points are then analyzed, figure 4 represents the contour of the left ventricle with the 49 delineated points.

In the first step of the decision tree evaluation, it was found that evaluation of only one point, corresponding to the diastole phase, may discriminate between pathological and non-pathological records. On the contrary, in the second step of the decision tree all the evaluation was based on one point of the systolic phase. This finding reveals that the diastolic phase is of the utmost importance in the discrimination between a pathologic and non-pathologic case. In pathology differentiation, the systolic phase is revealed to be more useful.

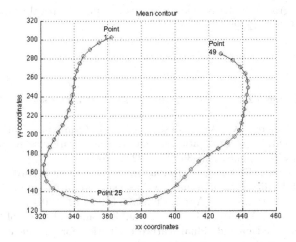

Fig. 4. Left ventricle contour mean, delineated by the 49 points

4   Conclusions

Automatic computer aided decision systems are welcome auxiliary means to help in the diagnosis of cardiovascular disease. The present study implemented a network classifier to differentiate non pathological from pathological records. It classifies three classes in two steps: 1. identifies a new record as pathological or not; 2. if pathological, the record will follow to a new classifier branch, where it will be classified as dilated or hypertrophic cardiomyopathy. The classifier, a decision tree, uses one input variable, velocity, sampled at 49 points in the left ventricular wall. Since the same variable is used in the two steps, a classifier for the three classes was tested, but the result performance decreases. This latter result is expected, because the classifier has to find more specificity in the data. Therefore, a binary decision is easier for the classifier.

Velocity is revealed to be the variable with best discriminatory power, in accordance with the physiological description of heart diseases: if our heart is sick the velocity at which it contracts and dilates is not normal, and therefore there will be differences with the non pathological records. The dilation phase (diastole) is useful in differentiation between pathological and non-pathological. On the contrary, the contraction phase (systole) gives information about the type of pathology.

Acknowledgments. This work was supported by the European Regional Development Fund (FEDER) through the COMPETE programme and by the Portuguese Government through the FCT, Fundação para a Ciência e a Tecnologia, in the scope of the project PEst-OE/EEI/UI0127/2014 (IEETA/UA), and Cloud Thinking (CENTRO-07-ST24-FEDER-002031, co-funded by QREN,

Mais Centro program). S. Brás acknowledges the postdoctoral grant from FCT (ref. SFRH/BPD/92342/2013).

References

1. World Health Organization: The top 10 causes of death, Fact sheet n° 310 (updated June 2011)
2. Fernandéz-Caballero, A., Vega-Riesco, J.M.: Determining heart parameters through left ventricular automatic segmentation for heart disease diagnosis. Expert Systems with Applications 36, 2234–2249 (2009)
3. Akdemir, B., Gunes, S., Oran, B., Karaaslan, S.: Prediction of cardiac end-systolic and end-diastolic diameters in m-mode values using adaptive neural fuzzy inference system. Expert Systems with Applications 37, 5720–5727 (2010)
4. Arauzo-Azofra, A., Aznarte, J.L., Benitez, J.M.: Empirical study of feature selection methods based on individual feature evaluation for classification problems. Expert Systems with Applications 38, 8170–8177 (2011)
5. Chykeyuk, K., Clifton, D.A., Alison Noble, J.: Feature extraction and wall motion classification of 2D stress echocardiography with relevance vector machines. In: 2011 IEEE International Symposium on Biomedical Imaging: From Nano to Macro, pp. 677–680. IEEE (2011)
6. Mansor, S., Alison Noble, J.: Local wall motion classification of stress echocardiography using a hidden markov model approach. In: 2008 IEEE International Symposium on Biomedical Imaging: From Nano to Macro, pp. 1295–1298. IEEE (2008)
7. Punithakumar, K., Ayed, I., Islam, A., Goela, A., Ross, I.G., Chong, J., Li, S.: Regional heart motion abnormality detection: An information theoretic approach. Medical Image Analysis 17(3), 311–324 (2013)
8. Pirat, B., Khoury, D.S., Hartley, C.J., Tiller, L., Rao, L., Schulz, D.G., Nagueh, S.F., Zoghbi, W.A.: A novel feature-tracking echocardiographic method for the quantitation of regional myocardial function validation in an animal model of ischemia-reperfusion. Journal of American College of Cardiology 51(6), 651–659 (2008)
9. Mor-Avi, V., Lang, R.M., Badano, L.P., Belohlavek, M., Cardim, N.M., Derumeaux, G., Galderisi, M., Marwick, T., Nagueh, S.F., Sengupta, P.P., Sicari, R., Smiseth, O.A., Smulevitz, B., Takeuchi, M., Thomas, J.D., Vannan, M., Voigt, J.-U., Zamorano, J.L.: Current and evolving echocardiographic techniques for the quantitative evaluation of cardiac mechanics: Ase/eae consensus statement on methodology and indications. Journal of American Society of Echocardiography 24, 277–313 (2011)
10. Cui, Z.: Two New Alternative Smoothing Methods in Equating: the Cubic B-Spline Presmoothing Methods and the Direct Presmoothing Method. ProQuest Information and Learning Company, UMI: 3229654 (2006)
11. Konrad, A.M., Prasad, P., Pringle, J.K.: Handbook of Workplace diversity. Sage Publications (2006)
12. Kantardzic, M.: Data Mining concepts, models, methods and algorithms. IEEE, Wiley & Sons (2011)

Optimal Elbow Angle for MMG Signal Classification of Biceps Brachii during Dynamic Fatiguing Contraction

Mohamed R. Al-Mulla[1], Francisco Sepulveda[2], and Mohammad Suoud[1]

[1] College of Computer Science and Engineering, Kuwait University
mrhalm@sci.kuniv.edu.kw, suoud@cs.ku.edu.kw
[2] School of Computer Science and Electronic Engineering, University of Essex,
Colchester CO4 3SQ, UK
fsepulv@essex.ac.uk

Abstract. Mechanomyography (MMG) activity of the biceps muscle was recorded from thirteen subjects. Data was recorded while subjects performed dynamic contraction until fatigue. The signals were segmented into two parts (Non-Fatigue and Fatigue), An evolutionary algorithm was used to determine the elbow angles that best separate (using DBi) both Non-Fatigue and Fatigue segments of the MMG signal. Establishing the optimal elbow angle for feature extraction used in the evolutionary process was based on 70% of the conducted MMG trials. After completing twenty-six independent evolution runs, the best run containing the best elbow angles for separation (fatigue and non-fatigue) was selected and then tested on the remaining 30% of the data to measure the classification performance. Testing the performance of the optimal angle was undertaken on eight features that where extracted from each of the two classes (non-fatigue and fatigue) to quantify the performance. Results show that the elbow angles produced by the Genetic algorithm can be used for classification showing 80.64% highest correct classification for one of the features and on average of all eight features including worst performing features giving 66.50%.

1   Introduction

Localised muscle fatigue has mainly been researched using two techniques, mechanomyography (MMG) and surface electromyograhhy (sEMG). MMG has also been utilised in studies on muscle activity [1] and prosthetic control [2]. MMG is a mechanical measure of the changes within the surface of a contracting muscle i.e., where the vibration of the muscle fibres that move are recorded orizio. MMG can be recorded using a variation of sensors, such as hydrophones, condenser microphones, piezoelectric contact sensors, laser distance sensors orizio,orizio2, sonomyography [2], accelerometers [3] and goniometer [3] among others.

F. Ortuño and I. Rojas (Eds.): IWBBIO 2015, Part I, LNCS 9043, pp. 303–314, 2015.

Studying the effect of the joint angle on muscle fatigue has been a topic that has been touched upon in research using both sEMG and MMG. Several of these studies are fairly dated, although their findings are similar. The muscle length will affect both the force signals and the MMG signals, and this will vary for the different joint angles in the different muscles in our body [1]. The joint angle affects the muscle length, which again affects the localised fatigue in the contracting muscle. Djordjevic et al. [2] found that fatigue occurred faster and at a higher rate in long muscles, which means that the fatigue occurs quicker at small elbow angles. Mamghani et al. studied signals from sustained isometric contraction recorded with both MMG and sEMG in time and frequency domains in various upper limb muslces, including the biceps brachii [3]. Results showed that the shortest endurance time occurred at the longest muscle length (smallest angle). They also found that the MPF (mean power frequency) and the RMS (Rooot mean square) value of the EMG and MMG signal changed according to the joint angle. The effect of elbow angle was different in BB than in the other muscles, where the RMS value was higher at bigger elbow angle and decreased as the elbow angle decreased. Various studies have linked fatigue occurrence to muscle length (which is dependent upon the joint angle) based on force loss in the muscle [4, 5]. In addition, other researchers have found that the rate of fatigue in long muscles is dependent upon the rate of changes in the signal for both sEMG [6, 7] and in the MMG signal [8].

Jaskolska et al [9] studied the RMS value of the MMG signal in different muscles during isometric elbow extensions in young and old women. They discovered that for the BB the RMS value increases when the muscles shortens (bigger elbow angle) for both age groups, and that the optimal average elbow angle values for the young women was smaller ($83.0 +/- 10.0°$) than that of the older women ($90.3 +/- 14.1°$) ($P > 0.05$). These findings are similar to research done on knee extensions [10].

There has been little research on classification of the fatigue content in MMG signals. Most research on classification of MMG signals in muscle activity are used for prosthetic control. Xie et al. [11] proposed a method using the classification results of Short-Time Fourier Transform (STFT), Stationary Wavelet Transform (SWT) and S-Transform (ST) combined with Singular value decomposition (SVD), to find the highest classification accuracy of hand movements based on MMG signals, which gave a classification performance of 89.7% between the two classes (wrist flexion and wrist extension). Another research has also classified the MMG signal emanating from muscle activity for prosthetic control, getting classification accuracy of 70% between the two classes (flexion and extension). However, that research [12] did not use wavelet transform for classification purposes, but rather used RMS (Root Mean Square) as a feature for classification.

Various research has used different classification techniques for sEMG signals in localised muscle fatigue [13], which may also be applicable to research on MMG signals. These include genetic programming and genetic algorithms [14–17], statistical analysis [18–20], as well as classification methods to predict fatigue by using neural networks [21] or linear discriminant analysis (LDA) [22]. A variation of these techniques have been adapted in this research to evolve a pseudo-wavelet for classifying fatigue content in the MMG signal.

In this study a GA was used to find the optimal elbow angle to be used in classification for localised muscle fatigue from fatiguing dynamic contractions based on MMG signals. The MMG signal was extracted using a previously developed pseduo-wavelet [23]. The GA utilised 26 evolutionary runs to find the optimal (best separate Non-Fatigue and Fatigue) joint angle window where the MMG signal is best classified. Then the DBI was used to determine the separation between the two classes (Non-fatigue and Fatigue). The classification performance of the GA was compared using eight commonly used feature extraction methods.

2   Methods

In the first part of this research MMG signals were recorded emanating from the biceps brancii during fatiguing dynamic contractions. The GA was used to determine the optimal elbow angle for fatigue classifications utilising a pseudo-wavelet as a feature extraction method. Finally, the classification performance of the GA was compared using eight commonly used feature extraction methods.

2.1   Data Recording and Pre-processing

Thirteen athletic, healthy male subjects (mean age 27.5 +/- 3.6 yr) volunteered for this research. The study was approved by the University of Essex's Ethical Committee and all subjects signed an informed consent form prior to taking part in the study.

The participants, all non-smokers, were seated on a 'preacher' biceps curl machine to ensure stability and biceps isolation while performing biceps curl tasks. The participants reached physiological fatigue and was encouraged during the trial to reach the complete fatigue stage (unable to continue the exercise).

To evaluate the Maximum Dynamic Strength (MDS) percentage for each participant we used the average of three 100% MDS measurements on three different days to ensure correct estimation. The 100% MDS measurements for each subject were determined by the one-repetition maximum (1RM), where the subjects managed to keep the correct technique while executing the repetition with the heaviest possible load on a preacher biceps curl machine. In other words 100% MDS is equal to 1RM. Determining each subject's 100% MDS allowed estimating the correct loading MDS (40% MDS and 70% MDS) across subjects when conducting the trials.

After establishing the MDS for each subject the trials where carried out. After the warm-up period, all the thirteen participants carried out 3 trials of non-isometric exercises with 40% Maximum Dynamic Strength (MDS) and 3 trials of 70% MDS with a one week resting period between trials to ensure full recovery from the biceps fatigue, giving a total of 104 trials. Only one trial was performed per day for each subject in order to avoid injury.

The MMG signal was recorded using a 3-axis accelerometer (Biometrics, ACL300 (range +/- 10G). The accelerometer was placed on the muscle belly of the biceps Brachii, without covering the end plate zone or getting too close to the musculotendinous region [24]. A flexible electrogoniometer (Biometrics Ltd.) was placed on the lateral side of the arm to measure the elbow angle and arm oscillations.

2.2   Labelling the Signals

The recorded MMG signals were grouped into Fatigue and Non-Fatigue epochs. Initial recordings in the first few repetitions – when the subjects felt "fresh"– were considered 'Non-Fatigue', and when the subject was unable to perform the sustained task the epochs were labelled as 'Fatigue, as per [25]. In the signal analysis, the first repetition was therefore labelled as Non-Fatigue, while the last repetition was labelled as Fatigue. This information was then used to train and test the classifier.

2.3   Genetic Algorithms

Genetic Algorithms (GA) can be used for solving linear and nonlinear problems [26]. The GA used a previously evolved pseudo-wavelet [23] as a feature extraction method to find the optimal elbow angle that could best separate between Non-Fatigue and Fatigue segments of the MMG signal.

The parameter settings for the GA runs are shown in Table 1.

Table 1. Parameter settings for the GA runs

Parameter	Value
Independent runs	26
Population size	5000
Maximum number of generations	20
Mutation probability	10%
Crossover probability	90%
Selection type	Tournament, size 5
Termination criterion	Maximum number of generations

The fitness function in the GA is used to find the optimal elbow angle in the search space. The modified Davies Bouldin Index (DBI) was selected in this study in the fitness function as it is a simple and effective index. Data cluster linear overlap was calculated applying the modified DBI [27] by deciding the proportion of intracluster spread to intercluster centroid distance. A good class separation was expressed by smaller DBI values. Normally the fitness function works by maximisation, using a hill climbing method; however, in this research the DBI was changed into negative numbers, letting the fitness function use the hill climbing method by trying to bring the (now) negative DBI closer to zero.

2.4   Evolved Elbow Angle Selection

The GA used a pseudo-wavelet as a feature extraction to select the most optimal angle for fatigue occurrence in the elbow joint. The pseudo-wavelet has been developed in previous research [23]. A window of the MMG signal was selected based on the starting elbow angle and the ending elbow angle. The starting and ending elbow angle (a window) was determined by the evolutionary process of the GA through testing the separation (DBI) of the two stages of fatigue (Fatigue and Non-Fatigue). This helped the evolutionary processes by intending to minimise the DBI, which allowed the fitness function to increase the separation between the two classes. Usually the fitness function operates by maximisation, utilising a hill climbing technique. This was enabled by the DBI being transformed into negative numbers, allowing the fitness function to use the hill climbing method by attempting to bring the (now) negative DBI closer to zero. The starting and ending angles that gave best MMG signal separation were used at a later stage in the classification.

Figure 1a shows a single rep when the muscle was fresh (Non-Fatigue) indicating the starting and ending joint angles for one of the subject trials using the best GA run. Figure 1b displays the same joint angles, but for a fatiguing rep.

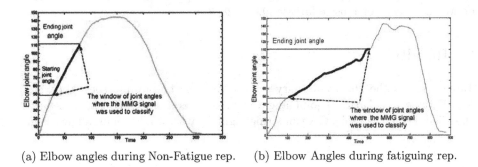

(a) Elbow angles during Non-Fatigue rep.   (b) Elbow Angles during fatiguing rep.

Fig. 1. Optimal window of joint angles selected by the GA showing where the MMG signal was used for classification

2.5   Classification

Establishing the optimal elbow angle for feature extraction used in the evolutionary process was based on 70% of the conducted MMG trials for training purposes. The best run containing the best separation value was selected and then tested on the remaining 30% of the data to measure the classification performance.

For a comparison between the evolved pseudo-wavelet and other feature extraction methods, LDA (linear discriminant analysis) was chosen due to its simplicity, being well established and light on computational resources. The eight features extracted from the MMG signal was the input for the training and testing phase of the LDA classifier.

2.6 Feature Extraction Techniques

A variation of parameters were used to test the optimal angle for the fatigue content in the MMG signal. These are utilised for comparison purposes to measure the classification performance. The parameters used for comparison purposes are:

- Higher-order statistics (HOS) (HO2 and HO3 were used as they gave the best results.)
- Mean Frequency (MF)
- Median Frequency (MDF)
- Power Spectrum Density (PSD)
- Root Mean Square (RMS)
- Mexican Hat (Mex H)
- Evolved Pseudo-wavelet (PW)

There are common parameters that have been applied in a variety of research on MMG signals for muscle fatigue [9, 23, 28–30].

3 Results

Table 2 shows the 26 evolutionary runs in finding the optimal (best separate Non-Fatigue and Fatigue) joint angle window where the MMG signal is best classified. From the table, GA run 9 (highlighted) with window joint angles from

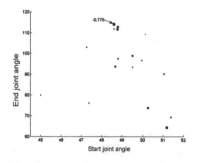
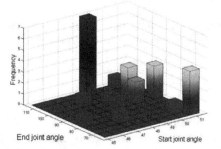

(a) A scatter plot of elbow angles selected by the GA. (larger dots indicate better separation)

(b) 3D histogram of elbow angles selected by the GA

Fig. 2. A scatter plot and a 3D histogram of elbow angles that was selected by 26 evolutionary runs

Table 2. GA runs with the best DBI

Ga Run	Elbow Joint 1	Elbow Joint 2	Dbi
1	48.61	111.75	-0.846
2	51.05	90.20	-0.836
3	49.53	93.42	-0.849
4	51.39	69.30	-0.843
5	49.96	96.66	-0.855
6	45.00	79.96	-0.860
7	48.67	93.71	-0.808
8	48.81	97.47	-0.823
9	50.26	73.90	-0.803
10	48.61	114.03	-0.775
11	48.80	112.78	-0.801
12	48.79	111.59	-0.789
13	51.19	64.35	-0.783
14	48.77	111.82	-0.809
15	51.39	69.30	-0.844
16	51.05	90.20	-0.836
17	48.81	112.87	-0.836
18	49.96	96.66	-0.855
19	49.53	93.42	-0.849
20	47.27	102.90	-0.853
21	51.39	69.30	-0.843
22	48.81	112.87	-0.839
23	50.13	109.22	-0.882
24	49.52	98.85	-0.804
25	47.38	76.10	-0.868
26	51.05	90.20	-0.836
Average	49.45	93.95	-0.83
St.Dev	1.51	16.05	0.03

48.61 to 114.03 gave best DBI between Fatigue and Non-Fatigue, giving -0.775 separation index.

Figure 2 shows a scatter plot 2a of the results obtained in Table2. Moreover, 2b shows a histogram of the joint elbow angles complementing Figure 2a showing the frequency of joint angles that was selected by the GA.

Table 3 shows the classification performance of all 13 subjects with 8 different extraction features. These results indicate that the joint angles found by the GA enabled a classification of the two classes (Non-Fatigue and Fatigue) with exceptional performance depending on the feature used ranging from 52.74% and up to 80.63% classification performance. It can also be seen that the pseudo-wavelet was the feature extraction method that gave best classification average, with the lowest standard deviation. The average classification accuracy for the pseudo-wavelet was much higher than the average classification accuracy for the other parameters. The second highest average classification by the GA used Mexican Hat (wavelet) as a feature extraction technique.

Table 3. Classification performance

Subjects	Ho2	Ho3	Mean freq	Median Freq	Psd	RMS	Mexican Hat	PW
Subject 1	39.01	58.24	57.69	68.68	46.15	47.80	59.20	87.36
Subject 2	67.42	71.35	91.01	70.79	80.34	89.33	81.12	83.92
Subject 3	38.46	63.74	56.59	69.23	66.48	62.09	58.62	79.89
Subject 4	40.62	60.93	81.46	60.93	78.15	79.47	78.73	81.34
Subject 5	44.76	33.47	34.27	52.02	72.18	29.03	76.13	77.78
Subject 6	32.98	70.21	75.53	77.66	74.47	75.53	70.45	70.45
Subject 7	39.64	71.07	60.71	72.86	43.93	45.36	82.59	92.31
Subject 8	49.02	72.06	75.00	66.18	75.00	25.00	52.28	70.56
Subject 9	84.34	89.16	90.36	89.16	85.54	85.54	86.67	88.00
Subject 10	64.66	74.44	58.65	70.68	53.38	54.14	87.30	84.92
Subject 11	65.31	64.29	69.39	64.29	60.20	59.18	75.58	74.42
Subject 12	49.64	46.04	55.76	53.96	56.83	53.60	60.23	67.18
Subject 13	69.16	69.16	71.03	76.64	58.88	66.36	75.82	90.11
Average	52.69	64.93	67.50	68.70	65.50	59.42	72.67	80.63
st.dev	15.74	13.73	15.78	9.87	13.33	19.99	11.55	8.11

Figure 3 shows graphical representation of the classification performance shown in Table 3.

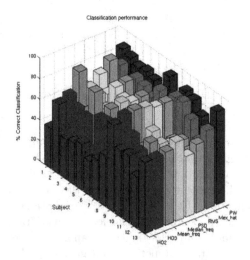

Fig. 3. Graphical representation of the classification performance

4   Discussion

The classification performance of the various feature extraction methods performed less well than the classification based on the GA using the pseudo-wavelet as a feature. The worst performance of the GA was when RMS was utilised as the feature extraction method. RMS is a common parameter used in feature extraction methods for both sEMG and MMG signals, however, for MMG RMS is often used in research on isometric contraction. The poor performance of the RMS parameter may be due to the fact that these results are based on MMG signals from dynamic contraction, while most research on joint angle has utilised fatiguing isometric contractions when analysing the signal with common feature extraction methods. Various researchers has recommended the use of wavelets for studies on MMG signals from dynamic contraction[29–31] due to the stochastic nature of the signal in these contractions. Previous research by Al-mulla et al. [23] also found that wavelets, and in particular the pseudo-wavelet, gave the best classification results for MMG signals emanating from fatiguing dynamic contractions. Results in this study show that the two best classification performance of the optimal elbow angle was both wavelet-based (Mexican Hat and evolved pseudo-wavelet), which may suggest that wavelets perform better at fatigue classification of MMG signals emanating from dynamic contraction.

In this study the optimal elbow angle for MMG classification was determined by the GA. The GA utilised a window that best separate between Non-fatigue and Fatigue at various elbow angles. As the contractions were dynamic, with constant movement of the elbow angle, a window of optimal elbow angle was selected rather than one specific elbow angle degree. Finding the optimal elbow angle of the biceps brachii during fatiguing dynamic contraction can be difficult. For isometric contraction, the optimal elbow angle is perceived to be 90°, however, this differs between subjects and the age of the subjects [9]. The various GA runs used different elbow angles to best separate the two classes, but run 9 gave the best separation index (DBI). This run has the widest range of the starting and ending joint angle, however, this is not a consistent finding between the different GA runs. Some of the runs that has a smaller range in joint angles also gave good separation between fatigue and non-fatigue.

This study was able to produce good classification results for MMG classification for fatiguing contractions from the biceps brachii. It is important to note that the various muscles in our body respond differently when it comes to endurance and fatigue occurrence at different joint angles [1]. While for most muscles it was found that a small joint angle, which gives a longer muscle length, fatigue would occur more quickly [2, 3]. However, for the biceps brachii it was found that the RMS value of the MMG signal increased with increases in the elbow angle. Other research has claimed that endurance time was not dependent upon the elbow angle for the biceps brachii muscle [32, 33]. Therefore, the findings in this study for the biceps brachii may not be applicable to other muscle groups. More research is needed to determine the optimal joint angle classification performance of the GA on other muscles using MMG signals.

5 Conclusion

This research shows that it is possible to find the optimal elbow angle for fatigue classification based on MMG signals from dynamic contractions. The GA, using a pseudo-wavelet as a feature extraction method, was able to separate between the two classes of fatigue (Non-Fatigue and Fatigue). Using the pseudo-wavelet proved to provide better classification results that other common parameters.

References

1. Toma, K., Honda, M., Hanakawa, T., Okada, T., Fukuyama, H., Ikeda, A., Nishizawa, S., Konishi, J., Shibasaki, H.: Activities of the primary and supplementary motor areas increase in preparation and execution of voluntary muscle relaxation: an event-related fMRI study. J. Neurosci. 19(9), 3527–3534 (1999)
2. Djordjevic, S., Tomazic, S., Zupancic, G., Pisot, R., Dahmane, R.: The influence of different elbow angles on the twitch response of the biceps brachii muscle between intermittent electrical stimulations
3. Mamghani, N.K., Shimomura, Y., Iwanaga, K., Katsuur, T.: Mechanomyogram and Electromyogram Responses of Upper Limb During Sustained Isometric Fatigue with Varying Shoulder and Elbow Postures. J. Physiol. Anthropol. 21(1), 29–43 (2002)
4. Fitch, S., McComas, A.: Influence of human muscle length on fatigue. J. Physiol (Lond.) 362, 205–213 (1985)
5. Sacco, P., McIntyre, D.B., Jones, D.A.: Effects of length and stimulation frequency on fatigue of the human tibialis anterior muscle. J. Appl. Physiol. 77(3), 1148–1154 (1994)
6. Doud, J.R., Walsh, J.M.: Muscle fatigue and muscle length interaction: effect on the EMG frequency components. Electromyogr. Clin. Neurophysiol. 35(6), 331–339 (1995)
7. Weir, J.P., McDonough, A.L., Hill, V.J.: The effects of joint angle on electromyographic indices of fatigue. Eur. J. Appl. Physiol. Occup. Physiol. 73(3-4), 387–392 (1996)
8. Weir, J.P., Ayers, K.M., Lacefield, J.F., Walsh, K.L.: Mechanomyographic and electromyographic responses during fatigue in humans: influence of muscle length. Eur. J. Appl. Physiol. 81(4), 352–359 (2000)
9. Jaskolska, A., Kisiel, K., Brzenczek, W., Jaskolski, A.: EMG and MMG of synergists and antagonists during relaxation at three joint angles. Eur. J. Appl. Physiol. 90(1-2), 58–68 (2003)
10. Ebersole, K.T., Housh, T.J., Johnson, G.O., Evetovich, T.K., Smith, D.B., Perry, S.R.: The effect of leg flexion angle on the mechanomyographic responses to isometric muscle actions. Eur. J. Appl. Physiol. Occup. Physiol. 78(3), 264–269 (1998)
11. Xie, H.B., Zheng, Y.P., Guo, J.Y., Chen, X., Shi, J.: Estimation of wrist angle from sonomyography using support vector machine and artificial neural network models. Medical Engineering and Physics 31, 384–391 (2009)
12. Silva, J., Heim, W., Chau, T.: A self-contained, mechanomyography-driven externally powered prosthesis. Arch. Phys. Med. Rehabil. 86(10), 2066–2070 (2005)
13. Al-Mulla, M.R., Sepulveda, F., Colley, M.: Semg techniques to detect and predict localised muscle fatigue. EMG Methods for Evaluating Muscle and Nerve Function, 978–953 (2012)

14. Kattan, A., Al-Mulla, M.R., Sepulveda, F., Poli, R.: Detecting localised muscle fatigue during isometric contraction using genetic programming. In: IJCCI, pp. 292–297 (2009)
15. Al-Mulla, M.R., Sepulveda, F., Colley, M., Kattan, A.: Classification of localized muscle fatigue with genetic programming on sEMG during isometric contraction. In: Annual International Conference of the IEEE Engineering in Medicine and Biology Society EMBC, September 2-6, pp. 2633–2638 (2009)
16. Al-Mulla, M.R.: Evolutionary computation extracts a super semg feature to classify localized muscle fatigue during dynamic contractions. In: 2012 4th Computer Science and Electronic Engineering Conference (CEEC), pp. 220–224 (2012)
17. Al-Mulla, M.R., Sepulveda, F., Colley, M.: Evolved pseudo-wavelet function to optimally decompose sEMG for automated classification of localized muscle fatigue. Med. Eng. Phys. 33(4), 411–417 (2011)
18. Al-Mulla, M.R., Sepulveda, F., Colley, M., Al-Mulla, F.: Statistical class separation using sEMG features towards automated muscle fatigue detection and prediction. In: International Congress on Image and Signal Processing, pp. 1–5 (2009)
19. Al-Mulla, M.R., Sepulveda, F.: A Novel Feature Assisting in the Prediction of sEMG Muscle Fatigue Towards a Wearable Autonomous System. In: Proceedings of the 16th IEEE International Mixed-Signals, Sensors and Systems Test Workshop (IMS3TW 2010), France (2010)
20. Al-Mulla, M.R., Sepulveda, F.: Novel feature modelling the prediction and detection of semg muscle fatigue towards an automated wearable system. Sensors 10(5), 4838–4854 (2010)
21. Al-Mulla, M.R., Sepulveda, F.: "Predicting the time to localized muscle fatigue using ANN and evolved sEMG feature. In: IEEE International Conference on Autonomous and Intelligent Systems (AIS 2010), Povoa de Varzim, Portugal, pp. 1–6 (2010)
22. Al-Mulla, M.R., Sepulveda, F., Colley, M.: An autonomous wearable system for predicting and detecting localised muscle fatigue. Sensors (Basel) 11(2), 1542–1557 (2011)
23. Al-Mulla, M.R., Sepulveda, F.: Novel pseudo-wavelet function for mmg signal extraction during dynamic fatiguing contractions. Sensors 14(6), 9489–9504 (2014)
24. Vedsted, P., Blangsted, A.K., Søgaard, K., Orizio, C., Sjøgaard, G.: Muscle tissue oxygenation, pressure, electrical, and mechanical responses during dynamic and static voluntary contractions. European Journal of Applied Physiology 96, 165–177 (2006)
25. Kumar, D.K., Pah, N.D., Bradley, A.: Wavelet analysis of surface electromyography to determine muscle fatigue. IEEE Trans. Neural Syst. Rehabil. Eng. 11, 400–406 (2003)
26. Michalewicz, Z.: Genetic algorithms + data structures = evolution programs. Springer, New York (1996)
27. Sepulveda, F., Meckes, M., Conway, B.A.: Cluster separation index suggests usefulness of non-motor eeg channels in detecting wrist movement direction intention. In: IEEE Conference on Cybernetics and Intelligent Systems, pp. 943–947. IEEE Press (2004)
28. Tarata, M.T.: Mechanomyography versus electromyography, in monitoring the muscular fatigue. Biomed. Eng. Online 2, 3 (2003)
29. Beck, T.W., Housh, T.J., Cramer, J.T., Weir, J.P., Johnson, G.O., Coburn, J.W., Malek, M.H., Mielke, M.: Mechanomyographic amplitude and frequency responses during dynamic muscle actions: a comprehensive review. Biomedical Engineering Online 4, 67 (2005)

30. Tarata, M.: Noninvasive Monitoring of Neuramuscular Fatigue: Techniques and Results. University of Pietisti Scientific Bulletin: Electyronics and Computer Science 11(1), 201–243 (2011)
31. Ryan, E.D., Cramer, J.T., Egan, A.D., Hartman, M.J., Herda, T.J.: Time and frequency domain responses of the mechanomyogram and electromyogram during isometric ramp contractions: A comparison of the short-time fourier and continuous wavelet transforms. Journal of Electromyography and Kinesiology 18(1), 54 (2008)
32. Caldwell, L.S.: Relative muscle loading and endurance. J. Eng. Psychol. 2(4), 155–161 (1963)
33. J. S. Petrofsky and C. A. Phillips, "The effect of elbow angle on the isometric strength and endurance of the elbow flexors in men and women," *J Hum Ergol (Tokyo)*, vol. 9, no. 2, pp. 125–131, Dec 1980.

From Single Fiber Action Potential to Surface Electromyographic Signal: A Simulation Study

Noureddine Messaoudi[1,2] and Raïs El'hadi Bekka[2]

[1] Physics Department, Sciences Faculty, University of Boumerdes,
35000 Boumerdes, Algeria
[2] Lab. LIS, Electronics Department, Faculty of Technology,
University of Sétif 1, 19000 Sétif, Algeria
{nor_messaoudi,bekka_re}@yahoo.fr

Abstract. The purpose of this investigation was to simulate surface electromyographic (EMG) signals generated in a cylindrical multilayer volume conductor constituted by bone (isotropic) muscle (anisotropic), fat (isotropic) and skin (isotropic) layers. This simulation was based on the distributions of the: MFs within each motor unit (MU), motor units (MUs) within the muscle, diameters of all activated MUs, conduction velocities of all activated MUs, lengths of all MFs, firing rates (FRs) of all recruited MUs, inter-spike intervals (ISIs) and the starting recruitment times of the activated MUs.

A MU is composed of an alpha motor neuron and connected MFs, thus the action potential generated in each MU (MUAP) is the sum of the action potentials generated from MFs (SFAPs) belonging to that particular MU. The simulation of surface EMG signal began first by simulating SFAP and then the simulation of the MUAP. The non uniform repetition of the MUAP with a firing rate gives the MUAP train (MUAPT). Finally, the surface EMG signal is the sum of non-synchronized MUAP trains of active MUs. Four filters were used to detect the surface EMG signals. Simulations results show that the amplitude and shape of surface EMG signals depend on the filter used for recording.

Keywords: Detection, Limb muscle, SEMG Simulation, spatial filter.

1   Introduction

Surface electromyography (EMG) is a technique in which the electrodes are placed on the skin overlying a muscle to record the electrical activity of the muscle [1]. Surface SFAP is the elementary component of surface EMG signal. Its simulation is important for the interpretation of the experimental recordings [2]. Several generation models of the SFAP signal were proposed [3, 4, 5, 6]. The difference between these models resides in the volume conductor description. The volume conductor was described as a space invariant system (the same shape of the potential in each point in the direction of source propagation) [3, 4, 5] or a non space invariant system [2], [7, 8, 9]. Analytical [3], [10] and numerical [11, 12, 13] approaches were proposed for the description of the electrical properties which separate the MFs and the detection point. Cartesian

F. Ortuño and I. Rojas (Eds.): IWBBIO 2015, Part I, LNCS 9043, pp. 315–324, 2015.

[4], and cylindrical [6], [14] coordinate systems were adopted for the mathematical description of the volume conductor.

The main steps to simulate surface SFAP are [11]: a) mathematical description of the volume conductor, b) modeling of the detection system and c) description of the generation, propagation and extinction of the intracellular action potential (IAP).

In the 2D spatial filtering techniques, the volume conductor and the detection system are described by transfer functions in the 2D spatial frequency domains [4].

The muscle is an ensemble of MUs and the MU is an ensemble of MFs. The surface EMG signal simulation was based first on the simulation of the SFAP. Secondly, the MUAP was simulated by summing the SFAPs belonging to the same MU. Then, the MUAPT was simulated on the basis of the repetition of the MUAP according to the motor neuron firing rate. Each MUAPT is the non synchronized repetition of the MUAP. Finally, surface EMG signal is the sum of the MUAPTs trains.

Our objective was to simulate surface EMG signal generated in a limb muscle (it can be the upper or the lower limb according to the given parameters). This muscle was represented with a multilayer cylindrical volume conductor composed of bone, muscle, fat and skin layers.

Four detection systems were used to record surface EMG signals. The findings of this study show that the amplitude and shape of surface EMG signals depend on the filter used for recording.

2   Methods

2.1   Simulation of the Surface SFAP

The SFAP was simulated using the analytical model describing the volume conductor as a cylindrical layered medium [6], [15, 14] and which was implemented as described by Farina et al., [6]. The volume conductor which represents the limb muscle is composed of four layers (bone, muscle, fat and skin as shown in Fig. 1). Parameters of simulation corresponding to the volume conductor are described in the table 1. These parameters have already been used in other studies [6], [16, 17]. The SFAPs were simulated taking into account the generation, propagation, and extinction of the intracellular action potential (IAP), at the neuromuscular junction, along the fiber and at the tendons respectively. These phenomena are described by progressive generation, propagation and extinction of the first derivative of the IAP [5]. The current density source is obtained from the second derivative of the IAP described analytically by Rosenfalck [18].

2.2   Simulation of the MUAP

Within the muscle layer there is a muscle of elliptical shape (Fig. 1) with short and long axes of 26mm and 54mm respectively [23]. The elliptical muscle is composed of 120 MUs uniformly distributed. The centers of all MUs cannot exceed the territory of the ellipse [23], [24]. Each MU contains a number of MFs according to the density of these fibers within the muscle (the density of MFs within the muscle was 2fibres/mm²) and the radius of each MU. The diameters of the MUs are distributed according to the Poisson law in the interval [2-8mm]. The fibers were uniformly distributed within the MUs. The number of repetition of the MUAP is related to the peak firing rate of each MU. The firing rates of the 120 MUs are also distributed according to the Poisson law [22], [25]. The MUAPT is the non-synchronized repetition of the MUAPs. Finally, surface EMG signal is the sum of the 120 MUAPTs [24].

Table 1. Selected parameters to simulate SFAP generated in a multilayer cylindrical volume conductor constituted by bone, muscle, fat and skin layers [6], [16, 17]

Parameter	Description	Value
a	Radius of the bone	$20\,mm$
b	Radius of bone + muscle compartment	$50\,mm$
c	Radius of bone + muscle + fat	$51\,mm$
d	Radius of the volume conductor	$52\,mm$
σ_{mz}	Longitudinal conductivity of the muscle	$0.5\,S/m$
$\sigma_{m\theta}$	Angular conductivity of the muscle	$0.1\,S/m$
σ_b	Conductivity of the bone	$0.02\,S/m$
σ_f	Conductivity of the fat	$0.05\,S/m$
σ_s	Conductivity of the skin	$1\,S/m$

Table 2. Selected parameters to simulate surface EMG signal [22, 23, 24, 25]

Parameter	Value	Description
Fiber diameter	$46\,\mu m$	For all fibers
Muscle fibers lengths	$[40-60\,mm]$	Gaussian $(mean=80,\ SD=1\,mm)$
MUs diameters	$[2-8\,mm]$	Poisson $(mean=6\,mm)$
MUs firing rates	$[8-42\,Hz]$	Poisson distribution
MUs conduction velocities	$[2.5-5.5\,m/s]$	Gaussian $(mean=4\ SD=0.75\,m/s)$
Recruitment starting times	$St=i\,(83.33/n)$	Uniform
MUs shape	-	Circular
MUs distribution	-	Uniform
MUs Number	120	-
Fibers distribution	-	Uniform
NMJs	$5\,mm$	Uniform
Tendons regions	$5\,mm$	Uniform
Muscle dimensions	$27-54\,mm$	Elliptic
Muscle fibers density	$2\,fibers/mm^2$	-

2.3   The Detection System

The detection system is modeled as a two-dimensional spatial filter rotated by an angle with respect to the fibers direction and it is constituted of 1D and 2D spatial filters according to the weights given to the electrodes. The investigated filters are the LDD, NDD, IB2 and the MKF (Fig. 3). Fig. 3 shows the electrodes arrangement with respect to the centre of the elliptical muscle and the electrodes weights for each spatial filter. Electrodes arrangement is adapted to the volume conductor shape [6]. The transverse direction of electrodes arrangement is along the angular direction. The detection point corresponds to the point of intersection of the central row and column [6].

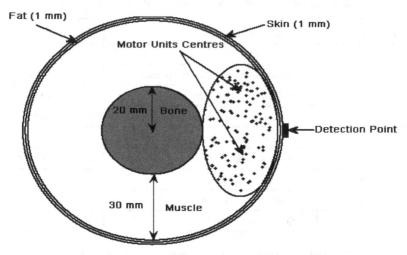

Fig. 1. Multilayer cylindrical volume conductor model constituted by bone (isotropic) muscle (anisotropic), fat (isotropic) and skin (isotropic) layers. The muscle layer includes a muscle of elliptical shape with semi short and semi large axes of *13mm* and *27 mm*, respectively. The elliptical muscle contains 120 motor units uniformly distributed. The other parameters are described in tables 1 and 2, respectively.

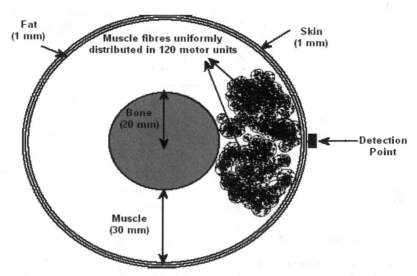

Fig. 2. This figure is the same as Fig. 1, but here, we have shown the distribution of 120 motor units within the elliptical muscle (uniform distribution) and the distribution of the muscle fibers within the territory of each motor unit (uniform distribution) with a density of *2fibres/mm²*

The parameters described in tables 1 and 2 can be changed according the studied limb (upper right or left limb or lower right or left limb).

$$LDD = \begin{vmatrix} 0 & 1 & 0 \\ 0 & -2 & 0 \\ 0 & 1 & 0 \end{vmatrix} \qquad MKF = \begin{vmatrix} 8 & 3 & 2 \\ -21 & -3 & 0 \\ 7 & 2 & 2 \end{vmatrix}$$

$$NDD = \begin{vmatrix} 0 & 1 & 0 \\ 1 & -4 & 1 \\ 0 & 1 & 0 \end{vmatrix} \qquad IB^2 = \begin{vmatrix} 1 & 2 & 1 \\ 2 & -12 & 2 \\ 1 & 2 & 1 \end{vmatrix}$$

fibres direction $\uparrow$

Fig. 3. A schematic representation of the spatial filter configurations on a nine-electrode grid arranged with respect to the muscle fiber direction (a). The abbreviations of these spatial filter configurations are: LDD: Longitudinal Double Differential; NDD: Normal Double Differential; IB2: Inverse Binomial of order two ([19], [20]) and MKF: Maximum Kurtosis Filter ([1], [21]).

3   Results

Fig. 2 shows the volume conductor in which the surface EMG signal is generated. In the muscle layer, there is a muscle of elliptic shape. This elliptic muscle was consisted, 120 MUs that were uniformly distributed. Inside each MU, the muscle fibers were uniformly distributed. The number of the muscle fibers within each motor unit is related to its diameter and also to the density of the fibers within the muscle area.

Fig. 4 shows simulated surface SFAPs generated in a multilayer cylindrical volume conductor as described in figure 1 and detected by longitudinal double differential (LDD) (Fig. 4a), normal double differential (NDD) (Fig. 4b), inverse binomial of order two (IB2) (Fig. 4c) and maximum kurtosis (MKF) (Fig. 4d) filters. The differences between these signals reside in the amplitude and the shape. The signal that has the lowest amplitude is the LDD and the signal that has the biggest amplitude is the MKF. This result is due to the configuration and the weights given to the surface electrodes.

Fig. 5 shows two simulated action potentials (left) generated in two motor units (right) and configured by LDD filter. The number of fibers within each motor unit was six. They were uniformly distributed. Positions of the fibers within the first motor unit were different with respect to their positions within the second motor unit (this is due to the random distribution of the fibers within each motor unit). The coordinates of the two motor units were also shown. The amplitude of the second MUAP was lower than the amplitude of the first motor unit (this is due to several parameters such as the depth of the motor unit within the muscle).

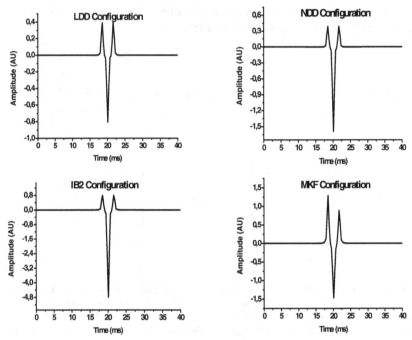

Fig. 4. Simulated surface single fiber action potentials generated in a multilayer cylindrical volume conductor model (Fig. 1) and detected by the LDD, NDD, IB2 and MKF spatial filters. Parameters of simulation are shown in tables 1 and 2, respectively.

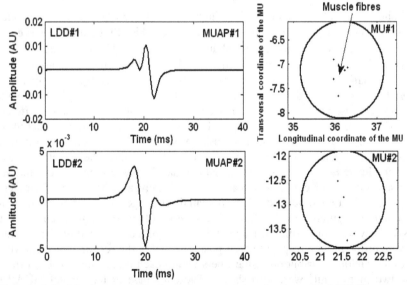

Fig. 5. At the left of the figure, we simulated the motor unit action potential (MUAP) generated in the first and the second motor units and detected by the LDD filter. At the right of the figure we simulated the two associated motor units. The muscle fibers were distributed uniformly in each motor unit with a density of *2fibres/mm²*. The first and the second motor unit contain 6 fibers.

Fig. 6 depicts the motor unit action potentials trains (MUAPTs) generated from the first and the second motor units, respectively (the left of this figure). On the right of the Fig. 6, we depict the same as shown on the right of the Fig 5. The first MUAPT has a firing rate of *9Hz* and the second has a firing rate of *12Hz*. These MUAPTs were corresponding to the MUAP shown in Fig. 5. The differences between the two simulated MUAPTs reside in the amplitude and in the number of repeated impulsions (this is due to the firing rate of each motor unit).

In Fig. 7, we show the same MUAPTs depicted in Fig. 6 (the left of the Fig. 6 and Fig. 7 are the same), however on the right of Fig. 7, we show the positions of the two motor units within the muscle and the distribution of the fibres within these two motor units (the axes of the fibres belonging to the same motor unit and the axes of the motor units belonging to the mauscle are uniformly distributed).

Fig. 8 depicts the simulated surface EMG signals generated in the multilayer cylindrical volume conductor as shown in figure 1 and configured by the LDD (Fig. 8a), NDD (Fig. 8b), IB2 (Fig. 8c) and MKF (Fig. 8d) spatial filters. The differences between these signals reside in the amplitude and the shape. The LDD signal has the lowest amplitude and the IB2 signal has the biggest amplitude (this result is due to the configuration of the electrodes in the filter mask (see Fig. 3) and also to the weights given to the electrodes). Parameters of simulation of these signals were resumed in tables 1 and 2 respectively.

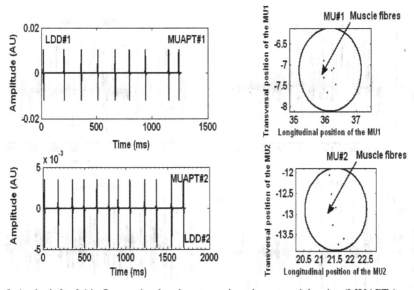

Fig. 6. At the left of this figure, simulated motor unit action potential trains (MUAPTs) generated in the first and the second motor units and detected by the LDD filter. The firing rates of the first and the second MUs were *9Hz* and *12Hz* respectively. At the right of the figure we simulated the two associated MUs. The muscle fibers were distributed uniformly in each motor unit with a density of *2fibres/mm^2*. Each MU contains 6 fibers with different positions.

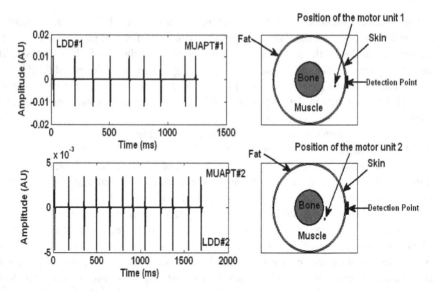

Fig. 7. At the left of this figure, simulated MUAPTs generated in the first and the second MUs and detected by the LDD filter. The firing rates of the first and the second motor units were *9Hz* and *12Hz* respectively. At the right of this figure, we showed the volume conductor model and positions of the two motor units within the elliptical muscle.

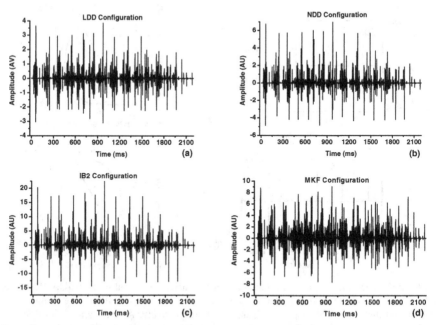

Fig. 8. Simulated surface EMG signal generated in a multilayer cylindrical volume conductor model constituted by bone, muscle, fat and skin layers (Fig. 1) and detected by LDD (a), NDD (b), IB2 (c) and MKF (d) spatial filters. The parameters of modeling are shown in tables 1 and 2, respectively.

4   Conclusion

Simulation of surface EMG signals is based on several distributions such as the distribution of the: MUs within the muscle (uniform distribution), the MFs within each MU (uniform distribution), MUs radiuses (Poisson distribution), MFs lengths (Gaussian distribution), firing rates of all MUs (Poisson distribution), ISIs of the MUAPTs (Gaussian distribution), conduction velocity of all MUs (Gaussian distribution) and starting times of all recruited MUs (uniform distribution).

Simulations results show that the amplitude and shape of surface EMG signals depend on the type of filters used for recording. In fact, the signal detected by the LDD filter has the lowest amplitude and the one recorded by IB2 has the biggest amplitude. These differences are due to the configuration of the electrodes in the filter mask and to the weights given to the electrodes.

References

1. Östlund, N., Yu, J., Roeleveld, K., Karlsson, J.S.: Adaptive spatial filtering of multichannel surface electromyogram signals. Med. Biol. Eng. Comput. 42, 825–831 (2004)
2. Mesin, L., Farina, D.: Simulation of surface EMG signals generated by muscle tissues with inhomogeneity due to fiber pinnation. IEEE Trans. Biomed. Eng. 51, 1521–1529 (2004)
3. Merletti, R., Lo Conte, L., Avignone, E., Guglielminotti, P.: Modelling of surface myoelectric signals – part I: model implementation. IEEE Trans. Biomed. Eng. 46, 810–820 (1999)
4. Farina, D., Alberto, R.: Compensation of the effect of sub-cutaneous tissue layers on surface EMG: a simulation study. Med. Eng. Physics. 21, 487–496 (1999)
5. Farina, D., Merletti, R.: A novel approach for precise simulation of the EMG signal detected by surface electrodes. IEEE Trans. Biomed. Eng. 48, 637–646 (2001)
6. Farina, D., Mesin, L., Martina, S., Merletti, R.: A surface EMG generation model with multilayer cylindrical description of the volume conductor. IEEE Trans. Biomed. Eng. 51, 415–426 (2004)
7. Mesin, L., Farina, D.: A model for surface EMG generation in volume conductors with spherical inhomogeneties. IEEE Trans. Biomed. Eng. 52, 1984–1993 (2005)
8. Mesin, L., Farina, D.: An analytical model for surface EMG generation in volume conductors with smooth conductivity variations. IEEE Trans. Biomed. Eng. 53, 773–779 (2006)
9. Mesin, L.: Simulation of surface EMG signals for a multi-layer volume conductor with triangular model of the muscle tissue. IEEE Trans Biomed Eng 53, 2177–2184 (2006)
10. Dimitrov, G.V., Dimitrova, N.A.: Precise and fast calculation of the motor unit potentials detected by a point and rectangular plate electrode. Med. Eng. Phys. 20, 374–381 (1998)
11. Farina, D., Mesin, L., Martina, S.: Advances in surface EMG signal simulation with analytical and numerical descriptions of the volume conductor. Med. Biol. Eng. Comput. 42, 467–476 (2004)
12. Lowery, M.M., Stoykov, N.S., Taflove, A., Kuiken, T.A.: A multiple-layer finite-element model of the surface EMG signal. IEEE Trans. Biomed. Eng. 49, 446–454 (2002)
13. Mesin, L., Joubert, M., Hanekom, T., Merletti, R., Farina, D.: A Finite Element Model for Describing the Effect of Muscle Shortening on Surface EMG. IEEE Trans. Biomed. Eng. 53, 593–719 (2005)

14. Blok, J.H., Stegeman, D.F., Van Oosterom, A.: Three-layer volume conductor model and software package for applications in surface electromyography. Ann. Biomed. Eng. 30, 566–577 (2002)

15. Gootzen, T.H.J.M.: Muscle fibre and motor unit action potentials. A biophysical basis for clinical electromyography. PhD. Dissertation, Univ. Nijmegen, Nijmegen, the Netherlands (1990)

16. Farina, D., Cescon, C., Merletti, R.: Influence of anatomical, physical and detection system parameters on surface EMG. Biol. Cybern. 86, 445–456 (2002)

17. Merletti, R., Bottin, A., Cescon, C., Farina, D., Gazzoni, M., Martina, S., Mesin, L., Pozzo, M., Rainoldi, A., Enck, P.: Multichannel surface EMG for the invasive assessment of the anal sphincter muscle. Oasis. Progress. Report. Digestion. 69, 112–122 (2004)

18. Rosenfalck, P.: Intra and extracellular fields of active nerve and muscle finres. Acta. Physiol. Scand. 32, 1–49 (1969)

19. Farina, D., Shulte, E., Merletti, R., Rau, G., Disselthorst-Klug, C.: Single motor unit analysis from spatially filtered surface electromyogram signals. Part I: Spatial Selectivity. Med. Biol. Eng. Comput. 41, 330-337 (2003)

20. Reucher, H., Silny, J., Rau, G.: Spatial filtering of non-invasive multi-electrode EMG: Part II-Filter performance in theory and modelling. IEEE. Trans. Biomed. Eng. 34, 106–113 (1987)

21. Disselhorst-Klug, C., Silny, J., Rau, G.: Improvement of spatial resolution in surface-EMG: A theoretical and experimental comparison of different spatial filters. IEEE. Trans. Biomed. Eng. 44, 567–574 (1997)

22. Wang, W., Stefano, A.D.E., Allen, R.: A Simulation Model of the Surface EMG Signal for Analysis of Muscle Activity during the Gait Cycle. Comput. Biol. Med. 36, 601–618 (2006)

23. Keenan, K.G., Farina, D., Meyer, F., Merletti, R., Enoka, R.M.: Sensitivity of the Cross-correlation between Simulated Surface EMGs for two Muscles to Detect Motor Unit Synchronization. Appl. Physiol. 102, 1193–1201 (2007)

24. Fuglevand, A.J., David, A., Winter, A., Patla, E.: Models of Recruitment and Rate Coding Organisation in Motor-Unit Pools. Neurophysiol. 70, 2470–2488 (1993)

25. Stashuk, D.W.: Simulation of Electromyographic Signals. Electromyo. Kinesiol. 3, 157–173 (1993)

26. Keenan, K.G., François, J., Valero, C.: Experimentally Valid Predictions of Muscle Force and EMG in Models of Motor-Unit Function Are Most Sensitive to Neural Properties. Neurophysiol. 98, 1581–1590 (2007)

27. Keenan, K.G., Farina, D., Maluf, K.S., Merletti, R., Enoka, R.M.: Influence of Amplitude Cancellation on the Simulated Surface Electromyogram. Appl. Physiol. 98, 120–131 (2004)

New Algorithm for Assessment of Frequency Duration of Murmurs Using Hilbert-Huang Transform

Tahar Omari[1] and Fethi Bereksi-Reguig[2]

[1] Department of Physics, Boumerdes University, Algeria
[2] Department of Biomedical Engineering, Tlemcen University, Algeria
tah.omari@gmail.com,
bereksif@yahoo.fr

Abstract. This paper presents a new algorithm for automatic assessment of frequency duration of murmurs developed in order to estimate the severity of aortic stenosis. The applied analysis method is based on Hilbert-Huang Transform (HHT). This technique can produce a significant time frequency distribution through Hilbert transform of different Intrinsic Mode Functions (IMF) obtained by the Empirical Mode Decomposition. In this work, the frequency duration of murmurs is computed using the instantaneous mean frequency produced via HHT. The algorithm is tested on 14 cases of heart murmurs with different degrees of severity. Those obtained results are compared with manual measurement through Short Time Fourier Transform (STFT). The obtained results show a very high correlation between the methods with a coefficient of correlation R = 0.93

Keywords: Hilbert-Huang Transform, Empirical mode decomposition, Heart murmurs, Instantaneous mean frequency, aortic stenosis severity estimation.

1   Introduction

The aortic stenosis (AS) is currently the most common valvular disease. It is characterized by a narrowing of the aortic valve opening, resulting resistance to blood flow from left ventricle to the aorta. Generally, aortic stenosis results from three conditions, whether a patient suffers from a congenital structure such as bicuspid aortic valve or catches a stenosis by secondary conditions such as rheumatic heart disease or idiopathic calcification of the aortic valve (Fig.1) [2]. This pathology can cause ventricular hypertrophy characterized by a myocardium thickening, which may lead to certain death without surgery. For most patients, the valvular replacement is the only effective remedy. Before taking such decision, the doctor must know with certainty the degree of severity of the aortic stenosis. Several diagnostic techniques exist to assess the valvular dysfunction and select the good timing for surgery, where Doppler echocardiography

F. Ortuño and I. Rojas (Eds.): IWBBIO 2015, Part I, LNCS 9043, pp. 325–336, 2015.
© Springer International Publishing Switzerland 2015

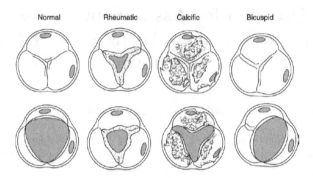

Fig. 1. Aortic stenosis aetiology: morphology of calcific AS, bicuspid valve, and rheumatic AS. Adapted from [2].

and catheterization are the most popular techniques. However, these techniques are relatively cumbersome and expensive and require a specialist in cardiology. Therefore, many researchers are trying to find new techniques easy, efficient and inexpensive to estimate the severity of aortic stenosis. Among the promising techniques is the numerical analysis of Phonocardiographic signal (PCG). This signal represents the recordings of heart sounds. Through an electronic stethoscope two sounds (S1 and S2) can be distinguished (Fig.2), produced respectively by the closure of atrio-ventricular valves (AV) and sigmoid valves. A third and a fourth sound (S3 and S4) may also exist. However, a variety of heart murmurs may also be present [3].

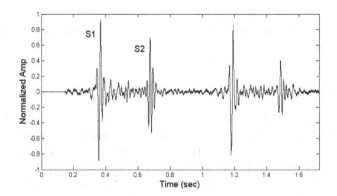

Fig. 2. PCG signal (healthy subject)

The detection of a murmur in PCG signal reflects an abnormal situation in blood circulation through valves heart. The identification of this anomaly is based on morphology and timing of murmur. In the case of aortic stenosis, the obstructed site produces a significant change in the blood velocity. The reduction of this site causes a compression of fluid, and consequently an increase in velocity. The increasing of velocity produces turbulence around the stenosis (Fig.3) causing the vibration of cardiac structures and producing a murmur [4-6]. The registered murmur has a crescendo-decrescendo evolution (diamond shape (Fig.4)), it reflects perfectly the opening and the closing of the valve, it often occupies the middle of the systole and it is called mesosystolic murmur. However, a protosystolic click is sometimes recorded; it indicates the conservation of some mobility of aortic valve and absence of calcified forms [7-8].

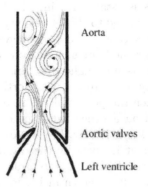

Fig. 3. Streamlines of flow through and distal to a stenotic aortic valve showing an asymmetric flow jet, recirculation regions and vortex shedding

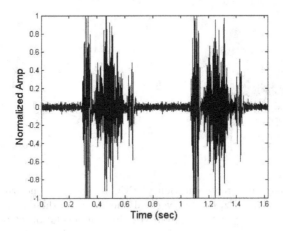

Fig. 4. PCG signal of subject with aortic stenosis

In the case of aortic stenosis (Fig.4), the increasing of severity has a direct influence on the shape and the timing of murmur [4], [9]. When the stenosis is tight:

- The intensity of murmur increases.
- The intensity of S1 decreases.
- The second sound S2 is frequently abolished.
- The peak of the murmur is telesystolic (It appears towards the end of systole).
- A high tone of murmur (frequency change).

The exploitation of these parameters can be very efficiency in the estimation of the severity of aortic stenosis. However, according to Gamboa et al [10], these are not convincing parameters. Nevertheless, Spectral properties of murmur seem more reliable, and a relationship between the murmurs frequency content and the severity of the stenosis is established in [11]. Furthermore, the dominate frequency of the murmur is related to the jet velocity of the stenosis [12] and the percentage of higher frequencies is related to transvalvular pressure differences [13]. In a recent work, using time frequency analysis D.Kim et al [1] found good correlation between the duration of higher frequency components and the peak pressure gradient. On fourty one subjects with aortic stenosis, they established a comparative study between the peak and the mean of transvalvular pressure gradient on one hand, and on another hand the evolution of the duration of murmur spectra at 200, 250 and 300 Hz. In their overall conclusion they found that the best correlated result is given by the duration of murmur spectra at 300Hz and the peak of transvalvular pressure gradient TPGmax measured by Doppler ultrasound. This result is shown in Figure 5 and 6. The approximate curve is an exponential generated by a high correlation coefficient R = 0.91 .

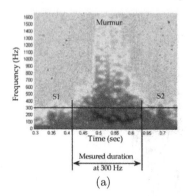

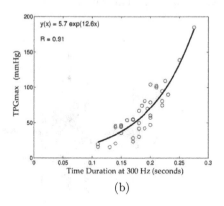

(a)   (b)

Fig. 5. (a) Time-frequency representation of PCG signals. The horizontal line indicates the frequency at 300 Hz, the two vertical lines indicate the duration of murmur measured at this frequency. Adapted from [1], (b): The evolution of the time duration at 300 Hz plotted against the Doppler echocardiogram-derived transvalvular peak pressure gradient. The line indicates the exponential regression curve. Adapted from [1].

Among limitations of this technique is the manual measurement of the frequency duration at 300Hz. In this work, we proposed a new algorithm able to surmount this limitation and provides automatically the frequency duration. The idea is to calculate the instantaneous mean frequency of the murmur. This approach is described in the next section.

2 Methodology

The proposed methodology involves the frequency duration measurement of murmurs at a given frequency using the instantaneous mean frequency. This last can be measured with different methods such as: short time Fourier transforms (STFT) algorithm, wavelet transform, Hilbert-Huang transform, and many others [14][15]. The proposed approach uses the Hilbert-Huang transform. Before the description of this approach, some mathematic backgrounds are summarized in following.

2.1 The Hilbert-Huang Transform

Before applying any mathematical technique, we must know the nature of the data. Generally, the biomedical signals are apparently random or aperiodic in time. Traditionally, the randomness in physiological signals has been ascribed to noise or interactions between very large numbers of constituent components. Therefore, the biomedical signals are most likely to be both nonlinear and non-stationary[16]. The non-stationary signals are those whose frequency contents change with time. However, the nonlinear systems are those which not follow the superposition principal. Traditional data-analysis methods are all based on linear and stationary assumptions. Only in recent years, where have been new methods introduced to analyze non-stationary and nonlinear data. For example, the wavelet analysis and the Wigner-Ville distribution were designed for linear but non-stationary data. However, various nonlinear time-series-analysis methods were designed for nonlinear but stationary and deterministic systems [17]. Therefore, this problematic was really a great challenge to the mathematical community to develop a new method able to analyse non-stationary and nonlinear data. The first solution was brought by Huang et al in [18]. Using Hilbert transform, they develop a new method which seems to be able to meet this challenge. This method took the name of Hilbert- Huang Transform (HHT). The HHT consists of two parts: empirical mode decomposition (EMD) and Hilbert spectral analysis (HSA). It is potentially viable for non-linear and non-stationary data analysis, especially for time-frequency-energy representations. It has been tested and validated in many projects. In all these projects, the HHT gave results much sharper than those from any of the traditional analysis methods in time-frequency-energy representations. Hilbert Transform (HT) is the easiest way to compute the instantaneous frequency. The HT of a real signal x(t) is defined as

$$\mathcal{H}[x(t)] = y(t) = \frac{P}{\pi} \int_{-\infty}^{+\infty} \frac{x(\tau)}{t - \tau} d\tau \qquad (1)$$

Where P indicates the Cauchy principal value.

From $y(t)$ it is possible to define the analytical signal $z(t) = x(t) + iy(t)$ or, in polar form, $z(t) = a(t)e^{i\theta(t)}$, in which

$$a(t) = \sqrt{x^2(t) + y^2(t)} \tag{2}$$

$$\omega(t) = \frac{d\theta(t)}{dt} \tag{3}$$

This method is efficient only when the data is constituted by single frequency component. However, most biomedical signal does not meet these criteria. At any given time, the data may involve more than one oscillatory mode. In this case the simple Hilbert Transform is unable to provide the full description of the frequency content in data. To solve this problem Huang et al [17] used the empirical mode decomposition method (EMD). Their idea is to decompose a non-stationary and nonlinear signal into a finite set of functions that have meaningful instantaneous frequencies defined by equation (3). These functions are called intrinsic mode functions (IMFs) in which each mode should be independent of the others and only one frequency component exists at a given time. Then the instantaneous frequencies can be computed by taking Hilbert transform of each IMF.

The decomposition of a signal $x(t)$ into a series of $IMFs$ is implemented with a so called sifting process. This process can be described generally on three steps: The first step is the detection of all picks and valleys present in the signal. After that, all pics will be connected together by a cubic interpolation to produce the upper envelope $e_u(t)$; the same operation will be repeated for the valleys to produce the lower envelope $e_l(x)$. The upper and lower envelopes should cover all the data between them. In the second step the mean between the envelopes will be extracted and designated as $m_1(t) = [e_u(t) + e_l(t)]/2$. The difference between the data and m_1 produce the first IMF, it is designated by h_1[19].

$$h_1(t) = x(t) - m_1(t) \tag{4}$$

An IMF is a function that satisfies two conditions:

1. The number of extrema and the number of zeros crossings may differ by no more than one, and
2. At any point, the average value of the envelope defined by the local maxima, and the envelope defined by the local minima, is zero.

If $h_1(t)$ taken as an IMF do not satisfy the conditions above, the first and the second steps will be repeated for k siftings in the third step until the resulting signal meet the criteria of an intrinsic mode; then,

$$h_{11} = h_1 - m_{11}$$
$$\vdots \tag{5}$$
$$h_{1k} = h_{1(k-1)} - m_{1k}$$

After repeating sifting operation in this manner, and h_{1k} satisfy the conditions above, the residue r can be extracted by the subtraction between the original signal $x(t)$ and c_1. Where h_{1k} is designated as c_1; then,

$$r_1(t) = x(t) - c_1(t) \qquad (6)$$

Finally, the residue r_1 will be treated as new data that are subject to the sifting, yielding the second IMF from $r_1(t)$. The procedure continues until finally a preset value for the residue is met or when the residue becomes monotonic function (non-oscillatory) from which no more IMF can he extracted. The original signal $x(t)$ can thus be expressed as follows

$$x(t) = \sum_{j=1}^{N} c_j(t) + r_N(t) \qquad (7)$$

Where N is the number of $IMFs$, $r_N(t)$ is the final residue which can be either the mean trend or a constant [19].

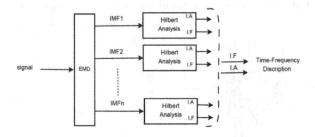

Fig. 6. HHT Diagram

Once all IFs are found, the instantaneous mean frequency of the signal will be computed to be used later in our algorithm.

2.2   The Frequency Duration Measurement Algorithm

The proposed algorithm is based on the extraction of the instantaneous mean frequency of murmur. However, before this operation the PCG signal must be segmented into different component to compute independently the murmur. After that, the instantaneous mean frequency of murmur is computed and thresholded with a given value to get finally frequency duration. The whole procedure is realized in five steps as presented in the following algorithm (Fig7).

For the first and the second step we used segmentation algorithm developed in [20]. This algorithm allows the extraction of heart sounds (S1,S2) and the systolic murmur with good precision. In the third step, the Hilbert-Huang Transform of the extracted murmur is computed. In our algorithm the first three decompositions (IMF1 to IMF3) are used. These last contain the principal higher frequencies in the murmur. The result of this operation is presented in figure 8.

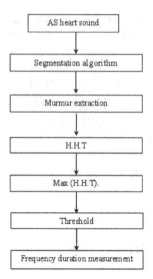

Fig. 7. The different steps of the measurment algorithm

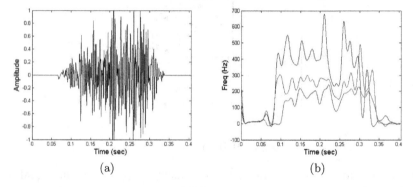

Fig. 8. (a) Aortic stenosis murmur, (b): Instantaneous frequencies (IF) obtained from the first three intrinsic mode functions (IMF1..IMF3)

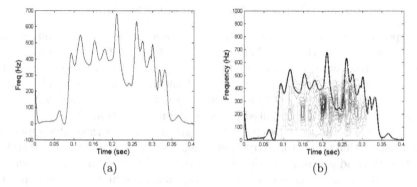

Fig. 9. (a) Max envelop of the instantaneous frequencies, (b): Time frequency representation plotted with max envelop

In order to detect the envelope of highest frequency in the signal, the maximum value found by the different IFs is took in every sample. The result is presented in figure 9(a). Figure 9(b) shows the perfect enveloping of the time frequency representation of the murmur obtained by STFT.

The last step in our algorithm is the application of a threshold to measure of the frequency duration. This last is the difference between the first and the last intersection of the threshold with the spectrum in figure 9(a).

3   Results

Fourteen aortic stenosis heart sounds with various degree of severity are used to evaluate the proposed method. These were collected from: `http://egeneral medical.com/listohearmur.html`, and from `www.med.umich.edu/lrc/psb/ heartsounds/index.htm` . After segmentation of the signals and extraction of the instantaneous frequency of murmur using HHT, a threshold is applied to measure the frequency duration. This last must be 300 Hz according to [1]. In performance test, a comparative study is done between the automatic measurements produced by our algorithm and the traditional manual measurements used in [1]. The sensitivity of the measure is calculated using equation(8). The result of the test is presented in table 1.

$$Sensitevity = \frac{max(\text{MM, AM})}{min(\text{MM, AM})} \times 100 \qquad (8)$$

Table 1. The measure found by the methods

cases	automatic measurement	manual measurement	Sensitivity
case 1	0,043	0,040	93,0%
case 2	0,078	0,075	96,2%
case 3	0,113	0,090	79,6%
case 4	0,100	0,070	69,7%
case 5	0,121	0,097	80,2%
case 6	0,106	0,110	96,7%
case 7	0,100	0,115	87,0%
case 8	0,136	0,130	95,6%
case 9	0,170	0,130	76,3%
case 10	0,097	0,137	70,8%
case 11	0,111	0,123	90,2%
case 12	0,133	0,145	91,7%
case 13	0,250	0,200	80,0%
case 14	0,248	0,225	90,8%
case 15	0,236	0,230	97,5%
		Median	90,2%
		standard deviation	9,5%

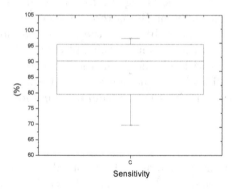

Fig. 10. Boxplot of sensitive results

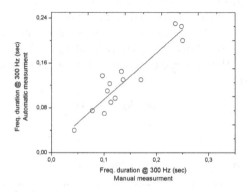

Fig. 11. Automatic measurement of frequency duration plotted against the manual measurement of frequency duration. The line indicates the linear regression curve.

Where, MM indicates manual measurement, and AM indicates automatic measurement.

form the table it can be observed three categories of results. The first category involves cases of sensitivity higher than 92%. From this value the measure is considered with high precision. In this category, 5 cases can be found(case {1,2,6,8,15}). The second category involves cases where their sensitivity is between 80% and 91%. The measures in this category is considered with medium precision. Six cases can be found in this category (case{5,7,11,12,13,14}). The third category involves cases where the sensitivity is lower than 80%. The measures in this category is considered with low precision. In this category, 4 cases can be found(case {3,4,9,10}). The median value of all categories is 90,2%. This last indicates a good precision in measurement, although the standard deviation is relatively high (Std = 9,5%). These results are presented in boxplot in figure 10. In this figure it can be observed perfectly the distribution of the results, where the half of population has a good sensitivity higher than 90%.

To confirm the above results, the correlation function between the methods is calculated (see figure 11). The result provides a linear curve with a high correlation coefficient (R= 0.93). This value indicates a good success of our algorithm in measurement of frequency duration at 300 Hz for most cases.

4 Conclusion

In this work, a completely automatic algorithm is elaborated in order to measure the frequency duration of murmurs. This parameter is used in the assessment of the severity of the aortic stenosis via phonocardiogram signal analysis. In this purpose, we developed a new algorithm to compute the instantaneous frequency using Hilbert-Huang transform. It is found that the frequency duration measured automatically by our algorithm correlates best with manual measurement with a high correlation coefficient R=0,93.

References

1. Kim, D., Tavel, M.E.: Assessment of severity of aortic stenosis through time-frequency analysis of murmur. Chest 124, 1638–1644 (2003)
2. Baumgartner, H., Hung, J., Bermejo, J., Chambers, J.B., Evangelista, A., Griffin, B.P., Iung, B., Otto, C.M., Pellikka, P.A., Quiones, M.: Echocardiographic Assessment of Valve Stenosis: EAE/ASE Recommendations for Clinical Practice. Journal of the American Society of Echocardiography 22(1), 1–23 (2009)
3. Visagie, J.C.: Screening of abdominal heart sounds and murmurs by implementing Neural Networks, Phd.Thesis, Stellenbosch Universety, South Africa, p. 5-40 (2007)
4. Ahlstorm, C.: Nonlinear Phonocardiographic Signal Processing, Phd. Thesis, Linkoping University, Sweden, pp.2–130 (2008)
5. Ask, P., Hok, B., Loyd, D., Terio, H.: Bio-acoustic signals from stenotic tube flow. Med. Bio. Eng. Comput. (1993)
6. Kolmogrov, A.N.: The local-structure of turbulence in incompressible viscous-fluid for very large reynolds-numbers. INIST 434(1890), 9–13 (1991)
7. Timmis, A.D., Nathan, A.: Cardiologie, 3rd edn., pp. 204–205 (2001)
8. Melly, L., Huber, C., Delay, D., Stumpe, F.: La valve aortique sous toutes ses formes. Forum Med. Suisse,Cardiovascular Surgery service, Sion, Swiss, p.p. 73- 78 (2009)
9. Kulbertus, H.E.: Smiologie des maladies Cardiovasculaires (1999)
10. Gamboa, R., Hugenholtz, P.G., Nadas, A.S.: Accuracy of the phonocardiogram in assessing severity of aortic and pulmonic stenosis. Circulation 30, 35–46 (1964)
11. Sabah, H.N., Stein, P.D.: Turbulent blood flow in humans: its primary role in the production of ejection murmurs. Circ. Res. 38, 513–525 (1976)
12. Donnerstein, R.L.: Continuous spectral analysis of heart murmurs for evaluating stenotic cardiac lesions. Am. J. Cardiol. 64, 625–630 (1989)
13. Nygaard, H., Thuesen, L., Hasenkam, J.M., Pedersen, E.M., Paulsen, P.K.: Assessing the severity of aortic valve stenosis by spectral analysis of cardiac murmurs (spectral vibrocardiography) Part I: Technical aspects. J. Heart Valve Dis. 2(4), 454–467 (1993)

14. Sun, L., Shen, M., Chan, F.H.Y.: A Method for Estimating the Instantaneous Frequency of Non-stationary Heart Sound Signals. In: IEEE Int. Conf. Neural Networks 8 Signal Processing (2003)
15. Blaska, J., Sedlacek, M.: Use of the Integral Transforms for Estimation of Instantaneous Frequency. Measurement Science Review 1(1), 169–172 (2001)
16. Fonseca-Pinto, R.: A New Tool for Nonstationary and Nonlinear Signals: The Hilbert-Huang Transform in Biomedical Applications. In: Biomedical Engineering Trends in Electronics, pp. 482–491 (2011)
17. Huang, N.E., Shen, S.S.P.: Hilbert-Huang Transform and Its Applications, vol. 5, p. 314. World Scientific Publishing (2005)
18. Huang, N.E., Shen, Z., Long, S.R., Wu, M.C., Shih, H.H., Zheng, Q., Yen, N.-C., Tung, C.C., Liu, H.H.: The empirical mode decomposition and the Hilbert spectrum for nonlinear and non-steady time series analysis. Proceedings of the Royal Society of London, A 454, 903–995 (1998)
19. Huang, N.E., et al.: The empirical mode decomposition and Hilbert spectrum for nonlinear and non-stationary time series analysis. Proc. Royal Society 454(1971), 903–995 (1998)
20. Kumar, D., Carvalho, P., Antunes, M., Henriques, J., Maldonado, M., Schmidt, R., Habetha, J.: Wavelet Transform And Simplicity Based Heart Murmur Segmentation. Computers in Cardiology 33, 173–176 (2006)

Heart Rate Regularity Changes in Older People with Orthostatic Intolerance*

Marcos Hortelano[1],**, Richard B. Reilly[2], Lisa Cogan[3], and Raquel Cervigón[1]

[1] Escuela Politécnica. DIEEAC. UCLM
Camino del Pozuelo sn. 16071, Cuenca, Spain
raquel.cervigon@uclm.es
[2] School of Engineering and School of Medicine
Trinity College Dublin
Dublin 2, Republic of Ireland
[3] St James's Hospital
Dublin 8, Republic of Ireland

Abstract. Orthostatic hypotension (OH) is an excessive fall in blood pressure when an upright position is assumed and is related symptoms associated with the occurrence of syncope. It presents as a heterogeneous group of diseases but commonly manifest with symptoms of orthostatic intolerance (OI). This study is focused to quantify the regularity in hemo-dynamic profile by Shannon Entropy (SE) hemodynamic measures to in older people with symptoms of OI undergoing an active stand and to investigate if their dynamic cardiovascular profile during a six-minute walk would be different to those of controls. The database included a total of 65 participants, aged over 70 years of age, of whom 65% were female. There was no significant differences in age and gender between symptomatic and asymptomatic participants 44.6% (n=29) had symptomatic OI and 55.4% (n=36) did not. SE measurement of HR showed differences during phase 2 in both groups with statistical signification (p=0.03), with 1.73 ± 0.56 in non-symptomatic OI and 2.05 ± 0.62 in symptomatic OI. These differences did not arrive to the statistical signification for the average of diastolic and systolic blood pressure. Moreover, HR entropy measures showed statistical significant differences between phases in the control group with a regularity increase during the physically active phase. The main conclusion is that orthostatic HR regularity response appears impaired during the exercise, especially in the symptomatic OI patients, with more irregular time series during the active phase.

Keywords: Orthostatic intolerance, heart rate, entropy.

* Technology Research for Independent Living Clinic in St James[TM] Hospital Dublin (www.trilcentre.org)
** Corresponding author.

F. Ortuño and I. Rojas (Eds.): IWBBIO 2015, Part I, LNCS 9043, pp. 337–346, 2015.
© Springer International Publishing Switzerland 2015

1   Introduction

The proportion of older people (aged 65 years and older) in the development countries is rising dramatically. Injuries and morbidity within this group represent a significant source of high cost for the health care sector in the present and future. A major health care cost incurred by older adults is the result of fall-related injuries. Cardiovascular impairments represent one intrinsic risk factor that can impact on individuals falling risk in various ways [1].

The prevalence of syncope rises with age and comorbidity, with the highest prevalence in older adults in long-term care [2]. It is suggested that between 2-10% of falls are the result of syncopal events [3], where transient blood pressure declines could act as a risk factor for falls in a larger proportion of individuals. Low blood pressure could affect cerebral perfusion, which is maintained through autoregulatory mechanisms [4].

Blood pressure decrease could potentially cause dizziness or loss of balance, thereby increasing falling risk. Cerebral autorregulation is normally preserved with healthy aging, but has been found to be different between genders and in different disease states [5,6]. It is important to understand the association between blood pressure impairments, and falling risk, because this discrimination impacts on treatment and management approaches. Orthostatic intolerance (OI) which is a clinical syndrome that is characterized by symptoms and loss of consciousness before impending syncope and that it has been reported that is caused by orthostatic hypotension (OH). Some previous studies suggest that the presence of OH may be a risk factor for falls [7,8]. OH has been found to increase risk of mortality and has been associated with a number of diseases including stroke, myocardial infarction, and coronary artery disease [9,10].

There are many tools and methods to assess and define blood pressure impairments in older adults [7,8,11,12,13,14,15]. Common tests asses cardiovascular risk in relation to falling measure changes in blood pressure in response to a change in posture, or orthostatic stress. Different measurements can effectively assess blood pressure responses to orthostatic stress. Variability in blood pressure and heart rate have been used to study the function of the cardiovascular control system [16,17,18,19,20]. Moreover, beat-to-beat fluctuations in heart rate has been widely believed to reflect changes in cardiac autonomic regulation [21]. In addition, complexity quantification of the physiologic signals, such as heart rhythm in health and disease was analyzed in previous studies [22,23].

To implement best clinical practice, effective tools are needed to properly assess and define orthostatic risk. This work presents entropy measure to provide a more accurate measure to aid with the assessment of orthostatic intolerance symptoms and its association with falls in a cohort of older adults.

2   Materials

The database included a total of 65 participants, aged over 70 years of age (70.11 ± 5.85), of whom 65% were female. There was no significant differences in age and gender between symptomatic and asymptomatic participants 44.6% (n=29) had symptomatic OI and 55.4% (n=36) did not. The measurement of the arterial pressure waveform was done at the finger with Finapress for Finger Arterial Pressure (Table 1). This method enabled for the first time a reliable measurement of the beat-to-beat blood pressure signal in a noninvasive manner. hemodynamic parameters were registered during three phases: Phase 1 a pre exercise standing by participates lasting 3 minutes, Phase 2 participates engaged in six minutes of walking and then phase 3 a post exercise standing upright which lasted 3 minutes.

The participants were assessed at the Technology Research for Independent Living (TRIL) Clinic in St James Hospital Dublin. All participants had a Mini-Mental State Examination (MMSE) score of $\geq$ 23 points, which is an optimal cut off when screening for dementia in an Irish setting [24]. None of the participants had parkinson diabetes mellitus, severe chronic renal failure (defined as Cockcroft Gault estimated Glomerular Filtration rate < 30m/min), vitamin B12 or folate deficiency or a cardiac pacemaker. The participants were not asked to stop any of their usual medications or fast before the assessment. All persons gave informed consent before their inclusion in the study.

Table 1. Patient Clinical Characteristics

Parameters	Symptomatic	Non-Symptomatic
Male (%)	13(36%)	10(34%)
Age (years)	70 ± 6	70 ± 6
Weight (kg)	77 ± 15	75 ± 12
Height (cm)	168 ± 8	168 ± 8

3 Methods

Many studies tried to demonstrate the nonlinear nature of the heart rate signal. This study is focused to quantify the regularity in hemodynamic profile by Shannon Entropy (SE) measures in older people with symptomatic and asymptomatic OI during active and passive stands.

3.1 Experiment

The participants underwent a lying to standing orthostatic test (active stand) with noninvasive beat-to-beat blood pressure monitoring system. Before standing, the participants were resting in a supine position for at least 10 minutes, and after standing, blood pressure was monitored for 3 minutes with participants standing still.

The blood pressure measured by the *FinometerPRO* device was calibrated at baseline (at least 2 minutes before the active stand) using the Return to Flow calibration system, which involves the use of an oscilometric pressure cuff on the ipsilateral upper arm for an individual calibration of the reconstruction of the finger pressure signal to brachial level [25]. Furthermore, the hydrostatic height correction system was used throughout the study to compensate for hand movements with respect to heart level. In our study participants were equipped, while walking, with a Portapress device. It is the ambulatory Finapres technology solution. The Portapres offers standard ambulatory blood pressure monitoring and displays hemodynamic parameters.

The 6 minutes walk test is a simple tool for the evaluation of functional exercise capacity, which reflects the capacity of the individual to perform activities of daily living [26]. It was conducted according to the following protocol:

- The participant stood for three minutes pre-exercise.
- A 30 meters flat, obstacle-free corridor was used marked out in 5 metes increments. The participants was instructed to walk at their fastest comfortable pace, turning 180° every 30 metes in the allotted time of 6 minutes.
- The participant stood for a further 3 minutes at the end of the test.

Three phases were captured: a pre-exercise stand lasting 3 minutes, a six minute walking phase and a post-exercise stand lasting 3 minutes. Hemodynamic parameters were exported with the Beatscope 1.1a software (Finapres Medical Systems) according to the 10-s average method outputted using Modelfow.

The following beat to beat derived hemodynamic measures were extracted:

1. Systolic blood pressure (SBP): as maximum pressure in arterial systole (mmHg).
2. Diastolic blood pressure (DBP): as low blood pressure just before the current upstroke (mmHg).
3. Heart rate (HR): pulse rate derived from the pulse interval (beats per minute) (pulse interval: time between the current and the next upstroke).

3.2 Shannon Entropy

Due to the huge irregularity of the correlation time series, it was applied entropy measures to evaluate whether if there were differences in the regularity in hemodynamic parameters between both groups.

Derived hemodynamic parameters from equal length intervals were extracted at each of the following phases of the active stand test. Shannon entropy (SE) was applied to the measured parameters with the following definition:

$$SE = -\sum_{i=1}^{M} p(i) \ln p(i), \tag{1}$$

SE is formally defined as the average value of logarithms of the probability density function, where M is the number of discrete values the considered variable can assume and $p(i)$ is the probability of assuming the ith value.

3.3   Statistical Analysis

Entropy from hemodynamic parameters were analyzed to test any significant differences regarding symptomatic and non symptomatic OI, unpaired t-tests and repeated measures ANOVA coupled with the Student-Newman-Keuls test were used. Results were considered to be statistically significant at $p < 0.05$.

4   Results

Results showed differences between phases in both groups. The main differences were found in entropy from HR and SBP. These parameters showed differences between different phases and between men and women in both groups.

Shannon entropy from HR values showed differences between phases in the non-symptomatic OI group, however not statistical significant differences between all phases were found in the symptomatic OI group (Table 2).

Table 2. HR Entropy along the three phases in both groups

Groups	Phases	Entropy HR Differences	Sig.(p)
Non-Symptomatic	$1-2$	$2,21 \pm 0,51 - 1,73 \pm 0,56$	$0,002$
	$1-3$	$2,21 \pm 0,51 - 2,78 \pm 0,54$	$< 0,001$
	$2-3$	$1,73 \pm 0,56 - 2,78 \pm 0,54$	$< 0,001$
Symptomatic	$1-2$	$2,31 \pm 0,48 - 2,05 \pm 0,62$	$0,125$
	$1-3$	$2,31 \pm 0,48 - 2,65 \pm 0,56$	$0,016$
	$2-3$	$2,05 \pm 0,62 - 2,65 \pm 0,56$	$< 0,001$

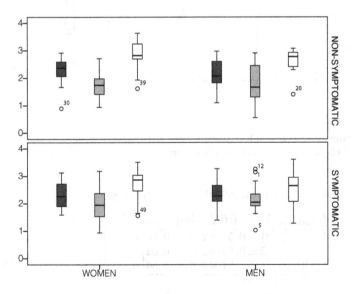

Fig. 1. HR entropy between genders during the three phases (phase 1: dark grey, phase 2: grey and phase 3: white) between both group

Table 3. HR Entropy along the three phases between genders in both groups

Groups	Genders	Phases	Entropy HR Differences	Sig.(p)
Non-Symptomatic	Men	$1-2$	$2,08 \pm 0,57 - 1,76 \pm 0,49$	$0,370$
		$1-3$	$2,08 \pm 0,57 - 2,61 \pm 0,49$	$0,027$
		$2-3$	$2,08 \pm 0,57 - 2,61 \pm 0,49$	$0,002$
	Women	$1-2$	$2,28 \pm 0,048 - 1,71 \pm 0,49$	$0,001$
		$1-3$	$2,28 \pm 0,48 - 2,67 \pm 0,56$	$< 0,001$
		$2-3$	$1,71 \pm 0,49 - 2,67 \pm 0,56$	$< 0,001$
Symptomatic	Men	$1-2$	$2,30 \pm 0,48 - 2,16 \pm 0,61$	$0,999$
		$1-3$	$2,30 \pm 0,48 - 2,53 \pm 0,68$	$0,603$
		$2-3$	$2,16 \pm 0,61 - 2,53 \pm 0,68$	$0,235$
	Women	$1-2$	$2,31 \pm 0,49 - 1,99 \pm 0,63$	$0,060$
		$1-3$	$2,31 \pm 0,49 - 2,71 \pm 0,49$	$0,009$
		$2-3$	$1,63 \pm 0,62 - 2,71 \pm 0,49$	$< 0,001$

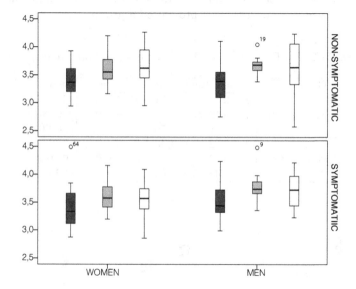

Fig. 2. SBP entropy between genders during the three phases (phase 1: dark grey, phase 2: grey and phase 3: white) between both group

Table 4. HR Entropy differences between groups

Phases	Entropy HR Control $-$ OI	Sig.(p)
1	$2,21 \pm 0,51 - 2,26 \pm 0,50$	$0,339$
2	$1,72 \pm 0,56 - 2,05 \pm 0,62$	$0,033$
3	$2,78 \pm 0,54 - 2,65 \pm 0,56$	$0,409$

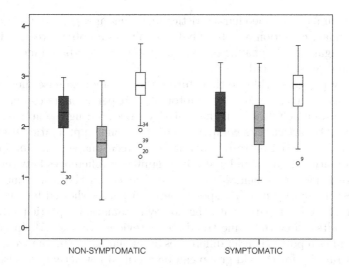

Fig. 3. HR entropy during the three phases (phase 1: dark grey, phase 2: grey and phase 3: white) between both group

In addition, HR entropy values showed differences between phases in women belong to non-symptomatic OI group, but not statistical significant differences between all the phases were found in the masculine group (Table 3). Moreover, in the symptomatic OI group not statistical significant gender differences between phases were found (Figure 1). HR entropy measure illustrated higher differences in women with more irregularity in non-symptomatic OI female human. Moreover, HR entropy measures showed statistical significant differences between phases in the control group with a regularity increase during the physically active phase.

Systolic blood pressure entropy measure showed statistical significant differences ($p=0.031$) in the phase 2 between men (3.78 ± 0.27) and women ($3,59 \pm 0.27$), and not statistical significant gender differences were found in SBP entropy measures in non symptomatic OI group were found. Moreover, blood pressure entropy differences between groups and phases were not statistically significant (Figure 2).

Entropy from HR measure during exercise showed statistically significant differences between both groups ($p=0,033$), moreover in the other phases were not found statistically significant differences (Table 4). In Figure 3 is possible to observe the difference in dispersion between both groups.

5   Conclusions

In this study we examined the hemodynamic response of 65 older people during a six minute walk functional exercise test. We analyzed the different profile of those participants with symptomatic OI and a non symptomatic group. Both groups had similar baseline characteristics including age, gender, Charlson Comorbidity Index, cognition and berg balance. The main objective of this study was to investigate hemodynamic parameters using entropy-based measures during active and passive stands.

This study presents evidence that during the phase of exercise the HR regularity response decreases more in symptomatic OI participants compared with the non symptomatic OI, with higher entropy values during the active stand in symptomatic OI participants compared with the non symptomatic OI group.

Moreover, higher HR entropy differences between phases were found in the non symptomatic OI group and statistical significant differences between all the phases were found in women, with lower differences in the masculine gender. Nevertheless, regularity in HR response between phases showed less differences in symptomatic OI compared with the non symptomatic OI participants.

These results follow the same trend that previous studies [27]. In addition, previous studies reported gender differences in repolarization inhomogeneity [28].

Lower regularity for the old group can be ascribed low activity levels in both groups. Our findings agree with comparative studies that predominantly looked exercise and passive tilt results [29].

As conclusion, this study highlights the important information heart rate regularity provide about hemodynamic changes that can occur during exercise in symptomatic OI patients. Identification of the mechanism that mediate OI is of clinical importance as OI is a recognized risk factor for falls and impairment of functional status, especially in women [30]. This would help focus therapies and interventions that potentially might enhance hemodynamic response.

Attention should also be drawn to entropy measure to detect changes in heart rate variability, elicited by orthostatic in older subjects. Entropy-based measures might provide useful indicators of pathological changes in cardiac activity during exercise. The proposed index could be used in treatment to provide information about hemodynamic changes in symptomatic OI patients during the exercise.

Acknowledgments. This work was supported by the Castilla-La Mancha Research Scheme (PPII-2014-024-P).

References

1. Carey, B.J., Potter, J.F.: Cardiovascular causes of falls. Age Ageing 30(suppl.4), 19–24 (2001)
2. Wöhrle, J., Kochs, M.: Syncope in the elderly. Z. Gerontol Geriatr. 36(1), 2–9 (2003)
3. Rubenstein, L.Z., Josephson, K.R.: Falls and their prevention in elderly people: what does the evidence show? Med. Clin. North. Am. 90(5), 807–824 (2006)
4. Schondorf, R., Stein, R., Roberts, R., Benoit, J., Cupples, W.: Dynamic cerebral autoregulation is preserved in neurally mediated syncope. Appl. Physiol (1985) 91(6), 2493–2502 (2001)
5. Mankovsky, B.N., Piolot, R., Mankovsky, O.L., Ziegler, D.: Impairment of cerebral autoregulation in diabetic patients with cardiovascular autonomic neuropathy and orthostatic hypotension. Diabet. Med. 20(2), 119–126 (2003)

6. Deegan, B.M., Serrador, J.M., Nakagawa, K., Jones, E., Sorond, F.A., Olaighin, G.: The effect of blood pressure calibrations and transcranial doppler signal loss on transfer function estimates of cerebral autoregulation. Med. Eng. Phys. 33(5), 553–562 (2011)

7. Gangavati, A., Hajjar, I., Quach, L., Jones, R.N., Kiely, D.K., Gagnon, P., Lipsitz, L.A.: Hypertension, orthostatic hypotension, and the risk of falls in a community-dwelling elderly population: the maintenance of balance, independent living, intellect, and zest in the elderly of boston study. Am Geriatr. Soc. 59(3), 383–389 (2011)

8. Heitterachi, E., Lord, S.R., Meyerkort, P., McCloskey, I., Fitzpatrick, R.: Blood pressure changes on upright tilting predict falls in older people. Age Ageing 31(3), 181–186 (2002)

9. Verwoert, G.C., Mattace-Raso, F.U.S., Hofman, A., Heeringa, J., Stricker, B.H.C., Breteler, M.M.B., Witteman, J.C.M.: Orthostatic hypotension and risk of cardiovascular disease in elderly people: the rotterdam study. Am. Geriatr. Soc. 56(10), 1816–1820 (2008)

10. Luukinen, H., Koski, K., Laippala, P., Airaksinen, K.E.J.: Orthostatic hypotension and the risk of myocardial infarction in the home-dwelling elderly. Intern. Med. 255(4), 486–493 (2004)

11. Kario, K., Tobin, J.N., Wolfson, L.I., Whipple, R., Derby, C.A., Singh, D., Marantz, P.R., Wassertheil-Smoller, S.: Lower standing systolic blood pressure as a predictor of falls in the elderly: a community-based prospective study. Am. Coll. Cardiol. 38(1), 246–252 (2001)

12. Tromp, A.M., Pluijm, S.M., Smit, J.H., Deeg, D.J., Bouter, L.M., Lips, P.: Fall-risk screening test: a prospective study on predictors for falls in community-dwelling elderly. Clin. Epidemiol. 54(8), 837–844 (2001)

13. Maurer, M.S., Cohen, S., Cheng, H.: The degree and timing of orthostatic blood pressure changes in relation to falls in nursing home residents. Am. Med. Dir. Assoc. 5(4), 233–238 (2004)

14. Poon, I.O., Braun, U.: High prevalence of orthostatic hypotension and its correlation with potentially causative medications among elderly veterans. Clin. Pharm. Ther. 30(2), 173–178 (2005)

15. Sorond, F.A., Galica, A., Serrador, J.M., Kiely, D.K., Iloputaife, I., Cupples, L.A., Lipsitz, L.A.: Cerebrovascular hemodynamics, gait, and falls in an elderly population: Mobilize boston study. Neurology 74(20), 1627–1633 (2010)

16. Casolo, G., Balli, E., Taddei, T., Amuhasi, J., Gori, C.: Decreased spontaneous heart rate variability in congestive heart failure. Am. J. Cardiol. 64(18), 1162–1167 (1989)

17. Saul, J.P., Arai, Y., Berger, R.D., Lilly, L.S., Colucci, W.S., Cohen, R.J.: Assessment of autonomic regulation in chronic congestive heart failure by heart rate spectral analysis. Am J. Cardiol. 61(15), 1292–1299 (1988)

18. Lipsitz, L.A., Mietus, J., Moody, G.B., Goldberger, A.L.: Spectral characteristics of heart rate variability before and during postural tilt. relations to aging and risk of syncope. Circulation 81(6), 1803–1810 (1990)

19. de Roquefeuil, M., Vuissoz, P.-A., Escanyé, J.-M., Felblinger, J.: Effect of physiological heart rate variability on quantitative t2 measurement with ecg-gated fast spin echo (fse) sequence and its retrospective correction. Magn. Reson. Imaging 31(9), 1559–1566 (2013)

20. Ledowski, T., Stein, J., Albus, S., MacDonald, B.: The influence of age and sex on the relationship between heart rate variability, haemodynamic variables and subjective measures of acute post-operative pain. Eur. J. Anaesthesiol. 28(6), 433–437 (2011)

21. Billman, G.E.: Heart rate variability - a historical perspective. Front Physiol. 2, 86 (2011)

22. Goldberger, A.L., Findley, L.J., Blackburn, M.R., Mandell, A.J.: Nonlinear dynamics in heart failure: implications of long-wavelength cardiopulmonary oscillations. Am. Heart J. 107(3), 612–615 (1984)

23. Goldberger, A.L., West, B.J.: Chaos and order in the human body. MD Comput. 9(1), 25–34 (1992)

24. Cullen, B., Fahy, S., Cunningham, C.J., Coen, R.F., Bruce, I., Greene, E., Coakley, D., Walsh, J.B., Lawlor, B.A.: Screening for dementia in an irish community sample using mmse: a comparison of norm-adjusted versus fixed cut-points. Int. J. Geriatr. Psychiatry 20(4), 371–376 (2005)

25. Guelen, I., Westerhof, B.E., Van Der Sar, G.L., Van Montfrans, G.A., Kiemeneij, F., Wesseling, K.H., Bos, W.J.: Finometer, finger pressure measurements with the possibility to reconstruct brachial pressure. Blood Press Monit. 8(1), 27–30 (2003)

26. Iwama, A.M., Andrade, G.N., Shima, P., Tanni, S.E., Godoy, I., Dourado, V.Z.: The six-minute walk test and body weight-walk distance product in healthy brazilian subjects. Braz. J. Med. Biol. Res. 42(11), 1080–1085 (2009)

27. García-Salmeron, F., Reilly, R.B., Cogan, L., Millet, R., Cervigon, J.: Haemodynamic parameters for assessment of orthostatic intolerance in older people. In: Computing in Cardiology (CinC) 2013, pp. 1111–1114 (September 2014)

28. Krauss, T.T., Mäuser, W., Reppel, M., Schunkert, H., Bonnemeier, H.: Gender effects on novel time domain parameters of ventricular repolarization inhomogeneity. Pacing Clin. lectrophysiol. 32(suppl. 1), S167–S172 (2009)

29. Tulppo, M.P., Hughson, R.L., Mäkikallio, T.H., Airaksinen, K.E., Seppänen, T., Huikuri, H.V.: Effects of exercise and passive head-up tilt on fractal and complexity properties of heart rate dynamics. Am. J. Physiol. Heart Circ. Physiol. 280(3), 1081–1087 (2001)

30. Ensrud, K.E., Nevitt, M.C., Yunis, C., Hulley, S.B., Grimm, R.H., Cummings, S.R.: Postural hypotension and postural dizziness in elderly women. the study of osteoporotic fractures. the study of osteoporotic fractures research group. Arch. Intern. Med. 152(5), 1058–1064 (1992)

Evolutionary Multiobjective Feature Selection in Multiresolution Analysis for BCI

Julio Ortega[1], Javier Asensio-Cubero[2], John Q. Gan[2], and Andrés Ortiz[3]

[1] Dept. of Computer Architecture and Technology, CITIC, University of Granada, Spain
[2] School of Computer Science and Electronic Eng., University of Essex, United Kingdom
[3] Dept. of Communications Engineering, University of Malaga, Spain
jortega@ugr.es, {jasens,jqgan}@essex.ac.uk, aortiz@ic.uma.es

Abstract. Although multiresolution analysis (MRA) may not be considered as the best approach for brain-computer interface (BCI) applications despite its useful properties for signal analysis in the temporal and spectral domains, some previous studies have shown that MRA based frameworks for BCI can provide very good performance. Moreover, there is much room for improving the performance of the MRA based BCI by feature selection or feature dimensionality reduction. This paper investigates feature selection in the MRA-based frameworks for BCI, proposes and evaluates several wrapper approaches to evolutionary multiobjective feature selection. In comparison with the baseline MRA approach used in previous studies, the proposed evolutionary multiobjective feature selection procedures provide similar or better classification performance, with significant reduction in the number of features that need to be computed.

Keywords: Brain-computer interfaces (BCI), Feature selection, Multiobjective optimization, Multiresolution analysis (MRA).

1 Introduction

Many high-dimensional pattern classification or modeling tasks require feature selection techniques in order to remove redundant, noisy-dominated, or irrelevant inputs. In particular, dimensionality reduction is very important to improve the accuracy and interpretability of the classifiers when the number of features is too large compared to the number of available training patterns, which is known as the curse of dimensionality.

Brain-computer interfacing (BCI) applications based on the classification of EEG signals pose the high-dimensional pattern classification problem [1], due to (1) the presence of noise or outliers (as EEG signals have a low signal-to-noise ratio); (2) the need to represent time information in the features (as brain signal patterns are related to changes in time); (3) the non-stationary of EEG signals, which may change quickly over time or within experiments. Moreover, the curse of dimensionality is usually present in the classification of EEGs as the number of patterns (EEGs)

F. Ortuño and I. Rojas (Eds.): IWBBIO 2015, Part I, LNCS 9043, pp. 347–359, 2015.

available for training is usually small and the number of features is usually much larger than the number of available patterns.

Feature selection is mandatory for BCI applications as it reduces the dimension of the input patterns making it possible to (1) decrease the computational complexity, (2) remove irrelevant/redundant features that would make it more difficult to train the classifier, and (3) avoid the curse of dimensionality [2]. Nevertheless, as the size of the search space depends exponentially on the number of possible features, an exhaustive search for the best feature set is almost impossible when the feature dimension is too high. Even for a modest number of features, feature selection procedures based on branch-and-bound, simulated annealing, and evolutionary algorithms have been proposed. Moreover, parallel processing could also be considered as an interesting alternative to take the advantage of high performance computer architectures for feature selection [3].

This paper describes several approaches for multi-objective feature selection in an MRA system for BCI [4]. An MRA system applies a sequence of successive approximation spaces that satisfy a series of constraints to reach a description as close as possible to the target signal [5], and thus it is useful whenever the target signal presents different characteristics in the successive approximation spaces. A specific example of MRA systems, the discrete wavelet transform (DWT), has been applied in [4] to characterize EEGs from motor imagery (MI). MI is a BCI paradigm that uses the series of amplifications and attenuations of short duration occasioned by limb movement imagination, the so called event related desynchronization (ERD) and event related synchronization (ERS). The task of ERD/ERS analysis is complex because they are weak and noisy and occur at different locations of the cortex, at different instants within a trial, and in different frequency bands. Moreover, there is no consistency in the patterns among subjects, and the patterns can even change within a session for the same subject, which may lead to high-dimensional patterns making the number of available patterns to conduct ERD/ERS analysis significantly less than the number of features. This constitutes a good scenario for evaluating the multiobjective feature selection approaches proposed in this paper.

This paper is organized as follows. Section 2 describes feature selection as a multiobjective optimization problem. Section 3 describes the MRA framework for BCI and the characteristics of the corresponding features. Section 4 describes several alternatives proposed for evolutionary multiobjective feature selection in MRA for BCI. Experimental results are presented and discussed in Section 5. Finally, the conclusions are given in Section 6.

2   Multiobjective Optimization in Supervised Feature Selection

In this paper we implement feature selection through a *wrapper* approach. A wrapper approach can be regarded as a search for the best feature set, which optimizes a cost function that evaluates the utility of the selected features for a given classification problem. In our case the cost function takes into account the classification performance obtained with the selected features. As the performance of a classifier is usually expressed not only by its accuracy for a given set of patterns, but also by other measures that quantify properties such as the generalization capability and computational efficiency. Using a multiobjective formulation for the feature selection problem, our method could be considered a powerful approach to feature selection.

A multiobjective optimization problem can be defined as finding a vector of decision variables $\mathbf{x}=[x_1,x_2,...,x_n]\in R^n$ that satisfies a restriction set, e.g., $g(\mathbf{x})\leq 0$, $h(\mathbf{x})=0$, and optimizes a function vector $\mathbf{f}(\mathbf{x})$, whose scalar values ($f_1(\mathbf{x})$, $f_2(\mathbf{x})$,..., $f_m(\mathbf{x})$) represent the objectives of the optimization. As these objectives are usually in conflict, instead of providing only one optimal solution, the procedures applied to multiobjective optimization should obtain a set of *non-dominated* solutions, known as Pareto optimal solutions, from which a decision agent will choose the most convenient solution in specific circumstances. These Pareto optimal solutions are optimal in the sense that in the corresponding hyper-area known as *Pareto front*, no solution is worse than the others when all the objectives are taken into account.

Feature selection as a multiobjective optimization problem can be for either supervised or unsupervised classifiers. A deep review on this topic is given by J. Handl and J. Knowles [6]. With respect to supervised classifiers, multiobjective feature selection procedures often take into account the number of features and the performance of the classifier [7, 8]. There are a lot of studies focusing on feature selection for unsupervised classification [6, 9, 10]. As the labels for the training and testing patterns are available in our BCI datasets, this paper deals with supervised multiobjective feature selection.

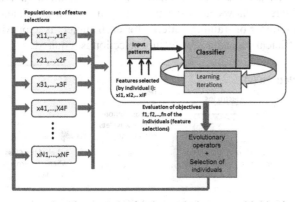

Fig. 1. Wrapper approach to feature selection by evolutionary multiobjective optimization

Figure 1 provides a scheme of multiobjective optimization for feature selection in a classification procedure. This scheme corresponds to the wrapper approach for feature selection implemented in our study. Each individual of the population encodes the features of the input patterns that are taken into account during the classifier training. An individual evaluation implies to train the classifier with the given input patterns and to determine the classifier performance by using several cost functions, as a multiobjective optimization procedure has been considered. From Figure 1 the usefulness of a multiobjective approach for feature selection is apparent as the classifier's behavior is not usually characterized by only one parameter. Besides the accuracy, there are other measures that quantify its performance. Among them we have measures to evaluate its generalization capabilities or the possible amount of overfitting the classifier could present.

3   Multiresolution Analysis of EEG for BCI

In this paper the dataset used to evaluate the proposed multiobjective feature selection procedures was recorded in the BCI Laboratory at the University of Essex. The dataset includes patterns that correspond to three different classes of imagined movements (right hand, left hand, and feet) from 10 subjects with ages from 24 to 50 (58% female, 50% naïve to BCI) and was recorded with a sampling frequency of 256 Hz during four different runs. There are a total of 120 trials for each class for each subject. More details about this dataset can be found in [4].

Each pattern was obtained from an EEG trial by the feature extraction procedure based on the MRA described in [4]. Thus, each signal obtained from each electrode contains several segments to which a set of wavelets detail and approximation coefficients are assigned. If there are S segments, E electrodes, and L levels of wavelets, each pattern is characterized by 2×S×E×L sets of coefficients (the number of coefficients in each level set depends on the level). In the Essex BCI dataset, S=20 segments, E=15 electrodes, and L=6 levels, therefore 3600 sets, with from 4 to 128 coefficients in each set are used to characterize each pattern (a total of 151200 coefficients). Figure 2 shows how the pattern features are generated. Taking into account that the number of training patterns for each subject is approximately 180, it is clear that an efficient procedure for feature selection is required.

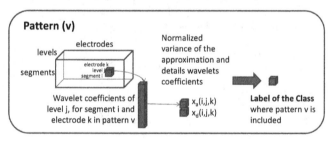

Fig. 2. Characterization of an EEG signal (pattern) in [4]

In [4] a simple approach to reduce the number of coefficients is applied, in which only one coefficient is assigned to each electrode and each level of approximation and detail. This coefficient is obtained by computing the second moment of the coefficient distribution (variance) and normalising the value between 0 and 1. This way, the number of coefficients for a given pattern is 2×S×E×L.

Another approach to reduce the number of features used in [4] is, instead of using only one classifier as shown in Figure 3, a set of LDA (linear discriminant analysis) classifiers are used, and the majority voting of all the LDA outputs is adopted as the final classification output. This way, a set of 2×S×L LDA classifiers with the number of inputs equaling the number of electrodes are adopted, as shown in Figure 4.

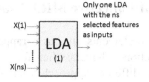

Fig. 3. EEG classification with only one LDA classifier

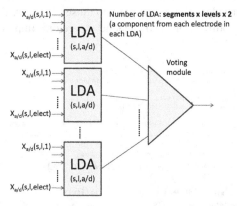

Fig. 4. EEG classification with multiple LDA classifiers based on majority voting, with one LDA classifier per segment per level [4]

In this paper we present several feature selection approaches for the classification framework shown in Figure 4. It is clear that it is possible to consider the selection of the LDAs in Figure 4, but there are also other possible arrangements of LDAs as the one shown in Figure 5. The feature searching space in the classifier structure of Figure 4 has a dimension of 2×S×L, while the classifier of Figure 5 has the same number of features as in the case of Figure 3, i.e., 2×E×L.

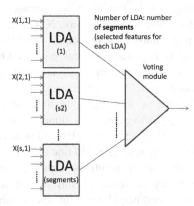

Fig. 5. EEG classification with multiple LDA classifiers based on majority voting, with one LDA classifier per segment

4   Evolutionary Multiobjective MRA Feature Selection

Figure 1 outlines the main elements of our wrapper approach to evolutionary multiobjective feature selection. The classifier for assigning the fitness to each individual is the one in the different approaches shown in Figures 3 to 5. In this section we provide the relevant characteristics of the multiobjective evolutionary algorithm proposed, which is briefly described in Figure 6. The main steps of the multiobjective evolutionary algorithm correspond to those of NSGA-II [11], including the specific individual codification and genetic operators implemented for the application at hand.

```
Input:
N       number of solutions in the population
P0      population of solutions (subsets of selected features)
M       number of objectives
D       set of training patterns
L       labels of the classes for training patterns
Output:
Pf      set of nondominantsolutions (selected subsets of features)
```

```
(1)      P0=initialize_population(P0);
(2)      (P0,f1..M(P0))=evaluate_population(P0,D,L);
(3)      (P,f1..M(P)) =NSGAII_nondomination_sort(P0,f1..M(P0));
(4)      repeat
(5)        (P',f1..M(P')) =NSGAII_tournament_selection(P,f1..M(P));
(6)        P" =genetic_operators(P');
(7)        (P",f1..M(P")) =evaluate_population(P",D,L);
(8)        (P*,f1..M(P*))=NSGAII_nondomination_sort(P",f1..M(P"),P,f1..M(P));
(9)        (P,f1..M(P)) =NSGAII_replace_chromosome(P*,f1..M(P*));
(10)     until end_condition
(11)     Pf = P
```

Fig. 6. Pseudocode for evolutionary multiobjective feature selection based on NSGA-II

In NSGA-II the fitness values of the individuals in the population are sorted accordingly to the different fronts of nondominated individuals (nondomination levels) where they belong, while the diversity among individuals in the same nondominated front is also preserved. This way, NSGA-II uses a fast procedure of $O(MN^2)$ complexity (M is the number of objectives and N the number of individuals in the population), and the storage requirements grow as $O(N^2)$ in order to sort the population according to the different levels of nondominance. This procedure is implemented by the function NSGAII_nondomination_sort() in lines (3) and (8) of the pseudocode of Figure 6, thus being applied once the evaluation of the individuals by evaluate_population() is done, either after the population is initialized by initialize_population()in line (1) or after new individuals appear due to the action of genetic_operators()in line (6).

To maintain the diversity among solutions with the same nondominance level, NSGA-II estimates the density of solutions surrounding a given solution through the average distance of the nearest neighbour solutions on either side of the considered solution for each dimension (objective) of the front. The density of solutions or crowding distance can be computed with complexity $O(MNlogN)$ as it implies to sort, at most, N individuals along the dimensions corresponding to m objectives.

This crowding distance avoids the need of setting a sharing parameter. Once each individual in the population has a nondomination rank and a crowding distance, a selection process chooses solutions with lower nondomination ranks and solutions with higher crowding distances (less crowded regions) in case of similar nondomination ranks. Thus, the complexity of NSGA-II is determined by the population sorting step, i.e., $O(mN^2)$. This task has been implemented, respectively in lines (5) and (9) of the pseudocode of Figure 6, through the functions NSGAII_tournament_ selection() and NSGAII_replace_chromosome().

To configure NSGA-II for our wrapper procedure for feature selection, the fitness evaluation is done by a function that implements the learning procedure for the corresponding classifier (Figures 3 to 5) by using a set of labelled patterns. The corresponding procedure is implemented through evaluate_population() in lines (2) and (7). Thus, three main issues should be considered for a whole specification of the evolutionary multiobjective search procedure: the codification of the individuals in the population, the genetic operators, and the cost functions.

Codification of the Individuals. If the number of features is very high, it is not useful to codify an individual with binary components, with each component corresponding to a possible feature. In our approach, each individual is codified by a set of vectors, with each vector corresponding to one of the features included in the selection codified by the individual. The components of the vector correspond to the dimensions that characterize each input pattern. For example, in the case described in Figure 2, each feature is codified by a vector with four components: segment, level, electrode, analysis/detail wavelet component. Each component will have an integer value between 1 and S for the segment component, between 1 and L for the level component, between 1 and E for the electrode component, and 0 or 1 for the analysis/detail component. This strategy provides efficient codification and allows to extract information about the characteristics of the feature selection (segments, levels, electrodes, or coefficients more frequently selected, etc.).

Genetic Operators. Two different operators can be applied to the individuals. The first one is a crossover operator that randomly selects two parents and a subset (also randomly selected) of features for each parent. These subsets are interchanged by the parents. The mutation operator applies changes to a subset of features randomly chosen among the features codified by the individual to be mutated, which is also randomly selected. The changes in the values of the components of the feature to be mutated are normally distributed with means and deviations that change with generations.

Cost Functions. To characterize the performance of the classifier while it has been trained or adjusted for a given set of features (an individual of the population), it is important not only to take into account the accuracy obtained for the training set but also to its behaviour for unseen instances, i.e., its generalization capabilities. Thus, two cost functions are considered to evaluate the feature selection. The first one is the kappa index [12], which provides an accurate description of the classifier performance. It can be considered even better than the classification ratio as it takes into account the per class error distribution. The other cost function evaluates aspects such as the generalization capability or the classifier overfitting. In our case, 10-fold cross-validation analysis to the training patterns was applied to obtain the cost function values.

5   Experimental Results

This section presents the experimental results obtained by the evolutionary multiobjective feature selection approaches described in Section 4. The experiments have been performed by using the dataset recorded in the BCI Laboratory at the University of Essex. For each subject, there is one data file named x1## with data recorded in two runs for training and another data file named xe1## with data recorded in two runs for evaluation. Each data file contains about 180 labelled patterns with data from 20 segments (S=20), six levels (L=6) of approximation or detail coefficients (a/d=2), and 15 electrodes (E=15). The labels correspond to three imagined movements of right hand, left hand, and feet.

The baseline method for comparison (OPT0) corresponds to the MRA framework depicted in Figure 4, where all the possible S×L×2 LDAs are considered for voting the class to which the corresponding pattern belongs. The number of inputs to each LDA is the same as the number of electrodes. The input from a given electrode is the normalized second moment of the wavelet coefficients of the signal in this electrode for the s-th segment, the l-th level, and approximation/detail type of the corresponding LDA. In OPT0 there is no feature selection.

The three alternative methods proposed in this paper (OPT1, OPT2, and OPT3) use different multiobjective feature selection approaches. OPT1 is based on the classification scheme shown in Figure 3 and the multiobjective feature selection procedure is applied on S×L×E×2 features corresponding to the possible segments, levels, variances of the approximation and detail coefficients, and electrodes. With the BCI dataset of the University of Essex there are 20×6×15×2=3600 possible features for selection.

OPT2 is based on the classifier structure shown in Figure 4. In this case, the feature selection problem is to select among the S×L×2 possible LDAs used for voting in the OPT0 method. This means the dimension of the search space is 20×6×2=240. OPT3 is based on the classification scheme shown in Figure 5. In this case, the number of LDAs used for voting is equal to the number of segments, and each LDA can have up to 2×L×E different features as inputs. With the BCI dataset, there are 20 LDAs, each with up to 2×6×15=180 inputs. There are 2×S×L×E features, but they are structured and no more than 2×E×L features are available for selection for a given LDA.

The evolutionary multiobjective feature selection procedures have been executed with populations of 20, 30, and 50 individuals respectively, and with 20, 30, and 50 generations respectively to determine the minimum number of individuals and generations that provide competitive results compared with OPT0. OPT1, OPT2, and OPT3 have been executed by using 50 individuals and 50 generations so that the amount of searching work in all the approaches is similar. Executions with different number of iterations and individuals would provide a fairer comparison among the different approaches. We will consider this at the end of this section.

We have used simulated binary crossover with a crossover probability of 0.5, a mutation probability of 0.5, and distribution index of 20 for crossover and mutation operators. It is worth mentioning that no work on tuning the parameters of the evolutionary multiobjective feature selection options to optimize their behavior has been considered, as our aim here is to analyze whether multiobjective optimization is able to provide some improvements on MRA approaches for BCI.

Tables 1 and 2 compare the Kappa indexes obtained by the 4 approaches (OPT0 to OPT3). The columns labeled as "Kappa index (x#)" in Table 1 correspond to the Kappa index values obtained by the trained classifier when it was evaluated by using the same training patterns. The columns labeled as "Kappa index (xe#)" in Table 2 provide the Kappa index values obtained by the 4 approaches when the testing patterns were used to evaluate the classifier performance for each subject.

Table 1. Comparison of different feature selection and classification methods for the University of Essex BCI data files (Kappa values evaluated with the training patterns)

	OPT0	OPT1	OPT2	OPT3
Subject	Kappa index (x#)	Kappa index (x# mean,std)	Kappa index (x# mean,std)	Kappa index (x# mean,std)
101	0.790	0.738±0.022*	**0.828±0.011^**	0.797±0.023
102	**0.857**	0.747±0.017*	**0.855±0.011**	**0.853±0.022**
103	**0.757**	0.665±0.012*	**0.747±0.013**	0.695±0.024*
104	0.899	0.819±0.013*	**0.902±0.010**	0.882±0.022*
105	**0.757**	0.646±0.023*	**0.751±0.013**	0.695±0.031*
106	0.774	0.604±0.023*	**0.776±0.014**	0.754±0.024*
107	0.857	0.816±0.018*	**0.880±0.011^**	0.845±0.022
108	**0.774**	0.508±0.021*	0.745±0.023*	0.683±0.032*
109	**0.790**	0.597±0.021*	0.770±0.015*	0.730±0.030*
110	0.883	0.821±0.016*	0.877±0.026	**0.897±0.015^**

Table 2. Comparison of different feature selection and classification methods for the University of Essex BCI data files (Kappa values evaluated with the test patterns)

	OPT0	OPT1	OPT2	OPT3
Subject	Kappa index (xe#)	Kappa index (xe# mean,std)	Kappa index (xe# mean,std)	Kappa index (xe# mean,std)
101	**0.438**	0.393±0.046*	**0.437±0.033**	0.367±0.032*
102	**0.455**	0.302±0.074*	**0.429±0.023**	0.382±0.044*
103	0.279	0.249±0.046	0.325±0.017^	**0.356±0.024^**
104	**0.564**	0.510±0.056	0.545±0.035	0.563±0.034
105	**0.287**	0.191±0.040*	**0.240±0.031**	0.227±0.023*
106	**0.321**	0.193±0.070*	**0.319±0.028**	0.246±0.036*
107	0.631	0.560±0.041*	**0.634±0.019**	0.603±0.027*
108	**0.254**	0.088±0.036*	0.184±0.027*	0.184±0.028*
109	**0.388**	0.207±0.071*	0.333±0.026*	0.321±0.037*
110	**0.648**	0.450±0.036*	0.605±0.041*	0.578±0.027*

The results shown in Table 1 and Table 2 for OPT1, OPT2, and OPT3 are the average (mean and standard deviation) over 15 executions of each approach for each subject. The statistical analysis has been conducted by applying a Kolmogorov-Smirnov test first to determine whether the obtained values of the Kappa index follow a normal distribution or not. If the experimental results do not have normal distribution, a non-parametric Kruskal-Wallis test has been used to compare the means of the different algorithms. A confidence level of 95% has been considered in the statistical tests. If the mean values of the Kappa index in OPT1, OPT2, and OPT3 are statistically significantly different from those in OPT0 according to the Kruskal-Wallis test, they are marked with either ^ or *, with ^ indicating performance improvement and * indicating performance loss.

Tables 3 and 4 show the best values of the Kappa index achieved by each approach over 15 executions. Compared to OPT0, feature selection (OPT1, OPT2, and OPT3) is able to provide competitive results, with the advantage that they require a smaller number of features. In some cases, OPT2 provides even better classification performance than OPT0. According to the statistical tests, except for subjects x108, x109, and x110, OPT2 is able to obtain the same or even better results than OPT0, with fewer features, which is highly valuable when designing online BCI systems.

Table 3. Comparison of different feature selection and classification methods for the University of Essex BCI data files: maxima Kappa values evaluated with the training patterns

Subject	OPT0 Kappa index (x#)	OPT1 Kappa index (x# max)	OPT2 Kappa index (x# max)	OPT3 Kappa index (x# max)
101	0.790	0.782	**0.849**	0.832
102	0.857	0.782	**0.883**	**0.883**
103	0.757	0.690	**0.765**	0.723
104	0.899	0.849	0.925	**0.933**
105	0.757	0.673	**0.774**	0.740
106	0.774	0.648	**0.799**	**0.799**
107	0.857	0.849	**0.899**	0.874
108	0.774	0.539	**0.782**	0.732
109	0.790	0.623	**0.799**	**0.799**
110	0.883	0.841	0.899	**0.920**

Table 4. Comparison of different feature selection and classification methods for the University of Essex BCI data files: maxima Kappa values evaluated with the test patterns

Subject	OPT0 Kappa index (xe#)	OPT1 Kappa index (xe# max)	OPT2 Kappa index (xe# max)	OPT3 Kappa index (xe# max)
101	0.438	0.472	**0.489**	0.430
102	0.455	0.405	**0.463**	0.447
103	0.279	0.329	0.354	**0.413**
104	0.564	0.589	**0.614**	0.606
105	**0.287**	**0.287**	**0.287**	**0.287**
106	0.321	0.338	**0.381**	0.292
107	0.631	0.631	**0.665**	0.656
108	**0.254**	0.170	0.245	0.237
109	**0.388**	0.346	0.371	**0.388**
110	0.648	0.530	**0.673**	0.639

With respect to the execution time required by each approach, the mean execution time is 4533±45s for OPT1, 13353±1031s for OPT2, and 1159±56s for OPT3. Taking into account these differences in the running time, it is possible to argue that the comparison among different approaches may not be fair as OPT1 and OPT3 could probably achieve better results when more generations (with more individuals) are adopted with execution time similar to that required by OPT2. Table 5 shows the results obtained, for two subjects only (x104 and x107), with OPT1 using a population of 100 individuals and 60 generations and OPT3 using 200 individuals and 90 generations. When 100 individuals and 60 generations are used, OPT1 requires a mean execution time of 11042 ± 162s, which is in a similar order as the one required by OPT2. In the case of OPT3, 200 individuals and 90 generations still consume less execution time (8408 ± 144s) than OPT2 with 50 individuals and 50 generations.

The results in Table 5 show improvements on the performance of OPT1 and OPT3 as the number of individuals in the population and generations are increased. In the values of the Kappa index obtained by using the training patterns (x# columns), the Kruskal-Wallis test shows that the differences are statistically significant for OPT1 ($p=0.00$ and $p=0.005$ for x104 and x107, respectively). However, when the evaluation was done by using the test patterns (xe# columns), these differences among OPT1 with 100 and 50 individuals are not significant ($p=0.75$ and $p=0.25$ for x104 and x107, respectively). In the case of OPT3, the situation is similar for the values of Kappa index obtained by using the training patterns ($p=0.04$ and 0.014 for x104 and x107, respectively). With respect to the values of Kappa index obtained by using the testing patterns, the results are not statistically significant for x104 ($p=0.63$) but are statistically significant for x107 ($p=0.005$). It can be seen that OPT3 with a population of 200 individuals and 90 iterations achieved better performance than OPT2 (evaluation with the test patterns, i.e., xe# columns). Nevertheless, the Kruskal-Wallis test only shows statistical significance for x107 ($p=0.005$ for x107, and $p=0.0997$ for x104).

Table 5. Comparison of Kappa indexes for OPT1 with 100 individuals and 60 generations, and OPT3 with 200 individuals and 90 generations, with respect to OPT1, OPT2, and OPT3 with 50 individuals and 50 generations (x#: evaluation was done with training patterns; xe#: evaluation was done with test patterns)

	OPT1 (100,60)	OPT1 (50,50)	OPT2 (50, 50)	OPT3 (200,90)	OPT3 (50,50)
Subject	Kappa index (x# mean,std)	Kappa index (x# mean,std)	Kappa index (x# mean,std)	Kappa index (x# mean,std)	Kappa index (x# mean,std)
x104	0.841 ± 0.018	0.819 ± 0.013	**0.902 ± 0.010**	0.897 ± 0.019	0.882 ± 0.022
x107	0.836 ± 0.016	0.816 ± 0.018	**0.880 ± 0.011**	0.865 ± 0.018	0.845 ± 0.022
Subject	Kappa index (xe# mean,std)	Kappa index (xe# mean,std)	Kappa index (xe# mean,std)	Kappa index (xe# mean,std)	Kappa index (xe# mean,std)
x104	0.515 ± 0.047	0.510 ± 0.056	0.545 ± 0.035	**0.573 ± 0.032**	0.563 ± 0.034
x107	0.580 ± 0.052	0.560 ± 0.041	0.634 ± 0.019	**0.644 ± 0.022**	0.603 ± 0.027

6   Conclusions

Procedures such as the one described in [4] provide approaches to cope with the curse of dimensionality based on the composition of multiple classifiers. Each classifier receives only a subset of pattern components (features) in such a way that the number of patterns is much higher than the number of features used as inputs. The problem with these approaches is that the number of classifiers to train and to accomplish the classification is usually very high (as finally, all the features should be taken into account). Beyond these approaches do not provide information about the most relevant features, the need to compute such a high number of features and to train a lot of classifiers could be a significant drawback to satisfy real-time requirements of many applications. In this context, the contribution of this paper is twofold. On the one hand, we provide a multiobjective approach to cope with the feature selection in LDA classification based on two cost functions that evaluate the classifier accuracy and its generalization capability. On the other hand, we have proposed several classification structures (OPT1 to OPT3) to take advantage of our multiobjective approach to feature selection. The experimental results show that evolutionary multiobjective feature selection is able to provide classification performance similar to that of using all the possible LDAs with all the possible feature inputs (OPT0). Thus the proposed approaches lead to simpler classification procedures with fewer features. Besides the analysis of the characteristics of the features selected for obtaining some knowledge about important electrodes and segments, etc., there are other issues that can be considered to improve the performance of multiobjective feature selection in BCI applications. It is clear that improving the cost function that evaluates the generalization capability is an important issue. Moreover, the implementation of cooperative coevolutionary approaches able to cope with problems with a large number of decision variables (features) is also an interesting topic.

Aknowledgements. This work has been partly funded by projects TIN2012-32039 (Spanish "Ministerio de Economía y Competitividad" and FEDER funds), P11-TIC-7983 (Junta de Andalucía, Spain), and Research Programme 2014 of the University of Granada.

References

1. Lotte, F., Congedo, M., Lécuyer, A., Lamarche, F., Arnaldi, B.: A review of classification algorithms for EEG-based brain-computer interfaces. Journal of Neural Engineering 4(2) (2007), doi:10.1088/1741-2560/4/2/R01
2. Raudys, S.J., Jain, A.K.: Small sample size effects in statistical pattern recognition: Recommendations for practitioners. IEEE Transactions on Pattern Analysis and Machine Intelligence 13(3), 252–264 (1991)
3. Zao, Z., Zhang, R., Cox, J., Duling, J.D., Sarle, W.: Massively parallel feature selection: an approach based on variance preservation. Machine Learning 92, 195–220 (2013)
4. Asensio-Cubero, J., Gan, J.Q., Palaniappan, R.: Multiresolution analysis over simple graphs for brain computer interfaces. Journal of Neural Engineering 10(4) (2013), doi:10.1088/1741-2560/10/4/046014
5. Daubechies, I.: Ten Lectures on Wavelets. SIAM, Philadelphia (2006)
6. Handl, J., Knowles, J.: Feature selection in unsupervised learning via multi-objective optimization. Intl. Journal of Computational Intelligence Research 2(3), 217–238 (2006)
7. Emmanouilidis, C., Hunter, A., MacIntyre, J.: A multiobjective evolutionary setting for feature selection and a commonality-based crossover operator. In: Proceedings of the 2000 Congress on Evolutionary Computation, pp. 309–316. IEEE Press, New York (2000)
8. Oliveira, L.S., Sabourin, R., Bortolozzi, F., Suen, C.Y.: A methodology for feature selection using multiobjective genetic algorithms for handwritten digit string recognition. International Journal of Pattern Recognition and Artificial Intelligence 17(6), 903–929 (2003)
9. Kim, Y., Street, W.N., Menczer, F.: Evolutionary model selection in unsupervised learning. Intelligent Data Analysis 6(6), 531–556 (2002)

10. Morita, M., Sabourin, R., Bortolozzi, F., Suen, C.Y.: Unsupervised feature selection using multi-objective genetic algorithms for handwritten word recognition. In: Proceedings of the Seventh International Conference on Document Analysis and Recognition, pp. 666–671. IEEE Press, New York (2003)

11. Deb, K., Agrawal, S., Pratab, A., Meyarivan, T.: A fast elitist Non-dominated Sorting Genetic Algorithms for multi-objective optimisation: NSGA-II. In: Deb, K., Rudolph, G., Lutton, E., Merelo, J.J., Schoenauer, M., Schwefel, H.-P., Yao, X. (eds.) PPSN 2000. LNCS, vol. 1917, pp. 849–858. Springer, Heidelberg (2000)

12. Cohen, J.: A coefficient of agreement for nominal scales. Educ. Psychological Meas. 20, 37–46 (1960)

Dose Calculation in a Mouse Lung Tumor and in Secondary Organs During Radiotherapy Treatment: A Monte Carlo Study

Mahdjoub Hamdi[1], Malika Mimi[1], and M'hamed Bentourkia[2,*]

[1] Department of Electrical Engineering, University of Mostaganem, Algeria
[2] Department of Nuclear Medicine and Radiobiology, Université de Sherbrooke, Canada
mhamed.bentourkia@usherbrooke.ca

Abstract. Radiotherapy in cancer treatment always affects surrounding tissues and even deposits doses in distant tissues not traversed by the radiation beams. In the present work, we report energy transfer and absorbed dose in a target tumor and in other distant organs in a digital mouse by Monte Carlo simulations. We simulated a selection of X-rays beams with seven energies, 50, 100, 150, 200, 250, 350 and 450 keV each oriented in seven irregularly incremented angles, and we computed the dose and the energy deposit as a function of photon interaction types. The results show that the absorbed dose increased with increasing energy even in the secondary organs not receiving the radiation beam, and that the lowest dose was obtained with 100 keV beam. The spinal cord, of comparable size to the tumor and excluding the spinal bones, which was not directly irradiated by the beams, received a dose representing in average 1% of that of the tumor, while the spinal bone received doses of 6.6 and 0.12 times those in the tumor at 50 and 450 keV, respectively. Such Monte Carlo simulations could be necessary to select the appropriate beam energy and beam angles to efficiently treat the tumor and to moderately reduce the impact of the radiations in the other organs.

Keywords: Dosimetry, Monte Carlo, Small animal, Tumor, X-rays, Photon interaction.

1   Introduction

With Intensity Modulated Radiation Therapy (IMRT), the external beam is adjusted in intensity and cross section to target the malignant tumor while preserving the normal tissues. The provided 3D anatomical images of structures with Computed Tomography (CT) serve as a guide for the radiation oncologist to precisely adjust the beam on the tumor and to select other beam trajectories. The beam flux, intensity, energy and orientation are then estimated to optimize the penetration of the beam until the target tissue. Monte Carlo simulations (MCS) are generally used to calculate dose distributions. MCS are accepted as the most accurate method for dose calculation, but

---

* Corresponding author.

F. Ortuño and I. Rojas (Eds.): IWBBIO 2015, Part I, LNCS 9043, pp. 360–367, 2015.

simulating a large number of particles and tracking their interactions in the medium need high memory and fast clusters of computers in conjunction with improved algorithms [1, 2].

Experimental preclinical small animal digital models are used to investigate tumor response to radiation therapy, and they are generally created from high resolution anatomical imaging such as magnetic resonance imaging, CT or from digital atlases [3-5].

The aims of the present work were to assess the radiotherapy dose deposit in a primary tumor and in other secondary organs of a mouse phantom [3] using multiple beam angles and energies targeting the tumor. To achieve these calculations we used Geant4 Applications for Tomographic Emission (GATE) based on Monte Carlo simulation code [6]. We also report the statistics of the transferred energy in the tumor and in the other tissues.

2   Materials and Methods

2.1   Heterogeneous Mouse Phantom

We used a digimouse based 3D image of a 28 g normal nude male mouse in format of micro-CT image provided by Digimouse [3] as shown in Fig. 1. Since the Digimouse data were obtained as gray scale intensities, we converted them to a voxelized phantom with density in Hounsfield Units (HU) [7, 8]. In fact, GATE utilizes HU, and within GATE, they are converted to mass densities. To convert mouse image from gray scale levels to HU, we chose two extreme volumes of interest (VOI) in air and in spinal bone in order to define the linearity between image intensities and tissues densities [7, 9, 10]. We evaluated the mean intensity values of each region as 0 for air and 254 for spinal bone, and we established their corresponding densities of 0.120×10^{-2} g/cm^3 and 1.85 g/cm^3, respectively. For calculation of HU values, we extracted the mass attenuation coefficients for these two materials from the tables of Photon Cross Sections and Attenuation Coefficients (http://atom.kaeri.re.kr/cgi-bin/w3xcom, the densities of the media are also given therein) by supposing 30 keV mono-energetic photon beam [3].

In the mouse phantom, we considered a lung tumor having 60 HU [11] with a spherical volume of 1.4 mm in diameter [12] (Fig. 1). The mean HU difference between the tumor and lung tissue was 625 HU. AMIDE Software [13] was used for image display and manipulation. For dosimetry analyses, we used 7 radiation beams focusing on the tumor and we evaluated photon interactions, energy transfer and the related absorbed dose in 8 volume regions including the tumor (Fig. 1). Volume 1 (V1) was around the tumor, V2 was in lung tissue, V3 was in lung tissue intercepted by beam 1, V4 and V5 were in the heart, V6 was located in the lung at the level of the tumor but horizontally translated by 1.8 mm, V7 was manually drawn around the spinal bone including bone marrow and spinal cord, and V8 was located around the spinal cord excluding the bone. The volumes of V7 in the spine and V8 in the spinal cord were respectively 7.32 and 1.09 times greater than the tumor volume (V1). In addition to the actual volume of the tumor, i.e. 1.4 mm of diameter (Fig. 1), the tumor volume was defined by two other regions of different diameters to assess the energy transfer and absorbed dose within the beams and at larger volumes than intercepted by the beams, at diameters of 1.6 mm, V9, and 1.8 mm, V10. These assessments were made to estimate the dose to the boundaries of the tumor and to nearest tissues as the real limits of the tumor are not always accurately known from the images in clinical situations. All these regions were identified by means of the indications on the Digimouse atlas [3] (http://neuroimage.usc.edu/neuro/Digimouse_Download).

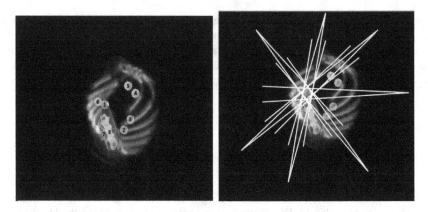

Fig. 1. Left: Mouse phantom with 8 tissue regions with region 1 being the targeted tumor. The circles around the digits indicate the actual spherical shape of the regions with 1.4 mm in diameter except for regions 7 and 8 which were manually drawn. Right: Same image repeated with 7 radiotherapy beams focused on the tumor only.

2.2   Simulation Procedure

A monoenergetic conical X-rays beam was simulated from a point source intercepting the tumor. Seven beams were used around the mouse phantom at 7 angles to avoid irradiating bone marrow and the spinal cord [14] (Fig. 1). To investigate the influence of the X-rays energies on the absorbed dose within the tumor, in its surrounding tissues and in the neighboring organs, the X-rays beams spanned the following energies of 50, 100, 150, 200, 250, 350 and 450 KeV with a total of $7x10^8$ photons simulated for each beam angle and energy. This number of photons corresponds to the expected absorbed dose, around 2 Gy, in a small volume of a mouse tissue, using 8 beams with a current of 50 mAs at 120 kVp [15, 16]. The simulations were conducted on a supercomputer having 2464 CPUs and 308 SGI XE320 compute nodes each with 2 Intel Xeon E5462 four core processors at 2.8 GHz and 16 to 32 Gbytes of memory per node (http://www.calculquebec.ca/en/resources/compute-servers/mammouth-serie-ii). A single beam with $7x10^8$ photons simulated in the mouse took around 5 min, and the whole beams for a single simulated energy were distributed on 40 nodes.

The radiation dose was calculated in each volume of the ten regions (Fig. 1). Also we concentrated our efforts here on energy transfer and absorbed dose in specific volume regions instead of lines of isodoses [17].

2.3   Data Analysis

The results of particle tracking were obtained in ROOT file format [18], then the data, i.e. photon identifier number, type of interaction, position of interaction, and energy transferred were extracted for further analysis. In addition, two 3D matrices (0.1 mm resolution) storing the energy transferred and the related absorbed dose in the whole mouse were provided in two files in Analyze format [8, 19].

For each VOI, the relative importance for elementary interactions (Compton and photoelectric), the energy spectra, the total transferred energy and its related absorbed dose were also calculated.

3   Results

The energy transfer and dose were assessed with the 7 X-rays beams for each of the 7 energies. The total energy transferred in the 10 VOIs and for each of the 7 beam energies are reported in Table 1. Although low energy photons are more prone to transfer energy such as 50 keV in comparison to those at 450 keV (Table 1), the absorbed dose appeared high at higher energies (450 keV) than at lower energies (Table 2). Note the heart (V4 and V5) and the spine (V7 and V8) were not traversed by the beams although at some energies they received high doses. The regions V9 and V10 which included V1 also showed higher energy deposit.

Table 1. Total energy transferred in the 10 VOIs and for the 7 beam energies. Values for the tumor volume V1 are in MeV and the other VOIs have % values of those of V1 respectively for each beam energy.

VOIs	Beam energy (keV)						
	50	100	150	200	250	350	450
V1	6211	9947	17010	24810	32781	48656	63812
V2	0.64	0.27	0.18	0.15	0.12	0.1	0.08
V3	10.27	9.58	9.21	9.03	9.18	9.17	9.14
V4	1.14	1.03	0.47	0.28	0.16	0.1	0.07
V5	3.54	4.68	2.13	1.16	0.64	0.34	0.23
V6	0.95	0.51	0.35	0.3	0.26	0.22	0.18
V7	5.96	7.08	3.1	1.65	0.92	0.48	0.3
V8	0.1	0.09	0.05	0.04	0.03	0.02	0.02
V9	115.7	115.39	115.48	115.32	115.42	115.34	115.4
V10	128.85	129.33	129.43	129.27	129.39	129.3	129.5

Table 2. Absorbed dose in Gray in the 10 VOIs and for the 7 beam energies. The dose in V2 to V10 are % of their respective in V1 for a beam energy.

VOIs	Beam energy (keV)						
	50	100	150	200	250	350	450
V1	0.9123	1.4391	2.4430	3.5582	4.6796	6.8127	8.5674
V2	0	0	0	0	0	0	0
V3	0.11	0.11	0.1	0.1	0.1	0.1	0.11
V4	0.76	0.31	0.24	0.23	0.22	0.22	0.23
V5	0.06	0.04	0.03	0.03	0.03	0.03	0.03
V6	0.01	0.01	0	0	0	0	0.01
V7	6.6	2.89	0.97	0.45	0.27	0.15	0.12
V8	0.02	0.02	0.01	0.01	0	0	0
V9	1.4	1.39	1.4	1.39	1.39	1.37	1.35
V10	1.85	1.85	1.86	1.84	1.83	1.79	1.76

4   Discussion

MCS in small animal radiotherapy are accessible and handy tools to help for better designing treatment planning, not only to the target tumors and to the tissues along the beam paths, but also to distant organs of interest. It has been reported that doses of more than 5 Gy could be considered as lethal in mice [20, 21]. It is recognized that exposure to repeated radiations during follow-up studies can have biological effects on the animal models and thus can affect the experimental results. MCS can therefore help in designing the appropriate radiation energies, beam directions and intensity and duration of experiment. It is preferable to study each beam direction separately in order to determine its optimal parameters. Instead of using monoenergetic photons in the simulations, an energy spectrum resembling the one produced by an X-ray tube can be generated reproducing also its shape and intensity.

The radiation beams used in this simulation were based on X-rays tube voltage and current of 120 kVp and 50 mAs. The doses obtained in these simulations at lower energy (50 – 100 keV) were in the same range as those for micro-CT/radiotherapy or for radiotherapy previous studies at an effective energy around 50 keV (up to 2 Gy) [15, 16, 22, 23].

The variation of the dose in the heart and spine regions was a function of the beam energy, except for the 100 keV beam. Moreover, the volumes determined within the spinal cord received around 1% of the absorbed dose in the tumor, knowing that this volume was not on the path of any beam, and it was surrounded by bones. The spinal bones, however, and including the spinal cord, received higher doses with respect to spinal cord and heart VOIs. In fact, V7 received 6.6 times the dose of the tumor at 50 keV and 0.12 at 450 keV. This high ratio at 50 keV could be due to the high scattering in the mouse at this energy and high attenuation in the bones. The high doses calculated in V9 and V10 and in V1 at energies above 150 keV were due to the high number of generated photons. In real treatments, the number of photons is also governed by beam application duration among other parameters.

The two extra volumes around the lung tumor directly received the beam from some angles, but they were not directly exposed to the beams for some other beam angles. Also, the tissue density of these extra volumes was less than that of the tumor. Despite these differences, photons interactions in these volumes V9 and V10 varied very slightly with beam energies in comparison to V1, and their average ratios of absorbed dose (average over beam energies) were $V9/V1 = 1.3841 \pm 0.0194$ Gy and $V10/V1 = 1.8265 \pm 0.0372$ Gy. Apart from uncertainty on tumor volume definition, the X-rays focal spot on the anode could also cause a penumbra that can have an impact on tumor neighboring tissues [15, 16].

5   Conclusions

The goal of this study was to assess the photon energy transfer and absorbed dose in tissues distant from or partially intercepted by the radiation beams as a function of beam energy. The spinal cord, even protected by the spinal bones and not intercepted by the beams, received in average 1% of the dose to the tumor, which, in turn, received 0.9 Gy at 50 keV and 8.56 Gy at 450 keV. The lowest dose to the secondary organs was found at the 100 keV beam. The dose to tumor surrounding tissue was shown to be independent of beam energy and had ratios of 1.38 and 1.83 for volumes of diameters 1.6 mm and 1.8 mm with respect to the lung tumor of diameter 1.4 mm.

Acknowledgements. The authors wish to thank M. HuiZhong Lu for his kind assistance in software installation on the supercomputer Mammouth at Université de Sherbrooke.

Computations were made on the supercomputer Mammouth MS2, Université de Sherbrooke, managed by Calcul Québec and Compute Canada. The operation of this supercomputer is funded by the Canada Foundation for Innovation (CFI), Nano-Québec, RMGA and the Fonds de recherche du Québec - Nature et technologies (FRQ-NT).

Conflict of interest disclosure. the authors declare that they have no conflict of interest.

References

1. Leal, A., Sanchez-Doblado, F., Perucha, M., Carrasco, E., Rincon, M., Arrans, R., Bernal, C.: Monte Carlo Simulation of Complex Radiotherapy Treatments. Computing in Science and Engg. 6(4), 60–68 (2004)
2. Wang, H., Ma, Y., Pratx, G., Xing, L.: Toward Real-Time Monte Carlo Simulation Using a Commercial Cloud Computing Infrastructure. Physics in Medicine and Biology 56(17), N175–N181 (2011)
3. Dogdas, B., Stout, D., Chatziioannou, A.F., Leahy, R.M.: Digimouse: a 3D whole body mouse atlas from CT and cryosection data. Phys. Med. Biol. 52(3), 577–587 (2007)
4. Segars, W.P., Tsui, B.M.W.: MCAT to XCAT: The Evolution of 4-D Computerized Phantoms for Imaging Research. Proceedings of the IEEE 97(12), 1954–1968 (2009)

5. Mauxion, T., Barbet, J., Suhard, J., Pouget, J.-P., Poirot, M., Bardiès, M.: Improved realism of hybrid mouse models not be sufficient to generate reference dosimetric data. Medical Physics 40(5), 052501 (2013)

6. Jan, S., Santin, G., Strul, D., Staelens, S., Assié, K., Autret, D., Avner, S., Barbier, R., Bardiès, M., Bloomfield, P.M., Brasse, D., Breton, V., Bruyndonckx, P., Buvat, I., Chatziioannou, A.F., Choi, Y., Chung, Y.H., Comtat, C., Donnarieix, D., Ferrer, L., Glick, S.J., Groiselle, C.J., Guez, D., Honore, P., Kerhoas-Cavata, S., Kirov, A.S., Kohli, V., Koole, M., Krieguer, M., van der Laan, D.J., Lamare, F., Largeron, G., Lartizien, C., Lazaro, D., Maas, M.C., Maigne, L., Mayet, F., Melot, F., Merheb, C., Pennacchio, E., Perez, J., Pietrzyk, U., Rannou, F.R., Rey, M., Schaart, D.R., Schmidtlein, C.R., Simon, L., Song, T.Y., Vieira, J., Visvikis, D., Van de Walle, R., Wieërs, E., Morel, C.: GATE - Geant4 Application for Tomographic Emission: a simulation toolkit for PET and SPECT. Phys. Med. Biol. 49(19), 4543–4561 (2004)

7. Schneider, W., Bortfeld, T., Schlegel, W.: Correlation between CT numbers and tissue parameters needed for Monte Carlo simulations of clinical dose distributions. Phys. Med. Biol. 45(2), 459–478 (2000)

8. Sarrut, D., Bardies, M., Boussion, N., Freud, N., Jan, S., Letang, J.M., Loudos, G., Maigne, L., Marcatili, S., Mauxion, T., Papadimitroulas, P., Perrot, Y., Pietrzyk, U., Robert, C., Schaart, D.R., Visvikis, D., Buvat, I.: A review of the use and potential of the GATE Monte Carlo simulation code for radiation therapy and dosimetry applications. Med. Phys. 41(6), 064301 (2014)

9. Chow, J.C.L., Leung, M.K.K.: Treatment planning for a small animal using Monte Carlo simulation. Medical Physics 34(12), 4810–4817 (2007)

10. Mah, P., Reeves, T.E., McDavid, W.D.: Deriving Hounsfield units using grey levels in cone beam computed tomography. Dentomaxillofac Radiol. 39(6), 323–335 (2010)

11. Suryanto, A., Herlambang, K., Rachmatullah, P.: Comparison of tumor density by CT scan based on histologic type in lung cancer patients. Acta Med. Indones 37(4), 195–198 (2005)

12. Larsson, E., Ljungberg, M., Strand, S.-E., Jönsson, B.-A.: Monte Carlo calculations of absorbed doses in tumours using a modified MOBY mouse phantom for pre-clinical dosimetry studies. Acta Oncologica 50(6), 973–980 (2011)

13. Loening, A.M., Gambhir, S.S.: AMIDE: a free software tool for multimodality medical image analysis. Mol. Imaging 2(3), 131–137 (2003)

14. Kirkpatrick, J.P., van der Kogel, A.J., Schultheiss, T.E.: Radiation Dose–Volume Effects in the Spinal Cord. International Journal of Radiation Oncology*Biology*Physics 76(suppl. 3), S42–S49 (2010)

15. Rodriguez, M., Zhou, H., Keall, P., Graves, E.: Commissioning of a novel microCT/RT system for small animal conformal radiotherapy. Phys. Med. Biol. 54(12), 3727–3740 (2009)

16. Zhou, H., Rodriguez, M., van den Haak, F., Nelson, G., Jogani, R., Xu, J., Zhu, X., Xian, Y., Tran, P.T., Felsher, D.W., Keall, P.J., Graves, E.E.: Development of a MicroCT-Based Image-Guided Conformal Radiotherapy System for Small Animals. Int. J. Radiat. Oncol. Biol. Phys. 78(1), 297–305 (2010)

17. Chow, J.C.L.: Dosimetric impact of monoenergetic photon beams in the small-animal irradiation with inhomogeneities: A Monte Carlo evaluation. Radiation Physics and Chemistry 86(0), 31–36 (2013)

18. Brun, R., Rademakers, F.: ROOT — An object oriented data analysis framework. Nuclear Instruments and Methods in Physics Research Section A: Accelerators, Spectrometers, Detectors and Associated Equipment 389(1-2), 81–86 (1997)

19. Visvikis, D., Bardies, M., Chiavassa, S., Danford, C., Kirov, A., Lamare, F., Maigne, L., Staelens, S., Taschereau, R.: Use of the GATE Monte Carlo package for dosimetry applications. Nuclear Instruments and Methods in Physics Research Section A: Accelerators, Spectrometers, Detectors and Associated Equipment 569(2), 335–340 (2006)
20. Ford, N.L., Thornton, M.M., Holdsworth, D.W.: Fundamental image quality limits for microcomputed tomography in small animals. Med. Phys. 30(11), 2869–2877 (2003)
21. Bartling, S.H.: Small Animal Computed Tomography Imaging. Current Medical Imaging Reviews 3, 45–59 (2007)
22. Taschereau, R., Chow, P.L., Chatziioannou, A.F.: Monte Carlo simulations of dose from microCT imaging procedures in a realistic mouse phantom. Med. Phys. 33(1), 216–224 (2006)
23. Graves, E.E., Zhou, H., Chatterjee, R., Keall, P.J., Gambhir, S.S., Contag, C.H., Boyer, A.L.: Design and evaluation of a variable aperture collimator for conformal radiotherapy of small animals using a microCT scanner. Medical Physics 34(11), 4359–4367 (2007)

Towards a More Efficient Discovery of Biologically Significant DNA Motifs

Abdulrakeeb M. Al-Ssulami and Aqil M. Azmi[*]

Department of Computer Science, King Saud University,
Riyadh 11543, Saudi Arabia
aqil@ksu.edu.sa

Abstract. DNA motifs are short recurring patterns which are assumed to have some biological function. Most of the algorithms that solve this problem are computationally prohibitive. In this paper we extend a recent work that discovered identical string motifs. In the first phase of our three phase algorithm we report all the string motifs of all sizes. In the next phase we filter out those motifs which fail to meet our constraints, and in the last phase the motifs are ranked using a combination of stochastic techniques and p-value. Our method outperforms other motif discovery algorithms including some well-known ones such as MEME and Weeder on benchmark data suites.

Keywords: Significant DNA motifs, sequence analysis, algorithm.

1 Introduction

Nucleotide motifs are short recurring patterns in DNA, and are significant for understanding the mechanism behind regulating gene expressions. DNA motifs are sometimes termed signals, e.g. regulatory sequences. Motifs are short biological sequences (about 30 nucleotides) whereas the regulatory sites within which they reside have a much longer size. The automatic motif discovery problem is defined as a multiple sequence local alignment. There are two different approaches for the motif finding problem: pattern-driven and sequence-driven [9]. In the former, the objective is to search for pattern of length L which has the highest occurring frequency in the sequence, indicating the motif as being the most significant pattern of that length. On the other hand, the sequence-driven is used to identify profile models with no pre-assumption on the statistical distribution of the patterns in the sequence. The pattern driven approach is guaranteed to find optimal solutions in the restricted space, but it becomes computationally expensive for $L > 10$ [15].

Mutations is a common phenomenon and a number of algorithms have been proposed to find motifs with mutations: PROJECTION [5], Weeder [14], Motif Enumerator [17], MEME [2,10], VAS [6], Seeder [8], MOGAMOD [12] and [20].

---

[*] Corresponding author.

F. Ortuño and I. Rojas (Eds.): IWBBIO 2015, Part I, LNCS 9043, pp. 368–378, 2015.

However, mutations complicates the situation further and so these algorithms suffer from an exponential time complexity deeming them impractical on any large sized sequence. In [13] they generated approximately 17.88 billion patterns of length 10 over the IUPAC alphabet. As the search space was huge, the authors had to develop a set of techniques to carefully prune it. The remaining patterns were scored and those below a certain threshold were thrown out, and we are left with motifs. The results were encouraging but the overall exponential time complexity makes it impractical for motifs of length over 10. Karci [11] proposed a quadratic time algorithm to find all the identical *string* motifs that occur at least twice in the input sequence. Azmi and Al-Ssulami [1] were able to improve the work in [11] reaching a practical complexity that is linear in time and space.

Yet the problem is far from over. Tompa et al [18] provided a benchmark data set to be used for assessing motif discovery algorithms. In [18] they proposed a scheme to measure the accuracy of the reported motifs for the benchmark dataset. Using the above setting they evaluated 13 different motif discovery algorithms and reported that none scored satisfactorily. Sandve et al [16] proposed an improved benchmark dataset that was designed to differentiate between the performance of motif discovery algorithms and the popular motif models, e.g. position weight matrices. Their supplementary web service allows users to visualize their score of submitted predictions of the benchmark datasets.

In this paper we extend the work in [1], so instead of reporting string motifs it will report biologically significant motifs. In [1] the authors defined a threshold and their algorithm reported all the string motifs of all sizes occurring in a single input sequence S provided that the motifs satisfied the threshold. Our adaption includes modifying the original algorithm so it works on multiple input sequences. Our objective is to discover identical motifs that are biologically significant. As we believe that significant biological exact motifs are a subset of the string motifs, we devise a kind of a filtering mechanism. The filtering will assess the significance of the string motifs, throwing those that fail to meet our criteria. This is followed by the ranking of the remaining motifs. At the end we pick the first few motifs with the highest ranking. And these will be our biologically significant motifs. The time complexity of our algorithm is $O(nL - NL^2 + z \log z)$, where N is the number of input sequences, n is the total length of N input sequences, L is the length of the largest discovered motif, and z is the size of the output. Though our algorithm only yields identical motifs, it scored higher on the benchmark datasets [16] compared to MEME [2] and Weeder [14].

2 Preliminaries

Let Σ denote the set of finite symbols. In our case the alphabet $\Sigma = \{A, C, G, T\}$, the four bases for the DNA sequence. For any integer $k > 0$, we define $\Sigma^k = \Sigma^{k-1} \circ \Sigma$, where Σ^0 is the empty string. The $\circ$ is the concatenation operation. The set of all strings formed using the symbols in Σ is denoted $\Sigma^\star$. Let $S_1, S_2, \ldots, S_N$ be the set of N strings each of different size. We define the composite string $S = S_1 S_2 \cdots S_N$. Let us denote the nucleotide at position i

$(0 \leq i < |S|)$ using $S[i]$, also let $S[i..j]$ denote the subsequence (substring) starting with the symbol at position i and ending with the symbol at position j, for any $0 \leq i \leq j < |S|$. A substring of length k is called k–mer. In this paper we will be using the terms (sub)string(s) and (sub)sequence(s) interchangeably.

Problem definition: given the set of strings $S_1, S_2, \ldots, S_N$ over the alphabet {A, C, G, T} with a total combined length equals n. Find all the biologically significant identical motifs of lengths up to L ($2 \leq L < \min_{i \leq N}\{|S_i|\}$), where there is no motif of length $L+1$ in any of the strings.

3   Our Proposed Algorithm

Our three phase algorithm first finds all the strings motifs, this is followed by filtering before it undergoes a ranking process. The algorithm in [1] was designed to report all the string motifs in a single sequence, we however, will combine the first two phases and modify the algorithm so that it will do the filtering prior to any output.

3.1   Phases I–II

We introduce two parameters: $max\_hits\_per\_sequence$ and $min\_sequences$ for the filtering. Respectively, these specify the maximum number of times a motif can occur in a single sequence, and the least number of different sequences a motif must occur in. For example, suppose we have four sequences ($N = 4$), and we want to report only the motifs that occur no more than twice in a sequence in three or more sequences then we just set the parameters $max\_hits\_per\_sequence = 2$ and $min\_sequences = 3$. Of course, in each of the three sequences it must occur at least once but no more than twice. The data structure used by the algorithm is shown in Figure 1. In the subsequent discussion the string S is the composite string $S = S_1 S_2 \cdots S_N$.

The main algorithm is listed in Algorithm 1. We use the term CanMotif (candidate motif), a term we borrowed from [11]. Any substring is a candidate motif, and if there is at least one more copy of the same substring then it is a "string" motif. We will go over the algorithm using the example input of three DNA sequences $S_1 = $ ATAGACAG, $S_2 = $ TGTATATACGCT and $S_3 = $ GACATTGCAG. Assume we execute the algorithm with the setting $max\_hits\_per\_sequence = 1$ and $min\_sequences = 2$. This basically mean to report all identical string motifs that occur exactly once in two or more of the input strings. Our composite string $Seq = $ ATAGACAGTGTATATACGCTGACATTGCAG. The algorithm starts by initializing the array T to the encoding of each character in the string Seq, its starting position and the sequence number. Without loss of generality we use the following encoding {A$\rightarrow$ 0, C$\rightarrow$ 1, G$\rightarrow$ 2, T$\rightarrow$ 3}. Any other encoding is also acceptable as long as each character in the alphabet Σ is assigned a unique consecutive number starting from 0. Next we sort the auxiliary array T on the .$NucVal$ component saving the result in array U(step 4). We use the counting sort algorithm [7], a linear sorting algorithm which works only if the keys were known beforehand.

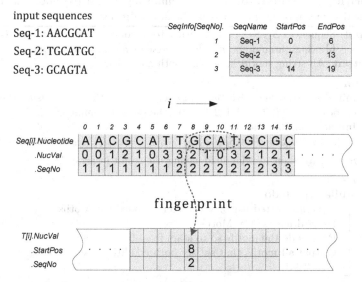

Fig. 1. An example illustrating the basic data structure as used by the algorithm. The array *Seq[ ].Nucleotide* is the composite string holding the three input sequences Seq-1, Seq-2, and Seq-3. Initially *.NucVal* holds the encoding of single nucleotides {A→ 0, C→ 1, G→ 2, T→ 3}, later it will hold the encoding of equi-length substrings. In this example *T[i].NucVal* holds the fingerprint of the substring GCAT, while *T[i].StartPos* and *T[i].SeqNo* holds this substring's starting position and the sequence number it belongs to (respectively). Note that all the indices start from zero.

The basic idea behind the algorithm is to start with a single nucleotide and successively discover larger motifs by augmenting each motif with a single nucleotide to its immediate right. We throw out some of the entries which violate the condition set forth. The entire process is repeated as long as we are successful in discovering motif(s) at the current iteration. The main while-loop (Algorithm 1, steps 6–11) is where we augment the substrings one character to their right in each iteration. At iteration k we generate a sorted list of $k + 1$-mer encoded CanMotifs. The sorting will let all the identical substrings to be next to each other. This simplifies the task of eliminating some of the useless CanMotifs. See Figure 2 for the entries marked for elimination. We eliminate two kind of entries. Those substrings that span multiple sequences; and those violating the condition *min_sequences*. We however, do not delete entries if they violate the condition *max_hits_per_sequence*. The reason should be obvious. Once an entry is deleted the information is lost. However, it is possible an entry that currently violate the condition *max_hits_per_sequence* will satisfy it in a subsequent iteration, i.e. when k is larger. In the next step we re-encode the motifs. More on that later. The final step is to output all the string motifs that satisfy both conditions. The whole process is repeated, going over for longer motifs.

Algorithm 1. STRINGMOTIF – the main algorithm to report all the identical string motifs that satisfies both conditions

Input: Sequences $S_1, S_2, \ldots, S_N$ with a total combined length n, and two filtering parameters *max_hits_per_sequence* and *min_sequences*

Output: All string motifs satisfying both conditions

```
1  begin
2      Initialize array Seq with the input sequence S₁, S₂, ... Sₙ, using the
        encoding (A, C, G, T → 0, 1, 2, 3), and the sequence number
3      Initialize auxiliary array
        T[i].{NucVal, StartPos, SeqNo} ← Seq[i].{NucVal, i, SeqNo} for all i < n
4      Sort array T on .NucVal field saving the result in array U
5      k ← 1
6      while |U| ≠ 0 do
7          Generate sorted list of k + 1-mer encoded CanMotifs
8          Discard useless CanMotifs
9          Re-encode the motifs so to start from 0
10         Output all motifs that satisfy both conditions
11         k ← k + 1
```

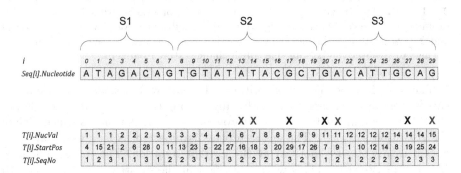

Fig. 2. The input sequence $Seq[\ ].Nucleotide$ made up of three sequences S_1, S_2 and S_3. Below it is the auxiliary array T holding the fingerprints for all substrings of length 2 (case $k = 1$). Note that the fingerprints are sorted in increasing order. For example, $T[0].NucVal = 1$ is the fingerprint for the substring of length 2 that starts at position 4, that is: AC. We have two more substrings AC, one in sequence S_2 starting at position 15, and another in sequence S_3 starting at position 21. The useless entries which we delete are marked X. We delete those that span multiple sequences. For example, the entry with $T[i].StartPos = 7$ is deleted because it spans S_1 and S_2. We also delete the entry $T[i].NucVal = 6$ since it violates the condition *min_sequences* $= 2$. We do not check for entries violating the condition *max_hits_per_sequence*.

Fingerprinting. It is one of the major operations in [1]. The idea is to simplify the task of comparing between two equal length substrings akin to comparing between two integers, as we are only interested in knowing whether or not they are equal. Given two equi-length substrings A and B, we want,

$$A = B \Leftrightarrow \mathcal{F}(A) = \mathcal{F}(B), \tag{1}$$

where $\mathcal{F}()$ is the fingerprint function. It is computed as follows. Suppose $\alpha \in \Sigma$, and let $encoding(\alpha) \in \{0, 1, \ldots, |\Sigma|-1\}$. Let $S[i .. i + k]$ be a substring of length $k + 1$. The fingerprint is recursively defined by,

$$\mathcal{F}(S[i .. i + k]) = \begin{cases} encoding(S[i]), & \text{if } k = 0, \\ |\Sigma| \cdot \mathcal{F}(S[i .. i + k - 1]) + \\ \quad encoding(S[i + k]) & \text{otherwise.} \end{cases} \tag{2}$$

Re-encoding the Motifs. Numerically, the fingerprint tends to grow exponentially. As we do not wish to impose any limit on the size of the motifs we need to hinder the growth of the fingerprint. We resort to the same trick as used in [1], that is to re-encode the fingerprint at each iteration so it always start from zero (Algorithm 1, step 9). Different fingerprint will be mapped into a different number.

Generating Sorted List of Encoded Motifs. This pertains to step 7 in Algorithm 1 where we used k–mer motifs to generate the list of $k + 1$–mer CanMotifs. The input is a sorted list of encoded k–mer substrings and the output is also a sorted list of encoded $k + 1$–mer substrings. From Eq. 2 it should be obvious we are augmenting the string of length k by a single nucleotide to its immediate right thereby forming an encoded $k + 1$–mer substrings. Figure 3 illustrates how we generate a sorted list of encoded $k + 1$–mer substrings in a single pass. If we treat the encoded k–mer substrings that have the same value as a group, then the encoded $k + 1$–mer substrings will be in the same group. To put them in order we can resort to counting sort, a linear sorting scheme. The cost of step 7 in Algorithm 1 is linear in the length of the input auxiliary array. In the first pass we look for the boundary of the groups, i.e. those entries with the same fingerprint. And in the second pass we go over each group and compute the fingerprint of the larger substring (Eq. 2) and then use counting sort to sort the content of each group.

List of Outputted Motifs. At iteration k, Algorithm 1 step 10 outputs the k–mer string motifs which satisfy both constraints *max_hits_per_sequence* and *min_sequences*. In Figure 3(top picture) we note that only four 2–mer string motifs will be outputted. These are motifs whose fingerprint (*.NucVal* = 0, 4, 5 and 7, corresponding to motifs AC, GA, GC and TG respectively. Motif with fingerprint 1 (corresponds to AG) violates *max_hits_per_sequence* as it occurs twice in string

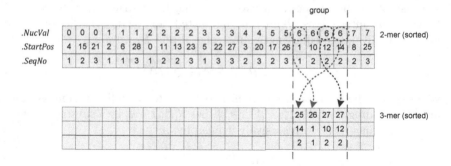

Fig. 3. How we generate a sorted list of encoded 3–mer CanMotifs out of sorted list of encoded 2–mer motifs. The input is the auxiliary array in Figure 2 after removing the entries marked X and $T[]$.$NucVal$ has been re-encoded, e.g. $1 \to 0, 2 \to 1, \ldots, 12 \to 6$, and $14 \to 7$. Note that 2–mer motifs with an a fingerprint value of 6 generates 3–mer CanMotifs with a fingerprint values of 24, 25, 26 and 27 only, which corresponds to the augmented nucleotide to its immediate right being A, C, G or T respectively.

S_1, and so it is not in the output list. The same is true for motifs with fingerprint 2, 3 and 6.

Complexity of First Two Phases. Assuming we start with N sequences whose total size is n, then the cost for steps 1–3 in Algorithm 1 is $O(n)$. Each of the steps in while-loop (steps 7–10) can be done in time linear to the length of the auxiliary array. Initially the size of the auxiliary array is n but as we discard useless CanMotifs (step 8) the size of the auxiliary array shrinks. It is difficult to estimate the size of the shrinkage but it does by at least N. This is because we can have N k–mer substrings spanning two adjacent sequences. Thus the size of the auxiliary array at iteration k is at most $n - (k-1)N$. The time complexity of phases I–II is $O(n) + \sum_{k=1}^{L} O(n - (k-1)N) = O(nL - NL^2)$, where L is the length of the longest discovered string motif.

3.2   Phase III

In this phase we rank the resultant string motifs out of phase II to determine their significance (see Algorithm 2). We believe that the top ranked string motifs are likely the biologically significant motifs we are after. Each input motif is assigned an initial score which simply is its occurring frequency. Next we pick the top twenty initial scored motifs, throwing out the rest. Compute the p-value of all picked motifs. It was computed using the standalone version of the algorithm in [3].[1] Finally we return the motifs with the best p-values. The reason for picking

---

[1] Download from, http://favorov.bioinfolab.net/ahokocc/zip/ahokocc.zip

only 20 highest scoring motifs as opposed to all the input is to save time since computing p-value is computationally expensive. The time complexity of this phase is $O(z \log z)$, where z is the number of motifs we receive out of phase II.

Algorithm 2. RANKSTRINGMOTIFS – the last phase of the algorithm to discover significant motifs. Here we rank the string motifs out of phase II

Input: All the string motifs out of phase II, and rank cutoff ρ
Output: The top ρ motifs

1 **begin**
2 | Assign an initial score to each input motif
3 | Let $Q = \{$the top 20 motifs based on initial score$\}$
4 | Assign a secondary score to each motif in Q
5 | Return the top ρ motifs based on secondary score

---

4 Evaluation and Discussion

Implementation wise, phases I–II (Algorithm 1) was implemented in MS C#, while phase III (Algorithm 2) was implemented in Activestate Perl.

Sandve et al [16] proposed a benchmark suite which we will use to evaluated our algorithm. The authors generated different datasets, which were generated either by extracting transcription factor binding site original neighborhood from their respective genomes or by implanting their occurrences into a background generated from a third order Markov model. The benchmark makes the performance analysis of the algorithms more useful as we can factor out the effect of motif model from the algorithm. We will use their 'algorithm' suite to evaluate our system. This suite in turn is divided into two parts: 'algorithm real' and 'algorithm Markov'. The latter contains true binding sites from the TRANS-FAC database [19] embedded into mock backgrounds generated by third order Markov. While in the real suite the binding sites are not tinkered with. Altogether there are 50 datasets in each part.

While assessing different tools for the discovery of transcription factor binding sites, Tompa et al [18] devised different statistical measures to assess these tools at the nucleotide level. Let nTP, nTN, nFP and nFN are the number of True/False Positive/Negative predicted nucleotides. We will use the following four measures [18]: Sensitivity nSn, Positive Predictive Value nPPV, Specificity nSP, and the Performance Coefficient nPC. The sensitivity gives the fraction of known site nucleotides that are predicted, and the positive predictive value gives the fraction of predicted site nucleotides that are known. The performance coefficient, initially devised by [15], is a single statistic that in some sense averages the above mentioned quantities. However, as [18] noted, there is no single

Table 1. Various statistics on the benchmark algorithm real and Markov suites. Each suite consists of 50 datasets. For our predictions we used motifs of length 8 (real suite) and 10 (Markov suite). We boldfaced the best result in each suite. See text for the definition of individual statistical measures.

Method	real suite				Markov suite			
	nSn	nPPV	nSP	nPC	nSn	nPPV	nSP	nPC
MEME	.1034	.0923	.9816	.0626	.1151	.1072	.9844	**.0775**
Weeder	**.2017**	.0714	.9605	.0552	**.1333**	.0427	.9590	.0324
Our	.1074	**.1497**	**.9924**	**.0770**	.0789	**.1255**	**.9939**	.0583

statistic which captures correctness perfectly. The four measures are given by,

$$nSn = nTP/(nTP + nFN),$$
$$nPPV = nTP/(nTP + nFP),$$
$$nSP = nTN/(nTN + nFP),$$
$$nPC = nTP/(nTP + nFN + nFP).$$

(3)

The authors in [16] set an online service[2] where users can upload their predictions and the system calculates various statistics along with that of Weeder [14] and MEME [2]. Both Weeder and MEME are well-known methods with proven performance in an assessment by [18]. So, it was reasonable to include them in the statistics. The study in [18] showed that Weeder outperformed 12 other tools in most of the measures. For our predictions we submitted two sets of predictions, one for algorithm real suites and another for algorithm Markov suites (Table 1). To prepare our predictions we used the setting: $max\_hits\_per\_sequence = 1$, $min\_sequences = 2$ (in Algorithm 1); and rank cutoff $\rho = 3$ (in Algorithm 2). Going over the Table 1 we see that none of the various statistical values indicate a good performance, however, judging from nPC our scheme had a lead over the other two schemes for algorithm real suite and ranked second in algorithm Markov suite.

To give a perspective of the execution speed, we timed Weeder 2.0 (http://159.149.160.51/modtools/), MEME v4.9.1 (http://meme.nbcr.net/meme/meme-download.html) and our software all running on the same environment and under the same condition. These are the latest version of both softwares. The largest dataset in [16] is M01011. The time to find all the motifs of size 10 was: 19.15 sec (Weeder), 36.07 sec (MEME) and 0.79 sec (our algorithm). While for the smallest dataset, M00622 the time to find all motifs of size 10 was: 4.17 sec (Weeder), 1.34 sec (MEME) and 0.23 sec (our). Overall, it took our three phase algorithm less than two minutes to prepare both predictions.

---

[2] At, http://tare.medisin.ntnu.no/index.php

5   Conclusion

This work is an attempt to get on biologically significant DNA motifs which happen to be conserved. The main philosophy behind our approach is that the exact match DNA motifs are a subset of string motifs. We solve this problem using a three phase algorithm where the input is a set of different sequences of different sizes. The first phase generates string motifs of different sizes. The second phase filters the string motifs allowing only for motifs that satisfy two different constraints. And in the third phase we rank the motifs using a combination of stochastic and p-value. Once the motifs are ranked we pick the top ones. The two constraints we used in phase two defines the upper bound on the number of motifs per sequence, and the lower bound on in how many sequences a motif must be in. We tested our algorithm on a benchmark dataset suites. In one benchmark suite, our algorithm outperformed both MEME and Weeder and in the other benchmark suite it was better than Weeder but lagged MEME.

Acknowledgments. This work is supported by the Research Center of the College of Computer & Information Sciences (CCIS) at King Saud University grant number RC-131010.

References

1. Azmi, A.M., Al-Ssulami, A.: A linear algorithm to discover exact string motifs. PLoS ONE 9(5), e95148 (2014)
2. Bailey, T.L., Elkan, C.: Unsupervised learning of multiple motifs in biopolymers using expectation maximization. Mach. Learning 21, 51–80 (1995)
3. Boeva, V., Clement, J., Regnier, M., Roytberg, M.A., Makeev, V.J.: Exact p-value calculation for heterotypic clusters of regulatory motifs and its application in computational annotation of cis-regulatory modules. Algo. Mol. Biol. 2, 13 (2007)
4. Burset, M., Gulg, R.: Evaluation of gene structure prediction programs. Genomics 34, 353–367 (1996)
5. Buhler, J., Tompa, M.: Finding motifs using random projections. In: Proc. 5th Annual Int. Conf. on Comput. Biol. (RECOMB 2001), Montreal, Canada, pp. 69–76 (2001)
6. Chin, F., Leung, H.: An efficient algorithm for string motif discovery. In: Proc. 4th Asia-Pacific Bioinfor. Conf (APBC 2006), Taipei, Taiwan, pp. 79–88 (2006)
7. Cormen, T.H., Leiserson, C.E., Rivest, R.L., Stein, C.: Introduction to Algorithms, 2nd edn. MIT Press (2001)
8. Fauteux, F., Blanchette, M., Strmvik, M.V.: Seeder: discriminative seeding DNA motif discovery. Bioinfor. 24, 2303–2307 (2008)
9. GuhaThakurta, D.: Computational identification of transcriptional regulatory elements in DNA sequence. Nucleic Acids Res. 34, 3585–3598 (2006)
10. Hu, J., Li, B., Kihara, D.: Limitations and potentials of current motif discovery algorithms. Nucleic Acids Res. 33, 4899–4913 (2006)
11. Karci, A.: Efficient automatic exact motif discovery algorithms for biological sequences. Expert Sys. With App. 36, 7952–7963 (2009)

12. Kaya, M.: MOGAMOD. Multi-objective genetic algorithm for motif discovery. Expert Sys. With App. 36, 1039–1047 (2009)
13. Marschall, T., Rahmann, S.: Efficient exact motif discovery. Bioinfor. 29, i356–i364 (2009)
14. Pavesi, G., Mereghetti, P., Mauri, G., Pesole, G.: Weeder Web: discovery of transcription factor binding sites in a set of sequences from co-regulated genes. Nucleic Acids Res. 32, W199–W203 (2004)
15. Pevzner, P.A., Sze, S.H.: Combinatorial approaches to finding subtle signals in DNA sequences. In: Proc. Int. Conf. Intel. Sys. Mol. Biol., vol. 8, pp. 269–278 (2000)
16. Sandve, G.K., Abul, O., Walseng, V., Drabls, F.: Improved benchmarks for computational motif discovery. BMC Bioinfor. 8, 163 (2007)
17. Sze, S.H., Zhao, X.: Improved Pattern-driven Algorithms for Motif Finding in DNA Sequences. In: Eskin, E., Ideker, T., Raphael, B., Workman, C. (eds.) RECOMB 2005. LNCS (LNBI), vol. 4023, pp. 198–211. Springer, Heidelberg (2007)
18. Tompa, M., Li, N., Bailey, T.L., Church, G.M., Moor, B.D., Eskin, E., Favorov, A.V., Frith, M.C., Fu, Y., Kent, W.J., Makeev, V.J., Mironov, A.A., Noble, W.S., Pavesi, G., Pesole, G., Regnier, M., Simonis, N., Sinha, S., Thijs, G., van Helden, J., Vandenbogaert, M., Weng, Z., Workman, C., Ye, C., Zhu, Z.: Assessing computational tools for the discovery of transcription factor binding sites. Nat. Biotech. 23, 137–144 (2005)
19. Wingender, E., Dietze, P., Karas, H., Knuppel, R.: TRANSFAC: A database on transcription factors and their DNA binding sites. Nucleic Acids Res. 24, 238–241 (1996)
20. Yu, Q., Huo, H., Vitter, J.S., Huan, J., Nekrich, Y.: StemFinder: An efficient algorithm for searching large motif stems over large alphabets. In: Proc. IEEE Int. Conf. Bioinfor. and Biomed. (BIBM), Shanghai, China, pp. 473–476 (2013)

Improved Core Genes Prediction
for Constructing Well-Supported Phylogenetic
Trees in Large Sets of Plant Species

Bassam AlKindy[1,2], Huda Al-Nayyef[1,2], Christophe Guyeux[1],
Jean-François Couchot[1], Michel Salomon[1], and Jacques M. Bahi[1]

[1] FEMTO-ST Institute, UMR 6174 CNRS, Department of Computer Science
for Complex Systems (DISC), University of Franche-Comté, France
[2] Department of Computer Science, University of Mustansiriyah, Baghdad, Iraq
{bassam.al-kindy,huda.al-nayyef,christophe.guyeux,jean-francois.couchot,
michel.salomon,jacques.bahi}@univ-fcomte.fr

Abstract. The way to infer well-supported phylogenetic trees that precisely reflect the evolutionary process is a challenging task that completely depends on the way the related core genes have been found.
In previous computational biology studies, many similarity based algorithms, mainly dependent on calculating sequence alignment matrices,
have been proposed to find them. In these kinds of approaches, a significantly high similarity score between two coding sequences extracted
from a given annotation tool means that one has the same genes. In
a previous work article, we presented a quality test approach (QTA)
that improves the core genes quality by combining two annotation tools
(namely NCBI, a partially human-curated database, and DOGMA, an
efficient annotation algorithm for chloroplasts). This method takes the
advantages from both sequence similarity and gene features to guarantee that the core genome contains correct and well-clustered coding
sequences (*i.e.*, genes). We then show in this article how useful are such
well-defined core genes for biomolecular phylogenetic reconstructions, by
investigating various subsets of core genes at various family or genus levels, leading to subtrees with strong bootstraps that are finally merged in
a well-supported supertree.

Keywords: Quality test, Phylogenetic tree, Bootstrap, RAxML, Core
genome, Core genes, Supertree.

1 Introduction

Given a collection of genomes, it is possible to define their core genes as the
common genes that are shared among all the species, while pan genome is all
the genes that are present in at least one genome (*all* the species have each core
gene, while a pan gene is in *at least one* genome).

The key idea behind identifying core and pan genes is to understand the
evolutionary process among a given set of species: the common part (that is, the

F. Ortuño and I. Rojas (Eds.): IWBBIO 2015, Part I, LNCS 9043, pp. 379–390, 2015.
© Springer International Publishing Switzerland 2015

core genome) is of importance when inferring the phylogenetic relationship, while accessory genes of pan genome explain in some extend each species specificity. We introduced in a previous study [1] two methods for discovering core and pan genes of chloroplastic genomes using both sequence similarity and alignment based approaches. Later, we presented in another study [2] the quality test approach as a method to find the core genes for chloroplast species. This article is an extended version of [1, 2] focusing on how the quality core genes will affect the phylogenetic inference, and also a performance analysis in terms of execution time and memory consumption.

Chloroplasts is one of many types of organelles in the plant cell. They are considered to have originated from cyanobacteria through endosymbiosis, when an eukaryotic cell engulfed a photosynthesizing cyanobacterium, which remained and became a permanent resident in the cell. The term of chloroplast comes from the combination of plastid and chloro, meaning that it is an organelle found in plant cell that contains the chlorophyll. Chloroplast has the ability to take water, light energy, and carbon dioxide (CO_2) to convert it in chemical energy by using carbon-fixation cycle [3] (also called *Calven Cycle*, the whole process being called photosynthesis). This key role can explain why chloroplasts are at the basis of most trophic chains and thus responsible for evolution and speciation. Moreover, as photosynthetic organisms release atmospheric oxygen when converting light energy into chemical energy and simultaneously produce organic molecules from carbon dioxide, they originated the breathable air and represent a mid to long term carbon storage medium. Consequently, exploring the evolutionary history of chloroplasts is of great interest and therefore further phylogenetic studies are needed.

A key idea in phylogenetic classification is that a given DNA mutation shared by at least two taxa has a larger probability to be inherited from a common ancestor than to have occurred independently. Thus shared changes in genomes allow to build relationships between species. Homologous genes are genes derived from a single ancestral one. These genes are divided in two types, namely paralogous and orthologous. Paralogous genes arise from ancestral gene duplication while the orthologous genes are products of speciation. In the case of chloroplasts, an important category of genomes changes is the loss of functional genes, either because they become ineffective or due to a transfer to the nucleus. Thereby a small number of genes lost among species may indicate that these species are close to each other and belong to a similar lineage, while a large lost means distant lineages. Phylogenies of photosynthetic plants are important to assess the origin of chloroplasts and the modes of gene loss among lineages. These phylogenies are usually done using a few chloroplastic genes, some of them being not conserved in all the taxa. This is why selecting core genes may be of interest for a new investigation of photosynthetic plants phylogeny.

To determine the core of chloroplast genomes for a given set of photosynthetic organisms, bioinformatics investigations using sequence annotation and comparison tools are required, and therefore various choices are possible. The purpose of our research work is precisely to study the impact of these choices on the obtained results. A state of the art for core genome discovery studies is detailed in Section 2, whereas a general presentation of the approaches we propose is provided in Section 3. To make this paper standalone, a closer examination of the approaches is given in Section 4, where we will present coding sequences clustering method based on sequence similarity, and quality test method based on quality genes. Information regarding computation time and memory usage is provided in Section 5, while an application example in the field of phylogeny is illustrated in Section 6. This research work ends with a conclusion section summarizing our investigations and giving suggestions for future work.

2   State of the Art

An early study of finding the common genes in chloroplasts was realized in 1998 by *Stoebe et al.* [4]. They established the distribution of 190 identified genes and 66 hypothetical protein-coding genes (*ysf*) in all nine photosynthetic algal plastid genomes available (excluding non photosynthetic *Astasia tonga*) from the last update of plastid genes nomenclature and distribution. The distribution reveals a set of approximately 50 core protein-coding genes retained in all taxa. *Grzebyk et al.* [5], for their part, have studied in 2003 the core genes among 24 chloroplastic sequences extracted from public databases, 10 of them being algae plastid genomes. They broadly clustered the 50 genes from *Stoebe et al.* into three major functional domains: (1) genes encoded for ATP synthesis (*atp* genes); (2) genes encoded for photosynthetic processes (*psa* and *psb* genes); and (3) housekeeping genes that include the plastid ribosomal proteins (*rpl* and *rps* genes). The study shows that all plastid genomes were rich in housekeeping genes with one *rbcLg* gene involved in photosynthesis.

In 2014, *De Chiara et al.* [6] aligned all of the 97 sequenced genomes to a reference, the complete genome of the *Haemophilus influenza* strain 86-028NP, using the Nucmer alignment program [7]. They generated a list of polymorphic sites with these alignments. This list was then filtered to include only the polymorphic sites in the core genome of NTHi, *i.e.*, the regions of the reference strain that could be aligned against all other strains, yielding a set of 149,214 SNPs. A clustering algorithm has been finally used on these SNPs to achieve core genes extraction. Remark that most of these studies used only a low amount of plant genomes to extract the core genome.

3   An Overview of the Pipeline

In previous work [1], an annotation based method has been presented in a pipeline for core genomes discovery. It is based on an Intersection Core Matrix (ICM) using gene features like gene names. The produced core tree has then been compared to the phylogenetic one. However, working with gene features alone does not lead to accurate core genomes. This is because of two reasons: first, gene name does not necessary point to the same sequence among different genomes. Second, gene features in the absence of gene sequences cannot provide information such as starting and ending codons, mutation rate, proteins, and so on. Such limitations in core genomes confidence is the main reason explaining why we are investigating a new direction. This new proposal consists of a pre-processing step using a Needleman-Wunch global alignment, before taking into account gene features, see Figure 1.

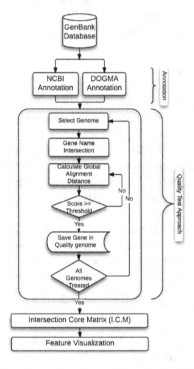

Fig. 1. An overview of the pipeline [2]

As a starting point, annotations (DNA coding sequences with gene names and locations) must be provided on the set of chloroplastic genomes under considerations. Obviously, this annotation stage must be of quality if we want to obtain acceptable core and pan genomes. Such a stage necessitates a DNA sequences database like NCBI's GenBank , the European *EMBL* database [8], or the Japanese *DDBJ* one [9]. The annotations can be directly downloaded from these websites, however it is preferable to launch an *ad hoc* annotation tool on complete downloaded genomes, like the DOGMA one [10].

Using such annotated genomes, we will employ two general approaches detailed in previous studies for extracting the core genome, which represent the second stage of the pipeline: the first approach involves similarity values that are computed on predicted coding sequences, while the second one takes benefits from all the information provided during the annotation stage.

Instead of considering only gene sequences taken from NCBI or DOGMA, we consider to use an improved quality test process provided in [2] in this new proposal. It works with gene names and sequences, to produce what we call "quality genes". Remark that such a simple general idea is not so easy to realize, and that it is not sufficient to only consider gene names returned by such tools. Providing good annotations is an important stage for extracting gene features. Indeed, gene features here could be considered as: gene names, gene sequences, protein sequences, and so on. We will subsequently propose methods that use gene names and sequences for extracting core genes and producing chloroplast evolutionary tree.

An extra step named *feature visualization* [1], will be added as a final stage of our pipeline. The construction of core tree and/or phylogenetic tree is done by taking the advantage of information produced during the core and pan genomes search. This feature visualization stage will then be used to encompass phylogenetic tree construction using core genes, genes content evolution illustrated by core trees, functionality investigations, and so on.

For illustration purposes, we have considered 99 genomes of chloroplasts downloaded from GenBank database. These genomes cover eleven types of chloroplast families (see [1]). Furthermore, two kinds of annotations will be considered in this work, namely the ones provided by NCBI on the one hand, and the ones by DOGMA on the other hand.

4   Core Genes Extraction

In this section, we consider the gene prediction approach based on sequence similarity presented in [1, 2]. This method starts with genomes annotated, either from NCBI or DOGMA, and uses a distance on genes coding sequences $d : N = \{A, T, C, G\}^* \times \{A, T, C, G\}^* \to [0, 1]$, where A^* is the set of words on alphabet N, to group similar alleles in a same cluster.

Let us now present the proposed quality test improvement. The inputs are genomes annotated twice, by NCBI and DOGMA respectively. To extract the common genes, a post-treatment of these annotations must first be achieved. On the NCBI side, due to the large variety of annotation origins (being produced either by human or by various automatic tools), we have to compute an edit similarity distance on gene names. The same name is then set to sequences whose names are close according to this edit distance. This stage is not required in the DOGMA side, as names are provided by an unique algorithm. However, DOGMA investigates the six reading frames when extracting coding sequences [10], and it sometimes produces various fragments for one given gene. So a gene whose name is present at least twice in the file is either a duplicated gene or a fragmented one. Obviously, these issues must be fixed and "fragmented" genes have to be defragmented before the DNA similarity computation stage (such defragmentation has normally already been realized on NCBI website). As the orientation of each gene fragment is given in output file, this defragmentation consists in concatenating all the possible permutations, and only keeping the permutation with the best similarity score in comparison with other sequences having the same gene name: this score has to be larger than a given predefined threshold.

The risk is now to merge genes that are different but whose names are similar (for instance, ND4 and ND4L are two different mitochondrial genes, but with similar names). To fix such a flaw, the sequence similarity, for intersected genes in a genome, is compared too in a second stage (with a Needleman-Wunsch global alignment) after selecting a genome accession number, and the genes correspondence is simply ignored if this similarity is below a predefined threshold. We call this operation, which will result in a set of quality genes, a *quality test*. These genes will then constitute the quality genomes. A list of generated quality genomes based on specific threshold is then produced. It is used to construct the intersection core matrix, which will generate the core genes, core tree, and phylogenetic tree after choosing an appropriate outgroup. In this work, to improve the confidence put in the core genes, we have discarded the paralogous genes. fragment is in this file with the same gene name.

5   Implementation

All algorithms have been implemented using Python language version 2.7, on a personal computer running Ubuntu 12.04 32 bits with 6 GByte memory, and a quad-core Intel Core i5 processor with an operating frequency of 2.5 GHz.

5.1   Construction of Quality Genomes

The first step in producing annotated genomes is to find the set of common genes, that is, genes sharing similar names and sequences, by using various annotation tools and following the method described previously. Figure 2a presents the original amount of genes based on NCBI and DOGMA annotations. Two quality test routines then take place to produce "quality genomes" by: (1) selecting all common genes based on gene names and (2) checking the similarity of sequences, which must be larger than or equal a predefined threshold (see Figure 2a). Note that predefined threshold is not used to determine the ortholog genes, it is used to ensure that core genes from NCBI and DOGMA annotations are identical. We also calculate the correlation function to see with whom the common genes have good relation (*e.g.* with NCBI or Dogma)? We found that the correlation value based on the number of genes produced by two annotation algorithms is 0.57. The correlation value based on the number of genes between the produced quality genomes and NCBI genomes is 0.6731, and 0.9664 between produced quality genomes and Dogma genomes. Note that gene differences between such annotation tools can affect the final core genome, if the naming and the functionality of these genes are well defined.

5.2   Core and Pan Genomes

Figure 2b represents the amount of genes in the computed core genome of 98 species. In this figure, two methods are used and compared using the same sample

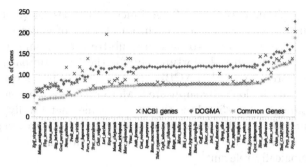

(a) Amount of genes based on NCBI and DOGMA w.r.t
quality common genes. DOGMA gives the larger number
of genes.

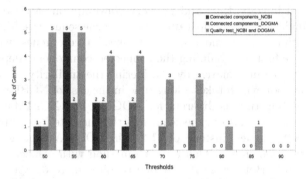

(b) Core genomes sizes w.r.t. threshold. A maximal number of core genes does not mean a good core genomes: we
are looking for genes meeting biological requirements.

Fig. 2. (a) Genes coverage for a threshold of 60% and (b) core genomes sizes

of genomes: in the first one, the gene prediction approach presented in [1, 2] has
been used on genomes annotated by NCBI, while on the other one the quality
test approach [2] has been applied on genomes annotated by DOGMA. Different
thresholds have been examined for both approaches. The amount of final core
genes within the two approaches is low, as the species considered here are highly
divergent. However even in that particular situation, it is obvious that the quality
test approach outperforms the other one at each tested threshold. Compare to
Coregenes [11], our approach allows to deal with a large class of genomes (98
species) whereas this tool is limited to six genomes. As stated previously, the
main goal is to find the largest number of core genes compatible with biological
background related to chloroplasts. In the quality approach case, one genome
(*Micromonas pusilla*, with accession number NC_012568.1) has been discarded
from the sample, as we observed that this genome always has the minimum
number of common genes with its correspondents. That can be explained by

two reasons: (1) either it consists of non-functional genes, or (2) the diversity value is too high. With quality approach, an absence of genes in rooted core genome means that we have two or more sub-trees of organisms completely divergent among each other. Unfortunately, for the first approach with NCBI annotation, the core genes within NCBI cores tree did not provide a distribution of genomes that are biologically supported. More precisely, *Micromonas pusilla* (accession number NC_012568.1) is the only genome that totally destroys the final core genome with NCBI annotations, for both gene features and gene quality methods. Conversely, in the case of DOGMA annotation, the distribution of genomes is biologically relevant[1].

5.3   Execution Time and Memory Usage

In computational biology, time and memory consumptions are two important factors due to high throughput operations among gene sequences. Figure 3 shows the amount of time and memory needed to extract core genes using the two approaches: in the first one, building the connected components depends on the construction of a distance matrix by considering the similarity scores from the global alignment tool, which takes a long time in the case of NCBI and DOGMA genomes. Calculation time is different for DOGMA and NCBI due to the size of genomes and the amount of gene sequences that need to be compared: NCBI genomes have 8,992 genes, instead of 11,242 in DOGMA genomes. Figure 3a presents the execution time needed for each method with respect to thresholds in range [50 − 100]. But the DOGMA one requires more computational time (in minutes) for sequence comparisons, while gene quality method needs a low execution time to compare quality genes. However, once the "quality genomes" have been constructed, this method takes only 1.29 minutes to extract core genes.

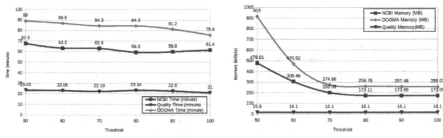

(a) Time needed to execute each method.   (b) Memory usage (MB unit) (sizes usually available on personal computers).

Fig. 3. Execution time and memory usage w.r.t. threshold

---

[1] Core tree is available on http://members.femto-st.fr/christophe-guyeux/en/chloroplasts.

The second important factor is the amount of memory used by each methodology, this one is highlighted by Figure 3b. The low values show that the gene quality method based on gene sequence comparisons presents the most reasonable memory usage (when constructing quality genomes). It also depends on the size of genomes. Determining which method to choose depends on the user preferences: if we search for a fast and semi-accurate method, then the second approach should be chosen. Otherwise, if an accurate but relatively slow approach is desired, then the first method with DOGMA annotations should be preferred.

6   Phylogenetic Study Illustration

To show the relevance of the obtained quality core genes, we will use all or a proportion of them to build a phylogenetic tree. Indeed, thanks to our approach we can precisely identify the common genes of a group of species and thus use the corresponding core genome to deduce their phylogenetic relation. The objective is to find the most well-supported phylogenetic tree (with high bootstrap values). In practice, to find a such tree, the popular program RAxML [12] is employed to compute the phylogenetic maximum-likelihood (ML) function with the following setup: the General Time Reversible model of nucleotide substitution with the Γ model of rate heterogeneity and the hill-climbing optimization method, while the *Prochlorococcus marinus* (NC_009091.1) Cyanobacteria species is chosen as outgroup due to the supposed cyanobacteria origin of chloroplasts. The tree representation is obtained with Geneious [13] based on the RAxML information.

The first experiments are done using all five genes in the core genome of 98 species. Thus, in order to find a well supported phylogenetic tree from all core genes, which reflects a real evolutionary scenario, we have achieved high level calculations of bootstrapping values for 120 trees, by considering all permutations (using *itertools* package) of gene orders[2]. Among all these trees, we have then selected the tree with the largest value of its lowest bootstrap, this one is denoted as the most accurate tree (MAT) in what follows, after having verified that gene order has no effect on the supports. The MAT has a lowest bootstrap equal to 32 and to improve this value, we have investigated in a second stage of experiments whether some core genes are homoplasic ones. In fact, when the core is large enough, it is possible to remove a few of them that obviously break the supports according to the maximum likelihood inference. After having removing systematically 1, 2, 3, and 4 genes, the best phylogenetic tree, having its lowest bootstrap value equal to 35, was obtained after one gene loss.

The low improvement observed previously when removing some core genes suggests that their number is not sufficient to produce a well-supported phylogenetic tree. Therefore we decided for the next experiments to split the set of species in two and to work with the core genome of the largest subset: 52

---

[2] Five core genes: 5! = 120 phylogenetic trees.

genomes lead to a core genome of 16 genes[3] (Core_81 in the core tree available online). As expected, working with this large core genome allows to really improve the lowest bootstrap value[4], since by removing randomly 1, 2, 3, and 4 genes the resulting MAT has 55 for lowest bootstrap value. Figure 4 presents this best tree obtained after removing one gene (*i.e. atpI*). Let us notice that for large core genomes such an approach is intractable in practice, due to the dramatic number of core genes combinations to calculate.

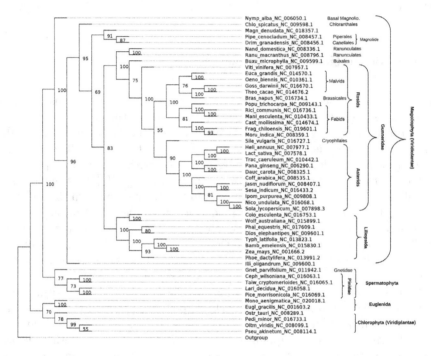

Fig. 4. Core_81 phylogenetic tree with 15 core genes (1 gene removed randomly)

Finally, the support of the best phylogenetic tree can be improved again by using the whole knowledge inherited by all the previously constructed trees. *SuperTriplets* [14] is one of the methods that can infer a supertree from a collection of bootstrapping phylogenetic trees. This tool[5] receives a file that stores all bootstrap values. In this last experiment, phylogenetic trees with 1, 2, 3, and 4 random gene loss have been concatenated in one file and transmitted to *SuperTriplets*. The obtained supertree with all taxa is provided in Figure 5. It can be seen that the minimum bootstrap has been further improved to 64.

---

[3] Core genes in Core_81: *psbE, psbD, petG, psbF, psbA, psbC, rpl36, psbN, psbI, psbJ, atpH, psaJ, atpI, atpA, psaA*, and *psaC*.

[4] The lowest bootstrap value for 16 core genes is 15.

[5] Available on http://www.supertriplets.univ-montp2.fr/index.php

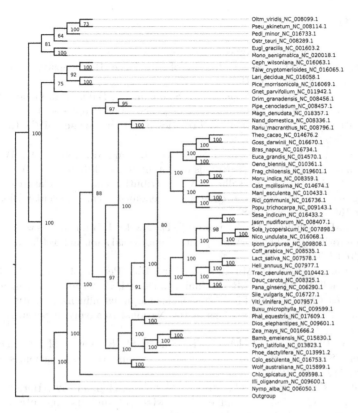

Fig. 5. Supertree for Core_81 from 248 bootstrap phylogenetic trees after removing 1, 2, 3, or 4 genes randomly

7   Conclusion

We have employed a "quality test approach" from previous study to extract core genes from a large set of chloroplastic genomes, and we compared it with the gene prediction approach developed also in our previous studies. A two stage similarity measure, on names and sequences, has thus been proposed for clustering DNA sequences in genes, which merges best results provided by NCBI and DOGMA. Results obtained with this quality control test have finally been deeply compared with our previously obtained results.

Phylogenetic trees have finally been generated to investigate the distribution of chloroplasts and core genomes. High computations are made to produce the highest bootstrap values tree by generating all trees by considering gene orders and through removing randomly some genes from core genome. A supertree is then generated, showing a highly accurate phylogenetic tree for a large amount of plant species.

Computations have been performed on the supercomputer facilities of the Mésocentre de calcul de Franche-Comté.

References

1. Alkindy, B., Couchot, J., Guyeux, C., Mouly, A., Salomon, M., Bahi, J.M.: Finding the core-genes of chloroplasts. Journal of Bioscience, Biochemistry, and Bioinformatics 4(5), 357–364 (2014)
2. Alkindy, B., Guyeux, C., Couchot, J., Salomon, M., Bahi, J.M.: Gene similarity-based approaches for determining core-genes of chloroplasts. In: 2014 IEEE International Conference on Bioinformatics and Biomedicine, BIBM (2014) 978-1-4799-5669-2/14/
3. Chaffey, N., Alberts, B., Johnson, A., Lewis, J., Raff, M., Roberts, K., Walter, P.: Molecular biology of the cell. Annals of Botany 91(3), 401–401 (2003)
4. Stoebe, B., Martin, W., Kowallik, K.V.: Distribution and nomenclature of protein-coding genes in 12 sequenced chloroplast genomes. Plant Molecular Biology Reporter 16(3), 243–255 (1998)
5. Grzebyk, D., Schofield, O., Vetriani, C., Falkowski, P.G.: The mesozoic radiation of eukaryotic algae: The portable plastid hypothesis1. Journal of Phycology 39(2), 259–267 (2003)
6. De Chiara, M., Hood, D., Muzzi, A., Pickard, D.J., Perkins, T., Pizza, M., Dougan, G., Rappuoli, R., Moxon, E.R., Soriani, M., Donati, C.: Genome sequencing of disease and carriage isolates of non typeable haemophilus influenzae identifies discrete population structure. Proceedings of the National Academy of Sciences 111(14), 5439–5444 (2014)
7. Kurtz, S., Phillippy, A., Delcher, A.L., Smoot, M., Shumway, M., Antonescu, C., Salzberg, S.L.: Versatile and open software for comparing large genomes. Genome Biology 5(2), R12 (2004)
8. Apweiler, R., ODonovan, C., Martin, M.J., Fleischmann, W., Hermjakob, H., Moeller, S., Contrino, S., Junker, V.: Swiss-prot and its computer-annotated supplement trembl: How to produce high quality automatic annotation. Eur. J. Biochem. 147, 9–15 (1985)
9. Sugawara, H., Ogasawara, O., Okubo, K., Gojobori, T., Tateno, Y.: Ddbj with new system and face. Nucleic Acids Research 36(suppl. 1), D22–D24 (2008)
10. Wyman, S.K., Jansen, R.K., Boore, J.L.: Automatic annotation of organellar genomes with dogma. Bioinformatics 20(17), 3252–3255 (2004)
11. Zafar, N., Mazumder, R., Seto, D.: Coregenes: A computational tool for identifying and cataloging. BMC Bioinformatics 33(1), 12 (2002)
12. Stamatakis, A.: Raxml version 8: A tool for phylogenetic analysis and post-analysis of large phylogenies. Bioinformatics (2014)
13. Kearse, M., Moir, R., Wilson, A., Stones-Havas, S., Cheung, M., Sturrock, S., Buxton, S., Cooper, A., Markowitz, S., Chris, Duran, o.: Geneious basic: an integrated and extendable desktop software platform for the organization and analysis of sequence data. Bioinformatics 28(12), 1647–1649 (2012)
14. Ranwez, V., Criscuolo, A., Douzery, E.J.: Supertriplets: a triplet-based supertree approach to phylogenomics. Bioinformatics 26(12), i115–i123 (2010)

Comparative Analysis of Bivalent Domains in Mammalian Embryonic Stem Cells

Anna Mantsoki and Anagha Joshi

Division of Developmental Biology, The Roslin Institute and Royal (Dick) School of
Veterinary Studies, University of Edinburgh, Edinburgh, United Kingdom
{anna.mantsoki,anagha.joshi}@roslin.ed.ac.uk

Abstract. Bivalent promoters are defined by the presence of both activating
(H3K4me3) and repressive (H3K27me3) chromatin marks. In this paper, we
first identified high confidence bivalent promoters in murine ES cells integrating data across eight studies using two methods; peak-based and cutoff-based.
We showed that peak-based method is more reliable as promoters are more enriched for developmental regulators than the cutoff-based method. We further
identified bivalent promoters in human and pig using the peak-based method to
show that the bivalent promoters conserved across species were highly enriched
for embryonic developmental processes.

Keywords: ChIP sequencing, embryonic stem cells, chromatin, H3K4me3,
H3K27me3, bivalent promoters, developmental genes, comparative genomics.

1 Introduction

The key cellular processes determining the fate of each cell type during development
and differentiation are thought to be controlled by gene regulation (Pearson et al.
2005). Genomic regulatory elements such as promoters receive and execute transcriptional signals, dependent on their epigenetic state and chromatin accessibility, controlling the expression of key developmental factors(Wilson et al. 2010). Apart from the
transcription control at the promoters and enhancers, gene expression is also controlled epigenetically, by post-translational histone modifications, which transform
the chromatin structure and thereby control gene expression (Bannister & Kouzarides
2011).

To unravel key developmental transitions that lead to different types of cell identi-
ties, embryonic stem cells (ESCs) offer a valuable model (Thomson et al. 1998) as
they have an unlimited potential to self-renew as well as to differentiate in specific
lineage when suitable external stimuli are provided. In ESCs, the majority of promoters with high CG content are un-methylated. During differentiation though, some of
them become methylated, assisting to the acquisition of their final cell identity (Mohn
et al. 2008). Azuara et al., 2006 proposed that particular histone modifications and
chromatin structure (Voigt et al. 2013; Thomson et al. 1998) are characteristic of ES
cells. Two of the most commonly studied histone modifications related to activation

F. Ortuño and I. Rojas (Eds.): IWBBIO 2015, Part I, LNCS 9043, pp. 391–402, 2015.
© Springer International Publishing Switzerland 2015

and repression of chromatin respectively are H3K4me3 and H3K27me3(Bannister & Kouzarides 2011). Polycomb (PcG) and Trithorax (TrxG) group proteins catalyze H3K27me3 and H3K4me3 respectively, regulating genes involved in development and differentiation (Ringrose & Paro 2004). Bernstein et al., 2006 observed activating (H3K4me3) and repressing (H3K27me3) chromatin signals in promoters of several developmentally regulated genes in murine ES cells. These activating and repressive marks were previously thought to be mutually exclusive and therefore the promoters marked with both modifications were named 'bivalent'. Mikkelsen et al., 2007 used ChIP sequencing technique to examine the bivalent status and construct chromatin state maps across three cell types: mESCs, neural progenitor cells (NPCs) and mouse embryonic fibroblasts (MEFs). Their study showed for the first time, that bivalent domains also exist in cells of restricted potency and 8-43% of bivalent domains retained their bivalent mark during the differentiation (Mikkelsen et al. 2007). More-over, Mohn et al., 2008 indicated that bivalent genes that are not present in the pluripotent cells may arise in reduced potency cells.

Bivalent genes were detected also in human ESCs (Pan et al. 2007; Zhao et al. 2007) and the majority of them was shared with bivalent genes in mouse ESCs. Spe-cifically, in two out of three studies there were 2,157 common bivalent genes (Mikkelsen et al. 2007; Pan et al. 2007; Sharov & Ko 2007; Zhao et al. 2007). In agreement with the studies in mice, human ESCs bivalent genes are functionally en-riched with developmental transcription factors and genes and most of them lose the repressive H3K27me3 mark during differentiation (Pan et al. 2007; Zhao et al. 2007).

ES cells employ various mechanisms to avoid losing their pluripotency. For exam-ple they manage to prevent DNA methylation that would silence important genes indefinitely. Bivalent genes belong to an important category of genes full of develop-mental factors that need to be poised for activation or repression at the right moment during the differentiation process (Voigt et al. 2013). The bivalent state preserves the plasticity of the developmental genes until certain environmental cues lead to proper differentiation.

Though bivalent genes have been identified across multiple species in ES cells as well as differentiated cells, there is no study so far collecting multiple data sets to build a high-confidence bivalent gene set. We therefore collected genome wide bind-ing patterns of H3K4me3 and H3K27me3 in murine ES cells from eight different studies. We then used two complementary approaches; peak-based and cutoff-based approach to define high confidence bivalent promoters. The high confidence bivalent promoters detected by the peak-based method were more enriched for developmental genes than the cutoff based. Finally, we collected data to identify bivalent promoters in human ES cells and pig induced pluripotent cells to study the evolutionary conser-vation of bivalency. By performing the comparative analysis of bivalent domains across three species we highlighted the functional relevance of coexistence of these marks on the developmental promoters.

2   Materials and Methods

Data Collection and Processing: Murine ChIP sequencing data for H3K4me3 and H3K27me3 histone marks in ESCs was obtained in fastq format from Gene Expression Omnibus (GEO) database (Barrett et al. 2013). Accession numbers for mouse are: SRX001923, SRX085431, SRX122633, SRX172569, SRX266814, SRX266815, SRX305921, SRX305922, SRX001921, SRX185810, SRX122629, SRX172574, SRX266816, SRX266817, SRX305910, and SRX305911. Human ChIP-Seq data (fastq format) for H3K4me3 and H3K27me3 histone marks in hESCs was obtained from Roadmap to epigenomics (Bernstein et al. 2010) and Gene Expression Omnibus (GEO). Accession numbers for human are: SRX006237, SRX012501, SRX27864, SRX007385, SRX019896, SRX006262, SRX006874, SRX012368, SRX007379, SRX019898, SRX003845, SRX064486, SRX027487, SRX189253, SRX027865, SRX056719, SRX003843, SRX064487, SRX027484, SRX189254, SRX040598, SRX056700. ChIP sequencing data for H3K4me3 and H3K27me3 in pig (Sus Scrofa) induced pluripotent stem cells (iPSCs) was downloaded from a published study with accession number GSE36114 (Xiao et al. 2012). After downloading all the raw sequence files for all the experiments, each technical and biological replicate of the samples was imported in FastQC 0.10.1 (S. Andrews 2010) for quality control. Alignment of reads was done using Bowtie 0.12.9 (Langmead et al. 2009) using reference genomes mm10 for mouse, hg19 for human and susScr3 for pig. For all the species we used single end alignment, seed length=28. We then performed the bowtie execution using custom bash scripts and the samtools (Li et al. 2009) pipeline to convert directly sam format file to a bam format file for each sample. The bam files that belonged to the same experiment (technical replicates) were merged into a common bam file in order to proceed with the further analysis. The biological replicates of each experiment were not merged. We downloaded the Gencode (Harrow et al. 2012) genes for human (Gencode 19) and mouse (Gencode M2). We filtered out and kept only the genes from the initial gtf files. Also, we created bed files for the promoter regions, keeping the areas that were (-1000 bp, +2000 bp) from the Transcription Start Site (TSS). For mouse there were 38,922 promoter regions and for human 57,818. Since there was not a Gencode file available for pig, we downloaded the ensembl gene file available from Biomart (Haider et al. 2009). After doing the same procedure as mentioned above in order to keep only the promoter regions, we got 21,116 regions for pig promoters.

Peak Calling Method: We used SICER (Zang et al. 2009), a tool that is recommended for enrichment analysis of histone modification data, since it outperforms every other tool in its category for peak calling. The input controls were used when they were provided with the samples. When input was available, the SICER parameters were: for H3K4me3, window=200 and gap size=200. For H3K27me3, window=200 and gap size=2x300, since this histone mark is found covering wider chromatin domains. The rest of the parameters (same for both H3K4me3 and H3K27me3) were effective genome fraction =0.7, false discovery rate (FDR) = 0.01, redundancy threshold = 1 and fragment size = 150. When the control library was unavailable, the FDR value parameter was replaced by the E-value parameter equal to

100. We intersected the resulting files after peak calling with the promoter files using the intersect command from BEDtools (Quinlan & Hall 2010).

Cutoff Method: We obtained the read density only at the regions we were interested in, the promoters. Using custom scripts and the coverageBed (BEDtools) (Quinlan & Hall 2010) command, we created bed files for each sample. In the resulting bed files, the column that we kept was the one that contained the number of reads in the promoter regions. We applied logarithmic scale to the read densities of all samples, followed by quantile normalization for H3K4me3 and H3K27me3 samples separately to define a threshold that would reveal the real enrichment for H3K4me3 and H3K27me3 and even out the variability across samples. We generated scatterplots (Figure 1) of the same histone modification samples against each other to examine what type of normalization to choose. To further increase the accuracy of the cutoff method, we created promoter files with sliding windows. Every promoter region was divided in windows of 200 bp, with a sliding step of 50 bp. For all the window regions corresponding to the initial promoter region, the maximum coverage value was chosen as the representative for this region. The distribution pattern of H3K4me3 reads is very close to the bimodal distribution. Following that, we used the mixtools package (Benaglia et al. 2009) in order to fit the bimodal distribution to all of our samples, both for H3K4me3 and H3K27me3. For most of the cases of H3K4me3 bimodal distribution was fitted successfully. In contrast, most H3K27me3 samples were not close to follow the bimodal distribution. For the successfully fitted H3K4me3 samples we kept the mean and standard deviation of the second curve of each distribution. After subtracting each standard deviation value from its respective mean value, we obtained the initial threshold values for each sample. The final threshold value for all the H3K4me3 samples was the average of all the initial values. In the case of H3K27me3 distributions, since we had no successful fitting bimodal distribution, we chose empirically 3 different thresholds and chose the one that would give results best matching to previous studies. The final threshold values used were 4.57 for H3K4me3 and 3.00 for H3K27me3. We used the study of Mikkelsen et al., 2007 to compare peak-based and cut off based method to a published study.

Functional Enrichment Analysis: We conducted gene ontology functional analyses for the bivalent promoters for both approaches, using DAVID (Dennis et al. 2003).

Overlap between Species: To obtain a list of common bivalent, expressed and repressed genes between the species, we used only the orthologous genes that mouse and pig share with human (18,255 genes). We got the common list of genes for all three species between them, but also for each combination by two (human-mouse, human-pig, mouse-pig).

P value Calculation: To calculate if the overlap of two gene lists can happen due to random chance, we used hypergeometric test. Specifically, to compare two lists we used the phyper function in R. When we were comparing more than two lists we used random permutation of the rows and columns of the results table (species in columns, genes in rows) simulated for 1000 times. We used the permatfull function from the vegan package (Oksanen et al. 2013) in R. Then we compared the mean of all the simulations with our result of common genes in order to find if there is significant difference between them.

3   Results and Discussion

3.1   Peak-Based Method to Detect High-Confidence Bivalent Promoters

Bivalent promoters are defined by the presence of both active (H3K4me3) and repressive (H3K27me3) chromatin modifications. In ES cells, they are highly enriched for developmental genes and therefore the identification of high confidence bivalent promoters might lead to discovery of novel developmental regulators. With this rationale, we set to look for high confidence bivalent marked promoters in murine ES cells and collected data for paired (H3K4me3 and H3K27me3) ChIP sequencing samples from eight studies from GEO (methods for details). The samples varied in their or nothing read length, ranging from 27 bp to 115 bp and the total number of mapped reads to the mouse genome assembly mm10, were ranging from 14 million to 200 million reads per sample. We called peaks using SICER (Zang et. al., 2009), the best suited algorithm for peak detection in histone modification data. For eight samples of H3K4me3, between 16 thousand and 66 thousand peaks were identified while for H3K27me3, between 9 thousand and 26 thousand peaks were identified. To check if this variation in peak number can be attributed to the variability in total number of reads across samples, we calculated Pearson's correlation coefficient between number of reads and number of peaks detected across eight samples and found a high correlation. The correlation coefficient for H3K4me3 was 0.84 while for H3K27me3 was 0.75. As the only way to adjust for the sequencing depth for peak based method is to consider the only 7 million reads for peak calling but it suffers a major drawback of not being able to use most of the available data, we defined an approach complementary to peak-based approach - cutoff-based method (described in detail in the following section). We then collected 38,922 transcribed units (genes) in mouse from GENCODE (Harrow et al. 2012) and defined promoters as -1kb and +2kb region around the transcription start site of each transcribed unit. We then intersected these promoters with the H3K4me3 and H3K27me3 peaks. Despite the large variance in the number of H3K4me3 peaks identified in individual samples, the number of peaks within promoters was very consistent across samples ranging from 18 thousand to 20 thousand H3K4me3 marked promoters. This suggests that most promoters have a high peak height of H3K4me3 and therefore H3K4me3 is a distinguishing mark for promoters. In contrast, the number of H3K27me3 promoter peaks showed a large variance ranging from 3 thousand to 9 thousand peaks. The Pearson's correlation coefficient value between the total H3K27me3 peaks and the fraction of these in promoters was 0.5. This suggests that H3K27me3 does not show preference to promoters and therefore is not a distinguishing mark for promoters. The number of bivalent marked promoters varied between 2 thousand and 7 thousand across eight samples. Pearson's correlation coefficient between the number of H3K4me3 promoters and bivalent promoters was 0.58 while between H3K27me3 promoters and bivalent promoters was 0.98. This shows that the classification of a promoter as a bivalent promoter highly depends upon identification of H3K27me3 modification rather than H3K4me3 modification.

In order to identify the high confidence bivalent promoters, we calculated cumulatively the number of promoters identified with the H3K4me3 modification in 'n' or more samples. Over 20 thousand promoters were H3K4me3 marked in at least one sample, while about 15 thousand promoters were H3K4me3 marked in all eight samples. This demonstrates that H3K4me3 modification on promoters across samples is quite stable (Table 1). On the contrary, Over 11 thousand H3K27me3 promoters were detected in at least one sample of which only about 2 thousand were H3K27me3 marked in all samples (Table 1). The rate of decrease in the number of bivalent promoters (ratio of six or more to one or more) was 0.44, in H3K4me3 promoters was 0.81 and in H3K27me3 promoters was 0.37 in 'n' or more samples. This again demonstrates that the number of high confidence bivalent promoters is dependent on the H3K27me3 histone mark. We noted that consistently over 80% of H3K27me3 promoters were marked bivalent. This means that most H3K27me3 marked promoters also have H3K4me3 modification present. This demonstrates that the co-existence of these two chromatin modifications on promoters initially thought as a surprise, is rather a rule than exception. ChIP enrichment signals can be missed during peak calling procedure or by experimental error in an individual sample. Peaks detected in all samples are likely to miss true bivalent promoters. As the ratio of bivalent to H3K27me3 marked promoters was highly consistent when 4, 5 or 6 or more samples are taken into account, we decide to use an arbitrary cut off of six or more to define high confidence bivalent promoters. This resulted into identification of 16,885 high confidence H3K4me3 marked, 4,239 high confidence H3K27me3 marked and 3,740 high confidence bivalent promoters.

We then checked if the high confidence detection was biased towards any individual study or were true representative of all eight studies. About 50% of high confidence peaks were present in individual H3K4me3 samples while the fraction of high confidence H3K27me3 peaks in individual sample varied between 40 and 70%. This again demonstrates that H3K4me3 is consistent while H3K27me3 varies on the promoters.

Table 1. Cumulative count of three categories of promoters in mESCs with the peak based method. The cells with bold font (6 or more) represent the high confidence cut off chosen.

Samples	MOUSE (WITH PEAK CALLING METHOD)			
	H3K4me3	H3K27me3	Bivalent	Biv./H3K27me3
1 or more	20761	11610	8515	0.73
2 or more	19980	8931	7252	0.81
3 or more	19358	7413	6343	0.85
4 or more	18523	6198	5458	0.88
5 or more	17848	5175	4679	0.90
6 or more	**16885**	**4239**	**3740**	**0.88**
7 or more	16062	3287	2764	0.84
8 or more	14720	2236	1555	0.69

3.2   Cutoff-Based Method to Detect High-Confidence Bivalent Promoters

As the peak calling method is highly sensitive to the sequencing depth, we defined another independent method to identify enriched genomic regions for a specific histone modification, henceforth called cutoff-based method. We calculated the number of reads mapping to each promoter in each H3K4me3 and H3K27me3 samples by using custom scripts and BEDtools (Quinlan & hall 2010). In order to normalize the reads across multiple samples, the logarithmic scaled promoter read counts across all H3K4me3 and H3K27me3 experiments were quantile normalized separately (see Methods). The H3K4me3 normalized promoter read density followed a clear bimodal distribution separating H3K4me3 unmarked from marked promoters (Figure 1a). We noted further that the H3K4me3 positive and H3K4me3 negative sets were conserved across samples. On the contrary, the normalized promoter read density for H3K27me3 did not show a clear bimodal distribution making it hard to distinguish between the H3K27me3 positive and H3K27me3 negative sets (Figure 1b). Moreover, though H3K27me3 mark was coherent across samples, the distinction of two groups unlike H3K4me3 was not clear (Figure 1). We fitted a bimodal distribution to H3K4me3 log scaled promoter read densities and consistently obtained a cut off of 4.57 to distinguish between H3K4me3 positive and negative promoters (Figure 1a). On the other hand, as bimodal distribution failed to fit, we defined an arbitrary cut off of 3.00 to distinguish between H3K27me3 positive and negative promoters (Figure 1b). The cutoff based method identified consistently about 7 thousand H3K27me3 marked promoters and about 13 thousand H3K4me3 marked promoters.

a   **b**

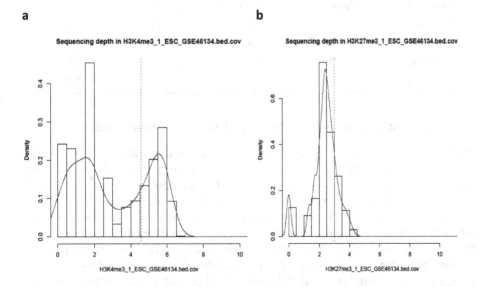

Fig. 1. Representative histograms and density plots for a) H3K4me3 and b) H3K27me3 samples in mouse ESCs. The vertical dotted red line marks the threshold used.

To identify high confidence bivalent promoters using the cutoff method, we calculated cumulatively the number of promoters identified with a given modification in 'n' or more samples. Both H3K4me3 and H3K27me3 marks showed a large variability across samples. Over 15 thousand promoters were H3K4me3 marked in at least one sample while only about 11 thousand promoters were H3K4me3 marked in all eight samples. Similarly, Over 16 thousand H3K27me3 promoters were detected in at least one sample from which only about 3 thousand were H3K27me3 marked in all samples (Table 2). The ratio levels were not as consistent as in the case of the peak calling method but for most of the cases (except for the extremes) more than 50% of the bivalent promoters were part of the repressed ones. Similar to the peak calling procedure, we used a threshold of six or more to define high confidence bivalent promoters. This resulted into the identification of 13,034 high confidence H3K4me3 marked, 4,660 high confidence H3K27me3 marked and 2,396 high confidence bivalent promoters.

Table 2. Cumulative count of three categories of promoters in mESCs with the cutoff based method. The cells with bold font (6 or more) represent the high confidence cutoff chosen.

	MOUSE (WITH CUTOFF BASED METHOD)			
Samples	**H3K4me3**	**H3K27me3**	**Bivalent**	**Biv/H3K27me3**
1 or more	15668	16624	7711	0.46
2 or more	14895	10327	5428	0.52
3 or more	14389	7748	4400	0.56
4 or more	13942	6419	3685	0.57
5 or more	13478	5479	3027	0.55
6 or more	**13034**	**4660**	**2396**	**0.51**
7 or more	12378	3846	1708	0.44
8 samples	11190	2829	945	0.33

3.3   Systematic Comparison of Peak-Based and Cutoff-Based Method

We performed a systematic comparison of the peak-based and cutoff-based method. Across individual samples, the variability in the total number of peaks identified by cutoff-based method was much lower compared to the peak-based method for both H3K4me3 and H3K27me3 data sets. Though cutoff-based method showed high consistency across samples for both modifications, there was more variability for the cutoff-based method when the cumulative analysis was performed (Table 1 & 2). We then compared the high confidence bivalent promoters obtained by both methods by defining the same threshold of six or more samples. The cutoff-based method concluded that only about 50% of H3K27me3 marked promoters were bivalent whereas peak-based method predicted this fraction to be over 80%. The peak-based method results are thus in agreement with literature. This was expected since the peak-based method is the most widely used approach in the literature. Over 80% of bivalent peaks detected by the cutoff method were also found by the peak method. The peak-based method is therefore able to identify high confidence bivalent promoters missed by the cutoff method (Figure 2a). Finally we calculated functional enrichment for bivalent promoters using both methods. Though both promoter sets were enriched for developmental categories such as anatomic structure development, and developmental process the enrichment was higher with the peak method than for the cutoff method (Figure 2b). Taken together, the peak-based method was more reliable in detecting high confidence bivalent promoters.

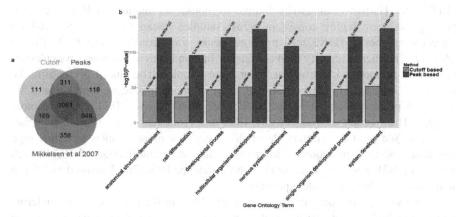

Fig. 2. a) Common bivalent promoters between the cutoff based method, the peak based method and from Mikkelsen et al., 2007 b) Functional enrichment values (-log10Pvalue) for the most enriched gene ontology terms for the two methods (P-value indicated on top of each bar)

3.4 Comparison of Serum-Grown High-Confidence Bivalent Promoters with 2i

Having established that peak detection method predicts reliable high confidence bivalent promoters, we used the bivalent promoters detected by the peak-based method for further analysis. Murine ES cells can be maintained in two distinct culture conditions in vitro, 2i (with inhibitors of two kinases Mek and GSK3) and serum. All eight samples used for high confidence bivalent promoter detection were grown in serum culture condition. Marks et al., 2012 identified 1,014 bivalent genes in murine ES cells grown under 2i media and 2,936 bivalent genes grown in serum and stated that the identification of fewer bivalent genes in '2i' was in agreement with the postulated naïve ground state of ES cells grown in '2i' and not in serum. If this were the case, the high confidence bivalent promoters should have higher overlap with 2i grown bivalent genes than bivalent genes detected in a serum grown sample. 76% of 2i-grown bivalent genes and 68% of serum-grown bivalent genes overlapped with our high confidence bivalent promoters respectively. 2i grown show higher overlap than serum-grown suggesting 2i might be more similar to naïve ground state. A fraction of 2i-grown bivalent genes were not identified bivalent in any of the ten samples grown in serum. This suggests that there are genes specifically bivalent marked in 2i and not in serum culture condition.

3.5   Identification of Bivalent Regions in Other Mammalian Species

In order to check if the high confidence bivalent regions are more conserved across species, we collected genome wide binding profiles for H3K4me3 and H3K27me3 in human ES cells and pig induced pluripotent cells. We collected 11 paired samples in humans from six studies with reads ranging from 13 million to 60 million in individual samples. We used the peak based method to call peaks in individual samples. These peaks were then mapped to promoters of 57,818 transcribed units defined by GENCODE (Harrow et al. 2012). Similar to mouse, the number of H3K4me3 promoter peaks were highly consistent across samples (mean 19,219.73, SD 462.88) while the number of H3K27me3 promoter peaks was variable (mean 8,035.73, SD 2,626.27). In order to identify high confidence human bivalent promoters we calculated bivalent promoters identified in 'n' or more samples. The rate of decrease in the number of bivalent promoters (ratio of eight or more to one or more) was 0.39, H3K4me3 promoters was 0.89 and H3K27me3 promoters was 0.31 in 'n' or more samples. The fraction of bivalent to H3K27me3 promoters was consistently higher than 80%. We used an arbitrary threshold of eight or more samples to define high confidence bivalent promoters. This resulted into the identification of 18,744 high confidence H3K4me3 marked, 5,841 high confidence H3K27me3 marked and 5,116 high confidence human bivalent promoters.

In pig (Sus Scrofa), only one study was available in the public domain hindering detection of high confidence bivalent promoters. Using 21,116 promoter regions we detected 8,383 H3K4me3 marked, 2,816 H3K27me3 marked and 1,561 bivalent marked promoters again demonstrating that over half of H3K27me3 marked promoters also contain an H3K4me3 modification.

Table 3. Cumulative count of three categories of promoters in mESCs with the peak based method. The cells with bold font (8 or more) represent the high confidence cut off chosen.

	HUMAN (WITH PEAK CALLING METHOD)			
Samples	**H3K4me3**	**H3K27me3**	**Bivalent**	**Biv./H3K27me3**
1 or more	21167	18701	13206	0.70
2 or more	20275	12066	9778	0.81
3 or more	19865	9825	8236	0.83
4 or more	19602	8560	7308	0.85
5 or more	19341	7789	6713	0.86
6 or more	19123	7102	6177	0.86
7 or more	18944	6480	5660	0.87
8 or more	**18744**	**5841**	**5116**	**0.87**
9 or more	18489	5171	4505	0.87
10 or more	18189	4087	3495	0.85
11 samples	17678	2771	2202	0.79

3.6   Comparative Analysis of Bivalent and Promoters Across Three Species

Finally, we computed the overlap of bivalent promoters across three species by considering only one-to-one mapping orthologs. The bivalent promoters were less conserved across three species compared to the active promoters (Figure 3a and b). Specifically less than 10% of human bivalent promoters were conserved across three species while over 25% of H3K4me3 marked promoters were conserved across three species. The functional enrichment of common bivalent genes resulted in development processes more specific to embryogenesis, such as pattern specification process, embryonic morphogenesis and embryonic organ development, suggesting that the three species have more commonalities during embryonic development.

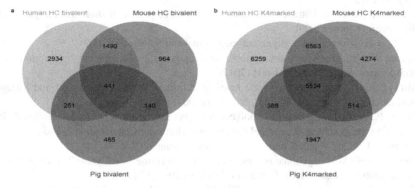

Fig. 3. Venn diagram of a) bivalent and b) K4marked promoters between human, mouse and pig using the peak calling method

4   Conclusion

In summary, we identified high confidence bivalent domains in murine ES cells by integrating data across eight studies using two methods; peak-based and cutoff-based and demonstrated that the peak-based method is more reliable. We then identified bivalent promoters in human and pig and performed a multi-species comparative analysis of bivalent promoters to show that the conserved bivalent promoters were highly enriched for embryonic developmental processes.

Acknowledgements. A.J. is a Chancellors Fellow at the University of Edinburgh. This work was supported by the Roslin Institute Strategic Grant funding from the BBSRC.

References

1. Azuara, V., et al.: Chromatin signatures of pluripotent cell lines. Nature Cell Biology 8(5), 532–538 (2006)
2. Bannister, A., Kouzarides, T.: Regulation of chromatin by histone modifications. Cell Research 21(3), 381–395 (2011)

3. Barrett, T., Wilhite, S.E., Ledoux, P.: NCBI GEO: archive for functional genomics data sets—update (2013)
4. Benaglia, T., et al.: mixtools: An R Package for Analyzing Finite Mixture Models. Journal of Statistical Software 32(6), 1–29 (2009)
5. Bernstein, B., et al.: A bivalent chromatin structure marks key developmental genes in embryonic stem cells. Cell 125(2), 315–326 (2006)
6. Bernstein, B.E., et al.: The NIH Roadmap Epigenomics Mapping Consortium. Nat. Biotech. 28(10), 1045–1048 (2010)
7. Dennis, G., et al.: DAVID: Database for Annotation, Visualization, and Integrated Discovery. Genome Biology, 4(5), P3 (2003)
8. Haider, S., et al.: BioMart Central Portal—unified access to biological data (2009)
9. Harrow, J., et al.: GENCODE: The reference human genome annotation for The ENCODE Project (2012)
10. Langmead, B., et al.: Ultrafast and memory-efficient alignment of short DNA sequences to the human genome (2009)
11. Li, H., et al.: The sequence alignment/map format and SAMtools (2009)
12. Marks, H., et al.: The transcriptional and epigenomic foundations of ground state pluripotency. Cell 149(3), 590–604 (2012)
13. Mikkelsen, T., et al.: Genome-wide maps of chromatin state in pluripotent and lineage-committed cells. Nature 448(7153), 553–560 (2007)
14. Mohn, F., et al.: Lineage-specific polycomb targets and de novo DNA methylation define restriction and potential of neuronal progenitors. Molecular Cell 30(6), 755–766 (2008)
15. Oksanen, J., et al.: vegan: Community Ecology Package (2013)
16. Pan, G., et al.: Whole-genome analysis of histone H3 lysine 4 and lysine 27 methylation in human embryonic stem cells. Cell Stem Cell 1(3), 299–312 (2007)
17. Pearson, J., Lemons, D., McGinnis, W.: Modulating Hox gene functions during animal body patterning. Nature reviews. Genetics, 6(12), 893–904 (2005)
18. Quinlan, A.R., Hall, I.M.: BEDTools: a flexible suite of utilities for comparing genomic features. Bioinformatics 26(6), 841–842 (2010)
19. Ringrose, L., Paro, R.: Epigenetic regulation of cellular memory by the Polycomb and Trithorax group proteins. Annual Review of Genetics 38, 413–443 (2004)
20. Andrews, S.: FastQC A Quality Control tool for High Throughput Sequence Data (2010)
21. Sharov, A.A., Ko, M.S.: Human {ES} cell profiling broadens the reach of bivalent domains. Cell Stem Cell 1(3), 237–238 (2007)
22. Thomson, J., et al.: Embryonic stem cell lines derived from human blastocysts. Science 282(5391), 1145–1147 (1998)
23. Voigt, P., Tee, W.-W., Reinberg, D.: A double take on bivalent promoters. Genes & Development 27(12), 1318–1338 (2013)
24. Wilson, N., et al.: Combinatorial transcriptional control in blood stem/progenitor cells: genome-wide analysis of ten major transcriptional regulators. Cell Stem Cell 7(4), 532–544 (2010)
25. Xiao, S., et al.: Comparative epigenomic annotation of regulatory DNA. Cell 149(6), 1381–1392 (2012)
26. Zang, C., et al.: A clustering approach for identification of enriched domains from histone modification ChIP-Seq data. Bioinformatics 25(15), 1952–1958 (2009)
27. Zhao, X.D., et al.: Whole-genome mapping of histone H3 Lys4 and 27 trimethylations reveals distinct genomic compartments in human embryonic stem cells. Cell Stem Cell 1(3), 286–298 (2007)

Finding Unknown Nodes in Phylogenetic Graphs

Luis Evaristo Caraballo, José Miguel Díaz-Báñez, and Edel Pérez-Castillo

Dpto de Matemática Aplicada II, Universidad de Sevilla, Spain
{luisevaristocaraballo-ext,dbanez}@us.es, edepercas@alum.us.es

Abstract. A phylogenetic tree estimates the "historical" connections between species or genes that they carry. Given a distance matrix from a set of objects, a phylogenetic tree is a tree whose nodes are the objects in the set and such that the distance between two nodes in the tree corresponds to the distance in the matrix. However, if the tree structure does not match the data perfectly then new nodes in the graph may be introduced. Such nodes may suggest "ancestral living beings" that can be used for phylogeny reconstruction. In general, finding these ancestral nodes on a phylogenetic graph is a difficult problem in computation and no efficient algorithms are known. In this paper we present an efficient algorithm to compute unknown nodes in phylogenetic trees when the similarity distance can be reduced to the L_1 metric. In addition, we present necessary conditions to be fulfilled by unknown nodes in general phylogenetic graphs that are useful for computing the ancestral nodes.

Keywords: Phylogenetic trees, ancestral sequences, algorithms, L_1 metric.

1 Introduction

Phylogenetic trees were conceived for applications in evolutionary biology for the purposes of describing and visualizing evolutionary relationships that exist between members of a group of biological organisms [7]. However, more recently phylogenetic methods have been applied to a wide variety of cultural objects [2]. We emphasize here that the musical domain has also been considered as an application of phylogenetic methods [9,3,1,4]. In fact, this similarity analysis could be applied in investigations of the origins of musical features that interpret them as living beings that evolve over time under the influence of social parameters such as preferences and fashions. Following this metaphor, bioinformatic techniques have been used in several contexts, putting the data into a similarity matrix in a phylogenetic tree. For example, the phylogenetic tree of a collection of rhythms provides a compelling visualization of the various relationships that exist between all the rhythms, as well as of their possible evolutionary phylogeny [3].

The bioinformatics and computational biology literatures are filled with a wide variety of different approaches for constructing phylogenetic trees. Distance methods assume that a distance (or dissimilarity) matrix is available containing the distance between every pair of objects being studied. From these distance

F. Ortuño and I. Rojas (Eds.): IWBBIO 2015, Part I, LNCS 9043, pp. 403–414, 2015.
© Springer International Publishing Switzerland 2015

matrices, the algorithms construct phylogenetic trees in such a way that the
minimum distance between every pair of rhythms, measured along the branches
in the tree (geodesic distances) approximates as closely as possible the corresponding distance entry in the distance matrix. Some of these methods have the
desirable property that they produce graphs that are not trees, when the underlying proximity structure is inherently not tree-like. One notable example is
SplitsTree [8]. *SplitsTree* computes a plane graph embedding with the property
that the distance in the drawing between any two nodes reflects as closely as
possible the true distance between the corresponding two objects in the distance
matrix. However, if the tree structure does not match the data perfectly then
new nodes in the graph may be introduced, as for example in Figure 1, with the
goal of obtaining a better fit. Such nodes may suggest implied "ancestral living
being" from which their "offspring" may sometimes be derived. *SplitsTree* also
computes the *splitability index*, a measure of the goodness-of-fit of the entire
splits graph. This *fit* is obtained by dividing the sum of all the approximate distances in the splits graph by the sum of all the original distances in the distance
matrix [8].

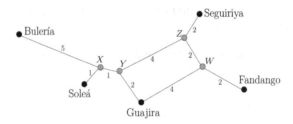

Fig. 1. SplitsTree with unknown nodes X, Y, Z, W using the Chronotonic distance
between flamenco rhythms in [3]

By using extant sequences and the phylogenetic relationships among them,
it is possible to infer the most plausible ancestral sequences from which they
have been derived when an estimation paradigm is used [10]. However, computing ancestral nodes on a phylogenetic graph based on a similarity distance is
a difficult problem in computation and is still under investigation [4]. In this
paper, we present an efficient algorithm that computes all unknown nodes in a
phylogenetic tree (without cycles) when the similarity distance can be reduced
to the L_1 metric after some suitable rewriting of the sequence. In the case of
phylogenetic trees with cycles, we also show how to extract valuable information
on the coordinates of the ancestral nodes.

The rest of the paper is organized as follows: Section 2 presents the distances
considered in this paper and the statement of the problem. In Section 3, we state
the theoretical properties and the algorithms to solve the problem. In Section
4 we extend our approach for the Hamming distance between general strings.

In Section 5 we show several examples and the conclusions of our study are summarized in Section 6.

2   Problem Statement. Similarity Distances

Given two k-dimensional vectors $X = (x_1, x_2, \cdots, x_k)$ and $Y = (y_1, y_2, \cdots, y_k)$ representing objects, the dissimilarity between the objects may be measured by any of a large variety of metrics or more general measures. Perhaps the most well-known and frequently used metrics are two special cases of the L_p-metrics: the Euclidean distance ($p = 2$) and the city-block distance (also called Manhattan metric, $p = 1$). Some authors argue in favor of using the L_1 norm given by the simpler sum of the absolute values of the differences of each coordinate. The L_1 distance between the vectors X and Y is $\sum_{i=1}^{n} |x_i - y_i|$. Moreover, as we will see in this section, other useful metrics can be reduced to the city-block distance. Thus, the problem we are dealing with in this paper is the reconstruction of ancestral sequences in a given phylogenetic graph whose nodes are the objects in a set and such that the distance between two nodes in the graph corresponds to the L_1 distance between the objects.

2.1   Other Related Distances

We include here some distances that are inherently the L_1 metric. Standard techniques in the field of evolutionary biology consider the distance between two objects to be the minimum number of *operations* that must be made on one object in order to transform it to the other. The simplest one is the *Hamming distance* where the operation is the *substitution*. Obviously, the Hamming distance between two binary sequences of the same length is equivalent to the L_1 distance. Another easy operation that may be made in binary sequences is to move an "1" from its position to either of its two adjacent neighboring positions. Such a change is called a *swap*, and the minimum number of swaps required to transform one string to another is termed the *Swap distance* [3]. For instance, the Hamming distance between the sequences [001001010101] and [001000110101] is 2 but the Swap distance is 1. Notice that we can compute the Swap distance by using the following simple idea. Let us represent each string by a vector of the positions of the ones. It is easy to see that we obtain the minimum number of swaps needed by adding the absolute values of the differences between the elements of these auxiliary vectors. Then the Swap distance between two binary sequences can be reduced to the L_1 distance between these auxiliary vectors. Let us look at an example: the sequences $S = [101010010010]$ and $G = [100100101010]$ correspond to the rhythms of Seguiriya and Guajira, respectively in flamenco music [3]. Thus, the auxiliary vectors are $S' = [1, 3, 5, 8, 11]$ and $G' = [1, 4, 7, 9, 11]$, respectively (see Figure 2) and the distance is

$$d_{swap}(S, G) = |1 - 1| + |3 - 4| + |5 - 7| + |8 - 9| + |11 - 11| = 4$$

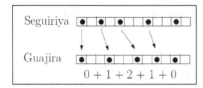

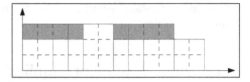

Fig. 2. The Swap distance   **Fig. 3.** Chronotonic distance

Another similarity distance is the *Chronotonic distance*, proposed in the area of *Phonetics* [5] to measure the similarity between two speech recordings (voice recognition). This distance has also been considered in the musical domain [3].

The chronotonic distance is based on histogram representation, specifically on the so-called TEDAS representation [5]. The vertical axis displays the inter-onset intervals, while the horizontal axis shows where the onsets occur. The result is a histogram representing the binary sequence that contains both pieces of information. The distance between two sequences is obtained by computing the area between the two given histograms. In the example above, if we superimpose the chronotonic representations of the sequences S and G, the area between them is 8 (see Figure 3).

Now, in order to reduce the chronotonic distance to the L_1 metric, we associate to each histogram the auxiliary vector $(f(1), f(2), ..., f(n))$ where $f(i)$ is the value of the chronotonic function in the interval $[i - 1, i]$. Thus, with this new annotation the sequences S and G can be rewritten as

$$S = (2, 2, 2, 2, 3, 3, 3, 3, 3, 3, 2, 2), \quad G = (3, 3, 3, 3, 3, 3, 2, 2, 2, 2, 2, 2).$$

Observe that the chronotonic distance between these rhythms is equivalent to the L_1 distance between the two auxiliary vectors,

$$d_{chrono}(A, B) = |2 - 3| + |2 - 3| + ... + |2 - 2| + |2 - 2| = 8$$

As a consequence, the Hamming, Swap and Chronotonic distances on binary sequences are examples that can be reduced to the L_1 metric.

3   Computing Ancestral Nodes. Properties

In this section we show some properties of the phylogenetic graphs that lead us to find the unknown nodes (if any) when the L_1 distance is used as similarity distance. Hereafter, we are given a phylogenetic graph with n nodes that are represented using k numerical components. Since we are interested in comput-ing the exact representation of the unknown nodes, we suppose the weights of the edges in the phylogenetic graph matches perfectly the L_1-distance matrix, that is, the *fit index* equal to 1. Otherwise, our solution can be interpreted as an approximation. We denote the nodes by capital letters $A, B, C, \ldots$ and the i-th component of a node by using the same letter with subscript i. For exam-ple, the i-th component of the node A is a_i and A can be represented by the

k-tuple $(a_1, a_2, \ldots, a_k)$. Unless otherwise is specified, we denote by $d(A, B)$ the L_1 distance between two known nodes A and B.

Lemma 1. *Let A, B and C be nodes in a phylogenetic graph, if B lies in the shortest path between A and C, then $a_i \leq b_i \leq c_i$ or $c_i \leq b_i \leq a_i$, $1 \leq i \leq k$.*

Proof. Using that B is in the shortest path between A and C we have

$$d(A, B) + d(B, C) = d(A, C) \tag{1}$$

From the triangle inequality we get that $|a_i - b_i| + |b_i - c_i| \geq |a_i - c_i|$ for $i = 1, 2, \ldots, k$. Adding these inequalities we obtain $|a_i - b_i| + |b_i - c_i| = |a_i - c_i|$ as a consequence of (1).

Let us suppose $a_i \leq c_i$. We analyze three cases: $b_i < a_i$, $c_i < b_i$ and $a_i \leq b_i \leq c_i$. If $b_i < a_i$ then $|c_i - b_i| = c_i - b_i > c_i - a_i = |a_i - c_i|$ and we get a contradiction. If $b_i > c_i$ then $|a_i - b_i| = b_i - a_i > c_i - a_i = |a_i - c_i|$ and we get a contradiction. Then we have $a_i \leq b_i \leq c_i$. The case $a_i \geq c_i$ is analogous and the result follows.

Definition 1. *An unknown node X in a phylogenetic graph is said to be <u>crossroad</u> if there exist at least three known nodes A, B and C such that the shortest paths between A and B, A and C, B and C, contain the node X. We say that A, B and C are <u>connected in crossing through</u> X.*

Lemma 2. *Let X be a crossroad node and A, B and C three nodes connected in crossing through X, then for all $i, 1 \leq i \leq k$, the value of x_i is the median element of the multi-set of numbers $\{a_i, b_i, c_i\}$.*

Proof. By Lemma 1, we have $\min(a_i, b_i) \leq x_i \leq \max(a_i, b_i), \min(a_i, c_i) \leq x_i \leq \max(a_i, c_i)$, and $\min(b_i, c_i) \leq x_i \leq \max(b_i, c_i)$. From the inequalities above we conclude that $\max(\min(a_i, b_i), \min(a_i, c_i), \min(b_i, c_i)) \leq x_i$ and $x_i \leq \min(\max(a_i, b_i), \max(a_i, c_i), \max(b_i, c_i))$.

Without loss of generality, let $a_i \leq b_i \leq c_i$. Then the inequality above become $b_i \leq x_i \leq b_i$, and the result follows.

The following property is a direct consequence from Lemma 2,

Corollary 1. *A crossroad node X can be computed in $O(k)$ time when three nodes connected in crossing through X are given.*

By construction, a phylogenetic graph is a connected graph (for every pair of nodes in the graph, there exists a path between them). Also, we can assume that all unknown nodes has degree at least 3, otherwise it has no sense to consider them in the phylogenetic graph.

Lemma 3. *Let X be an unknown node in a phylogenetic graph. If the resulting graph after removing X and the incident edges on X has at least three connected components, then X is crossroad.*

Proof. Let G be a phylogenetic graph and let $G-\{X\}$ be the resulting graph after removing X and the incident edges on X. Note that every connected component in $G - \{X\}$ contains known nodes because the phylogenetic software (e.g., the SplitTree) adds an unknown node only if it is necessary for a path between two known nodes. Let A, B and C be three known nodes from three distinct connected components of the graph $G - \{X\}$. Any path between A and B in G passes throw X, because they are in different connected components in $G-\{X\}$. Then X is in the shortest path between A and B. Applying the same argument to B and C, and C and A we conclude that X is crossroad and A, B and C are nodes connected in crossing through X.

The following results are straightforward,

Corollary 2. *An unknown node that is not in a cycle is crossroad.*

Corollary 3. *In a phylogenetic network with no cycles (phylogenetic tree), all the unknown nodes are crossroad.*

As a consequence of Corollaries 1 and 3, all unknown nodes in a tree can be computed in polynomial time. Our algorithm uses two stages. First, by using the necessary conditions stated in the properties above, the candidate representation for each unknown node is calculated. Note that this representation is unique by Corollary 1. After that, we have to check if the computed unknown nodes fit the weights of the adjacent edges, that is, the distance between the candidate X and every known node N matches the sum of the weights in the shortest path from X to N. In this case, we say that the representation is *valid*. In the negative case, the corresponding node does not exist. Now we are ready to show the main result of this section,

Theorem 1. *In a phylogenetic network with no cycles (phylogenetic tree), the possible values for unknown nodes can be computed in $O(nk)$. Each possible value represents a candidate node for which the matching on the distances can be done in $O(nk)$ time.*

Proof. The first step is to find, for every unknown node X, a triple of nodes connected in crossing through X. Let G be the graph that represents the phylogenetic network and let T be a rooted tree on G whose root is an arbitrary known node A. We denote by $tag(N)$ the first known node in the subtree of T with root in N, thus

$$tag(N) = \begin{cases} N & \text{if } N \text{ is known} \\ tag(C_1) & \text{if } N \text{ is unknown, where } C_1 \text{ is one child of } N \text{ in } T. \end{cases}$$

Note that the leaves of T are known nodes, then using a bottom up analysis we can find $tag(N)$ for every node in T in $O(n)$ time. Since X has degree at least 3, it has at least two children C_1 and C_2. The nodes A (root of T), $tag(C_1)$ and $tag(C_2)$ are connected in crossing through X. Now, from the triple of nodes connected in crossing through X, the components of X can be computed in $O(k)$ time by using Corollary 1. Finally, for each candidate vector X we can check the distance to every node in $O(nk)$ time by using a DFS from X.

3.1   Finding Unknown Nodes in a Phylogenetic Network with Cycles

If the unknown node lies in a cycle, it can be non-crossroad. In this case, we are not able to exactly compute the coordinates of the node. However, we give an algorithm to obtain bounds of their coordinates. Also, we show that if a valid representation for a crossroad node exists, our algorithm computes the exact value for each of its components.

Let G be the phylogenetic graph containing cycles. We use $d_G(A, B)$ to denote the cost of the shortest path between two known nodes A and B in the graph. Given an unknown node X in G, the Algorithm 1 computes the vectors *lower* and *upper* containing the bounds, and if a valid representation for X exists, then $lower[i] \leq x_i \leq upper[i]$, $1 \leq i \leq k$. In the line 1 we compute the *shortest paths tree from X*[1], and we can use $d_G(X, A)$ for every node A in the graph. If X lies in the shortest path between the known nodes A and B (line 5), then $x_i \in [\min(a_i, b_i), \max(a_i, b_i)]$ and we update the lower and upper bounds (lines 6, 7 and 8). Is easy to see that Algorithm 1 spends $O(n^2 k)$ time.

Algorithm 1. Computing lower and upper bounds on the unknown node coordinates

---

Input: Graph G and one unknown node X
Output: Two k-tuples *lower*, *upper*
1: $Dijkstra(Graph : G, root : X)$
2: $lower \leftarrow [-\infty, -\infty, ...]$
3: $upper \leftarrow [+\infty, +\infty, ...]$
4: **for** every known nodes A, B **do**
5:     **if** $d_G(X, A) + d_G(X, B) = d(A, B)$ **then**
6:         **for** $i \in \{1, 2, ..., k\}$ **do**
7:             **if** $lower[i] < Min(a_i, b_i)$ **then** $lower[i] \leftarrow Min(a_i, b_i)$
8:             **if** $upper[i] > Max(a_i, b_i)$ **then** $upper[i] \leftarrow Max(a_i, b_i)$

Theorem 2. *Let G be a phylogenetic graph and X an unknown node. If X is crossroad, Algorithm 1 allows to compute the exact representation or guarantee that it is impossible to find its value. Otherwise, lower and upper bounds for each its components are calculated.*

Proof. Let $(A^{(1)}, B^{(1)}), (A^{(2)}, B^{(2)}), \ldots, (A^{(r)}, B^{(r)})$ be all the pairs of known nodes such that X lies in the shortest path between $A^{(j)}$ and $B^{(j)}$, $1 \leq j \leq r$. By Lemma 1 x_i is in $I_i = \bigcap_{j=1}^{r}[\min(a_i^{(j)}, b_i^{(j)}), \max(a_i^{(j)}, b_i^{(j)})]$. Suppose that using Algorithm 1, we obtain $lower[i] > upper[i]$ for some i, then I_i is empty and, therefore a valid representation for X does not exist. In the case $lower[i] = upper[i]$ $\forall i$ ($1 \leq i \leq k$) we obtain the only possible representation of X and in order to guarantee the node is correct, we must check the distance to every

---

[1] A spanning tree rooted at X such that for any node Y, the reversal of the Y to X path in the tree is a shortest path from X to Y.

known node in G, that is, to verify if it is a valid representation. The matching operation takes $O(nk)$ time.

Finally, if X is crossroad then there exist three nodes $A^{(*)}$, $B^{(*)}$ and $C^{(*)}$ connected in crossing through X. Thus $I_i \subseteq \{[\min(a_i^{(*)}, b_i^{(*)}), \max(a_i^{(*)}, b_i^{(*)})] \cap [\min(b_i^{(*)}, c_i^{(*)}), \max(b_i^{(*)}, c_i^{(*)})] \cap [\min(a_i^{(*)}, c_i^{(*)}), \max(a_i^{(*)}, c_i^{(*)})]\}$ and by Lemma 2, I_i is only one point or is empty, $1 \leq i \leq k$. Therefore we can compute the exact representation of X or guarantee that it does not exist.

4   The Hamming Distance for Strings

In this section we extend the previous results from numerical to symbolic sequences when the Hamming distance is considered. We present the version of Lemmas 1 and 2 for this scenario. With these technical results, the algorithm can be easily derived.

Lemma 4. *Let A, B and C be nodes in a phylogenetic graph obtained for n strings and the Hamming distance. If B lies in the shortest path between the strings A and C, then $a_i = b_i$ or $c_i = b_i$, $1 \leq i \leq k$.*

Proof. By the definition of Hamming distance, we have:

$$d(A,B) = \sum_{i=1}^{k} \delta(a_i, b_i), \; d(B,C) = \sum_{i=1}^{k} \delta(a_i, b_i), \; d(A,C) = \sum_{i=1}^{k} \delta(a_i, b_i)$$

where $\delta(x,y)$ is 0 if $x = y$ and 1 if $x \neq y$. Note that $\delta(a_i, b_i) + \delta(b_i, c_i) \geq \delta(a_i, c_i)$ and $d(A,B) + d(B,C) = d(A,C)$ because B lies in shortest path between A and C. As a consequence, we get $\delta(a_i, b_i) + \delta(b_i, c_i) = \delta(a_i, c_i)$.

We have two cases $a_i = c_i$ or $a_i \neq c_i$. If $a_i = c_i$ then $\delta(a_i, b_i) + \delta(b_i, c_i) = 0$ and $a_i = b_i = c_i$. If $a_i \neq c_i$ then $\delta(a_i, b_i) + \delta(b_i, c_i) = 1$ and $a_i = b_i$, $b_i \neq c_i$ or $a_i \neq b_i$, $b_i = c_i$, and this proves the Lemma.

Lemma 5. *Let X be a crossroad node and A, B and C three nodes connected in crossing through X. If a valid representation for X exists, then the value of $x_i, (1 \leq i \leq k)$, is the element that repeats more (the mode) in the multi-set $\{a_i, b_i, c_i\}$.*

Proof. First, we prove that at least two of a_i, b_i, c_i are equal. Suppose that a_i, b_i, c_i are different, then by applying Lemma 4 on the shortest path between A and B, we get two cases $x_i = a_i$ or $x_i = b_i$. If $x_i = a_i$, by Lemma 4 on the shortest path between B and C we arrive to $x_i = b_i$ or $x_i = c_i$ and this is a contradiction. The other case is analogous. Therefore, at least two of a_i, b_i, c_i are equal.

Now, let suppose that $a_i = b_i$. By Lemma 4 on the shortest path between A and B we have $x_i = a_i = b_i$. The cases $b_i = c_i$ or $a_i = c_i$ are similar and the proof is done.

The Algorithm 1 can be slightly modified to calculate the candidate coordinates for each crossroad node when the Hamming distance is used.

5   Examples

In this section we show some examples of unknown nodes in phylogenetic trees that can be exactly computed from extant sequences. We consider two examples from the musical domain and one example from Biology. Every discovery of ancestral nodes suggests a possible ethnomusicological or Biological research project to determine the exact nature of its influence. Possible discussions on the influence of theses ancestral nodes in the evolution of the species is outside the scope of this paper.

5.1   Example 1

In [9] a phylogenetic tree for a set of musical rhythms by using the Hamming distance was computed via SplitTree. Each rhythm has a representative binary string of sixteen bits, see Table 1. The matrix distance is illustrated in Table 2 and the SplitsTree software is used to obtain the phylogenetic tree. This tree has only one unknown node, denoted by X in Figure 4. It is easy to see that X is a crossroad node.

Table 1. Binary representation of rhythms

Rhythm	Representation
1. Shiko	1000101000101000
2. Son	1001001000101000
3. Soukous	1001001000110000
4. Rumba	1001000100101000
5. Bossa-Nova	1001001000100100
6. Gahu	1001001000100010

Table 2. Distances between rhythms

	1	2	3	4	5	6
1	0	2	4	4	4	4
2	2	0	2	2	2	2
3	4	2	0	4	2	2
4	4	2	4	0	4	4
5	4	2	2	4	0	2
6	4	2	2	4	2	0

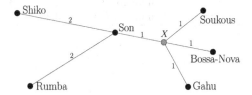

Fig. 4. SplitsTree using Hamming or L_1 distance in example 1

In order to find the unknown node X we consider the nodes Gahu, Bossa-Nova and Soukous as the nodes connected in crossing through X. Our algorithm assigned the value [1001001000100000] to the node X.

5.2   Example 2

By considering the Chronotonic distance between flamenco rhythms, a phylogenetic graph with four unknown nodes has been computed in [3], see Figure 5. The Chronotonic vector is obtained from the binary vector, subtracting the positions of successive one's. As noted in Section 2, we transform the
Chronotonic distance into L_1 by using the auxiliary vectors and it is easy to see
that the four unknown nodes in this phylogenetic network are crossroad. Therefore we can exactly compute their values, illustrated in Table 3. For example,
the node X was computed by using Soleá, Bulería and Guajira as the nodes
connected in crossing through it, and the computed values are valid because all
L_1 distances between the unknown and others nodes coincide with the length of
the corresponding shortest paths in the graph.

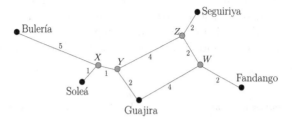

Fig. 5. The phylogenetic graph using the Chronotonic distance

Table 3. The found unknown flamenco rhythms

Unknown node	Connected in crossing	Auxiliary vector
X	Soleá, Bulería, Guajira	(2,2,3,3,3,3,2,2,2,2,2,1)
Y	Bulería, Guajira, Seguiriya	(2,2,3,3,3,3,2,2,2,2,2,2)
Z	Seguiriya, Fandango, Bulería	(2,2,3,3,3,3,3,3,3,3,2,2)
W	Seguiriya, Fandango, Guajira	(3,3,3,3,3,3,3,3,3,3,2,2)

5.3   Example 3

The following data were extracted from a research work on primate mitochondrial DNA evolution published in [6]. They consider 12 primates and the size of
each sequence is 898. For our example, we selected 5 strings with 60 characters
each. The taxa is illustrated in Table 4. The phylogenetic graph computed with
the SplitsTree software and Hamming distance is given in Figure 6. All unknown
nodes are crossroad. Then, if any, their values can be determined. However, since

Table 4. DNA of a group of primates

Primates	DNA
Homo_sapiens	CCACCCTCGTTAACCCTAACAAAAAAAACT CATACCCCCATTATGTAAAATCCATTGTCG
Pan	CCACCCTCATTAACCCTAACAAAAAAAACT CATATCCCCATTATGTGAAATCCATTATCG
Gorilla	CCACCTTCATCAATCCTAACAAAAAAAGCT CATACCCCCATTACGTAAAATCTATCGTCG
Pongo	CTACCCTCATTAACCCCAACAAAAAAAACC CATACCCCCACTATGTAAAAACGGCCATCG
Hylobates	CCACCCTTATTAACCCCAATAAAAAGAACT TATACCCGCACTACGTAAAAATGACCATTG

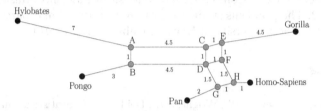

Fig. 6. The phylogenetic graph using Hamming distance and strings.

Table 5. The unknown nodes for phylogenetic reconstruction

Unknown	Connected in crossing	Value
A	Gorilla, Hylobates, Pongo	CCACCCTCATTAACCCCAACAAAAAAAACT CATACCCCCACTACGTAAAAACGACCATCG
B	Hylobates, Pongo, Pan	CCACCCTCATTAACCCCAACAAAAAAAACT CATACCCCCACTATGTAAAAACGACCATCG
G	Pongo, Pan, Homo-Sapiens	CCACCCTCATTAACCCTAACAAAAAAAACT CATACCCCCATTATGTAAAATCCATTATCG
H	Pan, Homo-Sapiens, Gorilla	CCACCCTCATTAACCCTAACAAAAAAAACT CATACCCCCATTATGTAAAATCCATTGTCG

the distance from the nodes C, D, E, F to the known nodes are not integer values, these nodes do not match the weights in any way. On the other hand, the distance from A, B, G, H to the known nodes are integer values and our algorithm is able to find their values. The Table 5 shows the values obtained by our algorithm. After finding their values we checked that the distances to the known nodes are correct, that is, the Hamming distance between every pair of nodes coincides with the distance on the phylogenetic graph.

6   Conclusions and Future Research

We have shown that the reconstruction of all ancestral nodes in a given phylogenetic tree can be done in polynomial time if the similarity distance can be reduced to the L_1 metric. Our algorithm runs in $O(mnk)$ if the tree has n nodes and m ancestral nodes in a k-dimensional space. Additionally, we proved some properties that must be satisfied by ancestral nodes in general graphs. By imposing these conditions, the unknown nodes can be obtained in many real examples, even for phylogenetic graphs with cycles. To our knowledge, this is the first efficient solution to the problem.

Acknowledgments. This research was partly funded by the Junta de Andalucía, project COFLA: Computational Analysis of Flamenco Music, FEDER-P12-TIC-1362. The problems studied here were introduced and partially solved during a visit to the University of Havana, Cuba, January 2014.

References

1. Brown, A.R., Towsey, M.W., Wright, S.K., Deiderich, J.: Statistical analysis of the features of diatonic music using Music software. In: Proceedings Computing ARTS 2001 Digital Resources for Research in the Humanities, Sydney, Australia: The University of Sydney, pp. 1–11 (2001)
2. Collard, M., Shennan, S.J., Tehrani, J.J.: Branching, blending, and the evolution of cultural similarities and differences among human populations. Evolution and Human Behavior 27, 169–184 (2006)
3. Díaz-Báñez, J.M., Farigu, G., Gómez, F., Rappaport, D., Toussaint, G.T.: El compás flamenco: a phylogenetic analysis. Bridges: Mathematical Connections in Art, Music, and Science, 61–70 (2004)
4. Guastavino, C., Gómez, F., Toussaint, G., Marandola, F., Gómez, E.: Measuring similarity between flamenco rhythmic patterns. Journal of New Music Research 38, 129–138 (2009)
5. Gustafson, K.: A new method for displaying speech rhythm, with illustrations from some Nordic languages. In: Prosody IV, N., Gregersen, K., Basboll, H. (eds.), pp. 105–114. Odense University Press (1987)
6. Hayasaka, K., Gojobori, T., Horai, S.: Molecular phylogeny and evolution of primate mitochondrial DNA. Molecular Biology and Evolution 5, 626–644 (1988)
7. Huson, D., Bryant, D.: Application of phylogenetic networks in evolutionary studies. Molecular Biology and Evolution 23, 254–267 (2006)
8. Huson, D.H.: SplitsTree: analyzing and visualizing evolutionary data. Bioinformatics 14, 68–73 (1998)
9. Toussaint, G.T.: Classification and phylogenetic analysis of African ternary rhythm timelines. In: Proceedings of BRIDGES: Mathematical Connections in Art, Music and Science, Granada, Spain, July 23–27, pp. 25–36 (2003)
10. Yang, Z.: PAML 4: phylogenetic analysis by maximum likelihood. Molecular Biology and Evolution 24, 1586–1591 (2007)

Supporting Bioinformatics Applications with Hybrid Multi-cloud Services

Ahmed Abdullah Ali[2], Mohamed El-Kalioby[2], and Mohamed Abouelhoda[1,2]

[1] Faculty of Engineering, Cairo University
Giza, Egypt
[2] Center for Informatics Sciences, Nile University
Sheikh Zaid City, Egypt
mabouelhoda@yahoo.com

Abstract. Cloud computing provides a promising solution to the big data problem associated with next generation sequencing applications. The increasing number of cloud service providers, who compete in terms of performance and price, is a clear indication of a growing market with high demand. However, current cloud computing based applications in bioinformatics do not profit from this progress, because they are still limited to just one cloud service provider. In this paper, we present different use case scenarios using hybrid services and resources from multiple cloud providers for bioinformatics applications. We also present a new version of the *elasticHPC* package to realize these scenarios and to support the creation of cloud computing resources over multiple cloud platforms, including Amazon, Google, Azure, and clouds supporting OpenStack. The instances created on these cloud environments are pre-configured to run big sequence analysis tasks using a large set of pre-installed software tools and parallelization techniques. In addition to its flexibility, we show by experiments that the use of hybrid cloud resources from different providers can save time and cost.

Keywords: Bioinformatics, High performance computing in Bioinformatics, Novel architecture, Cloud Computing, Sequence Analysis.

1 Introduction

Cloud Computing is no longer a new term that sounds strange to scientists in the life science domain. It is currently a usual accepted practice, due to its flexibility, efficiency, usability, scalability, and cost-effectiveness. This development has in fact been enabled 1) by companies offering cloud based services in the form of a *software as service* (SaaS) with a pay-as-you-go business model, and 2) by academic institutions providing open-source cloud-based software tools.

Prominent SaaS examples include BaseSpace (`https://basespace.illumina.com`) and IonReporter (`https://ionreporter.lifetechnologies.com`), which are developed by the two major manufacturers of Next Generation Sequencing (NGS) machines. Examples from academic institutes include CloVR [1] and QIIME [2] for metagenomics analysis. These are distributed in the form of virtual

F. Ortuño and I. Rojas (Eds.): IWBBIO 2015, Part I, LNCS 9043, pp. 415–425, 2015.

machine (VM) images that can run on private or public clouds. Other less recent examples include Crossbow [3] for NGS read alignment and SNP calling, RSD-Cloud [4] for comparative genomics, Myrna [5] for RNA-Seq data analysis, among others. These tools are used only in Amazon's cloud environment, after the user instantiates instances at own cost from the respective pre-configured machine images deposited in Amazon. To speed up computation, these cloud-based tools use some middleware packages to establish a computer cluster and analyze the data in parallel. These middleware packages include StarCluster [6], Vappio [7], Cloud-Man [8], and first version of our *elasticHPC* [9]. It is worth mentioning that these packages are limited to the Amazon cloud.

The cloud computing market (especially the IaaS one) is no longer a few players field, dominated by Amazon (AWS [10]). Very recently, significant entities, such as Google, Windows Azure [11], RackSpace, and IBM have joined the IaaS market, with very competitive performance and pricing models, as we will discuss below. Moreover, some academic clouds (such as Magellan [12] and DIAG [13]) started to offer services for researchers.

To profit from the current competition, it is important to extend current bioinformatics solutions to run on multiple cloud platforms. To exploit parallelization power, these solutions should enable two major scenarios: In the first, different separate clusters are automatically created and configured on different cloud environments (multiple non-federated cloud clusters). In the second, a hybrid cluster is created with machines from different providers (federated cloud cluster). Providing a package implementing these scenarios is not a straightforward task, because all of the above mentioned clouds are built with different architecture, usage scenarios, APIs, and business models.

In this paper, we present a new version of the *elasticHPC* package to support the creation and management of a computer cluster for bioinformatics applications over multiple clouds. The older version of *elasticHPC* was specific to Amazon but the new version supports also Google Cloud, Microsoft Windows Azure, and any other OpenStack-based cloud platform. The package supports different use case scenarios as well as job submission and data management to leverage the use of the multicloud for big data bioinformatics applications. To the best of our knowledge, it is the first package supporting multicloud (including recent commercial ones) for bioinformatics applications.

This paper is organized as follows: In the following section, we present use case scenarios for multicloud. In Section 3, we present the new features of *elasticHPC* supporting these scenarios. In Section 4, we discuss some implementation details. Sections 5 and 6 include experiments and conclusions, respectively.

2   Use Case Scenarios for Multicloud

To easily present different use case scenarios for multicloud, we use a simplified version of the variant analysis workflow based on NGS technology as an example. This workflow is used to identify mutations compared to a reference genome. The basic steps of this workflow are depicted in Figure 1. In the first step, the

Fig. 1. The variant analysis workflow: The tools BWA, Picard, GATK are usually used for the three steps of the workflow. On the arrows, we write the different file formats of the processed data.

NGS reads are mapped (aligned) to a reference genome. In the second step, the duplicate reads are marked to improve the subsequent variant calling step. The variant calling step involves the identification of mutations by analyzing the aligned reads in comparison to the reference genome. In biomedical practice, the program BWA [14] is usually used for the read mapping step, the package Picard [15] (http://picard.sourceforge.net) is used to mark duplicate reads, and GATK [15] is used to call the variants.

The currently existing scenario for running this workflow in the cloud includes the creation of a computer cluster composed of a number of nodes over one cloud computing platform. The parallelization is achieved by decomposing the set of input reads into blocks that are distributed over the cluster nodes.

In the case of executing this workflow over multiple clouds, there are different scenarios with different configurations to match different requirements. These scenarios can be broadly categorized into two basic ones: 1) Multiple clusters over multiple clouds, and 2) One cluster of federated cloud machines. (Combination of both scenarios, where one of the multiple clusters is composed of nodes from multiple clouds, is also possible.) Our *elasticHPC* package supports these scenarios as we will discuss later in the implementation and experiment sections.

2.1   Multiple Clusters Over Multiple Clouds

There are two different configurations of this scenario. The first configuration, which is represented in Figure 2(a), includes many computer clusters; each is created in one cloud environment. The data is partitioned and sent to each cluster for processing. After completion, the data is collected in one persistent storage. This scenario matches the case when the data partitions are independent from one another, as in the case of reads from different sample groups or from different chromosomes.

The decision of using this scenario depends on whether there is a *deadline* (time) constraint or not. In other words, there is a constraint to finish the computation before a certain time, while reducing the cost as much as possible. Assume that there is a deadline constraint and there is one cloud available free of charge (or with reduced price), but it has certain limitations on resource availability. Then, in this case, one would use whatever available of resources in the cheaper cloud to process as much as possible of the data and uses the other cloud

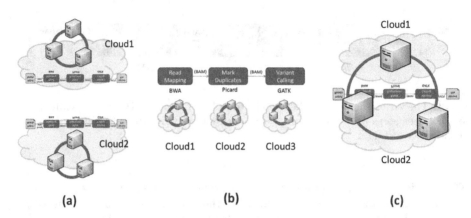

Fig. 2. Three general configurations for computer clusters over multiple clouds. Part(a): Multiple independent clusters over multiple clouds and each cluster processes part of the input data. Part(b): Each cluster is created in one cloud and solves a step of the workflow. Part(c): One cluster composed of different machines from different clouds.

to scale up the computation power to finish before the deadline. This saves cost compared to the case in which all the workflow runs in the more expensive cloud.

The second configuration, which is shown in Figure 2(b), includes the creation of the computer clusters in a staged fashion, where each cluster is dedicated to one step of the workflow. More precisely, one can select that the read mapping step of the variant analysis workflow runs in Google cloud, the mark duplicate step runs in Amazon, and the variant calling runs in Azure or on Google again. This configuration can be justified only in the case that there are some technical limitations (like the RAM size) preventing a step from running in one cloud, but the other steps can run in cheaper cloud(s). This is because the data transfer is costly and time consuming. In the experiments section, we show when this configuration can retrieve better results than the usual configuration including one cluster over a single cloud platform.

2.2   One Cluster of Federated Cloud Machines

In this use case scenario, a computer cluster whose nodes come from different cloud providers is created, as shown in Figure 2(c). This cluster includes one master job queue which dispatches the jobs among the nodes in different clouds. The job queue can be configured to schedule the jobs depending on the availability of resources for the task at hand and the performance of the nodes at each cloud site. The job manager can also consider extra cloud specific factors like the latency of transferring the data among cloud sites, access to storage, and the efficiency of the file system of each cloud.

This scenario provides more granularity than the previous ones, as it works on the job level rather than the whole (sub-) workflow level. This can be very useful in the case that the processing time differs from one job to another depending on the characteristics of the processed data. For example, the time of processing longer NGS read is more than that of processing shorter ones. In this case, the jobs associated with the longer reads might be assigned to some instances of certain computing capabilities in one specific cloud site and the jobs associated with the shorter reads might go to other nodes in another cloud site. Determining the affinity of a job to a cloud platform and its cluster nodes can be implemented by an additional module examining the data characteristics. The success of this scenario depends on fast Internet connection among the cloud sites and good management of input data according to its characteristics.

3   Package Features

elasticHPC supports the above-mentioned use case scenarios and related cluster configurations through the following set of features:

- *Creation of Multicloud Clusters*: The package can create virtual machines of different types on different cloud platforms. The nodes can be grouped into different clusters. Even within one cloud environment, we can have multiple clusters, where the nodes of each can have different specifications.
- *Management of Cluster*: The package supports addition (scaling up) and removal (scaling down) of nodes from running clusters. The type of the added nodes can be different from the currently used nodes in a cluster. There are also functions to retrieve the status of the allocated nodes.
- *Data Management Options*: The package allows mounting of virtual disks and establishment of a shared file system. (Default option is the NFS). The persistent storage can also be used as shared file system. In AWS, we use the library S3FS. The package has also a functionality to speed up data transfer among nodes. For data to be replicated over multiple nodes, we use peer-to-peer like protocol to reduce the total time transfer to $O(n \log n)$. Considering different data management scenarios is important to make the data available to the cluster nodes quickly and with reduced prices, noting that the user is charged for data transferred out of the cloud sites.
- *Job Submission Options*: *elasticHPC* automatically configures a job scheduler (including security settings among the different providers) when the cluster is started. The default job schedule is PBS Torque, but SGE is also supported. The automatic setup takes into consideration the use case scenarios and cluster configuration defined by the user.
- *Bioinformatics Tool Set*: The original package of *elasticHPC* includes about 200 sequence analysis tools coming from BioLinux, EMBOSS, and NCBI Toolkit. The new version of *elasticHPC* has been augmented with more tools for NGS data analysis tools including SHRiMP, Bowtie2, GATK, BWA, among others.

Table 1. Features of different clouds. The feature 'Max RAM per Core' is the result of dividing the RAM size by the number of cores. Amazon charges more for higher IO speed.

Resource	AWS	Azure	Google
Compute			
IaaS Service Name	EC2	Azure VM	GCE VM
Maximum Core VM	32	16	16
Maximum RAM VM	244 GB	112 GB	104 GB
Max RAM per Core	8.1	7	6.5
Hard Disk	EBS Volume	VHD Page Blob	Persistent HD
Maximum Volume/HDD	1 TB	1 TB	10TB
One HDD multiple mounts	No	No	Yes
VM Format	RAW,VMDK, VHD	VHD	RAW Format
Snapshot	Yes	Yes	Yes
IO speed	High(Variable)	High	NA
Storage			
Persistent Storage	S3	Page/Block Blobs	Buckets
Security			
Firewall	Security Groups	EndPoints	Firewall
Grouping VMs	Placement Group	Cloud Service	Instance Group
Business Model			
Min. Charge Unit	Hour	Minute	Minute
On-Demand Model	Yes	Yes	Yes
Spot Model	Yes	No	No
Reserved Model	Yes	No	No
Sustained use	No	No	Yes
Charge IO/Data Transfer	Yes	Yes	Yes (Cheaper)
Cost per IO speed	Yes	No	NA

4   Implementation of Multi-cloud *elasticHPC*

4.1   Quick Review of Major Cloud Platforms

Before introducing the implementation details of *elasticHPC*, we review the three major commercial providers Amazon, Azure, and Google. Table 1 compares the features among the three, including the business model. The following paragraphs include brief explanation.

Amazon Web Services (AWS). Amazon Web Services (AWS) is the part of Amazon responsible for cloud computing services. It includes many IaaS products, including virtual machines, storage, and networking. Amazon divides the virtual machines (VMs) into families depending on their power (in terms of CPU and RAM). For example, the highest CPU virtual machine of type c3.8xlarge in AWS has 32 Cores and 108 GB RAM and costs $1.68 per hour. (Prices are per

November 2014 and for Linux machines.) Persistent storage solutions in AWS include EBS volumes, which can be mounted to running instances as virtual hard disks, and S3 which functions as web-storage system.

Pricing models in the language of AWS include 'on demand', 'reserved', and 'spot'. On-demand model refers to pay-as-you-go, where users are charged by the time their machines are running. Minimum charging unit is one hour; i.e., a fraction of an hour is charged as an hour. Reserved model refers to machines allocated over long time period and there is a discount for that. In the spot model, the user bids on unused resources. Once the machine price gets below the user's bid, the machines are started. But amazon can terminate any spot instance, if the bid becomes less than the new spot price. AWS charges for IO operations and data transfer out of AWS sites to the Internet.

Microsoft Azure. Windows Azure [11] is a commercial cloud computing platform developed by Microsoft. The most powerful virtual machine (VM) of Azure is of type A9. It has 16 cores and 112 GB RAM and costs $4.47 per hour. A new VM can be instantiated from a supported image by Azure (with selected operating system), or from a previously created one stored on a virtual hard disk (VHD). To each VM, one can mount one or multiple VHD(s) as persistent disks including data that can persist, even after the VM is terminated. Azure provides two types of storage models: page blob and block blob. Page blob is used for file system storage with a maximum size of 1 TB. Block blob is used for object storage with maximum size of 200 GB[2] per object. VHDs and images use page blobs for storing files. To organize the Blobs, each storage account can include one or multiple containers, and each container includes the actual Blobs.

Azure pricing scheme includes only on-demand instances. Like Amazon, Azure also charges per data transfer out to the Internet. Unlike Amazon, Azure charges per minute of virtual machine use.

Google Cloud (GCE). Google has recently provided very competitive IaaS features in terms of performance and prices. Like EC2, Google Compute Engine is the interface to all infrastructure products, including VMs, hard disks, images, snapshots and networking. Depending on their capabilities, the VMs are categorized into families, such as 'high CPU' and 'high memory'. The most powerful machine in Google is of type n1-highmem-16. It has 16 cores and 104 GB RAM, and it costs $1.184 per hour. Google provides only one type of storage, which is object storage. It consists of buckets, and each bucket can contain unlimited number of objects. Buckets are unique within each project and cannot be nested.

Pricing model of Google includes pay-as-you-go for 'on-demand' instances, and charges for data transferred out of Google sites to the Internet. Google offers also discounts for long use of virtual machines with the "sustained use" model. Unlike Amazon, Google charges per minute of virtual machine use.

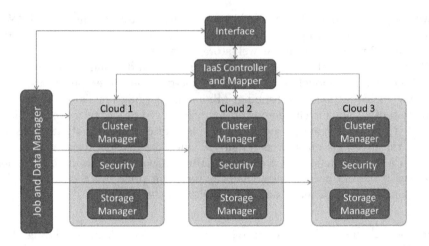

Fig. 3. Architecture Diagram of the multicloud *elasticHPC*

4.2   Implementation Details

Figure 3 shows the basic components of the *elasticHPC*. The *elasticHPC* follows a server client architecture. The user installs a client module at own desktop to control cloud resources. The interface component is responsible for handling all command lines submitted by the user. The command lines are then passed to the IaaS controller-and-mapper layer to translate the request to the corresponding APIs specific to each cloud platform.

Specifying the required resources in one or multiple clouds is achieved through a configuration file, including different clusters in different cloud platforms, number of nodes, machine types, storage, and security. The *elasticHPC* website includes detailed explaination of the configuration file.

For each cloud, there is a module that handles all functions related to the creation and management of clusters at that cloud site including security settings and storage. The reason having a module specific to each cloud is the lack of a standard among the providers and many details are involved. For example, one persistent virtual hard disks in Amazon (which defined as an EBS volume) cannot be mounted to multiple instances. To do this, one has to first copy the volume many times from a snapshot. In Google, however, it is possible to mount a persistent disk as read only for many machines. This fast mounting is very useful for bioinformatics applications, where the disk including tools, databases and indices can be quickly mounted without copy operations. Another example, is that multiple machine instances can be created in parallel in both Google and Amazon, but this is not always the case in Azure, depending on the Azure's cloud service.

The job and data related commands are implemented in the Job management module. This module uses the created cloud resources in one or multiple clouds.

Table 2. Running times in minutes. "MarkD" stands for the mark duplicate step. The numbers between brackets are the cost in USD.

	Azure			Google			Amazon		
	Nodes			Nodes			Nodes		
	4	8	16	4	8	16	4	8	16
BWA	21(1.4)	14(1.9)	12(3.2)	9 (0.3)	6 (0.36)	5 (0.6)	10 (2.2)	7 (4.5)	6 (8.9)
MarkD	284(19)	280(37)	281(75)	181 (5.5)	180 (10)	175 (22)	183 (6.8)	186 (14)	185 (28)
GATK	67(4.5)	35(4.7)	20(5.4)	47 (1.4)	26 (1.6)	15 (1.8)	46 (2.2)	24 (4.5)	13 (4.5)
Totals	372(25)	329(44)	313(84)	237 (7.2)	212 (12.8)	195 (24.4)	239(11.3)	217 (23)	204 (41)

5 Experiments

We demonstrate the new multicloud features of *elasticHPC* with two experiments: In the first experiment, we run the variant analysis workflow in Figure 1 three times on the three different clouds AWS, Azure and Google, and compare the performance and cost. In the second experiment, we use hybrid model in which parts of the workflow run on different clouds.

5.1 Experiment 1

For this experiment, we use an exome dataset from [15] of size $\approx$ 9 GB. (The exome is a set of NGS reads sequenced only from the whole coding regions of a genome.) We ran the three steps in Figure 1 using BWA for read mapping, Picard for marking duplicates, and GATK for variant calling. The workflow was executed three times independently on Google, AWS, and Azure clouds. In each cloud, the 9 GB input data is divided into blocks to be processed in parallel over the cluster nodes. We used nodes of types m3.2xlarge (8 Cores, 30 GB RAM, \$0.56/hour) for Amazon, n1-highmem-8(8 Cores, 52 GB RAM, \$0.452/hour) for Google, and Standard A7 (8 Cores, 56 GB RAM, \$1.00/hour) for Azure.

Table 2 summarizes the running times and the cost of each step using each cloud. It can be observed that Google and Amazon have similar performance and they are better than Azure. Regarding the cost, Google is the best as it charges per minute and not per hour as AWS. We observe that there is no improvement in the running time when adding more nodes for the mark duplication step. This is because Picard requires all the reads to be as one set to sort them.

5.2 Experiment 2

To improve the cost and running time of the previous workflow, one can use just one stronger machine for the mark duplicate step. The best machine is the Amazon c3.8xlarge, which has 32 cores and 108 GB RAM. It costs \$1.68. To further reduce the cost, we can still run steps 1 and 3 on Google cluster. This is equivalent to the configuration in Figure 2(b). This means that we will have hybrid scenario against the scenario based on a single cloud provider. As shown in Table 3, the time for running the mark duplicate step on such strong machine

Table 3. Running times in minutes using single provider and multicloud scenario of 2 providers. The numbers between brackets are the cost in USD.

	Amazon (4nodes)	Google (4nodes)	Google+Amazon (4nodes)
BWA	10 (2.24)	9 (0.27)	9 (0.27)
MarkD	183 (6.83)	181 (5.5)	58 (1.68)
GATK	46 (2.24)	47 (1.41)	47 (1.41)
Data Transfer	NA	NA	23 (2.0)
Totals	239(11.3)	237 (7.18)	138 (5.27)

is $\approx$58 minutes. The uni-directional data transfer took 10-15 minutes (depending on traffic) between Google and Amazon in US West site. The total cost of this scenario on 4 nodes is $\approx$ \$5 and it takes $\approx$2 hours. The key factor for this is that Amazon charges a fraction of an hour as an hour, while Google charges per minutes. That is, Google will always retrieve better cost when the parallelization leads to fractions of hour. So the best cost with comparable performance for these three steps workflow is when we use hybrid cloud of Amazon and Google.

6 Conclusions

In this paper, we have introduced the multicloud version of *elasticHPC*. It enables creation and management of computer clusters over multiple cloud platforms to serve bioinformatics applications. The cluster machines in any cloud is equipped with large set of sequence analysis tools. The package is available for free for academic use at www.elastichpc.org. On the web-page of the package, we have developed a web-interface from which users can start computer clusters in multiclouds and execute data analysis jobs.

Google and Azure currently offer "the charge per minute" pricing model. We expect Amazon will follow soon. With such dynamic market, *elasticHPC* enables the data analyst to use the cloud with the best offer at the time of analysis.

In future versions, we will include different ideas to use shared storage from multicloud as a shared file system, a research field referred to as cloud-backed file systems. This is useful in regions with fast Internet connectivity (as between Google and Amazon in US West).

It is also important to mention that the new multicloud features of *elasticHPC* opens the way for the development of more advanced layers for task scheduling and cost-time optimization.

Acknowledgments. This publication was made possible by NPRP Grant no. 4-1454-1-233 from the Qatar National Research Fund (a member of Qatar Foundation). The statements made herein are solely the responsibility of the authors. We also thank Hatem Elshazly for his support.

References

1. Angiuoli, S., Matalka, M., Gussman, A., et al.: CloVR: a virtual machine for automated and portable sequence analysis from the desktop using cloud computing. BMC Bioinformatics 12(1), 356+ (2011)
2. Gregory, J., Kuczynski, J., Stombaugh, J.: QIIME allows analysis of high-throughput community sequencing data. Nat. Meth. 7(5), 335–336 (2010)
3. Langmead, B., Schatz, M.C., Lin, J., Pop, M., Salzberg, S.L.: Searching for snps with cloud computing. Genome Biology 10(R134) (2009)
4. Wall, D.P., Kudtarkar, P., Fusaro, V.A., Pivovarov, R., Patil, P., Tonellato, P.J.: Cloud computing for comparative genomics. BMC Bioinformatics 11, 259 (2010)
5. Langmead, B., Hansen, K., Leek, J.: Cloud-scale rna-sequencing differential expression analysis with Myrna. Genome Biology 11(8), R83+ (2010)
6. StarCluster, http://web.mit.edu/stardev/cluster
7. Vappio, http://vappio.sf.net
8. Afgan, E., Coraor, N., Chapman, B., et al.: Galaxy CloudMan: delivering cloud compute clusters. BMC Bioinformatics 11(suppl. 12), S4+ (2010)
9. El-Kalioby, M., Abouelhoda, M., Krüger, J., et al.: Personalized cloud-based bioinformatics services for research and education: use cases and the elastichpc package. BMC Bioinformatics 13(S-17), S22 (2012)
10. AWS: Amazon Web Services, http://aws.amazon.com
11. Azure, W., http://www.microsoft.com/windowsazure
12. Magellan-a cloud for Science, http://magellan.alcf.anl.gov
13. DIAG-Data Intensive Academic Grid, http://diagcomputing.org
14. Li, H., Durbin, R.: Fast and accurate short read alignment with burrows and wheeler transform. Bioinformatics 25(14), 1754–1760 (2009)
15. DePristo, M., Banks, E., et al.: A framework for variation discovery and genotyping using next-generation DNA sequencing data. Nature Genetics 43(5), 491–498 (2011)

Relation between Insertion Sequences and Genome Rearrangements in *Pseudomonas aeruginosa*

Huda Al-Nayyef[1,3], Christophe Guyeux[1], Marie Petitjean[2], Didier Hocquet[2], and Jacques M. Bahi[1]

[1] FEMTO-ST Institute, UMR 6174 CNRS, DISC Computer Science Department
Université de Franche-Comté, 16, route de Gray, 25000 Besançon, France
[2] Laboratoire d'Hygiène Hospitalière, UMR 6249 CNRS Chrono-environnement,
Université de Franche-Comté, France
[3] Computer Science Department, University of Mustansiriyah, Iraq
{huda.al-nayyef,christophe.guyeux,marie.petitjean,
jacques.bahi}@univ-fcomte.fr, dhocquet@chu-besancon.fr

Abstract. During evolution of microorganisms genomes underwork have different changes in their lengths, gene orders, and gene contents. Investigating these structural rearrangements helps to understand how genomes have been modified over time. Some elements that play an important role in genome rearrangements are called insertion sequences (ISs), they are the simplest types of transposable elements (TEs) that widely spread within prokaryotic genomes. ISs can be defined as DNA segments that have the ability to move (cut and paste) themselves to another location within the same chromosome or not. Due to their ability to move around, they are often presented as responsible of some of these genomic recombination. Authors of this research work have regarded this claim, by checking if a relation between insertion sequences (ISs) and genome rearrangements can be found. To achieve this goal, a new pipeline that combines various tools has firstly been designed, for detecting the distribution of ORFs that belongs to each IS category. Secondly, links between these predicted ISs and observed rearrangements of two close genomes have been investigated, by seeing them with the naked eye, and by using computational approaches. The proposal has been tested on 18 complete bacterial genomes of *Pseudomonas aeruginosa*, leading to the conclusion that IS3 family of insertion sequences are related to genomic inversions.

Keywords: Rearrangements, Inversions, Insertion Sequences, Pseudomonas aeruginosa.

1 Introduction

The study of genome rearrangements in microorganisms has become very important in computational biology and bio-informatics fields, owing to its applications in the evolution measurement of difference between species [1]. Important

F. Ortuño and I. Rojas (Eds.): IWBBIO 2015, Part I, LNCS 9043, pp. 426–437, 2015.

elements in understanding genome rearrangements during evolution are called transposable elements, which are DNA fragments or segments that have the ability to insert themselves into new chromosomal locations, and often make duplicate copies of themselves during transposition process [2]. Indeed, within bacterial species, only cut-and-paste of transposition mechanism can be found, the transposable elements involved in such way being the insertion sequences. These types of mobile genetic elements (MGEs) seem to play an essential role in genomes rearrangements and evolution of prokaryotic genomes [3,4].

Table 1. P. aeruginosa isolates used in this study

Isolates	NCBI accession number	Number of genes
19BR	485462089	6218
213BR	485462091	6184
B136-33	478476202	5818
c7447m	543873856	5689
DK2	392981410	5871
LES431	566561164	6006
LESB58	218888746	6059
M18	386056071	5771
MTB-1	564949884	6000
NCGM2.S1	386062973	6226
PA1	558672313	5981
PA7	150958624	6031
PACS2	106896550	5928
PAO1	110645304	5681
RP73	514407635	5804
SCV20265	568306739	6190
UCBPP-PA14	116048575	5908
YL84	576902775	5856

In this research work, we questioned the relation between the movement of insertion sequences on the one hand, and genome rearrangements on the other hand, and tested whether the type of IS family influences this relation. Investigations will focus on inversion operations of rearrangement (let us recall that an inversion occurs within genomes when a chromosome breaks at two points, and when the segment flanked with these breakpoints is inserted again but in reversed order, this event being potentially mediated with molecular mechanisms [5,6]). To achieve our goal, we built a pipeline system module that combines existing tools together with the development of new ones, for finding putative ISs and inversions within studied genomes. We will then use this system to investigate the structure of prokaryotic genomes, by searching for IS elements at the boundaries of each inversion.

The contributions of this article can be summarized as follows. (1) A pipeline for insertion sequences discovery and classification is proposed. It uses unannotated genomes and then combines different existing tools for ORF predictions and clustering. It also classifies them according to an international IS database specific to bacteria. Involved tools in this stage are, among others, Prodigal [7], Markov Cluster Process (MCL) [8], and ISFinder[1] [9]. (2) We then use two

---

[1] www-is.biotoul.fr

different strategies to check the relation between ISs and genomic rearrangements. The first one used a well-supported phylogenetic tree, then genomes of close isolates are drawn together, while the questioned relation is checked with naked eye. In the second strategy, inversion cases are thoroughly investigated with ad hoc computer programs. And (3), the pipeline is tested on the set of 18 complete genomes of *Pseudomonas aeruginosa* provided in Table 1. After having checked left and right inversion boundaries according to different window sizes, the probability of appearance of each type of IS family is then provided, and biological consequences are finally outlined.

The remainder of this article is organized as follows. The proposed pipeline for detecting insertion sequences in a list of ORFs extracted from unannotated genomes is detailed in Section 2. Rearrangements found using drawn genomes of close isolates is detailed in Section 3, while a computational method for discovering inversions within all 18 completed genomes of *P. aeruginosa* and results are provided in Section 3.2. This article ends by a conclusion section, in which the contributions are summarized and intended future work is detailed.

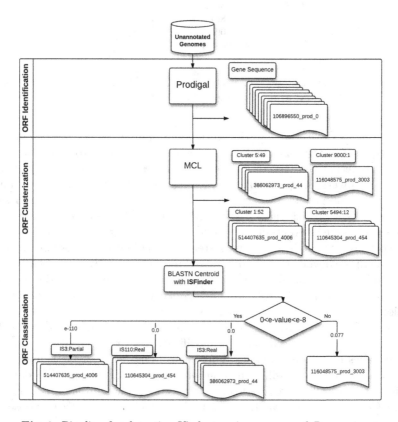

Fig. 1. Pipeline for detecting IS clusters in genomes of *P.aeruginosa*

2   Methodology for IS Detection

In a previous work [10], we have constructed a pipeline system that combines three annotation tools (BASys [11], Prokka [12], and Prodigal [7]) with OASIS [13], that detected IS elements within prokaryotic genomes. This pipeline produces various information about each predicted IS, each IS is bordered by an *Inverted-Repeat* (IR) sequence, number of ORFs in each IS family and group, etc. As we are now only interested in detecting which ORFs are insertion sequences, we have developed a new lightweight pipeline that focuses on such open reading frames.. This pipeline, depicted in Figure 1, relies on ISFinder database [9], the up-to-date reference for bacterial insertion sequences. The main function of this database is to assign IS names and to provide a focal point for a coherent nomenclature. This is also a repository for ISs that contains all information about insertion sequences such as family and group.

The proposed pipeline can be summarized as follows.

Step 1: ORF Identification. Prodigal is used as annotation tool for predicting gene sequences. This tool is an accurate bacterial and archaeal gene finding software provided by the Oak Ridge National Laboratory [7]. Table 1 lists the number of the predicted genes in each genome.

Step 2: ORF Clustering. The Markov Cluster Process (MCL) algorithm is then used to achieve clustering of detected ORFs [8, 14].

Step 3: Clusters Classification. The IS family and group of the centroid sequences of each cluster is determined with ISFinder database. BLASTN program is used here: if the e-value of the first hit is equal to 0, then the cluster of the associated sequence is called a "Real IS cluster". Otherwise, if the e-value is lower than 10^{-8}, the cluster is denoted as "Partial IS". At each time, family and group names of ISs that best match the considered sequence are assigned to the associated cluster. In Table 2 summarizes founded IS clusters found in the 18 genomes of *P. aeruginoza*.

Table 2. Summary of detected IS clusters

	No. of Clusters	Max. size of Cluster	Total no. of IS genes
Partial IS	94	57	362
Real IS	66	49	238
Total IS Cluster	160	-	600

3   Rearrangements in *Pseudomonas aeruginosa*

At the nucleotide level, genomes evolve with point mutations and small insertions and deletions [15], while at genes level, larger modifications including duplication, deletion, or inversion, of a single gene or of a large DNA segment, affect genomes by large scale rearrangements [16, 17]. The pipeline detailed previously investigated the relations between insertion sequences and these genome rearrangements, by using two different methods that will be described below.

3.1   Naked Eye Investigations

In order to visualize the positions of IS elements involved in genomic recombination that have occurred in the considered set of *Pseudomonas*, we have first designed Python scripts that enable us to humanly visualize close genomes. Each complete genome has been annotated using the pipeline described in the previous section, and the strict core genome has been extracted. This latter is constituted by genes shared exactly once in each genome. Thus polymorphic nucleotides in these core genes have been extracted, and a phylogeny using maximum likelihood (RAxML [18, 19] with automatically detected mutation model) has been inferred. (Figure 2).

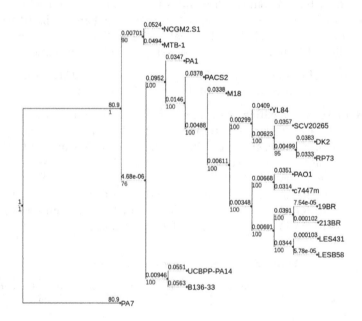

Fig. 2. Phylogeny of *P. aeruginosa* based on mutations on core genome

For each close isolates, a picture has then been produced using our designed Python script, for naked eye investigations. Real and Partial IS are represented with a red and green circles, respectively. Additionally, DNA sequences representing the same gene have been linked either with a curve (two same genes in the same isolate) or with a line (two same genes in two close isolates). Example of recombination events are given below.

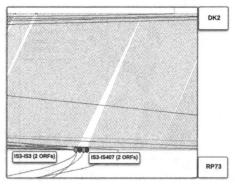

(a) Insertion events of IS sequences have occurred in this set of 18 *P. aeruginosa* species. For instance, when comparing DK2 and RP73, we have found that IS3-IS3 (2 ORFs) and IS3-IS407 (2 ORFs too) have been inserted inside RP73.

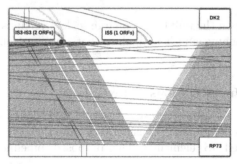

(b) Deletions of insertion sequences can be found too, IS5 (Partial IS) is present in the genome of DK2, while it is deleted in the close isolates RP73.

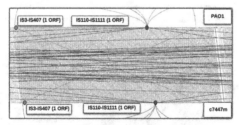

(c) A duplication occurs in the insertion sequence type IS110-IS1111 that contains one ORF (Real IS), as there are 6 copies of this insertion sequence in both PAO1 and C7447m genomes.

Fig. 3. Examples of genomic recombination events: Insertion, Deletion, and Duplication

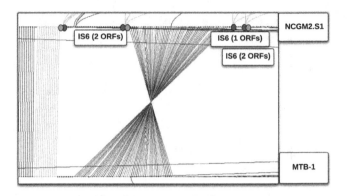

Fig. 4. The surrounding insertion sequences are within the same IS family (IS6) in the NCGM2.S1 genome. We have found too that insertion sequences are not always exactly at the beginning and end positions of the inversion, but they are overrepresented near these boundaries.

We will focus now on the link between large scale inversions and ISs as shown in Figure 4, by designing another pipeline that automatically investigate the inversions.

3.2   Automated Investigations of Inversions

The proposal is now to automatically extract all inversions that have occurred within the set of 18 genomes under consideration, and then to investigate their relation with predicted IS elements. The proposed pipeline is described in Figure 5.

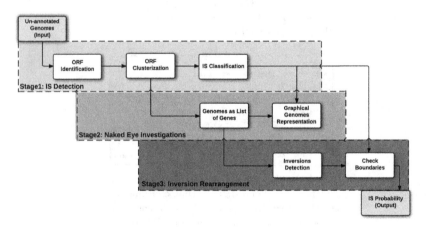

Fig. 5. Pipeline for detecting the role of ISs in inversions

Table 3. Small and large inversions detected from all genomes

Genome 1	Start	Stop	Genome 2	Start	Stop	Length (genes)
19BR	1001	1002	213BR	5094	5095	2
19BR	2907	2920	213BR	2933	2946	14
19BR	684	685	LES431	4689	4690	2
19BR	850	978	LES431	4393	4521	129
PAO1	997	998	c7447m	1977	1978	2
LESB58	4586	4587	LES431	2347	2348	2
DK2	2602	2603	RP73	2516	2517	2
DK2	1309	1558	RP73	3489	3738	250
DK2	260	261	SCV20265	3824	3825	2
DK2	2846	2852	SCV20265	3065	3071	7
M18	3590	3591	PACS2	1920	1921	2
M18	3194	3579	PACS2	2076	2461	386
MTB-1	5581	5582	B136-33	4742	4743	2
UCBPP-PA14	4820	4821	B136-33	2871	2872	2
NCGM2.S1	1053	1307	MTB-1	4507	4761	255
NCGM2.S1	1742	1743	MTB-1	4882	4883	2
PA1	95	96	B136-33	2691	2692	2
PA1	1334	1491	B136-33	1286	1443	158
PACS2	94	97	PA1	495	498	4
PACS2	970	1206	PA1	2220	2456	237
SCV20265	45	46	YL84	721	722	2
SCV20265	261	462	YL84	306	507	202
UCBPP-PA14	259	260	B136-33	3507	3508	2
YL84	721	722	M18	43	44	2
YL84	768	983	M18	5555	5770	216
YL84	721	722	PAO1	44	45	2
YL84	1095	1264	PAO1	5192	5361	170

- **Step1**: Convert genomes from the list of predicted coding sequences in the list of integer numbers, by considering the cluster number of each gene.
- **Step2**: Extract sets of inversion from all input genomes. 719 inversions have been found (see Table 3).
- **Step 3**: Extract IS clusters (Partial and Real IS) using the first pipeline, as presented in a previous section.
- **Step 4**: Investigate boundaries of each inversion (starting S and ending E positions), by checking the presence of insertion sequences within a window

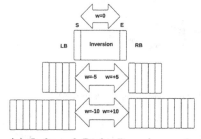

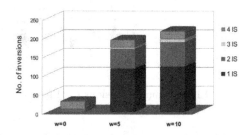

(a) Left and Right Boundary using window

(b) No. of inversions for three different window size

Fig. 6. Using different window size within all inversions ranging from $w = 0$ up to 10 genes. Between 0 and 4 insertion sequences have been found at the boundaries of each inversion, (Figure 6).

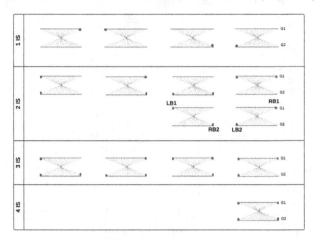

Fig. 7. Different cases of IS inversions

- **Step 5**: Finally, compute the presence probability for each IS families and groups near inversions. (Figure 7).

As presented in Figure 8, there is no major problem in dealing with small inversions because the small inversions having small ratio of increment as compared with big inversions (*i.e.*, during window size increment of inversion boundaries, the small inversions, which have length lower than 4 genes, have small increase ratios compared to large inversions).

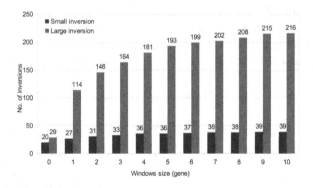

Fig. 8. Small inversions ($\leqslant$ 4 genes) vs. large inversions (> 4 genes)

Table 4 details the roles of IS in largest inversions found within two close isolates.

The IS family of type IS3 always have the most probability of appearance with left and right boundaries of inversions.(Figure 9)

Table 4. Summary of large inversion sets within closed genomes

Genome 1	Genome 2	Inversions no.	inversion inversion	IS family	Boundary	Window
19BR	213BR	9	14	IS110	LB1-RB2	w=0
PAO1	c7447m	2	2	IS3	(LB1-RB2)/(LB2-RB1)	w=0
LES431	LESB58	3	2	IS3	(LB1-RB2)/(LB2-RB1)	w=0
DK2	RP73	93	250	Tn3	LB1-RB2	w=3
UCBPP-PA14	B136-33	7	2	IS3	(LB1-RB2)/(LB2-RB1)	w=0
NCGM2.S1	MTB-1	91	255	IS5	LB1-RB2	w=5

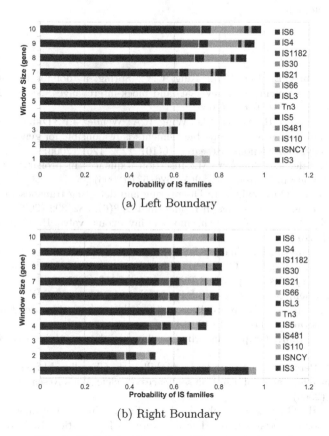

(a) Left Boundary

(b) Right Boundary

Fig. 9. IS distribution using different windows size

4   Conclusion

We designed a pipeline that detects and classify all ORFs that belong to IS. It has been done by merging various tools for ORF prediction, clusterization, and by finally using ISFinder database for classification.

This pipeline has been applied on a set of *Pseudomonas aeruginosa*, showing an obvious improvement in ORFs detection that belong to insertion sequences. Furthermore, relations between inversions and insertion sequences have been

H. Al-Nayyef et al.

emphasized, leading to the conclusion that the so-called IS3 family has the largest probability of appearance inside inversion boundaries.

In future works, we intend to investigate more deeply the relation between ISs and other genomic recombination such as deletion and insertion. We will then focus on the implication of other types of genes like rRNA (rrnA, rrnB, rrnC, rrnD) in *P. aeruginosa* recombination [20]. By doing so, we will be able to determine genes that are often associated with deletion, inversion, etc. The pipeline will be finally extended to eukaryotic genomes and to other kinds of transposable elements.

References

1. Lin, Y.C., Lu, C.L., Liu, Y.C., Tang, C.Y.: Spring: a tool for the analysis of genome rearrangement using reversals and block-interchanges. Nucleic Acids Research 34(suppl. 2), 696–699 (2006)
2. Hawkins, J.S., Kim, H., Nason, J.D., Wing, R.A., Wendel, J.F.: Differential lineage-specific amplification of transposable elements is responsible for genome size variation in gossypium. Genome Research 16(10), 1252–1261 (2006)
3. Siguier, P., File, J., Chandler, M.: Insertion sequences in prokaryotic genomes. Current Opinion in Microbiology 9(5), 526–531 (2006)
4. Bergman, C.M., Quesneville, H.: Discovering and detecting transposable elements in genome sequences. Briefings in Bioinformatics 8(6), 382–392 (2007)
5. Kirkpatrick, M.: How and why chromosome inversions evolve. PLoS Biology 8(9), e1000501 (2010)
6. Ranz, J.M., Maurin, D., Chan, Y.S., Von Grotthuss, M., Hillier, L.W., Roote, J., Ashburner, M., Bergman, C.M.: Principles of genome evolution in the drosophila melanogaster species group. PLoS Biology, 5(6), e152 (2007)
7. Hyatt, D., Chen, G.-L., LoCascio, P.F., Land, M.L., Larimer, F.W., Hauser, L.J.: Prodigal: prokaryotic gene recognition and translation initiation site identification. BMC Bioinformatics 11(1), 119 (2010)
8. Van Dongen, S.M.: Graph clustering by flow simulation. University of Utrecht (2000)
9. Siguier, P., Pérochon, J., Lestrade, L., Mahillon, J., Chandler, M.: Isfinder: the reference centre for bacterial insertion sequences. Nucleic Acids Research 34(suppl.1), D32–D36 (2006)
10. Al-Nayyef, H., Guyeux, C., Bahi, J.: A pipeline for insertion sequence detection and study for bacterial genome. Lecture Notes in Informatics (LNI) vol. 235, pp. 85–99 (2014)
11. Van Domselaar, G.H., Stothard, P., Shrivastava, S., Cruz, J.A., Guo, A., Dong, X., Lu, P., Szafron, D., Greiner, R., Wishart, D.S.: Basys: a web server for automated bacterial genome annotation. Nucleic Acids Research 33(suppl. 2), W455–W459 (2005)
12. Seemann, T.: Prokka: rapid prokaryotic genome annotation. Bioinformatics 30(14), 2068–2069 (2014)
13. Robinson, D.G., Lee, M.-C., Marx, C.J.: Oasis: an automated program for global investigation of bacterial and archaeal insertion sequences. Nucleic Acids Research 40(22), e174–e174 (2012)
14. Enright, A.J., Van Dongen, S., Ouzounis, C.A.: An efficient algorithm for large-scale detection of protein families. Nucleic Acids Research 30(7), 1575–1584 (2002)

15. Garcia-Diaz, M., Kunkel, T.A.: Mechanism of a genetic glissando: structural biology of indel mutations. Trends in Biochemical Sciences 31(4), 206–214 (2006)
16. Hurles, M.: Gene duplication: the genomic trade in spare parts. PLoS Biology 2(7), e206 (2004)
17. Proost, S., Fostier, J., Witte, D.D., Dhoedt, B., Demeester, P., Peer, Y.V.d., Vandepoele, K.: i-adhore 3.0 fast and sensitive detection of genomic homology in extremely large data sets. Nucleic Acids Research 40(2), e11 (2012)
18. Stamatakis, A.: Raxml version 8: A tool for phylogenetic analysis and post-analysis of large phylogenies. Bioinformatics 30(9), 1312–1313 (2014)
19. Alkindy, B., Couchot, J.-F., Guyeux, C., Mouly, A., Salomon, M., Bahi, J.M.: Finding the core-genes of chloroplasts. Journal of Bioscience, Biochemistery, and Bioinformatics 4(5), 357–364 (2014)
20. Stover, C.K., Pham, X.Q., Erwin, A.L., Mizoguchi, S.D., Warrener, P., Hickey, M.J., Brinkman, F.S.L., Hufnagle, W.O., Kowalik, D.J., Lagrou, M., et al.: Complete genome sequence of pseudomonas aeruginosa pao1, an opportunistic pathogen. Nature 406(6799), 959–964 (2000)

A Genetic Algorithm for Motif Finding Based on Statistical Significance

Josep Basha Gutierrez[1], Martin Frith[2], and Kenta Nakai[3]

[1] Department of Medical Bioinformatics,
Graduate School of Frontier Sciences, The University of Tokyo,
5-1-5 Kashiwanoha, Kashiwa-shi, Chiba-ken 277-8561, Japan
yusef@hgc.jp
[2] Computational Biology Research Center,
AIST Tokyo Waterfront Bio-IT Research Building,
2-4-7 Aomi, Koto-ku, Tokyo, 135-0064, Japan
[3] Human Genome Center, The Institute of Medical Science,
The University of Tokyo, 4-6-1 Shirokane-dai,
Minato-ku, Tokyo 108-8639, Japan
knakai@ims.u-tokyo.ac.jp

Abstract. Understanding of transcriptional regulation through the discovery of transcription factor binding sites (TFBS) is a fundamental problem in molecular biology research. Here we propose a new computational method for motif discovery by mixing a genetic algorithm structure with several statistical coefficients. The algorithm was tested with 56 data sets from four different species. The motifs obtained were compared to the known motifs for each one of the data sets, and the accuracy in this prediction compared to 14 other methods both at nucleotide and site level. The results, though did not stand out in detection of false positives, showed a remarkable performance in most of the cases in sensitivity and in overall performance at site level, generally outperforming the other methods in these statistics, and suggesting that the algorithm can be a useful tool to successfully predict motifs in different kinds of sets of DNA sequences.

Keywords: Motif finding, Genetic Algorithm, Transcription Factor Binding Site, Statistical significance.

1 Introduction

Sequence motifs are short nucleic acid patterns that are repeated very often and have some biological significance. Their function is usually to serve as sequence-specific binding sites for proteins such as transcription factors (TF). The discovery of these sequence elements in order to get a better understanding of transcriptional regulation is a fundamental problem in molecular biology research. Traditionally, the most common methods to determine binding sites were DNase footprinting, and gel-shift or reporter construct assays. Currently, however, the use of computational methods to discover motifs by searching for

F. Ortuño and I. Rojas (Eds.): IWBBIO 2015, Part I, LNCS 9043, pp. 438–449, 2015.

overrepresented (and/or conserved) DNA patterns in sets of functionally related genes (such as genes with similar functional annotation or genes with similar expression patterns) considerably facilitates the search. The existence of both computationally and experimentally derived sequence motifs aplenty, as well as the increasing usefulness of these motifs in the definition of genetic regulatory networks and in the decoding of the regulatory program of individual genes, make motif finding a fundamental problem in the post-genomic era.

One of such many strategies for motif discovery relies on the use of Genetic Algorithms (GA).

Genetic Algorithms. A genetic algorithm is a search heuristic that tries to imitate the process of natural selection in order to find exact or approximate solutions to optimization or search problems. The motivation for using genetic algorithms comes from the idea of reducing the number of searches in a high number of large DNA sequences.

The basic structure of a genetic algorithm consists of evolving a population of candidate solutions (individuals) in order to find the best solution or set of solutions possible. This is performed through an iterative process in which the population in each iteration will be considered a generation. In each one of these generations, the fitness (the score given to measure how good the individual is as a solution for the problem) of every individual in the population is evaluated. The fittest individuals are selected from the current population, and a new generation is created by crossover and mutation of these fit individuals. The new generation is then used in the next iteration of the algorithm. The algorithm will normally terminate when either a satisfactory solution has been found or a maximum number of generations has been reached. The main challenge therefore resides in successfully defining the population, and the fitness, crossover and mutation functions.

The most common approach for motif finding using Genetic Algorithms [1] assumes that every input sequence contains an instance of the motif, and relies on the following elements:

- Each individual is represented by a vector $P = \{p_1, p_2, ..., p_N\}$ storing the starting positions for each one of the TFBS instances for the given set of N sequences $S = \{S_1, S_2, ..., S_N\}$. Thus P represents a possible solution set $M = \{m_1, m_2, ..., m_N\}$, where each m_i is an instance with length w from sequence S_i.
- The fitness of each individual is computed using the similarity score of the consensus string produced by an individual, using the PWM (position weight matrix).

$$Fitness(M) = \sum_{i=1}^{w} f_{max}(i) \tag{1}$$

where M is a candidate motif, w is the motif length and $f_{max}(i)$ is the maximum frequency value in column i in the PWM.

Current Situation. As well as the standard GA, most of the existing motif finding methods deal with a set of DNA sequences in which all of the sequences (or at least most of them) are expected to contain at least one instance of each one of the transcription factor binding sites (TFBS) that the method will report, sustaining the search on the statistical enrichment of these TFBS in the set of sequences.

In the last years, new statistical properties of TFBS have been discovered [2]. By the use of this statistical information, this research explores the efficiency of the application of a different sort of genetic algorithm, with fewer restrictions about the input sequences, the presence of instances and the size of the datasets, and able to work with large datasets without consuming a great amount of time. In the method here described, we would like to find TFBS based on their statistical enrichment in any of the input sequences (not necessarily most of them). For that purpose, a new GA-based method was designed, in which the sequences are treated iteratively, creating random subsets of them in a random order and analyzing the statistical overrepresentation in several steps.

2   Methods

The method here proposed mixes a genetic algorithm with probabilistic methods, trying to integrate the advantages of both.

The main characteristics of the method are the following:

- There are no assumptions about the presence of the motifs in the input sequences. Unlike other methods, which assume that every sequence contains at least one instance of the motif, in this method the motifs can be distributed in any way in the given sequences, with the only assumption that they are overrepresented in at least a few of them.
- It is a heuristic algorithm. Thus, it may produce different results each time it is run
- Individual motifs are ungapped. Patterns with variable-length gaps might be predicted split into two or more separate motifs.
- The background set of sequences is generated dynamically throughout the process by shuffling the candidate motifs to analyze against the sequences instead of shuffling the sequences themselves.

2.1   Representation

To represent the candidate motifs, initially there were two possible options: either using a string with the sequence of nucleotides, or using a position in which the instance is located. As Vijayvargiya and Shukla [1] proved in their experiment, the approach with positions is more appropriate for a genetic algorithm (GA). Therefore, that was the approach chosen for our method.

However, in that standard algorithm, as it assumed there was one instance of the motif in each of the input sequences, each individual was represented by a vector with the beginning positions of each one of those instances. In the method here described, on the other hand, there are no assumptions about the presence of instances in every sequence, so the individuals are represented by a single position value in what we call the *supersequence*.

Supersequence. In order to deal with that, we needed to have a structure in which a single value for that position represents a unique individual for the whole set of sequences. For that purpose, all the input sequences are joined in a single supersequence before starting the algorithm. That way, each individual will be represented by a unique position in the supersequence. This position, at the same time, represents the motif given by the position itself, a fixed motif length and a maximum number of mutations allowed.

It is important to clarify that the supersequence serves only as a means of representation of the motifs during the algorithm process and there is no biological meaning in it. The final solutions will be represented by a position weight matrix, got by clustering all the predicted instances with a high level of similarity.

Subsequences. In order to discard unfit individuals faster to generate more diverse solutions, the supersequence is divided in subsequences of an arbitrary length regardless of the length of each sequence (defined by a parameter that by default has a value of 500 bp).

For each generation of the population, the fitness will be calculated against one of the subsequences. In other words, in each iteration the algorithm will search for overrepresentation of the motifs within the given subsequence. The purpose of creating the subsequences is only to simplify the fitness function and, as well as the supersequence, there is no biological meaning in it.

The order of the original sequences will be shuffled every S generations, being S the total number of subsequences.

Fig. 1 shows how the supersequence and the subsequences are created.

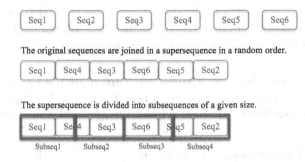

Seq1   Seq2   Seq3   Seq4   Seq5   Seq6

The original sequences are joined in a supersequence in a random order.

Seq1   Seq4   Seq3   Seq6   Seq5   Seq2

The supersequence is divided into subsequences of a given size.

Seq1   Seq4   Seq3   Seq6   Seq5   Seq2

Subseq1   Subseq2   Subseq3   Subseq4

Fig. 1. Creation of the supersequence and the subsequences

2.2 Evaluation and Selection

The algorithm does not have a typical fitness function, but a combination of different ones applied at different moments of the process.

Simple Overrepresentation. The first step to measure the quality of the solutions consists of checking if they are overrepresented in the given subsequence. In order to calculate it, the algorithm follows these steps for each individual of the population:

1. Getting the candidate motif i given by the position P of the individual and the fixed motif length l.
2. Shuffling the candidate motif to get a background motif $b(i)$.
3. For both the candidate motif and the background motif, counting the number of similar words $(S_W(x))$ in the given subsequence (those words with length l that are exactly the same as the motif except for as much as m mismatches, being m the number of mismatches allowed).
4. Storing both values $(S_W(i)$ and $S_W(b(i)))$ in vectors that are kept along with the individual.
5. Calculating the difference of similar words between the candidate motif and the background motif $(N(i))$:

$$N(i) = S_W(i) - S_W(b(i)) \tag{2}$$

Candidate Selection (First Fitness Measurement). In order to select the best candidate solutions as survivors and generate new individuals by crossover of these fit candidates, the *Fluffiness Coefficient* $(F_N(i))$, inspired by the "fluffy-tail test" proposed by Abnizova et al. [3], is used for every individual in every generation.

$$F_N(i) = \frac{N(i) - \mu_s}{\delta_s} \tag{3}$$

where μ_s and δ_s are, respectively, the mean and the standard deviation of the Simple Overrepresentation values $(N(i))$ for the set of solutions. The individuals with the lowest Fluffiness value are eliminated from the population.

Solution Selection (Second Fitness Measurement). Once an individual has survived for at least 10 generations, a new fitness test is performed in order to decide if the candidates are final solutions to the problem or not. For that purpose, two different coefficients are used.

– **Thinness Coefficient.** This coefficient is inspired by the "thin-tail test" proposed by Shu and Li [4]

$$T_N(i) = \frac{k_0 - 2\epsilon}{4\epsilon} \text{ where } \epsilon = 2\sqrt{\frac{6}{M}} \text{ and } k(i) = \frac{M(f_Z(i) - \mu_s)^4}{(M-1)\delta_s^4} - 3 \tag{4}$$

M is the total number of individuals in the population, $k(i)$ is the kurtosis of the individual i and $f_Z(i)$ is the number of individuals in the population with the same fitness value $N(i)$ as i. The individuals with a Thinness coefficient above 0.6 are eliminated from the population.

- **Mann-Whitney U Test.** The Mann-Whitney U test (also known as Wilcoxon rank-sum test) is a nonparametric test that, for two populations, measures if one of them tends to have larger values than the other.

$$U_1 = n_1 n_2 + \frac{n_1(n_1 + 1)}{2} - R_1 \tag{5}$$

where n_1 is the sample size for the sample 1, and R_1 is the sum of ranks in the sample 1.

In our algorithm, the first sample corresponds to the vector with the values stored for the number of similar words for the candidate motif in each generation, and the second sample is formed by the same values for the background motif.

If the probability of both data samples coming from the same population is lower than 0.05, then the motif is considered as a possible final solution. The final fitness value will be given by the following formula:

$$FV(i) = p_{val}(i) \times w \tag{6}$$

where $p_{val}(i)$ is the Mann-Whitney U test p-value and w is the motif width.

2.3 Genetic Operators

Crossover. The crossover function is a one-point crossover in which a child is generated by the parts of both parents joined in reversed order. The parents are the motifs given by the position of the individuals, and the position of the newborn child will be the position of the most similar word in the supersequence.

Mutation. Mutation will happen randomly, according to a parameter that defines the frequency. It will also be applied to random individuals. The mutated individual will slightly change its position by a random offset between 1 and the motif length

2.4 Post Processing

Filtering and Clustering of Solutions. After running the algorithm for every given motif width, the solutions are filtered and clustered to generate the final solutions, each one of them formed by a combination of instances (preliminary solutions given by the GA) with a high level of similarity. The fitness of the clustered solution will be the maximum of the fitnesses of the motifs that are part of the cluster.

In order to devise the similarity, the algorithm measures the distance between motifs.

The distance between motifs is defined by the sum of the distances between their IUPAC symbols. Each IUPAC symbol represents a subset of the symbols {A, G, C, T}. The distance between two symbols s_1 and s_2 is calculated as follows:

$$d(s_1, s_2) = 1 - 2\frac{|s_1 \cap s_2|}{|s_1| + |s_2|} \tag{7}$$

So the values will be always between 0 and 1, being 0 for identical symbols and 1 for disjoint symbols.

The distance between two motifs x and y with sizes m and n respectively is calculated as follows:

$$D(x, y) = \sum_{i=1}^{min(m,n)} d(x_i, yi) + w_u u \tag{8}$$

where u is the number of unpaired bases ($|m - n|$) and w_u is the weight assigned to them (0.6 by default). All possible shifts aligning the motifs are considered and the final distance will be the minimum.

Two motifs are considered similar (and clustered in the same motif) if:

$$\frac{D(x, y)}{mean(|x|, |y|)} \leq MS \tag{9}$$

where MS is the Maximum Similarity, a parameter that can be adjusted and that, by default, will be 0.5.

3   Results

Assessment. The tool was tested using the assessment provided by the study performed by *Tompa et al.* [5] to compare the accuracy of motif finding methods. This assessment provides a benchmark containing 52 data sets of four different organisms (fly, human, mouse and yeast) and 4 negative controls. The data sets are of three different types: the real promoter sequences in which the sites are contained (Type Real), random promoter sequences from the same genome (Type Generic) and synthetic sequences generated by a Markov chain of order 3 (Type Markov). The assesment compared the efficiency of 14 methods, to which we compared our method as well. The eight statistics that will define the accuracy of each tool are the following ones:

- nSn (Sensitivity, nucleotide level), gives the fraction of known site nucleotides that are predicted:

$$nSn = \frac{nTP}{nTP + nFN} \tag{10}$$

- $nPPV$ (Positive Predicted Value, nucleotide level), gives the fraction of predicted site nucleotides that are known:

$$nPPV = \frac{nTP}{nTP + nFP} \tag{11}$$

- nSp (Specificity):

$$nSp = \frac{nTN}{nTN + nFP} \tag{12}$$

- nPC (Performance Coefficient, nucleotide level) [6]:

$$nPC = \frac{nTP}{nTP + nFN + nFP} \tag{13}$$

- nCC (Correlation Coefficient) [7]:

$$nCC = \frac{nTP \times nTN - nFN \times nFP}{(nTP + nFN)(nTN + nFP)(nTP + nFP)(nTN + nFN)} \tag{14}$$

- sSn (Sensitivity, site level), gives the fraction of known sites that are predicted:

$$sSn = \frac{sTP}{sTP + sFN} \tag{15}$$

- $sPPV$ (Positive Predicted Value, site level), gives the fraction of predicted sites that are known:

$$sPPV = \frac{sTP}{sTP + sFP} \tag{16}$$

- sASP (Average Site Performance) [7]:

$$sASP = \frac{sSn + sPPV}{2} \tag{17}$$

Where TP refers to the number of true positives, FP refers to the number of false positives, TN refers to the number of true negatives, and FN refers to the number of false negatives. The n before each one of these measures refers to *nucleotide level* and the s refers to *site level*.

Tests. Our algorithm was run 3 times for each data set, using different motif lengths, and then all the results combined in the post processing stage.

- Motif width 8, allowing 2 mismatches
- Motif width 10, allowing 3 mismatches
- Motif width 12, allowing 4 mismatches

The parameters with which the algorithm was run for all of the data sets are the following:

- Population size: 200
- Number of generations: 100
- Maximum number of solutions: 100
- Mutation rate: 0.1
- Subsequence size: 500bp
- Maximum Similarity: 0.7

Fig. 2 summarizes the average values of the mentioned statistics got by each one of the 14 methods of the assessment and our own method regardless of the organism and the type of data set. Fig. 3 shows the average values grouped by organisms. The procedure to calculate the average in every case is as follows: The values of nTP, nFP, nFN, nTN, sTP, sFP and sFN of the different data sets are added, and then the given statistic is computed as if the summed values corresponded to a unique large data set.

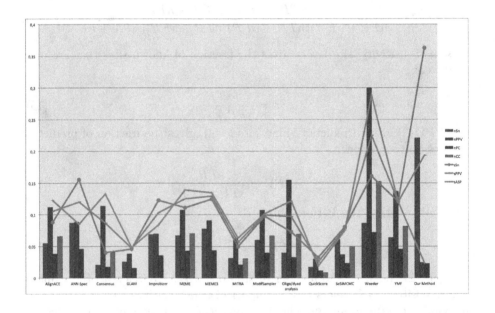

Fig. 2. Average statistical values for all 56 data sets

4   Discussion

First of all, as the authors of the assessment [5] in which our tests are based explain, these statistics should not be taken as an absolute measurement of the quality of the methods. There are many factors that affect the results:

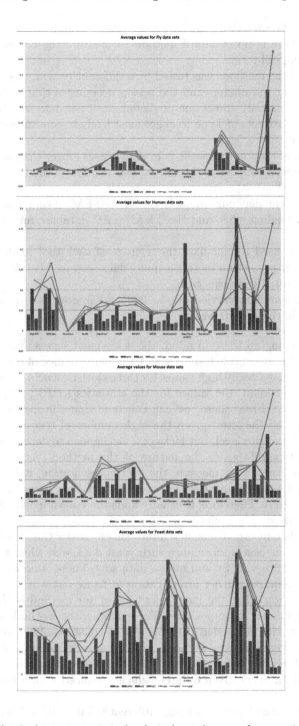

Fig. 3. Average statistical values depending on the organism

- Obviously, the underlying biology is yet to be completely understood, and therefore there is no standard method to measure the correctness of each tool in biological terms.
- Each one of the methods was tested by a different person, who made human choices for the parameters and the post processing of that method.
- The assessment only allows one predicted motif per data set (or none), even though the data sets of Type Real are most likely to contain several different binding sites.
- The length of the known motifs is in many cases longer than 30 bp, and our method, as well as most of the others, was run for motifs no longer than 12 bp.
- The assessment depends uniquely on TRANSFAC [8] for the definition of the known binding sites, and the TRANSFAC database might also contain errors.
- The method used to calculate the average of each tool tends to favor the methods that predict no motif for many data sets, as 0 is taken as the value for all the statistics in this case.
- The assessment was carried out in 2005, so it does not include methods developed in the last 10 years.

However, keeping all these in mind, the assessment serves as a powerful tool to infer some important conclusions about the performance of each method.

Our method shows really high values for three statistics: nSn, sSn, and $sASP$. But, on the other hand, the values for the statistics $nPPV$, nSp, and $sPPV$ are generally poor. From these, we can conclude that the method succeeds in predicting many of the sites, given the high number of true positives both at nucleotide level and site level, but lacks of a mechanism to detect false positives. This is understandable given the nature of the method. As it predicts sites according to statistics that measure the overrepresentation, it is very likely to happen that it reports many sites that are not actual instances of the motif but are very similar to it. Therefore, it usually finds the known motif, but with more instances than it actually contains.

As for the different organisms, it is interesting to notice that most of the other methods offer their best performance with yeast data sets, whereas our method gives its best results with fly and mouse data sets. There is no apparent reason for this, and it requires further investigation to figure out why this happens.

It is quite obvious that the main drawback of our method is the absence of a mechanism to detect false positives. For example, the method that gives the best overall statistics is Weeder. But this is, to a considerably extent, due to the fact that it was run in a cautious mode, predicting no motif in most of the cases. Our method, however, failed to detect the negative controls and predicted at least one motif for every given dataset, which produced a high number of false positives.

Even though there have been many different studies about DNA motif finding, it still remains as one of the most complicated challenges for researchers. Several different approaches have been recently developed and there has been an important progress in this area. However, the task of comparing the performances of different motif finding tools have proven to be quite a struggle, given that each tool is designed based on an algorithm and on motif models that are too diverse and complex. This happens basically because we do not have yet a clear understanding of the biology of regulatory mechanisms. Therefore, it is not possible for us to define a standard to measure the quality of tools.

As many studies comparing the performance of different tools suggest [9] (and as researchers' experiments corroborate), the best option when trying to find motifs is using a few complementary tools in combination (selecting the top predicted motifs of each one), instead of simply relying on a single one.

According to this, we believe that our Statistical GA approach has proven to be suitable for being one of those complementary tools that can be used in addition to other ones to successfully predict motifs in any kind of set of DNA sequences. There is still work to do to improve the method, especially the addition of a mechanism to detect false positives, but we think that the method can be useful for researchers and that it might offer new ways for future development of computer-based motif finding methods.

References

1. Vijayvargiya, S., Shukla, P.: Identification of Transcription Factor Binding Sites in Biological Sequences Using Genetic Algorithm. International Journal of Research & Reviews in Computer Science 2(2) (2011)
2. Tanaka, E., Bailey, T. L., Keich, U.: Improving MEME via a two-tiered significance analysis. Bioinformatics, btu163 (2014)
3. Abnizova, I., te Boekhorst, R., Walter, K., Gilks, W.R.: Some statistical properties of regulatory DNA sequences, and their use in predicting regulatory regions in the Drosophila genome: the fluffy-tail test. BMC Bioinformatics 6(1), 109 (2005)
4. Shu, J.J., Li, Y.: A statistical thin-tail test of predicting regulatory regions in the Drosophila genome. Theoretical Biology and Medical Modelling 10(1), 11 (2013)
5. Tompa, M., Li, N., Bailey, T.L., Church, G.M., De Moor, B., Eskin, E., Favorov, A.V., Frith, M.C., Fu, Y., Kent, W.J., et al.: Assessing Computational Tools for the Discovery of Transcription Factor Binding Sites. Nat. Biotechnol. 23137–23147 (2005)
6. Pevzner, P.A., Sze, S.H.: Combinatorial approaches to finding subtle signals in DNA sequences. ISMB 8, 269–278 (2000)
7. Burset, M., Guigo, R.: Evaluation of gene structure prediction programs. Genomics 34(3), 353–367 (1996)
8. Wingender, E., Dietze, P., Karas, H., Knüppel, R.: TRANSFAC: a Database on transcription factors and their DNA binding sites. Nucleic Acids Res. 24, 238–241 (1996)
9. Das, M.K., Dai, H.K.: A survey of DNA motif finding algorithms. BMC Bioinformatics 8(Suppl. 7), S21 (2007)
10. Lenhard, B., Wasserman, W.W.: TFBS: Computational framework for transcription factor binding site analysis. Bioinformatics 18(8), 1135–1136 (2002)

Identification and *in silico* Analysis of NADPH Oxidase Homologues Involved in Allergy from an Olive Pollen Transcriptome

María José Jiménez-Quesada[1], Jose Ángel Traverso[1,3], Adoración Zafra[1], José C. Jimenez-Lopez[1], Rosario Carmona[1,2], M. Gonzalo Claros[2], and Juan de Dios Alché[1]

[1] Estación Experimental del Zaidín (CSIC), Granada, Spain
{mariajose.jimenez,dori.zafra,josecarlos.jimenez,
juandedios.alche}@eez.csic.es
[2] Departamento de Biología Molecular y Bioquímica, Universidad de Málaga, Málaga, Spain
{rosariocarmona,claros}@uma.es
[3] Departamento Biología Celular. Universidad de Granada, Granada, Spain
traverso@ugr.es

Abstract. Reactive oxygen species generated by pollen NADPH oxidases are present in numerous allergenic pollen species. The superoxide generated by this enzyme has been suggested as a key actor in the induction of allergic inflammation. However, this enzyme has been characterized in *Arabidopsis thaliana* pollen only, where two pollen-specific genes (*RbohH* and *RbohJ*) have been described. The olive (*Olea europaea* L.) pollen is an important source of allergy in Mediterranean countries. We have assembled and annotated an olive pollen transcriptome, which allowed us to determine the presence of at least two pollen-specific NADPH oxidase homologues. Primers were designed to distinguish between the two homologues, and full-length sequences were obtained through a PCR strategy. Complete *in silico* analysis of such sequences, including phylogeny, 3-D modeling of the N-terminus, and prediction of cellular localization and post-translational modifications was carried out with the purpose of shed light into the involvement of olive pollen-intrinsic NADPH oxidases in triggering allergy symptoms.

Keywords: allergy, NADPH oxidase, pollen, Rboh, ROS, superoxide.

1 Introduction

One of the most important causes of seasonal respiratory allergy in the Mediterranean area is olive pollen [1]. Allergens are proteins significantly expressed in pollen, many of them showing essential roles in pollen physiology. In the case of olive, twelve allergens have been identified and characterized to date [2].

Together with the classic view of the allergic inflammatory response triggered by pollen allergens, several studies have shown that pollen is also able to induce IgE-independent mast cell degranulation mediated by reactive oxygen species (ROS) [3].

F. Ortuño and I. Rojas (Eds.): IWBBIO 2015, Part I, LNCS 9043, pp. 450–459, 2015.
© Springer International Publishing Switzerland 2015

Moreover, pollen-released soluble molecules such as superoxide, nitric oxide and nitrite seem to contribute to the allergic response [4-7].

NADPH oxidases (NOX or DUOX [dual NADPH oxidases] in animals, also called respiratory burst oxidase homologs [Rbohs] in plants [8]), are transmembrane enzymes which catalyze the generation of superoxide radical $O_2^{\cdot-}$ in the apoplast [9], leading to extracellular ROS increase [8]. This protein family shares 6 transmembrane central domains, two heme-binding sites, and a long cytoplasmic C-terminal owning FAD and NADPH binding domains. In addition, plant NOXs as well as animal NOX5 and DUOX possess Ca^{2+}-binding EF-hands motifs in the N-terminus [10]. In *Arabidopsis thaliana*, ten Rboh are encoded [11, 12], with two forms (RbohH and RbohJ) specifically expressed in pollen [13]. Rboh activity is essential for proper pollen tube growth to the female gametophyte, leading to fertilization [14]. Although pollen NOXs are expected to be located in the plasma membrane, a cytoplasmic localization has also been described [15].

NADPH oxidase enzymes are present in a wide number of allergenic pollens [4]. Superoxide generated by this enzyme activity locates in particular pollen compartments depending on the plant family: surface, cytoplasm, the outer wall or even in subpollen particles (SPPs) released during hydration [7]. These SPPs owning NADPH oxidase activity can reach the lower airways because their size is in the respirable scale [5] and this activity has been involved in the activation of dendritic cells leading to induction of adaptive immune responses [16].

Boldogh *et al.* [4] demonstrated that pollen intrinsic NADPH oxidase activity generates oxidative stress in the airway-lining fluid and epithelium, which in turns facilitates pollen antigen-induced allergic airway inflammation. The authors suggest that this ROS induction is independent of the adaptative immune response. Therefore, in the proposed '2-signal model', pollen NADPH oxidase activity promotes allergic inflammation induced by pollen antigens. Regarding allergic conjunctivitis, NOX-produced ROS from ragweed pollen were able to intensify the responsiveness of patients [17]. However, it has been recently suggested that this ROS-producing activity in pollen is not involved in either sensitization to pollen or in the allergic airway disease [18]. On the basis of the presence of contradictory results, further studies are necessary to determine how pollen intrinsic NADPH oxidase-produced ROS are acting in the allergic response.

An olive pollen transcriptome has been recently generated from cDNA by using a 454/Roche Titanium+ platform (Carmona et al., in preparation). The annotated information is highly valuable to determine the presence of putative new allergens, to elucidate the mechanisms governing the allergy process, and the implication in the olive pollen metabolism through germination, stigma receptivity, and the interaction pollen-stigma. We have screened such transcriptome to identify the presence of Rbohs in the olive pollen, and used the retrieved sequences for further sequence confirmation and bioinformatic analysis.

2   Materials and Methods

2.1   Screening and Identification of NADPH Oxidase Transcripts in the Olive Transcriptome

Different strategies were defined to select such transcripts. Searches were performed by definition using GO, EC, KEGG and InterPro terms and codes, orthologues and gene names in the annotated transcriptome. BLAST searches were also performed using heterologous sequences available in public databases and well- established bibliography resources as TAIR and GenBank.

2.2   Olive Pollen Rboh Cloning and Sequencing

Olive pollen transcriptome retrieved sequences corresponding to *Rboh*-homologues (partial, internal sequences) were used to design primers in order to amplify the known sequence from pollen cDNA. Plant material was obtained according to [19]. Total RNA from mature pollen was extracted using the RNeasy Plant Total RNA kit (Quiagen, U.S.A.) and first-strand cDNA was synthesized with oligo(dT)19 primer and M-MLV reverse transcriptase (Fermentas). Standard PCR were carried out using Taq polymerase (Promega) and Pfu (Promega) to obtain or confirm nucleotides sequences. Full sequences were obtained by means of both 3'- and 5'-RACE (smarter RACE, Clontech), following manufacturer's specifications. pGEMT-easy (Promega) was used for cloning purposes. Sanger sequencing was achieved in the facilities of the EEZ-CSIC institute in Granada.

2.3   *In silico* Analysis of the Sequences

Nucleotide sequences were aligned using CLUSTAL OMEGA multiple alignment tool with default parameters [20]. Phylogenetic trees were constructed with the aid of the software Seaview [21] using the maximum likelihood (PhyML) method and implementing the most probable nucleotide substitution model (GTR) previously calculated by JmodelTest2 [22]. The branch support was estimated by bootstrap resampling with 100 replications.

For the prediction of protein cell localization, the software Plant-mPloc [23] was used. PhosPhAt [24] and PlantPhos [25] were used to predict serine, threonine and tyrosine phosphorylation. GPS-SNO 1.0 was used for prediction of S-nitrosylation sites [26]. TermiNator was used to predict N-terminal methionine excision, acetylation, myristoylation or palmitoylation [27].

Structure prediction of OeNOX1 N-terminal region was modeled by using the fold recognition-based Phyre2 server [28], with c3a8rB [29] as the template and a 100% confidence. Ligand binding sites were predicted with the server 3DLigandSite [30].

3   Results

3.1   Retrieval and Alignment of NADPH Oxidase Sequences from the Olive Pollen Transcriptome

Transcriptome mining led us to pick seven entries identified as internal, partial sequences corresponding to NADPH oxidases (Table 1) by BLASTing each individual sequences against the GenBank database.

Table 1. Output   sequences identified as NADPH oxidase after screening. The lengths of the assembled sequences are indicated.

Sequence No.	Sequence name	Lenght (bp)
1	input_mira_c5325	2032
2	input_mira_c12648	463
3	HVRC56C02JI550_RL3	459
4	input_mira_c8822_split_1	610
5	HVRC56C02GU1BZ_RL1	370
6	input_mira_c5061	643
7	input_mira_c8822_split_0	281

Figure 1 displays a simplified representation of a multiple sequence alignment between the transcripts obtained and well-characterized plant NOXs available at public databases. Such alignment, together with the phylogenetic analysis displayed in Fig. 2, led us to conclude that the 7 outputs identified in the annotated transcriptome belong to at least two different coding sequences, that we named OeNOX1 and OeNOX2.

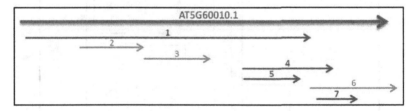

Fig. 1. Sequences alignment using *AT5G60010.1* (*Arabidopsis thaliana* pollen-specific *RbohH*) as the reference (blue arrow).   Green arrows: OeNOX1; red arrows: OeNOX2.

3.2   Cloning of OeNOX1 and OeNOX2 and Phylogenetic Analysis

PCR amplification of cDNA enabled us to obtain the whole sequence of the coding region of OeNOX1 (2502bp) and a large proportion of the coding region of OeNOX2 (2094bp). The nucleotide sequences of both genes were subjected to a phylogenetic analysis including a representation of NADPH oxidases identified in different taxonomical groups (Fig. 2). Different clusters are easily differentiated. Ancestral-type NOXs are presents in humans (NOX1-4) as well as in fungus (*Fusarium graminearum*: NOXA and NOXB) (yellow cluster). Animal NOX5 and DUOX1-2 share EF-hands motifs with plants NADPH oxidases, and all of them are considered calcium-dependents (green cluster). OeNOX1 and OeNOX2 proteins belong to a family of highly related proteins in plants (blue cluster) with ten orthologues in *Arabidopsis*. Olive pollen NADPH oxidases are gathered in a subgroup together with the three pollen specific NOXs identified up to now (AtRbohH and J from *Arabidopsis* and NtNOX from *Nicotiana*) (purple cluster) [31, 32].

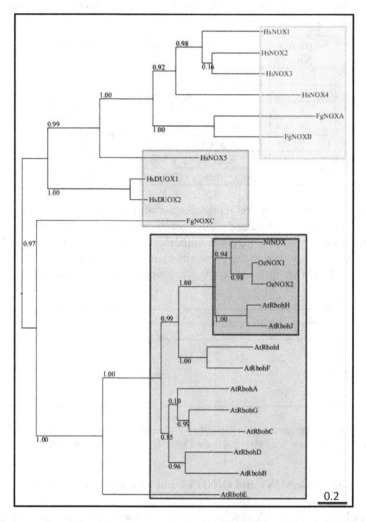

Fig. 2. Phylogenetic relationships between olive pollen NADPH oxidases OeNOX1 and 2 and their homologues in vegetative and reproductive tissues from *Arabidopsis* (AtRbohA-J) and other taxonomically distant NOX proteins.

3.3   *In silico* Prediction of Cellular Localization and Post-translational Modifications

Plant-mPloc software allowed to predict OeNOX1 and OeNOX2 cellular localization in plasma membrane, as it occurs with AtRbohH from *Arabidopsis*. Nevertheless, cytosolic localization was predicted for the *Arabidopsis* homologue AtRbohJ.

PhosPhAt and PlantPhos allowed us to predict 23 putative phosphorylation (Table 2). Potential S-nitrosylation sites were identified in 2 Cys-containing peptides by using the software GPS-SNO, as described in Table 3. Moreover, the N-terminus methionine of the mature protein was predicted to be acetylated with a 100% probability by using the prediction tools cited in material and methods. On the contrary, the presence of neither myristoylation nor palmitoylation sites was predicted.

Table 2. Amino acids prone to phosphorylation within the OeNOX1 sequence as predicted by PhosPhAt/PlantPhos

Serine	Threonine	Tyrosine
14	90	231
49	91	253
75	92	269
189	246	514
689	271	527
691	521	628
804	578	645
	625	736

Table 3. Cys-containing peptides prone to S-nitrosylation (B) within the OeNOX1 sequence as predicted by GPS-SNO

Position	Sequence
347	ALILLPVCRRTLTKL
815	KPLKQLCQELSLTS

3.4   3D Modeling of the Structure and *in silico* Analysis of the Function

The structure prediction served used confirmed that the corresponding translated amino acid sequences of OeNOX1 and OeNOX2 were consistent with available 3D models developed by using folding-recognition software. However such available models included partial structures only. N-terminus of OsNOX1/2 was modeled from Met73 to Leu241 with 100% confidence with the template corresponding to the (X-ray) structure of the N-terminal regulatory domain of a plant NADPH oxidase [29].

The most interesting feature was the presence of Calcium binding sites predicted at residues Asp181, Asn183, Asp185 and Met187, (Fig.3).

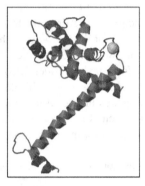

Fig. 3. 3D modeling of the N-terminus of OeNOX1/2. Predicted Ca-binding sites are displayed in blue and Calcium ion is displayed in green.

4   Discussion

In this paper we have carried out a bioinformatic screening of exome sequences of mature olive pollen. This allowed us to detect the existence of at least two *Rboh* genes, homologous to other pollen *Rboh* genes described previously in tobacco and *Arabidopsis*. Subsequent PCR and RACE experiments confirmed such expression in olive pollen and drove to the obtention of full sequence of one of the two genes, namely OeNOX1. The overall filtering of transcriptome outputs to identify the two NOX sequences obtained resulted to be a challenge due to the high homology between these two genes, which required not only a knowledge of bioinformatics tools but previous experience in these protein families. The phylogenetic analysis carried out clustered OeNOX1 and OeNOX2 within a pollen-specific NOXs subgroup, thus suggesting distinctive features for these Rboh homologous in pollen.

ROS production by RbohH and RbohJ is essential for proper pollen tube tip growth[14], acting as a speed control to dampen growth rate oscillations during polarized cell growth [15]. Furthermore, the activity of these enzymes has been demonstrated to be subjected to a complex regulation by multiple elicitors including Ca^{2+}, signaling phospholipids and Rac/Rop GTPases [33]. Although the key involvement of NADPH oxidases in olive pollen physiology has been determined as well [33], little is known as regard to the involvement of these enzymes in the development of allergy. Olive pollen generates superoxide which can be detected by using confocal microscopy in combination with the use of dihydroethidium [34]. Still many aspects have to be elucidated, for example putative release of sub-pollen particles from olive pollen upon hydration, superoxide dismutase inhibition of both pollen NOX activities, their presence in insoluble fractions which could facilitate the exposure of tissues to ROS generated by these enzymes etc. These many aspects may contribute to processes like sensitization and pathogenesis of allergic inflammation [7].

Concerning bioinformatic predictions of post-translational modifications, caution should be taken, considering the *in silico* nature of the outputs. However, some of the results agree with experimental data previously published in plants other than olive. This is the case of the highly conserved Cys815, which is a predicted candidate for S-nitrosylation. This residue was previously disclosed to be nitrosylated in *Arabidopsis thaliana*, leading to NADPH oxidase activity regulation during the immune response in plants [35]. Considering the proposed involvement of pollen-generated nitric oxide (NO) in the allergic response as well [6] we also suggest another possible interconnection of these events with pollen-induced allergy via NOX activity. Unfortunately, the available 3D templates did not enable us modeling the protein portion including such cysteine liable to be S-nitrosylated. Concerning other modifications, NADPH oxidases in plants have been previously described to be modified by phosphorylation, probably in a synergistic way with Ca^{2+} [36, 37]. OeNOX1 has been predicted to be acetylated at the N-terminus, as it occurs for almost 80% of eukaryotic proteins [27]. Finally, the expected plasma membrane localization reported is also in good agreement with determinations made by other authors [15].

Here we show for the first time the presence of two NADPH oxidase forms in a highly allergenic pollen like that of *Olea europaea*. Predictions of sequence, structure, location and post-translational modifications have been made in order to initiate a valuable characterization of this enzyme which help to further assess its implication in allergy.

5   Conflict of Interest

The authors confirm that this article content has no conflicts of interest.

Acknowledgments. This work was supported by ERDF-cofunded projects BFU2011-22779, P2010-AGR6274, P10-CVI-6075, P2010-CVI5767 and P2011-CVI-7487. This research was also partially supported by the European Research Program MARIE CURIE (FP7-PEOPLE-2011-IOF), under the grant reference number PIOF-GA-2011-301550 to JCJ-L and JDA. We also thank the I.F.A.P.A. "Venta del Llano" (Mengíbar, Jaén, Spain) for providing the plant material.

References

1. Liccardi, G., D'Amato, M., D'Amato, G.: Oleaceae pollinosis: a review. International Archives of Allergy and Immunology 111, 210–217 (1996)
2. Villalba, M., Rodriguez, R., Batanero, E.: The spectrum of olive pollen allergens. From structures to diagnosis and treatment. Methods (San Diego, Calif.) 66, 44–54 (2014)
3. Speranza, A., Scoccianti, V.: New insights into an old story: pollen ROS also play a role in hay fever. Plant Signal Behav. 7, 994–998 (2012)
4. Boldogh, I., Bacsi, A., Choudhury, B.K., Dharajiya, N., Alam, R., Hazra, T.K., Mitra, S., Goldblum, R.M., Sur, S.: ROS generated by pollen NADPH oxidase provide a signal that augments antigen-induced allergic airway inflammation. J. Clin. Invest. 115, 2169–2179 (2005)

5. Dharajiya, N., Boldogh, I., Cardenas, V., Sur, S.: Role of pollen NAD(P)H oxidase in allergic inflammation. Current Opinion in Allergy and Clinical Immunology 8, 57–62 (2008)
6. Bright, J., Hiscock, S.J., James, P.E., Hancock, J.T.: Pollen generates nitric oxide and nitrite: a possible link to pollen-induced allergic responses. Plant Physiol. Biochem. 47, 49–55 (2009)
7. Wang, X.L., Takai, T., Kamijo, S., Gunawan, H., Ogawa, H., Okumura, K.: NADPH oxidase activity in allergenic pollen grains of different plant species. Biochem. Biophys. Res. Commun. 387, 430–434 (2009)
8. Lamb, C., Dixon, R.A.: The oxidative burst in plant disease resistance. Annual Review of Plant Physiology and Plant Molecular Biology 48, 251–275 (1997)
9. Lambeth, J.D.: NOX enzymes and the biology of reactive oxygen. Nature Reviews. Immunology 4, 181–189 (2004)
10. Bedard, K., Lardy, B., Krause, K.H.: NOX family NADPH oxidases: not just in mammals. Biochimie 89, 1107–1112 (2007)
11. Dangl, J.L., Jones, J.D.: Plant pathogens and integrated defence responses to infection. Nature 411, 826–833 (2001)
12. Torres, M.A., Onouchi, H., Hamada, S., Machida, C., Hammond-Kosack, K.E., Jones, J.D.: Six Arabidopsis thaliana homologues of the human respiratory burst oxidase (gp91phox). Plant J. 14, 365–370 (1998)
13. Sagi, M., Fluhr, R.: Production of reactive oxygen species by plant NADPH oxidases. Plant Physiol. 141, 336–340 (2006)
14. Kaya, H., Nakajima, R., Iwano, M., Kanaoka, M.M., Kimura, S., Takeda, S., Kawarazaki, T., Senzaki, E., Hamamura, Y., Higashiyama, T., Takayama, S., Abe, M., Kuchitsu, K.: Ca2+-activated Reactive Oxygen Species production by Arabidopsis RbohH and RbohJ is essential for proper pollen tube tip growth. The Plant Cell Online (2014)
15. Lassig, R., Gutermuth, T., Bey, T.D., Konrad, K.R., Romeis, T.: Pollen tube NAD(P)H oxidases act as a speed control to dampen growth rate oscillations during polarized cell growth. The Plant Journal (2014)
16. Pazmandi, K., Kumar, B.V., Szabo, K., Boldogh, I., Szoor, A., Vereb, G., Veres, A., Lanyi, A., Rajnavolgyi, E., Bacsi, A.: Ragweed Subpollen Particles of Respirable Size Activate Human Dendritic Cells. PLoS ONE 7, e52085 (2012)
17. Bacsi, A., Dharajiya, N., Choudhury, B.K., Sur, S., Boldogh, I.: Effect of pollen-mediated oxidative stress on immediate hypersensitivity reactions and late-phase inflammation in allergic conjunctivitis. The Journal of Allergy and Clinical Immunology 116, 836–843 (2005)
18. Shalaby, K.H., Allard-Coutu, A., O'Sullivan, M.J., Nakada, E., Qureshi, S.T., Day, B.J., Martin, J.G.: Inhaled birch pollen extract induces airway hyperresponsiveness via oxidative stress but independently of pollen-intrinsic NADPH oxidase activity, or the TLR4-TRIF pathway. Journal of Immunology 191, 922–933 (2013)
19. Alché, J.D., M'rani-Alaoui, M., Castro, A.J., Rodríguez-García, M.I.: Ole e 1, the major allergen from olive (Olea europaea L.) pollen, increases its expression and is released to the culture medium during in vitro germination. Plant and Cell Physiology 45, 1149–1157 (2004)
20. McWilliam, H., Li, W., Uludag, M., Squizzato, S., Park, Y.M., Buso, N., Cowley, A.P., Lopez, R.: Analysis Tool Web Services from the EMBL-EBI. Nucleic Acids Research 41, W597–W600 (2013)
21. Gouy, M., Guindon, S., Gascuel, O.: SeaView Version 4: A Multiplatform Graphical User Interface for Sequence Alignment and Phylogenetic Tree Building. Molecular Biology and Evolution 27, 221–224 (2010)
22. Darriba, D., Taboada, G.L., Doallo, R., Posada, D.: jModelTest 2: more models, new heuristics and parallel computing. Nat. Meth. 9, 772–772 (2012)

23. Chou, K.C., Shen, H.B.: Plant-mPLoc: a top-down strategy to augment the power for predicting plant protein subcellular localization. PLoS One 5,e11335 (2010)
24. Heazlewood, J.L., Durek, P., Hummel, J., Selbig, J., Weckwerth, W., Walther, D., Schulze, W.X.: PhosPhAt: a database of phosphorylation sites in Arabidopsis thaliana and a plant-specific phosphorylation site predictor. Nucleic Acids Research 36, D1015–D1021 (2008)
25. Lee, T.-Y., Bretana, N., Lu, C.-T.: PlantPhos: using maximal dependence decomposition to identify plant phosphorylation sites with substrate site specificity. BMC Bioinformatics 12, 261 (2011)
26. Xue, Y., Liu, Z., Gao, X., Jin, C., Wen, L., Yao, X., Ren, J.: GPS-SNO: Computational Prediction of Protein S-Nitrosylation Sites with a Modified GPS Algorithm. PLoS ONE 5, e11290 (2010)
27. Martinez, A., Traverso, J.A., Valot, B., Ferro, M., Espagne, C., Ephritikhine, G., Zivy, M., Giglione, C., Meinnel, T.: Extent of N-terminal modifications in cytosolic proteins from eukaryotes. Proteomics 8, 2809–2831 (2008)
28. Kelley, L.A., Sternberg, M.J.: Protein structure prediction on the Web: a case study using the Phyre server. Nature Protocols 4, 363–371 (2009)
29. Oda, T., Hashimoto, H., Kuwabara, N., Akashi, S., Hayashi, K., Kojima, C., Wong, H.L., Kawasaki, T., Shimamoto, K., Sato, M., Shimizu, T.: Structure of the N-terminal regulatory domain of a plant NADPH oxidase and its functional implications. J. Biol. Chem. 285, 1435–1445 (2010)
30. Wass, M.N., Kelley, L.A., Sternberg, M.J.: 3DLigandSite: predicting ligand-binding sites using similar structures. Nucleic Acids Research 38,W469–W473 (2010)
31. Boisson-Dernier, A., Lituiev, D.S., Nestorova, A., Franck, C.M., Thirugnanarajah, S., Grossniklaus, U.: ANXUR receptor-like kinases coordinate cell wall integrity with growth at the pollen tube tip via NADPH oxidases. PLoS Biology 11, e1001719 (2013)
32. Potocky, M., Jones, M.A., Bezvoda, R., Smirnoff, N., Zarsky, V.: Reactive oxygen species produced by NADPH oxidase are involved in pollen tube growth. New Phytol. 174, 742–751 (2007)
33. Potocky, M., Pejchar, P., Gutkowska, M., Jimenez-Quesada, M.J., Potocka, A., Alche, J.D., Kost, B., Zarsky, V.: NADPH oxidase activity in pollen tubes is affected by calcium ions, signaling phospholipids and Rac/Rop GTPases. J. Plant Physiol. 169, 1654–1663 (2012)
34. Zafra, A., Rodriguez-Garcia, M.I., Alché, J.D.: Cellular localization of ROS and NO in olive reproductive tissues during flower development. BMC Plant Biol. 10, 36 (2010)
35. Yun, B.W., Feechan, A., Yin, M., Saidi, N.B., Le Bihan, T., Yu, M., Moore, J.W., Kang, J.G., Kwon, E., Spoel, S.H., Pallas, J.A., Loake, G.J.: S-nitrosylation of NADPH oxidase regulates cell death in plant immunity. Nature 478, 264–268 (2011)
36. Drerup, M.M., Schlucking, K., Hashimoto, K., Manishankar, P., Steinhorst, L., Kuchitsu, K., Kudla, J.: The calcineurin B-like calcium sensors CBL1 and CBL9 together with their interacting protein kinase CIPK26 regulate the Arabidopsis NADPH oxidase RBOHF. Mol. Plant (2013)
37. Ogasawara, Y., Kaya, H., Hiraoka, G., Yumoto, F., Kimura, S., Kadota, Y., Hishinuma, H., Senzaki, E., Yamagoe, S., Nagata, K., Nara, M., Suzuki, K., Tanokura, M., Kuchitsu, K.: Synergistic activation of the Arabidopsis NADPH oxidase AtrbohD by Ca2+ and phosphorylation. J. Biol. Chem. 283, 8885–8892 (2008)

Identification of Distinctive Variants of the Olive Pollen Allergen Ole e 5 (Cu,Zn Superoxide Dismutase) throughout the Analysis of the Olive Pollen Transcriptome

Adoración Zafra[1], Rosario Carmona[1,2], José C. Jimenez-Lopez[1], Amada Pulido[3], M. Gonzalo Claros[2], and Juan de Dios Alché[1]

[1] Estación Experimental del Zaidín (CSIC), Granada, Spain
{dori.zafra,juandedios.alche,josecarlos.jimenez}@eez.csic.es
[2] Departamento de Biología Molecular y Bioquímica, Universidad de Málaga, Málaga, Spain
{rosariocarmona,claros}@uma.es
[3] Departamento de Fisiología Vegetal, Universidad de Granada, Granada, Spain
amadapulido@ugr.es

Abstract. Ole e 5 is an olive pollen allergen displaying high identity with Cu,Zn superoxide dismutases. Previous studies characterized biochemical variability in this allergen, which may be of relevance for allergy diagnosis and therapy. The generation of an olive pollen transcriptome allowed us to identify eight Ole e 5 sequences, one of them including a 24 nt deletion. Further *in silico* analysis permitted designing primers for PCR amplification and cloning from both cDNA and gDNA. A large number of sequences were retrieved, which experimentally validated the predictive NGS sequences, including the deleted enzyme. The PCR-obtained sequences were used for further scrutiny, including sequence aligment and phylogenetic analysis. Two model sequences (a complete sequence and a deleted one) were used to perform 3-D modeling and a prediction of the T- and B-cell epitopes. These predictions interestingly foreseed relevant differences in the antigenicity/allergenicity of both molecules. Clinical relevance of differences is discussed.

Keywords: allergen, deletion, Ole e 5, pollen, Superoxide Dismutase.

1   Introduction

Olive pollen is one of the most important causes of respiratory allergy in the Mediterranean area [1]. To date, twelve allergens have been identified in olive pollen, and one in olive mesocarp, named as Ole e 1 to Ole e 13 [2,3]. Several studies have shown that olive pollen allergens posses a relatively high level of microheterogeneity in their sequence, which has been extensively characterized for allergens Ole e 1, Ole e 2 and

F. Ortuño and I. Rojas (Eds.): IWBBIO 2015, Part I, LNCS 9043, pp. 460–470, 2015.
© Springer International Publishing Switzerland 2015

Ole e 7 [3,4,5]. Moreover, olive pollen contains a complex mix of allergen forms, which is highly dependent of the genetic background (olive cultivar) used for the analysis [4].

Allergen polymorphism is a relevant issue concerning allergy diagnosis and treatment. Heterogeneity of vaccines used for specific immunotherapy (SIT) is considered a challenging factor their efficacy or even worse, it may represent a cause for secondary sensitization of patients or the development of undesirable reactions. Therefore, accurate identification and quantification of the different allergenic forms present in the extracts and its reactivity is a major concern nowadays.

Ole e 5 is considered a minor allergen in olive pollen, with prevalence around 35%, which has been identified as a Cu,Zn superoxide dismutase (SOD). Due to its ubiquity, it is considered as a cross-reactive allergen in the pollen-latex-fruit syndrome [6]. Studies carried out over Ole e 5 show that it is a 16 kDa protein with a high identity to Cu,Zn SODs from other species. In olive pollen, the presence of up to five isoforms has been detected [4, 7-9]. Concerning the biological function of this protein in pollen, SODs catalyse the dismutation of the superoxide anions into molecular oxygen and hydrogen peroxide, therefore acting as an antioxidant able to remove reactive oxygen species (ROS) [10]. ROS are produced in both unstressed and stressed cells. Under unstressed conditions, the formation and removal of O_2 are in or close to balance. Within a cell, the SODs constitute the first line of defence against ROS [11] and they have been localized in cytosol, chloroplasts, peroxisomes and the apoplast [12,13]. ROS may also act as signalling molecules. Thus, the interaction pollen-pistil is mediated by the accumulation or ROS in the stigma, where Ole e 5, is considered one of the main actors [14,15]. Under stress conditions, a clear rise of Cu,Zn-SOD has also been detected [16].

Actual olive tree transcriptomes mainly rely on libraries from vegetative tissues using different NGS strategies. The peculiarity of the reproductive tissues, and the widely reported presence of numerous tissue-specific transcripts in pollen grains, made us to attempt a similar approach in the later. Hence, reproductive cDNA libraries were prepared to be sequenced using 454/Roche Titanium+ platform. This information is highly valuable to elucidate the mechanisms governing the allergy process, as well as its implication in the olive pollen metabolism through germination, stigma receptivity, and the interaction pollen-stigma.

2   Materials and Methods

2.1   Construction of the Olive Pollen Transcriptome

For the construction of the pollen transcriptome, mature pollen obtained from dehiscent anthers of the olive cultivar 'Picual' was used. The samples were thoroughly grinded with a pistil and liquid N_2 followed by the extraction of the total RNA as RNeasy Plant Mini Kit (Qiagen) manual instructions recommends. RNA integrity was checked by formaldehyde gel analysis [17]. The mRNAs were purified using the Oligotex mRNA mini kit (Qiagen). The concentration and quality of the mRNAs were determined by the Ribogreen method (Quant-it RiboGreen RNA Reagent and kit) and the Agilent RNA 6000 Pico assay chip (Bioanalizer 2100). The isolated mRNAs were subjected to 454/Roche Titanium+ sequencing.

2.2   Pre-processing

All reads obtained were pre-processed using the SeqTrimNext pipeline, in order to remove unsuitable reads, such as low quality, low complexity, linkers, adapters, vector fragments, polyA/polyT tails and contaminated sequences [18].

2.3   Assembling

The assembling strategy relies on the combination of two completely different algorithms to compensate assembling biases. After several assembling attempts, the best assembler combination was MIRA3 (based on overlap-layout-consensus algorithm) with Euler-SR (based on de Bruijn graphs) and a final reconciliation with CAP3.

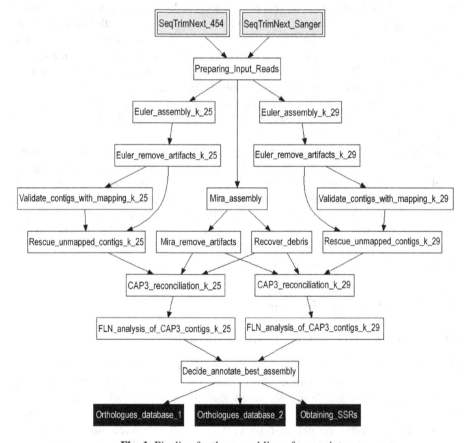

Fig. 1. Pipeline for the assembling of transcripts

2.4   Functional Annotation

A preliminary unigene analysis was performed using Full-LengtherNext. It provided a gene description, full-length unigenes, putative start and stop codons, the putative protein sequence, putative ncRNAs, an unknown unigenes. This gives a quick preview of the olive pollen transcriptome content. Full annotation was carried out using Sma3 to provide another gene description.

2.5   Selection of Cu,Zn-SOD Transcripts

Different strategies were defined to select such transcripts. First, EC codes of *A. thaliana* orthologous to olive sequences were extracted from annotations and subjected to analysis into the *Arabidopsis* reactome database of plant biological pathways. Finally, annotations were manually screened for specific enzymes selected from well-established bibliography resources.

2.6   Cloning of full Cu,Zn-SODs Sequences on the Basis of the Retrieved Sequences

Genomic DNA was obtained from the olive leaves of 7 cultivars and the extraction was carried out with the REDExtract-N-Amp Plant PCR (Sigma). Total RNAs corresponding to olive pollen from 17 olive cultivars were obtained using the RNeasy Plant Mini kit (Qiagen) and subjected to reverse transcription using the SuperScript II reverse transcriptase kit (Invitrogen). The gDNAs and cDNAs obtained were amplified with the following primers:

5′ ATG GTG AAG GCC GTA ACA GTC 3′ (forward) and
5′ TCA ACC CTG AAG GCC AAT G   3′ (reverse).

The PCR products were analyzed electrophoretically, and the amplified bands were excised, and purified with the MBL-Agarose Quikclean kit (Dominion). The purified PCR products were cloned in pGEM-Teasy vector system (Promega). *E. coli* DH5α were transformed with the vector followed by a selection of the colonies: the plasmidic DNAs were extracted as indicated in the Real mini-prep turbo kit (Real) manual instruction. After double-checking the correct insertion of the fragment, it was sequenced in the Sequencing Service of the Institute of Parasitology and Biomedicine "López-Neyra" (IPBLN-CSIC, Granada, Spain). At least 3 clones of each cultivar were sequenced.

2.7   Alignment and *in silico* Analysis of the Sequences

The alignment of all the nucleotidic sequences obtained was performed by using the Clustal W software (http://www.ebi.ac.uk/Tools/clustalw/). The WinGene 1.0 software (Henning 1999) was used to generate the translation to aminoacidic sequences, which were aligned using the Clustal W software as well. The ScanProsite software (http://www.expasy.org/tools/scanprosite) was used to identify potential post-translational modifications and SOD consensus sequences.

2.8   Phylogenetic Tree

Phylogenetic trees were constructed with the aid of the software Seaview [19] using the maximum likelihood (PhyML) method implemented with the JTT model of the most probable amino acid substitution calculated by the ProtTest 2.4 server [20]. The branch support was estimated by bootstrap resampling with 100 replications.

2.9   3D-Modeling

The two most distinctive forms of Cu,Zn SODs were subjected to 3D reconstruction (http://swissmodel.expasy.org/workspace/)[21,22,23,24] by using the 2q21 template (annotated as a dimer) available as PDB by means of the DeepView v3.7 software.

2.10   Identification of B- and T-cell Epitopes

To determine linear B- and T-cell, as well as conformational B-cell epitopes we made use of a set of tools as described for other olive pollen allergens like Ole e 12 [25].

3   Results

3.1   Screening of the Olive Pollen Transcriptome for Cu,Zn SOD Sequences

A total of eight inputs corresponding to Cu,Zn SOD were obtained from the database generated (Table1). The sequences were named OePOlee5-1 to OePOlee5-8 (*Olea europaea* Pollen allergen Ole e 5-1 to 8).

Table 1. Selected sequences from the olive pollen transcriptome identified as Cu,Zn SOD. Lenghts from the sequences obtained after the assembly and from the codificant protein are indicated, as well as the missing aminoacids. Putative cellular localization is also shown.

Identifier	Nucleotides (NGS)	Aa (no UTRs)	Seq. description	Subcellular location
OePOlee5-1	677	128	Lacking N-terminal(24A)	Cytosol
OePOlee5-2	532	150	Lacking N-terminal (2aa)	Cytosol
OePOlee5-3	450	144	Deleted sequence (8 aa)	Cytosol
OePOlee5-4	621	152	Complete sequence	Cytosol
OePOlee5-5	587	192	Lacking C-terminal (36aa)	Chloroplast
OePOlee5-6	615	155	Complete sequence	Cytosol
OePOlee5-7	407	39	Lacking N-terminal (189aa)	Chloroplast
OePOlee5-8	758	152	Complete sequence	Cytosol

3.2   Alignment of the NGS-Retrieved Sequences of Cu,Zn SODs to the GenBank Database and Phylogenetic Analysis

BLAST query of the individual sequences against the GenBank database confirmed all of them as Cu,Zn SODs, either highly homologous to Cu,Zn SODs previously described in olive (Sequences OePOlee5-1,3,4,6, and 8), or in other tree species (sequences OePOlee5-2,5, and 7). Sequences alignment allowed the detection of microheterogeneities among the sequences, affecting several positions, however a high level of identity was the main characteristic observed (data not shown). The most distinctive feature observed was the presence of one relevant deletion (24 nucleotides) in one of the sequences (OePOlee5-3). The phylogenetic tree generated by including most relevant BLAST scores showed the presence of two differentiated clusters, putatively corresponding to those previously described as chloroplastidic and cytosolic proteins, respectively (Figure 2). However, not all the sequences used for the phylogeny were annotated as chloroplastidic or cytosolic by the describing authors.

3.3   Alignment of the Cloned Sequences from Different Olive Cultivars

The nucleotide sequences experimentally obtained from cDNA (51 sequences) and gDNA (21 sequences) were aligned separately (not shown). In both cases, most of the sequences were 456 nt long. Three exceptions were found where the length of the sequences were 432 nt in all cases, with a deletion of 24 nt positioned between the nt 252-276. These three exceptions corresponded to clones obtained from different olive cultivars and origins ('Arbequina' and 'Empeltre' from cDNA; 'Loaime' from gDNA).

All 72 aminoacidic sequences were aligned in order to identify postraslational modifications (glucosylations and phosphorylations), cysteines putatively involved in 3D structure, Cu-binding histidines, and consensus sequences for SODs (figure not shown). The deletion observed in three of the sequences was positioned between the aa 85-92. All experimentally obtained sequences of Cu,Zn SOD clustered coincidentally with sequences OePOlee5-1,3,4 and 8 (Figure 2).

3.4   3D-Modeling of the Complete and Deleted Forms of Cu,Zn SODs and Putative Involvement of Modifications in Allergenicity

Sequences OePOlee5-4 and OePOlee5-3 were used as representative forms of the complete and deleted forms of Cu,Zn SOD, respectively. The modeling approaches performed confirmed that the missing part of the protein in the deleted form (8 aa) matched to an external loop which does not form part of the active centre neither binds to the Cu,Zn atoms (Figure 3).

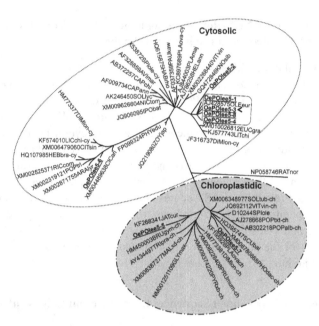

Fig. 2. Phylogenetic tree including olive Cu,Zn SODs sequences obtained either by NGS (underlined) or cloning strategies (which are situated inside the box -not fully listed-), and most relevant heterologous forms from other species (identified by their accession number, shortened name, and -ch or -cy if identified as a chloroplastidic or cytosolic form respectively). Two clusters, corresponding to chloroplastidic and cytosolic sequences are clearly defined. Deleted sequence OePOlee5-3 is pointed out with an arrowhead. The names of the experimentally obtained sequences are not fully displayed. These 72 sequences cluster with OePOlee5-1, 3, 4 and 8 within the rectangle inside the cytosolic cluster. A sequence from rat was included for rooting purposes.

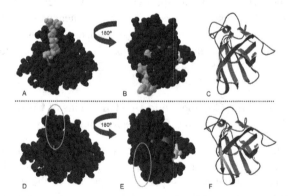

Fig. 3. 3D-modelling of the complete (A-C) and deleted forms (D-F) of the enzyme. The missing sequence is coloured in yellow.

The use of predictive software allowed us to identify the presence of T- and B- epitopes along the two representative forms of SOD mentioned above (OePOlee5-4 and -3). The short sequence (aa 85-92), which is absent from the deleted form, takes part of a defined T-epitope (aa 88-95), a predicted B-linear epitope (aa 87-102), and a conformational T-epitope in the complete sequence. Moreover, the 3-D restructuration occurring in the deleted form when compared to the complete form slightly alters the distribution of other T- and B- epitopes not directly involving the missing loop (results not shown).

4   Discussion

The transcriptomic approach used here greatly helped to define further strategies to identify the variability of the Ole e 5 (Cu,Zn SOD) pollen allergen both within and among olive cultivars, by allowing the identification of the most common variants, which included a deleted form of the enzyme. Such strategies comprised the design of accurate primers for PCR amplification. The presence among pollen transcripts of an even larger number of variants than those expected on the basis of transcriptome analysis was further confirmed by standard cloning procedures, which allowed an experimental validation of the predicted sequences generated by NGS assembly.

Nucleotide polymorphism in the olive pollen allergen Ole e 5 (Cu,Zn SOD) is however relatively lower than that of other olive pollen allergens like Ole e 1 [26] and the considered "highly conserved" allergen Ole e 2 [27] (profilin). Microheterogeneities observed in the sequence cannot be considered artefacts as the result of the miss-incorporation of nt by the polymerase as they have been detected in Cu,Zn SODs at much higher rate than the described for the enzyme. The observed substitutions do not affect key amino acids for the structure and/or function of the enzyme, like Cys involved in disulphide bridges, His residues involved in Cu-binding, of those motifs described as consensus for Cu,Zn SODs. Therefore, it is likely that the resulting gene products may represent active enzymes. The deleted forms of Cu,Zn SOD observed in both the sequences assembled after NGS as well as after experimental cloning developed in this work are likely not to be greatly affected in their functionality, as neither the reading frame nor the presence of key amino acids were disturbed. However, it is conceivable the presence of slight modifications either in the molecular weight, the isoelectric point of other properties of the protein. Moreover, subtle changes in the activity of the protein, including its kinetics could occur, and should be experimentally tested further.

As regard to the antigenicity/allergenicity of the deleted form of Cu,Zn SOD, predictive tools clearly report a modification of the overall ability of the protein to be recognized as a potential antigen/allergen. T-cell epitopes play a crucial role in immune responses for the induction of cytotoxic T-cell responses and in providing help to B cells for the development of antibody responses, whereas the identification of B-cell epitopes contributes to improve our understanding of structural aspects of allergens and of the pivotal role they play in the induction of hypersensitivity reactions against small molecules and allergens [28]. The observed differences in the distribution of T- and B-epitopes among the complete and deleted sequences of Cu,Zn SODs may help to develop biotechnological approaches aimed to improve allergy vaccines. These approaches include the generation of hypoallergenic variants including isoforms and folding variants and the engineering of vaccines combining B- or T-cell epitopes. Most recent studies at this regard, resulting in an increased safety profile and reduced side-effects compared with allergen extracts have been recently reviewed [28].

5   Conflict of Interest

The authors confirm that this article content has no conflicts of interest.

Acknowledgments. This work was supported by ERDF-cofunded projects BFU2011-22779 (MINECO), and P2010-AGR-6274, P10-CVI-6075 and P2011-CVI-7487 (Junta de Andalucía). A. Zafra thanks the Agrifood Campus of International Excellence ceiA3 for grant funding. This research was also partially supported by the European Research Program MARIE CURIE (FP7-PEOPLE-2011-IOF), under the grant reference number PIOF-GA-2011-301550 to JCJ-L and JDA. We also thank the I.F.A.P.A. "Venta del Llano" (Mengíbar, Jaén, Spain) for providing the plant material.

References

1. Busquets, J., Cour, P., Guerin, B., Michel, F.B.: Busquets et al 1984 Allergy in the Mediterranean area_ I_ Pollen counts and pollinosis (1984)
2. Rodríguez, R., Villalba, M., Monsalve, R.I., Batanero, E.: The spectrum of olive pollen allergens. Int. Arch. Allergy Immunol. 125, 185–195 (2001)
3. Jimenez-Lopez, J.C., Rodriguez-Garcia, M.I.: Olive tree genetic background is a major cause of pofilin (Ole e 2 allergen) polymorphism and functional and allergenic variability. Commun. Agric. Appl. Biol. Sci. 78, 213–219 (2013)
4. Tejera, M.L., Villalba, M., Batanero, E.: Mechanisms of allergy Identification, isolation, and characterization of Ole e 7, a new allergen of olive tree pollen. Mech. Allergy 797–802 (1999)
5. Villalba, M., Rodríguez, R., Batanero, E.: The spectrum of olive pollen allergens. From structures to diagnosis and treatment. Methods 66, 44–54 (2014)
6. Batanero, E., Rodríguez, R., Villalba, M.: Olives and Olive Oil in Health and Disease Prevention. Elsevier (2010)
7. Boluda, L., Alonso, C., Fernández-Caldas, E.: Purification, characterization, and partial sequencing of two new allergens of Olea europaea. J. Allergy Clin. Immunol. 101, 210–216 (1998)
8. Butteroni, C., Afferni, C., Barletta, B., Iacovacci, P., Corintia, S., Brunetto, B., Tinghino, R., Ariano, R., Panzani, R.C., Pini, C., Di Felice, G.: Cloning and expression of the olea europaea allergen Ole e 5, the pollen Cu/Zn Superoxide Dismutase (2005)

9. Alché, J.D., Corpas, F.J., Rodríguez-García, M.I., Del Rio, L.A.: Identification and immunulocalization of SOD isoforms in olive pollen. Physiol. Plant 104, 772–776 (1998)

10. Parida, A.K., Das, A.B.: Salt tolerance and salinity effects on plants: a review. Ecotoxicol. Environ. Saf. 60, 324–349 (2005)

11. Alscher, R.G., Erturk, N., Heath, L.S.: Role of superoxide dismutases (SODs) in controlling oxidative stress in plants. J. Exp. Bot. 53, 1331–1341 (2002)

12. Corpas, F.J., Fernández-Ocaña, A., Carreras, A., Valderrama, R., Luque, F., Esteban, F.J., Rodríguez-Serrano, M., Chaki, M., Pedrajas, J.R., Sandalio, L.M., del Río, L.A., Barroso, J.B.: The expression of different superoxide dismutase forms is cell-type dependent in olive (Olea europaea L.) leaves. Plant Cell Physiol. 47, 984–994 (2006)

13. Zafra, A., Traverso, J.A., Corpas, F.J., Alché, J.D.: Peroxisomal localization of CuZn superoxide dismutase in the male reproductive tissues of the olive tree. Microsc. Microanal. 18, 2011–2012 (2012)

14. Alché, J.D., Castro, A.J., Jiménez-López, J.C., Morales, S., Zafra, A., Hamman-Khalifa, A.M., Rodríguez-García, M.I.: Differential characteristics of olive pollen from different cultivars: biological and clinical implications. J. Investig. Allergol. Clin. Immunol. 17(Suppl 1), 17–23 (2007)

15. Zafra, A., Rodríguez-García, M.I., Alche, J.D.: Cellular localization of ROS and NO in olive reproductive tissues during flower development. BMC Plant Biol. (2010)

16. Iba, K.: Acclimative response to temperature stress in higher plants: approaches of gene engineering for temperature tolerance. Annu. Rev. Plant Biol. 53, 225–45 (2002)

17. Sambrook, J., Rusell, D.: Molecular Cloning: a laboratory manual. Cold Spring harbor Laboratory Press, New York (2001)

18. Falgueras, J., Lara, A.J., Fernández-pozo, N., Cantón, F.R., Pérez-trabado, G., Claros, M.G.: SeqTrim: a high-throughput pipeline for pre-processing any type of sequence read. BMC Bioinformatics (2010)

19. Gouy, M., Guindon, S., Gascuel, O.: SeaView version 4: A multiplatform graphical user interface for sequence alignment and phylogenetic tree building. Mol. Biol. Evol. 27, 221–224 (2010)

20. Abascal, F., Zardoya, R., Posada, D.: ProtTest: selection of best-fit models of protein evolution. Bioinformatics 21, 2104–2105 (2005)

21. Guex, N., Peitsch, M.C.: SWISS-MODEL and the Swiss-PdbViewer: An environment for comparative protein modeling (1997)

22. Benkert, P., Biasini, M., Schwede, T.: Toward the estimation of the absolute quality of individual protein structure models. Bioinformatics 27, 343–350 (2011)

23. Arnold, K., Bordoli, L., Kopp, J., Schwede, T.: The SWISS-MODEL workspace: a web-based environment for protein structure homology modelling. Bioinformatics 22, 195–201 (2006)

24. Schwede, T.: SWISS-MODEL: an automated protein homology-modeling server. Nucleic Acids Res. 31, 3381–3385 (2003)

25. Jimenez-Lopez, J.C., Kotchoni, S.O., Hernandez-Soriano, M.C., Gachomo, E.W., Alché, J.D.: Structural functionality, catalytic mechanism modeling and molecular allergenicity of phenylcoumaran benzylic ether reductase, an olive pollen (Ole e 12) allergen. J. Comput. Aided. Mol. Des. 27, 873–895 (2013)

26. Hamman-Khalifa, A., Castro, A.J., Jiménez-López, J.C., Rodríguez-García, M.I., Alché, J.D.D.: Olive cultivar origin is a major cause of polymorphism for Ole e 1 pollen allergen. BMC Plant Biol. 8, 10 (2008)

27. Jimenez-Lopez, J.C., Morales, S., Castro, A.J., Volkmann, D., Rodríguez-García, M.I., Alché, J.D.D.: Characterization of profilin polymorphism in pollen with a focus on multifunctionality. PLoS One 7, e30878 (2012)
28. Palomares, O., Crameri, R., Rhyner, C.: The contribution of biotechnology toward progress in diagnosis, management, and treatment of allergic diseases. Allergy 69, 1588–1601 (2014)

A Computational Method for the Rate Estimation of Evolutionary Transpositions

Nikita Alexeev[1,2], Rustem Aidagulov[3], and Max A. Alekseyev[1,*]

[1] Computational Biology Institute, George Washington University,
Ashburn, VA, USA
[2] Chebyshev Laboratory, St. Petersburg State University, St. Petersburg, Russia
[3] Department of Mechanics and Mathematics, Moscow State University,
Moscow, Russia
maxal@gwu.edu

Abstract. Genome rearrangements are evolutionary events that shuffle genomic architectures. Most frequent genome rearrangements are reversals, translocations, fusions, and fissions. While there are some more complex genome rearrangements such as transpositions, they are rarely observed and believed to constitute only a small fraction of genome rearrangements happening in the course of evolution. The analysis of transpositions is further obfuscated by intractability of the underlying computational problems.

We propose a computational method for estimating the rate of transpositions in evolutionary scenarios between genomes. We applied our method to a set of mammalian genomes and estimated the transpositions rate in mammalian evolution to be around 0.26.

1 Introduction

Genome rearrangements are evolutionary events that shuffle genomic architectures. Most frequent genome rearrangements are *reversals* (that flip segments of a chromosome), *translocations* (that exchange segments of two chromosomes), *fusions* (that merge two chromosomes into one), and *fissions* (that split a single chromosome into two). The minimal number of such events between two genomes is often used in phylogenomic studies to measure the evolutionary distance between the genomes.

These four types of rearrangements can be modeled by 2-breaks [1] (also called DCJs [2]), which break a genome at two positions and glue the resulting fragments in a new order. They simplify the analysis of genome rearrangements and allow one to efficiently compute the corresponding evolutionary distance between two genomes.

Transpositions represent yet another type of genome rearrangements that cuts off continuous segments of a genome and moves them to different positions. In contrast to reversal-like rearrangements, transpositions are rarely observed and

---

* Corresponding author.

F. Ortuño and I. Rojas (Eds.): IWBBIO 2015, Part I, LNCS 9043, pp. 471–480, 2015.

believed to appear in a small proportion in the course of evolution (e.g., in Drosophila evolution transpositions are estimated to constitute less than 10% of genome rearrangements [3]). Furthermore, transpositions are hard to analyze; in particular, computing the transposition distance is known to be NP-complete [4]. To simplify analysis of transpositions, they can be modeled by 3-breaks [1] that break the genome at *three* positions and glue the resulting fragments in a new order.

In the current work we propose a computational method for determining the proportion of transpositions (modeled as 3-breaks) among the genome rearrangements (2-breaks and 3-breaks) between two genomes. To the best of our knowledge, previously the proportion of transpositions was studied only from the perspective of its bounding with the weighted distance model [5, 6], where reversal-like and transposition-like rearrangements are assigned different weights. However, it was empirically observed [7] and then proved that the weighted distance model does not, in fact, achieve its design goal [8]. We further remark that any approach to the analysis of genome rearrangements that controls the proportion of transpositions would need to rely on a biologically realistic value, which can be estimated with our method.

We applied our method for different pairs among the rat, macaque, and human genomes and estimated the transpositions rate in all pairs to be around 0.26.

2   Background

For the sake of simplicity, we restrict our attention to circular genomes. We represent a genome with n blocks as a graph which contains n directed edges encoding blocks and n undirected edges encoding block adjacencies. We denote the tail and head of a block i by i^t and i^h, respectively. A *2-break* replaces any pair of adjacency edges $\{x, y\}$, $\{u, v\}$ in the genome graph with either a pair of edges $\{x, u\}$, $\{y, v\}$ or a pair of edges $\{u, y\}$, $\{v, x\}$. Similarly, a *3-break* replaces any triple of adjacency edges with another triple of edges forming a matching on the same six vertices (Fig. 1).

Let P and Q be genomes on the same set S of blocks (e.g., synteny blocks or orthologous genes). We assume that in their genome graphs the adjacency edges of P are colored black and the adjacency edges of Q are colored red. The *breakpoint graph* $G(P, Q)$ is defined on the set of vertices $\{i^t, i^h | i \in S\}$ with black and red edges inherited from genome graphs of P and Q. The black and red edges in $G(P, Q)$ form a collection of alternating black-red cycles (Fig. 1). We say that a black-red cycle is an *ℓ-cycle* if it contains ℓ black edges (and ℓ red edges), and we denote the number of ℓ-cycles in $G(P, Q)$ by $c_\ell(P, Q)$. We call 1-cycles *trivial cycles*[1] and we call *breakpoints* the vertices belonging to non-trivial cycles.

---

[1] In the breakpoint graph constructed on synteny blocks, there are no trivial cycles since no adjacency is shared by both genomes. However, in our simulations below this condition may not hold, which would result in the appearance of trivial cycles.

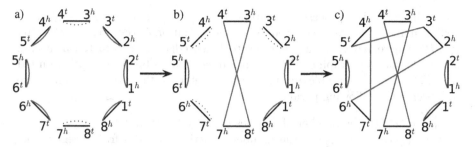

Fig. 1. a) The breakpoint graph $G(P, Q_0)$ of "black" genome P and "red" genome $Q_0 = P$, each consisting of a single circular chromosome $(1, 2, 3, 4, 5, 6, 7, 8)$. Here, $n = 8$, $b = 0$, and all cycles in $G(P, Q_0)$ are trivial. **b)** The breakpoint graph of "black" genome P and "red" genome $Q_1 = (1, 2, 3, -7, -6, -5, -4, 8)$ obtained from Q_0 with a reversal of a segment $4, 5, 6, 7$ (represented as 2-break on the dotted edges shown in a). Here we use $-i$ to denote opposite orientation of the block i. The graph consists of $c_1 = 6$ trivial cycles and $c_2 = 1$ 2-cycle, and thus $b = 2c_2 = 2$. **c)** The breakpoint graph of "black" genome P and "red" genome $Q_2 = (1, 2, -6, -5, 3, -7, -4, 8)$ obtained from Q_1 with a transposition of a segment $3, -7$ (represented as a single 3-break on the dotted edges shown in b). The graph consists of $c_1 = 3$ trivial cycles, $c_2 = 1$ 2-cycle, and $c_3 = 1$ 3-cycle; thus $b = 2c_2 + 3c_3 = 5$.

The 2-break distance between genomes P and Q is the minimum number of 2-breaks required to transform P into Q.

Theorem 1 ([2]). *The 2-break distance between circular genomes P and Q is*

$$d(P, Q) = n(P, Q) - c(P, Q) ,$$

where $n(P, Q)$ and $c(P, Q)$ are, respectively, the number of blocks and cycles in $G(P, Q)$.

While 2-breaks can be viewed as particular cases of 3-breaks (that keep one of the affected edges intact), from now on we will assume that 3-breaks change all three edges on which they operate.

3   Estimation for the Transposition Rate

In our model, we assume that the evolution represents a discrete Markov process, where different types of genome rearrangements (2-breaks and 3-breaks) occur independently with fixed probabilities. Let p and $1 - p$ be the rate (probability) of 3-breaks and 2-breaks, respectively. For any two given genomes resulted from this process, our method estimates the value of p as explained below. In the next section we evaluate the accuracy of the proposed method on simulated genomes and further apply it to real mammalian genomes to recover the proportion of transpositions in mammalian evolution.

Let the evolution process start from a "black" genome P and result in a "red" genome Q. It can be viewed as a transformation of the breakpoint graph $G(P,P)$, where red edges are parallel to black edges and form trivial cycles, into the breakpoint graph $G(P,Q)$ with 2-breaks and 3-breaks operating on red edges. There are observable and hidden parameters of this process. Namely, we can observe the following parameters:

- $c_\ell = c_\ell(P,Q)$, the number of ℓ-cycles (for any $\ell \geq 2$) in $G(P,Q)$;
- $b = b(P,Q) = \sum_{\ell \geq 2} \ell c_\ell$, the number of active (broken) fragile regions between P and Q, also equal the number of synteny blocks between P and Q and the halved total length of all non-trivial cycles in $G(P,Q)$;
- $d = d(P,Q)$, the 2-break distance between P and Q;

while the hidden parameters are:

- $n = n(P,Q)$, the number of (active and inactive) fragile regions in P (or Q), also equal the number of solid regions (blocks) and the halved total length of all cycles in $G(P,Q)$;
- k_2, the number of 2-breaks between P and Q,
- k_3, the number of 3-breaks between P and Q.

We estimate the rearrangement distance between genomes P and Q as $k_2 + k_3$ and the rate p of transpositions as

$$p = \frac{k_3}{k_2 + k_3} \, .$$

We remark that in contrast to other probabilistic methods for estimation of evolutionary parameters (such as the evolutionary distance in [9]), in our method we assume that the number of trivial cycles c_1 is not observable. While trivial cycles can be observed in the breakpoint graph constructed on homologous gene families (rather than synteny blocks), their interpretation as conserved gene adjacencies (which happen to survive just by chance) implicitly adopts the *random breakage model* (RBM) [10, 11] postulating that every adjacency has equal probability to be broken by rearrangements. The RBM however was recently refuted with the more accurate *fragile breakage model* (FBM) [12] and then the *turnover fragile breakable model* (TFBM) [13], which postulate that only certain ("fragile") genomic regions are prone to genome rearrangements. The FBM is now supported by many studies (see [13] for further references and discussion).

4   Estimation for the Hidden Parameters

In this section, we estimate hidden parameters n, k_2, and k_3 using observable parameters, particularly c_2 and c_3.

Firstly, we find the probability that a red edge was never broken in the course of evolution between P and Q. An edge is not broken by a single 2-break with the probability $\left(1 - \frac{2}{n}\right)$ and by a single 3-break with the probability $\left(1 - \frac{3}{n}\right)$. So, the probability for an edge to remain intact during the whole process of k_2 2-breaks and k_3 3-breaks is

$$\left(1 - \frac{2}{n}\right)^{k_2} \left(1 - \frac{3}{n}\right)^{k_3} \approx e^{-\gamma} ,$$

where $\gamma = \frac{2k_2 + 3k_3}{n}$.

Secondly, we remark that for any fixed ℓ, the number of ℓ-cycles resulting from occasional splitting of longer cycles is negligible,[2] since the probability of such splitting has order $\frac{b}{n^2}$. In particular, this implies that the number of trivial cycles (i.e., 1-cycles) in $G(P, Q)$ is approximately equal to the number of red edges that were never broken in the course of evolution between P and Q. Since the probability of each red edge to remain intact is approximately $e^{-\gamma}$, the number of such edges is approximated by $n \cdot e^{-\gamma}$. On the other hand, the number of trivial cycles in $G(P, Q)$ is simply equal to $n - b$, the number of shared block adjacencies between P and Q. That is,

$$n - b \approx ne^{-\gamma} . \tag{1}$$

Thirdly, we estimate the number of 2-cycles in $G(P, Q)$. By the same reasoning as above, such cycles mostly result from 2-breaks that merge pairs of trivial cycles. The probability for a red edge to be involved in exactly one 2-break is $\frac{2k_2}{n} \left(1 - \frac{1}{n}\right)^{2k_2 + 3k_3 - 1}$. The probability that another red edge was involved in the same 2-break is $\frac{1}{n} \left(1 - \frac{1}{n}\right)^{2k_2 + 3k_3 - 1}$. Since the total number of edge pairs is $n(n - 1)/2$, we have the following approximate equality for the number of 2-cycles:

$$c_2 \approx k_2 e^{-2\gamma} . \tag{2}$$

And lastly, we estimate the number of 3-cycles in $G(P, Q)$. As above, they mostly result from either 3-breaks that merge three 1-cycles, or 2-breaks that merge a 1-cycle and a 2-cycle. The number of 3-cycles of the former type approximately equals $k_3 e^{-3\gamma}$ analogously to the reasoning above. The number of 3-cycles of the latter type is estimated as follows. Clearly, one of the red edges in such a 3-cycle results from two 2-breaks, say ρ_1 followed by ρ_2, which happens with the probability about

$$\frac{2k_2(2k_2 - 2)}{2n^2} \left(1 - \frac{1}{n}\right)^{2k_2 + 3k_3 - 2} \approx 2\frac{k_2^2}{n^2} e^{-\gamma} .$$

One of the other two edges results solely from ρ_1, while the remaining one results solely from ρ_2, which happens with the probability about $\left(\frac{1}{n} e^{-\gamma}\right)^2$. Since there

---

[2] We remark that under the parsimony condition long cycles are never split into smaller ones. Our method does not rely on the parsimony condition and can cope with such splits when their number is significantly smaller than the number of blocks.

are about n^3 ordered triples of edges, we get the following approximate equality for the number of 3-cycles:

$$c_3 \approx k_3 e^{-3\gamma} + \frac{2k_2^2}{n} e^{-3\gamma} . \tag{3}$$

Fig. 2 provides an empirical evaluation of the estimates (2) and (3) for the number of 2-cycles and 3-cycles in $G(P, Q)$, which demonstrates that these estimates are quite accurate.

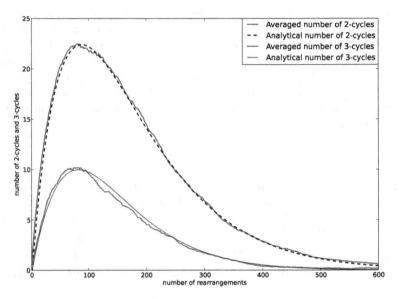

Fig. 2. Empirical and analytical curves for the number of 2-cycles and 3-cycles averaged over 100 simulations on $n = 400$ blocks with proportion of 3-breaks $p = 0.3$

Below we show how one can estimate the probability p from the (approximate) equations (1), (2), and (3).

We eliminate k_2 from (3), using (2):

$$c_3 \approx k_3 e^{-3\gamma} + \frac{2c_2^2}{n} e^{\gamma} .$$

Now we consider the following linear combination of the last equation and (2):

$$2e^{-\gamma}(c_2 - k_2 e^{-2\gamma}) + 3\left(c_3 - (k_3 e^{-3\gamma} + \frac{2c_2^2}{n} e^{\gamma}) \right) \approx 0 .$$

It gives us the following equation for γ and n:

$$\gamma e^{-3\gamma} \approx \frac{1}{n} \left(2c_2 e^{-\gamma} + 3c_3 - \frac{6c_2^2 e^{\gamma}}{n} \right) .$$

Using (1), we eliminate n from the last equation and obtain the following equation with respect to a single indeterminate γ:

$$\gamma e^{-3\gamma} \approx \frac{1 - e^{-\gamma}}{b} \left(2c_2 e^{-\gamma} + 3c_3 - \frac{6c_2^2 e^{\gamma}(1 - e^{-\gamma})}{b} \right) . \tag{4}$$

Solving this equation numerically (see Example 1, Section 5.1), we obtain the numerical values for γ^{est}, n^{est}, k_2^{est} and k_3^{est}, and, finally,

$$p_{est} = \frac{k_3^{est}}{k_2^{est} + k_3^{est}} .$$

5 Experiments and Evaluation

5.1 Simulated Genomes

We performed a simulation with a fixed number of blocks $n = 1800$ and variable parameters p and γ. In each simulation, we started with a genome P and applied a number of 2-breaks and 3-breaks with probability $1 - p$ and p, respectively, until we reached the chosen value of γ. We denote the resulting genome by Q and estimate p with our method as p_{est}. We observed that the robustness of our method mostly depends on p and γ, and it becomes unstable for $p_{est} < 0.15$

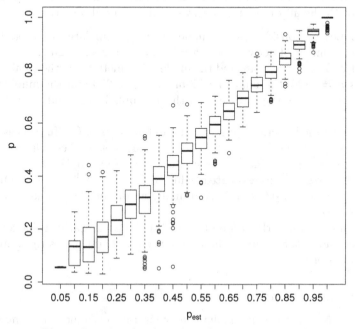

Fig. 3. Boxplots for the value of p as a function of p_{est}

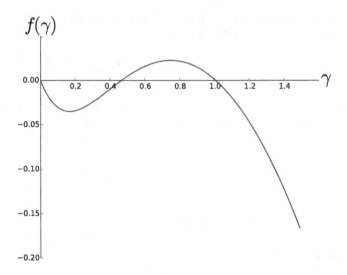

Fig. 4. Typical behavior of $f(\gamma) = \gamma e^{-3\gamma} - \frac{1-e^{-\gamma}}{b}\left(2c_2 e^{-\gamma} + 3c_3 - \frac{6c_2^2 e^{\gamma}(1-e^{-\gamma})}{b}\right)$, the difference between right and left hand sides of (4), where $b = 716$, $c_2 = 107$, $c_3 = 48$

(Fig. 3). So in our experiments we let p range between 0.05 and 1 with step 0.05 and γ range between 0.2 and 1.2 with step 0.1.

In Fig. 3, we present boxplots for the value of p as a function of p_{est} cumulative over the values of γ. These evaluations demonstrate that p_{est} estimates p quite accurately with the absolute error below 0.1 in 90% of observations.

Example 1. Let us consider the example from our simulated dataset. In this example, the number of active blocks $b = 716$, the number of 2-cycles $c_2 = 107$, the number of 3-cycles $c_3 = 48$, and the hidden parameters are: the total number of blocks is $n = 1800$, the number of 2-breaks $k_2 = 279$ and the number of 3-breaks is $k_3 = 114$. So, the value of p in this example is 0.29 and the value of γ is 0.5.

At first, using the bisection method, one can find roots of (4). In this case there are two roots: $\gamma = 0.466$ and $\gamma = 1.007$ (See Fig.4). Let us check the root 0.466 first. Then, using (1), one finds the estimated value of n: $716/(1-e^{0.466}) \approx 1922$. Equation (2) gives us the estimated value of k_2: $107e^{2\cdot 0.466} \approx 272$. One can estimate k_3 as $(\gamma n - 2k_2)/3 \approx 117$. And finally we obtain the estimated value of p: $117/(117+272) \approx 0.3$.

In this example, using the second root of (4) yields a negative value for k_3, so we do not consider it. So, our method quite accurately estimates the value of p, and also values of γ and n.

5.2   Mammalian Genomes

We analyzed a set of three mammalian genomes: rat, macaque, and human, represented as sequences of 1,360 synteny blocks [14, 15]. For each pair of genomes, we circularized[3] their chromosomes, constructed the breakpoint graph, obtained parameters b, c_2, c_3, and independently estimated the value of p. The results in Table 1 demonstrate consistency and robustness with respect to the evolutionary distance between the genomes (e.g., the 2-break distance between rat and human genomes is 714, while the 2-break distance between macaque and human genomes is 106). The rate of transpositions for all genome pairs is estimated to be around 0.26. Numerical experiments suggest that the 95% confidence interval for such values is [0.1, 0,4] (Fig. 3).

Table 1. Observable parameters b, c_2, c_3 and estimation p_{est} for the rate of evolutionary transpositions between circularized rat, macaque, and human genomes

Genome pair	b	c_2	c_3	p_{est}
rat-macaque	1014	201	85	0.27
rat-human	1009	194	79	0.26
macaque-human	175	45	17	0.25

6   Discussion

In the present work we describe a first computational method for estimation of the transposition rate between two genomes from the distribution of cycle lengths in their breakpoint graph. Our method is based on modeling the evolution as a Markov process under the assumption that the transposition rate remains constant. The method does NOT rely on the random breakage model [10, 11] and thus is consistent with more prominent fragile breakage model [12, 13] of chromosome evolution. As a by-product, the method can also estimate the true rearrangement distance (as $k_2 + k_3$) in the evolutionary model that includes both reversal-like and transposition-like operations.

Application of our method on different pairs of mammalian genomes reveals that the transposition rate is almost the same for distant genomes (such as rat and human genomes) and close genomes (such as macaque and human genomes), suggesting that the transposition rate remains the same across different lineages in mammalian evolution.

In further development of our method, we plan to employ the technique of stochastic differential equations, which may lead to a more comprehensive description of the c_ℓ behavior. It appears to be possible to obtain equations, analogous to (2) and (3), for c_ℓ with $\ell > 3$. This could allow one to verify the model and estimate the transposition rate more accurately.

---

[3] While chromosome circularization introduces artificial edges to the breakpoint graph, the number of such edges (equal to the number of chromosomes) is negligible as compared to the number of edges representing block adjacencies in the genomes. For subtle differences in analysis of circular and linear genomes see [16].

Acknowledgements. The authors thank Jian Ma for providing the synteny blocks for mammalian genomes.

The work is supported by the National Science Foundation under the grant No. IIS-1462107. The work of NA is also supported by grant 6.38.672.2013 of SPbSU and RFFB grant 13-01-12422-ofi-m.

References

1. Alekseyev, M., Pevzner, P.: Multi-break rearrangements and chromosomal evolution. Theoretical Computer Science 395(2), 193–202 (2008)
2. Yancopoulos, S., Attie, O., Friedberg, R.: Efficient sorting of genomic permutations by translocation, inversion and block interchange. Bioinformatics 21(16), 3340–3346 (2005)
3. Ranz, J., González, J., Casals, F., Ruiz, A.: Low occurrence of gene transposition events during the evolution of the genus Drosophila. Evolution 57(6), 1325–1335 (2003)
4. Bulteau, L., Fertin, G., Rusu, I.: Sorting by transpositions is difficult. SIAM Journal on Discrete Mathematics 26(3), 1148–1180 (2012)
5. Bader, M., Ohlebusch, E.: Sorting by weighted reversals, transpositions, and inverted transpositions. Journal of Computational Biology 14(5), 615–636 (2007)
6. Fertin, G.: Combinatorics of genome rearrangements. MIT Press (2009)
7. Blanchette, M., Kunisawa, T., Sankoff, D.: Parametric genome rearrangement. Gene 172(1), GC11–GC17 (1996)
8. Jiang, S., Alekseyev, M.: Weighted genomic distance can hardly impose a bound on the proportion of transpositions. Research in Computational Molecular Biology, 124–133 (2011)
9. Lin, Y., Moret, B.M.: Estimating true evolutionary distances under the DCJ model. Bioinformatics 24(13), i114–i122 (2008)
10. Ohno, S.: Evolution by gene duplication. Springer, Berlin (1970)
11. Nadeau, J.H., Taylor, B.A.: Lengths of Chromosomal Segments Conserved since Divergence of Man and Mouse. Proceedings of the National Academy of Sciences 81(3), 814–818 (1984)
12. Pevzner, P.A., Tesler, G.: Human and mouse genomic sequences reveal extensive breakpoint reuse in mammalian evolution. Proceedings of the National Academy of Sciences 100, 7672–7677 (2003)
13. Alekseyev, M.A., Pevzner, P.A.: Comparative Genomics Reveals Birth and Death of Fragile Regions in Mammalian Evolution. Genome Biology 11(11), R117 (2010)
14. Ma, J., Zhang, L., Suh, B.B., Raney, B.J., Burhans, R.C., Kent, W.J., Blanchette, M., Haussler, D., Miller, W.: Reconstructing contiguous regions of an ancestral genome. Genome Research 16(12), 1557–1565 (2006)
15. Alekseyev, M., Pevzner, P.A.: Breakpoint graphs and ancestral genome reconstructions. Genome Research, gr–082784 (2009)
16. Alekseyev, M.A.: Multi-break rearrangements and breakpoint re-uses: from circular to linear genomes. Journal of Computational Biology 15(8), 1117–1131 (2008)

Genome Structure of Organelles Strongly Relates to Taxonomy of Bearers

Michael Sadovsky, Yulia Putintseva, Anna Chernyshova, and Vaselina Fedotova

Institute of Computational Modelling of SB RAS,
Akademgorodok, 660036 Krasnoyarsk, Russia
msad@icm.krasn.ru, kinomanka85@mail.ru,
{anna12651,vasilinyushechka}@gmail.com
http://icm.krasn.ru

Abstract. We studied the relations between the triplet frequency dictionaries of organelle genome, and the phylogeny of their bearers. The clusters in 63-dimensional space were identified through K-means, and the clade composition of those clusters has been investigated. Very high regularity in genomes distribution among the clusters was found, in terms of taxonomy. The strong synchrony in evolution of nuclear and organelle genomes manifests through this correlation: the proximity in frequency space was determined over the organelle genomes, while the proximity in taxonomy was determined morphologically. Similar effect is also found in the ensembles of other (say, yeast) genomes.

Keywords: frequency, triplet, order, cluster, similitude, elastic map, morphology, evolution, synchrony.

1 Introduction

Statistical properties of nucleotide sequences may tell a lot to a researcher. The patterns observed in sequences correlate to functions encoded in a sequence, or to a taxonomy of a bearer of that latter. Here we shall study those correlations between the structure, and the taxonomy.

A variety of patterns in a nucleotide sequence is tremendous. Here a consistent and comprehensive study of frequency dictionaries answers some questions concerning the statistical and information properties of DNA sequences. A frequency dictionary, whatever one understands for it, is rather multidimensional entity. That latter is supposed to be the simplest structure. Further, we shall concentrate on the study of the frequency dictionaries [8–10] of the thickness $q = 3$; in other words, the triplet composition only will be taken into consideration. Here we studied the *structure − taxonomy* relations for mitochondrion vs. host genomes, and chloroplast vs. host genomes.

Let now introduce more strict definitions and issues. Consider a continuous symbol sequence of the length N (total number of symbols in it) from four-letter alphabet $\aleph = \{A, C, G, T\}$; such sequence represents some genetic entity (genome, chromosome, etc.). We stipulate that no other symbols or gaps in the

F. Ortuño and I. Rojas (Eds.): IWBBIO 2015, Part I, LNCS 9043, pp. 481–490, 2015.
© Springer International Publishing Switzerland 2015

sequence take place. Any coherent string $\omega = \nu_1\nu_2\ldots\nu_q$ of the length q makes a word. A set of all the words occurred within a sequence yields the support of that latter. Counting the numbers of copies n_ω of the words, one gets a finite dictionary; changing the numbers for the frequency

$$f_\omega = \frac{n_\omega}{N}$$

one gets the frequency dictionary W_q of the thickness q. This is the main object of our study.

Further, we shall concentrate on frequency dictionaries W_3 (i. e., the triplet composition) only. Thus, a frequency dictionary W_3 calculation converts any genetic entity into a point in (formally) 64-dimensional metric space. Obviously, two genetic entities with identical frequency dictionaries $W_3^{(1)}$ and $W_3^{(2)}$ are mapped into the same point in the space. On the contrary, the absolute congruency of two frequency dictionaries $W_3^{(1)} = W_3^{(2)}$ does not guarantee a complete coincidence of the original sequences. Nonetheless, such two sequences are indistinguishable from the point of view of their triplet composition.

Definitely, few entities may have very proximal frequencies of all the triplets, but few others may have not, thus making a distribution of the points in 64-dimensional space inhomogeneous. So, the key question here is what is the pattern of this distribution of mitochondrion genomes in that space? Are there some discrete clusters, and if yes is there a correlation to a phylogeny of the genome bearers and clusters? Some results preliminary answering this question could be found in [8, 5, 6].

To address the questions, we have implemented an unsupervised classification of both mitochondrion and chloroplast genomes, in (metric) space of frequencies of triplets. There were implemented a series of clusterizations, for two, three, four, ..., eight clusters, for both types of genomes. Then, the taxa composition of the classes has been studied; moreover, the relation between the clusters was specially studied, when we changed a clusterization in K clusters for that one in $K-1$ clusters. Besides, a considerable correlation in taxa composition was found, for the observed clusterizations. Briefly speaking, these correlations prove the high synchrony in the evolution of two (physically) independent genetic systems: somatic one, and the organelle one.

This paper presents the evidences of the strong synchrony in evolution of mitochondrion genomes and nuclear ones, as well as the synchrony in evolution of chloroplast genomes vs. nuclear ones.

2   Material and Methods

2.1   Genetic Sequences

All the sequences were retrieved from EMBL–bank. **"Junk" symbol agreement**: some entries contain the "junk" symbols (those that fall beyond the original alphabet $\aleph$). All such symbols have been omitted furthered with the concatination of the obtained fragment into an entity.

Originally, the release used to retrieve mitochondria genomes containes $\sim$ 6.4×10^3 entries. The final database used in our study enlists 3721 entries. Similarly, the full list of chloroplast genomes at the release used for them exceeded five hundred entries; the study on chloroplast clusterization has been carried out with 251 genomes. This discrimination comes from the (not obvious) constraint: we had to eliminate from the study the entries which represent rather highly ranked clade solely, as the single species in the clade. A highly order clade presented with a single genome (that is a single species) yields a "signal" strong enough to deteriorate a general pattern, but weak one to produce distinguishable details in the distribution pattern. Thus, we enlist into the final databases the entries representing an order (and higher clades) with five species or more, for mitochondria. Similar cut-off number for chloroplast genomes was equal to 3 species.

Table 1. Mitochondria database structure; M is the abundance of the clade

Order	M	Order	M	Order	M	Order	M
Actinopterygii	1181	*Amphibia*	151	*Anthozoa*	16	*Arachnida*	10
Aves	197	*Bivalvia*	34	*Cephalopoda*	45	*Cestoda*	28
Chromadorea	5	*Gastropoda*	16	*Homoscleromorpha*	14	*Insecta*	350
Malacostraca	33	*Mammalia*	1457	*Reptilia*	172	*Trematoda*	13

The stricture of the final mitochondrion database is shown in Table 1. Since the total number of chloroplast genomes under consideration is significantly less, in comparison to the mitochondria list, the clade composition of that former seems to be less apparent; meanwhile, the chloroplast database includes 157 broadleaf species against 94 conifer ones.

2.2   Clusterization Methods

We implemented unsupervised classification by K-means to develop classes (see details and a lot of examples in [3, 2, 11, 7]). To cluster, we had to reduce the data space dimension to 63: the reduction results from the equality to 1 of the sum of all frequencies. Formally speaking, any triplet could be excluded from the data set; practically, we excluded the triplets (specific, for mitochondria, and for chloroplasts) that yield the least standard deviation, over the set of genomes under consideration. Evidently, such triplets make least contribution into the distinguishing the entities, in the space of frequencies.

A K-means implementation may be based on a number of distances; here we used Euclidean distance. Also, no class separability has been checked. All the results were obtained with *ViDaExpert* software by A. Zinovyev[1].

---

[1] http://bioinfo-out.curie.fr/projects/vidaexpert/

"Downward" vs. "Upward" Classification. Two versions of the K-means classification implementation could be developed: "downward" vs. "upward" ones, respectively. They both are based on a standard K-means technique, but the difference is in the mutual interaction between the clusterizations developed for different number of clusters.

"Downward" classification. This kind of classification is designed to follow the classical morphology based classification. It starts from the clusterization of the entire set of genomes (frequency dictionaries) into the minimal number $\mathfrak{M}_c$ of clusters with the given stability of the clusterization. That latter is understood as the given number of volatile genomes, i. e. genomes that may change their cluster attribution with any new clustering realization. Then each of the clusters is to be separated into the similar (i. e. minimal stable subclusters) set of subclusters, etc. The procedure is to be trunked at the given "depth" of the cluster separation, usually determined by the volatility of a significant part of genomes.

Table 2. Standard deviation figures, for mitochondria and chloroplast databases

chloroplasts				mitochondria			
GAC	0,000540	ACC	0,000672	GCG	0,001329	TCG	0,001712
GGC	0,000593	GTC	0,000731	CGT	0,001608	GTC	0,001715
GCC	0,000612	CGC	0,000748	CGA	0,001690	AGG	0,001726

Thus, a "downward" classification yields the structure that is a tree, so making it close to a standard morphological classification.

"Upward" classification. On the contrary, the upward classification consists in the separation of the entire set of genomes, sequentially, into the series of clusters

$$\mathfrak{C}_2, \mathfrak{C}_3, \mathfrak{C}_4, \ldots, \ \mathfrak{C}_{K-1}, \mathfrak{C}_K .$$

Here we assume that the clusterization $\mathfrak{C}_2$ is stable. Again, the series is to be trunked at the given number K; we put $K = 8$ in this study with chloroplasts.

The key question here is the mutual relation between the members of a cluster $\mathfrak{C}(l)_j$ from $\{\mathfrak{C}(i)_j\}$ $(1 \leqslant i \leqslant j)$ clusterization with the clusters from $\{\mathfrak{C}(m)_{j-1}\}$ $(1 \leqslant m \leqslant j-1)$ clusterization. Here the index l enlists the clusters at the $\{\mathfrak{C}(i)_j\}$ clusterization. There could be (roughly) three options:

- A cluster $\mathfrak{C}(n)_j$ is entirely embedded into the cluster $\mathfrak{C}(l)_{j-1}$, with some l and j;
- The greater part of the members of a cluster $\mathfrak{C}(n)_j$ is embedded into the cluster $\mathfrak{C}(l)_{j-1}$, but the minor part is embedded into the other cluster $\mathfrak{C}(m)_{j-1}$;
- A cluster $\mathfrak{C}(n)_j$ is almost randomly spread between the set of clusters $\mathfrak{C}(l)_{j-1}$, $l = 1, 2, \ldots, l^*$.

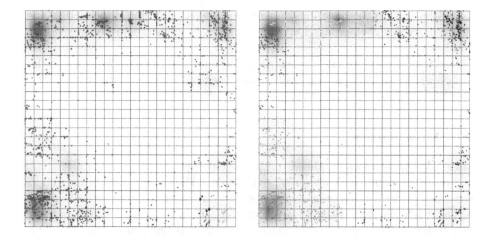

Fig. 1. Soft 25×25 elastic map for 3954 mitochondria genomes; left is the types distribution, right is the clades distribution for chordata type. See the text for details.

Thus, an upward classification yields a pattern that is a graph with cycles; at the worst case, the graph is fully connected, and here no essential structuredness is observed. If the graph has rather small number of cycles, then it reveals the relations between the clusters (determined through the proximity in frequency space), and the taxonomy (determined over the nuclear genome).

2.3   On the Stability of K-means Clusterization

Another essential point is the volatility of genomes (in K-means clusterization): that is equivalent to the clusterization stability. Speaking on stability, we stipulate some genomes always (or almost always) occupy the same cluster, for different starting distributions. Other genomes tend to change their cluster position, in a series of K-means implementations.

Thus, the former set of entities is supposed to be stable, while the latter gathers the unstable entities. Stability here could be evaluated through the portion of genomes always occupying the cluster together; volatile genomes, on the contrary, change their cluster occupation, for different implementations of K-means. Everywhere further we shall stipulate that stability in K-means clusterization means that the stable ensemble of genomes exhibits the same clusterization pattern in 0.85-part of the series of K-means implementations.

3   Results

We studied the relation between the structure defined in terms of a triplet composition of organelle genomes, and the taxonomy determined according to a morphology, for two types of organelles: chloroplasts and mitochondria. Everywhere below the upward classifications only are considered, both for mitochondria and chloroplasts. No class separation conditions has been checked, in both genomes databases.

Besides, the number of classes $\mathfrak{C}(n)$ varied form two to eight: $2 \leqslant \mathfrak{C}_K \leqslant 8$, for both databases. All these constraints have been put on mostly due to technical reasons.

3.1   Mitochondria

Mitochondria genomes database, unlike the chloroplasts one, is quite abundant; on the other hand, it is rather biased: Table 1 shows this fact. Some genera are overrepresented, in comparison to others, but others seem to be underrepresented. Such bias affects the K-means clusterization. In particular, it may result in a stability decay of the clusterization developed by K-means technique. Due to this discrepancy, we have used elastic map technique to figure out the clusters in triplet frequency space.

Elastic map is another powerful approach to visualize and analyze multidimensional data. This approach makes no way to establish either upward, or downward classification: it yields a distribution of genomes on a non-linear two-dimensional manifold (elastic map). We have used the detailed soft map of 25×25 size. Figure 1 shows the distribution of the entities over the map; left part of the figure presents the distribution of *Cordata* (the most abundant class) in pink ring labels, *Arthropoda* in red squares, *Mollusca* in green triangles, *Nematoda* are shown in brick-like colored diamonds, flat warms are whosn in sand-colored pentagons and finally *Porifera* are shown in dark-blue hexagons.

The right part of the figure shows the distribution of three main clades of *Chordata* type: *Mammalia* are shown in yellow diamonds, birds are shown in red pentagons, and fishes are shown in green triangles.

Color background indicates the average local density of the genomes in this map. One may see quite unexpected growth of the local density; that former is located at the map node $[5, 8]$, if the lowest left one is supposed to be the $[1, 1]$ node.

3.2   Chloroplasts

Figure 3 shows the graph of embedments for the clusters, where the number of these latter changed from 8 to 3. We developed the clusterizations for three, four, ..., eight clusters, and studied the composition of each cluster, at the each "depth". The key question was whether the species tend to keep together, when the number of clusters in a clusterization is decreased.

The Figure answers distinctly and apparently this question: the clades in the boxes correspond to genera, while the species (not shown in the figure) always make a solid group, when changing the number of clusters. Thus, boxes having two upright arrows showing the transfer of entities from $\mathfrak{C}_l$ clusterization to $\mathfrak{C}_{l-1}$ one contain two groups of species belonging the same genus (or family).

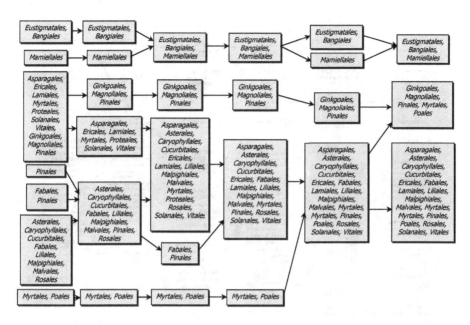

Fig. 2. The chain of clusters identified through K-means, with high stability level. Only orders are shown in the graph.

The point is that the scheme shown in Fig. 3 is just a part of a general pattern: the figure shows the genomes that exhibit reasonable stability level in the upward clusterization. There a sounding number of chloroplast genomes that exhibit a low stability, for various K-means clusterizations. It should be noticed that the ensemble of unstable genomes may change, as the number of clusters goes down from 8 to 3. Figure 2 shows this part of genomes. Careful examination of Fig. 2 has the following formula for cluster composition: 7, 6, 5, 4, 4, 3 clusters stably identified over the dataset. This formula comes from the degeneration of a stable cluster, when the clusterization over 8, 7, 6 and 5 clusters is carried out (on the contrary to Fig.3). In other words, an attempt to create an 8-class K-means clusterization yields seven stable clusters while the eighth one is opportunistic: it comprises various genomes that tend to occupy the different clusters, for different realizations of K-means.

The graph shown in Fig. 2 is not connected: the branch comprising the algae (both green and red ones) is isolated, at any depth of a classification (orders *Eustigmatales, Bangiales* and *Mamiellalis*.) Also, one can see that some orders occupy two clusters, at the depth 8 and some others. It means that this order is split: some species occupy other cluster than others.

An implementation of the clusterization through K-means for two classes yields the distinct and clear separation of algae from all other plants; the stability of such clusterization is not too high. A series of a thousand realizations of K-means with two-class separation exhibits about 680 realizations with discrete

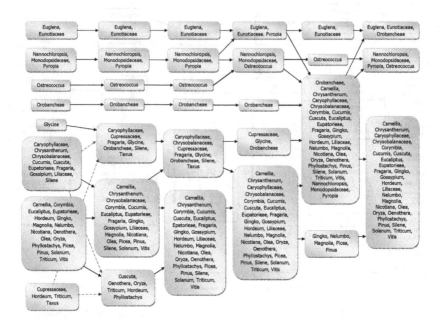

Fig. 3. Pattern of the "upward" embedment of various clades of plants in the unstable classification, developed over chloroplast genomes

isolation of algae from the other plants, while other part of realizations may combine some algae with higher plants.

Stable ensemble of chloroplast genomes differs from the unstable one in the relevant graph connectivity. Stable genomes exhibit significantly less complexity of the pattern: there is an isolated subgraph making the entire graph disconnected. Moreover, this subgraph have very simple structure (linear or almost linear) and is disjointed from the main body of that former. On the contrary, the graph representing the unstable genomes is connected: there are no isolated subgraphs or any other parts.

4 Discussion

To begin with, consider Table 2 in detail. Evidently, the figures for the standard deviation observed for mitochondria exceed the similar figures for chloroplasts in order. Nonetheless, this difference remains the same, for any subbase to be developed from the original one. Probably, such significant difference in the figures results from the fact that mitochondria exhibit the highest possible violation of Chargaff's parity rule, among all other genetic entities.

A study presented in this paper is done within the scope of population genomics methodology. The most intriguing result of the study is the very high correlation between the statistically identified clusters of genomes, and their taxonomy reference. The key point is that we used organelle genomes to derive the clusterization, while the taxonomy was determined traditionally, through morphology, which is ultimately defined by a nuclear genomes. There is no immediate interaction between the nuclear and organelle genomes. The study has been carried out for both main organelles: chloroplasts and mitochondria.

All mitochondria have the same function; same is true for chloroplasts. Thus, the impact of a function divergence was eliminated in our studies. Probably, a database structure is crucial in this kind of studies. We have used an unsupervised classification technique to develop a distribution of genomes into few groups. The results of such classification are usually quite sensitive to an original database composition [7]. Luckily, the genetic banks are rapidly enriched with newly deciphered genomes of organelles, so the stable and comprehensive results showing the reliable relation between structure and taxonomy could be obtained pretty soon. Moreover, a growth of genetic database may provide a comprehensive implementation of a "downward" classification.

The approach presented above looks very fruitful and powerful. One can expand the approach for the following problems to be solved:

- To study the clusterizations as described above for the database consisting of the genomes of mitochondria, and chloroplasts, of the same species. This study would unambiguously address the question on the relation between structure and function: since the organelle genomes under consideration would belong the same organisms, one may expect an elimination of the taxonomy impact, on the results. Meanwhile, this point should be carefully checked, since the results might be sensitive to the list of species involved into the study;
- To study the clusterization of the frequency dictionaries corresponding to the individual genes (or genes combinations) retrieved from the raw genomes of organelles. The clusterization of such genetic entities would address the question on the mutual interaction in a triadic pattern $structure-function-taxonomy$.

References

1. Bugaenko, N.N., Gorban, A.N., Sadovsky, M.G.: Open Systems & Information Dyn. 5, 265–281 (1998)
2. Gorban, A.N., Zinovyev, A.Y.: Int. J. of Neural Systems 20, 219 (2010)
3. Gorban, A.N., Kegl, B., Wünsch, D.C., Zinovyev A.Y. (eds.) Principal Manifolds for Data Visualisation and Dimension Reduction. LNCSE, vol. 58, p. 332. Springer, Heidelberg (2007)
4. Gorban, A.N., Popova, T.G., Zinovyev, A.Y.: In Silico Biology 3, 471 (2003)
5. Gorban, A.N., Popova, T.G., Sadovsky, M.G., Wünsch, D.C.: Information content of the frequency dictionaries, reconstruction, transformation and classification of dictionaries and genetic texts. In: Intelligent Engineering Systems through Artificial Neural Networks, 11 — Smart Engineering System Design, pp. 657–663. ASME Press, New York (2001)

6. Gorban, A.N., Popova, T.G., Sadovsky, M.G.: Open Systems & Information Dyn. 7, 1 (2000)
7. Fukunaga, K.: Introduction to statistical pattern recognition. 2nd edn., 591 p. Academic Press, London (1990)
8. Sadovsky, M.G., Shchepanovsky, A.S., Putintzeva Yu, A.: Theory in Biosciences 127, 69 (2008)
9. Sadovsky, M.G.: J. of Biol. Physics 29, 23 (2003)
10. Sadovsky, M.G.: Bulletin of Math. Biology 68, 156 (2006)
11. Shi, Y., Gorban, A.N., Yang, T.Y.: J. Phys. Conf. Ser. 490, 012082 (2014)

A Unified Integer Programming Model for Genome Rearrangement Problems

Giuseppe Lancia, Franca Rinaldi, and Paolo Serafini

Dipartimento di Matematica e Informatica,
University of Udine,
Via delle Scienze 206, (33100) Udine, Italy

Abstract. We describe an integer programming (IP) model that can be applied to the solution of all genome-rearrangement problems in the literature. No direct IP model for such problems had ever been proposed prior to this work. Our model employs an exponential number of variables, but it can be solved by column generation techniques. I.e., we start with a small number of variables and we show how the correct missing variables can be added to the model in polynomial time.

Keywords: Genome rearrangements, Evolutionary distance, Sorting by reversals, Sorting by transpositions, Pancake flipping problem.

1 Introduction

With the large amount of genomic data available today it is now possible to try and compare the genomes of different species, in order to find their differences and similarities. The model of sequence alignment is inappropriate for genome comparisons, where differences should be measured not in terms of insertions/deletions or mutations of single nucleotides, but rather of macroscopic events, affecting long genomic regions at once, that have happened in the course of evolution. These events occur mainly in the production of sperm and egg cells (but also for environmental reasons), and have the effect of rearranging the genetic material of parents in their offspring. When such mutations are not lethal, after a few generations they can become stable in a population and give rise to the birth of new species.

Among the main evolutionary events known, some affect a single chromosome (e.g., *inversions*, and *transpositions*), while others exchange genomic regions from different chromosomes (e.g., *translocations*). In this paper we will focus on the former type of evolutionary events. When an inversion or a transposition occurs, the fragment is detached from its original position and then it is reinserted, on the same chromosome. In an inversion, it is reinserted at the same place, but with opposite orientation than it originally had. In a transposition, similarly to a cut-and-paste operation in text editing, the fragment is pasted into a new position. Moreover, the fragment can preserve its original orientation, or it can be reversed (in which case we talk of an *inverted transposition*).

F. Ortuño and I. Rojas (Eds.): IWBBIO 2015, Part I, LNCS 9043, pp. 491–502, 2015.

Since evolutionary events affect long DNA regions (several thousand bases), the basic unit for comparison is not the nucleotide, but rather the gene. In fact, the computational study of rearrangement problems started after it was observed that many species share the same genes (i.e., the genes have identical, or nearly identical, DNA sequences), however differently arranged. For example, most genes of the mitochondrial genome of *Brassica oleracea* (cabbage) are identical in *Brassica campestris* (turnip), but appear in a completely different order. Much of the pioneering work in genome rearrangement problems is due to Sankoff and his colleagues [22].

Under the assumption that for any two species s and s' there is a closest common ancestror c in the "tree of life", there exists a set of evolutionary events that, if applied to s, can turn s back into c and then into s'. That is, there exists a path of evolutionary events turning s into s'. The length of this path is correlated to the so-called *evolutionary distance* between s and s'. The general genome comparison problem can then be stated as follows:

> *Given two genomes (i.e., two sets of common genes differently ordered), find which sequence of evolutionary events where applied by Nature to the first genome to turn it into the second.*

A general and widely accepted parsimony principle states that the solution sought is the one requiring the minimum possible number of events (a weighted model, based on the probability of each event, would be more appropriate, but these probabilities are very hard to determine). In the past years people have concentrated on evolution by means of some specific type of event alone, and have shown that these special cases can already be very hard to solve [2,3,9,11,18]. Only in a very few cases models with more than one event have been considered [1,20].

The ILP approach. When faced with difficult (i.e., NP-hard) problems, the most successful approach in combinatorial optimization is perhaps the use of Integer Linear Programming (ILP) [21]. In the ILP approach, a problem is modeled by defining a set of integer variables and a set of linear constraints that these variables must satisfy in order to represent a feasible solution. The optimal solution is found by minimizing a linear function over all feasible solutions. The solution of an ILP model can be carried on by resorting to a standard package, such as the state-of-the-art program CPLEX. The package solves the ILP via a branch-and-bound procedure employing mathematical programming ideas.

Since the solution algorithm is already taken care of, the main difficulty in the ILP approach is the design of the model, i.e., the definition of proper variables, constraints and objective function. Genome rearrangement problems have thus far defied the attempts of using ILP models for their solutions, since no direct ILP models seemed possible. In this paper we describe a very general new ILP model which can be used for all genome rearrangement problems in the literature.

2 Sorting Permutations

Two genomes are compared by looking at their common genes. After numbering each of n common genes with a unique label in $[n] := \{1, \ldots, n\}$ each genome is a permutation of the elements of $[n]$. Let $\pi = (\pi_1, \ldots, \pi_n)$ and $\tau = (\tau_1, \ldots, \tau_n)$ be two genomes. The generic genome rearrangement problem requires to find a shortest sequence of operators that applied to the starting permutation π yields τ as the final permutation. By possibly relabeling the genes, we can always assume that $\tau = \iota = (1, 2, \ldots, n)$, the identity permutation. Hence, the problem becomes *sorting* π by means of the allowed operators.

Each operator represents an evolutionary event that might happen to π so as to change the order of its genes. The most prominent such events are

1. **Reversal.** A fragment of the permutation is flipped over so that its content appears reversed. For instance, a reversal between positions 3 and 6 applied to

$$(2, 7, \boxed{4,\ 1,\ 5,\ 6}, 3)$$

   yields

$$(2, 7, 6, 5, 1, 4, 3)$$

   An interesting special case are *prefix reversals*, in which one of the pivoting points is always position 1, and therefore a prefix of the permutation is flipped over. For instance, a prefix reversal until position 4 applied to

$$(\boxed{2,\ 7,\ 4,\ 1}, 5, 6, 3)$$

   yields

$$(1, 4, 7, 2, 5, 6, 3)$$

2. **Transposition.** A fragment of the permutation is cut from its original position and pasted into a new position. The operator is specified by 3 indexes i, j, k where i and j specify the starting and ending position of the fragment, and k specifies the position where the fragment will be put (i.e. the fragment will end at position k after its placement). For instance a transposition with $i = 2$, $j = 3$ and $k = 6$ applied to

$$(2, \boxed{7,\ 4}, 1, 6, 5, \uparrow 3)$$

   yields

$$(2, 1, 6, 5, 7, 4, 3)$$

3. **Inverted transposition.** A fragment of the permutation is cut from its original position and pasted into a new position, but with the order of its elements flipped over. Of course this operator can be mimicked by a reversal followed by a transposition (or by a transposition followed by a reversal), but here we consider this as a single move and not two moves. Again, the operator is specified by 3 indexes i, j, k where i and j specify the stating and ending position of the fragment, and k specifies the position where the fragment

will be put (i.e. the fragment will end at position k after its placement). For instance an inverted transposition with $i = 2$, $j = 3$ and $k = 6$ applied to

$$(2, \boxed{7, 4}, 1, 6, 5, {\uparrow} 3)$$

yields

$$(2, 1, 6, 5, 4, 7, 3)$$

To each of these operators there corresponds a particular sorting problem. Therefore we have (i) Sorting by Reversals (SBR); (ii) Sorting by Prefix Reversals (also knonw as the *Pancake Flipping Problem*, SBPR); (iii) Sorting by Transpositions (SBT); (iv) Sorting by Inverted Transpositions (SBIT).

Permutation sorting problems are all very challenging and their complexity has for a long time remained open. As for the solving algorithms, there are not may excact approaches for these problems in the literature. In particular, there are not many ILP approaches, since the problems did not appear to have simple, "natural"' ILP formulations.

Historically, the study of permutation sorting problems started with the pancake flipping problem (SBRP), a few years before the age of bioinformatics. This problem has later gained quite some popularity due to the fact that Bill Gates, the founder of Microsoft, co-authored the first paper on its study [15]. The complexity of SBPR has remained open for over 30 years and recently the problem was shown to be NP-hard [6]. Papadimitriou and Gates have proved that the diameter of the pancake flipping graph (i.e., the maximum prefix-reversal distance between two permutations) is $\leq \frac{5}{3}n$. This bound has remain unbeated until recently, and we now know that it is $\leq \frac{18}{11}n$ [12]. There is no published algorithm for exact solution of SBPR, an no ILP formulation of the problem. There is an approximation algorithm achieving an approximation factor of 2 [14].

The other permutations sorting problems have started being investigated in the early nineties for their potential use in computing evolutionary distances. A particular interest was put in sorting by reversals. The problem of SBR was shown to be NP-hard by Caprara [8]. The first approximation algorithm for SBR achieves a factor $\frac{3}{2}$ [13], later improved to 1.375 [5]. Integer Linear Programming has been successfully applied to SBR [10,11], but in an "indirect" way. Namely, the approach exploits an auxiliary maximum cycle-decomposition problem on a bicolored special graph (the *breaklpoint graph* [2]). In this modeling, neither the permutation nor the sorting reversals of the solution are directly seen as variables and/or costraints of the model. Simply, ILP is used for the breakpoint graph problem which yields a tight bound to SBR. The latter is not modeld as an ILP, but solved in a branch-and-bound fashion, while employing the cycle-decomposition bound.

For SBPR and SBT (let alont SBIT) there are not even exact algorithms in the literature. There are approximation algorithms for SBT[3] and SBIT[17,16]. Only recently SBT has been proven NP-hard [7], closing a longstanding open question.

Results. In this paper we describe an ILP model for genome rearrangement problems. Our ILP is very general and possesses some nice features, such as (i) it yields (for the first time) a model for rearranging under two or more operators (such as prefix reversals *and* transpositions, or reversals *and* transpositions, etc.; (ii) it can be used to model each type of rearrangement separately; (iii) we can easily incorporate limitations on some operators (e.g., we may allow only reversals of regions of some bounded length, or transpositions which cannot move a fragment too far away from its original position); (iv) we can easily incorporate different costs for the various operators.

3 The Basic Model

Assume we have fixed the type of operator (or operators) of the particular sorting problem. Let L be an upper bound to the number of operators to sort π (obtained, e.g., by running a heuristic). We will considered a layered graph G with L layers. In particular, there will by L layers of edges (numbered $1, \ldots, L$), and $L + 1$ levels of nodes (numbered $1, \ldots, L + 1$). Each level of nodes consists of n nodes, which represent the current permutation. The nodes of level k are meant to represent the permutation after $k - 1$ solution steps. Between each pair of consecutive levels there are n^2 arcs (i.e., the two levels form a complete bipartite graph). These arcs are meant to show "where each element goes" (i.e., an arc from node i of level k to node j of level $k+1$ means that the k-th operator has moved an element from i-th position to j-th position).

The nodes of the first level are labeled by the starting permutation π, while the nodes of level $L+1$ are labeled by the sorted permutation. Given a solution, if we apply it to π and follow how elements are moved around by the sequence of operators, we will see that, for each $i = 1, \ldots, n$, the element in position i in π must eventually end in position π_i in the sorted permutation. Hence, the solution individues n node-disjoint directed paths in the layered graph, one for each node i of level 1, ending in nodes π_i on level $L + 1$. In Figure 1 we show an example of SBPR, with $L = 5$, and a solution corresponding to the prefix reversals $(< 2 >, < 5 >, < 3 >, < 4 >)$, where by $< i >$ we mean "reverse the first i elements". In bold we can see the path followed by the element labeled 3 in its movements toward its final position.

The model we propose has variables associated to paths in G (it is therefore an exponential model, but we will show how column-generation can be carried out in polynomial time).

Let $\mathcal{P}(i)$ be the set of paths in the layered graph which start at node i of level 1, and end at node π_i of level $L + 1$. Note that there are potentially n^{L-1} such paths. From each node, exactly one path must leave, which can be enforced by constraints

$$\sum_{P \in \mathcal{P}(i)} x_P = 1 \qquad \forall i = 1, \ldots, n \tag{1}$$

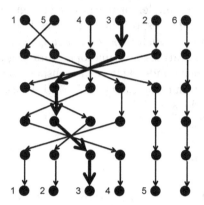

Fig. 1. Solving an instance of SBPR

Furthermore, in going from one level k to the next, the paths should only be allowed along a particular subset of arcs, i.e., the arcs corresponding to the k-th operator used.

Let us consider the complete bipartite graph $[n] \times [n]$, representing all the arcs from a level to the next. Each operator can be seen as a subset of these arcs, in particular, as a special perfect matching. To maintain the generality of the approach, we will just say that the sorting problem is defined by a set of perfect matchings $\mathcal{O}$ in $[n] \times [n]$. Each matching represents a legal operation that can be applied to the current permutation. We will assume that the identity matching $\mu_0 = \{(1,1),(2,2),\ldots,(n,n)\}$ always belongs to $\mathcal{O}$, so that $\mathcal{O} = \{\mu_0, \mu_1, \ldots, \mu_N\}$. The identity permutation specifies a special operation applied to π, i.e., the *no-operation*. It represents a null-event that should be used to fill the "extra" moves that the layered graph can fit with respect to the optimal solution (e.g., if $L = 10$ but the optimal solution uses only 7 moves, then there will be 3 no-operations in the solution using 10 moves).

We now introduce binary variables z_μ^k, for $k = 1, \ldots, L$ and $\mu \in \mathcal{O}$, with the meaning: "in going from level k to level $k+1$ of G, the subset of arcs that can be used by the paths is μ". Otherwise stated, "the k-th move of the solution is the operator (corresponding to) μ".

Since at each level we must use an operator (possibly, the null-operator), we have constraints

$$\sum_{\mu \in \mathcal{O}} z_\mu^k = 1 \qquad \forall k = 1, \ldots, L \tag{2}$$

We can see the setting of z_μ^k to 1 as the *activation* of a particular set of arcs at level k. Then, we have constraints stating that the paths can only use activated arcs, i.e.,

$$\sum_{i=1}^{n} \sum_{P \in \mathcal{P}(i) \,|\, e \in P} x_P \leq \sum_{\mu \in \mathcal{O} \,|\, (u,v) \in \mu} z_\mu^k \qquad \forall e = (u^k, v^{k+1}) \in E \qquad (3)$$

where E is the edge set of all arcs in the layered graph, and the generic arc $e \in E$ connects node u^k (with $u \in [n]$) of level k to node v^{k+1} (with $v \in [n]$) of level $k+1$.

Notice that the r.h.s. involves a number $O(|\mathcal{O}|)$ of variables. As we will soon see, $|\mathcal{O}|$ is polynomial for the specific sorting problems of interest in genomics, namely, $|\mathcal{O}| = O(n)$ for SBPR, $|\mathcal{O}| = O(n^2)$ for SBR, $|\mathcal{O}| = O(n^3)$ for SBT and SBIT.

As for the objective function, the parsimony model calls for minimizing the number of non-null operators used, i.e.

$$\min \sum_{k=1}^{L} \sum_{\mu \in \mathcal{O} - \mu_0} z_\mu^k \qquad (4)$$

Furthermore, we can add constraints in order to avoid different but equivalent solutions, that place the null operators at various levels of the layered graphs. In particular, we can enforce a "canonical" form of the solution in which all the true operators are at the beginning and then followed by the null operators. We obtain this via the constraint

$$z_{\mu_0}^k \leq z_{\mu_0}^{k+1} \qquad \forall k = 1, \ldots, L-1 \qquad (5)$$

The general IP model. Putting all together, we have the following ILP model for the "Sorting by X" problem, where X is a particular type of operator/s. Let $\mathcal{O}$ be the set of all legal moves under the operator/s X:

$$\min \sum_{k=1}^{L} \sum_{\mu \in \mathcal{O} - \mu_0} z_\mu^k \qquad (6)$$

subject to

$$\sum_{\mu \in \mathcal{O}} z_\mu^k = 1 \qquad \forall k = 1, \ldots, L \qquad (7)$$

$$\sum_{P \in \mathcal{P}(i)} x_P = 1 \qquad \forall i = 1, \ldots, n \qquad (8)$$

$$\sum_{i=1}^{n} \sum_{P \in \mathcal{P}(i) \,|\, e \in P} x_P \leq \sum_{\mu \in \mathcal{O} \,|\, (u,v) \in \mu} z_\mu^k \qquad \forall e = (u^k, v^{k+1}) \in E \qquad (9)$$

$$z_{\mu_0}^k \leq z_{\mu_0}^{k+1} \qquad \forall k = 1, \ldots, L-1 \qquad (10)$$

$$z_\mu^k \in \{0,1\}, \quad x^P \geq 0 \qquad \forall k = 1, \ldots, L, \forall \mu \in \mathcal{O}, \forall P \in \cup_i \mathcal{P}(i) \qquad (11)$$

Notice that only z variables need to be integer, since it can be shown that when the z are integer the x will be as well.

3.1 Solving the Pricing Problem

The above model has an exponential number of path variables. However, its Linear Programming (LP) relaxation can still be solved in polynomial time provided we can show how to solve the *pricing* problem for the path variables. The resulting approach is called *column generation* [4]. The idea is to start with only a subset S of the x variables. Then, given an optimal solution to the current $LP(S)$, we see if there is any missing x variable that could be profitably added to S (or, as usually said in the O.R. community, that could be *priced-in*).

Let $\gamma_1, \ldots, \gamma_n$ be the dual variables associated to constraints (8) and let λ_e, for $e \in E$, be the dual variables associated to constraints (9). To each primal variable y_P corresponds an inequality in the dual LP. The variable should be priced-in if and only if the corresponding dual constraints is violated by the current optimal dual solution. Assume P is a path in $\mathcal{P}(i)$. Then, the corresponding dual inequality for P is

$$\gamma_i - \sum_{e \in P} \lambda_e \leq 0 \tag{12}$$

If we consider λ_e as edge lengths, and define $\lambda(P) := \sum_{e \in P} \lambda_e$, we have that the dual inequalities, for all $P \in \mathcal{P}(i)$, are of type

$$\lambda(P) \geq \gamma_i \tag{13}$$

A path violates the dual inequality if $\lambda(P) < \gamma_i$. Notice that if the shortest path in $\mathcal{P}(i)$ has length $\geq \gamma_i$, then no path in $\mathcal{P}(i)$ can violate inequality (13). Hence, to find a path which violates (13), it is enough to solve n shortest-path problems, one for each $i \in [n]$. Being the graph layered, each shortest path can be found in time $O(m)$, where m is the actual number of arcs in the graph.

By the above discussion, we have in fact proved the following theorem:

Theorem 1. *The LP relaxation of (6)-(11) can be solved in polynomial time.*

4 A Compact Model

Alternatively to the exponential-size model with path variables, we describe a compact reformulation (see [19] for a review about compact optimization) based on a multi-commodity flow (MCF) model. The two models are equivalent, i.e., from a feasible solution of either one of the two we can obtain a feasible solution to the other which has the same objective-function value.

The compact model employs variables x_{ab}^{ki} for each level $k = 1, \ldots, L$, each "source" $i \in [n]$ (representing a node of level 1), and each pair $a, b \in [n]$ representing an arc of layer k (i.e., a is a level-k node and b is a level-$(k+1)$ node). The variable represent the amount of flow started from node i, traveling on the arc (a, b) and going to node π_i on level $L + 1$.

We have flow outgoing constraints:

$$\sum_{j=1}^{n} x_{ij}^{1i} = 1 \qquad \forall i = 1, \ldots, n \tag{14}$$

flow ingoing constraints

$$\sum_{j=1}^{n} x_{j,\pi_i}^{Li} = 1 \qquad \forall i = 1, \ldots, n \tag{15}$$

flow-conservation constraints:

$$\sum_{j=1}^{n} x_{aj}^{ki} - \sum_{j=1}^{n} x_{ja}^{k-1,i} = 0 \qquad \forall a = 1, \ldots, n, \quad \forall k = 2, \ldots, L \tag{16}$$

capacity constraints (i.e., arc activation):

$$\sum_{i=1}^{n} x_{ab}^{ki} \leq \sum_{\mu \in \mathcal{O} \,|\, (a,b) \in \mu} z_{\mu}^{k} \qquad \forall a, b \in [n], \quad k = 1, \ldots, L \tag{17}$$

The final MCF model is obtained from the path model by replacing constraints (8) and (9) with (14), (15), (16) and (17).

5   The Operators

Let us now describe the sets of perfect matchings corresponding to operators of interst for genomic rearrangement problems. We also add the prefix reversal operator, which, although not intersting in genomics, has received a lot of attention in the literature as the famous Pancake Flipping problem.

Reversals and Prefix Reversals. There are $\binom{n}{2}$ reversals, one for each choice of indexes $1 \leq i < j \leq n$. The generic reversal between positions i and j is identified (see Fig. 2(a)) by the perfect matching

$$\mu(i,j) := \{(1,1), \ldots, (i-1, i-1), (i,j), (i+1, j-1), \ldots, \\ (j,i), (j+1, j+1), \ldots, (n,n)\}$$

Let us call $\mathcal{R}$ the set of all reversals.

When in a reversal $\mu(i,j)$ it is $i = 1$, we talk of *prefix reversal*. The generic prefix reversal up to position j is identified (see Fig. 2(b)) by the perfect matching

$$\mu(j) := \{(1,j), (2, j-1), \ldots, (j,1), (j+1, j+1), \ldots, (n,n)\}$$

Let us call $\tilde{\mathcal{R}}$ the set of all prefix reversals.

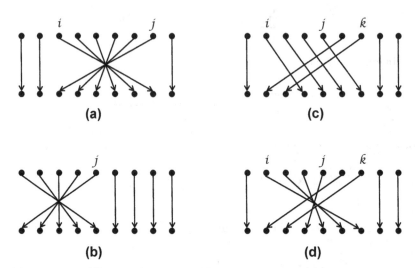

Fig. 2. (a) A reversal; (b) A prefix reversal; (c) A transposition; (d) An inverted transposition

Transpositions and Inverted Transpositions. Each transposition can be seen as the exchange of two consecutive blocks, where one of the two (but not both) can also have length 1. Let i be the position of the starting of the first block and k the position where the second block ends. Let j denote the position where the first block ends (inclusive). The transposition is identified by the triple

$$1 \leq i \leq j < k \leq n$$

Therefore, there are $\binom{n}{3} + \binom{n}{2}$ transpositions. The generic transposition is identified (see Fig. 2(c)) by the perfect matching

$$\begin{aligned}
\mu(i,j,k) := \ &\{(1,1),\ldots,(i-1,i-1)\} \cup \\
&\{(i,i+k-j),(i+1,i+k-j+1),\ldots,(j,k)\} \cup \\
&\{(j+1,i),(j+2,i+1),\ldots,(k,i+k-j-1)\} \cup \\
&\{(k+1,k+1),\ldots,(n,n)\}
\end{aligned}$$

Let us call $\mathcal{T}$ the set of all transpositions. Similarly, an inverted transposition in which the block from i to j is reversed and exchanged with the block from $j+1$ to k corresponds (see Fig.2(d)) to the perfect matching

$$\begin{aligned}
\mu'(i,j,k) := \ &\{(1,1),\ldots,(i-1,i-1)\} \cup \\
&\{(i,k),(i+1,k-1),\ldots,(j,i+k-j)\} \cup \\
&\{(j+1,i),(j+2,i+1),\ldots,(k,i+k-j-1)\} \cup \\
&\{(k+1,k+1),\ldots,(n,n)\}
\end{aligned}$$

Let us call $\mathcal{I}$ the set of all inverted transpositions.

The problems that we can model. Given the above definitions, there are many "sorting by X" problems that can be solved by our generic ILP model. The optimization of some of these problems has already been studied in the literature, while for some others (i.e., those considering more than one type of operator) optimization approaches have never been designed due to the problems hardness. In order to model a particular problem, all we need to do is specify which is the allowed set $\mathcal{O}$ of moves from one level to the next. Therefore, here are the problems that we can solve:

- **Sorting by Prefix Reversals (a.k.a. Pancake Flipping):** $\mathcal{O} := \{\mu_0\} \cup \tilde{\mathcal{R}}$.
- **Sorting by Reversals:** $\mathcal{O} := \{\mu_0\} \cup \mathcal{R}$.
- **Sorting by Transpositions:** $\mathcal{O} := \{\mu_0\} \cup \mathcal{T}$.
- **Sorting by Inverted Transpositions:** $\mathcal{O} := \{\mu_0\} \cup \mathcal{I}$.
- **Sorting by Reversals and Transpositions:** $\mathcal{O} := \{\mu_0\} \cup \mathcal{R} \cup \mathcal{T}$.
- **Sorting by Reversals and General Transpositions:.** $\mathcal{O} := \{\mu_0\} \cup \mathcal{R} \cup \mathcal{T} \cup \mathcal{I}$.

Furthermore, we can easily model situations in which some moves are forbidden, e.g., depending on the length of the fragment involved. For instance, if only fragments up to a maximum length of l can be reversed/transposed, (which is reasonable and has been studied in some computational biology papers) this can be easily modeled by redefing the above sets so as to contain only the allowed matchings.

References

1. Bader, M., Ohlebusch, E.: Sorting by weighted reversals, transpositions, and inverted transpositions. J. Comput. Biol. 14, 615–636 (2007)
2. Bafna, V., Pevzner, P.: Genome rearrangements and sorting by reversals. SIAM J. Comp. 25, 272–289 (1996)
3. Bafna, V., Pevzner, P.: Sorting by transpositions. SIAM J. Discr. Math. 11, 224–240 (1998)
4. Barnhart, C., Johnson, E.L., Nemhauser, G.L., Savelsbergh, M.W., Vance, P.H.: Branch-and-Price: Column Generation for Solving Huge Integer Programs. Op. Res. 46, 316–329 (1998)
5. Berman, P., Hannenhalli, S., Karpinski, M.: 1.375-Approximation algorithm for sorting by reversals. In: Möhring, R.H., Raman, R. (eds.) ESA 2002. LNCS, vol. 2461, pp. 200–210. Springer, Heidelberg (2002)
6. Bulteau, L., Fertin, G., Rusu, I.: Pancake Flipping is Hard. In: Rovan, B., Sassone, V., Widmayer, P. (eds.) MFCS 2012. LNCS, vol. 7464, pp. 247–258. Springer, Heidelberg (2012)
7. Bulteau, L., Fertin, G., Rusu, I.: Sorting by Transpositions Is Difficult. SIAM J. Discr. Math. 26, 1148
8. Caprara, A.: Sorting by reversals is difficult. In: 1st ACM/IEEE International Conference on Computational Molecular Biology, pp. 75–83. ACM Press (1997)
9. Caprara, A.: Sorting Permutations by Reversals and Eulerian Cycle Decompositions. SIAM J. on Disc. Math. 12, 91–110 (1999)

10. Caprara, A., Lancia, G., Ng, S.-K.: A Column-Generation Based Branch-and-Bound Algorithm for Sorting By Reversals. In: Mathematical Support For Molecular Biology. DIMACS Series in Discrete Mathematics and Theoretical Computer Science, vol. 47, pp. 213–226 (1999)
11. Caprara, L.G., Ng, S.K.: Sorting Permutations by Reversals through Branch and Price. INFORMS J. on Comp. 13, 224–244 (2001)
12. Chitturi, B., Fahle, W., Meng, Z., Morales, L., Shields, C.O., Sudborough, I.H., Voit, W.: An $18/11n$ upper bound for sorting by prefix reversals. Theor. Comp. Sc. 410, 3372–3390 (2009)
13. Christie, A.: A 3/2-approximation algorithm for sorting by reversals. In: 9th ACM-SIAM Symposium on Discrete Algorithms, pp. 244–252. ACM Press (1998)
14. Fischer, J., Ginzinger, S.W.: A 2-Approximation Algorithm for Sorting by Prefix Reversals. In: Brodal, G.S., Leonardi, S. (eds.) ESA 2005. LNCS, vol. 3669, pp. 415–425. Springer, Heidelberg (2005)
15. Gates, W., Papadimitriou, C.: Bounds for sorting by prefix reversal. Discr. Math. 27, 47–57 (1979)
16. Gu, Q.P., Peng, S., Sudborough, H.: A 2-approximation algorithm for genome rearrangements by reversals and transpositions. Theoret. Comput. Sci. 210, 327–339 (1999)
17. Hartman, T., Sharan, R.: A 1.5-approximation algorithm for sorting by transpositions and transreversals. J. Comput. Syst. Sci. 70, 300–320 (2005)
18. Kececioglu, J., Sankoff, D.: Exact and approximation algorithms for sorting by reversals, with application to genome rearrangement. Algorithmica 13, 180–210 (1995)
19. Lancia, G., Serafini, P.: Deriving compact extended formulations via LP-based separation techniques. 4OR 12, 201–234 (2014)
20. Meidanis, J., Walter, M.M.T., Dias, Z.: A Lower Bound on the Reversal and Transposition Diameter. J. Comput. Biol. 9, 743–745 (2002)
21. Nemhauser, G.L., Wolsey, L.A.: Integer and Combinatorial Optimization, 784 pages. Wiley (1999)
22. Sankoff, D., Cedergren, R., Abel, Y.: Genomic divergence through gene rearrangement. In: Molecular Evolution: Computer Analysis of Protein and Nucleic Acid Sequences, pp. 428–438. Academic Press, New York (1990)

Statistical Integration of p-values for Enhancing Discovery of Radiotoxicity Gene Signatures

Anna Papiez[1], Sylwia Kabacik[2], Christophe Badie[2],
Simon Bouffler[2], and Joanna Polanska[1]

[1] Silesian University of Technology, Institute of Automatic Control,
Akademicka 2A, 44-100 Gliwice, Poland
{anna.papiez,joanna.polanska}@polsl.pl
[2] Public Health England, Centre for Radiation, Chemical and Environmental Hazards
Chilton, Didcot, Oxfordshire, OX11 ORQ, United Kingdom
{sylwia.kabacik,christophe.badie,simon.bouffler}@phe.org.uk

Abstract. The aim of this study is to verify the effectiveness of a statistical integrative approach for merging expression data sets on the level of p-values. The data consist of two independent sets of expression levels for breast cancer patient lymphocyte tissue. The samples represent two groups: sensitive and resistant to the impact of ionizing radiation. Three approaches for integrating information derived from the two experiments to select a radiosensitivity gene signature were investigated: restrictive, non-statistical and integrative. Signature validity was assessed by verifying data separability using the support vector machine procedure and a logistic regression model selected using the likelihood ratio test to account for regularization. We demonstrated the value of additional information retained in the statistical method of gene selection based on combined p-values, which as the only logistic regression model in its optimal setting attained 100 % separability (AUC 86.2 % restrictive and 85.6 % non-statistical). Moreover, for the best support vector machine classifier, our p-value combination approach outperformed the restrictive and Arraymining methods (AUC 96.7 % versus 87.9 % and 94.6 % respectively). Further functional validation by signaling pathway research proved conjointly the biological relevance of the supreme gene signature.

Keywords: data integration, high-throughput, p-value combination, feature selection, radiosensitivity.

1 Introduction

The growing accessibility of high-throughput technology in molecular biology implies the constant increase of large amounts of data in the life sciences sector. The urge for transforming this information into knowledge incites the need for efficient and accurate data mining algorithms. On that account, the concept of combining various available data sets, with the aim of extracting material for additional conclusions, has become an important area of scientific research.

F. Ortuño and I. Rojas (Eds.): IWBBIO 2015, Part I, LNCS 9043, pp. 503–513, 2015.

In this study, we attempt to search for biomarkers of radiosensitivity in breast cancer patients based on gene expression measurements gathered in the course of two microarray experiments. It is commonly known, that people are exposed to small doses of radiation on an everyday basis and these may be the cause of an augmenting incidence rate of diseases of affluence, including various types of cancer. This phenomenon is related with the notion of radiosensitivity which is an individual feature reflecting the susceptibility of cells, tissues, organs or organisms to the harmful effect of ionizing radiation. This trait could play a key role in the dosimetry of radiotherapy in cancer treatment. This issue should be examined taking into account two aspects:

- radiosensitivity, where the subject of radiotherapy is fragile toward radiation exposure and should be treated with care and consideration of lower doses applied in longer time intervals,
- radioresistance, where the doses employed on the subject are insufficient to cure the patient with maximum efficacy.

The possibility to conduct a non-invasive prognostic test, simply by collecting a blood sample and testing how the lymphocytes respond to the small dose, environmental ionizing radiation, would be an undeniably enormous step toward the personalizing of radiotherapy.

In this study, we propose a statistical technique for merging gene expression sets which consists of data integration on the level of p-values obtained in the course of statistical tests. P-value combination methods exist in a wide variety, starting with Fisher's method [8], followed by Lancaster's modification [13], to Stouffer's procedure [20] based on the normal distribution probability density function and its numerous adjustments resulting in various weighted Z-score algorithms [15]. It is only recently, when joint p-values emerged in the field of biological high-dimensional data mining [4]. Likewise, we adopt the weighted Z-score transformation with weights reflecting experimental conditions such as sample size and dispersion [24].

2 Materials and Methods

2.1 Data

The research was carried out using two independent expression data sets. The microarray experiments were designed to identify potential biomarkers of radiosensitivity. In both settings blood samples were taken from breast cancer patients before radiotherapy. The group subjects were retrospectively assigned with one of either labels: radiosensitive or radioresistant, according to the occurrence of late adverse effects of radiotherapy. Lymphocyte RNA was extracted from these samples in order to conduct the microarray experiments. The first experiment [2] was performed using HuGene 1.0 ST Affymetrix oligonucleotide array platform whereas the second was carried out with custom Breakthrough 20K cDNA chips [7].

2.2   Data Integration

The expression data sets were preprocessed using quantile normalization appropriate for the type of microarray platform [18]. The HuGene 1.0 ST probes were annotated using custom chip description files established by Dai *et al.* [5]. Afterwards, the sets were transformed to a common space with ComBat batch effect filtration implemented in the SVA package in R [14]. Genes common for both microarray platforms were taken into account in further analysis. These data were examined for differential gene expression with a statistical inference approach with the application of two-sample t-tests, modified Welch's tests or U-Mann-Whitney test, according to population normality and variance homogeneity in the features. In order to combine information resulting from the two experiments, three approaches were adopted:

- Restrictive,
- Arraymining,
- Integrative.

Restrictive Approach. The two data sets were analyzed independently, resulting in lists of genes that were identified as differentially expressed ie. their p-value from statistical testing fell below the threshold of 0.05. Validation of the results from a single study was performed by taking the intersection of differentially expressed genes as the final gene list.

Arraymining. Data were analyzed independently with non-statistical algorithms available on the Arraymining webservice [10]. The gene signature was chosen based on the genes ranked as the most significantly differentiating in both experiments in the Ensemble of four methods: Empirical Bayes moderated t-test [16], Partial Least Squares cross-validation [11], Random Forest Mean Decrease in Accuracy [3] and Significance Analysis of Microarrays [21]. The Ensemble forms a gene list taking into account the sum of ranks for the individual algorithms.

Integrative Approach. This method is based on weighted Z-scores p-value combination [24]. The p-values from the two studies for each gene are joined after transformation with the inverse cumulative standard normal distribution function. The obtained Z-scores are combined with the weights set to the inverse standard error and transformed back to the form of a resulting p-value. This procedure is presented in Figure 1. The features with significant integrated p-values constitute the gene list.

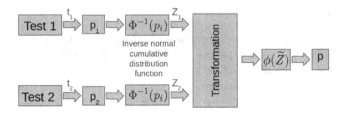

Fig. 1. Procedure for combining p-values using the weighted Z-scores method

2.3   Separability Validation

The effectiveness of three data integration approaches was measured with the separability of the data sets using obtained gene lists. The separability was assessed with a logistic regression model [1] and model selection was carried out by means of the likelihood ratio test for addressing the problem of regularization. Logistic regression models were chosen as an appropriate tool for signatures obtained in the course of statistical inference. As the Ensemble of methods available in the Arraymining service is not statistically-based, a classic Support Vector Machine was applied to juxtapose the logistic regression classifier. Model selection for the SVM was performed by means of minimizing the error rates. To ensure sufficient coverage of samples per variable, a space of maximum 20 gene features was examined. Initially, the regularization issue was controlled by means of the binomial test, yet this method, being very strict, produced a cutoff at one feature in all of the three studied cases. This ensued a large loss of information. Therefore, the minimal error approach remained the method of choice for feature selection.

The separability results were measured with Receiver Operating Characteristics and the Area under the Curve. The ROC curves were smoothed using the binormal algorithm. Additionally, metrics such as positive and negative predictive value (PPV & NPV) were investigated. The class separability thresholds were tuned based on the ROC curves Youden's index [23].

3   Results

3.1   Gene Lists

Restrictive Approach. The two experiments resulted in lists of genes of 471 and 927 differentially expressed genes at the significance level 0.05, respectively for the oligonucleotide and cDNA experiments. The intersection of these two sets consisted of 30 genes, as illustrated in Figure 2.

Arraymining. The algorithms implemented in the Gene Selection section of the Arraymining service returns a list of a predefined number of top-ranked genes. Therefore an intersection of the to 1000 genes for both experiments was considered. The gene set sizes are shown in Figure 3.

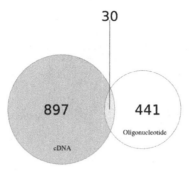

Fig. 2. Venn diagram for differentially expressed genes in two experiments

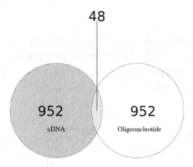

Fig. 3. Venn diagram for Arraymining top ranked genes in two experiments

Integrative Approach. The integrative approach gene list was established by the combination of p-values using the weighted Z-score method. Weights were assigned as the inverse standard error for the distribution of gene expression. This algorithm resulted in a list of 108 differentially expressed genes at the significance level of 0.05.

The proportions and overlap of the three gene lists are illustrated in Figure 4. A total of 12 genes were common for all three integration approaches.

3.2 Logistic Regression Model

The differentially expressed gene lists were utilized for building a logistic regression model for obtaining optimal separability of the two groups of patients (radioresistant and radiosensitive). For each of the three data merging strategies the best model was chosen with the use of the likelihood ratio test. The resulting models consisted of 6, 6 and 16 features for the restrictive, Arraymining and integrative approaches respectively. Only the signature obtained with the integrative

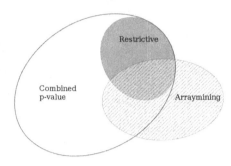

Fig. 4. Venn diagram for gene lists obtained in three integrative approaches [17]

approach provided a model with perfect separability. The comparison of model performance is illustrated by the Receiver Operating Characteristics (ROC) in Figure 5. Moreover, the decrease of error rates with regard to the number of features is presented (Figure 6).

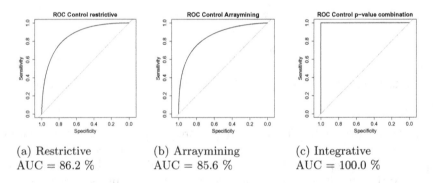

(a) Restrictive
AUC = 86.2 %

(b) Arraymining
AUC = 85.6 %

(c) Integrative
AUC = 100.0 %

Fig. 5. Receiver Operating Characteristic curves for logistic regression model separability

3.3   Support Vector Machine

As the logistic regression model is expected to have better performance with features selected using statistical tools, the assessment of support vector machine classifiers for the three gene signatures was carried out. The best model was chosen regarding the minimum error rate for a given number of features. The approaches resulted in models of 8, 19 and 19 features for the restrictive, Arraymining and integrative approaches respectively. The Receiver Operating

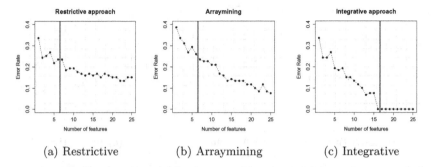

(a) Restrictive   (b) Arraymining   (c) Integrative

Fig. 6. Separability error rates subject to the number of features in the logistic regression model. The vertical line demonstrates the borderline, beyond which further feature addition does not provide a significant increase in model performance based on the likelihood ratio test.

Characteristics (Figure 7) indicate that again the supreme performance was obtained with the p-value combination signature. This model also provides the lowest error rates, as shown in Figure 8.

The efficiency of the models for group separability was also evaluated using the Positive and Negative Predictive Values. These statistics are summarized in Table 1.

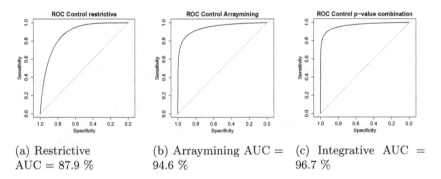

(a) Restrictive   (b) Arraymining AUC =   (c) Integrative AUC =
AUC = 87.9 %   94.6 %   96.7 %

Fig. 7. Receiver Operating Characteristic curves for support vector machine model separability

3.4 Signature Similarity

The gene signatures obtained in the course of the group separability study for the two classification algorithms were investigated for overlapping features. The results are presented in Figure 9. These signatures were also validated functionally with the use of Kyoto Encyclopaedia of Genes and Genomes signaling pathway database.

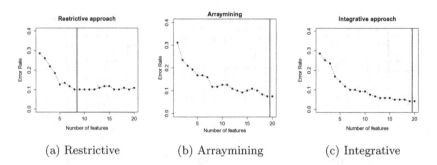

(a) Restrictive   (b) Arraymining   (c) Integrative

Fig. 8. Separability error rates subject to the number of features in the support vector machine model. The vertical line demonstrates the borderline separating the optimal number of feautres minimizing the error rate.

Table 1. Positive and negative predictive value for logistic regression model and support vector machine separability

	Logistic regression		Support vector machine	
	PPV [%]	NPV [%]	PPV [%]	NPV [%]
Restrictive	86.67	74.32	88.33	91.52
Arraymining	70.13	90.47	92.98	91.94
Integrative	100.00	100.00	98.18	93.75

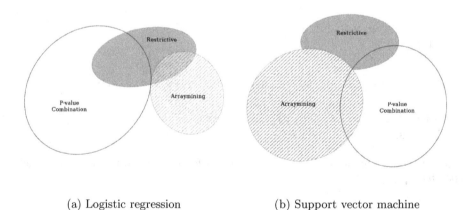

(a) Logistic regression   (b) Support vector machine

Fig. 9. Venn diagrams illustrating the proportion of overlapping genes in the logistic regression and support vector machine signatures for three integration approaches [17]

4   Discussion

Integrating expression data at the p-value level proves to be an adept approach for the improvement of gene signatures. Taking into consideration the true values referring to the test statistics enables a more precise incorporation of the differentiating strength of a particular gene feature, rather than setting a fixed p-value threshold on the individual studies. The latter case often leads to a binary decision where an arbitrarily set significance level provokes the rejection of potential genes of interest.

The expansion of the gene signature results not only in a quantitative increase, but also enhances the quality in terms of classification. The separability study showed the integrative approach model to be the only case of signature that allowed for a perfect split of the data groups in the logistic regression model selected by means of the likelihood ratio test for regularization handling. The best chosen model for the support vector machine is derived from the integrative approach method as well, yet not providing perfect separability. Furthermore, if the decrease of error rate is considered, the integrative approach yields the lowest error rate values. Moreover, also the PPV & NPV remain unsurpassed for both the logistic regression and the support vector machine algorithm.

The study of the gene signature relation to signaling pathways revealed pathway terms for the integrative approach gene set that were not available using the restrictive or Arraymining approach. These included pathways relevant to the issue of radiation exposure and susceptibility or cancer, such as Jak-STAT [6], T cell receptor [22], Cytokine-cytokine receptor interaction [12], Fc epsilon RI [9] and Natural killer cell mediated cytotoxicity [19].

5   Conclusions

The proposed strategy for gene signature selection involving statistical combination of p-values provides the best results in terms of radioresistant and radiosensitive patient separability. This was demonstrated using two methods: building a logistic regression model and a support vector machine classifier. The applied technique allows for the decrease of error rates and increase of positive and negative predictive values. Moreover, the obtained optimal gene signature presents links to radiosensitivity and cancer-related processes occurring in relevant signaling pathways. Although this work was intended to verify the effectiveness of the integrative approach in the simple setup of separating the two groups of patients, these findings imply the future investigation of gene signatures on their performance in more complex classification problems such as multiple random validation.

Acknowledgments. This work was financially supported by internal grant of Silesian University of Technology for young researchers BKM/524/RAU1/2014/ 17 (AP) and by the National Science Center grant Harmonia 4 DEC-2013/08/M /ST6/00924 (JP). Additionally, AP is a holder of the DoktoRIS scholarship for

Innovative Silesia. Computations were carried out using European Union grant GeCONiI infrastructure POIG.02.03.01-24-099/13.

Thank you to the anonymous reviewers for valuable comments and suggestions.

References

1. Antoniadis, A., Lambert-Lacroix, S., Leblanc, F.: Effective dimension reduction methods for tumor classification using gene expression data. Bioinformatics 19(5), 563–570 (2003)
2. Baumann, M., Hölscher, T., Begg, A.C.: Towards genetic prediction of radiation responses: Estro's genepi project. Radiotherapy and Oncology 69(2), 121–125 (2003)
3. Breiman, L.: Random forests. Machine Learning 45(1), 5–32 (2001)
4. Chen, Z.: Is the weighted z-test the best method for combining probabilities from independent tests? Journal of Evolutionary Biology 24(4), 926–930 (2011)
5. Dai, M., Wang, P., Boyd, A.D., Kostov, G., Athey, B., Jones, E.G., Bunney, W.E., Myers, R.M., Speed, T.P., Akil, H., et al.: Evolving gene/transcript definitions significantly alter the interpretation of genechip data. Nucleic Acids Research 33(20), e175 (2005)
6. Ding, M., Zhang, E., He, R., Wang, X.: Newly developed strategies for improving sensitivity to radiation by targeting signal pathways in cancer therapy. Cancer Science 104(11), 1401–1410 (2013)
7. Finnon, P., Kabacik, S., MacKay, A., Raffy, C., A'Hern, R., Owen, R., Badie, C., Yarnold, J., Bouffler, S.: Correlation of in vitro lymphocyte radiosensitivity and gene expression with late normal tissue reactions following curative radiotherapy for breast cancer. Radiother. Oncol. 105(3), 329–336 (2012)
8. Fisher, R.A.: Statistical methods for research workers. Genesis Publishing Pvt Ltd. (1925)
9. Fox, D.A., Chiorazzi, N., Katz, D.H.: Hapten specific ige antibody responses in mice v. differential resistance of ige and igg b lymphocytes to x-irradiation. The Journal of Immunology 117(5, Pt. 1), 1622–1628 (1976)
10. Glaab, E., Garibaldi, J.M., Krasnogor, N.: Arraymining: a modular web-application for microarray analysis combining ensemble and consensus methods with cross-study normalization. BMC Bioinformatics 10(1), 358 (2009)
11. Hall, M.A.: Feature selection for discrete and numeric class machine learning (1999)
12. Herok, R., Konopacka, M., Polanska, J., Swierniak, A., Rogolinski, J., Jaksik, R., Hancock, R., Rzeszowska-Wolny, J.: Bystander effects induced by medium from irradiated cells: similar transcriptome responses in irradiated and bystander k562 cells. International Journal of Radiation Oncology* Biology* Physics 77(1), 244–252 (2010)
13. Lancaster, H.: The combination of probabilities: an application of orthonormal functions. Australian Journal of Statistics 3(1), 20–33 (1961)
14. Leek, J.T., Johnson, W.E., Parker, H.S., Jaffe, A.E., Storey, J.D.: The sva package for removing batch effects and other unwanted variation in high-throughput experiments. Bioinformatics 28(6), 882–883 (2012)
15. Liptak, T.: On the combination of independent tests. Magyar Tud. Akad. Mat. Kutato Int. Kozl. 3, 171–197 (1958)
16. Lönnstedt, I., Speed, T.: Replicated microarray data. Statistica Sinica 12(1), 31–46 (2002)

17. Micallef, L., Rodgers, P.: eulerape: Drawing area-proportional 3-venn diagrams using ellipses. PLoS ONE (2014, to appear)
18. Papiez, A., Finnon, P., Badie, C., Bouffler, S., Polanska, J.: Integrating expression data from different microarray platforms in search of biomarkers of radiosensitivity. In: Proceedings IWBBIO 2014 (2014)
19. Son, C.-H., Keum, J.-H., Yang, K., Nam, J., Kim, M.-J., Kim, S.-H., Kang, C.-D., Oh, S.-O., Kim, C.-D., You-Soo, Park, o.: Synergistic enhancement of nk cell-mediated cytotoxicity by combination of histone deacetylase inhibitor and ionizing radiation. Radiation Oncology 9(1), 49 (2014)
20. Stouffer, S.A., Suchman, E.A., DeVinney, L.C., Star, S.A., Williams Jr., R.M.: The american soldier: adjustment during army life (studies in social psychology in world war ii), vol. 1 (1949)
21. Tusher, V.G., Tibshirani, R., Chu, G.: Significance analysis of microarrays applied to the ionizing radiation response. Proceedings of the National Academy of Sciences 98(9), 5116–5121 (2001)
22. Witek, M., Blomain, E.S., Magee, M.S., Xiang, B., Waldman, S.A., Snook, A.E.: Tumor radiation therapy creates therapeutic vaccine responses to the colorectal cancer antigen gucy2c. International Journal of Radiation Oncology* Biology* Physics 88(5), 1188–1195 (2014)
23. Youden, W.J.: Index for rating diagnostic tests. Cancer 3(1), 32–35 (1950)
24. Zaykin, D.V.: Optimally weighted z-test is a powerful method for combining probabilities in meta-analysis. Journal of Evolutionary Biology 24(8), 1836–1841 (2011)

A Pseudo de Bruijn Graph Representation
for Discretization Orders for Distance Geometry

Antonio Mucherino

IRISA, University of Rennes 1, Rennes, France
antonio.mucherino@irisa.fr

Abstract. Instances of the distance geometry can be represented by a simple weighted undirected graph G. Vertex orders on such graphs are discretization orders if they allow for the discretization of the K-dimensional search space of the distance geometry. A pseudo de Bruijn graph B associated to G is proposed in this paper, where vertices correspond to $(K+1)$-cliques of G, and there is an arc from one vertex to another if, and only if, they admit an overlap, consisting of K vertices of G. This pseudo de Bruijn graph B can be exploited for constructing discretization orders for G for which the consecutivity assumption is satisfied. A new atomic order for protein backbones is presented, which is optimal in terms of length.

1 Introduction

The ordering associated to the atoms of a given molecule plays a fundamental role in the discretization of Molecular Distance Geometry Problems (MDGPs) [15,19]. The MDGP is the problem of finding suitable three-dimensional conformations for a given molecule by exploiting the information concerning known distances between atom pairs. A simple weighted undirected graph $G = (V, E, d)$ can be formally used for representing an MDGP instance, where vertices u and $v \in V$ represent atoms, and there is an edge $(u, v) \in E$ between u and v if the corresponding distance is known. The weights associated to the edges provide the numerical values for such distances. These values can be either exact or represented by a real-valued interval. The MDGP basically asks whether the graph G can be embedded in dimension $K = 3$. Notice, however, that the same problem can be defined for any dimension $K > 0$.

The discretization of the MDGP allows for reducing the search conformational space of the problem to a tree [16]. While atoms can generally take any position in a continuous portion of the space (e.g. a (hyper)sphere containing the entire molecule), the discretization makes it possible to consider a discrete and finite subset of possible positions for each atom of the molecule. This space reduction does not decrease the problem complexity (which is NP-hard [21]), but it allows for the development of ad-hoc algorithms on search trees for discovering one solution (or even several solutions) to the problem [14].

A *discretization order* is an order given to the vertices of the graph G that allows for the discretization [11]. In previous works, discretization orders have been either handcrafted [7,13] or automatically generated [11,18]. When handcrafted, the orders have

F. Ortuño and I. Rojas (Eds.): IWBBIO 2015, Part I, LNCS 9043, pp. 514–523, 2015.

been particularly designed for an important class of molecules: the proteins. These molecules, in fact, perform several (often vital) functions in living beings. They are chains of *amino acids*, which are bonded to each other through peptide bonds. The simple examination of the known chemical structure of the 20 amino acids involved in the protein synthesis (which implies knowledge on distances) allowed for the identification of discretization orders for the protein backbone [13] and its side chains [7]. These handcrafted orders can also be seen as sequences of overlapping cliques of atoms (see Section 2 for more details): the possible positions for an atom u can be computed by using the information related to the atoms that immediately precede u in the order (these "reference" atoms belong to a common clique). This class of discretization orders satisfies the so-called *consecutivity assumption*, because all reference atoms are consecutive and they immediately precede u in the order.

The problem of finding a discretization order satisfying the consecutivity assumption is NP-complete [4]. When this assumption is relaxed, so that an atom u can have, as a reference, atoms that are not its immediate predecessors in the order, then the problem of finding the order has polynomial complexity [11]. A greedy algorithm for an automatic detection of discretization orders that do not necessarely satisfy the consecutivity assumption was proposed in [11,18].

When the consecutivity assumption is satisfied, it is possible to verify in advance whether the discretization distances (that are, for a given atom, grouped in the same clique) are compatible and are able to provide a finite number of positions for an atom of the molecule. For each u, since the reference atoms belong to a common clique, all relative distances are a priori known, so that their feasibility can be verified. This is not true anymore when the assumption is not satisfied: not all distances, necessary for the feasibility verification, may be available. On the one hand, therefore, orders satisfying the consecutivity assumption should be favored; on the other hand, however, the problem of identifying such orders is NP-complete.

In this work, the problem of finding discretization orders with the consecutivity assumption is studied, and, to this purpose, a *pseudo de Bruijn* graph representation [2] for cliques contained in MDGP instances is proposed. This novel representation allows in fact for an easier search for this kind of discretization orders. In this representation, cliques of G are vertices of a directed graph $B = (V_B, A_B)$, where there is an arc from the vertex b to the vertex $c \in V_B$ when the two corresponding cliques overlap. As a consequence, a discretization order can be seen as a *path* on the graph B, such that every atom of G appears at least once in the sequence of cliques. Orders induced from these paths on B are discretization orders satisfying the consecutivity assumption.

By exploiting the proposed pseudo de Bruijn graph representation, a new discretization order for protein backbones was identified. In comparison to the order previously proposed in [13], this new ordering contains fewer atomic repetitions, and it is optimal in terms of length. The de Bruijn graph representation provides a support for the identification of this particular class of orders. When no ordering can be found by exploiting this representation, then orders that the greedy algorithm in [11,18] is able to identify can be considered as valid alternatives, even if this algorithm cannot ensure that the consecutivity assumption be satisfied.

Two different graphs will be considered throughout the paper. The graph $G = (V, E, d)$ represents an instance of the MDGP, where vertices u, v, etc., are *atoms* and weighted edges are *distances*. The edge set E is partitioned in E' and its complement $E \setminus E'$, where E' only contains edges referring to exact distances. The graph $B = (V_B, A_B)$ represents the pseudo de Bruijn graph containing cliques of G, where the arc from the vertex b to the vertex c indicates that the two corresponding cliques overlap (see Section 2 for the rigorous definitions).

The rest of the paper is organized as follows. In Section 2, the pseudo de Bruijn graph representation for cliques in MDGP instances is presented and commented in details. By exploiting this novel representation, new discretization orders for the MDGP where the consecutivity assumption is satisfied can be found. Section 3 will present one of such orders for the protein backbones, that will result to be optimal in length. Section 4 will conclude the paper with a discussion.

2   de Bruijn Graph Representation

Graphs of *de Bruijn* are widely employed for formalizing problems related to DNA assembly [5,6,8,9]. New generation technologies are able to provide researches with subsequences of DNA (named *reads*) that need to be successively assembled into one unique sequence, which is the final DNA molecule. The best way to formalize this problem is to consider a graph where vertices represent reads, i.e. the subsequences, and where there is arc from a vertex to another when the ending of the former coincides with the beginning of the latter (there is an overlap).

The graph B considered in this work is an extension of the classical de Bruijn graph [2] which is used in the DNA application. If G represents an instance of the MDGP, the vertices of the *pseudo* de Bruijn graph $B = (V_B, A_B)$ are $(K + 1)$-cliques of the graph G, where K is the dimension of the search space. A vertex $b \in V_B$ can be seen as a subsequence of $K + 1$ atoms admitting an internal ordering.

In the standard de Bruijn graph, there is an arc from b to c if there is an overlap. In other words, if the ending of the subsequence b coincides with the beginning of the subsequence c, then the arc (b, c) is added in A_B. In this application, since the vertices in V_B cannot be considered as static objects (the internal order of their atoms is not constant), the standard definition of de Bruijn graph needs to be extended. Consider for example that $c \in V_B$ is a $(K + 1)$-clique composed by exact edges (all distances are exact): in this case, the $K + 1$ atoms in the clique can be reordered $(K + 1)!$ times. If instead $b \in V_B$ contains one interval distance, there are $2(K - 1)!$ permutations of the atoms that allow the extremes of the interval distance to be the first and the last atom in the clique (see Def. 2.5). In this application, it is necessary for the overlap to have length equal to K. Notice that, even if the main application of this work is to biological molecules, the theory presented in the following holds for any dimension $K > 0$.

Definition 2.1. *There is a K-overlap from the vertices b to the vertex c of V_B if there exists an internal order for the atoms in b and an internal order for the atoms in c for which the K-suffix of b coincides with the K-prefix of c.*

Algorithm 1. An algorithm for constructing an induced order r for the vertices of G from a total K-valid path on the pseudo de Bruijn graph B associated to G.

```
1: find_induced_order  in: P = {p₁,p₂,...,pₙ}  out: order r
2: i = 1
3: for all u ∈ p₁ in the internal order do
4:    rᵢ = u; i++
5: end for
6: for (j = 2, n) do
7:    u = last vertex in internal order of pⱼ
8:    rᵢ = u; i++
9: end for
```

Notice that this definition applies to any kind of clique (either consisting of exact distances, or containing interval data).

The interest in constructing the graph $B = (V_B, A_B)$ from the graph G is evident. When a set of coordinates has already been assigned to its first K atoms (in a given internal order), each $(K+1)$-clique allows for computing a finite set of possible positions for its last atom [12]. When all the distances in the clique are exact, there are only 2 possible positions for the atom; when the distances between the first and the last atom is represented by a real-valued interval, the positions for the last atom lie on two arcs, which can be discretized [13]. Each clique in the suitable *path* on B gives therefore the necessary information for computing a finite set of possible positions for each atom of the molecule. A path of K-overlapping $(K+1)$-cliques naturally implies a sorted sequence of atoms, i.e. an order for the vertices of the graph G.

Definition 2.2. *A K-valid path $P = \{p_1, p_2, \ldots, p_n\}$ on B is a sequence of K-overlapping cliques p_i where the internal order of each clique is preserved when referring to p_{i-1} and p_{i+1}. When every atom $u \in V$ is included in at least one clique p_i, then the path is said "total".*

Notice that the condition on the clique internal order is not necessary when standard de Bruijn graphs are concerned.

A total K-valid path on B implies the definition of an order $r : \mathbb{N}_+ \longrightarrow V \cup \{\clubsuit\}$ with length $|r| \in \mathbb{N}$ (for which $r_i = \clubsuit$ for all $i > |r|$) for the vertices of G. Alg. 1 is a sketch of the simple algorithm that is necessary to apply to this purpose. Let P be a total K-valid path on B. The first K labels are assigned to the atoms of $p_1 \in P$ (the internal order of the clique has to be preserved). Then, for all other p_j, with $j \geq 2$, the last atom of the clique p_j, in the internal order, is added to the induced order.

Proposition 2.3. *Any order r constructed by Alg. 1 from a total K-valid path P on B is a discretization order for which the consecutivity assumption is satisfied.*

Proof. By construction.   □

A simple verification for the existence of a total K-valid path on B is to check its connectivity. Naturally, if B is not connected, no total paths can be constructed. But even when B is connected, a total path on B may not exist, as it is the case for the protein backbone, even if all its $(K+1)$-cliques are considered. To overcome this issue, *auxiliary cliques* can be added in B.

Definition 2.4. *An auxiliary $(K+1)$-clique is a clique*

$$\{v_1, v_2, \ldots, v_K, v_1\}$$

where $\{v_1, v_2, \ldots, v_K\}$ is a K-clique of V having edges in E' (all distances are exact).

It is important to remark that several auxiliary cliques can be generated from one K-clique, depending on the selected internal order of its atoms. The set of vertices $\{v_1, v_2, \ldots, v_K, v_1\}$ evidently form a clique, because the distances between the duplicated v_1 and all other vertices are known. Moreover, the distance between the first and second copy of v_1 is exact and equal to 0. When deadling with protein backbones, the identification of a total K-valid path on B is only possible when auxiliary cliques are included in the pseudo de Bruijn graph B (see Section 3).

One immediate consequence in using auxiliary cliques is that atoms may be repeated one (or even several times) in the induced orders. The auxiliary clique allows for locally reordering a given subset of atoms, so that a K-overlap can become possible with other cliques. Every time an auxiliary clique is involved, an atom is repeated in the atomic sequence, exactly K places after its previous copy. This kind of orders were previously formalized in [13] and named *re-orders*. Recall that E' is the subset of E containing exact distances only.

Definition 2.5. *A repetition order (re-order) is a function $r : \mathbb{N}_+ \to V \cup \{\clubsuit\}$ with length $|r| \in \mathbb{N}_+$ (for which $r_i = \clubsuit$ for all $i > |r|$) such that:*

- *$G[\{r_1, r_2, \ldots, r_K\}]$ is a clique with edge set in E',*
- *$\forall i \in \{K+1, \ldots, |r|\}$ and $\forall j \in \{i-K+1, \ldots, i-1\}$, $(r_j, r_i) \in E'$,*
- *$\forall i \in \{K+1, \ldots, |r|\}$, either $(r_{i-K}, r_i) \in E$ or $r_{i-K} = r_i$.*

Since every re-order is a discretization order where the consecutivity assumption is satisfied, the following proposition holds.

Proposition 2.6. *Induced orders from total K-valid paths P on pseudo de Bruijn graphs B generated from G (with or without auxiliary cliques) are re-orders for the vertices of the graph G.*

3   Discretization Orders for Protein Backbones

Proteins are important molecules that perform vital functions in the bodies of living beings. They are chains of smaller molecules named *amino acids*, whose order is a priori known (in other words, every amino acid is known with its rank/position in the chain). The *protein backbone* is defined by this chain, and basically contains, in sequence for each amino acid, a nitrogen N, a carbon C_α and another carbon C, plus some additional atoms chemically bonded to them. Only 20 different amino acids can be involved in the protein synthesis. A group of atoms attached to the carbon C_α makes the 20 amino acids different from each other. Since this latter group of atoms looks like "hanging"

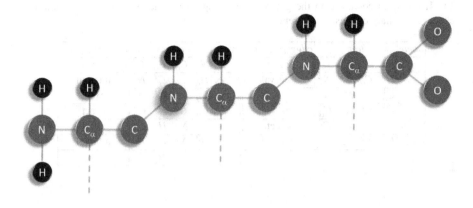

Fig. 1. The chemical structure of the considered 3-amino acid protein backbone. Some atoms are omitted because they can be positioned uniquely once the considered ones have been placed. Side chains may be attached to the atoms C_α through the bonds represented by the dashed gray lines.

from the protein backbone, it is said this is the *side chain* of the amino acid. Due to the complexity of the problem of identifying protein conformations, many proposed methods focus on protein backbones, and the information about the side chains is either approximated or neglected. In fact, once a suitable conformation for a protein backbone has been identified, there are methods that attempt the positioning of the side chains [3,10].

As in previous publications about discretization orders [13,18], a small 3-amino acid backbone will be considered in the following. Since the chemical structure of protein backbones is repetitive (no side chains ⇒ no difference among the 20 amino acids), the identification of an order for a small molecule with 3 amino acids is sufficient, because this order can be trivially extended to protein backbones of any length.

Figure 1 shows the chemical structure of the considered 3-amino acid backbone. For every chemical bond (light gray lines in the picture), there is a known exact distance that can be considered for the discretization. Moreover, the relative distance between atoms bonded to another common atom is known, and can also be considered as exact. Finally, every quadruplet of consecutive bonded atoms form a torsion angle, from which a lower and an upper bound can be obtained for the distance between the first and the last atom of the quadruplet. Since peptide bonds, which chemically connect consecutive amino acids, give a rigid configuration to a part of the backbone structure, some of the distances derived from torsion angles can be considered as exact [17].

Table 1 shows the (non-auxiliary) cliques that can be found in the 3-amino acid backbone. Only information deduced from its chemical structure are considered in the table: the distances derived from experiments of Nuclear Magnetic Resonance (NMR) [1] are here not considered. In fact, the interest is in finding orders that are suitable for every protein backbone, so that only instance-independent distances are used for defining the 4-cliques of the pseudo de Bruijn graph B.

Table 1. 4-cliques contained in the graph representing an instance related to a 3-amino acid backbone. Auxiliary cliques are not reported.

name	atoms	edge $\{r_{i-3}, r_i\}$	name	atoms	edge $\{r_{i-3}, r_i\}$
c_1	N^1 C_α^1 H_α^1 C^1	exact	c_7	N^2 C_α^2 H_α^2 C^2	exact
c_2	H_α^1 C_α^1 C^1 N^2	interval	c_8	H_α^2 C_α^2 C^2 N^3	interval
c_3	C_α^1 C^1 N^2 H^2	exact	c_9	C_α^2 C^2 N^3 H^3	exact
c_4	C_α^1 C^1 N^2 C_α^2	exact	c_{10}	C_α^2 C^2 N^3 C_α^3	exact
c_5	C^1 N^2 H^2 C_α^2	exact	c_{11}	C^2 N^3 H^3 C_α^3	exact
c_6	H^2 N^2 C_α^2 H_α^2	interval	c_{12}	H^3 N^3 C_α^3 H_α^3	interval
			c_{13}	N^3 C_α^3 H_α^3 C^3	exact

In [13], a discretization order for the protein backbones was previously proposed. This order was handcrafted and satisfies the consecutivity assumption (it is a re-order, see Def. 2.5). Since then, it was generally used for discretizing MDGPs, as for example in [20], where real NMR data were considered for the first time when working with a discrete approach to distance geometry. In terms of de Bruijn graph, the handcrafted order corresponds to the following total 3-valid path in dimension 3:

$$
\begin{aligned}
&\textit{(first amino acid)} && \Diamond \to c_1 \to c_2 \\
&\textit{(second amino acid)} && \to c_4 \to c_5 \to \Diamond \to \Diamond \to c_6 \to c_7 \to \Diamond \to c_8 \to \Diamond \\
&\textit{(third amino acid)} && \to c_{10} \to c_{11} \to \Diamond \to \Diamond \to c_{12} \to c_{13} \, .
\end{aligned}
\tag{1}
$$

The symbol $\Diamond$ indicates that an auxiliary clique is used in the order. The de Bruijn graph representation of the handcrafted order starts with the auxiliary clique $(C_\alpha^1, N^1, H^1, C_\alpha^1)$. Notice that the two hydrogens bonded to the nitrogen atom N^1 of the first amino acid, as well as the two oxygens bonded to the carbon C^3 of the last amino acid, are here omitted. In fact, positions for these atoms can be calculated at the end of the computation, when a position has already been assigned to all other atoms. In the path (1), there are 7 auxiliary cliques; in general, for a protein backbone consisting of n_{aa} amino acids, $1 + 4 \cdot (n_{aa} - 2) + 2$ auxiliary cliques are necessary for constructing this path. Notice that the second amino acid can be repeated as many times as necessary in a protein backbone formed by $n_{aa} > 3$ amino acids.

The following is another possible path for the 3-amino acid backbone:

$$
\begin{aligned}
&\textit{(first amino acid)} && c_1 \to c_2 \\
&\textit{(second amino acid)} && \to c_3 \to c_5 \to \Diamond \to c_6 \to \Diamond \to c_7 \to c_8 \\
&\textit{(third amino acid)} && \to c_9 \to c_{11} \to \Diamond \to c_{12} \to \Diamond \to c_{13} \, .
\end{aligned}
\tag{2}
$$

In this case, there are two auxiliary cliques in second amino acid, and other two auxiliary cliques in the third one. As a consequence, two atoms are duplicated in each amino acid in the corresponding induced re-order. In general, for n_{aa} amino acids, $2 \cdot (n_{aa} - 1)$ repetitions are necessary. The internal order of the starting clique c_1 is: $N^1, H^1, C_\alpha^1, C^1$. Naturally, this is only one possible path that can be identified on the pseudo de Bruijn graph B. It requires fewer auxiliary cliques than the handcrafted order. However, in order to verify whether there are other possible paths for which the number of necessary auxiliary cliques is smaller (implying therefore fewer repetitions), one could attempt the

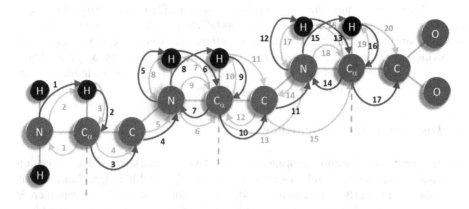

Fig. 2. An optimal (in terms of length) re-order for the protein backbone (in green). In the background, in light red, a previously proposed handcrafted order.

construction of all possible 3-valid total paths on the graph B by an exhaustive search. Naturally, even if an exhaustive search might be feasible for small instances, this is not an advisable procedure. For the considered 3-amino acid backbone, it is possible to prove that the discretization order induced from the path (2) is optimal in terms of length.

Theorem 3.1. *Let G be a graph representing an MDGP instance related to a protein backbone. For every amino acid in the protein backbone with rank greater than 2, every re-order for its atoms requires at least 2 repetitions.*

Proof. In a path starting with c_2 (see Table 1), the 4th place in the induced order (refer to Alg. 1) can be either for H_α^1 or for N^2, because of the constraint on the internal orders for the interval clique c_2 (refer to Def. 2.2). However, in order to construct a path to c_6 (and not to c_1), it is necessary to choose the internal order where N^2 is in position 4. At this point, the clique c_2 admits a 3-overlap with both cliques c_3 and c_4, and whichever the chosen clique is, the clique c_5 can follow either c_3 or c_4. The position of the atom H^2 in the induced order is the 5th (when c_3 is chosen) or the 6th (when c_4 is chosen). In order to add c_6 immediately after c_5, the atom H^2 should be instead in position 4, which is taken by N^2. However, the position 4 was fixed by c_2 at the beginning of the path. An auxiliary clique is therefore necessary for adjusting the internal order of c_5 and for making it possible to have a 3-overlap with the clique c_6. Naturally, the use of an auxiliary clique before c_6 might be avoided if a different path is rather constructed, where auxiliary cliques needs to be however involved earlier. This implies that at least one auxiliary clique is necessary for constructing a path on B from c_2 to c_6.

Similarly, it is possible to prove that at least one auxiliary clique is needed to step from the clique c_6 to the clique c_8. Because of the repetitive structure of protein backbones, the theorem is proved.   □

Fig. 2 graphically shows the re-order induced from path (2), in green. Since this path is basically a sequence of 15 cliques, $4 + 14$ atoms (all atoms contained in the first clique + one atom for all others) are included in this order (repetitions are also counted). Fig. 2 also shows the order induced from path (1), in light red. In this case, there are more repetitions: there are 18 cliques in total, and therefore there are $4 + 17$ atoms in the induced order. The order induced by path (2) is optimal, as Theorem 3.1 shows.

4   Discussion and Conclusion

Given a graph G representing an instance of the MDGP, the existence of a discretization order allows to make the search space discrete and to employ ad-hoc algorithms, such as the Branch & Prune (BP) algorithm [13,14], for its solution. If the discretization order satisfies the consecutivity assumption (as it is the case for the re-orders), it is possible to verify in advance whether all atoms in the molecule admit a finite number of positions. This advantage motivated this work on the pseudo de Bruijn graph representation of discretization orders.

The problem of finding a discretization order satisfying the consecutivity assumption is NP-complete [4]. It is expected therefore that the complexity of any possible algorithm designed for the solution of this problem grows exponentially with its size. In fact, the exploration of all possible total K-valid paths on the pseudo de Bruijn graph presented in this paper can be rather expensive in general.

This exploration can, however, still be feasible when considering small molecules, such as the 3-amino acid backbone considered in this work or the 20 side chains belonging to the 20 amino acids that can form a protein. For the 3-amino acid backbone (see Section 3), this was not necessary, because it was possible to prove that path (2) is an optimal one (see Theorem 3.1 in Section 3). For the side chains, instead, an exhaustive search on all possible paths on the pseudo de Bruijn graph could be performed. Once an optimal order, in terms of length, will be identified for each of them, the discretization order for an entire protein can be constructed by combining all found orders, including the optimal backbone order induced by path (2). The final order will depend on the amino acid sequence of the considered protein.

This procedure is obviously not applicable to large molecules that cannot be separated in relevant parts, such as backbone and side chains. The benefits in using the pseudo de Bruijn graph B and exploring the total paths on B still have to be investigated for this kind of instances. As already remarked in the Introduction, a current valid alternative is the greedy algorithm proposed in [11] and extended to interval data in [18]. This algorithm is able to provide discretization orders (where the consecutivity assumption is however not ensured) in polynomial time. One possible direction for future research can be the following. Is it possible to deduce a discretization order with consecutivity assumption from a generic order provided by the greedy algorithm in [18] ?

References

1. Almeida, F.C.L., Moraes, A.H., Gomes-Neto, F.: An Overview on Protein Structure Determination by NMR. Historical and Future Perspectives of the Use of Distance Geometry Methods. In: [19], pp. 377–412 (2013)
2. de Bruijn, N.G.: A Combinatorial Problem, Koninklijke Nederlandse Akademie v. Wetenschappen 49, 758–764 (1946)
3. Canutescu, A.A., Shelenkov, A.A., Dunbrack Jr., R.L.: A Graph-Theory Algorithm for Rapid Protein Side-Chain Prediction. Protein Science 12, 2001–2014 (2003)
4. Cassioli, A., Günlük, O., Lavor, C., Liberti, L.: Discretization Vertex Orders in Distance Geometry. To appear in Discrete Applied Mathematics (2015)
5. Chikhi, R., Rizk, G.: Space-Efficient and Exact de Bruijn Graph Representation based on a Bloom Filter. Algorithms for Molecular Biology 8(22), 9 (2013)
6. Compeau, P.E.C., Pevzner, P.A., Tesler, G.: How to Apply de Bruijn Graphs to Genome Assembly. Nature Biotechnology 29, 987–991 (2011)
7. Costa, V., Mucherino, A., Lavor, C., Cassioli, A., Carvalho, L.M., Maculan, N.: Discretization Orders for Protein Side Chains. Journal of Global Optimization 60(2), 333–349 (2014)
8. Drezen, E., Rizk, G., Chikhi, R., Deltel, C., Lemaitre, C., Peterlongo, P., Lavenier, D.: GATB: Genome Assembly & Analysis Tool Box. To appear in Bioinformatics. Oxford Press (2014)
9. Ellis, T., Adie, T., Baldwin, G.S.: DNA Assembly for Synthetic Biology: from Parts to Pathways and Beyond. Integrative Biology 3, 109–118 (2011)
10. Kim, D.-S., Ryu, J.: Side-chain Prediction and Computational Protein Design Problems. Biodesign 2(1), 26–38 (2014)
11. Lavor, C., Lee, J., Lee-St. John, A., Liberti, L., Mucherino, A., Sviridenko, M.: Discretization Orders for Distance Geometry Problems. Optimization Letters 6(4), 783–796 (2012)
12. Lavor, C., Liberti, L., Maculan, N., Mucherino, A.: The Discretizable Molecular Distance Geometry Problem. Computational Optimization and Applications 52, 115–146 (2012)
13. Lavor, C., Liberti, L., Mucherino, A.: The interval Branch-and-Prune Algorithm for the Discretizable Molecular Distance Geometry Problem with Inexact Distances. Journal of Global Optimization 56(3), 855–871 (2013)
14. Liberti, L., Lavor, C., Maculan, N.: A Branch-and-Prune Algorithm for the Molecular Distance Geometry Problem. International Transactions in Operational Research 15, 1–17 (2008)
15. Liberti, L., Lavor, C., Maculan, N., Mucherino, A.: Euclidean Distance Geometry and Applications. SIAM Review 56(1), 3–69 (2014)
16. Liberti, L., Lavor, C., Mucherino, A., Maculan, N.: Molecular Distance Geometry Methods: from Continuous to Discrete. International Transactions in Operational Research 18(1), 33–51 (2011)
17. Malliavin, T.E., Mucherino, A., Nilges, M.: Distance Geometry in Structural Biology: New Perspectives. In: [19], pp. 329–350 (2013)
18. Mucherino, A.: On the Identification of Discretization Orders for Distance Geometry with Intervals. In: Nielsen, F., Barbaresco, F. (eds.) GSI 2013. LNCS, vol. 8085, pp. 231–238. Springer, Heidelberg (2013)
19. Mucherino, A., Lavor, C., Liberti, L., Maculan, N. (eds.): Distance Geometry: Theory, Methods and Applications, 410 pages. Springer (2013)
20. Mucherino, A., Lavor, C., Malliavin, T., Liberti, L., Nilges, M., Maculan, N.: Influence of Pruning Devices on the Solution of Molecular Distance Geometry Problems. In: Pardalos, P.M., Rebennack, S. (eds.) SEA 2011. LNCS, vol. 6630, pp. 206–217. Springer, Heidelberg (2011)
21. Saxe, J.: Embeddability of Weighted Graphs in k-Space is Strongly NP-hard. In: Proceedings of 17th Allerton Conference in Communications, Control and Computing, pp. 480–489 (1979)

Using Entropy Cluster-Based Clustering for Finding Potential Protein Complexes

Viet-Hoang Le and Sung-Ryul Kim

AIS Lab, Department of Internet Multimedia
Konkuk University, South Korea
LHViet88@gmail.com, kimsr@konkuk.ac.kr

Abstract. Many researches have studied the complex system today because protein complexes, formed by proteins that interact with each other to perform specific biological functions, play a significant role in the biological area. And a few years ago, E. C. Kenley and Y. R. Cho introduced an algorithms which uses the entropy of graph for clustering in [2,3] based on protein-protein interaction network.

In our study, we extend the works to find potential protein complexes while overcoming existing weaknesses of their algorithms to make the results more reliable. We firstly clean the dataset, build a graph based on protein-protein interactions, then trying to determine locally optimal clusters by growing an initial cluster combined of two selected seeds while keeping cluster's entropy to be minimized. The cluster is formed when its entropy cannot be decreased anymore. Finally, overlapping clusters will be refined to improve their quality and compare to a curated protein complexes dataset. The result shows that the quality of clusters generated by our algorithm measured by the average cluster size considering f1-score is spectacular and the running time is better.

Keywords: cluster entropy, graph clustering, protein-protein interaction.

1 Introduction

Protein Complex is a group of proteins in which proteins work together to do a particular function. For example, the Hemoglobin protein complex, which is formed by subunit proteins Hb-α1 and Hb-α2 and Hb-β, is often responsible for carrying oxygen from the respiratory organs to the rest of the body. And because of its significant role, today there are many researches are trying to apply clustering algorithms to figure out these protein complexes. However protein complexes own some attributes such as overlapping allowance or the number of kinds of protein in one complex is small. Hence, clustering algorithms should be configured to satisfy these requirements also.

In 2011, E. C. Kenley and Y. R. Cho introduced Entropy-Based Graph Clustering algorithm which applied to Biological and Social Networks in [1]. They proposed a new method to determine potential protein complexes based on calculating graph entropy of mapped protein-protein interaction network. Assume that proteins are vertex nodes, and the interactions between them are edges, the algorithm builds a graph

F. Ortuño and I. Rojas (Eds.): IWBBIO 2015, Part I, LNCS 9043, pp. 524–535, 2015.

representing a protein-protein interaction dataset. Based on that graph, first it selects one seed, which has biggest vertex degree and form an initial cluster from selected seed's neighbors. The second step removes one by one the added neighbor vertices of seed node if removing can decrease entropy of the graph. Thirdly, it one by one adds neighbor nodes of initial cluster if adding they can decrease the entropy of the graph. Finally, the cluster is formed when graph entropy cannot be reduced anymore. The algorithm produces a good result when comparing to other algorithms such as Markov Clustering (MCL) in [7], MCode in [8], and CNM in [9] when assessing by f1-score. However, the algorithm then committed some issues: is it quickly fall into the local minimum, and the cluster size distributes different from the ground truth.

In 2012, Y. R. Cho and T. C. Chiam continue strengthening the algorithm with a modified version in which they slightly change the seed selection and cluster growing stages to overcome falling into the local minimum problem, and finally apply an over-lapping refinement process to increase the cluster's quality in [2]. The modified algorithm now instead of adding neighbor nodes one by one while considering graph entropy minimization, it selects a clique of three seeds as an initial cluster and at each growing step it's going to add all neighbor nodes of the cluster, and remove one by one to decrease the entropy of entire graph. After all, an overlapping refinement process attempts to assess overlapping cluster to add or remove cluster's node to form the new refined cluster. This algorithm looks better than the old one in terms of f1-score, however, it in turn committed new issues about running time cost and still did not solve cluster size problem. The basic principles, strategies and limitations of two original algorithms are presented in details in section 2.1.

Using entropy of a graph is a good idea to figure out the potential protein complexes from protein-protein interaction network, but calculating the entropy of entire graph at each step in the growing cluster iteration indeed consumes an abundant of computing power. In addition, we realized that finding the global minimum entropy does not really help to figure out good potential complexes, especially in the modified version of the original algorithm in [2] that led to the case a meta-cluster which contains more than 4000 nodes, leading to cannot figure out other potential complexes. Therefore, in our algorithm, to overcome these issues we introduce a concept about the entropy of cluster that is calculated based on the cluster and its neighbor. Since the algorithm tries to find the local optimized entropy, it generates clusters that have average size is small, and the distribution of cluster's size is closer to the ground truth's distribution when comparing to the original algorithms. To avoid quickly falling into the local minimum problem, we start with an initial cluster formed by two seeds and iteratively add its neighbor if the adding can decrease entropy of the cluster. To refine the overlapping result clusters, we apply the same method and measurement in the original algorithm but slightly change to drop the refined cluster, which has only one node (in the scenario that we are finding potential protein complexes formed from two different types of proteins in protein-protein dataset). We introduced the details of entropy cluster-based algorithm in section 2.2.

2   Entropy Cluster-Based Clustering

2.1   Graph Entropy

Graph Entropy H(G,P) is an information theoretic functional on a graph G based on probability distribution P on its vertex set in [6]. Assume that we have an undirected, unweight graph G(V,E). The graph G(V,E) contains a set of clusters, and a cluster is considered as a sub-graph $G'(V', E')$ of G.

Vertex Entropy
Suppose that we have a cluster $G'(V', E')$, entropy $e(v)$ of a vertex v will be calculated by formula (1)

$$e(v) = -p_i(v)log_2p_i(v) - p_o(v)log_2p_o(v) \qquad (1)$$

In which, $p_i(v)$ denotes the percentage (distribution, probability) of neighboring vertices n that is placed inside of graph G' over the total number of neighbor nodes $|N(v)|$ of node v as formula (2)

$$p_i(v) = \frac{n}{|N(v)|} \qquad (2)$$

Similarly, $p_o(v)$ denotes the rate (distribution, probability) of neighboring vertices n of node v that is placed outside of graph G' as formula (3).

$$p_o(v) = 1 - p_i(v) \qquad (3)$$

Graph Entropy
The entropy $e(G)$ of a graph G is the sum of all vertices' entropy in G calculated by the formula (4):

$$e(G) = \Sigma_{v \in V} e(v) \qquad (4)$$

Entropy-Based Graph Clustering
Based on the entropy theory, in 2011, E. C. Kenley and Y. R. Cho proposed a clustering algorithm to figure out clusters based on a given graph in [1] as described below:

1. Select a random seed vertex. Forms a seed cluster including the selected seed and its neighbors.
2. Iteratively remove any of the seed neighbors to minimize graph entropy.
3. Iteratively add vertices on the outer boundary of the cluster to minimize graph entropy.
4. Output the cluster. Repeat steps 1, 2 and 3 until all vertices have been clustered.

In step 1, a set of seed candidates is managed and when a cluster is generated, the vertices in the cluster are removed from the seed candidate set. In step 2-3, each removing or adding action is greedily checked if it can decreases graph entropy. The steps 1, 2 and 3 detect an optimal cluster with the lowest graph entropy.

These three steps are repeated to generate result clusters and stop until all vertex nodes are processed. The algorithm allows the vertices in a prior cluster can be members of the subsequent cluster.

The quality of a cluster can be evaluated by the connectivity; higher quality as denser intra-connections and sparser interconnections. A graph with lower entropy indicates the vertices in the cluster have more inner links and less outer links. In (a) of fig. 1., given a cluster with a, b, c, and d, entropy of the graph is 1.81. However, in (b), if the vertex f is added into the cluster, graph entropy increases to 1.92 because the cluster quality decreases.

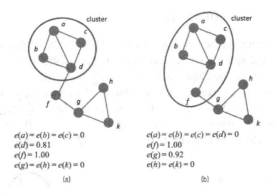

$e(a) = e(b) = e(c) = 0$
$e(d) = 0.81$
$e(f) = 1.00$
$e(g) = e(h) = e(k) = 0$

(a)

$e(a) = e(b) = e(c) = e(d) = 0$
$e(f) = 1.00$
$e(g) = 0.92$
$e(h) = e(k) = 0$

(b)

Fig. 1. Example of vertex entropy and graph entropy measurement

Modified Entropy-Based Graph Clustering
In 2012, Y. R. Cho and T. C. Chiam introduced a modified version of entropy-based graph clustering algorithm to improve the quality of result clusters in [2] to avoid quickly falling into the local minimum problem of the earlier version. It changes the way to select and grow the cluster, and concentrates to refine overlapping clusters. Below are 5 basic steps in their algorithm:

1. Select a clique of size 3 as an initial cluster.
2. Add all neighboring nodes of the cluster.
3. Remove nodes added on the step (2) iteratively to decrease graph entropy until it is mini-mal.
4. Repeat the steps (2) and (3) until the step (3) removes all nodes added on the step (2).
5. Output the cluster, and repeat the steps from (1) to (4) until no seed candidate remain.

From result clusters, it finds out overlaps of clusters and assess them by overlapping measurements to figure out what overlapping clusters need to be refined.

An overlapping cluster c_i will be selected and all elements from other clusters which overlapping the selected cluster c_i will be added to c_i. Then, each node in c_i will be remove if it overlapping value is less than a threshold to form the final cluster c_i. If the refined cluster c_i is not redundant, adding c_i to the result cluster set.

2.2   Cluster Entropy

Cluster Entropy

Assume that we have a cluster, a sub-graph, G' (V', E') and the set of all neighbors V_n of cluster G'. V_n contains all vertices placed outside of G' and they have links to all vertices placed inside of G'. The cluster entropy is defined as the sum of the entropy of all vertices in G' and its neighbors as formula (5):

$$e(G') = \sum_{v \in (V \cup V_n)} e(v) \tag{5}$$

The quality of a cluster can be evaluated by the connectivity; higher quality as denser intra-connections and sparser interconnections. It means the quality of the cluster is high when the entropy of the cluster is low.

Fig. 2. (a) illustrates an example of calculating cluster entropy, in which, given a cluster G' (V', E') where V'= {a, b, c, d, e} and its neighbors V_n = {f, g, h, i, k}. We can obtain entropy of vertices: e(a) = e(b) = e(c) = 0; e(e) = 1.00; e(d) = 0.92; e(f) = 0.81; e(g) = 0.92; e(h) = 0.81; e(i) = 0.81; e(k) = 0.72; Then entropy of cluster G' is e(G') = 5.99

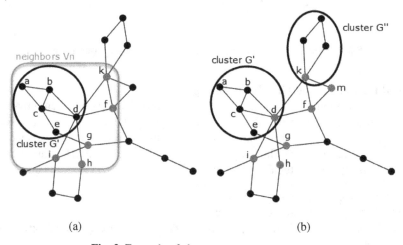

(a)   (b)

Fig. 2. Example of cluster entropy measurement

In Fig. 2. (b) We show the example about the differences between calculating cluster entropy and entire graph entropy. Suppose that graph G has two cluster G' and G''. Cluster G' has neighbors vertices f, g, h, i and k. Cluster G'' has neighbors vertices d, f and m. Cluster entropy of G' is e(G') = 5.99, and the cluster entropy of G'' is e(G'') = e(k)+e(m)+e(f)+e(d) = 0.97 + 1.00 + 0.81 + 0.65 = 3.43.

For graph entropy, the vertex f in (b) has two inner links and two outer links therefore, the entropy of vertex f is e(f) = 1.00, and the entropy of graph G is e(G) = 7.43.

Entropy Cluster-Based Clustering Algorithm

Investigating properties of protein complexes and protein-protein interaction dataset which only represent the interaction between proteins but not the portion or construction, we believe that finding a local optimum is a good way to figure out potential protein complexes. Therefore, like hill climbing problem, using cluster entropy can help us to implement the algorithm, and a slight change in the way of selecting seeds and forming an initial cluster can prohibit quick falling local optimal problem.

The algorithm is described in 4 steps below:

1. Select a clique of 2 seeds as an initial cluster, and add all neighboring nodes to the initial cluster.
2. Iteratively remove any neighbor node added in first step if removing will decrease cluster entropy.
3. Iteratively add any node from neighbor nodes of cluster if adding will decrease cluster entropy
4. Output the cluster. Remove cluster's nodes from set of seed candidates; and repeat the steps (1-3) until no seed candidate remain; it means all nodes are in clusters.

In fig. 3. we show the progress of growing a cluster based on cluster entropy.

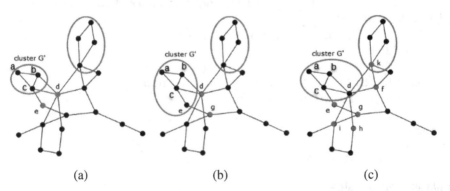

(a)   (b)   (c)

Fig. 3. Example of growing cluster based on cluster entropy

Suppose that cluster G' in fig. 3. (a) already has three nodes, then entropy of cluster G' is e(G') = e(a) + e(b) + e(c) + e(d) + e(e) = 0 + 0.92 + 1 + 0.92 + 1 = 3.84.

In (b), if the cluster adds neighbor vertex e into G', the cluster entropy now is

e(G') = e(a)+e(b)+e(c)+e(d)+e(e)+e(g) = 0 + 0.92 + 0.81 + 0.92 + 1 + 0.92 = 4.57.

The entropy of cluster in (b) is not smaller than (a), so we do not add new node into cluster as (b) but keep the original cluster (a).

In (c), we then add the neighbor vertex d into G' in (a), the cluster entropy now is

e(G') = e(a) + e(b) + e(c) + e(d) + e(e) + e(f) + e(k) + e(h) + e(i) = 0 + 0 + 1 + 0.92 +

$1 + 0.81 + 0.72 + 1 + 0.81 = 6.26$. The entropy of cluster in (c) is not smaller than in (a) also, hence we keep the original cluster in (a).

In step 1 of proposed algorithm, we setup a seed candidate set instead of random seed selection to strengthen the algorithm. The set contains all sorted nodes by vertex degree, and the algorithm will select the biggest vertex degree node to process first, because that node is a hub which has more chance to be a core of protein complexes. The second seed in step 1 of proposed algorithm will be the neighbor node of the first node which has smallest vertex degree value, because of ensuring minimized cluster entropy.

After step 4, we continue refining overlapping clusters as described in [2] in section 2.1. And because the protein-protein interaction data present only interactions between proteins so it's meaningless to get cluster which has only one node. Hence, we need to eliminate that kind of cluster after refining. The below pseudo code show how we treat the overlapping clusters:

```
find out all overlap clusters
for each overlap cluster
   evaluate if the cluster need to be refined
   if cluster need to be refined
     add all nodes from other overlap clusters of selected
cluster into it
      for each node in selected cluster
       if the node has overlapping value is lower than a
given threshold
          remove the node from cluster
        endif
     end for
     if the cluster has more than one node
       output refined cluster
      endif
   endif
end for
```

Dataset Preprocessing

In mif25 format PPI dataset, the id values of interactors and interactions do not make sense for comparison between result clusters and ground truth, therefore, to compare result clusters with protein complexes, we have to compare their elements directly. We obtain the universal id from xref values of each interactor and use it as the unique identity of the protein. For example, protein ufd4_yeast has id is 731427, but both its name and id are not stable; they are changed over different dataset from different providers. Only its universal ids in xref are not changed, so in our research, we choose xref from uniprotkb as its unique id to represent Ubiquit In Fusion Degradation Protein 4. The universal id will be formed in format:

(db name)_(id in that db), e.g. uniprotkb_P33202.

According the research article led by Ian M Donaldson in [5], the interaction records have only one interactor represent multimeric complexes, which contain one subunit type that is present in two or more copies. For example, in PPI dataset acquired from IntAct, interaction ImexId IM-18882-4 has only one participant is protein P09938 (UniProt). In fact, the interaction has two participants, subunit P09938 and subunit P21524, but these two subunits are derived from protein "Ribonucleoside-diphosphate reductase small chain 1" so the interaction describes only one protein. Similar to the gold standard, i.e., complex Syp1 dimer consists only one protein is P25623 (Suppressor of yeast profiling deletion - UniProt). The sole existence of only one protein in an interaction, which is unlike binary interaction, does not help to build the graph of proteins, so we remove them from the input dataset.

We also remove all interactions which have more than two interactors in the participant list since it increases the complexity, but does not offer good result. In addition, there is no exact description what model is correct for connections between these participants. For example, these proteins can be connected as a ring, or star, where one protein is a hub which connects other proteins.

Finally, the input data set contains only binary interactions and protein's information.

2.3   Clustering Accuracy Measurement

F1-score is a combination of precision and recall of a cluster when comparing the cluster to real protein complexes. It is calculated in formula (6) below:

$$f = 2 \cdot \frac{precision \cdot recall}{precision + recall} \tag{6}$$

In which, precision, also called positive predictive value, is the fraction of relevant instances that are retrieved. It is calculated in formula (7) by the ratio of common proteins in cluster c and complex p_i to the number of proteins in cluster c.

$$precision = \frac{|c \cap p_i|}{|c|} \tag{7}$$

And recall, also known as sensitivity, is a measure of relevance which is calculated in formula (8) by the ratio of common proteins of cluster c and complex p_i to the number of proteins in p_i.

$$recall = \frac{|c \cap p_i|}{|p_i|} \tag{8}$$

However, when assessing two algorithms that produce two sets of clusters, which have number of cluster and average size of cluster are much different, we realized that the f1-score cannot show a reliable result because it does not concern the average size of cluster. In table 1, we show the example of input and output data of two different algorithms. In that case, even algorithm B obtains a better f1-score value, it does not mean that quality of result clusters is better than result generated by the algorithm A, since the number of clusters of algorithm B is lower than expected, and the average size of them is far different from the average size of ground truth cluster. Besides, the too big cluster will limit the possibility of generating new cluster because the possible seeds were occupied by that big cluster. Therefore, in this case, algorithm A should be more suitable and better than algorithm B.

Table 1. Data of input and output of different algorithm

	Ground truth	Algorithm A	Algorithm B
Number of node	15	25	
Number of cluster (overlap allowance)	5	8	2
Cluster average size	3.4	3.75	10.0
Average f1-score	1	0.45	0.62

Because of the above reasons, we use a modified version of f1-score, that consider the ground truth average size by formula in (9). In which, f is original f1-score, $avg(g)$ is the average size of the ground truth cluster, and $avg(c)$ is the average size of cluster. If the similarity is low, the value of f' is low; if the similarity is higher, the value of f' will increase to get the same value of the original f1-score.

$$f' = f \cdot \frac{avg(g)}{avg(g) + |avg(g) - avg(c)|} \tag{9}$$

3   Result and Discussion

3.1   Data Source

We apply algorithms on a curated protein-protein interaction dataset of yeast named Saccharomyces Cerevisiae (Baker's yeast) in [3] and evaluate the result with the gold standard in [4]. The datasets are updated on 2014 July 18th by IntAct Molecular Interaction Database, The European Bioinformatics Institute. The Protein-protein interaction (PPI) dataset contains 6,184 proteins, and 16,133 interactions. The gold-standard dataset is a Protein complex dataset of Saccharomyces cerevisiae (strain ATCC 204508 / S288c – scientific name) which contains 306 protein complexes combined from 955 proteins, and there is 9 proteins do not appear in PPI dataset.

We show the distribution of size of curated protein complexes in fig. 4 to make clear our assessment based on it. In the complex dataset we can recognize that there are 131 complexes that have only two proteins, according 42.81%, and 58 complexes that have three proteins, 18.95%, it leads to the average size of protein complexes is 4.072.

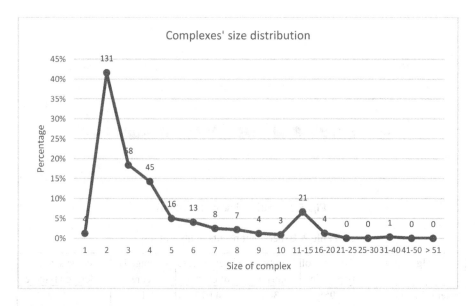

Fig. 4. Size distribution of protein complexes (ground truth)

3.2   Experiments and Result

We configure the three algorithms in its best condition and arguments:

- Entropy-Based Graph clustering [1]: select interactor which has biggest degree as seed from the candidate set first
- Modified Entropy-Based Graph Clustering [2]: select first seed has biggest degree and random select the second and the third seed. Using minCov = minCons = 0.4, and minCss = 0.8 for refining overlapping clusters.
- The Entropy Cluster-Based Clustering: select first seed has biggest degree and second seed from the neighbor node which has lowest degree. Using minCov = 0.4, minCons = 0.4, and minCss = 0.8 for refining overlapping clusters.

The chart in fig. 4 shows the execution time of the three algorithms, in which the proposed algorithm run in only 115 seconds while the original algorithm consumes about 1177 seconds. It runs longer because of greedily calculating all clusters to maintain the lowest graph entropy. The modified graph entropy algorithm even needs about 9.5 hours because steps (2-4) require much time to find the graph entropy for each iteration.

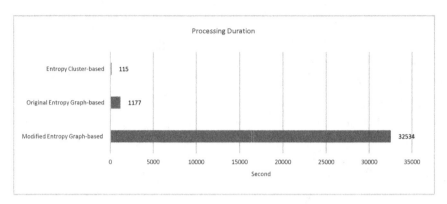

Fig. 5. Execution time of algorithms

Table 2. Experiment results of algorithms

Algorithm	Number of protein	Number of cluster	Average size of cluster	Average f1 score	Average size considering f1-score
Protein Complexes dataset(ground truth)	955	306	4.072	1	1
Original Entropy-based Graph		385	14.613	0.243	0.068
Modified Entropy-based Graph	6,184	174	51.006	0.625	0.050
Proposed entropy cluster-based		1540	3.771	0.404	0.377

To evaluate the effectiveness of each algorithm, we assess the average size considering f1 score when comparing to real protein complexes in table 2. From the dataset of 6,184 proteins, the original entropy-based graph clustering gives out 385 clusters with an average f1 score is 0.274. The modified entropy-based graph clustering algorithm offered a better result with clusters' quality is 0.625, however the number of clusters and its average size are not reasonable for predicting protein complexes, the number of clusters are too small and the average size of cluster is too big, the existence of some cluster which has up to 4,000 proteins will limit the possibility to generate other clusters. Therefore, even the algorithm improved its quality, we do not think that it is affordable to apply on finding potential protein complexes. In addition, our algorithm offers a reasonable result with 1,540 potential complexes from 6,184 proteins, and the average size of cluster is 3.771. Using the average size considering f1-score metric, apparent that the proposed algorithm offered a spectacular result, with 33 clusters that match the gold-standard protein complexes directly; conversely, other algorithms obtain less than 10 clusters that are matched directly.

4   Conclusions

There are various graph-based clustering algorithms that identify potential functional organizations such as protein complexes from protein-protein interaction networks. Among them, entropy-based clustering is a good competitor when offering a novel approach to reveal protein complexes from Protein-protein interaction data. In our study, we proposed an application to use the entropy concept in the graph clustering algorithm, which is effective when applying on the biological PPI dataset. The proposed algorithm may use with other network datasets which can be mapped to graphs.

The proposed algorithm first considers to compute cluster's entropy and ignore graph entropy, that by this way it helps to improve the speed and give out more reasonable results. Secondly, we refined the strategy of selecting seed candidates and apply the overlapping cluster refinement method in [2] to improve the quality of result clusters. And finally, we assess the algorithms by a modified f1-score, which considers the ground truth average size.

In addition, we also provided basic knowledge and detailed explanations which help persons who have not worked with PPI dataset before could understand what and why we process in that way to reach the final result.

References

1. Kenly, E.C., Cho, Y.-R.: Entropy-Based Graph Clustering: Application to Biological and Social Networks. In: 2011 IEEE 11th International Conference on Data Mining (ICDM), pp. 1550–4786 (December 2011), DOI =
   http://dx.doi.org/10.1109/ICDM.2011.64
2. Chaim, T.C., Cho, Y.-R.: Accuracy improvement in protein complex prediction from protein interaction networks by refining cluster overlaps. Proteome Sci. 10 (Suppl 1:S3)(Jun 21, 2012), doi:10.1186/1477-5956-10-S1-S3
3. IntAct Curated Yeast, Protein-protein Interaction Datasets,
   ftp://ftp.ebi.ac.uk/pub/databases/intact/current/psi25/
   species/yeast.zip
4. IntAct Curated Yeast, Complexes Datasets,
   ftp://ftp.ebi.ac.uk/pub/databases/intact/complex/current/
   psi25
5. Razick, S., Magklaras, G., Donaldson, I.M.: iRefIndex: A consolidated protein interaction database with provenance. BMC Bioinformatics 9, 405 (2008), doi:10.1186/1471-2105-9-405.
6. Graph Entropy – A Survey. G. Simonyi,
   http://www.renyi.hu/~simonyi/grams.pdf
7. Van Dongen, S.: A new clustering algorithm for graphs, National Research Institute for Mathematics and Computer Science in the Netherlands, Tech. Rep. INS-R0010 (2000)
8. Bader, G.D., Hogue, C.W.: An automated method for finding molecular complexes in large protein interaction networks. BMC Bioinformatics 4, 2 (2003)
9. Clauset, A., Newman, M.E.J., Moore, C.: Finding community structure in very large networks. Physical Review E 70, 66111 (2004)

A Weighted Cramér's V Index for the Assessment of Stability in the Fuzzy Clustering of Class C G Protein-Coupled Receptors*

Alfredo Vellido**, Christiana Halka, and Àngela Nebot

Department of Computer Science, Universitat Politècnica de Catalunya,
Barcelona 08034, Spain
{avellido,christiana.halka,angela}@cs.upc.edu
http://www.cs.upc.edu/~avellido

Abstract. After decades of intensive use, K-Means is still a common choice for crisp data clustering in real-world applications, particularly in biomedicine and bioinformatics. It is well-known that different initializations of the algorithm can lead to different solutions, precluding replicability. It has also been reported that even solutions with very similar errors may widely differ. A criterion for the choice of clustering solutions according to a combination of error and stability measures has recently been suggested. It is based on the use of Cramér's V index, calculated from contingency tables, which is valid only for crisp clustering. Here, this criterion is extended to fuzzy and probabilistic clustering by first defining weighted contingency tables and a corresponding weighted Cramér's V index. The proposed method is illustrated using Fuzzy C-Means in a proteomics problem.

Keywords: Fuzzy clustering, K-Means, Clustering stability analysis, Cramér's V index, G Protein-Coupled Receptors.

1 Introduction

G protein-coupled receptors (GPCRs) are cell membrane proteins of interest due to their role in transducing extracellular signals after specific ligand binding. They have in fact become a core interest for the pharmaceutical industry, as they are targets for more than a third of approved drugs [1].

GPCR functionality is mostly investigated from its crystal 3-D structure. Finding such structure, though, is a difficult undertaking[1] and only in the last decade, a handful of GPCR structures has been found, most of them belonging to the class A of the GPCR superfamily [2]. Only in 2013, one receptor not belonging to class A but to the Frizzled class and two class B [3,4] were reported.

---

* This research was partially funded by Spanish MINECO TIN2012-31377 research project.
** Corresponding author.
[1] Rhodopsin was the first GPCR crystal structure to be determined, back in 2000.

F. Ortuño and I. Rojas (Eds.): IWBBIO 2015, Part I, LNCS 9043, pp. 536–547, 2015.

The first structures of the 7-trans-membrane (7TM) domains of two class C receptors were published in 2014 [5,6].

Our research focuses on class C GPCRs, which have become a key target for new therapies [7]. The alternative approach when the tertiary 3D crystal structure is not available, is the investigation of the receptor functionality from its primary structure, that is, directly from the amino acid (AA) sequences. The comparative exploration of the sequences of the seven different described subtypes of this class may constitute a first step in the study of the molecular processes involved in receptor signalling.

Most of the existing data-based research on primary receptor sequences resorts to their alignment [8], which enables the use of more conventional quantitative analysis techniques. Given that the length of the class C GPCR sequences varies from a few hundred AAs to well over a thousand, alignment risks the loss of relevant sequence information. Alternatively, as in this paper, we can resort to methods for the analysis of alignment-free sequences from their transformation according to the AA properties (for a review see [9]).

Previous exploration of the class C GPCR sequences through visualization-oriented clustering [10] and semi-supervised analysis [11] has shown that the existing formal characterization of this class into seven subtypes only partially corresponds to the natural data cluster structure according to unaligned sequence transformations. In the current study, we investigate this issue within a more general clustering framework.

Clustering analysis often works by assigning individual data instances to one out of several clusters according to their similarity to representative examples Such assignment is often of a dichotomous or *crisp* nature: the instance either belongs to or does not belong to a given cluster. Fuzzy and probabilistic clustering methods, instead, assign each data instance to each cluster with an estimated degree or probability of membership. As a result, the uncertainty of the assignment decision is explicitly taken into account in the model.

Over the last decades [12], K-Means has become a stalwart method for data clustering, spawning many variants while remaining a common choice, even if as a benchmark, in many real-world applications. K-Means is based on crisp cluster assignments, although variants such as Fuzzy C-Means (FCM)[13] have extended the model to account for partial degrees of membership. K-Means limitations are well studied and include the lack of a closed criterion for the choice of the number of clusters K and the fact that, under different initializations, the algorithm may yield very different solutions.

Recent experimental evidence [14] has shown that K-Means solutions that might be expected to be similar according to the final value of the objective function may in fact be quite dissimilar, and that this effect increases with the value of K. This suggests the convenience of using the objective function as a criterion of model optimality only in combination with some cluster stability criterion if we aim to achieve cluster partition reproducibility. One such combined criterion is the *Separation and Concordance* (SeCo) map, which joins the standard sum-of-squares (SSQ) error and Cramér's V stability index, a variation of Pearson's χ^2, which can also be used to inform the choice of K.

The calculation of Cramér's V index is based on the use of contingency tables, which are only suitable for crisp cluster assignments. In this study, we extend the SeCo criterion to fuzzy and probabilistic clustering by first defining weighted contingency tables and a corresponding weighted Cramér's V index. This should be a more faithful assessment of the clustering solution stability for fuzzy and probabilistic methods.

The proposed methods are employed to investigate a class C GPCR primary sequence data set extracted from a publicly available database. Two experimental settings for the clustering experiments are used. The first fixes the number of clusters to the number of formal subtypes in the class in order to investigate the level of correspondence between both, while the second relaxes this constraint in order to analyze the stability of the clustering solutions using SeCo maps.

2 Methods

2.1 Alignment-Free Sequence Transformation Methods

As previously mentioned, in this paper we resort to methods for the analysis of alignment-free sequences from their transformation according to the AA properties. Three transformations were used in our experiments:

- *Amino Acid Composition (AAC)*: This simple transformation reflects the AA composition of the primary sequence. The frequencies of the 20 sequence-constituting AAs are calculated for each sequence and, as a result, an $N \times 20$ data matrix is obtained, where N is the number of data instances.
- *Auto Cross Covariance (ACC)*: The ACC transformation aims to capture the correlation of the physico-chemical AA descriptors along the sequence. The method relies on the translation of the sequences into vectors based on the principal physicochemical properties of the AAs. Data are transformed into a uniform matrix by applying a modified autocross-covariance transform [15]. First, the physico-chemical properties are represented by means of the five z-scores of AA as described in [16]. Then the Auto Covariance (AC) and Cross Covariance (CC) are computed on this first transformation. They, in turn, measure the correlation of the same descriptor (AC) or the correlation of two different descriptors (CC) between two residues separated by a lag along the sequence. From these, the ACC fixed length vectors can be obtained by concatenating the AC and CC terms for each lag up to a maximum lag, l. This transformation generates an $N \times (z^2 \cdot l)$ matrix, where $z = 5$ is the number of descriptors.
- *Digram Transformation*: The digram transformation is a particular instance of the more general n-gram transformation. It considers the frequencies of occurrence of any given pair of AAs. The n-gram concept has previously been used in protein analysis [17]. This particular transformation generates an $N \times 400$ matrix.

2.2   Data Clustering Using K-Means and Fuzzy C-Means

The K-Means algorithm creates a partition of a set $\mathbf{X} = \{\mathbf{x}_1, \ldots, \mathbf{x}_N\}$ of observed data into a set of K clusters $\varGamma = \{\varGamma_1, \ldots, \varGamma_K\}$ by defining a fixed number K of data centroids or prototypes $\{\mu_1, \ldots, \mu_K\}$. It does so by assigning individual data observations to their closest prototypes according to a given similarity measure or distance. Such assignment is crisp in the sense that individual observations are associated to individual clusters with complete certainty. Fuzzy and probabilistic clustering techniques relax this approach by assigning, to each observation, a fuzzy degree (for instance in FCM) or a probability (for instance in mixture models) of membership to each cluster. Most approaches to K-Means opt for different forms of random initialisation of the prototypes (*seeds*). The algorithm's objective is finding the set $\varGamma$ that minimises a SSQ error in the form $\sum_{k=1}^{K} \sum_{\mathbf{x} \in \varGamma_k} \|\mathbf{x} - \mu_k\|^2$. FCM generalizes this objective function to become:

$$\sum_{c=1}^{C} \sum_{n=1}^{N} \omega_{nc}^{m} \|\mathbf{x}_n - \mu_c\|^2 \tag{1}$$

for C fuzzy clusters, fuzzy weights ω and fuzziness parameter m. It is well known that different initialisations of the algorithm make it converge to different local minima and that there is no guarantee of convergence to a global minimum of the objective function. In practical applications, K-Means and FCM are run with a sufficient number of random initialisations and the $\varGamma$ generating minimum error is chosen among the rest.

2.3   Clustering Stability Measures

Even if finding a minimum error $\varGamma$ is a central objective of K-Means, the *stability* of the clustering solution is also relevant. Solutions that are reproducible are required. That is, cluster partitions that do not change (much) under different initialisations, i.e., that are stable. This cluster stability is paramount in practical applications and can be quantified using different indices [18].

It was recently brought into attention [14] that K-Means partitions with similar errors might be greatly different from each other (thus unstable) and that this effect increases with the value of K. Assuming that solutions that strike a balance between low error and high stability ought to be sought, Lisboa *et al.* [14] proposed a framework based on the calculation of Separation/Concordance (SeCo) maps for settings using multiple random initializations of K-Means for different values of K. This entails the simultaneous display of a pair of values for each run of the algorithm, namely: The ΔSSQ, calculated as the total SSQ minus the within-cluster SSQ:

$$\sum_{k=1}^{K} \sum_{n=1}^{N} \|\mathbf{x_n} - \mu_k\|^2 - \sum_{k=1}^{K} \sum_{\mathbf{x_n} \in \varGamma_k} \|\mathbf{x_n} - \mu_k\|^2 \tag{2}$$

and a *concordance index* (CI) quantifying stability. In [14], the use of Cramér's V index is recommended as a basis for it. The CI is calculated as the median of the $(nin - 1)$ pairwise Cramér's V calculations for nin initializations. For two cluster partitions Γ and Γ' of, in turn, K and K' clusters, Cramér's V index is a variation of a χ^2 test, calculated as $V = \sqrt{\chi^2/N \cdot min(K-1, K'-1)}$, where

$$\chi^2 = \sum_{k=1}^{K} \sum_{k'=1}^{K'} (O_{kk'} - E_{kk'})^2/E_{kk'} \tag{3}$$

Here, $\mathbf{O}$ is an observed contingency table ($K \times K'$) matrix, whose values $O_{kk'}$ indicate the number of instances in $\mathbf{X}$ that have been assigned to cluster k in one run of the algorithm and to cluster k' in another run. The $K \times K'$ matrix $\mathbf{E}$ contains the corresponding expected values for independent cluster allocations, calculated as $E_{kk'} = \frac{1}{N}(\sum_{j=1}^{K'} O_{kj} \sum_{i=1}^{K} O_{ik'})$.

This use of contingency tables is suitable for crisp cluster assignments such as those provided by K-Means. For *soft* assignments such as those provided by FCM or Gaussian Mixture Models, instead, this use occludes the richness of the cluster solution by requiring the assignment of instances to clusters to be based on the highest degree of membership or probability.

In this paper, we propose a variation of contingency tables that better suits the characteristics of fuzzy and probabilistic models. Elements in what we call weighted observed ($w\mathbf{O}$) contingency tables will now be calculated, following the notation of Eq.1 for FCM, as $wO_{cc'} = sum_{n=1}^{N}\omega_{nc}\omega_{nc'}$; this is, for data instance n, the product of the degree of membership to cluster c in a first run of the algorithm and the degree of membership to cluster c' in a second run. Consequently, we can obtain a weighted expected ($w\mathbf{E}$) contingency table matrix whose elements are defined as $wE_{cc'} = (\sum_{j=1}^{C'} wO_{cj} \sum_{i=1}^{C} wO_{ic'})/N$. This leads to the definition of a new *weighted Cramér's V index*, where $\mathbf{O}$ is replaced by $w\mathbf{O}$ and $\mathbf{E}$ by $w\mathbf{E}$ in the calculation of χ^2 in Eq.3.

If FCM estimated that all instances had a degree of membership of 1 for a single cluster, the weighted Cramér's V index would reduce to its standard formulation. This is unlikely to happen, which means that the proposed index will lead to lower levels of CI in SeCo. This should, therefore, be not only a conservative concordance estimator, but also a more reliable clustering assessment tool, capable of distinguishing solutions with varying levels certainty.

Note that SeCo can be used as a flexible tool to chose adequate values of the K parameter (number of clusters). This is equally true when using the modified index, but, in this case, there should be no bias in favour of "over-optimistic" solutions.

3   Experiments

3.1   Materials and Experimental Setting

The data in the following experiments were extracted from GPCRDB[2] [19] (version 11.3.4 as of March 2011), a public database of G Protein-Coupled Receptor

---

[2] http://www.gpcr.org/7tm

(GPCR) protein primary sequences. The data set comprises a total of 1,510 class C GPCR sequences, belonging to seven subfamilies and including: 351 metabotropic glutamate (mG), 48 calcium sensing (CS), 208 GABA_B (GB), 344 vomeronasal (VN), 392 pheromone (Ph), 102 odorant (Od) and 65 taste (Ta). Their AA lengths vary from 250 to 1,995.

Previous research [20] investigated the supervised classification of these data sequences, from several of their alignment-free transformations, including AAC, digram, and ACC, among others. Here, we use K-Means and FCM to investigate to what extent the natural clustering structure of the data fits the subfamilies (classes) description. For that, we first report the results of experiments in which the number of clusters is fixed *a priori* to be the same as the number of class C GPCR subtypes. These will provide us with a preliminary evaluation of the level of natural subtype overlapping. We then proceed to relax that constraint and the SeCo framework, with the proposed modification of the concordance index in the case of FCM, is applied in a setting with 500 random initializations of the algorithms for different number of clusters. Following [14], only the best 10% $\triangle SSQ$ results are displayed in the SeCo plots.

3.2   Results and Discussion

The three transformed data sets were fed to the FCM and K-Means algorithms. The class (subtype) specificity for each cluster for each data set was measured and the results are provided in the following paragraphs along with class-entropy measures. This will inform us to what extent the clusters extracted by K-means and FCM algorithms correspond (or not) to the theoretically labeled subtypes. The class-entropy for a given cluster k is calculated as $E_k = -\sum_{j=1}^{C} p_{kj} ln p_{kj}$, where j is one of the $C = 7$ class C GPCR subtypes and $p_{kj} = m_{kj}/m_k$, where, in turn, m_k is the number of sequences in cluster k and m_{kj} is the number of subtype j sequences in cluster k.

For FCM, Figure 1 and Table 1 show that, for the AAC data transformation, almost none of the defined clusters show clear class (subtype) specificity. Only in cluster 1, the first subtype (mG) of GPCR achieves a specificity that is close to 60%, but even in this case, the third subtype (GB) reaches a non-negligible 30%. Several clusters show common specificity profiles: for instance, clusters 1 and 3 are predominantly a mixture of mG and GB, which means that they might truly be a single cluster with some substructure. Cluster 4 is a very mixed combination of Ph and VN, but clusters 2, 6 and 7 seem to be variations of this combination, again suggesting one main cluster with further substructure and important levels of overlapping. The ACC and Digram transformations (Figures 2 and 3, and, again, Table 1), instead, manage to separate some of these clusters to become more subtype-specific. mG and GB are now more clearly discriminated (clusters 1 plus 5 and cluster 3, in turn) with the rest of subtypes showing clear overlapping in some clusters but also high specificity in others (for instance, Ph in ACC cluster 6 and Digram in cluster 7). In any case, the more complex transformations (ACC and Digram) seem to make the FCM clustering model more class C GPCR subtype-specific.

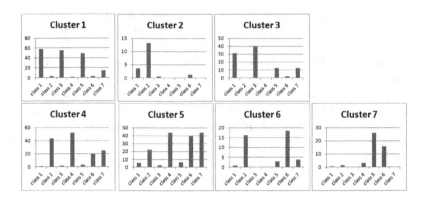

Fig. 1. Class specificity bar chart (with percentage values) for each FCM cluster of class C GPCR data set with the AAC transformation. Classes 1 to 7 are, in turn, mG, CS, GB, VN, Ph, Od and Ta.

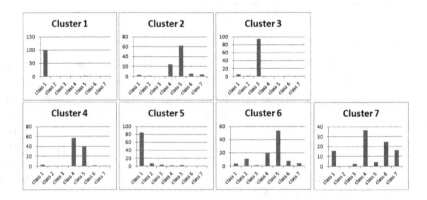

Fig. 2. As Figure 1, for the ACC transformation

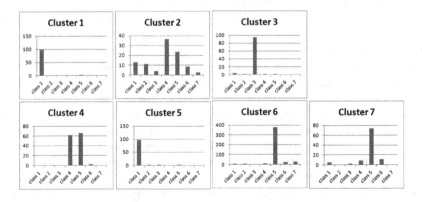

Fig. 3. As Figure 1, for the Digram transformation

Table 1. Number of GPCR sequences-per-cluster (♯) and cluster-specific (E_k) and total entropies for the FCM clustering of class C GPCR data with the three transformations

	AAC		ACC		Digram	
	♯	E_k	♯	E_k	♯	E_k
Cluster 1	245	1.45	107	0.13	112	0.12
Cluster 2	239	2.16	207	1.52	374	2.39
Cluster 3	200	1.34	202	0.33	200	0.37
Cluster 4	193	1.30	237	1.17	277	1.09
Cluster 5	67	1.97	199	0.89	179	0.26
Cluster 6	263	2.15	279	1.99	205	2.02
Cluster 7	303	1.95	279	2.20	163	1.28
Total Entropy	**1.77**		**1.34**		**1.29**	

The results of the K-means algorithm for the seven subtypes of class C GPCRs, for which, for the sake of brevity, we only report the entropy results in Table 2, are consistent with those of FCM. Again, almost none of the defined clusters show clear class (subtype) specificity with AAC data transformation. The ACC and Digram transformations, instead, manage to separate some of these clusters to become more subtype-specific. The similarity between the two algorithms is that mG and GB are more clearly discriminated than the rest of subtypes. Moreover, in the ACC transformation, Ph receptors can also be discriminated from the rest subtypes due to their high specificity in cluster 2. The remaining subtypes show clear overlapping in some of the clusters.

Table 2. Number of GPCR sequences-per-cluster (♯), together with cluster-specific (E_k) and total entropies for the K-Means clustering of class C GPCR data with the three transformations

	AAC		ACC		Digram	
	♯	E_k	♯	E_k	♯	E_k
Cluster 1	270	1.65	260	0.26	222	0.10
Cluster 2	406	2.17	136	0.72	398	1.47
Cluster 3	196	1.33	150	0.23	121	0
Cluster 4	189	1.24	188	1.07	284	1.10
Cluster 5	54	1.34	165	1.57	184	1.90
Cluster 6	56	1.74	379	1.79	197	1.95
Cluster 7	339	2.03	232	1.97	104	1.63
Total Entropy	**1.77**		**1.19**		**1.39**	

Comparing the results of both algorithms in terms of the total entropy measure, conclusions are not clear-cut. ACC and Digram show a clear advantage both in FCM and K-Means, but neither shows a clear advantage over the other.

We now move to the clustering stability analyses results, based on random algorithm initializations and varying number of clusters, using the SeCo maps. For each one of the transformed sets, three SeCo maps were created using:

- The K-means objective function and the standard Cramér's V index.
- The FCM objective function and the standard Cramér's V index.
- The FCM objective function and the novel weighted Cramér's V index proposed.

As previously mentioned, a threshold for the $\triangle SSQ$ values to select the 10% top values for each value of K is expected to allow the degeneracy of similar SSQ values to be resolved. The FCM 10% top results, as reported in Figs. 4 to 6, are very parsimonious (much more so than the complete ones, not reported here), revealing a high concentration of stability results around just a handful of median Cramérs V index values, in comparison with the still wide spread of K-Means. These results are also very consistent over data transformations. For K-Means, this effect does not necessarily increase as K increases for any of the data transformations. For FCM, though, an increase in spread as K increases is revealed. Overall, this indicates that FCM is much more resilient than its crisp K-Means counterpart to the variability introduced by random initializations.

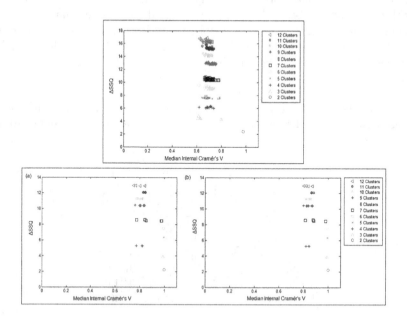

Fig. 4. Separation-Concordance maps for the AAC data set, including the 10% best results. Top: results for K-Means; bottom: results for FCM, a) with standard Cramér's V index; b) with proposed weighted Cramér's V index.

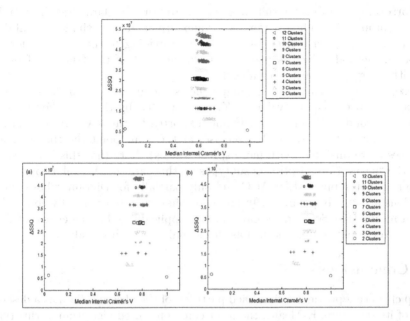

Fig. 5. Separation-Concordance maps for the ACC data set. Layout as in previous figure.

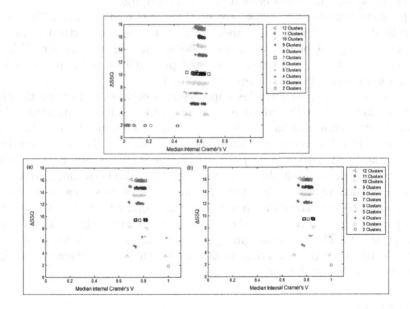

Fig. 6. Separation-Concordance maps for the *digram* data set. Layout as in previous figure.

Moreover, the stability results as measured by the standard Cramérs V index and the proposed weighted Cramérs V index are again very similar for all data transformations (even better for the latter, providing further support for the proposed method, which is a more faithful account of the true belief of the algorithm regarding cluster membership).

Also, the restricted 10% SeCo maps offer some guidance to make a decision about the most adequate value of K, as supported by the data. For FCM, a solution beyond 7 clusters is clearly not supported by the AAC transformation, as maximum stability suddenly decreases at the same point the cluster model becomes more unstable (with more spread values). Note that this is consistent with the "natural" description of subtypes for this data set. A similar conclusion, though, is not supported for ACC and only partially for Digram, whose low-K solutions are clearly polarized. In any case, these results are hardly conclusive, which means that SeCo maps have limited applicability for the choice of K in highly overlapping data sets such as those analyzed in this study.

4   Conclusions

Crisp clustering provides a simplified partition of the observed data as a description of its structure. K-Means, the most commonly used algorithm of this type, is known to be strongly dependent on initialization. Furthermore, recent studies suggest that different K-Means solutions with apparently similar error may in fact be quite dissimilar. In this context, it is recommended to base the choice of solution on a combination of error and stability criteria.

Separation-Concordance maps provide such combined criterion. They were originally developed for K-Means and, in this study, we have extended them to FCM by defining weighted contingency tables and a weighted Cramér's V index that could also be used in other fuzzy and probabilistic techniques in which cluster assignment is not of a *crisp* nature any longer.

We have experimented with this approach in a problem concerning the clustering of class C GPCRs, which have a pre-defined subtype structure. This sub-typology is known to have a highly overlapping structure from a clustering viewpoint, which has been confirmed in our experiments. The SeCo maps have revealed the FCM algorithm to yield much more stable results that K-Means under multiple random initializations, but they have also been shown to provide limited guidance for the choice of the K and C parameters, due to the highly overlapping nature of the data. In any case, the proposed weighted Cramér's V index provided consistent with and often better results than the standard one in our experiments. This is encouraging, given that the modified index is meant to reflect the nature of the clustering results more faithfully, something that might have revealed lower stabilities.

References

1. Rask-Andersen, M., Sällman-Almén, M., Schiöth, H.B.: Trends in the exploitation of novel drug targets. Nature Reviews Drug Discovery 10, 579–590 (2011)

2. Katritch, V., Cherezov, V., Stevens, R.C.: Structure-function of the G Protein-Coupled Receptor superfamily. Annual Review of Pharmacology and Toxicology 53, 531–556 (2013)

3. Hollenstein, K., et al.: Structure of class B GPCR corticotropin-releasing factor receptor 1. Nature 499, 438–443 (2013)

4. Siu, F.Y., et al.: Structure of the human glucagon class B G-protein-coupled receptor. Nature 499, 444–449 (2013)

5. Wu, H., et al.: Structure of a class C GPCR metabotropic glutamate receptor 1 bound to an allosteric modulator. Science 344(6179), 58–64 (2014)

6. Doré, A.S., et al.: Structure of class C GPCR metabotropic glutamate receptor 5 transmembrane domain. Nature 551, 557–562 (2014)

7. Kniazeff, J., Prézeau, L., Rondard, P., Pin, J.P., Goudet, C.: Dimers and beyond: The functional puzzles of class C GPCRs. Pharmacology & Therapeutics 130, 9–25 (2011)

8. Karchin, R., Karplus, K., Haussler, D.: Classifying G-protein coupled receptors with support vector machines. Bioinformatics 18, 147–159 (2002)

9. Liu, B., Wang, X., Chen, Q., Dong, Q., Lan, X.: Using amino acid physicochemical distance transformation for fast protein remote homology detection. PLoS ONE, 7 46633 (2012)

10. Cárdenas, M.I., Vellido, A., König, C., Alquézar, R., Giraldo, J.: Exploratory visualization of misclassified GPCRs from their transformed unaligned sequences using manifold learning techniques. In: Procs. of the IWBBIO 2014, pp. 623–630 (2014)

11. Cruz-Barbosa, R., Vellido, A., Giraldo, J.: The influence of alignment-free sequence representations on the semi-supervised classification of Class C G Protein-Coupled Receptors. Medical & Biological Engineering & Computing 53(2), 137–149 (2015)

12. Jain, A.K.: Data clustering: 50 years beyond K-means. Pattern Recognition Letters 31(8), 651–666 (2010)

13. Bezdek, J.C.: Pattern Recognition with Fuzzy Objective Function Algorithms. Kluwer Academic Pub. (1981)

14. Lisboa, P.J.G., Etchells, T.A., Jarman, I.H., Chambers, S.J.: Finding reproducible cluster partitions for the k-means algorithm. BMC Bioinformatics 14(S1) 8 (2013)

15. Gutcaits, A., et al.: Classification of G-protein coupled receptors by alignment-independent extraction of principal chemical properties of primary amino acid sequences. Protein Science 11(4), 795–805 (2002)

16. Sandberg, M., Eriksson, L., Jonsson, J., Sjöström, M., Wold, S.: New chemical descriptors relevant for the design of biologically active peptides. A multivariate characterization of 87 amino acids. Journal of Medicinal Chemistry 41(14), 2481–2491 (1998)

17. Caragea, C., Silvescu, A., Mitra, P.: Protein sequence classification using feature hashing. Proteome Science (10 Suppl.), S14 (2012)

18. Ben-Hur, A., Elisseeff, A., Guyon, I.: A stability based method for discovering structure in clustered data. Pacific Symposium on Biocomputing 7, 6–17 (2001)

19. Vroling, B., et al.: GPCRDB: information system for G protein-coupled receptors. Nucleic Acids Research, 39(suppl. 319) (2011)

20. König, C., Cruz-Barbosa, R., Alquézar, R., Vellido, A.: SVM-based classification of class C GPCRs from alignment-free physicochemical transformations of their sequences. In: Petrosino, A., Maddalena, L., Pala, P. (eds.) ICIAP 2013. LNCS, vol. 8158, pp. 336–343. Springer, Heidelberg (2013)

P3D-SQL: Extending Oracle PL/SQL Capabilities Towards 3D Protein Structure Similarity Searching

Dariusz Mrozek, Bożena Małysiak-Mrozek, and Radomir Adamek

Institute of Informatics, Silesian University of Technology,
Akademicka 16, 44-100 Gliwice, Poland
Dariusz.Mrozek@polsl.pl

Abstract. 3D protein structure similarity searching is one of the most popular processes performed in the structural bioinformatics. The process is usually performed through dedicated websites or desktop software tools, which makes a secondary processing of the search results difficult. One of the alternatives is to store macromolecular data in relational databases and perform the similarity searching on the server-side of the client-server architecture. Unfortunately, relational database management systems (DBMSs) are not designed for efficient storage and processing of biological data, such as 3D protein structures. In this paper, we present the P3D-SQL extension to the Oracle PL/SQL language that allows invoking protein similarity searching in SQL queries and perform the process efficiently against a database of 3D protein structures. Availability: P3D-SQL is available from P3D-SQL project home page at: http://zti.polsl.pl/w3/dmrozek/science/p3dsql.htm

Keywords: proteins, 3D protein structure, structural bioinformatics, similarity searching, structural alignment, databases, SQL, relational databases, query language.

1 Introduction

At the current stage of the development of structural bioinformatics, after a few decades of scientific research, nobody questions the necessity of possessing effective methods for 3D protein structure similarity searching, which is one of the core processes performed in this domain. The popularity of the process flows from the fact that it underlies other processes, such as protein classification, functional annotation, and plays a supportive role while verifying results of predictions of 3D protein structures. Scientists have access to a variety of tools by which they can carry out the 3D protein structure similarity searching. Most of the tools are dedicated web pages and there are a few desktop applications. However, performing the process through these tools leaves a very limited control on the process and makes it difficult to reprocess the data or obtained results.

For scientists studying structures and functions of proteins one of the serious alternatives in collecting and processing protein macromolecular data are relational databases [3]. Relational databases collect data in tables describing part of reality, where

F. Ortuño and I. Rojas (Eds.): IWBBIO 2015, Part I, LNCS 9043, pp. 548–556, 2015.

data are arranged in columns and rows. Modern relational databases also provide a declarative query language – SQL that allows retrieving and processing collected data. The SQL language gained a great power in processing regular data hiding details of the processing under a quite simple SELECT statement. This allows to move the burden of processing to the DBMS and leaves more freedom in managing the workload.

1.1   SQL for Manipulating Relational Data

SQL (Structured Query Language) is a query language that allows to retrieve and manage data in relational database management systems (RDBMS). It was initially developed at IBM in the early 1970s and later implemented by Relational Software, Inc. (now Oracle) in its RDBMS [3]. The great power of SQL lies in the fact that it is a declarative language. While writing SQL queries, SQL programmers and database developers are responsible just for specifying what they want to get, where the data are stored, i.e. in which tables of the database, and how to filter them, and this is the role of the database management system to build the execution plan for the query, optimize it, and perform all physical operations that are necessary to generate the final result. For example, a simple SELECT statement, which is used to retrieve and display data, and at the same time one of the most frequently used statement of the SQL, may have the following general form:

```
SELECT A₁, ..., Aₖ
FROM T
WHERE W;
```

In the query, the SELECT clause contains a list of columns $A_1, ..., A_k$ that will be returned and displayed in the final result, the FROM clause indicates the table(s) T to retrieve data from, and the WHERE clause indicates the filtering condition W that can be simple or complex. Other clauses, like GROUP BY for grouping and aggregating data, HAVING for filtering groups of data, ORDER BY for sorting result set, are also possible, but we will omit them in the considerations for the sake of clarity. The simple query retrieves the specified columns from table T and displays only those rows that satisfy the condition W. What is important for our considerations is that table T can be one of the tables existing in the database, can be result of other nested SELECT statement (the result of any SELECT query is a table), or can be a table returned by a table function that is invoked in the FROM clause. The last option was utilized in the P3D-SQL that we present in the paper.

1.2   Related Works

Advantages of a declarative processing of biological data with the use of the SQL were noticed in the last decade, which resulted in the development of various SQL extensions. For example, ODM BLAST [10] is a set of extensions to Oracle RDBMS that allows to align and match protein and DNA nucleotide sequences. The BioSQL [2], which incorporates modules of the BioJava project [8], provides extended capabilities by focusing on bio-molecular sequences and features, their annotation, a reference taxonomy, and ontologies. Several extensions to the SQL language, including PSS-SQL [6,7] and the query language developed by Hammel and Patel [4] and Tata et al. [11], allow searching on the secondary structure of protein sequences. All mentioned projects confirm that for bio-database developers, highly skilled users, also those working in the domain of structural bioinformatics, the SQL language became the communication interface, just like a web site is an interface for common users.

In this paper, we present the P3D-SQL extension to the Oracle PL/SQL language that allows invoking 3D protein structure similarity searching in SQL queries and perform the process against a whole database of 3D protein structures.

2   Algorithms for Protein Structure Similarity Searching

3D protein structure similarity searching refers to the process in which a given protein structure is compared to another protein structure or a set of protein structures collected in a database or any other collection. 3D protein structure similarity searching is usually done by alignment of protein structures. The alignment procedure finds fragments of protein structures that match to each other, i.e. indicate high similarity according to assumed scoring system and given objective function.

P3D-SQL realizes protein structure similarity searches by using two popular algorithms – CE [9] and FATCAT [12]. Both algorithms are publicly available through the Protein Data Bank (PDB) [1] website for those, who want to search for structural neighbors. FATCAT and CE perform structural alignments by combining so called Aligned Fragment Pairs (AFPs). Additionally, by entering twists, FATCAT allows for flexible alignments, eliminating drawbacks of many existing methods that treat proteins as rigid bodies. As a result, for the number of cases, FATCAT enables capturing actual homology that flows from the sequence similarity, which is impossible for other methods.

When developing the P3D-SQL, we integrated the CE and FATCAT algorithms included in BioJava libraries [8] into the Oracle PL/SQL. BioJava provides new, enhanced implementations of CE and FATCAT algorithms – jCE and jFATCAT. jFATCAT is delivered in two variants *rigid* and *flexible*, for rigid and flexible alignments. jCE performs a rigid-body alignment of protein structures, similar to jFATCAT-rigid. jCE also implements a *CE with Circular Permutations* (jCE-CP) variant, which solves the problem of handling of circular permutations. This problem is typical for many alignment algorithms, including CE and FATCAT, that compute sequence order-dependent alignments. For the purpose of P3D-SQL extension to Oracle PL/SQL several classes were added to BioJava and the whole package has been recompiled before it was registered in Oracle DBMS.

3   Implementation in Oracle PL/SQL

Oracle is a commercial object-relational database management system (RDBMS) that allows to store and process huge volumes of data. Oracle provides the PL/SQL language, which is a procedural language extension to SQL. Oracle RDBMS can store and execute stored procedures and functions implemented in the native PL/SQL or other languages, like Java. Although PL/SQL is a powerful language, some tasks can be more easily developed in low-level languages, especially, when they have a form of reusable libraries, like BioJava. Oracle Java Virtual Machine (Oracle JVM) is a part of the Oracle RDBMS that executes the Java code.

P3D-SQL, that we have developed, benefits from the ability of executing the Java code in Oracle. P3D-SQL provides a set of PL/SQL functions that are invoked from the SQL queries. The most important functions are CE_ALIGN and FATCAT_ALIGN. These two functions retrieve macromolecular data of proteins from relational tables, process parameters passed to them, schedule the process of similarity searching, and finally, they call appropriate methods from the BioJava libraries in order to perform the similarity searching of specified protein structures or to generate a detailed alignment for them.

Before a user can use both mentioned functions the BioJava bytecode must be loaded into Oracle Database instance and published to SQL. After loading the code, the executable form of BioJava classes is hold in the Library Cache in the System Global Area (SGA) of the database instance (Fig. 1).

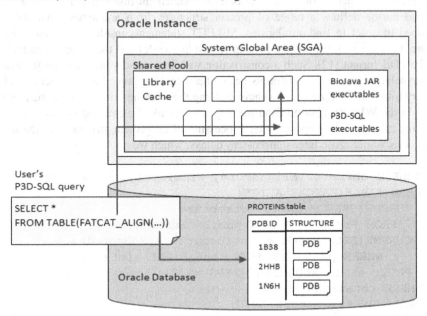

Fig. 1. Oracle database containing PDB repository, and Oracle memory architecture with BioJava executables and P3D-SQL extension

For each BioJava method that is called from PL/SQL we created an appropriate call specification, which exposes the top-level entry point of the method to Oracle Database. These call specifications are then used by the CE_ALIGN and FATCAT_ALIGN PL/SQL functions. Installation scripts for all these operations are available for the public use at the given P3D-SQL home web site (see Availability section at the end of the paper).

Macromolecular data of protein structures are stored as whole PDB files in a dedicated relational table (as CLOBs). The data must be loaded to the table prior to execution of any P3D-SQL statement. Our tests of P3D-SQL queries were performed on the Oracle 11g R2 database storing 93 043 protein structures from the Protein Data Bank (the size of the repository was 80GB).

4   Querying with P3D-SQL

Invocation of protein structure matching methods is nested in SQL queries. Sample query showing the invocation of the jFATCAT flexible method is presented in Listing 1. The query has a specific construction. Both matching methods (FATCAT_ALIGN and CE_ALIGN) need two cursors to be passed as arguments. The first cursor determines the query protein structure that will be compared to structures indicated by the second cursor. Query protein structure should be either present in a table in the database (*Proteins* table in presented example) or inserted to the table before it is used in the query. It is then retrieved by the SELECT statement inside the first cursor. The second cursor defines a range of protein structures from a database that will be scanned in order to find similarities. SELECT statements inside both cursors must return a unique identifier of the protein (e.g. *pdbid* code) and the protein structure(s) in the PDB format [12]. Such a construction with two cursors increases performance of the solution imposing constraints before the matching procedure is executed. This is very important feature of the syntax. Placing the filtering criteria for aligned structures in the WHERE clause of the main query (outside the cursors) would cause the filtering criteria to be imposed after processing of all protein structures in the database. This would cause huge unnecessary delays, which we now avoid.

```
SELECT dbPDBID, alignscore, similarity, totalRMSD
FROM TABLE(FATCAT_ALIGN(
  CURSOR(SELECT pdbid, structure
    FROM Proteins WHERE pdbid='1n6h'),
  CURSOR(SELECT pdbid, structure FROM Proteins
      WHERE pdbid BETWEEN '1n6a' AND '1n6z'),
  PRINT => 1, ALGORITHM_TYPE => 2))
WHERE totalRMSD < 4.0
ORDER BY alignscore DESC;
```

Listing 1. Sample P3D-SQL query displaying proteins (from the range of *pdbid* between 1N6A and 1N6Z) that are similar to the given protein (1N6H) with the total RMSD lower than 4.0 Å. Results are sorted by *score* measure in descending order.

Additional parameter PRINT=>1 allows generation of the HTML document showing a detailed alignment for each pair of query structure and a candidate database structure (Fig. 2). The parameter ALGORITHM_TYPE=>2 determines the use of flexible version of jFATCAT.

```
Align 1n6h.A.pdb 167 with 1n6a.ALL.pdb 235
Twists 2 ini-len 64 ini-rmsd 3,16 opt-equ 75 opt-rmsd 3,63 chain-rmsd 10,84 Score 57,15
P-value 8.26e-01 Afp-num 11285 Identity 4,26% Similarity 19,15%
Block  0 afp  3 score 44,61 rmsd  2,44 gap 25 (0,51%)
Block  1 afp  2 score 30,15 rmsd  2,57 gap  0 (0,00%)
Block  2 afp  3 score 32,85 rmsd  3,70 gap 28 (0,54%)


              .    :    .    :    .    :    .    :    .    :    .    :
Chain 1:  40 FVKGQFHEFQESTIGAAFLTQTVCLDDTTVKFEIWDTAGQERYHSLAPMYYR--GAQAAIVVYDITNEES
             111          1111111111   1111111111              111111111
Chain 2: 117 GVCWIYY-----PDGGSLVGEVNEDGEMTGEKIAYVYPDERTALYGKFIDGEMIEGKLATLMSTEEGRPH


              .    :    .    :    .    :    .    :    .    :    .    :
Chain 1: 108 FARAKNWVKELQRQASPNIVIALSGNKADLANKRAVDFQEAQSYADDNSLLFMETSAKTSMNVNEIFMAI
             2222222222222222    333333333333333333333                      3333
Chain 2: 182 -----------FELMPGNSVYHFDKSTS--SCISTNALLPDPYESERVYVAESLISS--------AGEG


Chain 1: 178 A
             3
Chain 2: 230 L

Note: positions are from PDB; the numbers between alignments are block index
```

Fig. 2. Detailed structural alignment for sample proteins (PDB ID: 1N6H) [14] *histone methyltransferase* (HMTase) from *Homo sapiens* and (PDB ID: 1N6A) [5] *human Rab5a* generated by P3D-SQL query from Listing 1

5   Efficiency Tests

Calculation of protein structure alignments is time-consuming, due to the complex construction of protein structures (thousands of atoms), the number of protein structures available and the exponential grows of the amount, and computational complexity of algorithms used. We tested the efficiency of the P3D-SQL queries for various numbers of structures that were aligned to the given query structure. During our tests the database contained 93 043 protein structures. Tests were performed on Oracle 11gR2 Enterprise Edition working on nodes of the virtualized cluster controlled by the HyperV hypervisor hosted on Microsoft Windows 2008 R2 Datacenter Edition 64-bit. The host server had the following parameters: 2x Intel Xeon CPU E5620 2.40 GHz, RAM 96 GB, 3x HDD 1TB 7200 RPM. Cluster nodes were configured to use 4 CPU cores and 4GB RAM per node, and worked under the Microsoft Windows 2008 R2 Enterprise Edition 64-bit operating system. Results are shown for tested sample molecule (PDB ID: 1ZNJ, chain A), which structure is shown in Fig. 3.

In Fig. 4 we show execution times for P3D-SQL queries that perform structural similarity searches against 50, 100, 500, 1000 protein structures from the database with the use of all variants of the implemented jCE and jFATCAT algorithms.

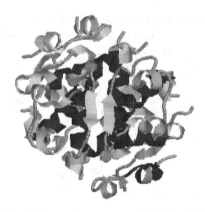

Fig. 3. Crystal structure of *Insulin* from *Homo sapiens* (PDB ID: 1ZNJ) used in efficiency tests

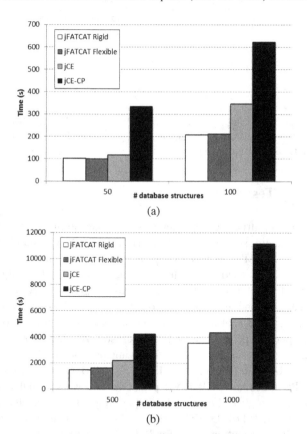

Fig. 4. Execution times for P3D-SQL queries that perform structural similarity searches against: a) 50 and 100, b) 500 and 1000 protein structures from the Oracle database with the use of jFATCAT Rigid, jFATCAT Flexible, jCE and jCE-CP algorithms

While analyzing the results of the efficiency tests we noticed that both variants of jFATCAT algorithm had similar execution times, e.g. the execution of P3D-SQL query against one hundred protein structures took 207 seconds for jFATCAT Rigid and 210 seconds for jFATCAT Flexible. For the same group of protein structures, the execution of the query with the use of jCE algorithm took 344 seconds. The slowest execution was noted for queries that used the jCE-CP variant of jCE algorithm – the execution took 621 seconds against the same set of one hundred proteins. Similar relationships between execution times of tested algorithms (invoked from P3D-SQL queries) were observed in all four groups of proteins. The execution time for a particular algorithm grows with the number of candidate structures that were compared with the given query structure. Average execution time per protein was different for various groups of database structures and various algorithms (varied between 2 seconds and 11 seconds). It depends on the size of the protein structures in the group.

6   Discussion and Concluding Remarks

P3D-SQL provides extension to Oracle PL/SQL, which allows to store, process, compare, align and match protein structures in the relational database management system. It can be especially beneficial for database developers and programmers working in the domain of structural bioinformatics. Implementation of 3D protein structure similarity searching on the database server-side gives several practical advantages. It increases the scalability and maintainability of the search process. Users may operate on their own repository of protein structures and submit similarity searches as simple P3D-SQL queries in batches, which is difficult to achieve through web pages without user interaction. In order to increase efficiency of P3D-SQL queries the database of protein structures can be partitioned on a cluster of many workstations. This solution allows to scale the search process on many nodes of the cluster or provides the possibility to publish the entire solution in one of the public clouds. Finally, P3D-SQL extension allows to carry out the calculations on the server hosting the database, which moves the processing from users to the server, reduces the load on the user's machine,   and reduces the network traffic between the user's application and the database (only results of similarity searches are transferred, not whole protein structures).

P3D-SQL is a successful example of such a DBMS-side processing. It joins a narrow group of bio-oriented SQL extensions, such as BioSQL, ODM BLAST, and PSS-SQL, complementing the group with the unique possibility of finding similarities among 3D protein structures. Thereby, P3D-SQL becomes a declarative, domain-specific query language for protein similarity searching and function identification.

Availability. P3D-SQL is free for scientific and testing purposes. It is available from P3D-SQL project home page at: http://zti.polsl.pl/w3/dmrozek/science/p3dsql.htm together with recompiled BioJava library that is required to work with Oracle.

References

1. Berman, H.M., Westbrook, J., Feng, Z., Gilliland, G., Bhat, T.N., Weissig, H., et al.: The Protein Data Bank. Nucleic Acids Res 28, 235–242 (2000)
2. BioSQL (accessed on January 13, 2015), http://biosql.org/
3. Date, C.: An introduction to database systems, 8th edn. Addison-Wesley, USA (2003)
4. Hammel, L., Patel, J.M.: Searching on the secondary structure of protein sequences. In: 28th International Conference on Very Large Data Bases, VLDB 2002, pp. 634—645, Hong Kong, China (2002)
5. Kwon, T., Chang, J.H., Kwak, E., Lee, C.W., et al.: Mechanism of histone lysine methyl transfer revealed by the structure of SET7/9-AdoMet. EMBO J 22, 292–303 (2003), http://dx.doi.org/10.1093/emboj/cdg025
6. Mrozek, D., Socha, B., Kozielski, S., Małysiak-Mrozek, B.: An efficient and flexible scanning of databases of protein secondary structures with the segment index and multithreaded alignment. J. Intell. Inf. Syst. (in press), http://dx.doi.org/10.1007/s10844-014-0353-0
7. Mrozek, D., Wieczorek, D., Małysiak-Mrozek, B., Kozielski, S.: PSS-SQL: Protein Secondary Structure - Structured Query Language. In: 32th Annual International Conference of the IEEE Engineering in Medicine and Biology Society, EMBS 2010, Buenos Aires, Argentina, vol. 2010, pp. 1073–1076 (2010)
8. Prlic, A., et al.: BioJava: an open-source framework for bioinformatics in 2012. Bioinformatics 28, 2693–2695 (2012)
9. Shindyalov, I., Bourne, P.: Protein structure alignment by incremental combinatorial extension (CE) of the optimal path. Protein Engineering 11(9), 739–747 (1998)
10. Stephens, S.M., Chen, J.Y., Davidson, M.G., Thomas, S., Trute, B.M.: Oracle Database 10g: a platform for BLAST search and Regular Expression pattern matching in life sciences. Nucl. Acids Res. 33(suppl. 1), D675-D679 (2005), http://dx.doi.org/10.1093/nar/gki114
11. Tata, S., Friedman, J.S., Swaroop, A.: Declarative querying for biological sequences. In: 22nd International Conference on Data Engineering, pp. 87–98. IEEE Computer Society, Atlanta (2006)
12. Westbrook, J., Fitzgerald, P.: The PDB format, mmCIF, and other data formats. Methods Biochem Anal 44, 161–179 (2003)
13. Ye, Y., Godzik, A.: Flexible structure alignment by chaining aligned fragment pairs allowing twists. Bioinformatics 19(2), 246–255 (2003)
14. Zhu, G., Liu, J., Terzyan, S., Zhai, P., Li, G., Zhang, X.C.: High resolution crystal structures of human Rab5a and five mutants with substitutions in the catalytically important phosphate-binding loop. J. Biol. Chem. 278, 2452–2460 (2003)

Evaluation of Example-Based Measures for Multi-label Classification Performance

Andrés Felipe Giraldo-Forero, Jorge Alberto Jaramillo-Garzón,
and César Germán Castellanos-Domínguez

Universidad Nacional de Colombia, sede Manizales
Km 7 vía al Magdalena, Manizales, Colombia
Instituto Tecnológico Metropoltano
Calle 54A 30-01, Medellín, Colombia
{afgiraldofo,jajaramillog,cgcastellanosd}@unal.edu.co

Abstract. This work presents an analysis of six *example*-based metrics conventionally used to measure the classification performance in multi-label problems. ROC curves are used for depicting the different trade-offs generated from each measure. The results show that measures diverge when performances decrease, which demonstrates the importance of selecting the right performance measure regarding to the application at hand. Hamming loss proved to be the wrong choice when sensitive classifiers are wanted, since it does not take into account the imbalance between classes. In turn, geometric mean showed a higher affinity to identify true positives. Additionally, the Matthews correlation coefficient and F-measure showed comparable results in most cases.

Keywords: Algorithm adaptation, Example-based, Multi-label learning, ROC curve.

1 Introduction

Multi-label classification is a frequent problem in machine learning, where a given sample can be associated to two or even more different categories. Multi-label framework is contrary to the conventional classification task, where each instance is associated with only one label among considered candidate classes. Real-world applications embracing multi-label classification include: text categorization, where a new article can cover multiple aspects of an event, thus being assigned with a set of multiple topics; in scene classification, one image can be tagged with a set of multiple words indicating the contents of the image; in bioinformatics, one gene sequence can be associated with a set of multiple molecular functions [1]; identification of emotions and others.

For solving multi-label classification tasks, it is commonly assumed that capturing all possible correlations or dependencies among classes should improve classification performance. Related algorithms relying on this hypothesis can be grouped into two categories: *problem transformation* and *algorithm adaptation* methods. The former algorithms break the multi-label learning problem into one or more single-label classification problems, while the latter strategy extends specific learning algorithms to handle multi-label data directly. In either case, measurement of the multi-label classification

F. Ortuño and I. Rojas (Eds.): IWBBIO 2015, Part I, LNCS 9043, pp. 557–564, 2015.
© Springer International Publishing Switzerland 2015

performance poses a challenge. In fact, it is essential to include multiple and contrasting measures because of the additional degrees of freedom that the multi-label setting introduces [2].

Therefore, the evaluation of methods that learn from multi-label data demands different measures than those used in the case of single-label data. Measures devoted to multi-label learning can be categorized in two types [2]: i) *label*-based measures that decompose the evaluation process into separate evaluations for each label, ii) *example*-based measures evaluating the performance based on the average differences of the actual and the predicted sets of labels over all examples of a given evaluation data set.

In the last years, several studies involving the comparison of measures have been published, but they are commonly focused on assessing the performance of given classifiers [3,4] instead of analyzing the predicton performance of each measure. A few studies like [5] have compared the performance of *example*-based measures against label-based yielding to the conclusion that *example*-based measures are more adequate in multi-label problems, since they consider all classes simultaneously. To the best of our knowledge, a critical analysis of the performance of *example*-based measures have been not presented yet in the literature.

In this paper, we use ROC curves constructed from *example*-based measures, for analyzing the performance of several multi-label measures and their applicability in several real-world problems. Moreover, besides of analyzing each measure, the results can also serve as basis for selecting the most adequate classifier depending on the classification problem at hand. Validation on real-word data bases shows that measures diverge when performances decrease, which demonstrates the need for selecting the right performance measure better fitting the application at hand. Hamming loss proved to be the wrong choice when sensitive classifiers are wanted since it does not take into account the imbalance among classes [6]. In turn, geometric mean showed a higher affinity to identify true positives. Additionally, the Matthews correlation coefficient and F-measure show comparable results in most cases.

2   Materials and Experimental Set-up

Algorithm adaptation Multi-label Classification Techniques:

ML-kNN: Provided an unseen instance, its k nearest neighbors in the training set are firstly determined. Then, the label set is inferred using the maximum a posteriori estimator of label sets of fixed neighboring instances [7]

TRaM: This strategy poses the multi-label learning as an optimization problem of estimating label concept compositions. They derive a closed-form solution to this optimization problem and propose an effective algorithm to assign label sets to the unlabeled instances [8].

RANKSVM: This technique makes use of a linear model that minimizes the ranking loss while having a low complexity controlled by the margin of the classifier [1].

LIFT: This method constructs specific features for each label by conducting clustering analysis on its positive and negative instances, and then performs training and testing by querying the clustering results [9].

Multi-label classifier performance measures: There exist two major tasks in supervised learning from multi-label data: multi-label classification (MLC) and label ranking (LR) [2]. Consequently, there are two groups of measures: *bipartition*-based and *ranking*-based. The former measures are calculated based on the comparison of the predicted relevant labels with the ground truth relevant labels and these are used in MLC. This group of evaluation measures is further divided into *example*-based (global), B_{global}, and *label*-based (micro and macro. B_{micro}, and B_{macro}). The *example*-based evaluation measures are based on the average differences of the actual and the predicted sets of labels over all examples of the evaluation dataset. On the other hand, the *label*-based evaluation measures, assess the predictive performance for each label separately and then average the performance over all labels. While LR is concerned with learning a model that outputs an ordering of the class labels according to their relevance to a query instance in this task is used the measures *ranking*-based. Since MLC is the problem to be further discussed, we only focus on the *bipartition*-based evaluation measures defined as follows [3]:

$$B_{global} = B\left(\sum_{i=1}^{N}|t_p|i, \sum_{i=1}^{N}|f_p|i, \sum_{i=1}^{N}|t_n|i, \sum_{i=1}^{N}|f_n|i\right) \tag{1a}$$

$$B_{micro} = B\left(\sum_{q=1}^{Q}|t_p|q, \sum_{q=1}^{Q}|f_p|q, \sum_{q=1}^{Q}|t_n|q, \sum_{q=1}^{Q}|f_n|q\right) \tag{1b}$$

$$B_{macro} = \frac{1}{Q}\sum_{q=1}^{Q}B\left(|t_p|q, |f_p|q, |t_n|q, |f_n|q\right) \tag{1c}$$

$B(\cdot)$ is a binary evaluation measure, t_p, f_p, t_n, and f_n are true positive, false positive, true negative, and false negative, respectively. N is the total number of samples tested and Q is the total number of classes. Notation $|\cdot|$ stands cardinality.

For our analysis, only the *example*-based evaluation measures are considered, because the *label-based* measurements are obtained separately and are unable to highlight dependencies between classes. Also, we use six measures, namely: Hamming loss (H_{loss}), Geometric mean (G_m), F-measure (F_m), Matthews coefficient correlation (M_{cc}), Sensitivity (S_n) and Specificity(S_p).

Table 1. Summary of the testing databases used for validation

Database	Domain	Labels	Instances	Features	Cardinality	Density	Distinct
Emotion	Music	6	593	72	1.869	0.311	27
Enron	Text	14	1679	250	2.787	0.199	306
Image	Image	5	2000	135	1.236	0.247	20
Molecular	Biology	10	2326	438	1.203	0.12	64
Scene	Image	6	2407	294	1.074	0.179	15
Yeast	Biology	13	2417	103	4.223	0.325	189

Experiments: Testing is carried out using six widely known multi-label classification benchmark problems relating to real-word applications and with different average number of labels per sample. Table 1 shows the description of the benchmark problems in terms of the application domain, total number of labels, number of instances, number of features, label cardinality (average number of labels associated to one sample), density (the relationship between the cardinality and the number of labels) and distinct (number of all possible groups formed by combination of labels that are associated with a instance). From the analysis, classes with less than 100 instances are discarded.

The *Scene* dataset scenes with the following six context annotations: beach, sunset, field, fall-foliage, mountain, and urban; the *Image* dataset contains images with labels: desert, mountains, sea, sunset, and trees. The number of images belonging to more than one class (sea+sunset) comprises over 22% of the data set. From the gene function prediction task, we have the *Yeast* and *Molecular* dataset. For *Yeast*, the instances are genes that can be associated to 13 biological functions of the *Saccharomyces cerevisiae*; it should be noted that the original dataset has an additional class. *Molecular* is a dataset for protein classification of *Embryophyte* organisms (land plants), where each protein may be assigned to at least one out of ten terms of the Gene Ontology. From the text analysis task, *Enron* was the chosen database. It has 1001 email messages belonging to 53 possible tags. A principal component analysis (PCA) was conducted and only the 250 principal components were extracted. Finally, in the *Emotions* dataset, each instance is a piece of music that can be labeled with six emotions: sad-lonely, angry-aggressive, amazed-surprised, relaxing-calm, quiet-still, and happy-pleased. All databases were obtained from "sci2s.ugr.es/keel/multilabel.php", excepting to *Molecular* which was obtained from [10] and *Image* available in "lamda.nju.edu.cn/Data.ashx"

The parameters of the different methods used in the classification step are fixed following the recommendations: for ML-kNN, the number of nearest neighbors considered is set to be 10; for Lift, the kernel lineal and a ratio $r=0.1$ are chosen; For RANKSVM, the kernel polynomial of degree eight; for TRAM the number of nearest neighbors considered is set to be 10. The parameters are fit over all the datasets. Validation of the results is obtained by 10-fold cross-validation.

3   Results and Discussion

For each database and classifier, one ROC curve is estimated. The points on the curve are obtained by varying step-wise the decision threshold with a step of 0.01 within the range from 0 to 1. For the classifiers with a different output range, a standardization process is carried out by mean of the following expression $y=x-x_{min}/(x_{max}-x_{min})$, where x_{max} and x_{min} are the highest and lowest values of x, respectively, being y the vector of the resulting values.

The results are presented through four groups of six figures. Each group represents the classification performance of a specific predictor: *Lift, MLkNN, Ranksvm* and *Tram*, over six different datasets: *Emotions, Enron, Image, Molecular, Scene* and *Yeast*. The best values of the performance measures are represented on the ROC curve, which are shown in the Figures 1, 2, 3 and 4. F-measure, geometric mean, hamming loss and Matthews correlation coefficient are denoted by ∘ □ ⋄ and △ respectively.

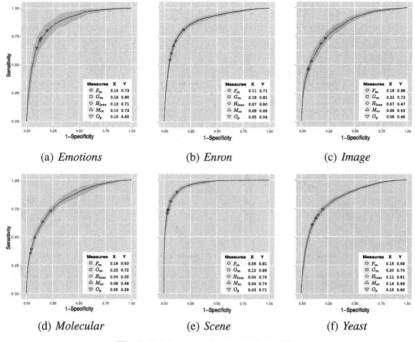

(a) *Emotions*   (b) *Enron*   (c) *Image*

(d) *Molecular*   (e) *Scene*   (f) *Yeast*

Fig. 1. ROC curves for the Lift clasifier

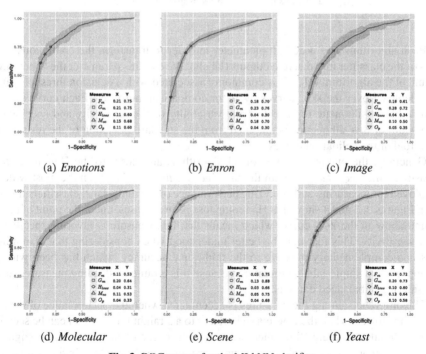

(a) *Emotions*   (b) *Enron*   (c) *Image*

(d) *Molecular*   (e) *Scene*   (f) *Yeast*

Fig. 2. ROC curves for the MLkNN clasifier

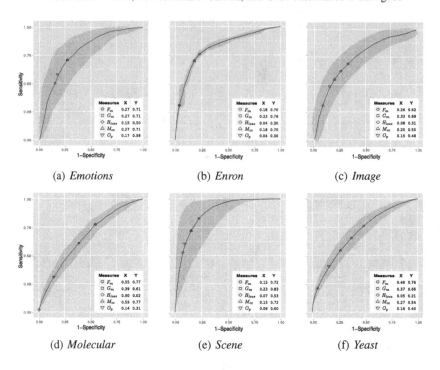

(a) Emotions   (b) Enron   (c) Image

(d) Molecular   (e) Scene   (f) Yeast

Fig. 3. ROC curves for the Ranksvm clasifier

Each algorithm uses a default parameter in order to transform the continuous outputs of the classifier into discrete outputs (labels). We call this parameter the "operating point" of the algorithm (O_p). It is commonly associated with a decision threshold, and it is depicted over the ROC curves by ▽ for comparison purposes. Each curve is surrounded by a silhouette representing the variability of the sensitivity and specificity. The width of this silhouette is computed from the standard deviation of the results for each fold of the validation process.

Generally, the operation points of all algorithms are closer to the Hamming loss measure, which is consistent with the fact that this measure has been the most widely employed for measuring and analyzing performance of the multi-label classifiers. However, Figures 2, 3 and 4 show that Hamming loss is, in all cases, under the point of maximum elbow in the ROC curve, which means that this measure is favoring specificity over sensitivity. This fact may be better understood when considering that Hamming loss is a complementary measure to the traditional accuracy, which has been widely disapproved in single-labeled literature because it is a quite misleading measure when considering unbalanced datasets.

The difference among the measures is accentuated when the performance of the algorithm is lower (when the ROC curve is closer to a straight line), which can be seen in Figures 3d and 3f, where all the measures are far from each other, contrarily to Figures 1a, 1e, 2e and others, where all the measures are more closely together. In this sense, in a perfect classifier all measures should be equally good. As there is no perfect classifier, a

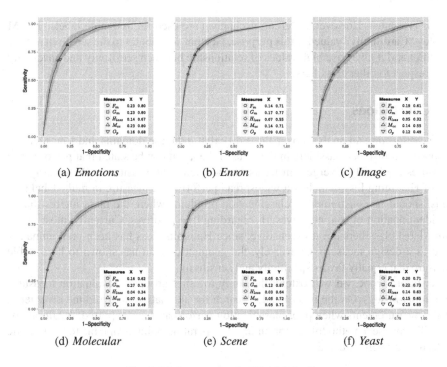

Fig. 4. ROC curves for the TRAM clasifier

carefull choise of the evaluation measure is needed for each specific application. On the other hand, the existence of a close relationship between the Matthews correlation coefficient and F-measure is viewed globally. Additionally, Further analysis is carried out, this analysis shows the influence of the relationship between density (cardinality vs the number of labels) and curvature on the location of the Matthews correlation coefficient and F-measure. Geometric mean, in turn, is equally favoring sensibility and specificity, which locates it on the ideal point in the ROC curve. Turning it, in the measure with the highest relation sensitivity-specificity. So it is ideal for measuring the performance of systems that require a high degree of sensitivity. For example, the detection of a pathology when it threatens human life (cancer) or the preventive identification of fault types of a machine, when a false alarm does not involve a high cost.

In the local analysis, we appreciate a low variability when LIFT and ML-kNN clasifiers are used. However, ML-kNN show greater divergence in performances than LIFT. Classification by RANKSVM show high variability, in the Figures 3a, 3c 3e indicated by width of the tape around the bend. One relationship is evidenced between "Distinct" value of the databaset with variability of the possible. In contrast of the previous figures, In Figures 3b, 3d and 3f, low variabilities are presented and "Distinct" values are increased in the range 50% to 70% approximately.

In some cases operation points may out of the ROC curve, due to way as the outputs of the algorithm are transformed into discrete output, this process is described in section 3. Likewise, In clasification by TRAM a gap between operating point and ROC curve also is presented. However, different mechanism are used for assigning labels, TRAM employ "Cardinality" value for this purpose, fact that contributes to the sensitivity of the classifier and it gives rise high values of relationship between sensitivity and specificity with respect to other classifiers.

4   Conclusions

After performing an analysis over three different database domains and four recently proposed multi-label classifiers in the state of the art, it can be stated that performance measures are highly divergent in terms of specificity and sensitivity. Particularly, the use of Hamming loss, which is a widespread performance measure in multi-label systems, should be constrained to classification problems where the main interest lie on the identification of true negatives, disregarding true positives. Currently, most of the multi-label classifiers are less sensitive, since Hamming loss has been predominantly used as a design measure. Conversely, the geometric mean has shown to be a stable measure that equally favors specificity and sensitivity.

As future work, we will extend the study to other groups of evaluation measures (e.g. *label-based*). On the other hand, a search for alternatives to avoid the divergence of the measurements when poor classification performance are reported. Additionally, we will propose a methodology for tuning the operating point, adaptable to the specific tasks that are being treated.

References

1. Elisseeff, A.: Kernel methods for multi-labelled classification and categorical regression problems. Advances in Neural Information Processing (2002)
2. Tsoumakas, G., Katakis, I., Vlahavas, I.: Mining multi-label data. In: Data Mining and Knowledge Discovery Handbook, pp. 667–685 (2010)
3. Madjarov, G., Kocev, D., Gjorgjevikj, D., Džeroski, S.: An extensive experimental comparison of methods for multi-label learning. Pattern Recognition 45(9), 3084–3104 (2012)
4. Yu, Y., Pedrycz, W., Miao, D.: Multi-label classification by exploiting label correlations. Expert Systems with Applications 41(6), 2989–3004 (2014)
5. Nowak, S., Lukashevich, H., Dunker, P., Rüger, S.: Performance measures for multilabel evaluation: a case study in the area of image classification. In: Proceedings of the international conference on Multimedia information retrieval, pp. 35–44. ACM (2010)
6. Giraldo-Forero, A.F., Jaramillo-Garzón, J.A., Ruiz-Muñoz, J.F., Castellanos-Domínguez, C.G.: Managing imbalanced data sets in multi-label problems: A case study with the SMOTE algorithm. In: Ruiz-Shulcloper, J., Sanniti di Baja, G. (eds.) CIARP 2013, Part I. LNCS, vol. 8258, pp. 334–342. Springer, Heidelberg (2013)
7. Zhang, M., Zhou, Z.: ML-KNN: A lazy learning approach to multi-label learning. Pattern Recognition 40(7), 2038–2048 (2007)
8. Kong, X., Ng, M., Zhou, Z.: Transductive Multi-Label Learning via Label Set Propagation. IEEE Transactions on Knowledge, 1–14 (2011)
9. Zhang, M.l.: LIFT: Multi-Label Learning with Label-Specific Features
10. Jaramillo-Garzón, J.A., Gallardo-Chacón, J.J., Castellanos-Domínguez, C.G., Perera-Lluna, A.: Predictability of gene ontology slim-terms from primary structure information in embryophyta plant proteins. BMC Bioinformatics 14(1), 68 (2013)

The MetaboX Library: Building Metabolic Networks from KEGG Database

Francesco Maiorano[1], Luca Ambrosino[2] and Mario Rosario Guarracino[1]

[1] Laboratory for Genomics, Transcriptomics and Proteomics,
Institute for High-Performance Computing and Networking,
National Research Council of Italy
[2] Dept. of Agricultural Sciences, University of Naples Federico II

Abstract. In many key applications of metabolomics, such as toxicology or nutrigenomics, it is of interest to profile and detect changes in metabolic processes, usually represented in the form of pathways. As an alternative, a broader point of view would enable investigators to better understand the relations between entities that exist in different processes. Therefore, relating a possible perturbation to several known processes represents a new approach to this field of study. We propose to use a network representation of metabolism in terms of reactants, enzymes and metabolites. To model these systems, it is possible to describe both reactions and relations among enzymes and metabolites. In this way, analysis of the impact of changes in some metabolites or enzymes on different processes are easier to understand, detect and predict.

Results. We release the MetaboX library, an open source PHP framework for developing metabolic networks from a set of compounds. This library uses data stored in the *Kyoto Encyclopedia for Genes and Genomes* (KEGG) database using its RESTful *Application Programming Interfaces* (APIs), and methods to enhance manipulation of the information retrieved from the KEGG webservice. The MetaboX library includes methods to extract information about a resource of interest (e.g. metabolite, reaction and/or enzyme) and to build reactants network, bipartite enzyme-metabolite and unipartite enzyme networks. These networks can be exported in different formats for data visualization with standard tools. As a case study, the networks built from a subset of the Glycolysis pathway are described and discussed.

Conclusions. The advantages of using such a library imply the ability to model complex systems with few starting information represented by a collection of metabolites KEGG IDs.

1 Background

In metabolomics applications it is often of interest to model relationships and interactions among compounds and enzymes, such as protein-protein interactions, metabolite pathways or pathway flows. One of the main challenges in metabolomics is to model these interactions in the form of networks [1,2]. Such networks make it easier to understand the topological and functional structure of

F. Ortuño and I. Rojas (Eds.): IWBBIO 2015, Part I, LNCS 9043, pp. 565–576, 2015.

molecules and their interactions. Furthermore, once the network has been built, various statistics can be obtained for characterization and comparison. Indeed, a network represents a convenient way to model objects and their relationships as complex systems. Modeling a network of metabolites provides several ways to further analyze types of interactions, to understand the role of each metabolite in a particular pathway and to detect changes. The problem we address is to build *reaction, unipartite enzymes* and *bipartite enzyme-metabolite* networks, starting from a list of metabolites and information on metabolism gathered from a database. These networks models were used in other research studies in order to identify so-called reporter metabolites [3]. In a metabolite reaction network, two metabolites are connected if they are known to react in the same reaction. In a unipartite enzyme network, two enzymes are connected if they share at least one metabolite in the reactions they catalyze. In a bipartite enzyme-compound network, each enzyme is connected to every metabolite that is present in the reactions it catalyzes.

There are several publicly available databases that store and distribute information on molecular compounds providing different access methods. Among these we cite MetaCyc [4], EcoCyc [5], HMDB [6], Lipid Maps [7], BioCyc [8], Reactome [9], PubChem [10], Chebi [11], ChemSpider [12], Meltin [13], IIMDB [14], and KEGG [15]. It is out of the scope of this paper to describe the characteristics of all these databases, and we focus our attention on the latter.

KEGG is a database containing the largest collection of metabolites, as well as enzymes, reactions and other information[16]. It is possible to query the database with a web interface using one compound, and obtain information on the reactions in which it is involved, the stoichiometric equations, enzymes that catalyze the reaction, and metabolic pathways in which these reactions are involved. Its website graphically displays the stored pathways, but there is no functionality to build networks with a custom topology. To overcome these difficulties, some software exist, and partially solve these problems.

KEGGgraph [17] represents an interface between KEGG pathways and graph objects. It parses KGML (KEGG XML) files into graph models. This tool only provides modeling for KEGG pathways and it is only available for R. MetaboAnalyst [18] provides a web-based analytical pipeline for metabolomic studies. Its web interface can be used to load data as a list, for statistical analysis, as well as pathway analysis. When queried with a list of compounds, it returns information on pathways taken from KEGG but, for reasons related to XML representation of KEGG pathways, information concerning reactions and substrates are partially lost. Therefore, the resulting metabolic network is often disconnected and it does not represent a good model for graph analysis. The source code is not available, neither it provides APIs of any form.

INMEX [19], introduces an integrative meta-analysis of expression data and a web-based tool to support meta-analysis. It provides a web interface to perform complex operations step-by-step. While it supports custom data processing, annotation and visualization, it does not provide any APIs to extend core functionalities and it cannot be deployed in a custom environment. MetaboLyzer [20] implements a workflow for statistical analysis of metabolomics data. It aims at both simplifying analysis for investigators who are new to metabolomics, and providing the flexibility to conduct sophisticated analysis to experienced investigators. It uses KEGG, HMDB, Lipid Maps, and BioCyc for putative ion identification. However, it is specifically suited for analysis of post-processed liquid chromatography-mass spectrometry (LCMS)-based metabolomic data sets. Finally, a tool that implements network construction is MetaboNetworks [21] which builds the networks using main reaction pairs and provide analyses for specific organisms. Although these software give the possibility to model a new network starting from a list of compounds, providing relevant tools for statistical and functional analyses, they miss the cabability to programmatically query metabolomics resources in order to develop novel applications and the software development choices made them suited for very specific environments raising difficulties to use them in production. With the aim to fill that gap, we introduce the MetaboX library which is a framework that enables investigators to extract information that is not visible at a first glance in KEGG. In fact, it is possible to retrieve many information available in the database with just a collection of KEGG IDs and then connect the gathered data in ways KEGG does not provide. The MetaboX library is written in PHP and it is platform independent. The latest version is available under the AGPL license on gitHub repository (https://github.com/Gregmayo/MetaboX-Library). On the other hand, it is possible to programmatically query the database, obtaining such information in the form of flat files. For large lists of input nodes it can be very difficult to manually gather information of interest to build a network, as well as other metadata useful to get a complete understanding of the biological system. With respect to the tools presented above, MetaboX provides a framework to model custom network layouts from a list of input nodes. Based on the nature of input nodes, we provide a set of classes to gather related information and programmatically build a network. The design of the MetaboX library is suited for web production environments, in fact it can be embedded in a custom webservice as it is released under the AGPL license. Therefore, the MetaboX library is an open source framework that aims to get a growing community of researchers and developers to support metabolomic analysis. With the MetaboX library, developers are able to model a network in different ways using the available methods to create a custom network layout that meets their needs. The library design is modular, with the aim to give developers the ability to implement different types of network builders from lists of compounds. Thus, gathering information and detecting interactions programmatically represents a benefit when working with large lists of metabolites. In the network construction process, MetaboX handles the following steps: *(i) connect* to the resource provider database using the PHP `libcurl` library. *(ii) query* a resource provider using methods to retrieve nodes and interactions. As KEGG does not provide a structured query response, we built a translation layer to extract information from flat files. *(iii) extract* requested resource attributes from returned data, parsing and storing them. This task is achieved using regular expressions. *(iv) cache* resource attributes to file using a convenient data structure for serialization and for sharing and processing purposes such as JSON. *(v) build* a consistent data structure with information about requested resources using all previous steps, in order to build a network from collected data. Results consist of a weighted edgelist and a list of network nodes. It also includes specific resource information. In the *connect* step, MetaboX currently supports HTTP, HTTPS, FTP, and ldap protocols. It also supports HTTPS certificates, POST, PUT and FTP uploading, which are natively available in the *curl* library. It is also possible to *query* the KEGG database with a list of resources of interest and then parse the response to firstly separate information about each one and then extract specific data. In the *extract* step downloaded data are parsed to produce new files ready for next steps. We locally *cache* data to load resource information every time it is requested again by a new process. We design the caching system for MetaboX to speed up computation and to produce a sustainable amount of requests to the resource provider system. It is possible to invalidate the cache in order to reload updated data, and to manually delete the cache so that the library can update it. Finally, the *build* step is intended to put together all gathered information and output the resulting network. Every build method connects two nodes differently in each network model, that is *Reactant Graph* implementation of the build method is different from both *Enzyme Unipartite Graph* and *Enzyme Bipartite Graph*. In the following section, we report the implementation of the MetaboX library, detailing how it provides easy access to KEGG database and data manipulation. We explain how to use the library to build the different networks proposed and how to export the result for visualization and analysis with external tools like Cytoscape [22], which we use to render the figures presented in this paper. Then, we provide a case study and discuss the results. Finally, we conclude providing details on future work directions and open problems.

2   Implementation

At the moment of this writing, KEGG only offers a RESTful API interface thus MetaboX is designed to query these in an appropriate manner. KEGG used to expose SOAP APIs to standard software but these were suppressed on 31st december 2012 and the toolbox does not work anymore (https://www.biocatalogue.org/announcements/37). KEGG returns plain text upon web-service calls, thus making it necessary to parse results and arrange them in a data structure. We query KEGG multiple times and store the gathered information to file. The file format we use is JSON which is a lightweight data-interchange format, human-readable and writable. JSON is a text format that is completely language independent but uses conventions that are familiar to C-family programmers. These properties make JSON an ideal data-interchange language. To limit requests to KEGG, the MetaboX library loads previously processed resources from local storage, if they are available. A sample request for a resource in KEGG can be achieved using the following url: http://rest.kegg.jp/<operation>/<argument>. For instance, http://rest.kegg.jp/get/cpd:C01290 can be used to retrieve metabolite C01290.

Network Construction. We deal with compounds, reactions and pathways. To handle such a variety of entities, the MetaboX library defines proper classes. *AbstractResourceLoader* is an abstract class that provides methods needed to load an entity. To model an entity with a new class, this has to extend the abstract class and implement the abstract *load* method. When an entity is instatiated, this method first checks for existing records in the cache. If the requested entity has not been processed previously, a new file is built upon KEGG response. This pattern is used to load metabolites, reactions, pathways and enzymes. We provide several helper methods to extract information about resources from plain text using regular expressions. When the entity has been successfully processed, we serialize it to file for further reference. The attributes that define a metabolite are: *id, formula, exact mass, molecular weight, reaction list, pathway list, enzyme list*. Lists of other entities of interest that are related to a metabolite, such as reactions, pathways and enzymes, are loaded with different API call. For instance, if the *load* of C01290 returns a list of 10 reactions, we use a RESTful url to instantiate each of these reactions. It is possible to query KEGG RESTful APIs using collections of metabolites, reactions, enzymes or pathways. Using this capability, we designed the MetaboX library to construct queries splitting the input collection in chunks of 10 items (as this is the maximum chunk size KEGG supports). For reactions, we collect *id, name, definition, equation, enzymes* and *pathways*. For data manipulation purposes and to conveniently organize reaction information of input metabolites, we process reaction equations and split reactants from products in a data structure. Cache directories can be set in a configuration file. Each resource is stored in a dedicated resource directory and files are named after resource id (e.g. {resource}/{resource_id}.json would result in compound/C00002.json). The configuration file 'config.ini' is divided in sections and it is possible to specify storage directories for entities (e.g. config->directory->compound or config->directory->reaction) as well as KEGG API urls (e.g. config->url->compound or config->url->reaction). This approach is helpful if the entities become available in different urls or from another resource provider. In the MetaboX library we provide an interface to build several networks. *AbstractGraphBuilder* is an abstract class that defines the general structure of the resulting network. Specific network builder classes must implement the abstract *build* method provided in the abstract builder which takes one optional parameter. This is a list of metabolites out of which a sub network has to be built. To create a new type of network, a builder class should provide the construction of a network involving input metabolites and others involved in common reactions, or other entities, such as enzymes. If the optional parameter is specified, the builder method should create a network with set of nodes given by input parameter. When the network-construction process is completed, *getGlobalGraph* and *getSubGraph* methods return a multidimensional array containing the list of nodes, a weighted edgelist, where the weight represents the times a reaction has been found, and the list of connected and not connected nodes, in the case of a sub network.

Reactants Network. A network of reactants $G = (V, E)$ is an undirected graph where each node represents a metabolite and two given nodes A and B in V interact with each other only if there is at least one reaction equation where A and B are involved as reactants. *ReactantsGraph* class builds a network out of a list of metabolites. To achieve this task, we first gather metabolites and reactions data from KEGG (Listing 1.1). We create a list of reactions that involve input metabolites and pass it to the class. In this case, the *build* method cycles through the list of reactions and, for each one, the list of substrates is extracted. We then connect each substrate to one another and when all direct network interactions have been built, we produce a weighted edgelist. Such edgelist represents a network including input compounds and all other compounds involved in processed reactions. We also save a weighted list of interactions that only include input compounds, this resulting in a smaller network which can be seen as a sub network of the global weighted interaction list. As shown in Fig. 1, the sub network is embedded in the global one. A builder class exposes methods to compute results and pass them to the graph writer classes in order to produce a file format that is suitable to the needs of further analysis, such as SIF and XML. The modeling of this class of networks allows to detect which compounds are directly connected, being reagents of the same reactions. It highlights what are the highly connected hubs in a network made up of the collection of metabolites under analysis. This information is useful for planning metabolic engineering strategies. It is clear that if we wish to modify a node of this type of network, it is crucial to know what are other reactants to be considered, so that the change can effectively impact on the metabolic system of the studied biological organism.

```
1  // Retrieve and collect compound information
2  foreach( $compounds as $compound ){ $_cpd_id = trim($compound);
3      $cpd_loader = new MetaboX\Resource\Loader\Compound($_cpd_id, $cpdLoaderConfig);
4      $_compounds[$_cpd_id] = $cpd_loader->load();
5  }
6
7  // Retrieve and collect reactions information
8  foreach($_compounds as $id => $compound){
9      $rn_list = $compound->reactionIdCollection;
10
11     if( $rn_list ){
12         foreach( $rn_list as $rn ){ $_rn_id = trim($rn);
13             $rn_loader = new MetaboX\Resource\Loader\Reaction($_rn_id, $rnLoaderConfig);
14             $_reactions[$_rn_id] = $rn_loader->load();
15     }   }   }
16
17 // Create reactants graph
18 $_graph = new MetaboX\Graph\ReactantsGraph($_reactions);
19 $_graph->build($compounds);
```

Listing 1.1. Loading Metabolites and Reactions metadata from KEGG

```
1  // Retrieve and collect reactions information
2  foreach($_compounds as $id => $compound){
3      $ec_list = $compound->enzymeIdCollection;
4
5      if( $ec_list ){
6          foreach( $ec_list as $ec ){
7              $_ec_id = trim($ec);
8              $ec_loader = new MetaboX\Resource\Loader\Enzyme($_ec_id, $ecLoaderConfig);
9              $_enzymes[$_ec_id] = $ec_loader->load();
10         }
11     }
12 }
```

Listing 1.2. Loading Enzymes metadata from KEGG

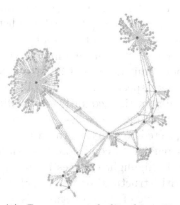

(1) A network of reactants obtained from the 11 input metabolites (darker nodes) selected from glycolysis pathway. This network shows 108 nodes and 151 edges.

(2) Enzyme-metabolite bipartite network: 342 nodes and 393 interactions. Darker nodes represent metabolites.

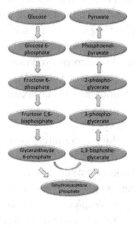

(3) Enzymes unipartite network: 297 nodes and 7705 interactions.

(4) A standard view of glycolysis.

Bipartite Enzyme-Metabolite Network. A network of enzymes and metabolites is a bipartite undirected graph $Z = (U, V, E)$ with set of nodes U representing metabolites and V representing enzymes. A metabolite node is connected to all the enzymes nodes that catalyze a reaction involving that metabolite, and an enzyme node is connected to all the metabolites that take part in the corresponding reaction. That is, if an enzyme F in V catalyzes a reaction where a metabolite M in U is a substrate, then an interaction between F and M exists in the network. We achieve this task using *EnzymeBipartiteGraph* class which parameters are: a metabolite collection, an enzyme collection and a reaction collection. We cycle through the list of metabolites and select the related enzymes. We search current metabolite M in the substrates of the reaction catalyzed by enzyme F. If we have a match, we connect nodes F and M. An enzymes network, both unipartite and bipartite, provides a kind of visualization that highlights some aspects that are not observable by a reactants network. If we are analyzing different time conditions with different concentration levels of some compounds, for instance, this class of networks would quickly identify which nodes are most affected, restricting the area of interest to the enzyme directly susceptible to a particular condition. Therefore, the construction of this type of graphs can help highlight changes in the enzymatic expression levels or to detect enzymes with structural or functional defects due to particular conditions of stress. An example of such a network is shown in Fig. 2.

Unipartite Enzymes Network. A unipartite network of enzymes is an undirected graph $G = (V, E)$ where nodes represent enzymes and two enzymes sharing a common compound in the corresponding reactions are connected to each other. The class used to model such a network is *EnzymeUnipartiteGraph*. This builder class is instantiated with a list of enzymes and a list of reactions. These lists are created collecting all reactions and enzymes that involve input metabolites (Listing 1.2). For each enzyme in the collection, we load data of the reaction catalyzed and select all substrates. We cycle through the enzyme collection comparing the current enzyme substrates to all others. Given two enzymes T and S in V, we connect them if the intersection between substrates in T and substrates in S is not empty. An example of such a network is shown in Fig. 3.

Data Export. In the MetaboX library, there are two classes that can be used to export the constructed network in other formats. As for the other components, a *AbstractGraphWriter* is an abstract class that exposes an abstract *write* method. The class constructor takes one parameter, that is a multidimensional array containing the node list and the weighted edgelist of the network. This can be set using *getGlobalGraph* or *getSubGraph* to export respectively a network or a sub network. The *write* method has two parameters: the name of the file to be written and the data that needs to be exported. If the output needs to be prepared or modified somehow, it is possible to call *prepareOutput* within the *write* method. This is the case of *CytoscapeGraphWriter* class where interactions are converted to string and then written to file. To work with the D3JS visualization library as well as D3py or NetworkX [25], the MetaboX library provides several classes to export a network in one of the formats accepted by other analysis tools. For instance, a *D3JSGraphWriter* converts the network to JSON and writes it to file.

3   Results and Discussion

In order to test MetaboX, we build a network starting from a set of eleven compounds, listed below: Glucose (C00031), Glucose 6-phosphate (C00668), Fructose

6-phosphate (C05345), Fructose 1,6-bisphosphate (C05378), Dihydroxyacetone phosphate (C00111), Glyceraldehyde 3-phosphate (C00118), 1,3-bisphospho-glycerate (C00236), 3-phosphoglycerate (C00197), 2-phosphoglycerate (C00631), Phosphoenolpyruvate (C00074), Pyruvate (C00022). These compounds belong to glycolysis, the metabolic pathway that leads, starting from glucose, to the production of pyruvic acid through several reactions. Fig. 1 shows the network of all reactions involving input metabolites. This results in a network that includes input metabolites as well as others related to them. Here we find 151 interactions between 108 metabolites, including all the input metabolites. An interaction in this graph means to be reactants of the same reactions, therefore each node is directly affected by a decrease in the concentration of one of its neighbors, not by an increase. In this second case, in fact, there would be an excess of one of the two reactants. For instance, this information is useful if we want to plan a change in the levels of a compound starting from other compounds already known. Glucose and Pyruvic acid, as well as representing the start and the end of the glycolysis, are also hubs, namely highly connected nodes, in the network shown in Fig. 1.The MetaboX library also outputs a reactants subnetwork which only contains the nodes and the edges of the input metabolites. In this case study, the subnetwork results in just few nodes and edges (3 nodes and 2 edges) because the compounds choosen from glycolysis represent a subset of the glycolysis pathway (map00010) shown in Fig. 4. Therefore, each metabolite is both a reactant and a product of its neighbors in the pathway. The networks built with the MetaboX library are not pathway representation of the input metabolites. A classic pathway view in the form of substrate-product flow is provided by many databases of metabolic data (KEGG, MetaCyc, MetaboAnalyst), and it is not implemented in the MetaboX library. Instead, we mainly focus our efforts to highlight other information, like the relationships between metabolites of the same reaction. In this case study, compounds do not follow the common representation of glycolysis. Indeed, they do not show the same connections in the reactants network built with the MetaboX library. Another example we provide is the construction of a unipartite and bipartite enzymes network. In the first case, in Fig. 3, each node of the network is an enzyme, and two of them are connected if they share at least one metabolite in the reactions they catalyze, with the constraint that the substrate of one enzyme is the product of another enzyme. We add this constraint to allow the user to easily have a view of the substrate-product flow, looking at enzymes instead of metabolites. In this way, we are able to build a unipartite enzymes network with 297 nodes and 7705 interactions. Finally, we build a bipartite enzyme-metabolite network, in Fig. 2. As already mentioned, this network consists of two sets of nodes: metabolites and enzymes. Nodes are connected alternatively, that is a metabolite to an enzyme or vice versa. Connections between two metabolites or two enzymes are not possible. In the resulting network, we are able to find 342 nodes (11 metabolites and 331 enzymes) and 393 interactions. Looking at this network, we can easily identify the hubs (namely glucose, pyruvate, glyceraldehyde 3-phosphate and phosphoenolpyruvate) and all the enzymes related to them. To change something in these hubs, the variables (enzymes) to take into account can be many, and the design of a subsequent experiment in metabolic engineering would be too complicated. A better solution is represented by focusing on compounds that show only few connections, in order to limit the analysis to few enzymes and to decrease the complexity of further analyses. Anyhow, we strongly believe that this type of network significantly simplifies the work of those who analyze metabolic pathways to understand metabolic disorders, to connect disease to enzyme defects, to design successful metabolic engineering strategies. MetaboNetworks provides analyses for specific organisms whereas in the MetaboX library we do not account for this feature, considering all available reactions in the first place. In such a way, a user can plan pipelines or methodologies from a compound-wise point of view, and not an organism-wise point of view. For instance, in soil remediation from copper, MetaboX provides access to all known reactions containing copper. A user can then find which organisms use copper within their metabolic pathways. In the enzyme API call, KEGG provides the *GENES* attribute containing a list of genes where that particular enzyme is involved. Each one of those genes is specific for an organism, enabling the user to filter organism specific reactions. In conclusion, if we start without any initial information about compounds concentration and we look at the topology of the network, the highly connected nodes are fundamental in the metabolism, and changes in these nodes would have probably led to the death of the organism. On the contrary, if we start from experimental data, it might be useful to correlate increases or decreases of the concentration of some compounds to a particular disease or to a particular disorder. Therefore highlighting compounds within a network should be useful in designing any strategy aimed at clarifying ways of occurrence of the disease, extracting from that network information like mumber of edges, enzymes, reactions, etc. The MetaboX library is a suitable tool created to solve both issues: a first preliminary view and a second in-depth analysis.

4   Conclusions

In this paper we describe the MetaboX library, a framework to build metabolic networks using information gathered from the KEGG database. The advantages of using such a library are: *(i)* the possibility to gather information from KEGG using a collection of KEGG IDs. *(ii)* the possibility to build a representation of the metabolic processes that can highlight how changes in metabolites or enzymes might affect other processes. *(iii)* the possibility to export the networks to other formats for visualization and analysis with standard software, such as Cytoscape, NetworkX, D3JS or D3py. Because of its extensibility, the MetaboX library may add support to other fields as in the construction of protein-protein interaction (PPI) networks for performing different topological and functional analyses [23]. In this case, the MetaboX library should use a resource provider that stores information about interactions between proteins such as STRING [24]. An organism filter can be implemented in the MetaboX library, as used in MetaboNetworks, in order to select specific reactions and build a sub network that enables an organism-wise network. This can be achieved building an organism list for each enzyme of each processed reaction.

The library has been built with the possibility to extend data information gathering, such as downloading these from databases other than KEGG or merge information collected from multiple databases. Indeed, in order to make the network construction as complete as possible, the MetaboX library will implement a merge process among different resource providers. As KEGG information is limited, it makes sense to gather data about metabolites, reactions, pathways and enzymes from other databases like MetaCyc.

The MetaboX library is the starting point of a three layer project involving a web service and a web application. "Web services provide a standard means of interoperating between different software applications, running on a variety of platforms and/or frameworks.". Following this vision, MetaboX library offers a framework to work with metabolic networks. On top of that, we will develop a web service to expose core functionalities to the web. The MetaboX webservice will work in a RESTful fashion, providing APIs to retrieve resources information, network construction options and job submission. Moreover, we will implement alternative ways to make data persistent, such as database storage.

Acknowledgements. This work has been partially funded by MIUR projects PON02_00619 and Italian Flagship project *Interomics*. We wish to thank LabGTP (Laboratory for Genomics, Transcriptomics and Proteomics) researchers from ICAR CNR for helpful discussion during the early stages of this project and for testing the MetaboX library giving an important feedback for improvements.

References

1. H.K., et al.: Metabolic network modeling and simulation for drug targeting and discovery. Biotechnol J., 30–42 (2011)
2. Cloots, L., et al.: Network-based functional modeling of genomics, transcriptomics and metabolism in bacteria. Curr Opin Microbiol, 599–607 (2011)
3. Raosaheb, K., et al.: Uncovering transcriptional regulation of metabolism by using metabolic network topology. PNAS, 2685–2689 (2005)
4. Krieger, C.J., et al.: MetaCyc: a multiorganism database of metabolic pathways and enzymes. Oxford Journals Nucl. Acids Res, 511–516 (2004)
5. Keseler, I.M., et al.: EcoCyc: fusing model organism databases with systems biology, Oxford Journals Nucl. Oxford Journals Nucl. Acids Res, D605-D612 (2013)
6. Wishart, D.S., et al.: HMDB: the Human Metabolome Database. Nucleic Acids Res, D521-D526 (2007)
7. Sud, M., et al.: LMSD: LIPID MAPS structure database. Oxford Journals Nucl. Acids Res, 527–532 (2007)
8. Karp, P.D., et al.: Expansion of the BioCyc collection of pathway/genome databases to 160 genomes. Nucl. Acids Res, 6083–6089 (2005)
9. Matthews, L., et al.: Reactome knowledgebase of human biological pathways and processes. Nucl. Acids Res, D619-D622 (2009)

10. Wang, Y., et al.: PubChem's BioAssay Database, Nucl. Nucl. Acids Res, D400-D412 (2012)
11. Hastings, J., et al.: The ChEBI reference database and ontology for biologically relevant chemistry: enhancements for 2013. Oxford Journals Nucl. Acids Res, 456–463 (2013)
12. Pence, H.E., et al.: ChemSpider: An Online Chemical Information Resource. J. Chem. Educ., 1123–1124 (2010)
13. Smith, C.A., et al.: METLIN: A Metabolite Mass Spectral Database Therapeutic Drug Monitoring. In: Proc. of the 9th ICTDM, pp. 747–751 (2005)
14. Menikarachchi, L.C., et al.: Silico Enzymatic Synthesis of a 400.000 Compound Biochemical Database for Nontargeted Metabolomics. J. Chem. Inf. Model., 2483–2492 (2013)
15. Kanehisa, M., et al.: KEGG for integration and interpretation of large-scale molecular data sets. Nucl. Acid Res. 14, D109-D114 (2011)
16. Altman, T., et al.: A systematic comparison of the MetaCyc and KEGG pathway databases. BMC Bioinformatics (2013)
17. Jitao, D.Z., et al.: KEGGgraph: a graph approach to KEGG PATHWAY in R and bioconductor. Bioinformatics, 1470–1471 (2009)
18. Xia, J., et al.: MetaboAnalyst 2.0 - a comprehensive server for metabolomic data analysis. Nucl. Acids Res., 1–7 (2012)
19. Xia, J., et al.: INMEXa web-based tool for integrative meta-analysis of expression data. Nucl. Acids Res., W63-70 (2013)
20. Mak, T.D., et al.: MetaboLyzer: A Novel Statistical Workflow for Analyzing Post-processed LCMS Metabolomics Data. Anal. Chem. Article ASAP, 506–513 (2013)
21. Posma, J.M., et al.: MetaboNetworks, an interactive Matlab-based toolbox for creating, customizing and exploring sub-networks from KEGG. Bioinformatics (2013)
22. Cline, M.S., et al.: Integration of biological networks and gene expression data using Cytoscape. Nat Protoc., 2366–2382 (2007)
23. Sharma, A., et al.: Rigidity and flexibility in protein-protein interaction networks: a case study on neuromuscular disorders, arXiv, arXiv:1402.2304v2 (2014)
24. Franceschini, A., Szklarczyk, D., et al.: STRING v9.1: protein-protein interaction networks, with increased coverage and integration. Nucleic Acids Res 41, D808-D815 (2013)
25. Hagberg, A.A., et al.: Exploring Network Structure, Dynamics, and Function using NetworkX. Proc. SciPy, 11-16 (2008)

Pseudoknots Prediction on RNA Secondary Structure Using Term Rewriting

Linkon Chowdhury[1] and Mohammad Ibrahim Khan[2]

[1] Department of Computer Science and Engineering
[2] Chittagong University of Engineering and Technology
Cuet Road, Zip Code-4349, Bangladesh
linkoncu@gmail.com, mohammad_khancuet@yahoo.com

Abstract. The presences of Pseudoknots generate computational complexities during RNA (Ribonucleic Acid) secondary structure analysis. It is a well known NP hard problem in computational system. It is very essential to have an automated algorithm based system to predict the Pseudoknots from billions of data set. RNA plays a vital role in meditation of cellular information transfer from genes to functional proteins. Pseudoknots are seldom repeated forms that produce misleading computational cost and memory. Memory reducing under bloom filter proposes a memory efficient algorithm for prediction Pseudoknot of RNA secondary structure. RNA Pseudoknot structure prediction based on bloom filter rather than dynamic programming and context free grammar. At first, Structure Rewriting (SR) technique is used to represent secondary structure. Secondary structure is represented in dot bracket representation. Represented secondary structure is separated into two portions to reduce structural complexity. Dot bracket is placed into bloom filter for finding Pseudoknot. In bloom filter, hashing table is used to occupy the RNA based nucleotide. Our proposed algorithm experiences on 105 Pseudoknots in pseudobase and achieves accuracy 66.159% to determine structure.

Keywords: Pseudoknot, Structure Rewriting (SR), Dot Bracket, Bloom Filter, Hashing Table.

1 Introduction

Deoxyribonucleic acid (DNA) contains the genetic information of all living organisms in it and transcribes them into Ribonucleic acid (RNA).RNA is translated into amino acid sequences from proteins.RNA forms the secondary structure by the complementary base pair A(Adenine),C(cytosine),G(guanine) and U(Uracil).Occurring widely stable base pair is Watson-Crick(A=C and G=U) and wobble (A=U).RNA secondary structure can be decomposed into different types of structure motifs[1]: stem, hairpin loop, bulge, internal loop, start sequence, multi-branched loop, Pseudoknot and external loop(Fig.1) which are described in section 3.1.

F. Ortuño and I. Rojas (Eds.): IWBBIO 2015, Part I, LNCS 9043, pp. 577–588, 2015.

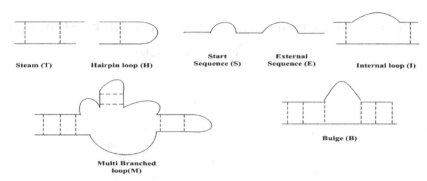

Fig. 1. RNA secondary Structure Motif in different patterns (Motif, Hairpin loop, Interior loop, Bulge, Multi branched loop)

Like many other proteins, RNA structure represents the distance base pairing within a sequence. This bonding within a nucleotide is the cause of structural folding known as Pseudoknot. These pseudoknots have drawn a considerable attention as they give a 3-D structure of molecule and a structure that will determine most of the cases for biological functionalities [8,9].A properly folded Pseudoknot is catalytically active in telomerase ribonucleic protein complexes [2].A noncoding RNA(ncRNA) are not translated into protein and involved into many cellular functionalities. Recently some databases have reveled over 212,000 ncRNAs [3] and more 1200 ncRNA families [4]. The huge scale of ncRNAs discovery and family indicates the importance of RNA identification. A number of researchers attempted to identify ncRNAs considering them as a stability structure of genome [5]. However, it is not efficient in random sequence with high GC-composition.

Here we have designed a technique for secondary structure and predict Pseudoknot considering time and space. At first, RNAs base pair sequence is converted into secondary grammatical structure using dot brackets technique. Then term rewriting technique is used to retrieve motif-motif interaction. Secondly, secondary grammatical structure motif is separated into two sub regions which can reduce the time and space cost. Finally, sub region motif structure is inserted into bloom filter in which hashing technique is used for randomly memory allocation.

2   Literature Review

NcRNA structure alignment can be classified into three types of interactions [1, 10]. Firstly align and fold concurrently. Most accurate algorithm is developed by David [11] for first type. But it needs expensively time complexity O (L^{3N}) and space complexity O (L^{2N}), where L and N are length and input sequence respectively. Variants of the David's algorithm have been proposed to reduce the computational time of multiple alignments, rather than Consan [12] and Stemloc [13].Second type of method   builds in first place a sequence then fold the alignment [14].Their Alignment structures are generated from pre-align structure method such as MULTIZ [16] and CastalW [15].This tool affects the alignment quality. In homologous NcRNA, only share alignment is similar and it is difficult to build meaningful sequence alignment. The third type of method folds the input sequence and conducts the alignment.

Different types of tools are used for different alignment algorithms. "Folds and align" tool is used for secondary structure alignment.

Secondary structure is encoded by using base pair probability matrix which is derived from McCaskill approach [17]. NcRNA secondary structure alignments are converted into base pair alignment matrix. But base pair alignment matrix is extremely resource insist .For a n length of base pair, Pcomp [7] takes memory $O(n^4)$ and $O(n^6)$ for operations. FLODALIGNM [18] and LocARNA[19] based on pruning technique and reduce time complexity applying various restriction. But these algorithms are more expensive than sequence alignment.Tree profiler algorithm such as RNAfoster[20] is used to represent the secondary structure alignment. Tree alignment is used pair wise and multiple alignment computation. Efficiency of tree alignment depends on depth and degree d of a tree node. For n structure of average size , pair wise algorithm time complexity $O(s^2 d^2)$ and memory $O(s^2 d)$.Tree alignment is more efficient than alignment matrix algorithm. CARNAC [21] is used stem graph theory to represent structure alignment. It can handle only 15 input sequences which is not sufficient for practical usage.

3   Approaches: System Architecture

Our research work is based on memory mapping technique based on bloom filter (Fig. 2) . Our model has four steps. In first step, RNA secondary structure prediction from RNA sequence. The Input sequences are data set of NcRNA which not translated into protein. Different types of NcRNA families make it possible to take data like input data set of NcRNA data stream. Data input file is read as fasta format. In Step 2, secondary structure is parsed by term rewriting for retrieving motif. Motif parsing is important for prediction Pseudoknot. In step 3, parsed motifs are separated into two substructures. Every secondary structure with Pseudoknot is separated into two simple Pseudoknots structure. Structural problem is divided into two alignment substructures between base pair of shorter sequence. In step 4, divided substructures are placed into bloom filter. Bloom filter [22] contains compact data structure for probabilistic representation to support the membership queries.

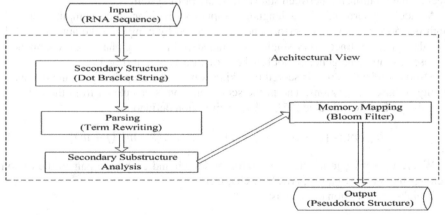

Fig. 2. Model for RNA secondary structure with Pseudoknot. Architectural view with four different phases (Secondary structure, Parsing, Secondary substructure, Memory Mapping)

3.1   RNA Secondary Structure Representation

The secondary structure includes the pure sequence that pairing among the base. A secondary structure decomposes into different loops and steam structures. The steams are the region of pair bases while loops are formed by the unpaired parts. A loop that is adjacent to only one steam is called a hairpin loop and a loop that is adjacent to two steams is known as internal loop. Loop is adjacent to more than two steams known as multi loop. The unpair region at the beginning and end of the chain is known as dangling end. RNA structure can be illustrated in different types of ways. The simplest type of RNA secondary structure prediction is Dot Bracket presentation. This presentation is used because it is easy and simple for sorting purpose. In the dot bracket presentation, base pairs are represented by round bracket '(' and ')' while unpair bases are represented by dots ('.').Pseudoknot which include cross pairing in stream regions are represented by square bracket ('['and '[').RNA structure is represented in dot bracket format (Fig.3).

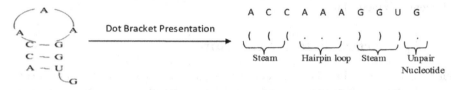

Fig. 3. RNA secondary structure presentation including different types of Motifs using dot bracket presentation

3.2   Parsing

Parsing is a term rewriting technique for logical secondary structure of RNA representation. Term rewriting is a logical formulation, which has a state and behavior of system idea. States of the system are represented by an element of algebraic data type and behaviors of the systems are represented by some rewrite rules. Rewrite rules declare the relationship between state and behavior of system.

Maude is a term rewriting language, supports rewriting logic computation with high performance. Its high performance is achieved by compiling the automata [23]. Maude can easily trace every single rewriting step. In parsing phase, RNA sequences are used as input and generate dot bracket structure as output (Fig.4).

Motif-motif interaction is scored by taking weight of base pair region and distances plenty of base pair regions. The in the score function is calculated from the physical chemistry of nucleotide acid [25]. A base pair region defined as:

$$Region = \{(i,j),(i-1,j+1), \ldots \ldots \ldots \ldots \ldots (i-m),(j+m)\} \quad (1)$$

Where i$\in$ motif1, j$\in$ motif2, i<j, m=region length and each base pair in the region belongs the set of {CG, GC, AU, UA, GU, UG}

Weight as the region defined as:

$$W_{region} = \sum_{(i,j)\in region} weight\,(i,j) \quad (2)$$

Where weight (CG/GC): weight (AU/UA): weight (GU/UG)=3:2:1.The weight are approximate value that indicates the trends in base pair energy.

The distance of the base pair is defined as:

$$Dis_{region}=i\text{-}j \tag{3}$$

Where (i,j) is the closest base pair of region and i<j. Score function can be defined as:

$$Score = Max(\alpha \times W_{region}) + (1 - \alpha) \times Dis_{region} \tag{4}$$

Where $\alpha \in[0,1]$ is the heuristic parameter and adjust the significance of W_{region} and Dis_{region} in score function. In our experiment α is set 0.7.

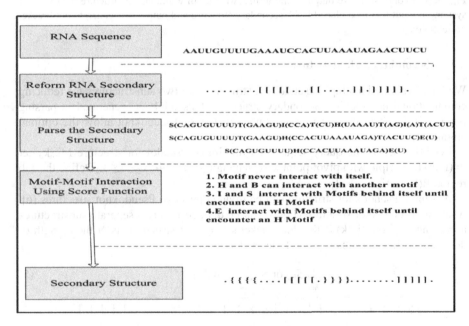

Fig. 4. Parsing for RNA Secondary Structure. In Parsing technique RNA sequences as Input and dot bracket presents as output.

The motifs with the highest score are considered as a Pseudoknot structures. RNA secondary structure is parsed in step 2.Secondary structure of RNA with Pseudoknot is parsed in this step. The Pseudoknot of final structure is labeled with "{" and "}".This structures is used for next substructures process. Our system generates the same output (Fig.5) like our described method.

.{ { { { ....[[[[[.} } } }........]]]]].

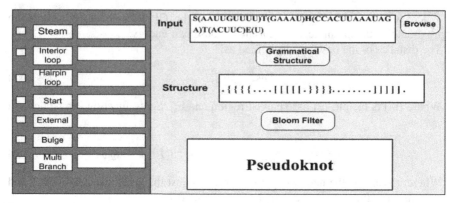

Fig. 5. Secondary structure output using term rewriting. In grammatical structure phase of our system, in input text area placed RNA secondary structure and output text generate dot bracket presentation.

3.3   Secondary Sub Structure

We have separated a simple Pseudoknot structure into two substructures for reducing computational complexity. Secondary structure of Pseudoknot, generated in parsing process is divided into two regions. Substructures of Pseudoknot allow the optimal alignment.

Let $S[i_0,....,k_0]$ is the query sequence with simple Pseudoknot structure Mi_0,k_0. Let $v=(i,j,k)$ be a triple with pivot point x_1, x_2 having $i_0 \leq i < x_1 \leq j < x_2 \leq k \leq k_0$. Define the sub region $R(S,v)=[i_0.....i] \vee [j.......k]$

A Simple Pseudoknot structure can be divided into two Pseudoknot structures [6]. A simple Pseudoknot structure (Fig.6) can be divided into two separate substructures according to triple (i,j,k). If the dot-bracket's length of structures is N then length of dot-brackets in each substructure, L defined as

$$L(S_{\text{left}} \text{ or } S_{\text{right}}) = \lfloor N/2 \rfloor \tag{5}$$

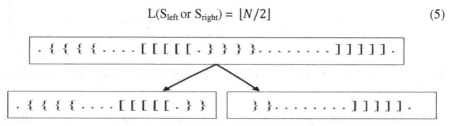

Fig. 6. N length of pseudoknot substructure divided into left and right substructures

3.4   Bloom Filter

A bloom filter [26] is a data structure used for testing the set of membership for every large set. It allows a small percentage of false positives for exchange for speed and space. For a given set S, a bloom filter uses m bits array and k hash functions to be applied to objects same time as the elements in S. Each hash function produces an integer value between 1 and m as an index. In filter setup phase, k hash functions are applied to each element of S and the bit indexed by each resulting value set to 1 in the array. We used feed forward bloom filter for RNA secondary matching (Fig.7).

When testing membership, the k hash functions are also applied for tested elements and bit indexed by the resulting values are checked. If they are all 1, the element is a member of set S. Otherwise, if one bit is 0, the element is not a part of the set S. The number of hash function is used and the size of the bit array determines the false positive rate of the bloom filter. For a set of n elements, asymptotic notation for false positive probability test is $(1-e^{-km/n})^k$.

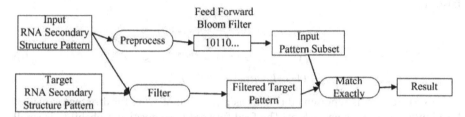

Fig. 7. Diagram of RNA secondary structure matching using Feed Forward Bloom Filter

Steps of Pseudoknot for the pattern matching using feed forward bloom filter.
Preprocessing:
  1. Find $F \epsilon P$, subset of most frequent patterns
  2. Choose l, the minimum size for the patterns for the bloom filter.
  3. Compute two substructures l1 and l_2, subset of l. $l=l_1 U l_2$.
  4. Build Feed Forward Bloom Filter (FFBF) from $P\backslash(F Ul_1 Ul_2)$
Filtering:
  1. $(T',P')\leftarrow FFBF(T)$
     with
$T' \epsilon T$ and $P' \epsilon P\backslash(FU l_1 U l_2))$
Exact matching:
  1. $T_1 \leftarrow exact\_match|FU l_1 U l_2)|(T)$
  2. $T_2 \leftarrow exact\_match|P'|(T')$
  3. Output $T_1 U T_2$
Assume for now that true positive (items that are actually match with target pattern) are rare. Let m be the number of bits in the first array, k are the number of hashes are used for every item search/insert in it and m' and k' are the corresponding parameters for second bloom filter. If $n=|S|$ is the number of symbol inserted, then the false positive probability in the first bloom filter.

$$P_{FP}^{Filter1} = function(n, m, k) \tag{6}$$

Then the false positive for the second filter will be

$$P_{FP}^{Filter2} = function(P_{FP}^{Filter1} * number\ of\ quries * m', k') \tag{7}$$

Assuming perfectly uniform hash function, after inserting n items in first bloom filter, the probability that any particular bit still 0 in the first array is $P_0 = (1 - \frac{1}{m})^{kn}$. The probability of searching false positive in the bloom filter is

$$P_{FP} = P_{FP}^{filter1} = (1 - p_0)^k = (1 - \left(1 - \frac{1}{m}\right)^{kn})^k \tag{8}$$

We begin ignore the true positive and consider $P_{hit} = P_{FP} + P_{TP} \approx P_{FP}$. The probability that a bit is 0 in second array, after w queries, by using the expectation for the number of false positive queries in the first filter (w, P_{FP}):

$$P_0 \approx (1 - \frac{1}{m})^{kw} \times P_{FP} \tag{9}$$

The probability that the item in S will be selected to be a part of S'(as a feed forward false positive):

$$P_{feed-fwdFP} = P_{FP}^{filter2} = (1 - P_0')^{k'} \approx$$
$$(1 - (1 - \frac{1}{m'})^{k'w'(1-e^{-kn/m})^k})^{k'} \approx \tag{10}$$
$$(1 - e^{-k'w/m(1-e-knm)k})k'$$

This expression is represented (Fig.8) as a function of w/m, for different values of k and m/n, where k'=k and m'=m. The feed forward false positive probability is small and it can be radically reduced by small adjustment to the feed forward bloom filter parameters.

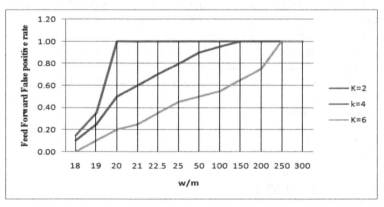

Fig. 8. The feed forward false positive rate as w/m and varying k. We consider k=k' and m=m'.

Factoring of True Positives:

It is not number of true positive tests (against the first bloom filter) that affect the feed forward false positive. For example, if only one item generates all the true positives, then at least k' will be set second bit array bit array. Assume that there are n' items from S that will generate true positive. Then the probability bit is 1 in the second bit array because of true positive test is:

$$P_{TP} = 1 - (1 - \frac{1}{m})^{kn} \tag{11}$$

The probability that a bit set in second array, after w test they are not true positives and any number of tests they are true positive is:

$$P_1 = P_{TP} + (1 - P_{TP})(1 - P_0) \tag{12}$$

where P_0 has the expression presented in the previous section.

The probability of a feed forward false positive becomes:

$$P_{feed-fwFP} = (P)^k \tag{13}$$

4   Result Analysis

For each nucleotide a_i in the sequence, a_j and a_k are the member of a_i in the predicted structure and reference structure respectively with $0 \leq j$, $k \leq n$ and $1 \leq i \leq n$. For a_i, our prediction have two possible results:

1. Correct if j=k.
2. Wrong if j≠k.

Accuracy can be defined as follow:

$$Accuracy = \frac{correct}{n} \times 100 \tag{14}$$

Our method was tested on 105 single-standard Pseudoknot which are chosen in randomly from PseudoBase with length varies 21 to 121. These Pseudoknots are classified into 9 classes. This classification shown in Table 1:

Table 1. RNA classification in PseudoBase showing different lengths. First column indicates the different types of RNA, the second column shows the number of pseudoknots considered and the third column reflects the range of sequence lengths [27]".

Classification	RNA	Length
1. Viral ribosomal frame shifting signals	15	39-73
2. Viral ribosomal read through signals	6	61-63
3. Other Viral-5´	1	35
4. Other Viral-3´	80	21-96
5. rRNA	3	46-51
6. mRNA	7	28-120
7. tmRNA	10	30-90
8. Ribozymes	3	73-89
9. Others	4	35-121

In Table 1 first column is the names of 9 classes of RNA pseudoknot in pseudoBase. Second column is the number of RNA pseudoknots in each class. The third column is the length range of RNA pseudoknot in each class. During procedure of pattern analysis of pseudoknots, the bases of the stems are separated according to length of stem in parsing phase. To effect of stems on pseudoknots, we have tested 105 pseudoknots by considering two phases: parsing and bloom filter. If the stem length is less than N, then stem is separated in substructure phase. Experiment results on the effect of stem-length are shown in Table 2.

Table 2. Effect on stem length and accuracy

Case	Length	Average accuracy (%)
a.	3	73.042
b.	3	73.345
c.	4	75.085
d.	4	76.225
e.	No	71.784

In Table 2 each case has (a)-(d) higher accuracy than case (e) in which no stem is separated. This indicates that small stems have effect on pseudoknots. Testing the 105 pseudoknots using stem length 4,our method achieves an average accuracy of 66.159%. 38 pseudoknots reach 100% accuracy, only 11 pseudoknots reach 40% less accuracy. The Fig.9 shows the accuracy distribution.

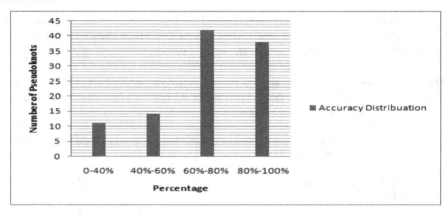

Fig. 9. Accuracy distribution using stem length

We compare our method with the algorithm introduced in Ruan Server [28]. Their method was implemented in their web server (http://cic.cs.wustl.edu/RNA/) which supports thermodynamic and comparative analysis for prediction of RNA secondary structure with pseudoknots. We tested 105 pseudoknots on this web server by using their default parameters. The result is described in Table 3.

In Table 3, there 105 pseudoknots in 9 classes.In Table 3,the first column is the classification of pseudoknots in pseudobase. The second column is the number of pseudoknots in each class. The third column is the accuracy of different method, including Mflod method [24], our algorithm and Raun server [28]. From Table 3,we can see that our method obtains higher average accuracy than both Mflod and Raun Server.Mflod specially designed to predict pseudoknot-knot free secondary structure and Ruan server is designed by using thermodynamic concept.

Table 3. Accuracy comparison between different pseudoknot prediction methods.

Classification	RNA	Accuracy(%)		
		Our method	Mflod[39]	Ruan Server[45]
1	15	64.163	54.983	56.317
2	6	38.3	31.388	51.89
3	1	100	71.429	37.143
4	80	89.276	66.94	31.822
5	3	54.364	51.161	36.676
6	7	59.94	51.362	40.66
7	10	62.03	56.316	36.356
8	3	57.67	45.02	36.94
9	4	69.69	65.715	46.248
Average Accuracy		66.159	54.923	41.561

5   Conclusion

In this paper, we propose a pseudobase RNA secondary structure based on term rewriting technique and bloom filter. We use Maude knowledge for term rewriting and generate RNA secondary structure. Generated secondary structure is divided into substructures which reduce the complexity cost..Bloom filter is a robust tool which reduces memory space in pattern matching. Our method has been tested on 105 pseudoknot in 9 classes and achieved average accuracy 66.159% compare to Mfold and Run server. We get maximum accuracy (89.276%) in class 4 which contains 80 types of RNA structure. The combination of Mfold and bloom filter appears to be a powerful tool for RNA Pseudoknot prediction.

References

1. Batey, R.T., Rambo, R.P., Doudna, J.: Tertiary Motifs in RNA Structure and Folding. Angew. Chem. Int. Ed. Engl 38(16), 2326–2343 (1999)
2. He, S., Liu, C., Skogerbo, G., Zhao, H., Wang, J., Liu, T., Bai, B., Zhao, Y., Chen, R.: Noncode v2.0: Decoding the Non-Coding.Nucleic Acids Research 36(Database), D170-D172 (2007)
3. Griffiths-Jones, S., Bateman, A., Marshall, M., Khanna, A., Eddy, S.R.: Rfam: An RNA Family Database. Nucleic Acids Research 31(1), 439–441 (2003)
4. Le, S.Y., Chen, J.H., Maizel, J.: Structure and Methods: Human Genome Initiative and DNA Recombination. Chapter Efficient Searches for Unusual Folding Regions in RNA Sequences 1, 127–130 (1990)
5. Rivas, E., Eddy, S.: Secondary Structure Alone is Generally Not Statistically Significant for the Detection of Noncoding RNAs. Bioinformatics 16(7), 583–605 (2000)
6. Thomas, K.F., Wong, Y.S., Chiu, T.W., Lam, S.M.: Memory Efficient Algorithms for Structural Alignment of RNAs with Pseudoknots. IEEE/ACM Transactions On Computational Biology And Bioinformatics 9(1), 161–168 (2012)
7. Hofacker, I.L., Bernhart, S.H.F., Stadler, P.F.: Alignment of RNA base pairing probability matrices. Bioinformatics 20(14), 2222–2227 (2004)

8. Lee, D., Han, K.: Prediction of RNA Pseudoknots – Comparative Study of Genetic Algorithms. Genome Informatics 13, 414–415 (2003)
9. Staple, D.W., Butcher, S.E.: Pseudoknots: RNA Structures with Diverse Functions. PLoS Biology 3(6), e213 (2005)
10. Pley, H.W., Flaherty, K.M., McKay, D.B.: Threedimensional structure of a hammerhead ribozyme. Nature 372, 68–74 (1994)
11. Sankoff, D.: Simultaneous Solution of the RNA Folding, Alignment and Protosequence Problems. SIAM Journal on Applied Mathematics 45(5), 810–825 (1985)
12. Dowell, R.D., Eddy, S.R.: Efficient pair wise RNA structure prediction and alignment using sequence alignment constraints. BMC Bioinformatics 7(400) (2006)
13. Holmes, I.: Accelerated probabilistic inference of rna structure evolution. BMC Bioinformatics 6(1), 73 (2005)
14. Knudsen, B., Hein, J.: Pfold: RNA secondary structure prediction using stochastic context-free grammars. Nucleic Acids Res 31(13), 3423–3428 (2003)
15. Thompson, J.D., Higgins, D.G., Gibson, T.J.: CLUSTAL W: improving the sensitivity of progressive multiple sequence alignment through sequence weighting, position-specific gap penalties and weight matrix choice. Nucl. Acids Res. 22(22), 4673–4680 (1994)
16. Blanchette, M., Kent, W.J., Riemer, C., Elnitski, L., Smit, A.F., Roskin, K.M., Baertsch, R., Rosenbloom, K., Clawson, H., Green, E.D., Haussler, D., Miller, W.: Aligning multiple genomic sequences with the threaded blockset aligner. Genome Res 14(4), 708–715 (2004)
17. Mccaskill, S.: The equilibrium partition function and base pair binding probabilities for rna secondarystructure. Biopolymers 29(6-7), 1105–1119 (1990)
18. Torarinsson, E., Havgaard, J.H., Gorodkin, J.: Multiple structural alignment and clustering of RNA sequences. Bioinformatics 23(8), 926–932 (2007)
19. Washietl, S., Hofacker, I.L., Stadler, P.F.: Fast and reliable prediction of noncoding RNAs. Proc Natl Acad Sci U S A 102(7), 2454–2459 (2005)
20. Hochsmann, M., Voss, B., Giegerich, R.: Pure multiple RNA secondary structure alignments: a progressive profile approach. IEEE/ACM Trans Comput Biol Bioinform 1(1), 53–62 (2004)
21. Touzet, H., Perriquet, O.: CARNAC:folding families of related RNAs. Nucl. Acids Res. 32(suppl. 2), W142–W145 (2004)
22. Bloom, B.: Space/time Trade-offs in Hash Coding with Allowable Errors. Communications of the ACM 13(7), 422–426
23. Eker, S.: Fast matching in combination of regular equational theories. In: Meseguer, J. (ed.) Proceedings First International Workshop on Rewriting Logic and its Applications. Electronic Notes in Theoretical Computer Science, vol. 4, pp. 90–109. Elsevier (1996)
24. Zuker, M.: Mfold web server for nucleic acid folding and hybridization prediction. Nucleic Acids Research 31(13), 3406–3415 (2003)
25. Rouzina, I., Bloomfield, V.A.: Heat Capacity Effects on the Melting of DNA.2. Analysis of Nearest-Neighbor Base Pair Effects. Biophysical Journal 77(6), 3252–3255 (1999)
26. Bloom Space/time, B.H.: trade-offs in hash coding with allowable errors. Communications of the ACM 13(7), 422–426 (1970)
27. Tabaska, J., Cary, R., Gabow, H., Stormo, G.: An RNA folding method capable of identifying pseudoknots and base triples. Bioinformatics 14(8), 691–699 (1998)
28. Ruan, J., Stormo, G.D., Zhang, W.: An Iterated loop matching approach to the prediction of RNA secondary structures with Pseudoknots. Bioinformatics 20(1), 58–66 (2004)

A Simulation Model for the Dynamics
of a Population of Diabetics with and without
Complications Using Optimal Control

Wiam Boutayeb[1,*], Mohamed E.N. Lamlili[1], Abdesslam Boutayeb[1],
and Mohammed Derouich[2]

[1] URAC04, Department of Mathematics Faculty of Sciences Boulevard Mohamed VI,
BP: 717, Oujda, Morocco
{wiam.boutayeb,mohamed.lamlili}@gmail.com,
x.boutayeb@menara.ma
[2] National School of Applied Sciences Univeristy Mohamed Premier Boulevard
Mohamed VI, BP: 717, Oujda, Morocco
mderouich2011@gmail.com

Abstract. In this paper, we simulate the evolution of diabetes from
the stage without complication to the stage with complications and vice
versa. Three scenarios are proposed according to the estimated level of
incidence of diabetes. Our model shows that the number of diabetics
with complications can be limited by an optimal control and hence the
overall burden of diabetes can be reduced.

Keywords: Diabetes, mathematical model, simulation, optimal control.

1   Introduction

Once known as a disease of the rich and associated with economic development,
diabetes is now sweeping the entire globe as a silent epidemic, affecting particularly low and middle-income countries. According to the last International
Diabetes Federation (IDF) report, about 80% of the 382 million diabetics live
in low- and middle-income countries; more than five million deaths were due
to diabetes; and more than 480 billion USD was spent on healthcare for diabetes in 2013 [1]. Due to its chronic nature with severe complications, diabetes
needs costly prolonged treatment and care, affecting individuals, families and
the whole society. Consequently, efficient and optimal strategies aiming to reduce the burden of diabetes are needed. During the last decades, a large number
of publications were devoted to mathematical modelling for diabetes as indicated
by recent reviews [2–6]. Following a previous mathematical model for the burden
of diabetes and its complications [7], we propose in this paper a mathematical
model simulating the evolution of the disease from diabetes without complication to diabetes with complications and vice versa. Three scenarios are proposed
according to the estimated level of incidence of diabetes and other scenarios may
be considered by varying the values of parameters.

F. Ortuño and I. Rojas (Eds.): IWBBIO 2015, Part I, LNCS 9043, pp. 589–598, 2015.
© Springer International Publishing Switzerland 2015

2   Formualtion of the Model

We consider the model developed by A. Boutayeb et al. [7]

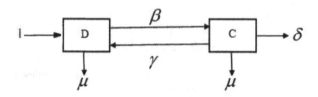

$$\frac{dD}{dt} = I - (\beta + \mu)D(t) + \gamma C(t)$$

$$\frac{dC}{dt} = \beta D(t) - (\gamma + \mu + \delta)C(t)$$

(1)

Where:
- $D = D(t)$ is the number of diabetics without complications
- $C = C(t)$ is the number of diabetics with complications
- I denotes the incidence of diabetes mellitus (assumed constant)
- μ : natural death rate,
- β : probability of developing a complication,
- γ : rate at which complications are cured,
- δ : death and severe disability rate due to complications

Suppose that $C = C(t)$ and $D = D(t)$ represent the numbers of diabetics with and without complications, respectively, and let $N = N(t) = C(t) + D(t)$. As indicated in th introduction section, the IDF estimated in 2013 that $N(2013) = 382$ million people were living with diabetes worldwide. The proposed model simulates the dynamics of the population of diabetics with and without complications in a period of 10 years in presence and absence of an optimal control, assuming that incidence is constant and that intial conditions are: $N(0) = D(0) + C(0) = 250 \times 10^6 + 132 \times 10^6 = 382 \times 10^6$.

The controlled model is given by the following system:

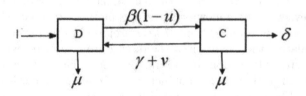

$$\frac{dD}{dt} = I - (\beta(1-u) + \mu)D(t) + (\gamma + v)C(t)$$

$$(2)$$

$$\frac{dC}{dt} = \beta(1-u)D(t) - (\gamma + \mu + \delta + v)C(t)$$

Where u and v are controls.

- The objective function is defined as:

$$J(u,v) = \int_0^T \left(C(t) + Au^2(t) + Bv^2(t)\right) dt$$

Where A and B are positive weights that balance the size of the terms. U is the control set defined by $U = \{u, v/u$ and v are measurable, $0 \le u(t) \le 1, t \in [0,T]\}$.

- The objective is to characterize an optimal control $(u^*, v^*) \in U$ satisfying:

$$J(u^*, v^*) = \min_{u,v \in U} J(u,v)$$

3   Existence of an Optimal Control Pair

3.1   Theorem

- Consider the control problem with system 2. There exists an optimal control pair $(u^*, v^*) \in U$ such that $J(u^*, v^*) = \min_{u,v \in U} J(u,v)$ • Given an optimal control pair (u^*, v^*) and solutions C^* and D^* of the corresponding state system, there exist adjoint variables λ_1 and λ_2 satisfying:

$$\lambda_1' = \lambda_1 \mu + (\lambda_1 + \lambda_2)(1 - u^*)\beta$$

$$\lambda_2' = -1 + \lambda_2(\mu + \delta) + (\lambda_2 - \lambda_1)(\gamma + v^*)\beta$$

With transversality conditions: $\lambda_1 = \lambda_2 = 0$ Moreover the optimal control is given by:

$$u^* = min(1, max(0, \frac{\lambda_2 - \lambda_1}{2A}\beta D^*(t)))$$

$$v^* = min(1, max(0, \frac{\lambda_2 - \lambda_1}{2B}C^*(t)))$$

3.2   Proof

1. The existence of optimal control can be proved by using the results from Fleming and Rishel [8] checking the following points:

- $\mathcal{J}(u,v) = \int_0^T \left(C(t) + Au^2(t) + Bv^2(t) \right) dt$ is convex

- The control space $U = \{u,v/u$ and v are measurable, $0 \leq u(t) \leq 1, t \in [0,T]\}$ is convex and closed by definition.

- The integrand in the objective functional, $\left(C(t) + Au^2(t) + Bv^2(t) \right)$ is clearly convex in U
- There exist constants a, b, c such that: $\left(C(t) + Au^2(t) + Bv^2(t) \right) \geq a + b|u|^\propto + c|v|^\propto$

The necessary conditions for the optimal control pair arise from the Pontryagins maximum principle [9].

2. The Hamiltonian is defined as follows: $H = C + Au^2 + Bv^2 + \lambda_1 f_1(D^*, C^*) + \lambda_2 f_2(D^*, C^*)$
where:

$$f_1(D,C) = I - (\beta(1-u) + \mu)D(t) + (\gamma + v)C(t)$$

$$f_2(D,C) = \beta(1-u)D(t) - (\gamma + \mu + \delta + v)C(t)$$

The optimal control can be determined from the optimality condition :

$$\frac{\partial H}{\partial u} = 0 \Rightarrow 2Au + \lambda_1 \beta D^*(t) - \lambda_2 \beta D^*(t) = 0$$

$$\frac{\partial H}{\partial u} = 0 \Rightarrow u^* = \frac{1}{2A}(\lambda_2 - \lambda_1)\beta D(t)^*$$

$$\frac{\partial H}{\partial v} = 0 \Rightarrow 2Bv + \lambda_1 C^*(t) - \lambda_2 C^*(t) = 0$$

$$\frac{\partial H}{\partial v} = 0 \Rightarrow v^* = \frac{1}{2B}(\lambda_2 - \lambda_1)C(t)^*$$

The adjoint variables λ_1 and λ_2 are obtained by the following system:

$$\lambda_1' = \frac{\partial H}{\partial D} = \lambda_1 \frac{\partial f_1}{\partial D} - \lambda_2 \frac{\partial f_2}{\partial D} = (\lambda_2 - \lambda_1)(1 - u^*)\beta + \mu\lambda_1$$

$$\lambda_2' = \frac{\partial H}{\partial C} = -1 - \lambda_1 \frac{\partial f_1}{\partial C} - \lambda_2 \frac{\partial f_2}{\partial C} = -1 - \lambda_1(\gamma + v^*) + \lambda_2(\mu + \gamma + \delta + v^*)$$

$$= -1 - (\lambda_1 - \lambda_2)(\gamma + v^*) + \lambda_2(\mu + \delta)$$

$$\Rightarrow$$

$$\lambda_1' = (\lambda_2 - \lambda_1)(1 - u^*)\beta + \mu\lambda_1$$

$$\lambda_2' = -1 - (\lambda_1 - \lambda_2)(\gamma + v^*) + \lambda_2(\mu + \delta)$$

$$\lambda_1(T) = \lambda_2(T) = 0$$

4   Numerical Discritization

Following Gumel et al., we use a Gauss-Seidel-like implicit finite-difference method [10]. The time interval $[t_0, T]$ is discritized with a step h (time step size) such that $t_i = t_0 + ih \;\; i = 0, 1, \cdots, n$ and $t_n = T$.

So at each point t_i we will note

$D_i = D(t_i), \;\; C_i = C(t_i),$

$\lambda_1^i = \lambda_1(t_i), \;\; \lambda_2^i = \lambda_2(t_i),$

$u_i = u(t_i)$ and $v_i = v(t_i)$

For the approximation of the derivative we used simultaneously forward difference for $\dfrac{dD(t)}{dt}$ and $\dfrac{dC(t)}{dt}$ and backward difference for $\dfrac{d\lambda_1(t)}{dt}$ and $\dfrac{d\lambda_2(t)}{dt}$.

So the derivatives $\dfrac{dD(t)}{dt}$ and $\dfrac{dC(t)}{dt}$ are approached by the following finite differences:

$$\left. \begin{aligned} \frac{dD_{i+1}}{dt} &\approx \frac{D_{i+1} - D_i}{h} \\[2mm] \frac{dC_{i+1}}{dt} &\approx \frac{C_{i+1} - C_i}{h} \end{aligned} \right\} \quad \text{for} \quad i = 0, \cdots, n-1.$$

Similarly, $\dfrac{d\lambda_1(t)}{dt}$ and $\dfrac{d\lambda_2(t)}{dt}$ are approached by finite differences.

Hence the problem is given by the following numerical scheme for $i = 0, \cdots, n-1$

$$\frac{D_{i+1} - D_i}{h} = I - (\mu + \beta(1 - u_i))D_{i+1} + (\gamma + v_i)C_i$$

$$\frac{C_{i+1} - C_i}{h} = \beta(1 - u_i)D_{i+1} - (\mu + \gamma + \delta + v_i)C_{i+1}$$

$$\frac{\lambda_1^{n-i} - \lambda_1^{n-i-1}}{h} = (\lambda_1^{n-i-1} - \lambda_2^{n-i})(1 - u_i)\beta + \mu\lambda_1^{n-i-1}$$

$$\frac{\lambda_2^{n-i} - \lambda_2^{n-i-1}}{h} = -1 - (\lambda_2^{n-i-1} - \lambda_3^{n-i})(\gamma + v_i) + \lambda_2^{n-i-1}(\mu + \delta)$$

Then we consider : $D_0 = D(0), C_0 = C(0), u_0 = 0, \lambda_1^n = 0$ and $\lambda_2^n = 0$

So for $i = 0, \cdots, n-1$

$$D_{i+1} = \frac{D_i + hI + h(\gamma + v_i)C_i}{1 + h(\mu + \beta(1 - u_i))}$$

$$C_{i+1} = \frac{C_i + h\beta(1 - u_i)D_{i+1}}{1 + (\mu + \gamma + \delta + v_i)h}$$

$$\lambda_1^{n-i-1} = \frac{\lambda_1^{n-i} + h\beta\lambda_2^{n-i}(1 - u_i))}{1 + h(\mu + \beta(1 - u_i))}$$

$$\lambda_2^{n-i-1} = \frac{\lambda_2^{n-i} + h\lambda_1^{n-i-1}(\gamma + v_i) + h}{1 + h(\mu + \gamma + \delta + v_i)}$$

$$M_1^{i+1} = \frac{1}{2A}\left[\beta D_{i+1}(\lambda_2^{n-i-1} - \lambda_1^{n-i-1})\right]$$

$$M_2^{i+1} = \frac{1}{2B}\left[C_{i+1}(\lambda_2^{n-i-1} - \lambda_1^{n-i-1})\right]$$

$$u_{i+1} = \min(1, \max(0, M_1^{i+1}))$$

$$v_{i+1} = \min(1, \max(0, M_2^{i+1}))$$

5　Simulation and Results

5.1　Parameter Estimation

Different simulations can be carried out using various values of parameters. In the present numerical approach, we use the following parameters:

• Incidence of diabetes:

The 2013 IDF report indicates that the number of people living with diabetes is expected to increase from 382 million in 2013 to 592 million in 2035 which gives an average incidence of diabetes around 10 million per year [1].

According to Centers for Disease Control and Prevention (CDC), from 2006 to 2011, crude and age-adjusted incidence of diabetes of the population aged 18-79 years has not changed significantly in USA. It was around 7 per 1000 people [11]. Based on these remarks, we assume that incidence of diabetes is constant over a 10 years period and we propose three scenarios according to a low(LI=8x10^6), medium(MI=10x10^6) and high (HI=20x10^6) level of incidence.

• Natural death rate:

For the natural death rate μ, we use the estimated value of 8 per 1000 people given for the year 2012 by the World Bank[12].

• Death rate caused by complications:

According to IDF, aroud 5 million deaths are caused by diabetes each year. Although some deaths may be caused by hypoglycemia, we assume that nearly five million deaths are caused by complications of diabetes and hence we propose 0.02 as an estimated value of δ.

• The rate at which complications are cured is very difficult to estimate. Complications at advanced stage are not curable and hence $\gamma = 0$, whereas for complications at a non advanced stage, γ is given the value 0.01.

• Finally, assuming that, every year a quarter of the population with diabetes develop some kinds of complications, the parameter β is given the value 0.25. The parameter values are summarised in Table1 below:

Table 1. Parameter values used in numerical simulation

Parameter	Value yr^{-1}
μ	0.008
γ	0.01
δ	0.02
β	0.25
LI	8×10^6
MI	10×10^6
HI	20×10^6

5.2   Results

A simulation is carried out on an interval time of T=10 years, subdivised into n subintervals of step h=0.01 so that:

$n = 1000; h = 0.01; T = nh = 10$ years $\lambda_1(n) = 0; \lambda_2(n) = 0;$

Since control and state functions are on different scales, the weight constant value is chosen as follows: : $A = 250106000; B = 100106000$

Using the parameter values given in Table1, a first simulation is carried out according to the three scenarios on incidence and without control. The results are given in Table2 and illustrated by Fig1, Fig2 and Fig3

Table 2. Results of simulation using parameter values given in Table1 without control

	LI	MI	HI
D	58×10^6	66×10^6	102×10^6
C	319×10^6	330×10^6	385×10^6
Total	377×10^6	395×10^6	487×10^6
$\frac{C}{Total}(\%)$	84.5	83	80

Similarly, a second simulation is used based on the parameter values given in Table1 according to the three scenarios on incidence and with control. The results are given in Table3 and illustrated by Fig1, Fig2 and Fig3.

Table 3. Results of simulation using parameter values given in Table1 with control

	LI	MI	HI
D	307×10^6	324×10^6	407×10^6
C	101×10^8	103×10^6	114×10^6
Total	408×10^6	427×10^6	521×10^6
$\frac{C}{Total}(\%)$	25	24	22

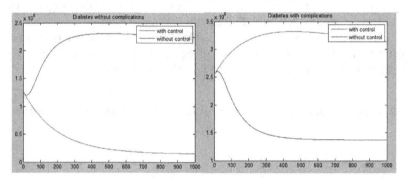

Fig. 1. Evolution of the population of diabetics with and without complications using the parameters given in Table 1 with low incidence of diabetes (LI=8×10^6)

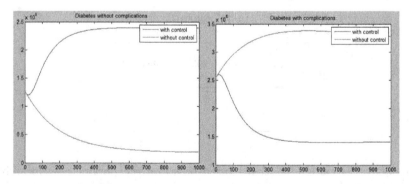

Fig. 2. Evolution of the population of diabetics with and without complications using the parameters given in Table 1 with medium incidence of diabetes (MI=10×10^6)

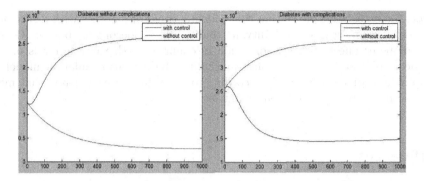

Fig. 3. Evolution of the population of diabetics with and without complications using the parameters given in Table 1 with high incidence of diabetes (HI=20x10^6)

6   Discussion

The results yielded by the present model show that, without any control, the population of diabetes will reach 377 million, 395 million and 487 million people in 10 years under the three scenarios: low incidence, medium incidence and high incidence of diabetes respectively. The proportions of diabetics with complications will be 84.5%, 83% and 80% respectively (Table2). Under the same scenarios, the optimal control will lead to a total population of respectively, 408 million, 427 million and 521 million of which diabetics with complications will represent 25%, 24% and 22% respectively (Table3). Under the three scenarios, the strategy with optimal control limits the proportion of diabetics with complications to less than 25% and hence reduces the rate of mortality caused by diabetes essentially due to complications (CVDs, kidney failure, amputations, blindness,). Moreover, beyond mortality, the reduction of complications reduces the socio-economic burden of diabetes, including direct and indirect costs as well as intangible and not quantifiable costs such as inconvenience, anxiety, pain and more generally lower quality of life[13]. In the real world, achievement of optimal control can be obtained by acting on behavioral factors like healthy diet, physical activity, smoking and alcohol reduction; and metabolic risks like overweight/obesity and hypertension. Early diagnosis and efficient control of blood glucose provided by an affordable regular healthcare will also be needed.

7   Conclusion

The incidence of diabetes and the probability of developing complications are the main parameters determining the burden of diabetes, including the size of the population living with diabetes, the rate of morbidity, the rate of mortalirty and socioeconomic costs. The present simulation model highlights the role of reducing incidence of diabetes and controlling the rate of evolution from diabetes without complications to the stage of diabetes with complications. Our simple model indicates ways of limiting the epidemic trend of diabetes with disastrous consequences in terms of mortality, morbidity and socioeconomic burden. Given that diabetes affects people worldwide in countries with variable levels of incidence, prevalence, mortality and morbidity, the present simulation model is adaptable to different contexts according to the values of parameters that may be used in strategies of different countries.

References

1. International Diabetes Federation: IDF Diabetes Atlas sixth edition. DF, Bressels, Belgium (2013)
2. Boutayeb, A., Chetouani, A.: A critical review of mathematical models and data used in diabetology. BioMedical Engineering Online 5, 43 (2006)
3. Makroglou, A.: Mathematical models and software tools for the glucose-insulin regulatory system and diabetes: an overview. Applied Numerical Mathematics 56, 559–573 (2006)
4. Li, J., Johnson, J.D.: Mathematical models of subcutaneous injection of insulin analogues: a mini-review. Discrete Continuous Dynamical Systems. Series B 12, 401–414 (2009)
5. Bergman, R.N., Finegood, D.T., Ader, M.: Assessment of Insulin Sensitivity in Vivo. Endicrine Reviews 6(1), 45–86 (1985)
6. Ajmera, I., Swat, S., Laibe, C., Le Novére, N., Chelliah, V.: The impact of mathematical modeling on the understanding of diabetes and related complications: A review. Phamacometrics and Systems Pharmacology 2(e54), 1-14 (2013)
7. Boutayeb, A., Twizell, E.H., Achouyab, K., Chetouani, A.: A mathematical model for the burden of diabetes and its complications. BioMedical Engineering Online 3, 20 (2004)
8. Fleming, W.H., Rishel, R.W.: Deterministic and Stochastic Optimal Control. Springer Verlag, New York (1975)
9. Pontryagin, L.S., Boltyanskii, V.G., Gamkrelidze, R.V., Mishchenko, E.: The mathematical theory of Optimal Process. Gordon and Breach Science Publishers 4 (1986)
10. Gumel, A.B., Shivakumar, P.N., Sahai, B.M.: A mathematical model for the dynamics of HIV-1 during the typical course of infection. In: proceeding of the 3d World Congress of Nonlinear Analysts, vol. 47, pp. 2073–2083 (2011)
11. Centers for Disease Control and Prevention: Diabetes Public Health Resource: http://www.cdc.gov/diabetes/statistics/incidence/fig2.htm
12. The World Bank: World Development Indicators:Population dynamics, http://wdi.worldbank.org/table/2.1
13. Boutayeb, W., Lamlili, M., Boutayeb, A., Boutayeb, S.: Direct and indirect cost of diabetes in Morocco. Journal of Biomedical Science and Engineering 6, 732–773 (2013)

Logical Modeling and Analysis of Regulatory Genetic Networks in a Non Monotonic Framework

Nicolas Mobilia[1], Alexandre Rocca[1], Samuel Chorlton[2], and Eric Fanchon[1], and Laurent Trilling[1]

[1] Laboratoire TIMC-IMAG, Université de Grenoble, France
[2] Department of Medicine, McMaster University, Hamilton, Canada

Abstract. We present a constraint based declarative approach for analyzing qualitatively genetic regulatory networks (GRNs) with the discrete formalism of R. Thomas. For this purpose, we use the logic programming technology ASP (Answer Set Programming) whose related logic is non monotonic. Our aim is twofold. First, we give a formal modeling of both Thomas' GRNs and biological data like experimental behaviors and gene interactions and we evaluate the declarative approach on three real biological applications. Secondly, for taking into account both gene interaction properties which are only **generally** true and automatic inconsistency repairing, we introduce an optimized modeling which leads us to exhibit new logical expressions for the conjunction of defaults and to show that they can be applied safely to Thomas' GRNs.

Keywords: computational systems biology, gene networks, discrete modeling, AI-oriented declarative approach, non monotonic logic, Answer Set Programming.

1   Introduction

Mathematical modeling and simulation tools may help to understand how complex genetic regulatory networks (GRNs), composed of many genes and their intertwined interactions, control the functioning of living systems. They provide a framework to unambiguously describe the network structure and to infer predictions of the dynamical behavior of the system.

The typical model building cycle starts with gathering existing knowledge on a biological system and formulating working hypotheses, on the basis of which a model formalism is chosen and the structure of the GRN is defined. The development of the dynamical model and its parametrization lead to an initial model, whose predictions are confronted with experimental data. This often reveals inconsistencies, and calls for a revision of the structure of the GRN and/or the parameter values of the model. The process is repeated iteratively until the validation step is considered satisfactory. The generate-and-test approach underlying the above-mentioned method demands a large number of simulations to be carried out and usually leads to the formulation of a unique model consistent with biological data. In this paper, we adopt an alternative method for the systematic construction and analysis of models of GRNs by means of an Artificial Intelligence oriented *declarative* approach. The models are developed using the formalism

F. Ortuño and I. Rojas (Eds.): IWBBIO 2015, Part I, LNCS 9043, pp. 599–612, 2015.

of R. Thomas [15], which offers an appropriate discrete description of GRNs, as most
available data on regulatory interactions are qualitative. Instead of instantiating the models as in classical modeling approaches, all possible knowledge on the network structure
and its dynamics (e.g., existence of cycles or stationary states, response of the network
to environmental or genetic perturbations) is formulated in the form of constraints, i.e.
logical formulae. Without resorting to numerous simulations, the compatibility of the
network structure with the biological constraints is determined and an *intensional* (implicit) representation of the set of consistent models is returned, in case all the constraints
are satisfied. This is well suited for biological data which are often incomplete. Futher-
more, in case of inconsistency, an automatic repairing can be applied. Also, for the profit
of biologists, all properties, expressed in a predefined language, that are common to all
consistent models can be deduced [8].

In this paper, we use for that purpose the logic programming technology ASP [2]
which is based on a non monotonic logic defined with *stable* models. The aim of this
paper is to show the benefits provided by ASP for the declarative approach of Thomas'
GRNs. We pay a special attention to modeling methods to take advantages of the ASP
non monotonic feature for tackling potential inconsistencies and for expressing gene
interaction rules that are **generally** true.

The paper is organized as follows. ASP is introduced in Section 2 and the logical
specifications of Thomas' GRNs together with biological data in Section 3. Then, three
illustrating applications are presented in Section 4. Finally an optimized modeling exemplifying the non monotonic aspects of ASP is described in Section 5.

2   Answer Set Programming (ASP)

Here is a short presentation based on [11] which proposes the *gringo* language for
expressing ASP programs.

A logical ASP program is a finite set of rules:
$a_0 \leftarrow a_1, \ldots, a_m, not\ a_{m+1}, \ldots, not\ a_n.$
where $0 \leq m \leq n$ and $\forall i \mid 0{\leq}i{\leq}n$, a_i is an atom. For any rule r, $head(r) = a_0$ is
the head of the rule, and $body(r) = \{a_1, \ldots, a_m, not\ a_{m+1}, \ldots, not\ a_n\}$ is the body
of the rule. If $head(r)$ is empty, r is called an *integrity constraint*. If $body(r)$ is empty,
r is a *fact*.

Let A be the set of atoms, $body^+(r) = \{a{\in}A \mid a \in body(r)\}$ and $body^-(r) = \{a{\in}A \mid not\ a \in body(r)\}$. A set $X \subseteq A$ is an *answer set* (AS) or stable model of a
program P if X is the minimal model [1] [2] of the *reduct*
$P^X = \{head(r){\leftarrow}body^+(r) \mid r{\in}P, body^-(r) \cap X = \varnothing\}.$

Example: let E be the following ASP program where $\leftarrow$ is represented by : - :
```
a :- not b,c.
c.
```
Let $X = \{a,c\}$. The corresponding reduct is $E^X = \{a \leftarrow c, c\}$ and its minimal
model is $\{a,c\}$. Then X is an AS of E.

---

[1] A logical model is minimal when removing an atom from it cannot provide a model. A reduct
has a unique minimal model.

[2] It is important to note that ASs also have been shown to be minimal (see Section 5.2).

Let $X' = \{a, b, c\}$. The corresponding reduct is $E^{X'} = \{c\}$, and its minimal model is $\{c\}$. Then X' is not an AS of E.

The first rule of this example is typical of a *default* rule [5]. It expresses that generally (in the absence of knowledge on b) a is implied by c. But if b holds because of additional knowledge, this rule is no longer applicable. For instance, by addition of the fact b. to this example, we get the AS $\{b, c\}$ exemplifying the non monotonic character of ASP: a does not hold any more, without leading to an inconsistency.

Rules in the *gringo* language are extended for accepting heads which are disjunctions of literals (exclusive unless both literals are proven) using the operator $|$. Furthermore, *gringo* provides logical variables and functional terms, in a limited way (so that the program can be transformed in an equivalent finite propositional one). It also provides cardinality constraints on the number of true literals. If we impose the constraint $u\{l\_1, \ldots, l\_n\}v$ we obtain only models such that the number of true literals l_i is bigger than (or equal to) u (0 by default) and smaller than (or equal to) v (n by default). Moreover, this formalism allows the expression of conditional enumerations of literals through the symbol ":", as conjunctions (resp. disjunctions) in the body (resp. head) of a rule. For example, in the following program:

```
dom(0). dom(1).    all_true :- p(X) : dom(X).
q(X) : dom(X) :- one_of.
at_least_one_true :- 1{p(X) : dom(X)}.
```

the first line expresses that all_true is deduced if p(0) and p(1) hold. This line is, therefore, equivalent to the rule: all_true :- p(0), p(1). and the second line equivalent to q(0) | q(1) :- one_of. The third line expresses that at_least_one_true is deduced if a least one among p(0) and p(1) holds.

Para-logic operators are also provided for maximizing or minimizing the number of atoms true among a set of atoms. Asserting #maximize{f_1,...,f_n} will produce only models with the highest number of f_i true.

The solver [11] we use proceeds in two steps to compute the ASs of a program P. First, a "grounder" substitutes the variables of the program by terms without free variables, and consequently produces a propositional program $\mathbf{P}$ corresponding to P. In the second step, a solver computes the ASs of $\mathbf{P}$. This motivates the programmer to reduce as far as possible the number of resulting Boolean variables and rules subject to a big expansion (see Section 5.1).

3 Thomas' GRNs and the Declarative Approach

In Section 3.1, we specify Thomas' GRNs. Biological constraints and typical queries for constructing and analyzing models of GRNs are described in Section 3.2.

3.1 Thomas' GRN Specification

A common representation of a GRN is to view it as an *interaction graph*, where nodes represent genes and arrows represent interactions between genes. An arrow $j \to i$ is labeled with the sign of the regulatory influence (to indicate whether the gene i is activated or inhibited by the product J of gene j), and with the index r of the threshold concentration θ_j^r above which the protein J controls the expression of gene i. A simple example of interaction graph for a two-gene network is shown in Fig. 1(a).

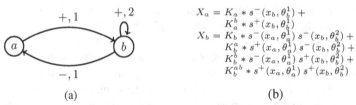

$$X_a = K_a * s^-(x_b, \theta_b^1) +$$
$$K_a^b * s^+(x_b, \theta_b^1)$$
$$X_b = K_b * s^-(x_a, \theta_a^1) s^-(x_b, \theta_b^2) +$$
$$K_b^a * s^+(x_a, \theta_a^1) s^-(x_b, \theta_b^2) +$$
$$K_b^b * s^-(x_a, \theta_a^1) s^+(x_b, \theta_b^2) +$$
$$K_b^{ab} * s^+(x_a, \theta_a^1) s^+(x_b, \theta_b^2)$$

(a)   (b)

Fig. 1. (a) Interaction graph corresponding to a GRN of two genes. The product of gene a stimulates the expression of gene b, while the product of gene b inhibits expression of a. In addition, b activates its own expression. The label $-, 1$ from gene b to a means that b inhibits a expression when b is above its threshold θ_b^1. (b) Focal equations relating a state $[x_a, x_b]$ and its focal state $[X_a, X_b]$.

The dynamic behavior of a GRN is represented in terms of an oriented graph called *state transition graph*, where each node represents a specific *state* of the system and the arrows represent *transitions* between a state and its possible successors. A state S of a network of n genes is represented by a vector of protein concentrations: $S = [x_1, ..., x_n]$. The concentrations take discrete values, each one representing an interval between two consecutive thresholds. For instance, $x_j = 0$ indicates a concentration of protein J lower than the lowest threshold of J, say θ_j^m, whereas $x_j = 1$ means that the concentration of J is lower than $\theta_j^m + 1$ and higher than (or equal to) θ_j^m.

A specific attractor value called *focal state*, $[X_1, ..., X_n]$, is associated to a given state S. It represents the expression levels toward which the genes tend to evolve (see precise definition with *focal equations* below). A successor $S' = [x_1', ..., x_n']$ of S in the graph is deduced from S by comparing the value of each variable x_i with that of its focal state. The transition of S to S' is assumed to be asynchronous, in the sense that at most one variable x_i is updated at a time. If the variable x_i is updated, the formal relationship between these states is expressed as follows: $x_i' = x_i + 1$ if $X_i > x_i$ and $x_i' = x_i - 1$ if $X_i < x_i$. If no logical variable x_i is updated then the focal state of S is equal to S and S is its own successor: in that case S is said to be *steady* (or stationary).

The focal state value X_i of gene i depends on the state S of the network and in particular, on a set of conditions regarding the presence or absence of activators and inhibitors of gene i. For the simple example in Fig. 1, the focal state value X_a of gene a depends on the influence of B (the product of b) on a, that is, if the concentration of B is below ($x_b = 0$) or above its first threshold value. Such interactions are expressed by means of products of step functions of the form: $s^+(x_j, \theta_j^r) = 1$ if $x_j \geq \theta_j^r$ else 0, $s^-(x_j, \theta_j^r) = 1$ if $x_j < \theta_j^r$ else 0.

We will call *cellular context* of gene i any set of states which are equivalent with respect to i for regulation purpose. For example, if i is influenced by only one gene j associated to the threshold θ_j^1, there are only two possible cellular contexts for i, depending on whether x_j is below or above θ_j^1. If i is the target of two interactions, four contexts have to be distinguished for i. More formally, let $[(j_1, \theta_{j_1}^{r_1}), ..., (j_p, \theta_{j_p}^{r_p})]$ be the ordered list of interactions acting on gene i. A cellular context for i is represented as a product $c_i(\sigma) = s^{\sigma_1}(x_{j_1}, \theta_{j_1}^{r_1}) * ... * s^{\sigma_p}(x_{j_p}, \theta_{j_p}^{r_p})$ defining a set of conditions, where

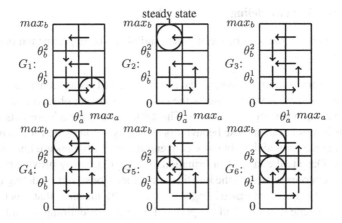

Fig. 2. Transition graphs $G_1, \ldots, G_6$ satisfying all the observability and additivity constraints associated to the example in Fig. 1, with $\theta_b^1 < \theta_b^2$. Arrows represent possible transitions between states represented by boxes. Each graph corresponds to a specific set of instantiated kinetic parameters.

$\sigma \in E^i = \{(\sigma_1, \ldots, \sigma_p) | k : 1..p, \sigma_k \in \{+, -\}\}$. In other words, a cellular context is identified by a p-tuple σ of signs, which specify the position of the region with respect to the p thresholds belonging to the interaction set. From the above definition it follows that the set of cellular contexts with respect to any gene i constitutes a partition of the state space. For the example of Fig. 1, let $[(a, \theta_a^1), (b, \theta_b^2)]$ be the list of interactions acting on b, then the cellular context $c_b((+, -))$ is $s^+(a, \theta_a^1) * s^-(b, \theta_b^2)$.

Each cellular context $c_i(\sigma)$ is associated to a *logical kinetic parameter* defining the focal value of gene i when the network state belongs to that context. We will denote such parameters by $K_i^{l(\sigma)}$, where $l(\sigma)$ is the set $\{j_k | k : 1..p, \sigma_k = +\}$. In this way, the cellular context associated to a logical parameter appears in its notation. For the simple example in Fig. 1, the logical parameters of gene b are: K_b, K_b^b, K_b^a and K_b^{ab}. They respectively are the focal value of b: when the concentrations of A and B are under their thresholds, when the concentration of B only is above its threshold, when the concentration of A only is above its threshold, and when both concentrations are above their thresholds.

Given the cellular contexts defined above and their associated logical parameters, X_i is specified by the following focal equation: $X_i = \sum_{\sigma \in E^i} K_i^{l(\sigma)} * c_i(\sigma)$.

In the declarative approach, some logical parameters may not be instantiated (i.e. unknown). When all these parameters are instantiated, the focal equations define a unique instantiated model. The focal equations corresponding to the simple network example are shown in Fig. 1(b). For the following values of logical parameters: $K_a = 1$, $K_a^b = 0$, $K_b = 0$, $K_b^b = 1$, $K_b^a = 0$, and $K_b^{ab} = 2$, we obtain the transition graph G_1 in Fig. 2. The state $[1, 1]$ belongs to the cellular context $c_a((+))$ and to the cellular context $c_b((+, -))$. The focal state of $[1, 1]$ is therefore $[K_b^b, K_b^a] = [0, 0]$. It follows that the successors of $[1, 1]$ are $[1, 0]$ and $[0, 1]$.

3.2 Biological Data Modeling

We focus here on interactions between genes and observed experimental behaviors of the network [3].

Behaviors. Experimental behavioral data are expressed using constraints on *paths* which are successions of states. This is the case for modeling observed steady states, cycles or repairing behavior due to stress. The declarative approach presents a decisive advantage as information on these behaviours is usually incomplete; for example, there could exist a cycle for which only some concentrations of proteins are known throughout the cycle. Despite the lack of information, this approach may provide biologically meaningful properties regarding the kinetic parameters. Constraint modeling of paths is described with the predicate species(N, V, I, P) meaning that V is the expression level of the gene N at step I of the path P. For example, defining a steady state can be done through the predicate statpath(P) (the two states of the path P of length 2 are equal):

statpath(P) :- path(P), length(2,P), succeg1(N,P):node(N).

where succeg1(N,P) is true if at the two first steps of the path P, the concentrations of the species N are equal. Enforcing the existence of the steady state ss can then be performed with the two facts and the integrity constraint which follow:

path(ss). length(2,ss). :- not statpath(ss).

Applied to the example of Fig. 1, this gives the models corresponding to the transition graphs $\{G_1, G_2, G_4, G_5, G_6\}$ of Fig. 2.

This expressive power provides significant benefits over well-known temporal logics like Computational Tree Logic (CTL), which have been proposed to check instantiations of Thomas networks [4]. For example, a query asking for the existence of a model admitting three different steady states (without knowing them beforehand), is easy to formulate as an extension of the above rules, but cannot be expressed in CTL.

Nevertheless, CTL is useful to express biological observations, typically with $EF\varphi$ formulas applied to a state, meaning that there exists at least a path originating from this state leading to a state with property φ. In the declarative framework, one can easily enforce such formulas. For the example in Fig. 1, enforcing the existence of a path respecting the CTL formula $(a = 0 \wedge b = 0) \wedge EF(a = 0 \wedge b = 2)$ (there exists a path beginning with a state respecting $a = 0$ and $b = 0$ and reaching a state where $a = 0$ and $b = 2$) is achieved with the following rules[4]:

path(p). length(5,p).
:- not species(a,0,1,p). :- not species(b,0,1,p).
exist_path :- species(a,0,I,p), species(b,2,I,p).
:- not exist_path.

The only models satisfying this formula are G_4 and G_6 (Fig. 2).

Enforcing universal CTL properties like $AF\varphi$, meaning that from the state to which it is applied all paths lead to a state with property φ, is not easily handled (see Section 6).

---

[3] Modeling other data like those obtained from *mutant* networks, resulting from suppression or over-expression of genes, can be found in [9].

[4] The length of p is set to 5 because it is the maximal length of a non looping path for this example.

However, such formulae are not appropriate for transcribing a biological experiment in CTL because the experiments usually include only a few trials. An AF formula would be too strong: some valid models could be unduly eliminated. However, if we change the context, and search for constructing robust networks in a synthetic biological perspective, it becomes crucial, as it is in computer programming, to ensure universal properties.

Interaction Signs. It is mandatory in a logical framework to define rigorously the meaning of the signs "+" or "−" on edges of an interaction graph (they may be loosely interpreted in the literature). We propose here a rather general and intelligible definition in the form of conditions called *observability constraints* and *additivity constraints* (not to be confused with the ASP integrity constraints).

A "+" (resp. "−") sign on an edge targeting a gene is understood as implying the existence of a couple of states $(s1, s2)$, with $s1$ just below the edge threshold, such that 1) $s2$ differs from $s1$ only by a $+1$ change in the value of the source gene, and 2) $s2$ has a greater (resp. lower) focal value for the target gene than $s1$. One may see why the transition graph G_4 (Fig. 2) respects the "+" label associated with the edge $a{\rightarrow}b$ (Fig. 1). The state $[0, 1]$ is such that the value of the source node a is lower than the threshold θ_a^1 of this edge. This state has a neighbouring state $[1, 1]$, which differs only in the value of a by a change of $+1$. Furthermore, this neighbour shows a positive tendency ($K_b^a = 2$) for b, indicating a future growth in expression level, while the state $[0, 1]$ shows a negative one ($K_b = 0$).

As all states of a cellular context have the same focal state, the existence of states $(s1, s2)$ is equivalent to the existence of cellular contexts $(c1, c2)$ of the target node which have the following properties for a "+"(resp. "−") sign: all states in $c1$ below the edge threshold and 1) $c2$ differs from $c1$ only by value of the source gene greater or equal than the edge threshold and 2) the focal value of the target gene in the context $c2$ has a greater (resp. lower) value than in context $c1$. In the transition graph G_4, considering again the positive interaction $a{\rightarrow}b$, such a couple of cellular contexts of b is for $c1$ the cellular context where $a < \theta_a^1 \wedge b < \theta_b^2$ holds and for $c2$ where $a \geq \theta_a^1 \wedge b < \theta_b^2$ holds.

Then, observability constraints for an interaction are expressed by a union of strict inequalities between kinetic parameters of the target of this interaction, just differing by one gene. For example, the observability constraint associated to the positive interaction $a{\rightarrow}b$ is $(K_b < K_b^a) \vee (K_b^b < K_b^{ab})$.

Additivity constraints are considered to indicate that **generally** no inhibition (resp. activation) can exist in case of a positive (resp. negative) interaction. For example, this means that in the general case for the positive interaction $a{\rightarrow}b$ where there is no proven inhibition (e.g. $(K_b > K_b^a) \vee (K_b^b > K_b^{ab})$ does not hold), then the negation of this inhibition is true (e.g. $(K_b \leq K_b^a) \wedge (K_b^b \leq K_b^{ab})$ holds). Solving this issue requires this ASP predicate:

```
addit(S,N1,N)  :- obs(S,N1,N), opposite_sign(S,Sp),
                        not obs(Sp,N1,N).
```

where `obs(S,N1,N)` means that if S=p (resp. S=m) then the edge N1$\rightarrow$N is an activation (resp. inhibition) and to introduce the integrity constraint:

```
:- addit(S, N1, N), opposite_sign(S, Sp),
   not -obs(Sp, N1, N).
```

where -obs(S,N1,N) is the (usual) negation of obs(S,N1,N). One should notice the default character of the rule defining addit. In the general case obs(Sp,N1,N) is not established, thus this rule is applied, with possible consequences. But if, due to the addition of new data, obs(Sp,N1,N) is established, there will be no inconsistency because of these consequences.

For reducing the number of ASs, and then increasing the number of properties deduced from them, a rather radical criterion (discussed in Section 5.2) can be applied by maximizing, with the para-logical operator #maximize, the number of addit atoms.

4   Applications

The three applications that are presented below illustrate the following advantages of the declarative approach: inconsistency checking and repairing, minimization of interaction and threshold numbers, and temporal series modeling.

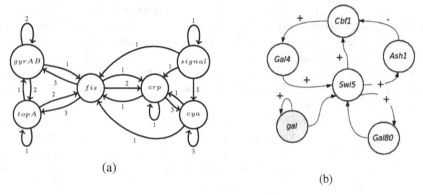

(a)

(b)

Fig. 3. (a) Interaction graph of the regulation of the carbon starvation response in *Escherichia coli* [8]. (b) Interaction graph of the IRMA network.

Carbon Starvation Response in *Escherichia Coli*. The declarative approach has been applied to the re-examination of the regulation network of the carbon starvation in *E. coli* presented in [13]. As long as environmental conditions are favourable, a population of *E. coli* bacteria grows quickly. The bacteria are in a state called exponential phase. Upon a nutritional stress due to carbon starvation, the bacteria are no longer able to maintain a fast growth rate. They enter in a stationary phase. Their response can be reversed as soon as the environmental conditions become favourable again. Modeling with the generate-and-test approach classically used for constructing GRN models led to a unique, instantiated and inconsistent model.

A declarative analysis of this network, using CLP and SAT solvers, has been presented in [8]. The network (Fig. 3 (a)) and biological observations on interactions, paths (stationary states and paths leading from the exponential phase to the stationary phase and vice-versa) and even characteristics of the shape of the DNA (*supercoiling*) were described using constraints. This analysis was resumed with an ASP implementation. We illustrate here the repairing of inconsistency.

Logical inconsistency was established. This showed rigorously the non-existence of alternative models, i.e. with a reasoning not based on the inconsistency of only one particular instantiated model. Repairing inconsistency was related to additivity constraints to the extent that they were not supported experimentally. Then repairing process proposed two solutions, that is to remove one constraint among $K_{gyrAB}^{fis} \leq K_{gyrAB}$ and $K_{topA} \leq K_{topA}^{fis}$. After biological investigations, it appeared that the first one should not be removed, but that the second could be, as it can be considered as not biologically plausible.

Computer performances stay very acceptable for solving such requests which require numerous recombination computations. For example, it is for determining the removable constraints that [8] reports the highest computer time (around 25min), with Prolog and a SAT solver loosely cooperating. This result was understandable because of the size of the solution space in this case. The same issue took 4s when solved by our ASP implementation (with a Core 2 Duo 3GHz, 4Go RAM).

Drosophila Embryo Gap Genes Network. This ASP approach has been applied in [7] to the regulatory network controlling the earliest steps of *Drosophila* embryo segmentation, i.e. the gap genes and their cross-regulations, under the additional control of maternal gene products [14,12,1]. Three kinds of data were considered: 1) published molecular genetic studies enabling the identification of the main actors (seven genes), as well as the establishment or the potentiality of cross-regulatory interactions, 2) qualitative information on the spatio-temporal expression profiles of the main genes involved in the process, giving seven regions with different stable states, 3) data available on the gap gene expression profiles for seven loss-of-function mutations, affecting maternal or gap genes. On the basis of this combination of interaction and gene expression constraints, the challenge was to identify the model(s) involving all established regulatory edges, along with a minimal set of potential ones, while minimizing the number of distinct thresholds.

In a first step, the consistency of the data was proven in 3338s, using a Linux PC with an Intel Core2Duo processor at 2.4GHz and 2.9GB of memory. Then, a unique regulatory network structure was obtained in 1016s which included only 2 potential interactions (on 11). Surprisingly, from this network, there was a unique instantiation of the thresholds minimizing the number of threshold values per component (obtained in 368s). Finally, some properties concerning the kinetic parameters were deduced: 52 parameters fixed (over 72), 12 inequalities connecting a threshold and a parameter, and 36 inequalities connecting two parameters.

IRMA Interaction Network. The IRMA (In vivo benchmarking of Reverse-engineering and Modelling Approaches) network [6] comprises five genes: Swi5, Ash1, Cbf1, Gal4 and Gal80, as well as one input (gal) and eight interactions (Fig. 3 (b)). These genes were chosen in the synthesis of the network so that different types of interactions were considered, including transcription regulation and protein-protein interaction, thereby capturing the behaviour of eukaryotic GRNs. In [6] Cantone et al. explored the dynamics of the IRMA network by measuring each gene's expression level in response to two different perturbations using qRT-PCR. In the first set of experiments, they shifted yeast cells from a glucose to galactose medium ('switch-on' experiments) and in the second set of experiments they shifted the cells from a galactose to glucose medium ('switch-off' experiments). The presence of galactose allows for increased transcription of Swi5 and is thus 'switch-on', while the opposite is true for the 'switch-off' experiments. From these data, two temporal series, composed of averaged gene expressions over five 'switch-on' and four 'switch-off' independent experiments, have been extracted.

Finding possible models of the IRMA network respecting these time series is a challenge proposed in [3]. The network is given in such terms that the order between the kinetic parameters is known. So the issue is to find a consistent order between thresholds and these parameters and between the thresholds themselves. Time series are formalized by CTL formulas of the form: $EF(prop_1 \wedge EF(...EF(prop_n)...))$ where $n = 12$ for the switch-off experiments and $n = 10$ for the switch-on experiments. A property $prop_i$ relates to the values of the components of a state and also to the derivative signs of these components. In [3] Batt et al. propose a modeling taking into account *singular states* (states admitting for a component a threshold value) leading to more states, together with the use of the model checking tool NuSMV. They claim, reviewing their work, that they provide more precise results and efficient coding.

When applying an ASP declarative approach to this problem (not yet published), we designed the appropriate constraints for expressing that a path satisfies a time series, while remaining in the Thomas framework, i.e. without singular states. The same number of parametrizations (64) were exhibited as in [3] [5], in 139s (compared to 885s, on a similar computer), showing that, at least for this analysis, taking account of singular states was not necessary.

5 Optimized Modeling and Additivity Constraints

Two issues arise for modeling additivity constraints: 1) escaping from a possible inconsistency that would result if these constraints would be imposed, 2) getting only the "most general" networks that is, intuitively, those which accept as many as possible additivity constraints compatible with the biological data. Recall that such a minimization of the number of resulting models leads to increasing the number of properties which can be deduced. The first issue is elegantly solved[6] with the default rule presented in Section 3.2. The second one can be solved with a naive modeling [9] which enumerates all models and uses the para-logical maximization operator of the *gringo* language as suggested in Section 3.2. However, both enumerating too many atoms and using para-logic operators are costly.

The refined modeling which follows attempts to reduce as far as possible these costs by taking advantage of the non monotonicity of ASP[7]. In Section 5.1 default rules [5] will be applied for lowering the enumeration of the models. In Section 5.2 an appropriate conjunction of defaults will be introduced for logically minimizing the number of resulting models.

---

[5] Parameterizations were found identical, except for two of them that were erroneous in [3].

[6] A more brutal para-logical approach with a maximization operator is proposed in [8].

[7] Note that this modeling allows also to associate additivity constraints even to edges that would not be labeled by any sign in the interaction graph but that would support nonetheless observability constraints as a result of the given behaviours.

5.1   Lowering the Enumeration of `kparam` Atoms

The predicate kparam(K,Ik) means that K is the value of Ik = k(N,CC) representing $K_N^{l(\sigma)}$ with CC being a term representing $l(\sigma)$. Its definition could be:
`1{kparam(K,k(N,CC)):val(N,K)}1 :- param(k(N,CC)).`
where val(N,K) means that K is a possible value of N and param(Ik) that Ik is a parameter identifier of N. Unfortunately, this definition leads to an exhaustive enumeration of all networks. To produce kparam atoms more savingly, the method consists to design specific rules related to the three origins of their production: observability constraints due to the interaction graph, additivity constraints and biological behaviors. All of these take advantage of the non monotonic property of ASP. Here, we will focuse on rules related to the second origin.

Producing correct couples of parameters is expressed via the rules:
`kparam(K, Ik) :- couple_kpr(K, Ik, _, _).`
`kparam(K_r, Ik_r) :- couple_kpr(_, _, K_r, Ik_r).`
The couple_kpr atoms due to additivity constraints are introduced by the default rule:
```
1{couple_kpr(K, Ik, K_r, Ik_r)
              : -param_obs(Sp, N1, N, K, K_r)}1
   :- neighboring_cell_cont(N1, N, Ik, Ik_r),
       addit(S, N1, N), opposite_sign(S, Sp).
```
where neighboring_cell_cont(N1, N, Ik, Ik_r) ensures that Ik and Ik_r identify two cellular contexts of N separated by the edge N1→N, Ik_r being Ik less the edge N1→N. The atom param_obs(S, N1, N, K, K_r) ensures that K and K_r, possible parameters of N, are in the right order regarding observability constraints with the sign S considering that K_r is associated to a cellular context that is the one associated to K less (with the above meaning) the edge N1→N. The predicate -param_obs is the (true) negation of param_obs. Note that expressing the logical conjunction (of inequalities) representing an additivity constraint requires every couple_kpr atoms associated to a couple of cellular contexts of N separated by the edge N1→N.

5.2   Conjunction of Defaults and Appropriate Use of the Para-Logical Maximization Operator

Defining the notion of "most general" networks regarding additivity constraints raise two different questions. The first is: "What is the logical definition of the answer sets satisfying observability constraints and behaviors (e.g. paths) to be retained regarding additivity constraints ?", the second: "Among the answer sets that result from the first question, are there still some to be eliminated ?". For minimizing the number of resulting models, there are two means: a logical one, coming from the minimality of the stable models and adapted to the first question, and a para-logical one via the maximization operator possibly useful for the second question.

For a set of parameters satisfying observability constraints and behaviors, it appears natural to ask for only answer sets having additivity constraints for all edges of all species, if such an answer set exists. If not, one would like to keep only the answer sets having additivity constraints for all edges of the species for which it is possible. For example, for the network of Fig. 1 with a behavior implying only $K_b^{ab} = 2$, there are the answer sets represented by the graphs $G_1, ..., G_6$ of Fig. 2 with additivity constraints for all edges of all species. But there are also other possible networks, for example with one edge of b with no additivity constraints. Unfortunately, the above modeling provides such undesirable networks, due to possible additivity constraints for one edge that implies the non additivity for some other edges. This is the case when having additivity constraints for the edge $a{\rightarrow}b$ ensured with the additional parameter values $K_b = 1$, $K_b^a = 1$ and $K_b^b = 0$ (thus $(K_b \leq K_b^a) \wedge (K_b^b \leq K_b^{ab})$ holds). These parameter values forbid additivity constraints for the edge $b{\rightarrow}b$ since $K_b \leq K_b^{ab}$ does not hold.

A simple program would help for illustrating this last point and for exhibiting a methodology to solve it. Let us consider the two following default rules which mimic cross influences between edges:

```
ad1_2 | ad1 :- not op_ad1.    ad2_1 | ad2 :- not op_ad2.
```
where the predicates prefixed by ad$i$ represent additivity constraints on edge i and the predicates op_ad$i$ represent the (rarely proven) conditions preventing additivity constraints on edge i. Influences between edges are modeled with the rules:
```
op_ad2 :- ad1_2.    op_ad1 :- ad2_1.
```
These four rules have three ASs: {ad1,ad2}, {ad1_2,op_ad2} and
{ad2_1,op_ad1}. The challenge is to transform these rules so that we only get the AS {ad1, ad2}. The methodology consists firstly in introducing the rules:
```
c :- op_ad1.    c :- op_ad2.
```
so that not c represents the case where both op_ad1 and op_ad2 are unknown or false, and secondly in completing the body of each original default rules with a kind of "guard" which is a tautological term provided with a default impact power:
```
ad1_2 | ad1 :- not op_ad1, 1{c, not c}1.
ad2_1 | ad2 :- not op_ad2, 1{c, not c}1.
```
The point is that when not c is true then ad1_2 (or ad2_1) cannot be deduced. But if the fact op_ad1. is added then two ASs are obtained: {op_ad1,ad2_1} and {op_ad1,ad2}.

Applied to our case, this methodology simply asks for the introduction of a guard of the following form into the body of the rule producing the additivity constraints:
```
1{not one_no_addit(N), one_no_addit(N)}1,
1{not one_no_addit, one_no_addit}1
```
where one_no_addit(N) means that one edge leading to the species N is not additive and one_no_addit means that one species has not all its entering edges additive.

Keeping only ASs with all possible additivity constraints may be prevented not only by behaviors but also by some parameter instantiations respecting observability constraints. Guarding in the same way the rule producing observability constraints solves this issue, in the case where no other behavior occur. This necessitates to prove that by doing so at least one AS is obtained. For this purpose, it has been necessary to show that the definition of interaction signs that we propose in Section 3.2 satisfies the following theorem: "Whatever are the interactions on a species N, one from each source, there exists an AS respecting all addivity constraints, in the absence of any other constraints on the parameters".

Meanwhile, there remain cases that generate further questions. For example, for the network of Fig. 1 with at least two stationary states, this new modeling provides nonetheless 3 ASs: one with the two edges of b being additive (graph G_6) and providing the stationary states $(0, 1)$ and $(0, 2)$, the two others respectively with one and not any of these edges being additive and providing the stationary states $(0, 1)$ and $(1, 0)$. The parameters values of these last ASs come from the stationary states ($K_b = 1$ and $K_b^a = 0$) and the observability constraints ($K_b^{ab} = 2$, $K_b^b = 0$ or $K_b^b = 1$). They are acceptable models from the "logical" point of view developed above. Consequently, discriminating some ASs among these three ASs can require para-logic standards like the one suggested in Section 3.2, i.e. the winners are those having in the whole the greatest number of additive edges, which eliminates these two possibly undesirable models.

6 Conclusion

We gave the main lines of an ASP modeling of Thomas' GRNs and illustrated the declarative approach interest with biological applications which make use of inconsistency repairing, minimization of interactions and temporal series representation. To take into account properties only generally true, we presented adequate default rules and an optimized modeling both reducing the number of used atoms (an efficient way for improving performances in computational logics), and keeping only most generally accepted models in a logical way (as opposed to a para-logical one). For this purpose, we were led to express conjunction of defaults with surprising and powerful ASP logical expressions, and to apply safely such expressions to Thomas' GRNs.

Few other teams use a declarative approach for analysing Thomas' networks. For this purpose, they use model checking tools and formalize paths with the temporal logic CTL, as in a seminal paper [4] on the subject. We showed advantages of our approach in Sections 3.2 and 4. It can be added that a constraint programming approach is well suited to avoid external processes to extract properties common to the set of consistent models, because these models are defined intensionaly as solutions of the constraints. However, as mentioned in Section 3.2, imposing (opposed to checking) in a convenient way CTL formulae of the form $AF\varphi$ with logic programming remains an issue. In [9] an acceptable solution taking advantage of the underlying non monotonicity of ASP is proposed. It based on the fact that a circular combination of rules like a1 :- a2. a2 :- a1. has only an empty AS while the corresponding logical formulae $a1 \Leftarrow a2.\quad a2 \Leftarrow a1.$ have also the model $\{a1, a2\}$ (which is not minimal).

Finally, it should be noted that ASP has also been applied successfully by other teams to the modeling of biological networks (see for example [16]), but not specifically, at our knowledge, to the modeling of Thomas' GRNs. The apparently closest ASP based work is reported in [10] but it deals only with simplistic instantiated deterministic Boolean networks, thus excluding issues coming from the multi-valued non deterministic Thomas'GRNs modeling in the declarative approach perspective, yet offered by the programming logic technology. Furthermore, it does not emphasize the non monotonicity offered by ASP.

Acknowledgments. This work was supported by Microsoft Research through its PhD Scholarship Programme. We acknowledge funding by the Agence Nationale de la Recherche through the CAPMIDIA project.

References

1. Alves, F., Dilao, R.: Modeling segmental patterning in drosophila: Maternal and gap genes. J. Theor. Biol. 241, 342–359 (2006)
2. Baral, C.: Knowledge Representation, Reasoning, and Declarative Problem Solving. Cambridge University Press, New York (2003)
3. Batt, G., Page, M., Cantone, I., Goessler, G., Monteiro, P., De Jong, H.: Efficient parameter search for qualitative models of regulatory networks using symbolic model checking. Bioinformatics 26(18), i603–i610 (2010)
4. Bernot, G., Comet, J.-P., Richard, A., Guespin, J.: Application of formal methods to biological regulatory networks: extending Thomas' asynchronous logical approach with temporal logic. Journal of Theoretical Biology 229(3), 339–347 (2004)
5. Besnard, P.: An Introduction to Default Logic. Springer (1989)
6. Cantone, I., Marucci, L., Iorio, F., Ricci, M.A., Belcastro, V., Bansal, M., Santini, S., di Bernardo, M., di Bernardo, D., Cosma, M.P.: A yeast synthetic network for in vivo assessment of reverse-engineering and modeling approaches. Cell 137(1), 172–181 (2009)
7. Corblin, F., Fanchon, E., Trilling, L., Chaouiya, C., Thieffry, D.: Automatic inference of regulatory and dynamical properties from incomplete gene interaction and expression data. In: Lones, M.A., Smith, S.L., Teichmann, S., Naef, F., Walker, J.A., Trefzer, M.A. (eds.) IPCAT 2012. LNCS, vol. 7223, pp. 25–30. Springer, Heidelberg (2012)
8. Corblin, F., Tripodi, S., Fanchon, É., Ropers, D., Trilling, L.: A declarative constraint-based method for analyzing discrete genetic regulatory networks. Biosystems 98, 91–104 (2009)
9. Farinas de Cerro, L., Inoue, K. (eds.): Logical Modeling of Biological Systems, pp. 167–206. Wiley, Chichester (2014)
10. Fayruzov, T., Janssen, J., Vermeir, D., Cornelis, C., Cock, M.D.: Modelling gene and protein regulatory networks with answer set programming. Int. J. Data Min. Bioinformatics 5(2), 209–229 (2011)
11. Gebser, M., Kaminski, R., Kaufmann, B., Ostrowski, M., Schaub, T., Thiele, S.: A user's guide to gringo, clasp, clingo, and iclingo (version 3.x) (October 2010)
12. Jaeger, J., Blagov, M., Kosman, D., Kozlov, K.N., Myasnikova, M.E., Surkova, S., Vanario-Alonso, C.E., Samsonova, M., Sharp, D.H., Reinitz, J.: Dynamical analysis of regulatory interactions in the gap gene system of drosophila melanogaster. Genetics 167, 1721–1737 (2004)
13. Ropers, D., de Jong, H., Page, M., Schneider, D., Geiselmann, J.: Qualitative simulation of the carbon starvation response in escherichia coli. Biosystems 84(2), 124–152 (2006)
14. Sánchez, L., Thieffry, D.: A logical analysis of the *Drosophila* gap-gene system. J. Theor. Biol. 211, 115–141 (2001)
15. Thomas, R., Kaufman, M.: Multistationarity, the basis of cell differentiation and memory. II. logical analysis of regulatory networks in terms of feedback circuits. CHAOS 11(1), 180–195 (2001)
16. Videla, S., Guziolowski, C., Eduati, F., Thiele, S., Gebser, M., Nicolas, J., Saez-Rodriguez, J., Schaub, T., Siegel, A.: Learning boolean logic models of signaling networks with ASP. Theoretical Computer Science (2014)

The Impact of Obesity on Predisposed People to Type 2 Diabetes: Mathematical Model

Wiam Boutayeb[1,*], Mohamed E.N. Lamlili[1], Abdesslam Boutayeb[1],
and Mohammed Derouich[2]

[1] URAC04, Department of Mathematics Faculty of Sciences Boulevard Mohamed VI,
BP: 717, Oujda, Morocco
{wiam.boutayeb,mohamed.lamlili}@gmail.com, x.boutayeb@menara.ma
[2] National School of Applied Sciences Univeristy Mohamed Premier Boulevard
Mohamed VI, BP: 717, Oujda, Morocco
mderouich2011@gmail.com

Abstract. Several mathematical models have been developed to simulate, analyse and understand the dynamics of β-cells, insulin and glucose. In this paper we study the effect of obesity on type 2 diabetes in people with genetic predisposition to diabetes. Equilibrium analysis and stability analysis are studied and the model shows three equilibrium points: a stable trivial pathological equilibrium point P_0, a stable physiological equilibrium point P_1 and a saddle point P_2. A simulation is carried out to understand the models behaviour.

Keywords: Type 2 diabetes, obesity, mathematical modeling, equilibrium analysis, stability analysis.

1 Introduction

According to the International Diabetes Federation (IDF) 2013, 8.3% of adults (382 million people) are living with diabetes all over the world with a particular growing trend of type 2 diabetes. [1]

Obesity is thought to be the primary cause of type 2 diabetes, especially for people having a genetic predisposition to the disease [2, 3]. Actually, an elevated level of Free Fatty Acids (FFA) leads to a chronic insulin resistance and thus β-cell apoptosis that consequently raises the blood glucose level [4].

Several studies have been carried out in order to understand the dynamics of insulin and glucose leading to diabetes. Bolie (1961) introduced a simple linear model, using ordinary differential equations in glucose and insulin [5]. Bergman et al. published the minimal model [6]. Diverse models based on the minimal model were published by different authors, including Derouich and Boutayeb (adding physical effort) [7], Roy and Parker dealt with the interaction between insulin, glucose and FFA [8]. Other authors introduced the dynamics of β-cells in the mechanisms leading to diabetes. Topp et al incorporate the β-cell mass,

---

* Corresponding author.

F. Ortuño and I. Rojas (Eds.): IWBBIO 2015, Part I, LNCS 9043, pp. 613–622, 2015.
© Springer International Publishing Switzerland 2015

insulin, and glucose kinetics [9]. Hernandez et al. proposed an extension of the Topp model by adding the surface insulin receptor dynamics [10]. Boutayeb et al. extended Topps model by stressing the effect of genetic predisposition to diabetes [11].

Our model is based on mathematical models published by Boutayeb et al[11], Roy et al[8] and Hernandez et al[10].

2   The Mathematical Model

In this model we assume, for glucose dynamics that the concentration of glucose in the blood is determined by a differential equation of the form:

$a - bG(t) - cI(t)R(t)G(t) + m_1(F(t) - F_b)$[9, 10].

Where $G(t)(g/l)$ is the concentration of glucose that increases by a rate a (in $mg/(dl.d)$) (glucose production by liver and kidneys) and decreases by a rate $bG(t)$ where b in (d^{-1})(independent of insulin) and a rate $cI(t)R(t)G(t)$ representing the glucose uptake due to insulin sensitivity c[10].We assume that the concentration of glucose increases by a rate $m_1(F(t) - F_b)$ where m_1 (in $l/d\mu mol$) which is the effect of FFA on glucose uptake.

Insulin dynamics is governed by the differential equation of the form: $\frac{d\beta(t)}{1+R(t)}\frac{G(t)^2}{e+G(t)^2} - fI(t) - fR(t)I(t)$, which has the same expression used by Henandez et al. Where $I(t)$($(\mu U)/ml$) is the plasma insulin concentration [10]. The dynamics of β-cell mass for predisposed people to type 2 to diabetes[2] as used in the model of Topp et al. takes the form: $(-g + hG(t) - iG(t)^2)$. Where $\beta(t)$ (mg) is the β-cell mass [10].

For the insulin receptors dynamics we keep the expression used by Hernandez et al.: $j(1 - R(t)) - kI(t)R(t) - lR(t)$. Where $R(t)$ is the insulin receptor [10]. The concentration of FFA increases by a rate $m_3(G(t) - G_b)$which represents the excess glucose used in lipogenesis and decreases by $m_2(F(t) - F(t)_b)$ which is the effect of the rate of insulin on FFA. Where $F(t)$ ($(\mu mol)/l$)

So, the model is written as follows:

$$\frac{dG(t)}{dt} = a - bG(t) - cI(t)R(t)G(t) + m_1(F(t) - F_b)$$

$$\frac{dI(t)}{dt} = \frac{d\beta(t)}{1 + R(t)}\frac{G(t)^2}{e + G(t)^2} - fI(t) - fR(t)I(t)$$

$$\frac{d\beta(t)}{dt} = (-g + hG(t) - iG(t)^2)$$

$$\frac{dR(t)}{dt} = j(1 - R(t)) - kI(t)R(t) - lR(t)$$

$$\frac{dF(t)}{dt} = -m_2(F(t) - F(t)_b) + m_3(G(t) - G_b)$$

3 Equilibrium Analysis

The steady state solutions are the solutions of the equations:

$$a - bG(t) - cI(t)R(t)G(t) + m_1(F(t) - F_b) = 0$$
$$\frac{d\beta(t)}{1 + R(t)}\frac{G(t)^2}{e + G(t)^2} - fI(t) - fR(t)I(t) = 0$$
$$-g + hG(t) - iG(t)^2 = 0$$
$$j(1 - R(t)) - kI(t)R(t) - lR(t) = 0$$
$$-m_2(F(t) - F(t)_b) + m_3(G(t) - G_b) = 0$$

This model has three equilibrium points:
$P_0(G_0, I_0, \beta_0, R_0, F_0), P_1(G_1, I_1, \beta_1, R_1, F_1)$ ans $P_2(G_2, I_2, \beta_2, R_2, F_2)$
• The first equilibrium point $P_0 = (G_0, I_0, \beta_0, R_0, F_0)$ is a trivial pathological point.
With:

$$G_0 = \frac{m_1 m_3 G_b - a m_2}{m_1 m_3 - m_2 b},$$
$$I_0 = 0,$$
$$\beta_0 = 0,$$
$$R_0 = \frac{j}{j+1},$$
$$F_0 = \frac{a m_3 - b m_3 G_b - m_1 m_3 F_b + m_2 b F_b}{-m_1 m_3 + m_2 b}$$

• The second equilibrium point $P_1 = (G_1, I_1, \beta_1, R_1, F_1)$ is a physiological point.

With:

$$G_1 = \frac{h - \sqrt{h^2 - 4ig}}{2i},$$
$$I_1 = \frac{-jG_b + aj + jm_1F_1^* - jm_1F_b + la + lm_1F_1^* - lm_1F_b}{ak - cjG_1^* + m_1kF_1^* - m_1kF_b},$$
$$\beta_1 = \frac{fI_1^*(R_1^* + 1)(e + G_1^{*2})}{dG_1^*},$$
$$R_1 = \frac{ak - cjG_1^* + km_1F_1^* - km_1F_b)}{(G_1^*(bk - cj - cl)},$$
$$F_1 = \frac{2m_2iF_b + hm_3 - m_3\sqrt{h^2 - 4ig} - 2im_3G_b}{2im_2}$$

The third equilibrium point $P_2 = (G_2, I_2, \beta_2, R_2, F_2)$
with:

$$G_2 = \frac{h + \sqrt{h^2 - 4ig}}{2i},$$

$$I_2 = \frac{-jG_b + aj + jm_1F_2^* - jm_1F_b + la + lm_1F_2^* - lm_1F_b}{ak - cjG_2^* + m_1kF_2^* - m_1kF_b},$$

$$\beta_2 = \frac{fI_2^*(R_2^* + 1)(e + G_2^{*2})}{dG_2^*},$$

$$R_2 = \frac{ak - cjG_2^* + km_1F_2^* - km_1F_b}{G_2^*(bk - cj - cl)},$$

$$F_2 = \frac{2m_2iF_b + hm_3 + m_3\sqrt{h^2 - 4ig} - 2im_3G_b}{2im_2}$$

The conditions of existence of the equilibrium points are presented in the following Table:

Table 1. Conditions of existence

	Coordinates	Conditions of existence
P_0	$(G_0, I_0, \beta_0, R_0, F_0)$	$\frac{m_1m_3}{m_2} < b,\ \frac{a}{b} > G_b$
P_1	$(G_1, I_1, \beta_1, R_1, F_1)$	$\frac{h - \sqrt{h^2 - 4ig}}{2i} < G_b$, $bk > c(j + l)$
P_2	$(G_2, I_2, \beta_2, R_2, F_2)$	$\frac{h + \sqrt{h^2 - 4ig}}{2i} > G_b,\ bk > c(j + l)$

4   Stability Analysis

The stability analysis based on variational principle is used. The variational matrix of the system at any point $P_i(i = 0, 1, 2)$ is written as:

$$\begin{pmatrix} -b - cIR & -cRG & 0 & -cIG & m_1 \\ \frac{2d\beta Ge}{(R+1)(e+G^2)^2} & -f - fR & \frac{dG^2}{(R+1)(e+G^2)} & \frac{-d\beta G^2}{(R+1)^2(e+G^2)} - fI & 0 \\ (h - 2iG)\beta & 0 & -g + hG - iG^2 & 0 & 0 \\ 0 & -kR & 0 & -j - kI - l & 0 \\ m_3 & 0 & 0 & 0 & -m_2 \end{pmatrix}$$

4.1   The Stability Analysis of the P_0

The eigenvalues of the variational matrix at P_0:

$$\lambda_1 = -j - l$$

$$\lambda_2 = -g + hG - iG^2$$
$$\lambda_3 = -\frac{1}{2}b - \frac{1}{2}m_2 + 1/2\sqrt{4m_1m_3 + b^2 - 2bm_2 + m_2^2}$$
$$\lambda_4 = -\frac{1}{2}b - \frac{1}{2}m_2 - 1/2\sqrt{4m_1m_3 + b^2 - 2bm_2 + m_2^2}$$
$$\lambda_5 = -f - fR$$

Since: $\lambda_1, \lambda_2, \lambda_3, \lambda_4, \lambda_5 < 0$ (following the conditions of existence in Table1) we conclude that the point P_0 is stable.

4.2   The Stability Analysis of the P_1:

The calculus of P_1s eigenvalues is computed using numerical approximation by Maple, showing that P_1 is a stable node.

4.3   The Stability Analysis of the P_2:

We put:

$$A = b + cIR$$
$$B = cRG$$
$$C = cIG$$
$$D = \frac{2d\beta Ge}{(R+1)(e+G^2)^2}$$
$$E = \frac{dG^2}{(R+1)(e+G^2)}$$
$$F = f + fR$$
$$J = \frac{d\beta G^2}{(R+1)^2(e+G^2)} + fI$$
$$K = kR$$
$$L = j + kl + l$$
$$M = (h - 2iG)\beta$$

Given the characteristic polynomial,

$\lambda^5 +$
$(n + L + F + A)\lambda^4 +$
$(m_1m_3 + m_2L + m_2F + nA - KJ + LF + LA + DB + FA)\lambda^3 +$
$(m_1m_3L - m_1m_3F - m_2KJ + m_2LF + m_2LA + m_2DB + m_2FA - KDC - KJA + LJA + LDB + LBD + LFA + MBE)\lambda^2 +$
$(m_1m_3KJ - m_1m_3LF - m_2KDC - m_2KJA + m_2LDB + m_2LFA + m_2MBE - KEMC + LMBE)\lambda +$
$m_2KEMC - m_2LMBE$

Following the conditions of existence of P_2 given in Table1: $m_2KEMC - m_2LMBE < 0$, whereas the coefficient of the highest order is positive. We conclude that P_2 is unstable since the necessary condition of routh Hurwitz is not satisfied.

5   Simulation

Our simulation is based on the parameters given in Table 2. [8–11].

Table 2. Parameters for an average healthy person

Param	Value	Units	Biological Interpretation
a	864	$\frac{mg}{dl\,d}$	glucose production rate by liver when G $=0$
b	1.44	d^{-1}	glucose clearance rate independent of insulin
c	0.85	$\frac{ml}{\mu U\,d}$	insulin induced glucose uptake rate
d	43.2	$\frac{\mu U}{ml\,d\,mg}$	β-cell maximum insulin secretory rate
e	20000	$\frac{mg^2}{dl^2}$	gives inflection point of sigmoidal function
f	216	d^{-1}	whole body insulin clearance rate
g	0.06	d^{-1}	β-cell natural death rate
h	0.572-3	$\frac{dl}{mg\,d}$	determines β-cell glucose tolerance range
i	0.252e-5	$\frac{dl^2}{mg^2 d}$	determines β-cell glucose tolerance range
j	2.64	$\frac{1}{d}$	insulin receptor recycling rate
k	0.02	$\frac{ml}{\mu U d}$	insulin dependent receptor endocytosis rate
l	0.24	d^{-1}	insulin independent receptor endocytosis rate
m_1	0.0864	$\frac{l}{d\mu mol}$	the effect of plasma FFA on glucose uptake
m_2	43.2	d^{-1}	the influence of insulin
m_3	97.92	ml^{-1}	the rate constant representing plasma FFA concentration
G_b	98	$\frac{md}{dl}$	the basal glucose concentration
F_b	380	$\frac{\mu mol}{l}$	the basal FFA concentration

Using the parameters giving in Table 2 yields the results presented in Table 3.

Table 3. Stability analysis using the values of parameters given in Table 2

equilibrium points (G, I, β, R, F)	Stability
(679, 0, 0, 0.9, 1698.2)	stable
(82, 12.65, 853.32, 0.85, 343.7)	stable
(145, 6.13, 211.25, 0.88, 486.6)	instable

In this model we considered the effect of obesity on type 2 diabetes. It was shown in the first point P_0 that an elevated rate of FFA has an impact on insulin secretion and insulin-resistance and hence on the development of type 2 diabetes.

The results of the simulation using parameters given in Table2 with (I(0)=6.5, β(0)=220, R(0)=0.87 and F(0)=580) are illustrated by Fig1, Fig2, Fig3, Fig4 and Fig5.

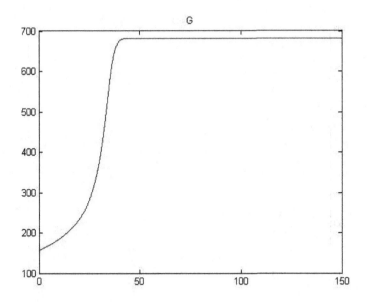

Fig. 1. Plot of the trajectory of G over 150 days

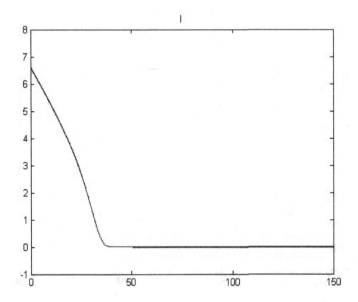

Fig. 2. Plot of the trajectory of I over 150 days

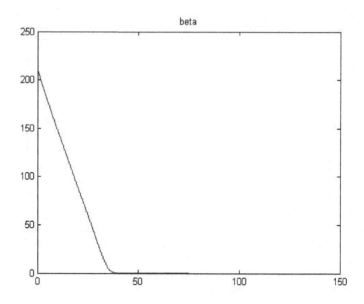

Fig. 3. Plot of the trajectory of β over 150 days

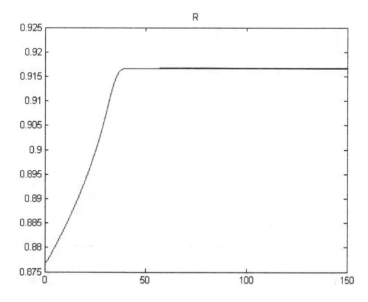

Fig. 4. Plot of the trajectory of R over 150 days

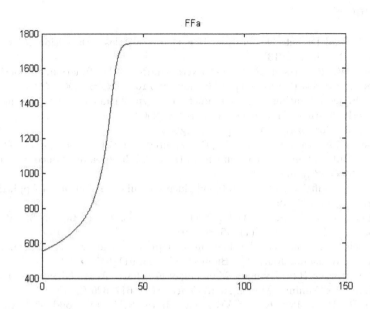

Fig. 5. Plot of the trajectory of FFA over 150 days

The mathematical model has three equilibrium points: a stable pathological point corresponding to an hyperglycemic state with zero level of β-cell mass and insulin $P_0(679, 0, 0, 0.9, 1698)$, and a high level of FFA, a stable physiological point with basal values of FFA, glycemia, insulin, insulin receptor and β-cell mass $P_1(82, 12.645, 853.32, 0.85, 343.7)$, and an unstable saddle point with intermediate values of FFA, glycemia, insulin, insulin receptor and β-cell mass $P_2(145, 6.13, 211.25, 0.88, 486.6)$.

6   Conclusion

In this model we considered the effect of obesity on type 2 diabetes in presence of pre-disposition to diabetes on the dynamics of β-cells, insulin, glucose, insulin receptors and Free Fatty Acids (FFA). It was shown that the pathological and physiological equilibrium points are stable and the saddle equilibrium point with intermediate values of Glucose, Insulin, β-cell mass, insulin receptors and FFA is unstable. An elevated rate of FFA, leads to an evolution towards the pathological point (G=679,I=0,β=0,R=0.9,FFA=1698.2). This model confirms that FFA has an impact on insulin secretion and insulin-resistance and hence on the development of type 2 diabetes for people with predisposition to diabetes.

References

1. International Diabetes Federation: IDF Diabetes Atlas fifth edition 2013 update. Brussels, Belgium (2013)
2. Poitout, V., Robertson, R.P.: Fuel Excess and -Cell Dysfunction. Endocrine Reviews 29, 351-366 (2008), http://dx.doi.org/10.1210/er.2007-0023
3. Beale, S.: Deteriorating Beta-Cell Function in Type 2 Diabetes a Long-Term Model. Quarterly Journal of Medicine, 96, 281-288 (2003), http://dx.doi.org/10.1093/qjmed/hcg040
4. Turpeinen, A., Takala, T., Nuutila, P.: Impaired Free Fatty Acid Uptake in Skeletal Muscle But Not in Myo-cardium in Patients With Impaired Glucose Tolerance. DIABETES 48 (June 1999)
5. Bolie, V.: Coefficients of normal blood glucose regulation. Journal of Applied Physiology 16, 783–788 (1961)
6. Bergman, R.N., Finegood, D.T., Ader, M.: Assessment of Insulin Sensitivity in Vivo. Endicrine Reviews 6(1), 45–86 (1985)
7. Derouich, M., Boutayeb, A.: The effect of physical exercise on the dynamics of glucose and insulin. Journal of Biomechanics 35, 911–917 (2002)
8. Roy, A., Parker, R.S.: Dynamic Modeling of Free Fatty Acids, Glucose, and Insulin: An Extended Minimal Model. Neurosiences 8(9), 617–626 (2006)
9. Topp, B., Promislow, K., De Vries, G., Miura, R.M., Finegood, D.T.A.: Model of -Cell Mass, Insulin, and Glucose Kinetics: Pathways to Diabetes. Journal of Theoretical Biology 99A, 605-619 (2000), http://dx.doi.org/10.1006/jtbi.2000.2150
10. Hernandez, R.D., Lyles, D.J., Rubin, D.B., Voden, T.B., Wirkus, S.A.: A Model of -Cell Mass, Insulin, Glucose and Receptor Dynamics with Applications to Diabetes. Tech-nical Report, Biometric Department, MTBI Cornell University (2001), http://mtbi.asu.edu/research/archive/paper/model-%CE%B2-cell-mass-insulin-glucose-and-receptor-dynamics-applications-diabetes
11. Boutayeb, W., Lamlili, M., Boutayeb, A., Derouich, M.: Mathematical Modelling and Simulation of β-Cell Mass, Insulin and Glucose Dynamics: Effect of Genetic Predisposition to Diabetes. J. Biomedical Science and Engineering 7, 330–342 (2014)

Gene Expression vs. Network Attractors

Gianfranco Politano, Alessandro Savino, and Alessandro Vasciaveo

Politecnico di Torino, Corso Duca degli Abruzzi 24, 10129 Torino, Italy
{firstname,lastname}@polito.it
www.sysbio.polito.it

Abstract. Microarrays, RNA-Seq, and Gene Regulatory Networks (GRNs) are common tools used to study the regulatory mechanisms mediating the expression of the genes involved in the biological processes of a cell. Whereas microarrays and RNA-Seq provide a snapshot of the average expression of a set of genes of a population of cells, GRNs are used to model the dynamics of the regulatory dependencies among a subset of genes believed to be the main actors in a biological process. In this paper we discuss the possibility of correlating a GRN dynamics with a gene expression profile extracted from one or more wet-lab expression experiments. This is more a position paper to promote discussion than a research paper with final results.

1   Objectives

Microarrays, RNA-Seq, and Gene Regulatory Networks (GRNs) are widely used to study and model gene expression status and mechanisms in a cell. Whereas microarrays and RNA-Seq techniques are the output of wet-lab experiments, GRNs are used for in-silico analysis and simulation of biological pathways [8-9]. GRN models, commonly known as "pathways", are available in several web repositories as Kegg, WikiPathway, Reactome [10-12]. Nevertheless, despite focusing on the same biological process, gene expression technologies and GRN are intrinsically different. The first big difference is in the type of information they can provide. Gene expression experiments provide an indication (more qualitative or quantitative depending on the technology) of the expression level of a large amount of genes in a particular physiological condition. Nevertheless they cannot provide any reliable information about causal relationships between genes (which gene activates which). On the contrary, GRNs are used to model regulatory relationships between genes, rather than focusing on the precise expression status of a single gene in a particular physiological condition.

A second important difference is the source of information. A GRN ideally models the regulatory network dynamics of a single cell, whereas gene expression technologies show the "average" expression status of all the cells present in the sample used in the experiment. This can be misleading and for this reason in certain wet-lab experiments cells are "synchronized" in order to increase the probability of having most of them in the same state.

F. Ortuño and I. Rojas (Eds.): IWBBIO 2015, Part I, LNCS 9043, pp. 623–629, 2015.
© Springer International Publishing Switzerland 2015

Despite these differences, GRNs and gene expression experiments are very often used together to try to elucidate several regulatory mechanisms involved in the life of a cell. In this case the obvious assumption is that the Microarray or the RNA-Seq show a "live" view of the regulatory mechanisms modeled by the GRN. The problem is that there is no measure to quantify how much a GRN model is "compliant" or "compatible" with the gene expression profiles result of a lab experiment. If they provide data referring to the same biological process, then there should be a way to correlate them. To our knowledge, in literature there is no formal way designed to understand if a gene expression experiment is actually showing expression data compatible with a GRN model. "Compatible" in this case means: "that it has a high probability of having been generated by a set of genes regulated as described by the GRN model".

The goal of this paper is therefore to discuss the following problem: "Is it possible to find a relation between a GRN model and the expression profile of its genes extracted from one or more lab experiments?" In this paper we are not presenting final and conclusive data (yet); we try instead to formalize the problem and explaining the methodology we are applying to find an answer.

2   Methods

The problem stated before requires subdividing the investigation into several steps. The first and most important one is to understand which data needs to be used, and how to make the data extracted from a GRN and from an expression experiment statistically comparable. The second step is to choose the correct statistical method to compare the data, and the third and final one is to understand if and which part of the network dynamics extracted from the GRN are correlated to the ones showing in the expression experiments.

Gathering the Correct Data

When running a gene expression experiment, especially on a tissue sample, the resulting data are not a clear image of the expression profile of the target genes. They are instead the quantification of the average expression of certain genes in the cells of the sample. In general, the activation of a particular gene in the sample is either static or dynamic. It is static if the expression of that gene does not change in time, for example in housekeeping genes. It is dynamic if the gene is involved in one or more biological processes that are active at the time of the experiment. In this case, the expression of the same gene in different cells is likely to be different, depending on the particular status of the cell. As an analogy, think about taking several different pictures of the same street always from the same position but at different times. In the street only two cars are present. One is blue and it is always parked (static expression gene), and one is red and always moving (dynamic expression gene). Imagine now overlapping all the pictures and trying to pinpoint the position of the blue and red car. The blue parked car will always be in focus and still, while the red moving car will appear in several different positions. It will not be possible to pinpoint its exact position, but only to have an "average" idea of its position.

Microarrays provide a less quantitative data than RNA-Seq [6]. In Microarrays the expression of a gene is converted into a real number corresponding to the intensity of the corresponding spot. Nevertheless, because of technological biases, the expression of a gene is not easily comparable with the expression of a different gene. For this reason Microarrays are usually the choice for differential expression experiments, where the focus is on the difference of expression of a set of genes between two different phenotypes [1-2]. RNA-Seq technology instead is able to provide a much more reliable quantitative value of the expression of a gene. Databases like Gene Atlas [3] store large sets of experiments that characterize the average expression of most known genes in different tissues and in different conditions (baseline and pathological). For these reasons, RNA-Seq data appears to be better suited for the study discussed in this paper.

Data concerning the network dynamics collectable from a GRN requires making a number of assumptions. To make the simulation of a realistic network (tens of nodes) computationally feasible in a reasonable time, it is possible to use a Boolean model to limit the possible states of each node/gene to only two possible values. Despite this simplification, this allows studying the network dynamics in terms of "steady-states" or "equilibrium-states" or, more, formally, "attractors". An attractor is a set of states (one or more) towards which the network tends to converge. Once an attractor is reached (and the inputs of the network remain steady) it is not possible to transition out of it, unless external perturbations are applied. In case of a point attractor, the system's state freezes whenever the network enters the attractor. Differently, cyclic attractors (most common in GRNs) show a cyclic behavior of the system: once a cell falls into one of the states belonging the attractor, the system keeps cyclically moving among the attractor's states. Attractors are also present in multi-valued or even continuous models (where the state of each gene is a real number), but their computation is computationally unfeasible for networks of more than a few nodes.

From the definition of "attractor", it is reasonable to assume that they represent high probable states for a cell [7], and therefore can be considered the states that contribute more to the expression of the genes. "Attractors" are therefore the network dynamics data that we consider more suitable to be compared with a gene expression profile.

Making Data Statistically Comparable

As we discussed before, if the pathway is correct (i.e. it reliably represents the real regulatory dynamics of the genes included in the model), then we can make the assumption that, without strong external perturbations, at any given moment a cell has a high probability of being into an attractor state. If this assumption holds, then the expression profile of a gene expression experiment should be correlated to the attractors in which these cells were during the experiment.

To prove this, we need to generate, from both the GRN model and the expression experiments, two statistically comparable datasets.

From the GRN model it is possible to compute the set of attractor states by running the Enhanced Boolean Network Simulator presented in [4][8]. The result is a set of n Boolean arrays. Each element of the arrays represent a gene in the network, and its value ('0' or '1') its expression value. These steps are those depicted in violet in Figure 1.

Given a subset of attractors 'a', for each gene (array position) it is possible to compute the Attractor Expression Frequency (AEF_a) as the number of '1' that the gene shows in the attractors set 'a'. Now, if (most of) the cells participating in the expression experiment are in one of the attractor states of 'a', then the average expression of a particular gene should be correlated with its AEF_a. Obviously, the set 'a' of attractors to be used to compute AEF has to be carefully selected. This step will be explained later.

From the gene expression experiment point of view, shown in orange in Figure 1, and for the particular purpose of this work, the normalized value of FPKM (Fragments Per Kilobase of transcript per Million mapped reads) available from the Expression Atlas [5] of the EMBL European Bioinformatics Institute currently appears to be the best choice, since it represents an "expression frequency" that can be statistically compared with the AEF measure extracted from the GRN model.

Statistically Comparing Data

Once these two datasets are obtained, we need to statistically compare them. To perform this comparison it is worth to consider that the normalized FPKM of each gene can be considered as a frequency of the appearance of the gene expressed in the specific experiment. The gene profile composed of FPKM values therefore represents a profile of **expected frequencies** for all genes. On the other hand the AEF represents a set of **observed frequencies** from the simulation of the GRN. In order to compare the two set of frequencies a chi-square test can be used to determine whether there is a significant difference between the expected frequencies and the observed frequencies. The acceptance of the test's null hypothesis would be a confirmation that the two set of frequencies are not statistically different, giving an indication that the AEFs are somehow linked to the gene expression profiles obtained in the lab.

Looking for a Correlation

If we compared the expression profile with the complete set of attractors of the target GRN, we would probably not obtain any significant result. From a theoretical point of view, this is intuitive: if the attractor set represents a set of the network possible states, it is very likely that the number of '1' and '0' in each position (and therefore the expression of the corresponding gene) will be very similar and therefore the AEF would be probably close to 0,5. But biologically, considering all attractors of the network does not make a lot of sense: of all the possible attractors of a network, probably only a subset is biologically valid and/or significant. Moreover, within this subset, not all attractors are equally probable. It is instead very likely that there are very probable attractor states, as well as very rare ones. This bias could affect the result of the statistical tests. It is therefore necessary to devise an algorithm able to find the best set of attractors that better correlates with the gene expression profile. We plan to solve the problem with an evolutionary algorithm (in green in Figure 1), since they are very efficient in analyzing very large solution spaces.

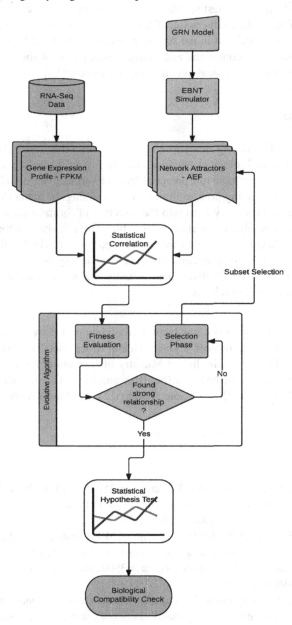

Fig. 1. Flow chart of the overall method

A simple algorithm consists in the evaluation of the statistical correlation between AEF and FKPM measures performed by a fitness function. If the correlation could be considered statistically strong (i.e. the Chi-square value is greater than a threshold, or a rho value between 0.5 and 1 if the correlation test is the Spearman's rank-order), then a statistical hypothesis test should be performed in order to confirm the statistical significance of the result. Otherwise, if no strong relationship is found, then a new attractors set is collected within the selection phase and new AEF measures are computed. The termination conditions and genetic operators (not shown in Figure 1 for brevity) are the same of classical evolutionary algorithms [13].

3   Conclusions

The main reason for writing this paper is to stimulate other researchers to look into this idea. The use of biological network already proved to be successful in several Life Sciences areas, but their full potential has not surfaced yet mainly because there is no way of understanding how well a GRN is modeling the actual regulatory mechanisms in the cell. This study goes into the direction of formalizing a methodology to investigate if making this correlation is possible. The possible outcomes are significant. First of all having a realistic measure of a pathway model "biological compatibility" would allow to dramatically increase the quality of the pathway network models, and, concurrently, making them actually usable for understanding the complex systems that regulate cell life.

Acknowledgments. This work has been partially supported by grants from Regione Valle d'Aosta (for the project: "Open Health Care Network Analysis" - CUP B15G13000010006), from the Italian Ministry of Education, University & Research (MIUR) (for the project MIND - PRIN 2010, and FIRB Giovani RBFR08F2FS-002 FO), from the Compagnia di San Paolo, Torino (DT), and from AIRC 2010 (IG 10104 DT).

References

[1] Seita, J., Sahoo, D., Rossi, D.J., Bhattacharya, D., Serwold, T., Inlay, M.A., Weissman, I.L.: Gene Expression Commons: an open platform for absolute gene expression profiling. PloS One 7(7), 40321 (2012), doi:10.1371/journal.pone.0040321

[2] McCall, M.N., Uppal, K., Jaffee, H.A., Zilliox, M.J., Irizarry, R.A.: The Gene Expression Barcode: leveraging public data repositories to begin cataloging the human and murine transcriptomes. Nucleic Acids Research 39(Database issue), 1011–1015 (2011), doi:10.1093/nar/gkq1259

[3] Robert, P., et al.: Expression Atlas update—a database of gene and transcript expression from microarray-and sequencing-based functional genomics experiments. Nucleic Acids Research 42(D1), D926-D932 (2014)

[4] Benso, A., Di Carlo, S., Politano, G., Savino, A., Vasciaveo, A.: An Extended Gene Protein/Products Boolean Network Model Including Post-Transcriptional Regulation. Theoretical Biology and Medical Modelling 11(Suppl 1), 1–17, ISSN: 1742-4682

[5] http://www.ebi.ac.uk/gxa/home (last visit, December 2014)

[6] Ugrappa, N., Waern, K., Snyder, M.: RNA-Seq: A Method for Comprehensive Transcriptome Analysis. Current Protocols in Molecular Biology, 4–11 (2010)

[7] Huang, S., Eichler, G., Bar-Yam, Y., Ingber, D.E.: Cell Fates as High-Dimensional Attractor States of a Complex Gene Regulatory Network. Physical Review Letters 94(12), 128701 (2005), doi:10.1103/PhysRevLett.94.128701

[8] Benso, A., Di Carlo, S., Rehman, H.U., Politano, G., Savino, A., Squillero, G., Vasciaveo, A., Benedettini, S.: Accounting for Post-Transcriptional Regulation in Boolean Networks Based Regulatory Models. In: International Work-Conference on Bioinformatics and Biomedical Engineering, IWBBIO 2013, pp. 397–404 (2013)

[9] Politano, G., Benso, A., Di Carlo, S., Savino, A., Ur Rehman, H., Vasciaveo, A.: Using Boolean Networks to Model Post-transcriptional Regulation in Gene Regula tory Networks. Journal of Computational Science 5(3), 332–344 (2014), ISSN 1877-7503

[10] Minoru, K., Goto, S.: KEGG: kyoto encyclopedia of genes and genomes. Nucleic Acids Research 28(1), 27–30 (2000)

[11] Thomas, K., et al.: WikiPathways: building research communities on biological pathways. Nucleic Acids Research 40(D1), D1301–D1307 (2012)

[12] David, C., et al.: Reactome: a database of reactions. Pathways and Biological Processes. Nucleic Acids Research (2010)

[13] Thomas, B., Fogel, D.B., Michalewicz, Z.: Evolutionary computation 1: Basic algorithms and operators, vol. 1. CRC Press (2000)

A Computational Domain-Based Feature Grouping Approach for Prediction of Stability of SCF Ligases

Mina Maleki[1], Mohammad Haj Dezfulian[2], and Luis Rueda[1]

[1] School of Computer Science,
[2] Department of Biological Sciences,
University of Windsor, 401 Sunset Avenue, Windsor, Ontario, N9B 3P4, Canada
{maleki,hajdez,lrueda}@uwindsor.ca

Abstract. Analyzing the stability of SCF ubiquitin ligases is worth investigating because these complexes are involved in many cellular processes including cell cycle regulation, DNA repair mechanisms, and gene expression. On the other hand, interactions of two (or more) proteins are controlled by their domains – compact functional units of proteins. As a consequence, in this study, we have analyzed the role of Pfam domain interactions in predicting the stability of protein-protein interactions (PPIs) that are known or predicted to occur involving subunit components of the SCF ligase complex. Moreover, employing the most relevant and discriminating features is very important to achieve a successful prediction with low computational cost. Although, different feature selection methods have been recently developed for this purpose, feature grouping is a better idea, especially when dealing with high-dimensional sparse feature vectors, yielding better interpretation of the data. In this paper, a correlation-based feature grouping (CFG) method is proposed to group and combine the features. To demonstrate the strength of CFG, two filter methods of χ^2 and correlation are also employed for feature selection and prediction is performed using different methods including a support vector machine (SVM) and k-Nearest Neighbor (k-NN). The experimental results on a dataset of SCF ligases indicate that employing feature grouping achieves significant increases of 10% for svm and 13% for k-NN, being more efficient than employing feature selection in identifying a set of relevant features

Keywords: protein-protein interaction, domain-domain interaction, prediction of complex stability, feature grouping.

1 Introduction

The largest class of E3 ligases that are responsible for the selection of up to 20% of the proteome for ubiquitin mediated degradation is SCF ubiquitin ligases [1]. This class of ligases is involved in many biological processes within a cell including cell cycle regulation, DNA repair mechanisms, gene expression, kinase/phosphotase cascades, RTK signaling, and angiogenesis. An SCF ligase has been shown to be comprised of four subunits of RBX1, CUL1, SKP1, and at least one F-Box protein and each subunit comprised on one or more domains.

Domains are compact structural and functional units of proteins that mediate the interactions of two or more proteins [2]. On the other hand, protein-protein interactions

F. Ortuño and I. Rojas (Eds.): IWBBIO 2015, Part I, LNCS 9043, pp. 630–640, 2015.

(PPIs) can be classified into two groups of obligate and non-obligate complexes based on their interaction stability. Obligate complexes are more stable and have high-affinity interactions than non-obligate ones [3]. As a consequence, prediction and analysis of PPI types and their relevant properties employing domain information has been studied from different perspective [4–8].

Moreover, applying feature selection to obtain more relevant features before running a classifier is important in order to reduce the dimensionality of the data by discarding redundant and/or irrelevant features, and, hence, reducing the prediction time, while improving the classification performance [9]. As a consequence, automatic feature selection methods have been studied for a long period of time in pattern recognition and have been successfully applied in many fields especially biological problems such as prediction of tyrosine sulfation [10] or lysine ubiquitination [11], prediction of protein-protein interactions [12, 13] or protein-nucleic acids interactions [14], and gene selection [15, 16].

However, defining the optimal number of features is a main challenge in feature selection. In other words, after scoring and ranking the features, the main question is: "how many features should be selected for classification?". Moreover, in many applications, the features are correlated to each other and tend to work well in groups [17]. For instance, it has been claimed in many biological studies that there are some regulatory relationships among genes that work together based on their functions [18]. In this case, feature grouping instead of feature selection helps achieve a better insight on the data and find these kinds of relations. On the other hand, grouping features before performing classification is also a good approach for sparse high-dimensional data to combine similar features together, which leads to reduce the sparsity and estimation variance, and improves the stability of feature selection methods and classifiers [19, 20]. Lasso and its variants [21], pursuit [22], and OSCAR [23] are some of the most well-known feature grouping methods.

A potential application of feature grouping is prediction of protein-protein interaction types of SCF-ligase complexes considering domain interactions as prediction properties. This is because (a) there are less interacting domains in each PPI, and hence, the final extracted feature vectors are too sparse, and (b) grouping features contributes to a better biological insight of PPIs and their interacting domains.

In this paper, which is an extended version of [4], the idea of grouping features based on their correlation scores, namely correlation-based feature grouping (CFG), is proposed. This method can find the best number of groups, dynamically. Considering a manually curated PPI dataset of SCF-ligase complexes, performances of various classification techniques, including support vector machines (SVM) and k-Nearest Neighbor (k-NN), are compared in terms of their accuracies employing different feature selection and grouping methods to demonstrate the strength of the proposed method. The classification results verify that employing CFG to obtain relevant features before running a classifier is effective and yields higher classification accuracies (10%-13% increase) than using feature selection methods. Moreover, CFG can find the number of groups dynamically for each dataset, which is a challenge in all feature selection methods.

2   Materials and Methods

To predict complex types, initially, the prediction properties (features) of each complex in the dataset are extracted. Then, after selecting the most powerful and discriminative features for prediction by employing a feature selection or feature grouping method, a classifier is applied on the selected features to predict the complex types. More explanations regarding the dataset, extracted features and also feature selection and classifier methods used in this paper and also proposed feature grouping method are discussed below.

2.1   The SCF-Ligases Dataset

As mentioned earlier, SCF-ligases are the largest class of E3 ligases which are minimally comprised of four subunits of RBX1, CUL1, SKP1 and an F-Box protein, as shown in Figure 1. RBX1 is responsible for the recruitment of the E2 ligase, CUL1 acts as a scaffold for the assembly of SCF ligase, SKP1 acts as an adaptor connecting CUL1 to the F-box protein, and the F-box protein dictates the target specificity of E3 through substrate selection [24].

Our manually curated SCF-ligase dataset contains 30 complexes. Of these, 21 complexes have strong interactions (obligate) and 9 complexes have weak interactions (non-obligate). The Protein Data Bank (PDB) IDs of these complexes and the interacting chains are shown in Table 1.

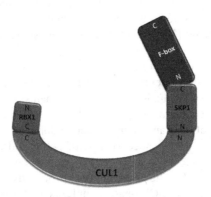

Fig. 1. A schematic view of an SCF-ligase

2.2   Prediction Properties

To extract domain-based prediction properties, first, the tertiary and quaternary structures of the complexes in the dataset were downloaded from the PDB [25]. After filtering and modifying the PDB files, the sequence domain content of each subunit was gleaned from the Pfam website [26] and mapped to the corresponding amino acids in the chain.

In the dataset, 27 unique Pfam domains present in the interface of at least one complex were identified. Of these, 17 domains were in the obligate complex class and 4 were in the non-obligate class, while the remaining 7 domains were both obligate and non-obligate complexes. A domain is considered to be in the interface, if it has at least one residue interacting with a domain in the other chain.

Table 1. Dataset of SCF-ligase complexes

Non-obligate Complexes (9)		
1NEX E:B	2AST C:D	2P1N B:C
1P22 C:A	2E33 A:B	3DB3 A:B
2AST B:D	2OVQ B:C	3OGK Q:B

Obligate Complexes (21)		
1FQV B:A	2ASS A:B	2P1M A:B
1LDK A:D	2ASS B:C	2QHO A:B
1LDK B:C	2E31 A:B	3MTN A:B
1LDK E:D	2HYE A:C	3NHE A:B
1NEX A:B	2HYE D:C	3OGK A:B
1P22 A:B	2HYE A:B	3OLM A:D
1U6G A:C	2OVP A:B	3PT2 A:B

To calculate the features for prediction, first, all pairs of interacting amino acids and their corresponding domains that are less than 7 Å apart from each other were extracted for each complex in the dataset. After that, the extracted amino acid pairs are grouped into two-domain, single-domain and also no-domain groups based on their corresponding domains. For instance, an amino acid pair is a member of the single-domain group if only one of the interacting amino acids belongs to at least one domain. To generate a domain-domain type (*DDT*) feature vector for each complex, all pairs of domains were considered. Since the order of the interacting domain pairs is not important, generated feature vectors for domain-domain type features contain $378 =_2^{27} C + 27$ values. The value of each domain pair in the *DDT* feature vector is the cumulative frequency across all occurrences of their corresponding amino acid pairs present in the two-domain group. To generate a single-domain type (*SDT*) feature vector for each complex, all 27 identified unique domains in the dataset were considered, individually. Each feature contains the sum of the frequencies for all amino acid pairs present in the single-domain group with the same domain-peptide chain interactions. The no-domain (*noD*) feature vector has only one feature that shows the number of amino acid pairs for each complex in the no-domain group.

Finally, after pre-processing and finding *DDT*, *SDT* and *noD* type feature vectors for all the complexes of the dataset, all zero-columns were removed, yielding only 41 features for the dataset.

2.3   Correlation-Based Feature Grouping (CFG)

To group the features, first, related features should be identified. Assuming N training inputs (samples) in a dataset $D = \{X_1, ..., X_N\}$, the feature vector of each sample X_i ($i = 1,..., N$) can be defined as an M-dimensional vector of $\{x_{i1}, x_{i2}, ..., x_{iM}\}$ where $j = 1,..., M$ in which M is the number of features. Although in this study, each x_{ij} is considered a binary value (0 or 1) of the jth feature for the ith sample in the dataset, the proposed method can be generalized to other types of data, e.g. real. Similarly, for each feature, a vector F_j can be defined as $\{x_{1j}, x_{2j}, ..., x_{Nj}\}$ and represents the values of the jth feature for all N samples in the dataset.

In this study, the correlation (Pearson's) scores between features and classes are considered as a measure to group the related features. The correlation score of a feature vector F_j with respect to class attribute Y is calculated as follows:

$$CorrelationScore(F_j, Y) = \frac{cov(F_j, Y)}{\sigma_{F_j}\sigma_Y}, \tag{1}$$

where $cov(F_j, Y)$ is the covariance of F_j and Y, and σ_{F_j} and σ_Y are the standard deviations of F_j and Y, respectively.

Then, the features whose differences of their correlation scores are less than a predefined threshold (δ) are placed in a group. In the CFG method, the number of groups can be found dynamically for each dataset based on a pre-defined threshold. For example, assume that there are five feature vectors with the following correlation scores: $\{0.12, 0.27, 0.25, 0.12, 0.25\}$. Considering $\delta = 0.01$, leads to three groups of features: $\{F_1, F_4\}, \{F_3, F_5\}, \{F_2\}$, while by assuming $\delta = 0.1$, two groups of features can be found: $\{F_1, F_4\}, \{F_2, F_3, F_5\}$.

After finding the related groups, the features are merged using a "combination" function. In this study, the features are combined in the following two steps:

1. In each group, any redundant features should be eliminated. A feature vector F_i is redundant if there is at least another feature vector F_j in the group that contains all the feature values of F_i. For example, if $F_i = \{1, 0, 0, 1\}$ and $F_j = \{1, 0, 0, 1\}$ or $F_j = \{1, 0, 1, 1\}$, then F_i is a redundant and can be removed.
2. All the remaining features in each group will be combined together to generate a new representative feature vector for the group, named F_{G_k}, where $k = 1,..., K$ with K being the number of groups. For instance, by considering the OR function, the combination vector of $\{1, 0, 0, 0\}$, $\{0, 0, 1, 0\}$, and $\{0, 0, 0, 1\}$ is $\{1, 1, 0, 1\}$. Any other functions can be used as a function to combine feature vectors.

2.4   Feature Selection

Applying feature selection before running a classifier is important to reduce the dimensionality of the data by discarding redundant and/or irrelevant features, and, thus, reducing the prediction time while improving the classification performance.

In this study, two filter methods, χ^2 and correlation scores, are used for feature selection. In filter methods, the features are scored and ranked independently of the classification algorithm by using some statistical criteria based on their relevance to each class, and then, the top ranked features can be selected for classification.

Chi2 Feature Selection (χ^2): The χ^2 method, which is an efficient feature selection method for numeric data, measures the degree of independence of each feature to the classes by computing the value of the chi2 statistic [29]. χ^2 discretizes numeric features and selects features as well. It keeps merging adjacent discrete statuses with the lowest value of χ^2 until all χ^2 values exceed their confidence interval. The χ^2 value of feature F_i with respect to class attribute Y is calculated as follows:

$$\chi^2(F_i, Y) = \frac{N \times (AD - CB)^2}{(A + C) \times (B + D) \times (A + B) \times (C + D)}, \tag{2}$$

where A is the number of times feature X and class Y co-occur, B is the number of times X occurs without Y, X is the number of times Y occurs without X, D is the number of times neither X and Y occurs, and N is the total number of samples. The ChiSquaredAttributeEval module of the Waikato Environment for Knowledge Analysis (WEKA) with default parameters is used for features selection based on χ^2 [28].

Correlation Feature Selection (CFS): The correlation (Pearson's) scores between features and classes can be calculated as defined in Eq. (1). In this study, the CorrelationAttributeEval module of WEKA with default parameters is used for ranking the features based on their correlation scores and selecting the top ranked features [28].

2.5 Prediction Method

There are a variety of methods that have been proposed for classification, of which SVM and k-NN are two of the most well-known ones, and which are used in this study. Details of the classifiers used in our model are discussed below.

Support Vector Machine: The main goal of the SVM is to find the support vectors, and derive a linear classifier, which ideally separates all the feature vectors into two regions. Using a linear classifier is inefficient in most cases when the data are not linearly separable. Thus, kernels, such as polynomial, radial basis function (RBF) and sigmoid, can be used to map the data onto a higher dimensional space in which the classification boundary can be found much more efficiently. The effectiveness of the SVM depends on the selection of the kernel and optimizing its parameters [27]. In addition, sequential minimal optimization (SMO) is a fast learning algorithm that has been widely applied in the training phase of a SVM classifier as one possible way to solve the underlying quadratic programming problem. In this work, the SMO module of WEKA with a normalized polynomial kernel, default parameter settings, and 10-fold cross-validation is used to perform classification via the SVM [28].

k-Nearest Neighbor: k-NN is one of the simplest classification methods, in which the class of each test sample can be easily found by a majority vote of the class labels of its neighbors. To achieve this, after computing and sorting the distances between the test sample and each training sample, the most frequent class label in the first "k" training samples (nearest neighbors) is assigned to the test sample. Determining the appropriate number of neighbors is one of the challenges of this method. In this study, the IBK module of WEKA with Euclidean distance, default parameter settings, and 10-fold cross-validation is used [28].

3   Results and Discussion

To evaluate our proposed grouping method and perform an in-depth analysis of the strength of feature grouping, different experiments have been conducted on the dataset and the extracted vectors. For different experiments, the OR function is employed to combine the feature vectors with a pre-defined threshold value of 0.001. Using 10-fold cross validation, the performances of the classification methods are compared in terms of their accuracies, which are computed as follows: $acc = (TP + TN)/N$, where TP and TN are the total numbers of true positive (obligate) and true negative (non-obligate) counters over the 10-fold cross-validation procedure, respectively, and N is the total number of complexes in the dataset.

3.1   Analysis of the Feature Selection

In this experiment, χ^2 and CFS methods are applied to score and rank the features, while SVM and k-NN are employed for classification. The selected features are all the same for the 10 folds because we employed classifiers with the 10-fold cross validation procedure after selecting features. The performance of SVM and k-NN using different numbers of selected features are shown in Table 2. From the table, it is clear that (a) the best obtained accuracy is 80.64% for both SVM and k-NN, (b) the k-NN performance improves almost 6% by using both χ^2 and CFS methods, and (c) achieving the same classification accuracies by using less features demonstrates the strength of the feature selection methods in selecting more powerful and discriminating features for classification.

3.2   Analysis of the Feature Grouping

For the SCF-ligase dataset, after grouping the features considering Δ less than 0.001 and combine the features in each group based on the CFG method, the final feature vector has 10 features.

In this experiment, CFG is compared with feature selection methods. For this, the performances of the SVM and k-NN classifiers using CFG along with the best accuracies obtained using feature selection methods (selecting 2 or 4 features) are shown in Table 3. The best accuracies of 90.32% for SVM and 93.55% for k-NN are achieved using features grouped by CFG which are significantly better than using the original

Table 2. Accuracies of SVM and k-NN using χ^2 and CFS for feature selection

FS method	# features	SVM	k-NN
no FS	41	80.64%	74.19%
χ^2	8	80.64%	74.19%
	4	80.64%	80.64%
	2	80.64%	80.64%
CFS	8	80.64%	80.64%
	4	80.64%	80.64%
	2	80.64%	80.64%

feature vectors or selected features. In other words, employing CFG helps the classifiers perform much better than employing feature selection methods. Moreover, in the CFG method, there is no need to pre-define the number of groups, while for feature selection methods defining the optimal number of features is still a challenge.

Table 3. Accuracies of SVM and k-NN using original, selected or grouped features

Feature Types	# features	SVM	k-NN
Original features	41	80.64%	74.19%
Selected features	2 (or 4)	80.64%	80.64%
Grouped features by CFG	10	90.32%	93.55%

3.3   Analysis of Interaction Types

As explained earlier, after identifying all unique domains present in the interface of at least one complex in the dataset, for each complex, the number of domain-domain interactions (*DDIs*), domain-peptide chain interactions (*DIs*) and also the number of interactions that none of the interacting amino acids belong to any domains (*noD*) were calculated to acheive a better insight of the complexes and their interacting domains. In Figure 2, the number and type of interactions for non-obligate (left) and obligate (right) complexes are shown in different colors: blue for *DDIs*, red for *DIs* and green for *noD* interactions.

From the histogram, it is clear that obligate complexes have more interactions and most of them are domain-domain interactions. In contrast, non-obligate complexes have less interactions in comparison to obligate ones and most of their interactions are single-domain (*DIs*). Also, most of the interactions, are mediated by at least one domain.

Similarly, the results of the average number of interactions for each obligate and non-obligate complexes of the SCF-ligase dataset categorized by their interaction types shown in Table 4, confirm the results demonstrated in Figure 2. This means that in both obligate and non-obligate complexes, less than 1% of the interactions are mediated by no domains. Also, the average number of interactions of obligate complexes (3,879 pairs) is approximately five times greater than the number of interactions of

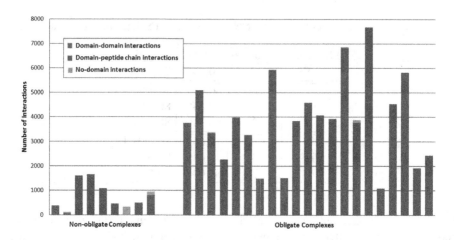

Fig. 2. Number and type of interactions for two groups of non-obligate (left) and obligate (right) SCF-ligase complexes

non-obligate complexes (794 pairs) in the dataset. In addition, more interactions of non-obligate complexes (greater than 86%) are *DIs*, while for obligate complexes, more than 63% of interactions are *DDIs*.

Table 4. A summary of the average numbers of interactions for obligate and non-obligate complexes of the SCF-ligase dataset categorized by their interaction types

Complex Type	Type of interactions			
	# DDIs	# DI	# noD	Total
Obligate	43	686	64	794
Non-obligate	2475	1394	10	3879

4   Conclusion

A correlation-based feature grouping method, namely CFG, has been proposed as an alternative for feature selection to obtain more relevant features for classification.

To demonstrate the strength of CFG, different feature selection methods of χ^2 and CFS are also used; the classification is performed using various techniques, including SVM and k-NN. The classification results on the SCF-ligase PPI dataset show significant increases of 10% for SVM and 13% for k-NN by grouping features based on the proposed CFG method in comparison with selecting features using different methods. Thus, it can be concluded that the CFG method has the ability to (a) group related features, (b) reduce number of features by eliminating redundant features and combining more relevant ones, (c) reduce the computational cost for classification, (d) improve the classification performance, and (e) define the optimal number of features (groups) dynamically. The approach proposed here can also be used for classification of different types of data and for different applications to verify its efficiency. Moreover, different threshold values (δ) and also different combination functions can be employed to find the groups and merge the features in each group.

Acknowledgements. This work has been partially supported by NSERC, the Natural Science and Engineering Council of Canada.

References

1. Dezfulian, M.H., Soulliere, D.M., Dhaliwal, R.K., Sareen, M., Crosby, W.L.: The skp1-like gene family of arabidopsis exhibits a high degree of differential gene expression and gene product interaction during development. PLOS One 7(11) (2012)
2. Chen, L., Wang, R., Zhang, X.: Biomolecular Networks: Methods and Applications in Systems Biology. John Wiley and Sons (2009)
3. Jones, S., Thornton, J.M.: Principles of protein-protein interactions. Proc. Natl. Acad. Sci., USA 93(1), 13–20 (1996)
4. Maleki, M., Rueda, L., Dezfulian, M.H., Crosby, W.: Computational Analysis of the Stability of SCF Ligases Employing Domain Information. In: 5th ACM Conference on Bioinformatics, Computational Biology, and Health Informatics (BCB 2014), pp. 625–626 (2014)
5. Maleki, M.H.M., Rueda, L.: Using desolvation energies of structural domains to predict stability of protein complexes. Journal of Network Modeling Analysis in Health Informatics and Bioinformatics (NetMahib) 2, 267–275 (2013)
6. Hall, M., Maleki, M., Rueda, L.: Multi-level structural domain-domain interactions for prediction of obligate and non-obligate protein-protein interactions. In: ACM Conference on Bioinformatics, Computational Biology and Biomedicine (ACM-BCB), Florida, USA, pp. 518–520 (October 2012)
7. Chandrasekaran, P., Doss, C., Nisha, J., Sethumadhavan, R., Shanthi, V., Ramanathan, K., Rajasekaran, R.: In silico analysis of detrimental mutations in add domain of chromatin remodeling protein atrx that cause atr-x syndrome: X-linked disorder. Network Modeling Analysis in Health Informatics and Bioinformatics 2(3), 123–135 (2013)
8. Lim, S., Peng, T., Sana, B.: Protein-protein interaction prediction using homology and inter-domain linker region information. In: Ao, S.-I., Gelman, L. (eds.) Advances in Electrical Engineering and Computational Science. LNEE, vol. 39, pp. 635–645. Springer, Heidelberg (2013)
9. Theodoridis, S., Koutroumbas, K.: Pattern Recognition, 4th edn. Elsevier Academic Press (2008)
10. Niu, S., Huang, T., Feng, K., Cai, Y., Li, Y.: Prediction of tyrosine sulfation with mRMR feature selection and analysis. J. Proteome. Res. 9(12), 6490–6497 (2010)
11. Cai, Y., Huang, T., Hu, L., Shi, X., Xie, L., Li, Y.: Prediction of lysine ubiquitination with mRMR feature selection and analysis. Amino Acids (2011)
12. Maleki, M., Aziz, M., Rueda, L.: Analysis of relevant physicochemical properties in obligate and non-obligate protein-protein interactions. In: IEEE International Conference in Bioinformatics and Biomedicine Workshops (BIBMW), pp. 345–351 (2011)
13. Liu, L., Cai, Y., Lu, W., Peng, C., Niub, B.: Prediction of protein-protein interactions based on PseAA composition and hybrid feature selection. Biochemical and Biophysical Research Communications 380(2), 318–322 (2009)
14. Yuan, Y., Shi, X., Li, X., Lu, W., Cai, Y., Gu, L., Liu, L., Li, M., Kong, X., Xing, M.: Prediction of interactiveness of proteins and nucleic acids based on feature selections. Mol. Divers. 14(4), 627–633 (2009)

15. Mundra, P., Rajapakse, J.: SVM-RFE with mRMR filter for gene selection. IEEE Transactions on Nanobioscience 9(1), 31–37 (2010)
16. Zhao, Y., Yand, Z.: Improving MSVM-RFE for multiclass gene selection. In: The Fourth International Conference on Computational Systems Biology (ISB 2010) (2010)
17. Yang, S., Yuan, L., Lai, Y., Shen, X., Wonka, P., Ye, J.: Feature grouping and selection over an undirected graph. In: Proceedings of the International Conference on Knowledge Discovery & Data Mining (KDD) (2012)
18. Li, C., Li, H.: Network-constrained regularization and variable selection for analysis of genomic data. Bioinformatics 24(9), 1175–1182 (2008)
19. Zhong, L.W., Kwok, J.T.: Efficient sparse modeling with automatic feature grouping. IEEE Transactions on Neural Networks and Leraning Systems 23(9), 1436–1447 (2012)
20. Suzuki, J., Nagata, M.: Supervised model learning with feature grouping based on a discrete constraint. In: Proceedings of the 51st Annual Meeting of the Association for Computational Linguistics, Sofia, Bulgaria (August 2013)
21. Tibshirani, R.: Regression shrinkage and selection via the lasso: A retrospective. Journal of the Royal Statistical Society: Series B (Statistical Methodology) 73(3), 273–282 (2011)
22. Shen, X., Huang, H.: Grouping pursuit through a regularization solution surface. Journal of the American Statistical Association 105(490), 729–739 (2010)
23. Bondell, H., Reich, B.: Simultaneous regression shrinkage, variable selection, and supervised clustering of predictors with OSCAR. Biometrics 64(1), 115–123 (2008)
24. Chen, B.B., Mallampalli, R.K.: F-box protein substrate recognition-a new insight. Cell Cycle 12(7), 1009–1010 (2013)
25. Berman, H.M., Kleywegt, G.J., Nakamura, H., Markley, J.L.: The Protein Data Bank at 40: reflecting on the past to prepare for the future. Structure 20(3), 391–396 (2012)
26. Punta, M., Coggill, P., Eberhardt, R., Mistry, J., Tate, J., Boursnell, C., Pang, N., Forslund, K., Ceric, G., Clements, J., Heger, A., Holm, L., Sonnhammer, E., Eddy, S., Bateman, A., Finn, R.: The Pfam protein families database. Nucleic Acids Res. 40(D1), D290–D301 (2012)
27. Duda, R., Hart, P., Stork, D.: Pattern Classification, 2nd edn. John Wiley and Sons, Inc., New York (2000)
28. Hall, M., Frank, E., Holmes, G., Pfahringer, B., Reutemann, P., Witten, I.H.: The WEKA data mining software: An update. SIGKDD Explorations 11(1), 10–18 (2009)
29. Liu, H., Setiono, R.: Chi2: Feature selection and discretization of numeric attributes. In: Proceedings of the Seventh International Conference on Tools with Artificial Intelligence, pp. 388–391 (1995)

A New Approach to Obtain EFMs Using Graph Methods Based on the Shortest Path between End Nodes

Jose Francisco Hidalgo Céspedes, Francisco De Asís Guil Asensio, and Jose Manuel García Carrasco

Grupo de Arquitectura y Computación Paralela Universidad de Murcia, Spain
{jhidalgo,fguil,jmgarcia}@um.es
http://www.um.es/gacop

Abstract. Genome-scale metabolic networks let us to understand the behavior of the metabolism in the cells of live organisms. The availability of great amounts of such data gives scientific community the opportunity to infer *in silico* new metabolic knowledge. Elementary Flux Modes (EFM) are minimal contained pathways or subsets of a metabolic network that are very useful to achieve the comprehension of a very specific metabolic function (as well as dis-functions), and to get the knowledge to develop new drugs. Metabolic networks can have large connectivity and, therefore, EFMs resolution faces a combinational explosion challenge to be solved. In this paper we propose a new approach to obtain EFMs based on graph methods and the shortest path between end nodes. Our method finds all the pathways in the metabolic network and it is able to prioritize the pathway search accounting the biological mean pursued. Our technique has two phases, the exploration one and the characterization one, and we show how it works in a well-known case study.

Keywords: Metabolic networks, graph theory, EFM, flux modes, pathways, systems biology.

1 Motivation

Cellular metabolism is the set of biochemical enzyme-catalyzed reactions involved in the generation of nutrients and energy necessary for the cells in living organisms. Those reactions are equations of metabolites with stoichiometric coefficients. All the reactions and metabolites used to be grouped in a stoichiometric matrix. A metabolic pathway of a cell is a piece of the network, that is, a sequence of some of its reactions. Metabolic pathways have been found quite useful in different domains such as personalized medicine, drug discovery techniques or genomic feature discovery. Therefore, many efforts have been lately made to find pathways experimentally or by inferring them computationally.

Several mathematical methods modeling metabolism are emerging that are able to incorporate datasets provided by different *omics* technologies. Many of these methods are encompassed within constraint-based models, in which a set of mathematical constraints are defined using a genome-scale metabolic

F. Ortuño and I. Rojas (Eds.): IWBBIO 2015, Part I, LNCS 9043, pp. 641–649, 2015.

network (GSMN) reconstruction as a starting point. Several curated GSMNs can be found in the literature [19]. However, being able to automatically characterize the biochemical reactions present in a particular metabolism through *omics* data truly constitutes a challenge [15].

The term constraint-based modeling (CBM) groups different approaches that analyze the metabolic behavior based on the stoichiometric relations between compounds participating in enzymatic reactions. CBM defines two constraints that pathways must fulfill. The first one is the steady-state condition that refers to the property of mass balance within the cell. In other words, the concentration of internal metabolites remains constant over the time. The second relevant constraint refers to thermodynamic feasibility, which restricts some fluxes from being non-negative (irreversibility constrain).

An elementary flux mode (EFM) [16] is a special type of metabolic pathway comprising a subset of reactions that meets the two aforementioned conditions plus the non-decomposability condition, that is, the pathway cannot be decomposed into smaller solutions (i.e., a subset of the pathway is not a feasible pathway as well). In other words, EFMs are solutions with the minimum support necessary to operate in stoichiometric steady-state balance with all reactions in the appropriate direction. EFMs are an effort to translate a complex network into a canonical expression of vector generators of a solution space.

In a typical metabolic network the number of reactions is higher than the number of metabolites, so that many possibilities can be found that are a solution to the system. As the metabolic network increases in size so do the amount of EFMs, which number explodes in a combinatorial fashion. Computing the full set of EFMs in large metabolic networks still constitutes a challenging issue.

Continuing with this effort, we have developed a new method to find systematically all the pathways from a metabolic network. In this paper we present our approach based on a novel strategy to find shortest pathways between end nodes in a graph representation of the network. Specifically, our approach exploits the well-known graph theory and tools to drive the search of EFMs prioritizing, if needed, the pathway search to account the biological mean quest. Our technique is composed of two phases, the exploration and the characterization one, and along the paper we describe the how the first phase works in a case study.

Unlike traditional Linear Programming (LP) approaches, our proposal avoids expensive floating-point calculations allowing us to speed-up the quest of all the available pathways in a certain metabolic network. Moreover, our approach is quite suitable to be developed in new commodity parallel architectures (such as multi- and many-cores and accelerators like GPUs), allowing shorter execution times and less energy consumption.

The rest of the paper is structured as follows. Section 2 gives some background on the constraint-based mathematical modeling. In Section 3 we show the method we have followed to design our technique. Section 4 presents a case study of our approach, and the paper concludes giving some related work in Section 5, and offering our conclusions and future work in Section 6.

2 Background

Constraint based modeling (CBM) starts with a stoichiometric matrix S where the values are the stoichiometric coefficients for metabolites (rows) on each reaction (columns). Every reaction is characterized by the reaction rate (also known as flux rate) which numerically gives the rate at which the substrate metabolites are converted to the product metabolites.

Be $\overrightarrow{r}$ a vector of flux rate that represents a pathway, therefore fulfilling the steady-state and the thermodynamic feasibility constrains. The steady-state condition means that internal metabolites are balanced and concentration remains constant ($S \cdot \overrightarrow{r} = \overrightarrow{0}$), and the feasibility constraint means that each irreversible reaction only participates with a positive rate ($\forall i, r_i \geq 0$) when it is part of the solution. Finally, $\overrightarrow{r}$ represents an EFM if the non-decomposability condition is met ($\overrightarrow{r}$ is not a lineal combination of other flux rate vectors).

The stoichiometric matrix S let us build an adjacency matrix that corresponds to the graph $G = (V, E)$, a non-weighted directed bipartite graph where V are both reactions and metabolites, and the edges E are directed attending the sign of the stoichiometric coefficients. A pathway is a sub-graph of G, $G' = (V', E')$, which is equivalent to $\overrightarrow{r}$ and vice versa.

A known drawback of graph exploration methods is that the flux rate vector is missing at the final of the process. In order to verify which ones of the pathways founds are EFMs, it is needed to do a final verification test using stoichiometry.

3 The Shortest Path Technique to Find EFMs

We propose a new CBM approach based on path-finding techniques. Our method consists of two phases, the exploration phase and the characterization one. The exploration phase consists of 3 stages for traversing the graph and finding the feasible pathways. In the first stage, we use the pathway distance metric approach (that is, the amount of reactions participants at the pathway) and take advantage of the fact that it should be biologically meaningful [1]. Therefore, the quest starts the graph exploration by building an axis between a source and a target of the network applying the Dijkstra's shortest path algorithm [2]. The choice of the path end nodes (source, target or both) comes from the biological problem we are dealing with.

At the end of this stage, an axis has been built using the Dijkstras algorithm that traverses the graph through metabolites and reactions from the source node to the target one using the shortest distance. Some of the reactions included in the axis can need metabolites that have not been included yet. We name this kind of metabolites as orphan metabolites.

The second stage goes back from the target to the source (bottom-up approach) to solve the orphan metabolite problem. This process traverses the inverted graph and it is done in a recursive way.

The third stage consists of the simulation of all the reactions that should occur due to the presence of the required metabolites produced by other reactions.

The third stage ends when the end nodes are connected by a complete graph of reactions without orphans nor non-consumed internal metabolites.

After the third stage, our approach has found systematically all the pathways in the axis formed by those end nodes in a metabolic network. This process should be iterative by every pair of interesting end nodes.

The characterization phase needs to check all the pathways produced to determine which of them are EFMs. The final pathways obtained seem minimal because none of the elements can be eliminated without sacrificing consistency. Moreover, pathways fulfill the necessary conditions to have the steady-state balance. However, it cannot be assured the steady-state condition because the stoichiometry is not playing a role during the run of our approach. Without stoichiometry, and depending on the network structure, it can be got a lot of false positives but also some other real positives. In terms of feasibility, the pathways are built fulfilling the necessary conditions to be feasible, that is, respecting the positive direction of every reaction, but the feasibility constrain is only met conditioned to the steady-state consistency.

This second phase is needed as the steady-state constraint has not been granted during the exploration phase of the graph and, therefore, it must be checked afterwards. Currently, we are developing some heuristics to properly select EFMs from the full set of feasible pathways produced.

4   Case Study

As mentioned before, our approach produces all possible pathways and, for certain cases, EFMs can inferred from those pathways. In simple networks like the EFMtool example published in [4] (6 metabolites and 12 reactions), once all the pathways has been found, the characterization phase has got the full list of EFMs easily discarding decomposable pathways. In addition, for this small and not complex network, the flux rate vector has easily been calculated for any found EFM.

Let us consider as an example the aforementioned network represented by the stoichiometric matrix S shown in the matrix 1. Note that the reactions $R2$ and $R8$ are reversible reactions. For the rest of the process these reactions need to be unfolded in $R2$, $R2\_rev$, $R8$ and $R8\_rev$ automatically. Unfolded reactions must be included in the matrix with individual columns in the new extended stoichiometric matrix that it is shown next. Therefore, all the reactions are from now irreversible.

$$S = \begin{array}{c} \\ A \\ B \\ C \\ D \\ E \\ F \end{array} \begin{array}{c} R1 \quad R2^r \quad R3 \quad R4 \quad R5 \quad R6 \quad R7 \quad R8^r \quad R9 \quad R10 \\ \left(\begin{array}{cccccccccc} 1 & 0 & 0 & 0 & -1 & -1 & -1 & 0 & 0 & 0 \\ 0 & 1 & 0 & 0 & 1 & 0 & 0 & -1 & -1 & 0 \\ 0 & 0 & 0 & 0 & 0 & 1 & 0 & 1 & 0 & -1 \\ 0 & 0 & 0 & 0 & 0 & 0 & 1 & 0 & 0 & -1 \\ 0 & 0 & 0 & -1 & 0 & 0 & 0 & 0 & 0 & 1 \\ 0 & 0 & -1 & 0 & 0 & 0 & 0 & 0 & 1 & 1 \end{array} \right) \end{array} \quad (1)$$

Based on the extended S we build the graph $G = (V, E)$ represented graphically in the figure 1, where vertices V are both metabolites and reactions, and edges E are the incidence arcs following the direction of the reactions.

$$S = \begin{array}{c} \\ A \\ B \\ C \\ D \\ E \\ F \end{array} \begin{array}{cccccccccccc} R1 & R2 & R2\_rev & R3 & R4 & R5 & R6 & R7 & R8 & R8\_rev & R9 & R10 \\ \left(\begin{array}{cccccccccccc} 1 & 0 & 0 & 0 & 0 & -1 & -1 & -1 & 0 & 0 & 0 & 0 \\ 0 & 1 & -1 & 0 & 0 & 1 & 0 & 0 & -1 & 1 & -1 & 0 \\ 0 & 0 & 0 & 0 & 0 & 0 & 1 & 0 & 1 & -1 & 0 & -1 \\ 0 & 0 & 0 & 0 & 0 & 0 & 0 & 1 & 0 & 0 & 0 & -1 \\ 0 & 0 & 0 & 0 & -1 & 0 & 0 & 0 & 0 & 0 & 0 & 1 \\ 0 & 0 & 0 & -1 & 0 & 0 & 0 & 0 & 0 & 0 & 1 & 1 \end{array}\right) \end{array}$$

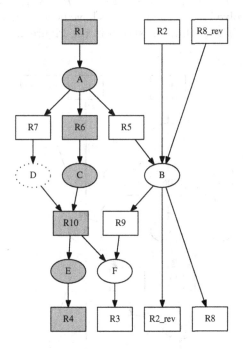

Fig. 1. Graph obtained after the first stage

Our technique starts in the exploration phase, which has three stages. In its first stage, the Dijkstra's shortest algorithm is run to build an axis for the foreseeable pathway. A shortest path for this example is shown in the Figure 1 with the participating nodes in gray. This shortest path is a route between $R1$ as input extreme of our metabolic network and $R4$ as output extreme. Obviously every pair of extreme points can be considered. Many times, the paths obtained in this stage could have the orphan metabolite problem. In the example we are considering, R10 needs that the metabolite D (dotted in the Figure) is also available in the cell to be part of the pathway.

The second stage has the objective to fix this inconvenience. Following with the example, this stage try to include the metabolite D in the axis {R1, A, R6, C, R10, E, R4} to form the axis {R1, A, R6, C, D, R10, E, R4}. Many solutions with different complexity can be developed for each found shortest path. In our case, this stage incorporates the reaction R7 to the pathway to supply D.

The third stage is responsible to assure that every metabolite produced by the pathway has consumer reactions, that is, it should be consumed inside the pathway. In our example, R10 produces the metabolite F but there is no consumer reaction for it. This stage looks for what reactions could occur with the metabolite F in order to be consumed. In this example, there is only one possibility (R3 reaction), and it will be incorporated to the pathway. After this three stages we have the pathway {R1, R6, R10, R4, R7, R3} with the metabolites {A, C, D, E, F} involved in it. The reactions have been shown in the same order they were obtained. The pathway is shown in the Figure 2.

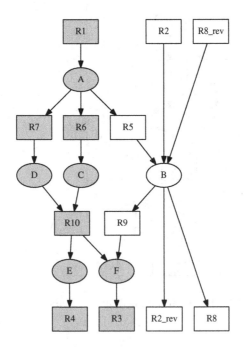

Fig. 2. Final pathway

Following with the example, in [4] the set of EFMs correspondent to the metabolic network are available. One of those EFMs is given by the *flux rate* $\vec{r} = (2, 0, 1, 1, 0, 1, 1, 0, 0, 1)$ and it corresponds with our pathway {R1, R6, R10, R4, R7, R3} (same non-zero and positive coefficients). Therefore, as there is not possible to have other EFMs with the same non-zero and positive rates, our approach has obtained the same EFM that `efmtool`.

In the end, $\overrightarrow{r}$ can be translated into the stoichiometric sub-matrix S' by maintaining columns and rows correspondent to those nodes of G'. It is important to note that G', S' and $\overrightarrow{r}$ are fully equivalent. It can be proven that if $\overrightarrow{r}$ is an EFM, then $S' \cdot \overrightarrow{r} = \overrightarrow{0}$.

$$
S' = \begin{array}{c} \\ A \\ C \\ D \\ E \\ F \end{array}
\begin{array}{c} R1 \quad R3 \quad R4 \quad R6 \quad R7 \quad R10 \end{array}
\left(\begin{array}{cccccc}
1 & 0 & 0 & -1 & -1 & 0 \\
0 & 0 & 0 & 1 & 0 & -1 \\
0 & 0 & 0 & 0 & 1 & -1 \\
0 & 0 & -1 & 0 & 0 & 1 \\
0 & -1 & 0 & 0 & 0 & 1
\end{array}\right)
$$

Finally, our approach is currently being checked against other small network like the network of *E. coli core model* [3] with 95 reactions and 72 metabolites, and it is obtaining some promising results.

5   Related Work

The advantages of analyzing metabolic networks based on EFMs have been shown in different works [5] [14]. However, their use has been limited because enumerating them is computationally demanding. Algorithms have been developed to enumerate all the EFMs in medium-size metabolic networks [20][21] [17]. However, despite the development of novel methods using state of the art computational techniques expediting their application in larger networks [7], this family of algorithms fails on GSMNs using standard computers, because of the combinatorial explosion in the number of EFMs [9]. In this light, several methods have been recently proposed to determine a subset of EFMs in GSMNs [6] [12] [13].

Computational approaches to metabolic pathways can be classified in two groups: stoichiometric approaches and path-finding approaches [10]. Summarizing, the first ones use the stoichiometric data to do calculations during the process. Linear Programming and Null-Space Algorithm [8] are some of the mathematical strategies applied to find pathways, mainly solving the system of linear equations propose by the stoichiometric matrix. Stoichiometric approaches have the quality of impose biochemically meaningful stoichiometric constraints to the solutions but at the cost of intense floating point calculations.

The second ones translate the network into a directed graph to explore it. Path-finding approaches are considered to constitute some advance with respect to stoichiometry approaches mainly because they rest on the well-known graph theory and let the use of techniques based on distance metric, revealed as biologically relevant [1]. Because of the combinatorial nature of the search, some proposals only find a subset of all feasible pathways, whereas other approaches get the full set of feasible pathways [18]. The major drawback of path-finding approaches is that the lack of use of stoichiometry during the exploration process cannot assure that the solution has biological meaning and meets all the constraints. Therefore, an extra stage is needed to determine if a found pathway meets the constraints and it constitutes an EFM.

Finally, some other authors combine both approaches trying to build on strengths of each and avoid respective drawbacks and computational expenses [11].

6   Conclusions and Future Work

In this paper we propose a new approach to obtain EFMs based on graph methods and the shortest path between end nodes. The novel approach we have presented here constitutes an advance with respect to previous approaches as it relies on a three-stage method based on the Dijkstra's shortest path algorithm, and an extra heuristic and mathematical phase that can produce systematically candidates to EFM.

Our method finds all the pathways in the metabolic network and it is able to prioritize the pathway search accounting the biological mean pursued. Our technique has two phases, the exploration one and the characterization one, and we show how it works in a well-known case study.

Unlike traditional Linear Programming (LP) approaches, our proposal avoids expensive floating-point calculations allowing us to speed-up the quest of all the available pathways in a certain metabolic network. We realize that the fact of the combinatorial explosion while exploration of the graph is a common problem to path-finding approaches (loops and the increasing size of the networks worsen the problem), so we foresee that the parallelization of this process could give us a lot of benefits. Our approach is quite suitable to be developed in new commodity parallel architectures (such as multi- and many-cores and accelerators like GPUs), allowing shorter execution times and less energy consumption.

As for future work, the characterization phase of the EFMs from the set of pathways obtained is still immature and more work should be done in relation with it, as developing some heuristics from artificial intelligence techniques like ants colony. Another direction of future work is the parallelization of all of the stages of our method using HPC commodity architectures, as multicore processors and accelerators (like GPUs or Xeon Phi).

Acknowledgments. This work was jointly supported by the Fundación Séneca (Agencia Regional de Ciencia y Tecnología, Región de Murcia) under grant 15290/PI/2010 and the Spanish MEC and European Commission FEDER under grant TIN2012-31345.

References

1. Croes, D., Couche, F., Wodak, S.J., et al.: Metabolic PathFinding: Inferring relevant pathways in biochemical networks. Nucleic Acids Res. 2005(33), W326–W330
2. Dijkstra, E.W.: A note on two problems in connexion with Graphs. Numerische Mathematik 1, 269–271 (1959)
3. Roman, M.T., Flemming, B.O.: Palsson. Reconstruction and use of microbial metabolic networks: The core Escherichia coli metabolic model as an educational guide. In: Escherichia coli and Salmonella: Cellular and Molecular Biology, Chapter 10.2.1, Washington, DC (2010)
4. Elementary Flux Mode Tool, http://www.csb.ethz.ch/tools/efmtool
5. De Figueiredo, L.F., et al.: Can sugars be produced from fatty acids? A test case for pathway analysis tools. Bioinformatics 24, 2615–2621 (2008)
6. De Figueiredo, L.F., et al.: Computing the shortest elementary flux modes in genome-scale metabolic networks. Bioinformatics 25, 3158–3165 (2009)
7. Hunt, K.A., et al.: Complete enumeration of elementary flux modes through scalable, demand-based subnetwork definition. Bioinformatics (2014) (in press)
8. Jevremovic, D., Boley, D., Sosa, C.P.: Divide-and-Conquer Approach to the Parallel Computation of Elementary Flux Modes in Metabolic Networks. In: IEEE International Symposium on IPDPS 2011, pp. 50–511 (2011)
9. Klamt, S., Stelling, J.: Combinatorial complexity of pathway analysis in metabolic networks. Mol. Biol. Rep. 29(1-2), 233–236 (2002)
10. Planes, F.J., Beasly, J.E.: A critical examination of stoichiometric and path-finding approaches to metabolic pathways. Briefings in Bioinformatics 9, 422–436 (2008)
11. Pey, J., Prada, J., Beasley, J.E., Planes, F.: Path finding methods accounting for stoichiometry in metabolic networks. Genome Biol. 12(5), 49 (2011)
12. Pey, J., Planes, F.J.: Direct calculation of Elementary Flux Modes satisfying several biological constraints in genome-scale metabolic networks. Bioinformatics (2014) (in press)
13. Rezola, A., et al.: Selection of human tissue-specific elementary flux modes using gene expression data. Bioinformatics 29, 2009–2016 (2013)
14. Rezola, A., et al.: Advances in network-based metabolic pathway analysis and gene expression data integration. Bioinform in press, Brief (2014)
15. Schmidt, B.J., et al.: GIM3E: condition-specific models of cellular metabolism developed from metabolomics and expression data. Bioinformatics 29, 2900–2908 (2013)
16. Schuster, S., Hilgetag, C.: On elementary flux modes in biochemical reaction systems at steady state. J. Biol. Syst. 2, 165–182 (1994)
17. von Kamp, A., Schuster, S.: Metatool 5.0: Fast and flexible elementary modes analysis. Bioinformatics 22(15), 1930–1931 (2006)
18. Seo, H., Lee, D.-Y., Park, S., Fan, L.T., Shafie, S., Bertk, B., Friedler, F.: Graph-theoretical identification of pathways for biochemical reaction. Biotechnology Letters 23, 1551–1557 (2001)
19. Thiele, I., Palsson, B.: A protocol for generating a high-quality genome-scale metabolic reconstruction. Nat. Protoc. 5, 93–121 (2010)
20. Terzer, M., Stelling, J.: Large-scale computation of elementary flux modes with bit pattern trees. Bioinformatics 24, 2229–2235 (2008)
21. Urbanczik, R., Wagner, C.: An improved algorithm for stoichiometric network analysis: theory and applications. Bioinformatics 21, 1203–1210 (2005)

Identifiability of Nonlinear ODE Models in Systems Biology: Results from Both Structural and Data-Based Methods

Maria Pia Saccomani

Department of Information Engineering, University of Padova, Padova, Italy
pia@dei.unipd.it
http://www.dei.unipd.it/persona/0A7F06976D2D39F73AEDD095D97CF91A

Abstract. The aim of this paper is to show that the two methodologies of testing identifiability of models described by nonlinear differential equations, the *structural identifiability*, which does not require the use experimental data and the *practical non-identifiability*, which is a data-based method, can be integrated in the overall identification analysis of the model. The two methodologies are compared on two much quoted biological models, one describing the Escherichia coli system, the other the dynamic behavior of the tumor-suppressor p53.

Keywords: Structural identifiability, Practical non-identifiability, Modeling and identification, Tumor model, Escherichia coli model, Bioinformatics, Systems biology, Nonlinear system, ODE model.

1 Introduction

Knowledge of the kinetics of biomolecular interactions is important for facilitating the study of cellular processes and underlying molecular events, and is essential for quantitative study and simulation of biological systems. Ordinary nonlinear differential equations (ODE) are commonly used as mathematical models in quantitative molecular biology and biotechnology with applications from metabolic engineering to cancer therapy. The first relevant question is whether the parameters of the model can be (uniquely) determined, at least for suitable input functions, assuming that all observable variables are error free. This is a mathematical property called *a priori* or *structural identifiability* of the model which can (and should) in principle be checked before collecting experimental data and guarantee the uniqueness of the parameter solution, which is a prerequisite for the parameter estimation problem to be well-posed. If the postulated model is not identifiable, the parameter estimates which could, nevertheless, be obtained by some numerical optimization algorithms, would be unreliable.

Obviously, although necessary, structural identifiability is not sufficient to guarantee an accurate identification of the model parameters from real, possibly noisy, input/output data. The answer obtained from structural identification

F. Ortuño and I. Rojas (Eds.): IWBBIO 2015, Part I, LNCS 9043, pp. 650–658, 2015.

algorithms may be very sensitive to noise and a measure of this sensitivity can be important in applications.

Also many currently studied models in systems biology are rather large networks containing many states and parameters and in some cases, checking structural identifiability can be prohibitively complex. The nature of the algorithms for checking structural identifiability imposes restrictions on the size and complexity of systems that so far the methods can handle. Situations of this kind can be approached by semi-empirical techniques which are essentially based on simulations and on the study of the level curves of a cost function. The level curves about the minimum (assuming that the minimization should yield a *unique* parameter value) can provide a measure of the sensitivity alluded at before. The study of sensitivity is called *practical non-identifiability* in the literature [9]. Checking practical non-identifiability should and can be done on more realistic models which explicitly involve noise in the measurements. Sometime authors may also talk about *practical identifiability* but we shall not use this term in this paper. It should, in our opinion, be kept in mind that "practical" data-based (or simulation-based), identifiability tests cannot provide a mathematically rigorous answer to the uniqueness problem.

2 Structural vs. Practical Non-identifiability

This section provides the reader with the definitions which are necessary to set the notations used in the paper. Consider a nonlinear dynamic system described in state space form

$$\dot{\mathbf{x}}(t) = \mathbf{f}(\mathbf{x}(t), \mathbf{p}) + \sum_{i=1}^{m} \mathbf{g}_i(\mathbf{x}(t), \mathbf{p}) u_i(t) \tag{1}$$

$$\mathbf{y}(t) = h((\mathbf{x}(t), \mathbf{u}(t), \mathbf{p}) \tag{2}$$

with state $\mathbf{x}(t) \in \mathbb{R}^n$, input $\mathbf{u}(t) \in \mathbb{R}^q$ ranging on some vector space of piecewise smooth (infinitely differentiable) functions, output $\mathbf{y}(t) \in \mathbb{R}^m$, and the constant unknown parameter vector $\mathbf{p}$ belonging to some open subset $\mathcal{P} \subseteq \mathbb{R}^p$. Whenever initial conditions are specified, the relevant equation $\mathbf{x}(0) = \mathbf{x}_0$ is added to the system. The essential assumption here is that the functions $\mathbf{f}, \mathbf{g}_1, \ldots, \mathbf{g}_m$ and $\mathbf{h}$ are vectors of *rational functions* in $\mathbf{x}$. Also we assume that there is no feedback, so that $\mathbf{u}$ is a free variable not depending on $\mathbf{y}$. The affine structure in $\mathbf{u}$ is not essential and could be relaxed.

Let $\mathbf{y} = \psi_{\mathbf{x}_0}(\mathbf{p}, \mathbf{u})$ be the input-output map of the system (1,2) started at the initial state $\mathbf{x}_0$ (we assume that this map exists).

Definition 1. *The system (1,2) is a priori globally (or uniquely) identifiable from input-output data if, for at least a generic set of points $\mathbf{p}^* \in \mathcal{P}$, there exists (at least) one input function $\mathbf{u}$ such that the equation*

$$\psi_{\mathbf{x}_0}(\mathbf{p}, \mathbf{u}) = \psi_{\mathbf{x}_0}(\mathbf{p}^*, \mathbf{u}) \tag{3}$$

has only one solution $\mathbf{p} = \mathbf{p}^*$ for almost all initial states $\mathbf{x}_0 \in X \subseteq \mathbb{R}^n$.

A weaker notion is that of local identifiability.

Definition 2. *The system (1, 2) is locally identifiable at* $\mathbf{p}^* \in \mathcal{P}$ *if there exists (at least) one input function* $\mathbf{u}$ *and an open neighborhood* $U_{\mathbf{p}^*}$ *of* $\mathbf{p}^*$, *such that the equation (3) has a unique solution* $\mathbf{p} \in U_{\mathbf{p}^*}$ *for all initial states* $\mathbf{x}_0 \in X \subseteq \mathbb{R}^n$.

According to these definitions, for a system which is not even locally identifiable, equation (3) has generically an infinite number of solutions for all input functions $\mathbf{u}$. This is commonly called *nonidentifiability* or *unidentifiability* [1,5].

In cases when the initial condition is known or some a priori information on the initial condition is available, one should study *identifiability from input-state-output data*.

Remark 1. (Global) identifiability is a *system-related* concept and should in principle hold irrespective, i.e. for all, possible initial conditions. It happens frequently in the applications that the property holds only generically, i.e. except for a "thin" set of initial conditions. In these situations the system is (incorrectly but forgivably) nevertheless declared to be (global) identifiable, excluding certain subsets of initial states.

In this paper the structural identifiability based on differential algebra and on the companion software DAISY (Differential Algebra for Identifiability of SYstems) [2] is used. The reader is referred to [1] and [10] for a detailed documentation of the theory behind the software tool and to [2] for the algorithm.

To check *practical non-identifiability* the method proposed in [7] is adopted.

The principal goal of this paper is to discuss and compare the relative merits of the two different methods on two significant examples. We would like also to effectively facilitate the adoption of the structural identifiability software system in modeling studies for biosystems, by showing with the examples that DAISY requires very little information from the user, just the model structure (equations) and input-output experimental configuration. Thus high-level programming languages, mathematics and computer algebra will not be prerequisites for using the software DAISY.

3   Identifiability of the Model of the $KdpD/KdpE$ System of Escherichia Coli

The mathematical model of the Escherichia coli system, proposed in [7] and successively discussed in [8], is here considered. A sensor kinase ($KdpD$) and a response regulator ($KdpE$) regulate the expression of the $KdpFABC$ operon encoding the high affinity $K+$ uptake system of Escherichia coli. A mathematical model for the $KdpD/KdpE$ two-component system was developed and the parameters where identified from the available in vitro and in vivo experimental data. The detailed description of the biochemical processes underlying the model is reported in the referenced papers.

We report below the algebraic-differential equations describing the model in [7], where a cancelation of one equation has already been done, as suggested by the authors. The symbol $mRNA$ represents the concentration of messenger RNA, $KdpD_0$ the total concentration of the sensor kinase, $KdpD^P$ the concentration of the phosphorylated $KdpD$, $KdpE_0$ the total concentration of the response regulator, $KdpE^P$ the concentration of the phosphorylated $KdpE$, $KdpFABC$ the concentration of the protein complex and $KdpE_f^p$ the concentration of the unbound response regulator. The vector $\mathbf{y}$ denotes the in vitro and in vivo measurements vector.

$$
\begin{cases}
\frac{dmRNA}{dt} & = k_{tr}\frac{DNA_f}{K}(1 + \frac{(KdpE_f^p)^2}{\alpha K_a}) - \\
& \quad (k_z + \mu)mRNA \\
\frac{dKdpD_0}{dt} & = k_{tl}mRNA - (k_d + \mu)KdpD_0 \\
\frac{dKdpE_0}{dt} & = k_{tl2}mRNA - (k_d + \mu)KdpE_0 \\
\frac{dKdpE^P}{dt} & = k_2KdpD^P(KdpE_0 - KdpE^P) - \\
& \quad (k_d + (k_{3f} + k_{-2})KdpD)KdpE^P \\
\frac{dKdpFABC}{dt} & = k_{tl3}mRNA - (k_{d2} + \mu)KdpFABC
\end{cases}
\tag{4}
$$

$$
\begin{cases}
0 = KdpE^P - KdpE_f^p - \\
\quad 2\frac{(KdpE_f^p)^2 DNA_f}{K_a}(1 + \frac{1}{\alpha K}) \\
0 = DNA_0 - DNA_f(1 + \frac{1}{K}) - \\
\quad \frac{(KdpE_f^p)^2 DNA_f}{K_a}(1 + \frac{1}{\alpha K})
\end{cases}
\tag{5}
$$

$$
\begin{cases}
y_1(t) = mRNA \\
y_2(t) = KdpD_0 \\
y_3(t) = KdpE_0 \\
y_4(t) = KdpE^P \\
y_5(t) = KdpFABC
\end{cases}
\tag{6}
$$

The vector of parameters is:

$$\mathbf{p} = [k_{tr}, DNA_f, K, KdpE_f^p, \alpha, K_a, k_z, \mu, k_{tl}, k_d, k_{tl2}, k_2, KdpD^P,$$

$$k_{3f}, k_{-2}, KdpD, k_{tl3}, k_{d2}, DNA_0]$$

The question to be addressed is whether the unknown parameter vector $\mathbf{p}$ is globally identifiable from the experiment. In the recent literature the identifiability of this model has been analyzed at a local level, about a nominal parameter value and (to our knowledge) not by differential algebra techniques. In [7] the authors find that the full set of parameters is not uniquely identifiable with the available data; they check the local identifiability of the model by using a *FIM-rank calculation* method with a *best set of parameters*. In particular parameter

k_2 has the lowest sensitivity index while $k_t r$ and DNA_0 have the highest sensitivity. Parameter μ presents high correlations with other parameters. Other pairs of parameters show also high correlation among them but they could still be identified. The authors thus modify the model structure by fixing the nonidentifiable parameters to known values and arrive at a second formulation of the model that fits the experimental data equally well.

Next the structural identifiability of model 4 is checked in an analytic way, using the differential algebra-based method implemented by DAISY. This allows to obtain an exact answer, not depending on nominal parameter values nor on indices that require the choice of a threshold. DAISY automatically checks the identifiability of each parameter. In agreement with the previous local identifiability results, the model is found nonidentifiable, but this method gives, without any approximation, the analytical relations among the dependent parameters. In particular, only parameters k_{tl}, k_{tl2}, k_{tl3}, are globally identifiable, while the other parameters are nonidentifiable. The result directly suggests how to proceed with the model reduction. Thus if, for example, parameter μ is fixed, as in [7], then k_z, k_d, k_{d2} become globally identifiable. If also K and DNA_f are fixed, then parameters $\alpha, K_a, k_{tr}, DNA_0$ become globally identifiable. This reveals that parameters k_{tr}, DNA_0 having a high sensitivity index are structurally nonidentifiable. One can also distinguish among parameters which are structurally nonidentifiable and parameters that cannot be estimated with precision due to problems related to the choice of a nominal parameter value or with the particular set of data, as for example k_2, μ. Hence DAISY provides, in about 3-4 seconds, the identifiability results holding in all the parameter space, with neither using experimental data nor requiring additional measurements. Note that using DAISY does not require any expertise on mathematical modeling to the experimenter.

4 Identifiability of the Tumor-Suppressor p53 Dynamics Model

The p53 tumor suppressor is thought to act as a surveillance molecule, detecting harmful insults to cells such as DNA damage, and initiating a cascade of events designed to stem the damage. Mutations in the p53 protein are found at a high frequency in a variety of tumors (e.g., 50% of breast cancers), and abnormal p53 activity has been linked to tumor growth. Thus mathematically modeling the regulation of p53 is of particular interest in cancer research. By using equations based on actual biophysical mechanisms, models can be created that represent hypotheses of how the relevant proteins interact. To identify the unknown parameters of these models from the available experimental data, provides a better understanding of how p53 is regulated, information critical for determining what goes wrong in cancer and designing rational therapies. Here the mathematical description of a recently proposed nonlinear model [6] describing the dynamic behavior of the tumor-suppressor p53 is reported. In particular the model has 4 ODEs, 4 measurement equations and 23 unknown parameters. For a detailed explanation of the model the reader is referred to [6].

$$\begin{cases}
\dot{x}_1 = p_1x_4 - p_3x_1 - p_4((x_1^2/(p_5 + x_1))(1 + p_6u_1/(p_7 + u_1))) \\
\dot{x}_2 = p_8 - p_9x_2 - p_{10}x_1x_2/((p_{11} + x_2)(1 + p_{12}u_1/(p_{13} + u_1))) \\
\dot{x}_3 = p_{14} - p_{15}x_3 - p_{16}x_1x_3(1 - p_{18}u_1)/(p_{17} + x_3) \\
\dot{x}_4 = (p_{20} - p_{21}(1 - p_{24})(1 - p_{25})/(p_{22}^4 + 1)) - p_{20}x_4 + \\
\quad p_{21}x_3^4(1 + p_{23}u_1)(1 - p_{24}x_1)(1 - p_{25}x_2)/(p_{22}^4 + x_3^4) \\
y_1 = x_1 \\
y_2 = x_2 \\
y_3 = x_3 \\
y_4 = x_4
\end{cases} \tag{7}$$

Even if the model has a large number (23) of parameters, DAISY provides the structural identifiability results in about 2-3 seconds: all parameters except p_{22} are globally identifiable; p_{22} is locally identifiable with four solutions (this was obvious given that p_{22} appears only in the single quartic p_{22}^4 term, in the fourth equation).

In practice, to check the global identifiability of this model with DAISY, the user has to write the input file in a given format. In the following the input file is reported:

Input File of DAISY

```
WRITE "Tumor-Suppressor p53 Dynamics Model"$
% B_ IS THE VARIABLE VECTOR
B_:={u1,y1,y2,y3,y4,x1,x2,x3,x4}$

 FOR EACH EL_ IN B_ DO DEPENDEL_,T$
%B1_ IS THE UNKNOWN PARAMETER VECTOR
B1_:={p1,p3,p4,p5,p6,p7,p8,p9,p10,p11,p12,p13,p14,p15,p16,\\
p17,p18,p20,p21,p22,p23,p24,p25}$
%NUMBER OF INPUTS
NX_:=1$
%NUMBER OF STATES
NX_:=4$
%NUMBER OF OUTPUTS
NY_:=4$
%MODEL EQUATIONS
c_:={df(x1,t)=p1*x4-p3*x1-p4*((x1*x1/(p5+x1))*(1+p6*u1/(p7+u1))),
df(x2,t)=p8-p9*x2-p10*(x1*x2/(p11+x2)*(1+p12*u1/(p13+u1))),
df(x3,t)=p14+p15*x3-p16*x1*x3*(1-p18*u1)/(p17+x3),
df(x4,t)=(p20-p21*(1-p24)*(1-p25)/(p22^4+1))-\\
p20*x4+p21*(x3^4)*(1+p23*u1)*(1-p24*x1)*(1-p25*x2)/(p22^4+x3^4),
    y1=x1,
    y2=x2,
    y3=x3,
    y4=x4}$
```

```
SEED_:=131$
DAISY()$
END$
```

Due to space limitations the output file is not reported here but the reader can directly run the above input file and see the required structural identifiability answer in just 2-3 seconds. Note that DAISY does not require expertise on mathematical modeling by the experimenter.

However, the experimenter may well be interested in parameter identifiability for just the actual data she has at hand. In this case, by moving to the data-based identifiability, she can easily check if p_{22} has only one nonnegative real and therefore feasible solution. If this is the case, she can conclude that this model, around a nominal point, is globally identifiable. Testing a different numerical point for $p*$ might lead to a different number of nonnegative real solutions.

Furthermore, the experimenter may well be interested in checking practical identifiability. In fact, if by evaluating the sensitivity of the above globally identifiable parameters, a parameter turns out to be practically non-identifiable, she can assess that the intrinsic reason of the apparent non-identifiability is not due to the model structure, but only to the paucity of the experimental data or to noise. By performing only data-based identifiability test it should be impossible to obtain this useful information.

5   Discussion

Data-based methods seem to be the only choice when collected data, perhaps obtained in a unrepeatable or very expensive experiment, are already at hand. Then the experimenter may want to check if the model parameters can in fact be recovered uniquely from the given data. There are some caveats:

1. In principle, for a large family of models, structural identifiability can be checked by suitable mathematical procedures directly on the model, without the need of collecting experimental data. This may avoid waste of resources for doing useless experiments, given the high costs, not only in economic terms, of biological experiments.

2. In case of nonidentifiability, structural methods calculate a unique parametrization of the model and thus provide guidelines to simplify the model structure or indicate, before performing the real experiment, when more information (measured outputs) is needed to guarantee unique identifiability [1]. Sometimes a model may be redundant and involve non-observable parameters. It is hence of great interest to check for *minimality* (i.e. non redundancy) of the model from the input-output configuration. Since when dealing with biological/physiological systems, severe constraints exist on experiment design it is also valuable to describe an experimental setup by a minimal model. By checking structural identifiability one can also check if the number of inputs and outputs are necessary and sufficient to guarantee a priori unique identifiability [4].

On the other hand one should also be aware of the limitations of the analytic procedures for checking structural identifiability, in particular the following:

1. When the model is very complex, with complicated nonlinearities and a large number of states or unknown parameters, and/or few measurement equations, most algorithms take a very long time to terminate or may even not terminate at all due to computational complexity problems.
2. Algorithms based on differential algebra, like the one employed in the above examples, in general, require the model differential equations to be of polynomial or rational form.
3. The experimenter may well be interested in parameter identifiability for just the actual data she has at hand.

6   Conclusion

The goal of this paper was to discuss the relations between structural and practical or data-based (non-)identifiability studies. We have discussed the two approaches both in general and by a detailed identifiability analysis of two biological models. For these two examples we have shown that the results of the data-based approach can also be obtained in an analytical way, by using a differential algebra based software tool [2]. This test can actually provide some additional information helpful for experiment design. Our suggestion would be to apply first the structural and successively the data-based approach to provide an integrated information. This is essential to make the model identification process more rigorous and reliable.

Acknowledgments. The author wishes to express her gratitude to Profs. Giuseppina Bellu, Stefania Audoly and Leontina D'Angiò (University of Cagliari, Italy) who offered invaluable knowledge and assistance.

References

1. Audoly, S., Bellu, G., D'Angiò, L., Saccomani, M.P., Cobelli, C.: Global identifiability of nonlinear models of biological systems. IEEE Trans. Biomed. Eng. 48(1), 55–65 (2001)
2. Bellu, G., Saccomani, M.P., Audoly, S., D'Angiò, L.: DAISY: A new software tool to test global identifiability of biological and physiological systems. Comp. Meth. Prog. Biom. 88, 52–61 (2007)
3. Chis, O., Banga, J.R., Balso-Canto, E.: Structural identifiability of systems biology models: a critical comparison of methods. Plos One 6(11), e27755 (2011)
4. Cobelli, C., Saccomani, M.P.: Unappreciation of a Priori Identifiability in Software Packages Causes Ambiguities in Numerical Estimates. Letter to the Editor. Am. J. Physiol. 21, E1058–E1059 (1990)
5. Ljung, L., Glad, S.T.: On global identifiability for arbitrary model parameterizations. Automatica 30(2), 265–276 (1994)
6. DiStefano III, J.J.: Dynamic systems biology modeling and simulation. Academic Press/Elsevier (2013)

7. Kremling, A., Heermann, R., Centler, F., Jung, K., Gilles, E.D.: Analysis of two-component signal transduction by mathematical modeling using the KdpD/KdpE system of Escherichia coli. BioSystems 78, 23–37 (2004)

8. Rodriguez-Fernandez, M., Rehberg, M., Kremling, A., Banga, J.R.: Simultaneuos model discrimination and parameter estimation in dynamic models of cellular systems. BMC System Biology 7(76), 1–14 (2013)

9. Raue, A., Kreutz, C., Maiwald, T., Bachmann, J., Shilling, M., Klingmüller, U., Timmer, J.: Structural and Practical Identifiability Analysis of Partially Observed Dynamical Models by Exploiting the Profile Likelihood. Bioinformatics 25, 1923–1929 (2009)

10. Saccomani, M.P., Audoly, S., D'Angiò, L.: Parameter identifiability of nonlinear systems: The role of initial conditions. Automatica 39, 619–632 (2004)

Non-canonical Imperfect Base Pair Predictor: The RNA 3D Structure Modeling Process Improvement

Jacek Śmietański

Faculty of Mathematics and Computer Science,
Jagiellonian University, Kraków, Poland
jacek.smietanski@ii.uj.edu.pl

Abstract. RNA is a large group of macromolecules involved in many essential cellular processes. They can form complex secondary and three-dimensional structures, and their biological functions highly rely on their forms. Therefore a high quality RNA structure determination is a key process to address RNA functions and roles in molecular pathways. However, in many cases the structure cannot be experimentally solved or the process is too expensive and laborious. This problem can be avoid using bioinformatics methods of computational RNA structure prediction. Such applications have been developed, however the quality of predictions, especially for large RNA structures, still remains too low.

One of the most important aspects in RNA 3D model building is the intramolecular interactions identification and validation. In this work I propose a method which can improve this stage of model building, and should result in creation of better final three-dimensional RNA models.

In my work I constructed a predictor that can identify both canonical and non-canonical base pair interactions within a given structure. The main advantages of this predictor are: 1) the ability to work on incomplete input structures, and 2) the ability to correctly predict base pair type even for imperfect (fuzzy) input atoms coordinates.

The predictor is based on the set of SVM multi-class classifiers. For each input base pair the classifier chooses one of 18 recognized pair types. The predictor was trained on the experimental high quality data and tested on different, imperfect and incomplete (coarse-grained) structures. The average quality of predictor for tested fuzzy nucleotide pairs is at the level about 96% of correct recognitions.

Keywords: RNA, base pairs, structure modeling, interaction prediction, classification, SVM, machine learning.

1 Introduction

RNA is a family of biological molecules that are capable of performing a wide range of biological functions. Carrying genetic information and protein synthesis is its the best known, but the not only function. RNA may also detect the presence of ions or

F. Ortuño and I. Rojas (Eds.): IWBBIO 2015, Part I, LNCS 9043, pp. 659–668, 2015.
© Springer International Publishing Switzerland 2015

small molecules, regulate gene expression, catalyze chemical reactions or have other complex regulatory roles in cells [3],[10]. The RNA biological functions are performed thanks to their ability to form complex secondary and three-dimensional structures, and those functions strongly depends on such structures. Therefore a high quality RNA structure determination is a key process to address its functions and roles in molecular pathways. However, in many cases the structure cannot be experimentally solved or the process is too expensive and laborious.

This problem can be overcome using bioinformatics methods of computational RNA structure prediction. A number of such applications have been developed to date, for example FARNA [7], iFoldRNA [29], MC-Fold MC-Sym Pipeline [22], NAST [8], RNAComposer [25] and ModeRNA [27] (to mention only most popular and easy available tools). Those methods are based on different concepts, like conformational space search, discrete molecular dynamics, knowledge-based, coarse-grained refinement, template-based and force-field-based approaches. For more details see [6] and [26]. However, in spite of many improvements towards the RNA structure prediction, the quality of them, especially for large RNA structures, still remains unsatisfactory [30].

Automatic RNA structure prediction is a difficult and complex process. The algorithms are typically composed of several consecutive steps, for example in homology based methods we must first identify the template, then align both (template and target) sequences, construct a frame, fill it with all atoms positions, optimize, and verify the correctness of the model. The result of each step strictly depends on the output of the previous step. So, the tool, which can cope with non-perfect data from an one step output and correct them to a good next-step input, is needed.

One of the most important aspects in RNA 3D model building is the intramolecular interactions identification and validation. In this work I propose a method which can improve this stage of model building, and should result in creation of better final three-dimensional RNA models.

The most important interaction type within RNA structures is a base pairing. When RNA molecule folds, hydrogen bonds between spatially contacting residues are formed. Canonical (cis Watson-Crick A-U and C-G) and wobble (cis Watson-Crick G-U) base pairs are the most popular ones and those base pairs determine the secondary structure of RNA molecule. However, nucleotide base pairs may also be folded into non-canonical forms and those interactions play essential role in 3D structure stabilization [11][15][16].

Among the methods developed to perform automatic assignment of residue pairs from atomic coordinates of RNA 3D structures, the three: MC-Annotate [9], FR3D [28] and RNAView [33] and are the most notable. They do not always agree about non-canonical pairs and to get around this inconvenience the ClaRNA classifier was recently developed as a consensus between those three classifications [31].

According to the nomenclature proposed by Leontis and Westhof [18], 12 geometric families were considered, which give us 18 classes of interaction pairs to recognize.

In presented work a predictor that can identify both canonical and non-canonical base pair interactions within a given structure is presented. The predictor has two important advantages over other existing classifiers. The first is that it works even on incomplete input data, that means coarse-grained representation. Only three atoms per residue are needed for it to work properly.

Second: it was shown that the classifier works well also on imperfect input data. To test this feature, the strict coordinates from validation data set were randomly moved in a random direction (each atom independently from others) and validation was made for really fuzzy data.

2   Materials and Methods

2.1   Training Data Sets

Training data were prepared based on data used to train the ClaRNA [31] classifier. As it was mentioned in previous section, ClaRNA functions as a consensus between the all other top base pair detection services. Therefore such data are the most valuable and then used to train our classifier.

Data set was downloaded from supplementary materials available on ClaRNA on-line web service (http://iimcb.genesilico.pl/clarna/supp-data/ref/ref.zip, downloaded on 2014, December).

From the given set, 6 disjoint subsets were selected. Each for different nucleotide pair (AC, AG, AU, CG, CU and GU). The AC subset contains only those base pairs, where adenine contacts with cytosine, AG is for adenine – guanine contacts, etc. I neglect reverse pairs, for example CA subset for cytosine – adenine contacts is not created because it is symmetrical to AC subset (if you need to predict pairing type for CA residue pair, the system will do it just for AC residue pair and reverse the result, for example if for AC the result is cWH (cis Watson-Crick/Hoogsteen) pair type, it is obvious that for corresponding CA residues the pair type should be cHW (cis Hoogsteen/ Watson-Crick)).

In this version the predictor cannot correctly recognize pairing types for identical residue pairs (AA, CC, GG and UU) yet. The problem here is that we do not know, which residue should be considered as the first in the pair and which as the second. The significance of this issue is clarified in the Figure 1. Thus the subsets for AA, CC, GG and UU were not created.

Each subset contains a set of experimentally determined (collected from the PDB database [1]) and proved (by ClaRNA) exemplars of interactions for different pairing types.

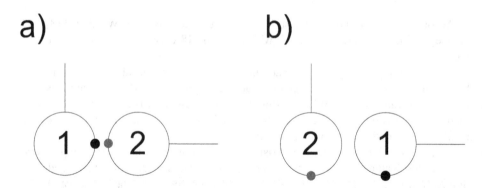

Fig. 1. The importance of nucleotide order within base pair. In both situations a) and b) identical interaction between the same nucleotides is presented. The only difference is the order in which those nucleotides are analyzed. Assume than we have to measure distance between black and gray points. The first (black) is the point on nucleotide number "1", 90 degrees right from the side chain. The second (gray) is the point on nucleotide number "2", opposite to the side chain. In the first case (a), where the left nucleotide is treated as number "1", the measured distance is several times smaller than in the second (b) case, where we set the right nucleotide as number "1".

The classification is based on the Leontis-Westhof nomenclature [18]. They proposed geometric families as combinations of three possible nucleotide edges, for each nucleotide, where the H-bonding interactions are possible and the relative orientations of the glycosidic bonds of the two bases.

Let us assume the following notation: W – Watson-Crick edge, H – hoogsten edge, S – sugar edge, c – cis orientation, t – trans orientation. The base pair type is denoted by the three letters notation, where the first (c or t) determines orientation, and next two – interacting edges (W, H or S). Combining all the possibilities give us 18 possible pair types: cWW, tWW, cWH, tWH, cWS, tWS, cHW, tHW, cHH, tHH, cHS, tHS, cSW, tSW, cSH, tSH, cSS, tSS.

Not always all 18 pair types are theoretically possible for given nucleotide pair. Also the frequency of possible nucleotide pairs differs. So, for some pair types there are no exemplars in training data (if pairing is not possible then classifier does not recognize such pair) and for other, the number of exemplars differs. The Table 1. presents a detailed summary for the numbers of exemplars in each training set. Each individual training set has a total of 334 – 599 base pairs and overall training size is 3189 base pairs.

To create final training sets, a base3 reduced representation (more details are given in section 2.3) was selected. It is a coarse-grained model, in which only three atoms from each nucleotide are taken into consideration. Then the distances between all atom pairs were calculated. Finally each exemplar in training set consists of 9 distances (for each of 3 atoms from the first base we measure distances to all 3 considered atoms from the second base). Those 9 distances will serve as features used to train all classifiers.

Table 1. Training sets statistics. Each row presents a number of exemplars for given base pair type.

pair type	set 1: AC	set 2: AG	set 3: AU	set 4: CG	set 5: CU	set 6: GU
cWW	50	50	50	50	50	50
tWW	50	16	50	50	50	50
cWH	18	50	4	50	1	2
tWH	14	25	12	50	4	24
cWS	50	50	50	50	50	50
tWS	50	50	3	50	1	50
cHW	6	50	50	5	31	50
tHW	50	26	50	8	3	11
cHH	3	4	8	5	0	0
tHH	50	50	50	50	19	0
cHS	11	50	50	6	50	0
tHS	50	50	50	2	50	1
cSW	50	11	50	50	19	50
tSW	42	11	50	41	4	50
cSH	50	1	2	0	1	50
tSH	50	1	1	1	1	50
cSS	5	50	5	50	0	50
tSS	0	50	0	20	0	50
total	599	595	535	538	334	588

2.2   Testing Data Sets

Testing data sets were created in a similar way as training data sets. As the predictor has to be tested in two modes: strict and fuzzy, the two different sets were used.

First – for strict validation, contains distances calculated directly upon the PDB coordinates data. The size of each validate set is 634 – 2117 base pairs, and overall size of all validate sets for thin mode is 8491 base pairs.

Second – for fuzzy validation. For each considered base pair from the PDB database (there were the same pairs as for creation strict validation set) 100 copies were created and for each copy its atoms were randomly moved within the distance up to 0,5Å and random direction. Each atom was moved independently from others. So each element from the previous (strict) validate set was replaced by other 100 elements. Technical details about this fuzzification process are described below, in section 2.4. Finally the validation sets for this mode were 100 times larger than the previous ones.

2.3   Base3 Coarse-Grained Scheme

Our predictor's classification is constructed upon only three atoms for one base. Such simplification enables predictions for even incomplete models and makes calculations faster. The concept is based on SimRNA3 coarse-grained model [2]. Since in the interaction mainly nucleotide bases are involved, phosphate and ribose groups are neglected, so we take into consideration atoms: N1, C2 and C4 for pyrimidines and N9, C2, C6 for purines. This model is presented on Figure 2.

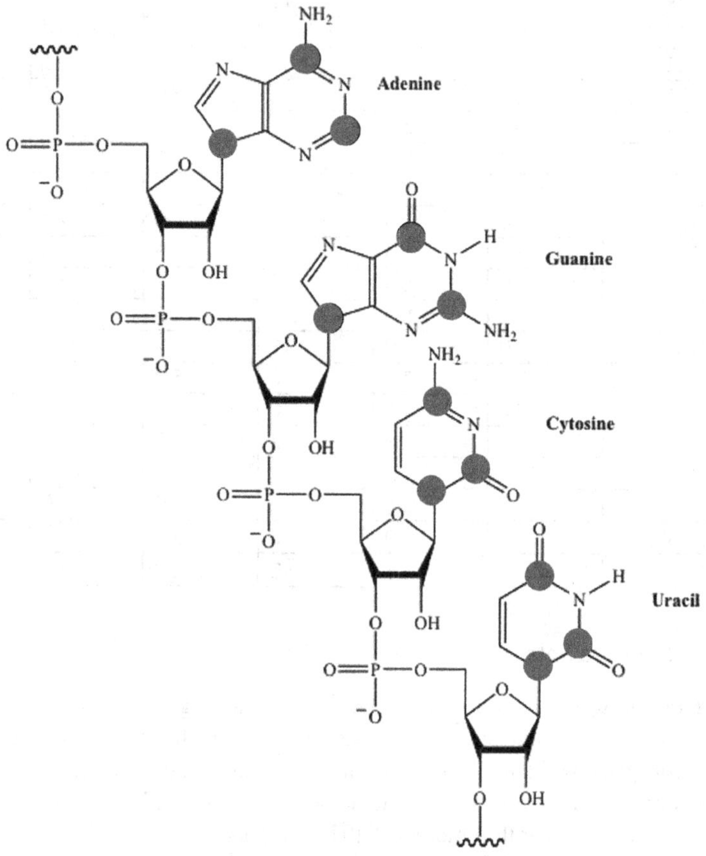

Fig. 2. Coarse-grained base3 RNA representation

2.4   Fuzzification Module

In the fuzzification process, for each atom the starting Cartesian coordinates (x,y,z) are transformed into spherical coordinate system (r,θ,φ) [20], then translated with vector (r', θ',ψ'), where r', θ' and ψ' are randomly generated. r' is a distance for which the atom is moved and is upper bounded: $r' \leq 0.5\text{Å}$, θ' and ψ' represents the direction (angle) in which translation is done. After translation, new coordinates are back transformed to Cartesian system (x',y',z').

Such transformation is performed independently for each atom.

2.5   Classifier Algorithm

The Support Vector Machine (SVM) algorithm [5] was used. SVM is one of the most successful algorithms for classification and has been widely applied to many areas of bioinformatics, including gene expression data classification, transcription initiation site prediction, protein functional site recognition, subcellular localization, secondary structure prediction, fold prediction, etc. [13],[19],[34].

As originally developed for binary problems, the SVM method can be successfully extended also for multi-class classification tasks. The one-versus-one [14] ,[24] model was used here for the assembly of the multi-class classifier from binary classifiers. For a classification problem of N class, it trains every pair-wise binary (sub-) classifier. This gives a total of $\frac{1}{2}*N*(N-1)$ sub-classifiers and the final classification is done by a max-wins voting strategy, in which every sub-classifier assigns the instance to one of the two classes, then the vote for the assigned class is increased by one vote, and finally the class with the most votes determines the final classifier decision.

For all classifiers (sub-classifiers, to be strict with the above notation), the radial basis function (RBF) kernel was used and the cost (c) and gamma (γ) parameters [32] optimized.

For each nucleotide-pair type tested (AC, AG, AU, CG, CU, GU) the separate multi-class classifier was trained. Each classifier has to recognize up to 18 classes (sometimes less, because for particular nucleotide-pairs not all theoretically possible pair types are allowed [17]). The number of classes corresponds to the exemplars in training data sets (Table 1).

2.6   Software

The software was entirely written in Python, version 3.4, using free additional libraries: NumPy, SciPy [21] (as necessary dependencies for the following), biopython [4] (Bio.PdbParser module [12] for parsing PDB data), and scikit-learn [23] (machine learning SVM algorithm).

2.7   Hardware

As calculations are not a time- and memory- consuming, the standard PC is enough for complete them. Python support multiplatform, so it can be done on any operating system.

Both training and testing experiments were performed on a ordinary home PC. The complete calculation cycle – train all classifiers and validate them in both – strict and fuzzy mode – takes about 2 minutes and 40 seconds.

3   Results

The quality of constructed predictor was tested in two independent modes. The first was strict validation. In this mode the distances within validation sets were based upon atom coordinates taken directly from the PDB. For each classifier appropriate validation set was used. The quality assessment was based on percentage of correct classifications measure. Results are presented in Table 2. The average predictor quality for strict data is 96,8 %.

Table 2. Results for strict validation

base pair	number of exemplars	correctly classified pairs	incorrectly classified	percentage of correct classifications
AC	1524	1491	33	97.8 %
AG	2117	2097	20	99.1 %
AU	2064	2035	29	98.6 %
CG	974	944	30	97.0 %
CU	634	569	65	89.7 %
GU	1178	1158	20	98.3 %

The second mode – fuzzy validation was performed using imperfect atomic coordinates. Results for this mode are presented in table 3. The average predictor quality for imperfect data is 95,9 %.

Table 3. Results for fuzzy validation

base pair	number of exemplars	correctly classified pairs	incorrectly classified	percentage of correct classifications
AC	152400	148855	3545	97.7 %
AG	211700	209159	2541	98.8 %
AU	206400	203407	2993	98.5 %
CG	97400	94574	2826	97.1 %
CU	63400	54149	9251	85.4 %
GU	117800	115438	2362	98.0 %

4   Conclusions

In this work I proposed a novel predictor for canonical and non-canonical base pairs identification based on – even imperfect and incomplete – RNA structures. It uses a machine learning approach, in which classification is performed on trained multi-class SVM classifiers. The main advantages of presented solution to the existing classifiers are: the ability to classify using only coarse-grained base3 model and to predict using imperfect input data. The algorithm is fast and effective.

So far it works well only for pairs formed by non-identical nucleotides. Avoiding this issue will be the goal of the future work.

Acknowledgements. I would like to thank Janusz Bujnicki from Laboratory of Bioinformatics and Protein Engineering, Warsaw, for inspiration and valuable comments.

References

1. Berman, H.M., Westbrook, J., Feng, Z., et al.: The Protein Data Bank. Nucleic Acids Res. 28, 235–242 (2000)
2. Boniecki, M.J., Łach, G., Tomala, K., et al.: SimRNA: A program for RNA folding simulations. In: SocBiN/BIT13 Book of Abstracts, Torun, Poland, June 26-29 (2013)
3. Clancy, S.: RNA functions. Nature Education 1(1), 102 (2008)
4. Cock, P.J., Antao, T., Chang, J.T., et al.: Biopython: freely available Python tools for computational molecular biology and bioinformatics. Bioinformatics 25(11), 1422–1423 (2009)
5. Cortes, C., Vapnik, V.: Support-vector networks. Mach. Learn. 20, 273–297 (1995)
6. Cruz, J.A., Blanchet, M.-F., Boniecki, M., et al.: RNA-Puzzles: A CASP-like evaluation of RNA three-dimensional structure prediction. RNA 18(4), 610–625 (2012)
7. Das, R., Baker, D.: Automated de novo prediction of native-like RNA tertiary structures. Proc. Natl. Acad. Sci. U.S.A. 104(37), 14664–14669 (2007)
8. Flores, S.C., Altman, R.B.: Coarse-grained modeling of large RNA molecules with knowledge-based potentials and structural filters. RNA 15(9), 1769–1778 (2010)
9. Gendron, P., Lemieux, S., Major, F.: Quantitative analysis of nucleic acid three-dimensional structures. J. Mol. Biol. 308, 919–936 (2001)
10. Gesteland, R.F. (ed.): The RNA World, 3rd edn. Cold Spring Harbor Laboratory Press (2005)
11. Halder, S., Bhattacharyya, D.: RNA structure and dynamics: a base pairing perspective. Prog. Biophys. Mol. Biol. 113, 264–283 (2013)
12. Hamelryck, T., Manderick, B.: PDB file parser and structure class implemented in Python. Bioinformatics 19(17), 2308–2310 (2003)
13. Jensen, L.J., Bateman, A.: The rise and fall of supervised machine learning techniques. Bioinformatics 27(24), 3331–3332 (2011)
14. Knerr, S., Personnaz, L., Dreyfus, G.: Single-layer learning revisited: A stepwise procedure for building and training neural network. In: Neurocomputing: Algorithms, Architectures and Applications. NATO ASI, Springer, Berlin (1990)
15. Lee, J.C., Gutell, R.R.: Diversity of base-pair conformations and their occurrence in rRNA structure and RNA structural motifs. J. Mol. Biol. 344, 1225–1249 (2004)
16. Leontis, N.B., Lescoute, A., Westhof, E.: The building blocks and motifs of RNA architecture. Curr. Opin. Struct. Biol. 16, 279–287 (2006)
17. Leontis, N.B., Stombaugh, J., Westhof, E.: The non-Watson-Crick base pairs and their associated isostericity matrices. Nucleic Acids Res 30(16), 3497–3531 (2002)
18. Leontis, N.B., Westhof, E.: Geometric nomenclature and classification of RNA base pairs. RNA 7, 499–512 (2001)
19. Mirmohammadi, S.N., Shishehgar, M., Ghapanchi, F.: Applications of ANNs, SVM, MDR and FR Methods in Bioinformatics. World Applied Sciences Journal 31(6), 1109–1117 (2014)
20. Moon, P., Spencer, D.E.: Spherical coordinates (r, θ, ψ). In: Field Theory Handbook, Including Coordinate Systems, Differential Equations, and Their Solutions, pp. 24–27. Springer, New York (1988)

21. Oliphant, T.E.: Python for Scientific Computing. Computing in Science & Engineering 9, 90 (2007)
22. Parisien, M., Major, F.: The MC-Fold and MC-Sym pipeline infers RNA structure from sequence data. Nature 452(1), 51–55 (2008)
23. Pedregosa, F., Varoquaux, G., Gramfort, A., et al.: Scikit-learn: Machine Learning in Python. JMLR 12, 2825–2830 (2011)
24. Platt, J.C., Cristianini, N., Shawe-Taylor, J.: Large margin DAGs for multiclass classification. In: Advances in Neural Information Processing Systems, vol. 12, pp. 547–553. MIT Press (2000)
25. Popenda, M., Szachniuk, M., Antczak, M., et al.: Automated 3D structure composition for large RNAs. Nucleic Acids Res. 40(14), 1–12 (2012)
26. Rother, K., Rother, M., Boniecki, M., et al.: RNA and protein 3D structure modeling: similarities and differences. J. Mol. Model. 17(9), 2325–2336 (2011)
27. Rother, M., Rother, K., Puton, T., Bujnicki, J.M.: ModeRNA: A tool for comparative modeling of RNA 3D structure. Nucleic Acids Res. 39(10), 4007–4022 (2011)
28. Sarver, M., Zirbel, C.L., Stombaugh, J., et al.: FR3D: finding local and composite recurrent structural motifs in RNA 3D structures. J. Math. Biol. 56, 215–252 (2008)
29. Sharma, S., Ding, F., Dokholyan, N.V.: iFoldRNA: Three-dimensional RNA structure prediction and folding. Bioinformatics 24(17), 1951–1952 (2008)
30. Sripakdeevong, P., Beauchamp, K., Das, R.: Why can't we predict RNA structure at atomic resolution? In: Leontis, N., Westhof, E. (eds.) RNA 3D Structure Analysis and Prediction, Nucleic Acids and Molecular Biology 27, 43–65 (2012)
31. Waleń, T., Chojnowski, G., Gierski, P., Bujnicki, J.M.: ClaRNA: a classifier of contacts in RNA 3D structures based on a comparative analysis of various classification schemes. Nucleic Acids Research 42(19), e151 (2014)
32. Wu, T.-F., Lin, C.-J., Weng, R.C.: Probability Estimates for Multi-class Classification by Pairwise Coupling. J. Mach. Learn. Res. 5, 975–1005 (2004)
33. Yang, H., Jossinet, F., Leontis, N., et al.: Tools for the automatic identification and classification of RNA base pairs. Nucleic Acids Res. 31, 3450–3460 (2003)
34. Yang, Z.R.: Biological applications of support vector machines. Brief. Bioinform. 5(4), 328–338 (2004)

Author Index